METHODS IN MOLECULAR BIOLOGY

Series Editor
John M. Walker
School of Life and Medical Sciences
University of Hertfordshire
Hatfield, Hertfordshire, AL10 9AB, UK

For further volumes:
http://www.springer.com/series/7651

Evolutionary Genomics

Statistical and Computational Methods

Second Edition

Edited by

Maria Anisimova

Institute of Applied Simulations, School of Life Sciences and Facility Management, Zurich University of Applied Sciences (ZHAW), Wädenswil, Switzerland

Swiss Institute of Bioinformatics, Lausanne, Switzerland

OPEN Humana Press

Editor
Maria Anisimova
Institute of Applied Simulations
School of Life Sciences and Facility Management
Zurich University of Applied Sciences (ZHAW)
Wädenswil, Switzerland

Swiss Institute of Bioinformatics
Lausanne, Switzerland

ISSN 1064-3745 ISSN 1940-6029 (electronic)
Methods in Molecular Biology
ISBN 978-1-4939-9073-3 ISBN 978-1-4939-9074-0 (eBook)
https://doi.org/10.1007/978-1-4939-9074-0

This Humana imprint is published by the registered company Springer Science+Business Media, LLC, part of Springer Nature.
The registered company address is: 233 Spring Street, New York, NY 10013, U.S.A.

Preface

This volume is a thoroughly revised second edition of *Evolutionary Genomics: Statistical and Computational Methods* published in 2012. Like the first edition, the new volume includes comprehensive reviews of the most recent and fundamental developments in bioinformatics methods for evolutionary genomics and related challenges associated with increasing data size, heterogeneity, and its inherent complexity.

Throughout the volume, prominent authors address the challenge of analyzing and understanding the dynamics of complex biological systems, and elaborate on some promising strategies that would bring us closer to the ultimate "holy grail" of biology—uncovering of the relationships between genotype and phenotype. Consequently, the presented collection of peer-reviewed articles also represents a synergy between theoretical and experimental scientists from a range of disciplines, working together towards a common goal. Once again, the revised volume reiterates the power of taking an evolutionary approach to study molecular data.

This book is intended for scientists looking for a compact overview of the cutting-edge statistical and computational methods in evolutionary genomics. The volume may serve as a comprehensive guide for both graduate and advanced undergraduate students planning to specialize in genomics and bioinformatics. Equally, the volume should be helpful for experienced researchers entering genomics from more fundamental disciplines, such as statistics, computer science, physics, and biology. In other words, the material presented here should suit both a novice in biology with strong statistics and computational skills and a molecular biologist with a good grasp of standard mathematical concepts. To cater to differences in reader backgrounds, *Part I* is composed of educational primers to help with fundamental concepts in genome biology (Chapter 1), probability and statistics (Chapter 2), and molecular evolution (Chapter 3). As these concepts reappear repeatedly throughout the book, the first three chapters will help the neophyte to stay "afloat". The exercises and questions offered at the end of each chapter serve to deepen the understanding of the material.

Part II of this volume focuses on sequence homology and alignment—from aligning whole genomes (Chapter 4) to disentangling orthologs, paralogs, and transposable elements (Chapters 5 and 6). *Part III* includes chapters on phylogenetic methods to study genome evolution. Chapter 7 presents multispecies coalescent methods for reconciling phylogenetic discord between gene and species trees. However, a mathematically convenient "binary tree" model does not always live up to scrutiny as numerous evolutionary processes act in reticulate (network-like) fashion, complicating the statistical description of evolutionary models and increasing computational complexity, often to prohibitive levels. One simplification is to assume that some molecular sequence units (genes, gene segments) still evolve in a treelike manner. If so, Chapter 8 describes one practical approach to meaningfully summarize the binary tree distributions for a set of genomes as a "forest of trees". Alternatively network-like phylogenetic relationships can be represented by graphs (Chapter 9). Dating methods for genome-scale data are discussed in Chapter 10, while Chapter 11 provides more examples of non-treelike processes in a comparative review of genome evolution in different breeding systems.

By disentangling different evolutionary forces acting on genomes, we hope to understand the origins of biological innovation, which is often thought to be coupled with natural selection. After all, how do we explain that, by the words of Darwin, "from so simple a beginning endless forms most beautiful and most wonderful have been, and are being, evolved"? This is the main topic of *Part IV* that discusses the methodology for evaluating selective pressures on genomic sequences (Chapters 12–14) and genomic evolution in light of protein domain architecture and transposable elements (Chapters 15 and 16). *Part V* of this book is dedicated to population genomics and other omics, with example applications to disease. Indeed, as evolution starts in populations, there is much interest in generating and studying population genome data for a wide range of species. Chapter 17 discusses models for genetic architectures of complex disease and genome-wide association studies for finding susceptibility variants. Chapter 18 reviews approaches to study ancestral population genomics. Chapters 19, 20 and 21 illustrate first principles of analyzing environmental sequences and applications to clinical trials and systems genetics. Finally, *Part VI* concludes the book by discussing current bottlenecks in handling and analyzing genomic data. Chapter 22 focuses on challenges and approaches for large and complex data representation and simultaneous querying of heterogeneous databases. Chapter 23 makes the case for using efficient high-performance computing strategies for computationally demanding phylogenetic analyses, in particular in the Bayesian framework. Solutions for scalable workflows and sharing programming resources are presented in Chapters 24 and 25.

On behalf of all authors, I hope that this book will become a source of inspiration and new ideas for our readers. Wishing you a pleasant reading!

Wädenswil, Switzerland
Lausanne, Switzerland

Maria Anisimova

Acknowledgements

This renewed edition of *Evolutionary Genomics: Statistical and Computational Methods* is a result of a dedicated effort by 94 co-authors of the book representing research institutions from nearly two dozen different countries. Special thanks go to almost 50 independent reviewers whose constructive and detailed comments have greatly contributed to improving the overall quality of the book chapters and the clarity of the presentation. As for the first edition of this book, the cover image was made by the author of Chapter 6 and a talented photography artist, Wojciech Makałowski, from the University of Münster, Germany.

By a mutual agreement between all authors of the book, all chapters are available *Open Access.* Swiss Institute of Bioinformatics (SIB) and Zurich University of Applied Sciences (ZHAW) have generously contributed to cover a part of the Open Access publication fees. Finally, I would like to thank my colleagues at the Institute of Applied Simulations and the School of Life Sciences and Facility Management of ZHAW (Zurich University of Applied Sciences) as well as my family for their support and encouragement.

Contents

Part IV Natural Selection and Innovation in Genomic Sequences

Part V Population Genomics and Omics in Light of Disease and Evolution

Part VI Handling Genomic Data: Resources and Computation

Contributors

ADRIAN M. ALTENHOFF • *Computer Science Department, ETH Zurich, Zurich, Switzerland; Swiss Institute of Bioinformatics, Lausanne, Switzerland*

CHRISTIAN ANDERSON • *Advantage Testing of Boston, Newton Centre, MA, USA*

MARIA ANISIMOVA • *Institute of Applied Simulation, School of Life Sciences and Facility Management, Zurich University of Applied Sciences (ZHAW), Wädenswil, Switzerland; Swiss Institute of Bioinformatics, Lausanne, Switzerland*

DANNY ARENDS • *Animal Breeding Biology and Molecular Genetics, Albrecht Daniel Thaer-Institute for Agricultural and Horticultural Sciences, Humboldt University zu Berlin, Berlin, Germany*

STÉPHANE ARIS-BROSOU • *Department of Biology, University of Ottawa, Ottawa, ON, Canada; Department of Mathematics and Statistics, University of Ottawa, Ottawa, ON, Canada*

DANIEL L. AYRES • *Center for Bioinformatics and Computational Biology, University of Maryland, College Park, MD, USA*

GUY BAELE • *Department of Microbiology and Immunology, Rega Institute, KU Leuven, Leuven, Belgium*

CANER BAĞCI • *Algorithms in Bioinformatics, Faculty of Computer Science, University of Tübingen, Tübingen, Germany*

ERIC BAPTESTE • *Sorbonne Universités, Institut de Biologie Paris-Seine, UPMC Université Paris 6, Paris, France*

NIKO BEERENWINKEL • *Department of Biosystems Science and Engineering, ETH Zurich, Basel, Switzerland*

SINA BEIER • *Algorithms in Bioinformatics, Faculty of Computer Science, University of Tübingen, Tübingen, Germany*

DOMINIQUE BELHACHEMI • *Life Technologies, Waltham, MA, USA*

SØREN BESENBACHER • *Department of Clinical Medicine (MOMA), Aarhus University, Aarhus, Denmark*

JOSEPH P. BIELAWSKI • *Department of Biology, Dalhousie University, Halifax, NS, Canada; Department of Mathematics & Statistics, Dalhousie University, Halifax, NS, Canada*

RAOUL J. P. BONNAL • *Istituto Nazionale Genetica Molecolare INGM Romeo ed Enrica Invernizzi, Milan, Italy*

AIDAN BUDD • *Structural and Computational Biology (SCB) Unit, European Molecular Biology Laboratory (EMBL), Heidelberg, Germany*

PHILIPPE COLSON • *Fondation Institut Hospitalo-Universitaire Méditerranée Infection, Pôle des Maladies Infectieuses et Tropicales Clinique et Biologique, Fédération de Bactériologie-Hygiène-Virologie, Centre Hospitalo-Universitaire Tione, Assistance Publique-Hôpitaux de Marseille, Marseille, France; Unité de Recherche sur les Maladies Infectieuses et Tropicales Emergentes (URMITE) UM63, CNRS 7278, IRD 198, INSERM U1095, Aix-Marseille University, Marseille, France*

EDUARDO COREL • *Sorbonne Universités, Institut de Biologie Paris-Seine, UPMC Université Paris 6, Paris, France*

MICHAEL R. CRUSOE • *Common Workflow Language Project, Vilnius, Lithuania*

TARCISIO MENDES DE FARIAS • *University of Lausanne, Lausanne, Switzerland; SIB Swiss Institute of Bioinformatics, Lausanne, Switzerland*

JOEP DE LIGT • *Department of Genetics, Center for Molecular Medicine, University Medical Center Utrecht, Utrecht University, Utrecht, The Netherlands*

CHRISTOPHE DESSIMOZ • *Swiss Institute of Bioinformatics, Lausanne, Switzerland; Department of Computational Biology, University of Lausanne, Lausanne, Switzerland; Center for Integrative Genomics, University of Lausanne, Lausanne, Switzerland; Department of Genetics, Evolution and Environment, University College London, London, UK; Department of Computer Science, University College London, London, UK*

COLIN N. DEWEY • *Department of Biostatistics and Medical Informatics, University of Wisconsin-Madison, Madison, WI, USA*

PAOLO DI TOMMASO • *Centre for Genomic Regulation (CRG), The Barcelona Institute for Science and Technology, Barcelona, Spain*

MARIO DOS REIS • *School of Biological and Chemical Sciences, Queen Mary University of London, London, UK*

JULIEN Y. DUTHEIL • *Department of Evolutionary Genetics, Max Planck Institute of Evolutionary Biology, Plön, Germany*

PETER EBERT • *Max Planck Institute for Informatics, Saarbrücken, Saarland, Germany*

SCOTT V. EDWARDS • *Department of Organismic and Evolutionary Biology & Museum of Comparative Zoology, Harvard University, Cambridge, MA, USA*

CHRISTOPHER FIELDS • *Institute for Genomic Biology, University of Illinois at Urbana-Champaign, Urbana, IL, USA*

SOFIA K. FORSLUND • *EMBL Heidelberg, Heidelberg, Germany; Max Delbrück Centre for Molecular Medicine, Berlin, Germany*

CLÉMENTINE M. FRANÇOIS • *Institut des Sciences de l'Evolution, UMR5554, Université Montpellier II, Montpellier, France*

NICOLAS GALTIER • *Institut des Sciences de l'Evolution, UMR5554, Université Montpellier II, Montpellier, France*

LAURENT GAUTIER • *DMAC, Center for Biological Sequence Analysis, Department of Systems Biology, Technical University of Denmark, Kongens Lyngby, Denmark*

MANUEL GIL • *ZHAW Zurich University of Applied Sciences, Winterthur, Switzerland; SIB Swiss Institute of Bioinformatics, Lausanne, Switzerland*

GEORGE GITHINJI • *KEMRI Wellcome Trust Research Programme, Kilifi, Kenya*

SYLVAIN GLÉMIN • *Institut des Sciences de l'Evolution, UMR5554, Université Montpellier II, Montpellier, France*

NATASHA M. GLOVER • *Swiss Institute of Bioinformatics, Lausanne, Switzerland; Department of Computational Biology, University of Lausanne, Lausanne, Switzerland; Center for Integrative Genomics, University of Lausanne, Lausanne, Switzerland*

JOSEFA GONZÁLEZ • *Institute of Evolutionary Biology (CSIC-Universitat Pompeu Fabra), Barcelona, Spain*

ANNA GÓRSKA • *Algorithms in Bioinformatics, Faculty of Computer Science, University of Tübingen, Tübingen, Germany*

VALER GOTEA • *National Human Genome Research Institute, National Institutes of Health, Bethesda, MD, USA*

NAOHISA GOTO • *Department of Genome Informatics, Genome Information Research Center, Research Institute for Microbial Diseases, Osaka University, Osaka, Japan*

LAIN GUIO • *Institute of Evolutionary Biology (CSIC-Universitat Pompeu Fabra), Barcelona, Spain*

Asger Hobolth • *Bioinformatics Research Center (BiRC), Aarhus University, Aarhus, Denmark*

Daniel H. Huson • *Algorithms in Bioinformatics, Faculty of Computer Science, University of Tübingen, Tübingen, Germany*

Ritsert C. Jansen • *Groningen Bioinformatics Centre, GBB, University of Groningen, Groningen, Netherlands*

Roel Janssen • *Department of Genetics, Center for Molecular Medicine, University Medical Center Utrecht, Utrecht University, Utrecht, The Netherlands*

Christopher T. Jones • *Department of Mathematics and Statistics, Dalhousie University, Halifax, NS, Canada*

Mateusz Kaduk • *Department of Biochemistry and Biophysics, Stockholm Bioinformatics Centre, Science for Life Laboratory, Stockholm University, Solna, Sweden*

Slim Karkar • *Sorbonne Universités, Institut de Biologie Paris-Seine, UPMC Université Paris 6, Paris, France; Department of Ecology, Evolution, and Natural Resources, School of Environmental and Biological Sciences, Rutgers, The State University of NJ, New Brunswick, NJ, USA*

Toshiaki Katayama • *Database Center for Life Science, Joint Support-Center for Data Science Research, Research Organization of Information and Systems, Chiba, Japan*

Eugene V. Koonin • *National Center for Biotechnology Information, National Library of Medicine, National Institutes of Health, Bethesda, MD, USA*

Sergei L. Kosakovsky Pond • *Institute for Genomics and Evolutionary Medicine, Temple University, Philadelphia, PA, USA*

Carolin Kosiol • *Centre of Biological Diversity, School of Biology, University of St Andrews, Fife, UK; Institut für Populationsgenetik, Vetmeduni Vienna, Wien, Austria*

Romain Lannes • *Sorbonne Universités, Institut de Biologie Paris-Seine, UPMC Université Paris 6, Paris, France*

Philippe Lemey • *Department of Microbiology and Immunology, Rega Institute, KU Leuven, Leuven, Belgium*

Michael Li • *Institute for Genomics and Evolutionary Medicine, Temple University, Philadelphia, PA, USA*

Liang Liu • *Department of Statistics, University of Georgia, Athens, GA, USA*

Philippe Lopez • *Sorbonne Universités, Institut de Biologie Paris-Seine, UPMC Université Paris 6, Paris, France*

Brittany Rife Magalis • *Institute for Genomics and Evolutionary Medicine, Temple University, Philadelphia, PA, USA*

Thomas Mailund • *Bioinformatics Research Centre, Aarhus University, Aarhus, Denmark*

Izabela Makałowska • *Institute of Anthropology, Adam Mickiewicz University, Poznań, Poland*

Wojciech Makałowski • *Institute of Bioinformatics, University of Muenster, Muenster, Germany*

Raphaël Méheust • *Sorbonne Universités, Institut de Biologie Paris-Seine, UPMC Université Paris 6, Paris, France*

Suparna Mitra • *Leeds Institute of Medical Research, University of Leeds, Microbiology, Old Medical School, Leeds General Infirmary, Leeds LS1 3EX, West Yorkshire, UK*

Steffen Möller • *Institute for Biostatistics and Informatics in Medicine and Ageing Research (IBIMA), Rostock University Medical Center, Rostock, Germany*

Megan K. Mulligan • *Department of Genetics, Genomics and Informatics, The University of Tennessee Health Science Center, Memphis, TN, USA*

Amit Pande • *Institute of Bioinformatics, University of Muenster, Muenster, Germany*
Jananan S. Pathmanathan • *Sorbonne Universités, Institut de Biologie Paris-Seine, UPMC Université Paris 6, Paris, France*
Dennis Pearl • *Department of Statistics, Pennsylvania State University, University Park, PA, USA*
Pjotr Prins • *Department of Genetics, Center for Molecular Medicine, University Medical Center Utrecht, Utrecht University, Utrecht, The Netherlands; Department of Genetics, Genomics and Informatics, The University of Tennessee Health Science Center, Memphis, TN, USA; Laboratory of Nematology, Department of Plant Science, Wageningen University, Wageningen, The Netherlands*
Pere Puigbò • *National Center for Biotechnology Information, National Library of Medicine, National Institutes of Health, Bethesda, MD, USA; Division of Genetics and Physiology, Department of Biology, University of Turku, Turku, Finland*
Andrew Rambaut • *Institute of Evolutionary Biology, University of Edinburgh, Edinburgh, UK*
Nicolas Rodrigue • *Department of Biology, Carleton University, Ottawa, ON, Canada; Institute of Biochemistry, Carleton University, Ottawa, ON, Canada; School of Mathematics and Statistics, Carleton University, Ottawa, ON, Canada*
Mikkel H. Schierup • *Bioinformatics Research Centre, Aarhus University, Aarhus, Denmark*
Stephen D. Shank • *Institute for Genomics and Evolutionary Medicine, Temple University, Philadelphia, PA, USA*
Juliane Siebourg • *Department of Biosystems Science and Engineering, ETH Zurich, Basel, Switzerland*
Ana Claudia Sima • *ZHAW Zurich University of Applied Sciences, Winterthur, Switzerland; University of Lausanne, Lausanne, Switzerland*
Geert Smant • *Laboratory of Nematology, Department of Plant Science, Wageningen University, Wageningen, the Netherlands*
Erik L. L. Sonnhammer • *Department of Biochemistry and Biophysics, Stockholm Bioinformatics Centre, Science for Life Laboratory, Stockholm University, Solna, Sweden*
Stephanie J. Spielman • *Institute for Genomics and Evolutionary Medicine, Temple University, Philadelphia, PA, USA*
Kurt Stockinger • *ZHAW Zurich University of Applied Sciences, Winterthur, Switzerland*
Francesco Strozzi • *Enterome Bioscience, Paris, France*
Marc A. Suchard • *Department of Human Genetics and Biomathematics, David Geffen School of Medicine, University of California, Los Angeles, CA, USA*
Edward Susko • *Department of Mathematics and Statistics, Dalhousie University, Halifax, NS, Canada*
Bjarni J. Vilhjálmsson • *Bioinformatics Research Centre, Aarhus University, Aarhus, Denmark*
Andrew K. Watson • *Sorbonne Universités, Institut de Biologie Paris-Seine, UPMC Université Paris 6, Paris, France*
Steven Weaver • *Institute for Genomics and Evolutionary Medicine, Temple University, Philadelphia, PA, USA*
Rob W. Williams • *Department of Genetics, Genomics and Informatics, The University of Tennessee Health Science Center, Memphis, TN, USA*
Scooter Willis • *Department of Computer & Information Science & Engineering, University of Florida, Gainesville, FL, USA*

Yuri I. Wolf • *National Center for Biotechnology Information, National Library of Medicine, National Institutes of Health, Bethesda, MD, USA*
Ricardo Wurmus • *BIMSB Scientific Bioinformatics Platform, Max Delbrück Center for Molecular Medicine, Berlin, Germany*
Ziheng Yang • *Department of Genetics, Evolution and Environment, University College London, London, UK*
Andrew Yates • *European Molecular Biology Laboratory, European Bioinformatics Institute, Wellcome Genome Campus, Hinxton, Cambridge, UK*
Noor Youssef • *Department of Biology, Dalhousie University, Halifax, NS, Canada*

Part I

Introduction: Bioinformatician's Primers

Chapter 1

Introduction to Genome Biology and Diversity

Noor Youssef, Aidan Budd, and Joseph P. Bielawski

Abstract

Organisms display astonishing levels of cell and molecular diversity, including genome size, shape, and architecture. In this chapter, we review how the genome can be viewed as both a structural and an informational unit of biological diversity and explicitly define our intended meaning of genetic information. A brief overview of the characteristic features of bacterial, archaeal, and eukaryotic cell types and viruses sets the stage for a review of the differences in organization, size, and packaging strategies of their genomes. We include a detailed review of genetic elements found outside the primary chromosomal structures, as these provide insights into how genomes are sometimes viewed as incomplete informational entities. Lastly, we reassess the definition of the genome in light of recent advancements in our understanding of the diversity of genomic structures and the mechanisms by which genetic information is expressed within the cell. Collectively, these topics comprise a good introduction to genome biology for the newcomer to the field and provide a valuable reference for those developing new statistical or computation methods in genomics. This review also prepares the reader for anticipated transformations in thinking as the field of genome biology progresses.

Key words Organism diversity, Viruses, Prokaryotes, Eukaryotes, Organelles, DNA, RNA, Protein, Regulatory DNA, Epigenetics, Plasmids, Transcription, Translation, DNA replication, Chromatin, Gene structure

1 Introduction

Following the introduction of the concept of the genome in 1920 [1], the field of genome science has grown to encompass a vast range of interconnected topics (e.g., nucleic acid chemistry, molecular structure, replication and expression biochemistry, mutational processes, evolutionary dynamics, and interactions with cellular processes). Although the notion of the genome as a fundamental biological unit has been with us for nearly a century, it is only within the last decade that genomics has emerged as a transformative discipline within biology and the health sciences [2]. Its rapid development was in large part due to advances in massively parallel next-generation sequencing [3], which yielded unprecedented levels of genomic data. Those data revealed extensive natural

Maria Anisimova (ed.), *Evolutionary Genomics: Statistical and Computational Methods*, Methods in Molecular Biology, vol. 1910, https://doi.org/10.1007/978-1-4939-9074-0_1,

variation in the way that genomes are structured and processed. This led modern biologists to reevaluate the fundamental definition of the genome.

The typical definition of the genome is often dualistic, referencing both structural features and its function to store and transmit biological information [4]. For example, the US National Institutes of Health (NIH) uses the following definition: "A genome is an organism's complete set of DNA, including all of its genes. Each genome contains all of the information needed to build and maintain that organism. In humans, a copy of the entire genome—more than three billion DNA base pairs—is contained in all cells that have a nucleus." This conception, as with many others, is structural with regard to physical features (viz., genes and DNA base pairs) and informational with regard to its role in carrying out cellular functions (viz., to build and maintain the organism). Through increased knowledge of genome diversity, the field has come to realize that both conceptions of the genome are sometimes insufficient [4]. We now understand that the physical structures of the genome can be transient and that the expression of information contained within a genome is often conditioned on non-genomic factors. The science of genome biology is entering a new era based on a deeper understanding of the relationship between genotype and phenotype [5].

The purpose of this review is to provide a condensed overview of genome biology and to anticipate transformations in thinking that will occur as the field progresses. The remainder of this article is structured into four parts, with the next section providing a brief overview of the diversity of organismal cell types. The two subsequent sections introduce the structural and informational aspects of genomes, respectively. In the final section, we reassess the definition of the genome through selected biological examples and conclude with an updated perspective on the nature of the genome as an informational entity.

2 Organism Diversity and Cell Types

Cells are the smallest living unit of an organism. All cells have three attributes in common: cell membrane, cytoplasm, and genome. Structurally, cells can be divided into two basic types: *prokaryotic* and *eukaryotic* cells. Eukaryotic cells tend to be more complex. They possess a nucleus and other membrane-bound *organelles*, which are specialized components in the cell that perform unique functions (e.g., nucleus, mitochondria, plastids). Conversely, prokaryotic cells lack membrane-bound organelles. Although similar in cell structure, prokaryotes include two fundamentally distinct domains: the eubacteria (true bacteria, often referred to simply as bacteria) and the archaea.

Cellular life is detected in almost every environment on Earth. As life has colonized and adapted to the vast number of niches, cells have evolved an incredible amount of diversity in regard to size [6], form [7], lifestyle [8, 9], and complexity [10]. Understanding the basis of such diversity remains one of the central aims of biology. Readers interested in the latest understanding of Earth's biodiversity, the unique characteristics of its organisms, and how both extant and extinct forms are related to each other are encouraged to explore the following resources: the University of California Museum of Paleontology "History of life through time" exhibit [11], the Tree of Life Web Project [12], the Encyclopedia of Life [13].

2.1 Viruses

Viruses are infectious agents of living cells that are unable to reproduce in the absence of a host. Viruses are not considered cellular entities since they lack two of the essential attributes that define a cell; they possess neither a cell membrane nor cytoplasm. The discovery of ***virophages***, viruses that parasitize other viruses, resurrected the debate on their classification as living organisms [14]. Some consider viruses to be living entities since they can be hosts to other viruses, with a virophage infection leading to the eventual death of the host virus, implying an initial "living" state [15]. The opposing view asserts that a virus' inability to reproduce outside of a cellular host makes them nonliving entities [16, 17]. Irrespective of their delineation as living or nonliving, viruses are relevant to this review as they possess genomes and are the most abundant biological replicators in the biosphere [18].

Outside of their host, viruses exist as viral particles (***virions***) consisting of a protein capsule that protects and encloses their genome. Once a virion has entered a host cell, it "hijacks" the host's cellular structures and processes to carry out the metabolically active phase of the viral life cycle. At this stage, the virus exhibits physiological properties reminiscent of living cells; they metabolize, grow, and reproduce. There is a wide range of viral lifestyles, with corresponding diversity in viral forms, sizes, hosts, and genomes [16]. The largest known virus, the mimivirus, was originally identified as an infectious agent of an amoeba [19] and can itself become a host for virophages [14]. To put this in context, the virion of a mimivirus can be larger than some prokaryotic cells [16]. At the other end of the scale are viruses such as the circoviruses, some of which have small genomes made up of less than 2000 nucleotides [20]. A more detailed account of viral diversity can be found at the ViralZone website [21].

2.2 Bacteria

The bacterial cell is prokaryotic, and it is relatively simple as compared to eukaryotic cells. It has no membrane-bound organelles, and the chromosome (usually one) is not separated from the other components of the cell. While predominantly unicellular, they often live in *biofilms*, a community of cells bound together by a secreted polymer matrix [22], displaying a range of cooperative behaviors [23]. They can also exhibit regulated differentiation into different cell types, where two cells with the same genome have different morphology and function [22, 24].

Only a very small fraction of bacterial diversity (less than 1%) can be cultured and grown in the laboratory [25]. The problem of uncultivable bacteria is a consequence of our limited knowledge of their physiological diversity and the interactions necessary for their growth [26]. To this end, efforts are being made to study bacteria in nature [27–29] but with limited progress given the immense metabolic diversity of bacteria. Even within the incomplete sampling of cultivable bacteria, there is considerable diversity in cell shape [30], mode of reproduction [9], and cell cycle regulation [31].

The bacterial *cell cycle* involves the coordination of genome replication and segregation of replicated copies into daughter cells, followed by cell division. In this way, the transmission of genetic material is "vertical" from one cell generation to the next. Under certain conditions, some bacteria, such as *E. coli*, can initiate a new round of genome replication prior to completion of cell division [32, 33], thereby resulting in an increase in the number of gene copies near the origin of replication as compared to loci replicated later [31]. Other bacteria, such as *Caulobacter*, maintain a tightly regulated cell cycle to ensure a single replication event per division [34]. Under optimal conditions, some species can complete their cell cycle every 20 min, implying that a single cell could produce more than a billion descendants in a mere 10 h. In addition to vertical transfer, genetic information can be transferred "horizontally" between unrelated cells via the processes of transformation, conjugation, or transduction [35]. An event that transfers gene(s) between different species (or cells) by any of these three processes is referred to as a *horizontal gene transfer* (*HGT*) event.

2.3 Archaea

Archaea are single-celled organisms that appear strikingly similar to bacteria under light and electron microscopes. Like bacteria they often have a single circular chromosome and lack a nucleus, and for a long period of time the archaea were wrongly categorized as bacteria. The first indication that the archaea might be a separate domain of life was obtained from phylogenetic analyses of the 16S rRNA gene [36]. Advancements in genome sequencing and analysis yielded further evidence of the evolutionary distinction between the bacterial and archaeal domains [37]. Despite their superficial cellular similarity to bacteria, the archaea have many molecular-level similarities to eukaryotes, leading researchers to hypothesize that the ancestor of the eukaryotes arose within the archaea [38].

Previously, archaea were assumed to be a minor group of organisms inhabiting extreme environments beyond the tolerance of bacteria (salt brines, hydrothermal vents, acidic and anoxic conditions, etc.). Through culture-independent methods, archaea were discovered to be much more widespread and metabolically diverse. Archaea are now known to inhabit the human gut, and through mutualistic community relationships, they play a key role in human health and metabolism [39–41]. There is increasing evidence for archaea playing a significant role in global nutrient cycling [42]. They contribute major mechanisms for anaerobic methane oxidation [42], ammonia oxidation [43], and other parts of the nitrogen cycle including nitrogen fixation [44]. The archaea also appear to be ecologically competitive with bacteria, as they make significant contributions to the microbial communities of non-extreme soil, aquatic, and marine environments [43, 45]. Although they can be highly abundant in such environments, archaeal diversity is greatest in the more extreme habitats [45].

Archaea possess an array of bacteria-like, eukaryote-like, and archaea-specific features. The archaeal cell wall is chemically and structurally diverse, yet they systematically lack a cell wall peptidoglycan, murein, that is ubiquitous among the bacteria [46, 47]. Their membrane lipids are chemically different from those found in either bacteria or eukaryotes [48], and they possess many novel enzymes that are required for the biosynthesis of their unique membranes [49, 50]. Consequently, most archeoviruses are unique to archaea [51]. Even structural appendages that initially appeared to be homologous to bacterial appendages are often structurally distinct and have different genetic basis than the bacterial counterparts [52–54]. At the biochemical level, the archaea use many sources of energy and are metabolically diverse, probably more so than either bacteria or eukaryotes [55].

2.4 Eukaryotes

All complex multicellular organisms are eukaryotes (animals, plants, fungi, red algae, and brown algae), as are many unicellular organisms [56, 57]. Eukaryotic cells are found in a wide diversity of sizes and shapes [58, 59]. They are generally larger and have a more complex internal organization than the bacteria and archaea. A key characteristic of the eukaryotic intracellular organization is the use of lipid membranes to separate their contents into different compartments [60, 61]. The bulk of the eukaryotic genetic material is surrounded by a nuclear envelope and is thus maintained in a separate organelle, the *nucleus*. This provides a fundamental perspective on how eukaryotic cells differ from bacterial and archaeal cells and has important consequences on the expression of eukaryotic genetic information.

In addition to the nucleus, other organelles (mitochondria and plastids) contain small genomes that encode additional genes. Both mitochondria and plastids originated from ancient endosymbiosis events between ancestral eukaryotic cells and bacterial organisms. Following these events, the invading bacteria underwent a process of genome reduction in which they transitioned from autonomous organisms to cell-dependent organelles [62].

Despite our familiarity with plants, animals, and fungi, the vast majority of eukaryotic diversity lies outside of those groups and is largely microbial [63]. These "other" eukaryotes are collectively called *protists.* They do not form a monophyletic group, i.e., protists do not from a phylogenetic group that is comprised of a shared common ancestor and all of its descendants [57, 64]. The term protist is used largely for convenience to classify all eukaryotes that are not plants, animals, or fungi. Protists embody extensive ecological and structural diversity and include several important groups of unicellular eukaryotes involved in human diseases [65]. For example, the unicellular apicomplexan eukaryote *Plasmodium* is the causative agent of malaria, which affects around 10% of the world population [65]. More positively, protist species are important primary producers and are an essential link in the ocean's biogeochemical cycles [66].

3 Genome Structure and Organization

The notion of the gene as the physical carrier of hereditary information existed years before its physical and chemical structures were known. In 1902, Sutton provided the first clear support for the chromosomal theory of inheritance, allocating genes to segments on chromosomes [67]. The modern view of the gene is more often focused on a particular chemical sequence of nucleic acids rather than a chromosomal locus, but the two are not independent. The genetic instructions encoded within an organism's nucleic acid molecules comprise the organism's *genotype.* The physical manifestation of such genetic information, which will depend on environmental interactions, comprises the organism's *phenotype.*

There are two types of nucleic acids: deoxyribonucleic acid (DNA) and ribonucleic acid (RNA). Both are polymers consisting of chains of nucleotides. Each *nucleotide* includes three components: a 5-carbon sugar, a phosphate group, and a nitrogenous base. A nitrogenous base together with the sugar (without the phosphate group) is called a *nucleoside.* The sugar component in RNA, ribose, is a normal sugar with one hydroxyl group (OH) attached to each carbon atom. Deoxyribose, the sugar present in DNA, differs only in the absence of one oxygen atom at the 2′ carbon atom (H instead of OH). This chemical difference is crucial for enabling enzymes to distinguish between RNA and DNA polymers. The 5′ sugar carbon carries a phosphate group and is referred to as the *5′ end* of the polynucleotide molecule (DNA or RNA). The *3′ end* has a free hydroxyl (OH) group that is available to form chemical bonds with other atoms. As a result, synthesis of DNA and RNA in the cell proceed through the addition of a nucleotide to a 3′ terminal hydroxyl group. The polynucleotides, therefore, exhibit directionality, and synthesis occurs in a 5′ to 3′ direction.

All living cells employ the double helical structure of DNA as a chemical means to store information. Each of the two longitudinal strands is an alternating sequence of phosphate and a 5-carbon sugar. At each sugar, the two strands are bridged by two nitrogenous bases, one purine molecule (of type adenine [A] or guanine [G]) and the other a pyrimidine molecule (of type cytosine [C], thymine [T], or uracil [U]). The chemical bridges between purine and pyrimidine molecules (called *base pairs*) are held together by hydrogen bonds. Each purine can be complemented by only one pyrimidine: A forms two hydrogen bonds with T (or U in RNA) and C forms three hydrogen bonds with G. These are referred to as the *canonical* or *Watson-Crick pairings*. Given this pairing pattern, the sequences of the double-stranded DNA are said to be *complementary*, and the sequence of one strand can be deduced from the sequence of its complementary strand. The order of the nitrogenous bases in DNA (or RNA) is what confers the meaning of the information encoded in the genome.

A vital feature of genetic information is its ability to be replicated and passed on to daughter cells. The core mechanisms that copy DNA are conserved in all three domains of cellular life: bacteria, archaea, and eukaryotes [68]. Accurate DNA replication is essential to produce viable offspring—too many alterations in the DNA impede the production of functional proteins, thereby increasing the chances of nonviable progeny. Therefore, most DNA replicates with high fidelity. However, mistakes do occur. In humans, on average one error occurs in 30 million bases copied per cell division [69]. The cells produced from these altered genes are called *mutants*.

Although all living things carry DNA, the processes through which genetic information is physically transferred from DNA to RNA (called *transcription*) and then used to create a polypeptide molecule with a unique sequence of amino acids (called *translation*) differ between domains of life. The lack of membrane-bound nuclei in prokaryotes permits the simultaneous occurrence of transcription and translation [70]. In eukaryotes, those processes are separated by the nuclear membrane; DNA is first transcribed to RNA in the nucleus, and the RNA product is subsequently translated to an amino acid sequence in the cytoplasm, ultimately leading to the construction of a protein.

Organisms from all domains of life, and many of the viruses that parasitize them, have a very large genome compared with the size of the cell or compartment to which it is confined. For instance, the human nuclear DNA consists of approximately three billion base pairs; when stretched out, it amounts to about 2 m of total DNA per cell. The average human cell size is merely 10 μm. The impressive ability to store DNA within the cell is possible through a process of *genome packaging*. In eukaryotes and some archaea, the DNA wraps around histone proteins to form *nucleosomes*. In humans, this results in a two-million-fold decrease in size, allowing the DNA to compact into the nucleus [68]. Prokaryotic DNA compaction is achieved using a combination of supercoiling, macromolecular crowding, and association with DNA-binding proteins [71]. The degree of the supercoiling used in prokaryotes varies considerably between different species.

Prokaryotic cells tend to have efficient genomes, with most of their genetic material composed of protein-coding regions. Archaeal genomes are, on average, more compact than bacterial genomes [72]. An increase in prokaryotic genome size is therefore often accompanied by an increase in the number of genes encoded. This trend is not evident in eukaryotes, for which there is little association between genome size and the number of protein-coding genes [73]. Consider the *E. coli* genome, more than 90% of its DNA encodes proteins. This is in stark contrast with the modest 2% protein-coding regions present in human DNA [74]. Most eukaryotic genomes are riddled with non-protein-coding regions (*see* Subheading 4.2 for an evolutionary mechanism). This results in them having larger genome sizes on average than prokaryotic cells [74].

3.1 Viral Genomes

Viruses use any combination of either RNA or DNA, either single- or double-stranded molecules, in either circular or linear forms, to encode their genetic instructions [75, 76]. The viral genetic material is typically referred to as *segments* rather than chromosomes. Viral genomes composed of multiple segments are referred to as *segmented*. When different strains of the same segmented viral species infect a cell, genomes from the different strains can mix to produce hybrids—a process known as *reassortment*. Hybrid flus such as the H1N1 swine influenza A virus originated in this way [77].

Viral strains package their genomes in various ways. Most DNA and RNA viruses with small genomes (<20 kb) employ energy-independent packaging systems where capsid assembly and genome condensation are coupled. One example is the RNA genome of the HIV retrovirus that, in the mature virion, forms a RNA-protein complex with one of the cleavage products of the Gag polyprotein [78]. Other viruses, such as the lambda bacteriophage, require ATP to pump their genome directly into a preassembled capsid [79]. The latter type of machinery is ubiquitous in bacterial viruses. Alternatively, large viruses package their genome using histone-like proteins that are critical for eukaryotic genome packaging [80]. For a review on genome packaging in viruses, *see* ref. 81.

3.2 Bacterial Genomes

Despite not being confined within a membrane-bound compartment, the prokaryotic genome will be unevenly distributed throughout the cell. It often clusters in an irregularly shaped viscous region known as the *nucleoid* that makes up about a quarter of the intracellular volume [82]. The organization and distribution of the nucleoid are dynamic and dependent on the growth rate and presence of antibiotics [83].

It was previously thought that all bacterial cells possessed a single circular chromosome. In 1989, the first linear bacterial chromosome was discovered in the spirochaete *Borrelia burgdorferi*, the causative agent of Lyme disease [84, 85]. Additionally, recent advancements have revealed that many cells retain multiple circular or linear chromosomes [86]. These often consist of a *primary chromosome*, which is larger and harbors a higher density of essential genes compared to the *secondary chromosome(s)* [87].

The replication of bacterial DNA initiates at a well-defined sequence, called the *origin of replication*. The proteins involved in replication bind to the origin site and DNA synthesis proceeds in both directions. Circular chromosomes require a single origin, and replication is terminated by either a stop signal or when the two replication forks meet [88]. Linear bacterial chromosomes typically have a central origin, and replication proceeds bidirectionally much as in circular chromosomes. However, replication enzymes are unable to synthesize new DNA at the ends of a linear chromosome, and this results in the gradual shortening of DNA after each replication event [89]. Linear chromosomes, therefore, require terminal structures known as *telomeres* to protect against DNA degradation. Telomeres are characterized by the presence of multiple tandem repeats of short noncoding nucleotide sequences.

Linear prokaryotic chromosomes have evolved two different types of telomeres [90]. The first, best understood in the streptomycetes, uses a terminal protein complex covalently attached to the 5′ end of the DNA molecules. During replication, DNA polymerase binds the first synthesized nucleotide directly to the terminal protein. This replication strategy allows for the complete duplication of the linear molecule with no loss of genetic information [91]. The second type, best studied in the spirochetes, involves the formation of closed hairpin structures at the termini [92]. Replication of the linear DNA proceeds as expected. Once duplication of each DNA strand is completed the newly synthesized DNA are temporarily still attached—forming a structure superficially resembling a circular chromosome. A specific enzyme is then recruited to separate the two linear strands and re-form the telomeres [93]. For an overview of telomeric structures, *see* ref. 94.

3.3 Archaeal Genomes

Archaeal genomes share features with both bacteria and eukaryotes. Archaea typically possess circular chromosomes reminiscent of bacteria genomes; some have a single chromosome and a single origin of replication, while other species have multiple chromosomes and multiple origins on each [95, 96]. Given that archaea have the prokaryote cell type that lacks membrane-bound organelles (and hence nuclei), they are similar to bacteria in permitting the simultaneous occurrence of transcription and translation. Nonetheless, there are fundamental differences from the bacteria in the processing of genomic information. The initiation of amino acid synthesis in archaea more closely resembles that used in the eukaryotic transcription process. Additionally, the core archaeal transcription machineries are more closely related to eukaryotes [97, 98]. Archaeal and eukaryotic DNA replication and repair systems have also been shown to have many features in common [99].

Relatively little is known about the structure of archaeal genomes [100], but some are packaged into chromatin via histone proteins. *Chromatin* is a compact and organized chromosome structure that consists of DNA in close association with proteins. Interestingly, this form of chromatin is present in all eukaryotes and missing from bacteria [101]. Among the archaea that use histones (i.e., *Thermoproteales* and *Euryarchaea*), the geometry of their histone-mediated chromatin is the same as in eukaryotes [102]. However, archaeal histones are often shorter than the eukaryotic histones [101]. Groups of archaea that lack histones (e.g., *Crenarchaea*) encode other DNA-binding proteins associated with the architecture of bacterial chromatin [100]. Another family of DNA-binding proteins called Alba (acetylation lowers binding affinity) is ubiquitous among archaea. They are abundant small proteins that facilitate genome compaction, play a key role in determining the architecture of archaeal chromatin, and regulate gene expression on a genomic scale [101]. Alba proteins have been detected in both histone-lacking and histone-containing archaea [103].

3.4 Eukaryotic Genomes

Eukaryotes sequester their linear chromosomes within a membrane-bound nucleus. Linear eukaryotic chromosomes have three essential structural elements: a centromere, a pair of telomeres, and origins of replication. The *centromere* is the attachment point for spindle microtubules—the filaments responsible for physically moving chromosomes during cell division. Telomeres are the protective ends of a linear chromosome. The origins of replication are the sites where DNA synthesis begins. Eukaryotes typically have multiple linear chromosomes, each with many origins of replication. The larger genome size and slower replication machinery in eukaryotes necessitate the need of multiple origins to speed up the replication process.

In eukaryotic cells, nuclear DNA compaction involves the association of DNA with the protein products of a family of genes, the histones, whose sequence variants provide for a variety of different functions. The eukaryotic chromosome is organized at the lowest level by wrapping the DNA around histones, forming nucleosomes. This structure constitutes the basic unit of the chromatin fiber, which is further organized into higher-order structures mediated by other proteins [104, 105]. Sequence variation in histones, in combination with posttranslational modification of the protein, affects the structural properties of chromosomal nucleosomes and gene expression.

Eukaryotic DNA consists of at least three types of sequences: unique-sequence DNA, moderately repetitive DNA, and highly repetitive DNA. *Unique-sequence DNA* are regions that are present only once or at most a few times in the genome. Most protein-coding regions fall within this category. Alternatively, more than half of the total DNA in all eukaryotic genomes is made up of repeated sequence motifs that are either moderately or highly repetitive [106]. *Moderately repetitive DNA* are sequences from 160 to 180 base pairs (bp) in length that are repeated thousands of times [106]. Some of these sequences perform important functions for the cell, such as coding for types of RNA [107]. *Highly repetitive DNA* are short sequences, less than 60 bp that are present in hundreds of thousands of copies repeated throughout the genome. Repeats that are 2–10 bp are known as *microsatellites*, whereas motifs that are 10–60 bp are termed *minisatellites* [108].

Most of the repetitive sequences arise through transposition (*see* Subheading 4.2). The repeated sequences can be found either in *tandem arrays*, i.e., appearing adjacent to each other, or interspersed throughout the genome. The evolution and maintenance of nonfunctional repeated sequences have spurred the interest of genome scientists, with some classifying these motifs as *selfish-genes* that reproduce to propagate themselves and provide no positive contribution to the organism's phenotype or fitness [106]. Repeats also represent technical challenges for bioinformaticians developing software for sequence alignment and genome assembly. From a computational perspective, repeats create ambiguities that are challenging to resolve. For a review on computational challenges and solutions, *see* ref. 108.

3.5 Auxiliary DNA Structures

Both prokaryotes and eukaryotes have secondary chromosomal structures. For eukaryotes, this refers to any form of DNA found outside of a nucleus—although the discovery of microDNA extends this classification [109]. Eukaryotic auxiliary DNA often contains essential genes that are necessary for normal cell production. For example, the DNA chromosome located within the mitochondrial organelle encodes genes that are involved in oxidative phosphorylation and the creation of different types of RNA [110]. For prokaryotes, auxiliary DNA refers to any DNA that is not associated with the primary chromosome, and unlike eukaryotes, the genes encoded in such DNA are often dispensable. For example, small circular chromosomes, called plasmids, often contain genes that allow the bacterium to survive various environmental conditions; however they are not usually essential for normal cell function [110].

3.5.1 Mitochondrial DNA

The mitochondrion is a double membrane-bound organelle that is ubiquitous in eukaryotic cells. There is only one known case of a eukaryotic cell able to survive without a mitochondrion [111]. Mitochondria are essential because they are the site of production for most of the cell's energy, which is produced as ATP by the oxidative phosphorylation metabolic pathway. Additionally, the mitochondrion is the site of iron-sulfur (Fe/S) cluster assembly. Fe/S clusters are protein cofactors that are essential for various extramitochondrial pathways [112]. The mitochondria-lacking eukaryote, a species of *Monocercomonoides*, is unique in that it lives only within the intestine of the chinchilla and has evolved different strategies for Fe/S cluster formation and obtaining energy absorbed from its environment [111].

Mitochondria are the derivatives of prokaryotic cells that were engulfed by a common ancestor of all eukaryotes. The DNA within these organelles are the remnants of the DNA genome of the ancestral prokaryotic endosymbiont. Thus, the mitochondrial DNA (mtDNA) more closely resembles a prokaryotic genome. For example, in most animals and fungi, mtDNA consists of a single circular chromosome. However, small linear mtDNA chromosomes with defined telomeres have been identified within various protists, animals, and fungi [113, 114]. Additionally, the architecture of mtDNA is not determined by histones but instead by a set of small DNA-binding proteins that induce structures analogous to the bacterial chromatin. Mitochondrial genomes have been categorized into six different types depending on shape, size, structure, and number (*see* ref. 115).

In humans, the mitochondrial genome encodes 13 of the 80 proteins that are directly involved in oxidative phosphorylation. The remaining proteins are encoded in the nuclear chromosomes [110]. The exact contribution from mitochondrial and nuclear genomes varies across eukaryotes. Nonetheless, in the vast majority of known eukaryotic species, the mtDNA is essential to produce important proteins involved in energy production, demanding that all cells have faithfully inherited the mtDNA.

3.5.2 Plastid DNA

Plastids are similarly derived from an endosymbiosis with a bacterium, with the organelle retaining remnants of that ancestral bacterial genome. Like the mitochondrion, the plastid is a double membrane-bound cytoplasmic organelle. Unlike the mitochondrion, plastids often contain pigment used in photosynthesis. Plastids are found in the cytoplasm of protists and all higher plants. Plastid DNA (ptDNA) is highly reduced relative to the genomes of extant photosynthetic bacteria. In part, the reduction in genome size is due to gene loss with some regions excised and incorporated into the host nuclear DNA [116]. The ptDNA encodes important proteins that are essential for cell viability [117]. Almost all plastids have circular DNA, with the alveolate *Chromera velia* being the single known case of linear ptDNA. The linear extrachromosomal ptDNA has a telomere arrangement resembling those of linear mtDNA [117, 118].

Genes encoded in ptDNA are involved in the synthesis and storage of various cellular components, including those necessary for photosynthesis. Plastids have diverged to carry out different functions with multiple types identified. For example, *chloroplasts* are specialized for carrying out photosynthesis; *chromoplasts* contain pigments that provide petal colors, whereas *amyloplasts* are used for bulk storage of starch [117].

3.5.3 Nucleomorph DNA

A nucleomorph is a vestigial eukaryotic nucleus found in cryptomonads and chlorarachniophytes, which are both plastid-containing algae. The nucleomorph is located in these organisms between the inner and outer membranes of the plastid and is believed to be derived from the nucleus of an endosymbiotic algal cell engulfed by a larger eukaryotic cell [119]. Thus, the plastid organelle in this case evolved from two endosymbiotic events: a prokaryote was engulfed by a eukaryote which thereby became photoautotrophic and that cell was then engulfed by another eukaryote. The nucleomorph genomes are extremely small compared to the typical nuclear genome, being comprised of mostly single-copy housekeeping genes and having no mobile elements. The nucleomorph genome of the cryptomonads suggests that it was derived from a red algal ancestor, whereas the nucleomorph genome of the chlorarachniophytes suggests a green algal ancestor [119].

3.5.4 Plasmid DNA

Plasmids are present in bacteria, archaea, and eukaryotes [120]. Most plasmids are circular, although linear plasmids have been identified [121]. The genes carried on plasmids tend to be associated with functions that enable or enhance survival and growth under specific conditions. They can be horizontally transferred between prokaryotic cells and represent an important vehicle for sharing genetic information [122]. For example, a plasmid that has evolved an antibiotic resistance gene(s) can be transferred to neighboring bacteria promoting their rapid adaptation to various stresses associated with an antibiotic environment.

The eubacteria *E. coli* is estimated to have more than 270 plasmids having different distributions among and within cells; some promote mating, while others contain genes that kill other bacteria. The number of plasmids known and sequenced is much higher in bacteria as compared to archaea, with the lowest number having been identified in eukaryotes [122]. In recent years, plasmids have been used extensively in genetic engineering as a means of introducing and modifying target genes [122, 123].

3.5.5 MicroDNA

In 2012, Shibata et al. discovered a new form of extrachromosomal DNA in eukaryotes, called microDNA [122]. In contrast with other auxiliary DNA, microDNA is derived from non-repetitive sequences that are often associated with functional genes. They are circular DNA between 200 and 400 bp and are found in the nuclei of mammalian cells [122]. microDNA is thought to be associated with the repair and maintenance processes of nuclear DNA. It is not yet clear if microDNA plays a functional role in these processes or if they are merely an unavoidable by-product. For the time being, detection of specific microDNA is being proposed as a screening measure to aid the successful eradication of tumors in humans and as a potential method for cancer diagnosis and prognosis [124].

4 Genomic Storage and Processing of Information

It was not possible to understand how hereditary information was encoded and transmitted across generations without first having knowledge of the structure of DNA. Knowledge of DNA structure led to a structure-oriented conception of genomes as linear sequences of ordered nucleotides. Once protein synthesis was linked to gene sequences, the structural view of the genome began to be supplanted by the informational view [125]. Genetic information was initially viewed as a static property belonging to the specific sequence of ordered subunits. However, others have argued that the static view of information is not satisfactory (e.g., [4, 125]). Barbieri [125] contends that "it is only when a sequence provides a guideline to a copymaker that it becomes information for it. It is only an act of copying, in other words, that brings organic information into existence." Based on Barbieri's viewpoint, information is not always a property of a specific structure (e.g., DNA or RNA); rather his view is that such molecules are information relevant only when they are used to perform a biological function. A DNA sequence, for example, is said to have information if it is transcribed or interacts with a protein in a biologically relevant way. Similarly, an mRNA transcript also encodes information as it is translated into a protein. Also then, a protein could be viewed as an informational entity in the sense that it is necessary to carry out a biological function. Therefore, under this new conception, as well as the static view, it is clear that biological information can be manifest in different biological molecules; an observation that has complicated the notion of the genome as the fundamental unit of biological information [4].

We now understand that storage of the genetic information required to sustain life does not need to be restricted to biological molecules. This was vividly illustrated in the laboratory when a bacterial genome was chemically sequenced, its information stored within a computer (a completely different medium composed of binary states), then resynthesized in the form of a new DNA chromosome, and that synthetic DNA ultimately used as the sole means to maintain a living cell [126]. Although the information required for life can be stored independently of the chemical structure of the DNA, it cannot be expressed in a biologically useful form without various proteins and RNA molecules. Thus, expression of information encoded within a genome (bringing that information into existence) is contingent on its cellular context. In this section, we examine different ways in which information may be contained within a genome and mechanisms that result in biologically useful expression of that information.

4.1 Gene Expression

Mere knowledge of the DNA sequence of a genome is often insufficient to predict phenotype. The amount and timing of gene expression play a key role. For example, human cells with a nucleus have copies of almost identical DNA sequences. Yet cells perform varying functions, and they organize to create the multiple organs that constitute the human body. Cells achieve this primarily by differentially regulating the rate of transcription and/or translation of genes.

DNA transcription and protein translation comprise elementary levels of information transfer from genotype to phenotype. Maintaining control of these processes is fundamental for all organisms. Genetic elements involved in regulating gene expression are referred to as *regulatory elements*. They often represent sequences found on the DNA or RNA. In this way, regulatory information can be encoded directly within the nucleic acid sequence. Direct structural proximity if often not necessary, as regulatory elements may be found proximal or distal to the genes they affect. In humans, approximately 8% of nuclear DNA is composed of elements involved directly in regulation such as promoters, enhancers, silencers, and insulators (defined in Subheading 4.1.1; [127, 128]).

If *all* genetic and regulatory information is encoded in the DNA sequence, why can't any cell with a complete genome be used to produce a viable organism? The specificity of cells suggests that additional regulatory markers also exist outside of the primary DNA sequence. This type of regulation is *epigenetic* (above the genes) and is essential for normal development. Epigenetic information is derived from chemical modifications of the chromosome (e.g., DNA methylation or histone modification) that do not change the primary sequence of chromosomal DNA and can be passed from one generation to the next [129, 130]. It is only through the collective actions of all cellular processes that gene products contribute to biochemical pathways and participate in the network of regulatory interactions to produce a complex organism or phenotype.

4.1.1 Transcriptional Regulation

DNA transcription is the chemical process through which information is transferred from DNA to RNA. The transcribed RNA may itself carry out some biological function or may be part of an intermediate information-carrying class of RNA known as messenger RNA (mRNA). mRNA along with other RNA molecules (tRNA and rRNA) are part of the machinery used to synthesize proteins. The flow of genetic information from DNA to RNA to protein is present in all forms of life. However, it is important to note that information transfer is not exclusively unidirectional. The enzyme *reverse transcriptase* can transfer genetic information from an RNA template into DNA.

The basic model of transcriptional regulation requires that regulatory proteins called *transcriptional factors* (TFs) bind specific DNA sequences in *regulatory modules* (RMs). TFs are protein products that are themselves subjected to regulation of gene expression. RMs are defined according to both the primary DNA sequence to which TFs bind and their role in the process of regulating gene expression. One type of RMs are *promoters*. They are specific motifs on DNA that are necessary regulatory elements for RNA transcription in prokaryotes and eukaryotes. They bind the basal transcriptional machinery, RNA polymerase and general TFs. *Enhancers* are RMs that bind activator proteins and enhance the affinity of RNA polymerase to the promoter region. They, therefore, result in an upregulation of transcription of a gene or set of genes. Enhancers often act by stabilizing RNA polymerase binding through structural histone modifications [131]. *Silencers* are regulatory elements that when bound to repressor proteins function to prevent gene transcription. Silencers and enhancers are often distance-independent, meaning that they can act on gene(s) that are proximal or distal to their location [132]. Enhancers can be thought of as *on*-switches for gene expression, whereas silencers are the *off*-switches.

4.1.2 Translational Regulation

The fate of all mRNAs, transcribed from protein-coding genes, is not the same. The mRNA is often subjected to translational regulation depending on cellular and environmental conditions. These regulatory mechanisms affect the rate of protein synthesis. In prokaryotes and eukaryotes, most translational regulation involves structural changes in the mRNA molecule that impact its accessibility [133, 134]. The mRNAs can be sequestered in stress granules or localized in specific regions of a cell's cytoplasm [135–137]. Another mechanism of translational regulation is

RNA interference (RNAi). This regulation strategy is common in eukaryotes and involves short noncoding RNAs—microRNA (miRNA) or small interfering RNA (siRNA)—that bind with imperfect complementarity to their target mRNA transcripts. The binding of miRNA (or siRNA) to mRNA destabilizes (or degrades) the target mRNA, thereby inhibiting its translation. The imperfect pairing allows a single RNAi molecule to affect the expression of multiple genes. In the human genome, almost 50% of mRNA transcripts are regulated by one or more miRNAs [138].

In prokaryotes, transcription and translation are more tightly coupled than in eukaryotes, and this allows prokaryotes to regulate their gene expression primarily by controlling the amount of transcription. Nevertheless, prokaryotes can still conduct translational regulation. They can employ fundamentally different types of translational regulatory machinery: the recently discovered CRISPR-Cas system. Although the CRISPR loci were first identified in prokaryotes in 1987 [139], it was only recently described as a bacterial immune defense system [140]. The CRISP-Cas system is most commonly known to target external DNA (viral or plasmid) and degrade it before it can be transcribed or translated. Recent advancement suggests that some CRISPR-Cas systems are more general and have the capacity to target RNA molecules. This was first discovered in *Pyrococcus furiosus* [141]; similar RNA targeting was later found in *Sufolobus solfataricus* [142]. Throughout these advancements, CRISPR-Cas system was still strictly viewed as an immune response to target and degrade external nucleic acid molecules. It was only in 2016 that a CRISPR-Cas system was discovered that targets cellular mRNAs and thereby participates in translational regulation [143].

4.1.3 Epigenetics

The term epigenetics was coined in 1942 by Waddington [144]. He defined it as changes in an organism's phenotype without an underlying alteration of its genome. It is now understood that epigenetic effects cause variation in phenotypes not associated with a change in the primary sequence but by chemical alterations of the DNA. Consider this analogy: throughout this review, whenever a word was being defined it was written in this *format*. If this chapter was rewritten with all bolds and italics removed, the informational content would be unaltered; however, the emphasis would be different. These "decorative" changes in font are akin to chemical epigenetic markers appended to the DNA. DNA methylation is a type of chemical decoration that is analogous to striking through a phrase. Specifically, it corresponds to the addition of a methyl group to parts of the DNA that results in gene silencing [145]. This additional information is not directly encoded within the primary DNA sequence but is manifested through chemical changes in nucleotides [145]. Thus, DNA methylation is one form of epigenetic control of gene expression. Epigenetic factors may also have an impact on regulation by changing protein-DNA binding. In eukaryotes, epigenetic factors may bind to consecutive histones moving them closer to each other. This results in local DNA compaction and prevents the expression of the gene(s) in this location.

Importantly, an organism's exposure to certain environmental conditions can impact the epigenetic markers on its genome. Because epigenetic mechanisms ultimately affect the physiological form of the chromosome, such environmental exposures can lead to heritable changes in gene expression with no change to the underlying DNA sequence. It was initially thought that these alterations are not heritable and that following fertilization all epigenetic markers are removed from the zygote genome. Accumulating evidence suggests that such erasure of epigenetic marks occurs for most but not all genes [129, 130].

4.2 Mobile Genetic Elements

Also known as transposons or jumping genes, mobile genetic elements are sequences that can move around within a genome independently of the complex networks which otherwise regulate gene expression [146]. Through their movement, transposons often cause mutations either by inserting into a gene and disturbing its function or by promoting DNA rearrangement. If a transposon is inserted within a protein-coding region, then it will undoubtedly affect the expression of this gene by altering the final protein product. Transposons may also be inserted into regulatory regions resulting in over- or under-expression of certain gene(s). The capability of these DNA sequences to produce new copies of themselves elsewhere in a genome is called *transposition*. The two types of transposition are:

Copy-and-paste (replicative) transposition: a new copy of the transposable element is inserted into a new site, while the old copy remains integrated into the original site [147]. This type of transposition requires transfer of information into an RNA intermediate (retrotransposons) and subsequent retrotranscription into DNA. This mechanism results in an increase in the number transposon copies.

Cut-and-paste (non-replicative or conservative) transposition: the transposable element is excised from the old site and is inserted into a new site in the genome. The number of transposons is not increased in this case [147].

Transposable elements are found in all cell types. The kinds of transposable elements vary within and between prokaryotes and eukaryotes. They are often viewed as genetic parasites since they rely on a host cell for information processing systems (replication, transcription, and/or translation). In humans, about 44% of the genome is comprised of sequences that are related to transposable elements [148]. These mobile genetic elements had an important impact on eukaryotic evolution [149, 150]. For example, siRNA regulation is believed to have evolved to regain control of the expression of transposable elements [151]. For a review of the regulatory mechanisms of transposable elements, *see* ref. 152.

5 The Role of the Genome as an Informational Entity in Biology

Although the information contained within a genome is necessary to maintain a living cell, it is not sufficient on its own. Expression of biologically useful information requires a complex network of cellular components for processing and regulation of the genome. This dependency on external cellular components permits considerable flexibility in how the information is stored. As we have seen, the information essential for eukaryotic life is partitioned between chromosomes located in nuclear and organelle compartments, with some nuclear-encoded proteins being transported to the organelle for assembly with other proteins synthesized within the organelle [110]. Thus, as long as the cellular mechanisms for expression and processing are in place, genomic information can be physically dispersed within the cell. The Cryptophytes have taken this to an extreme, having their genomic information distributed across four cellular compartments: the nucleus, nucleomorph, mitochondria, and plastids [153]. Clearly, the physical location of the genome is not a constraint to information storage and processing. Furthermore, the storage of that information need not remain in a particular physical location. In the case of temperate phages, genomic information is transferred, for a period of time, to the genome of its host where it is maintained by its host's replication processes [154]. These examples, and others (e.g., [126]), underscore the importance of viewing the genome foremost as an informational entity irrespective of its physical location.

In a well-argued critique of conventional notions of the genome, Goldman and Landweber [4] argue that viewing DNA as the sole source of information leads to additional difficulties. Recall that the NIH definition refers to the genome as containing *all of the information needed to build and maintain that organism.* We now understand that even the cell and its associated cytoplasm are not always sufficient for realization of all functional capabilities encoded within a genome. In other words, the genome, as conventionally defined, appears to be an incomplete informational entity [4]. Genome research has identified a variety of extracellular informational entities that can influence, and in some cases are even essential to, the creation and maintenance of an organism. Below we review selected examples of this phenomenon prior to reassessing the definition of the genome in light of modern genome science.

Marine cyanobacteria (*Prochlorococcus* and *Synechococcus*) are among the most abundant photosynthetic organisms in the world's oceans. The viruses that infect them (cyanophages) were discovered to possess copies of some of their hosts photosynthesis genes (e.g., PsbA and PsbD: [155, 156]). Through the process of HGT, the cyanophages acquired host genes, which they express after infection to optimize their own gene expression and broaden their host range [157]. As novel as this discovery was, it was completely unexpected that the cyanobacteria and their phages continued to exchange genetic variation through homologous recombination [157]. Through such exchanges, the PsbA and PsbD genes participate in gene pools that extend beyond the photosynthetic species boundaries [157]. Given that cyanobacteria contribute as much as 30% of carbon fixation worldwide, those findings suggest that viral gene pool dynamics have influenced the evolution of oceanic photosynthesis on a global scale. This case demonstrates that to fully understand the origin and distribution of photosynthetic diversity, one must be aware that relevant genetic information can reside outside of the genomes of the photosynthetic organisms.

The bacterial genus *Listeria* is comprised of ecologically divergent lineages that share gene pools through the process of homologous recombination [158, 159]. *Listeria monocytogenes* is a pathogen closely related to the nonpathogenic species *L. innocua*. *L. monocytogenes* evolved as a pathogen through the process of HGT [160] and then subsequently evolved into ecologically divergent lineages differing in population structure and ability to respond to environmental stress [161]. Among *Listeria*, recombination is frequent enough to permit natural selection to act independently of the variability present at unlinked loci, thereby promoting or impeding exchangeability of genes among species and ecotypes residing in different niches [159]. This is just one example of the "mosaic genome" model of prokaryotic genome evolution, where the combined effects of recombination, drift, and selection lead to genomes comprised of a mosaic of differentially extendible trans-species gene pools. A wide variety of bacterial species are now thought to have genome dynamics consistent with the mosaic genome model [159, 162–165]. In some cases, the process of genomic divergence can even become decoupled from the process of ecological divergence [159, 163]. Thus, the physical genomes of some species of prokaryotes are incomplete informational entities.

The single-celled stichotrichous ciliates *Oxytricha* and *Stylonychia* have two nuclei that store genomic information in very different forms [166]. One nucleus, called the *macronucleus*, contains information in the form required for growth and maintenance of a cell. Hence, the macronuclear DNA is often referred to as "active." The second nucleus, called the *micronucleus*, contains the same information in a "stored" form, which is used to produce the active form of the DNA in the next generation. However, information storage in the micronucleus is extremely complex. Protein-coding genes expressed by the macronucleus are partitioned into small segments, inverted, and scrambled among ~1 GB of other DNA sequences within the micronucleus. Furthermore, the production of a working macronucleus in the next generation cannot be accomplished without information contained within both small RNA molecules (piRNA) and long RNA templates (lncRNA), which are passed across generations via the cytoplasm of the maternal macronucleus [167, 168]. The piRNA are crucial to the elimination of DNA during the development of an active macronucleus, and the lncRNA mediate (1) unscrambling of the inactive micronuclear DNA, (2) regulation of gene dosage in the macronucleus, and (3) epigenetic transfer of somatic (macronuclear) alterations that are not found within the germ-line (micronuclear) DNA [167]. Thus, without those RNA molecules, the DNA genome of the stichotrichous ciliates is an incomplete informational entity [4]. Furthermore, emerging work on both *Oxytricha* and *Stylonychia* suggests that epigenetic modification of their DNA may play a role in the production of active macronuclear DNA [166, 169–171]

Complex microbial communities live in close association with the human body and have a strong impact on human health and disease. Host genetic variation is known to influence the composition of those communities [172], and, conversely, microbial variability is thought to influence various host disease states [173]. This association is so intimate that the microbiome has been referred to as an additional "human organ" [174], and substantial amounts of missing heritability associated with many complex human diseases are now being attributed, in part, to a failure to adequately account for microbial genetic variation [175]. Taking inflammatory bowel disease (IBD) as an example, host human genetic variation accounts for less than 50% of its estimated heritability [176]. This result implies that there exists undiscovered context dependence of human genetic variation for IBD. We have since come to understand that there is extensive inter-individual variation in the genetic composition of the gut microbiome and this metagenomic variation can influence healthy and dysregulated human immune responses [177] and is predictive of IBD patient outcomes [178]. Because the development of the IBD phenotype is related to gut microbiome variability, and because genetically similar human hosts can have different microbiomes, heritability estimates for human DNA variation will be impacted [175]. In other words, the expression of similar IBD phenotypes in humans is a function of both human and microbial genetics. Regardless of whether such interactions should be formally included within any future conception of the genome, this example illustrates how the human genome is also an incomplete informational entity with respect to prediction of healthy and disease states.

Goldman and Landweber [4] suggest that the notion of the genome should be reconceptualized in light of our modern, and deeper, understanding of genomic diversity and the mechanisms of information storage and processing. We agree and follow Goldman and Landweber [4] when they call for a "more expansive definition of the genome as an informational entity, often but not always manifest as DNA, encoding a broad set of functional capabilities that, together with other sources of information, produce and maintain the organism." At first glance, this appears to be consistent with the controversial idea that a collection of functionally integrated organisms, called a *holobiont*, is a fundamental unit of biological organization and their set of genomes, called a *hologenome*, is itself a unit subject to evolution by natural selection [179]. However, we cannot go this far. We expect that any hologenome composed of informational entities having even a little independence is analogous to intra-genomic epistasis with just a little recombination. In the latter case, adaptive coevolution is not very effective at moving the system on its fitness landscape via compensatory substitutions [180]. Further, when informational entities are largely independent, either through high recombination (as observed in *Listeria*) or through independent replication (as within the human gut microbiome), the process of genomic divergence can become decoupled from ecological dynamics. Thus, we cannot agree with the notion of the hologenome as a unit of selection. Rather, we view the genome as a potential mosaic of gene pools subject to different evolutionary dynamics, and we follow Goldman and Landweber [4] by considering it foremost as an informational entity, which may be incomplete and which does not have to manifest exclusively as the DNA within a species boundary.

References

1. Lederberg J, McCray AT (2001) Ome SweetOmics--a genealogical treasury of words. Scientist 15(7):8
2. Patra S, Andrew AA (2015) Human, social, and environmental impacts of human genetic engineering. J Biomed Sci 4:2
3. Behjati S, Tarpey PS (2013) What is next generation sequencing? Arch Dis Child Educ Pract Ed 98(6):236–238
4. Goldman AD, Landweber LF (2016) What is a genome? PLoS Genet 12(7):e1006181
5. Tyler-Smith C, Yang H, Landweber LF, Dunham I, Knoppers BM, Donnelly P et al (2015) Where next for genetics and genomics? PLoS Biol 13(7):e1002216
6. Mueller RL (2015) Genome biology and the evolution of cell-size diversity. Cold Spring Harb Perspect Biol 7(11):a019125
7. Kysela DT, Randich AM, Caccamo PD, Brun YV (2016) Diversity takes shape: understanding the mechanistic and adaptive basis of bacterial morphology. PLoS Biol 14(10): e1002565
8. Minelli A, Fusco G (2010) Developmental plasticity and the evolution of animal complex life cycles. Philos Trans R Soc Lond B Biol Sci 365(1540):631–640
9. Forster SC (2017) Illuminating microbial diversity. Nat Rev Microbiol 15(10):578

10. Carroll SB (2001) Chance and necessity: the evolution of morphological complexity and diversity. Nature 409(6823):1102
11. History of life through time UCMP. www.ucmp.berkeley.edu/exhibits/historyoflife.php
12. The tree of life web project. tolweb.org
13. The encyclopedia of life. eol.org
14. Claverie J, Abergel C (2009) Mimivirus and its virophage. Annu Rev Genet 43:49–66
15. Pearson H (2008) 'Virophages' suggests viruses are alive. Nature 454(7205):677
16. Forterre P (2010) Defining life: the virus viewpoint. Orig Life Evol Biosph 40 (2):151–160
17. Koonin EV, Starokadomskyy P (2016) Are viruses alive? The replicator paradigm sheds decisive light on an old but misguided question. Stud Hist Philos Biol Biomed Sci 59:125–134
18. Koonin EV (2010) The wonder world of microbial viruses. Expert Rev Anti-Infect Ther 8(10):1097–1099
19. Raoult D, Audic S, Robert C, Abergel C, Renesto P, Ogata H et al (2004) The 1.2-megabase genome sequence of Mimivirus. Science 306(5700):1344–1350
20. Finsterbusch T, Mankertz A (2009) Porcine circoviruses—small but powerful. Virus Res 143(2):177–183
21. Swiss Institute of Bioinformatics. ViralZone. http://www.expasy.org
22. Nadell CD, Drescher K, Foster KR (2016) Spatial structure, cooperation and competition in biofilms. Nat Rev Microbiol 14 (9):589–600
23. Rosenberg SM (2009) Life, death, differentiation, and the multicellularity of bacteria. PLoS Genet 5(3):e1000418
24. Flores E, Herrero A (2010) Compartmentalized function through cell differentiation in filamentous cyanobacteria. Nat RevMicrobiol 8(1):39
25. Lasken RS, McLean JS (2014) Recent advances in genomic DNA sequencing of microbial species from single cells. Nat Rev Genet 15(9):577–584
26. Stewart EJ (2012) Growing unculturable bacteria. J Bacteriol 194(16):4151–4160
27. Qin Y, Hou J, Deng M, Liu Q, Wu C, Ji Y, He X (2016) Bacterial abundance and diversity in pond water supplied with different feeds. Sci Rep 6:35232
28. Jovel J, Patterson J, Wang W, Hotte N, O'Keefe S, Mitchel T et al (2016) Characterization of the gut microbiome using 16S or shotgun metagenomics. Front Microbiol 7:459
29. Peterson BF, Scharf ME (2016) Metatranscriptome analysis reveals bacterial symbiont contributions to lower termite physiology and potential immune functions. BMC Genomics 17(1):772
30. Young KD (2006) The selective value of bacterial shape. Microbiol Mol Biol Rev 70 (3):660–703
31. Willis L, Huang KC (2017) Sizing up the bacterial cell cycle. Nat Rev Microbiol 15 (10):606–620
32. Haeusser DP, Levin PA (2008) The great divide: coordinating cell cycle events during bacterial growth and division. Curr Opin Microbiol 11(2):94–99
33. Thanbichler M (2010) Synchronization of chromosome dynamics and cell division in bacteria. Cold Spring Harb Perspect Biol 2 (1):a000331
34. Brown PJ, Hardy GG, Trimble MJ, Brun YV (2008) Complex regulatory pathways coordinate cell-cycle progression and development in Caulobacter crescentus. Adv Microb Physiol 54:1–101
35. Frost LS, Leplae R, Summers AO, Toussaint A (2005) Mobile genetic elements: the agents of open source evolution. Nat Rev Microbiol 3(9):722
36. Woese CR, Fox GE (1977) Phylogenetic structure of the prokaryotic domain: the primary kingdoms. Proc Natl Acad Sci U S A 74 (11):5088–5090
37. Hug LA, Baker BJ, Anantharaman K, Brown CT, Probst AJ, Castelle CJ et al (2016) A new view of the tree of life. Nat Microbiol 1:16048
38. Williams TA, Foster PG, Cox CJ, Embley TM (2013) An archaeal origin of eukaryotes supports only two primary domains of life. Nature 504:231–236
39. Eckburg PB, Lepp PW, Relman DA (2003) Archaea and their potential role in human disease. Infect Immun 71(2):591–596
40. Nakamura N, Lin HC, McSweeney CS, Mackie RI, Gaskins HR (2010) Mechanisms of microbial hydrogen disposal in the human colon and implications for health and disease. Annu Rev Food Sci Technol 1:363–395
41. Saengkerdsub S, Ricke SC (2014) Ecology and characteristics of methanogenic archaea

in animals and humans. Crit Rev Microbiol 40 (2):97–116
42. Jarrell KF, Walters AD, Bochiwal C, Borgia JM, Dickinson T, Chong JP (2011) Major players on the microbial stage: why archaea are important. Microbiology 157 (4):919–936
43. Prosser JI, Nicol GW (2008) Relative contributions of archaea and bacteria to aerobic ammonia oxidation in the environment. Environ Microbiol 10(11):2931–2941
44. Leigh JA (2000) Nitrogen fixation in methanogens: the archaeal perspective. Curr Issues Mol Biol 2:125–131
45. DeLong EF (1998) Everything in moderation: archaea as 'non-extremophiles'. Curr Opin Genet Dev 8(6):649–654
46. Kandler O, König H (1998) Cell wall polymers in archaea (archaebacteria). Cell Mol Life Sci 54(4):305–308
47. Vollmer W, Bertsche U (2008) Murein (peptidoglycan) structure, architecture and biosynthesis in Escherichia coli. Biochim Biophys Acta 1778(9):1714–1734
48. Kates M (1992) Archaebacterial lipids: structure, biosynthesis and function. Biochem Soc Symp 58:51–72
49. Sato T, Atomi H (2011) Novel metabolic pathways in archaea. Curr Opin Microbiol 14(3):307–314
50. Lombard J, López-García P, Moreira D (2012) Phylogenomic investigation of phospholipid synthesis in archaea. Archaea 2012:630910
51. Forterre P (2013) The common ancestor of archaea and eukarya was not an archaeon. Archaea 2013:372396
52. Zillig W (1991) Comparative biochemistry of archaea and bacteria. Curr Opin Genet Dev 1 (4):544–551
53. Moissl C, Rachel R, Briegel A, Engelhardt H, Huber R (2005) The unique structure of archaeal 'hami', highly complex cell appendages with nano-grappling hooks. Mol Microbiol 56(2):361–370
54. Szabo Z, Stahl AO, Albers SV, Kissinger JC, Driessen AJ, Pohlschroder M (2007) Identification of diverse archaeal proteins with class III signal peptides cleaved by distinct archaeal prepilin peptidases. J Bacteriol 189 (3):772–778
55. Siebers B, Schönheit P (2005) Unusual pathways and enzymes of central carbohydrate metabolism in archaea. Curr Opin Microbiol 8(6):695–705
56. Coelho SM, Peters AF, Charrier B, Roze D, Destombe C, Valero M, Cock JM (2007) Complex life cycles of multicellular eukaryotes: new approaches based on the use of model organisms. Gene 406(1):152–170
57. Adl SM, Simpson AG, Lane CE, Lukeš J, Bass D, Bowser SS et al (2012) The revised classification of eukaryotes. J Eukaryot Microbiol 59(5):429–514
58. Mathur J (2004) Cell shape development in plants. Trends Plant Sci 9(12):583–590
59. Mogilner A, Keren K (2009) The shape of motile cells. Curr Biol 19(17):R771
60. Fagarasanu A, Rachubinski RA (2007) Orchestrating organelle inheritance in Saccharomyces cerevisiae. Curr Opin Microbiol 10(6):528–538
61. Bornens M (2008) Organelle positioning and cell polarity. Nat RevMol Cell Biol 9(11):874
62. Dyall SD, Brown MT, Johnson PJ (2004) Ancient invasions: from endosymbionts to organelles. Science (New York, NY) 304 (5668):253–257
63. Corliss JO (2002) Biodiversity and biocomplexity of the protists and an overview of their significant roles in maintenance of our biosphere. Acta Protozool 41(3):199–220
64. Schlegel M, Hülsmann N (2007) Protists–a textbook example for a paraphyletic taxon. Organisms Divers Evol 7(2):166–172
65. Parfrey LW, Walters WA, Knight R (2011) Microbial eukaryotes in the human microbiome: ecology, evolution, and future directions. Front Microbiol 2:153
66. Caron DA, Alexander H, Allen AE, Archibald JM, Armbrust EV, Bachy C et al (2017) Probing the evolution, ecology and physiology of marine protists using transcriptomics. Nat Rev Microbiol 15(1):6–20
67. Sutton WS (1902) On the morphology of the chromoso group in Brachystola magna. Biol Bull 4(1):24–39
68. O'Donnell M, Langston L, Stillman B (2013) Principles and concepts of DNA replication in bacteria, archaea, and eukarya. Cold Spring Harb Perspect Biol 5(7):10
69. Dolgin E (2009) Human mutation rate revealed. Nat News. https://doi.org/10.1038/news.2009.864

70. Gowrishankar J, Harinarayanan R (2004) Why is transcription coupled to translation in bacteria? Mol Microbiol 54(3):598–603
71. Griswold A (2008) Genome packaging in prokaryotes: the circular chromosome of E. coli. Nat Educ 1(1):57
72. Koonin EV, Wolf YI (2008) Genomics of bacteria and archaea: the emerging dynamic view of the prokaryotic world. Nucleic Acids Res 36(21):6688–6719
73. Hou Y, Lin S (2009) Distinct gene number-genome size relationships for eukaryotes and non-eukaryotes: gene content estimation for dinoflagellate genomes. PLoS One 4(9): e6978
74. Cooper GM (2000) The cell: a molecular approach, 2nd edn. Sinauer Associates, Sunderland, MA
75. Gelderblom HR (1996) Structure and classification of viruses. In: Baron S (ed) Medical microbiology, 4th edn. The University of Texas Medical Branch at Galveston, Galveston, TX
76. Kay A, Zoulim F (2007) Hepatitis B virus genetic variability and evolution. Virus Res 127(2):164–176
77. Trifonov V, Khiabanian H, Rabadan R (2009) Geographic dependence, surveillance, and origins of the 2009 influenza A (H1N1) virus. N Engl J Med 361(2):115–119
78. Ganser-Pornillos BK, Yeager M, Sundquist WI (2008) The structural biology of HIV assembly. Curr Opin Struct Biol 18 (2):203–217
79. Sun S, Rao VB, Rossmann MG (2010) Genome packaging in viruses. Curr Opin Struct Biol 20(1):114–120
80. Thomas V, Bertelli C, Collyn F, Casson N, Telenti A, Goesmann A et al (2011) Lausannevirus, a giant amoebal virus encoding histone doublets. Environ Microbiol 13 (6):1454–1466
81. Chelikani V, Ranjan T, Kondabagil K (2014) Revisiting the genome packaging in viruses with lessons from the "Giants". Virology 466:15–26
82. Teif VB, Bohinc K (2011) Condensed DNA: condensing the concepts. Prog Biophys Mol Biol 105(3):208–222
83. Chai Q, Singh B, Peisker K, Metzendorf N, Ge X, Dasgupta S, Sanyal S (2014) Organization of ribosomes and nucleoids in Escherichia coli cells during growth and in quiescence. J Biol Chem 289 (16):11342–11352
84. Baril C, Richaud C, Baranton G, Saint Girons I (1989) Linear chromosome of Borrelia burgdorferi. Res Microbiol 140(7):507–516
85. Ferdows MS, Barbour AG (1989) Megabase-sized linear DNA in the bacterium Borrelia burgdorferi, the lyme disease agent. Proc Natl Acad Sci U S A 86(15):5969–5973
86. Rocha EP (2008) The organization of the bacterial genome. Annu Rev Genet 42:211–233
87. Cooper VS, Vohr SH, Wrocklage SC, Hatcher PJ (2010) Why genes evolve faster on secondary chromosomes in bacteria. PLoS Comput Biol 6(4):e1000732
88. Worning P, Jensen LJ, Hallin PF, Stærfeldt H, Ussery DW (2006) Origin of replication in circular prokaryotic chromosomes. Environ Microbiol 8(2):353–361
89. Nandakumar J, Cech TR (2013) Finding the end: recruitment of telomerase to telomeres. Nat Rev Mol Cell Biol 14(2):69–82
90. Cui T, Moro-oka N, Ohsumi K, Kodama K, Ohshima T, Ogasawara N et al (2007) Escherichia coli with a linear genome. EMBO Rep 8 (2):181–187
91. Hopwood DA (2006) Soil to genomics: the Streptomyces chromosome. Annu Rev Genet 40:1–23
92. Casjens S (1999) Evolution of the linear DNA replicons of the Borrelia spirochetes. Curr Opin Microbiol 2(5):529–534
93. Chaconas G, Kobryn K (2010) Structure, function, and evolution of linear replicons in Borrelia. Annu Rev Microbiol 64:185–202
94. Fulcher N, Derboven E, Valuchova S, Riha K (2014) If the cap fits, wear it: an overview of telomeric structures over evolution. Cell Mol Life Sci 71(5):847–865
95. Samson RY, Bell SD (2011) Cell cycles and cell division in the archaea. Curr Opin Microbiol 14(3):350–356
96. Samson RY, Bell SD (2014) Archaeal chromosome biology. J Mol Microbiol Biotechnol 24(5–6):420–427
97. Bell SD, Jackson SP (2001) Mechanism and regulation of transcription in archaea. Curr Opin Microbiol 4(2):208–213
98. Reeve JN (2003) Archaeal chromatin and transcription. Mol Microbiol 48(3):587–598

99. Kelman LM, Kelman Z (2003) Archaea: an archetype for replication initiation studies? Mol Microbiol 48(3):605–615
100. Bell SD, White MF (2010) Archaeal chromatin organization. Bacterial chromatin. Springer, New York, pp 205–217
101. White MF, Bell SD (2002) Holding it together: chromatin in the archaea. Trends Genet 18(12):621–626
102. Mattiroli F, Gu Y, Yadav T, Balsbaugh JL, Harris MR, Findlay ES et al (2017) DNA-mediated association of two histone-bound complexes of yeast Chromatin Assembly Factor-1 (CAF-1) drives tetrasome assembly in the wake of DNA replication. elife 6: e22799
103. Forterre P, Confalonieri F, Knapp S (1999) Identification of the gene encoding archeal-specific DNA-binding proteins of the Sac10b family. Mol Microbiol 32(3):669–670
104. Zlatanova J, Caiafa P (2009) CCCTC-binding factor: to loop or to bridge. Cell Mol Life Sci 66(10):1647–1660
105. Tark-Dame M, van Driel R, Heermann DW (2011) Chromatin folding--from biology to polymer models and back. J Cell Sci 124 (Pt 6):839–845
106. Biscotti MA, Olmo E, Heslop-Harrison JP (2015) Repetitive DNA in eukaryotic genomes. Chromosom Res 23(3):415–420
107. Rubin GM, Spradling AC (1982) Genetic transformation of drosophila with transposable element vectors. Science (New York, NY) 218(4570):348–353
108. Treangen TJ, Salzberg SL (2011) Repetitive DNA and next-generation sequencing: computational challenges and solutions. Nat Rev Genet 13(1):36–46
109. Shibata Y, Kumar P, Layer R, Willcox S, Gagan JR, Griffith JD, Dutta A (2012) Extrachromosomal microDNAs and chromosomal microdeletions in normal tissues. Science (New York, NY) 336(6077):82–86
110. Chen XJ, Butow RA (2005) The organization and inheritance of the mitochondrial genome. Nat RevGenet 6(11):815
111. Karnkowska A, Vacek V, Zubáčová Z, Treitli SC, Petrželková R, Eme L et al (2016) A eukaryote without a mitochondrial organelle. Curr Biol 26(10):1274–1284
112. Stehling O, Lill R (2013) The role of mitochondria in cellular Iron-sulfur protein biogenesis: mechanisms, connected processes, and diseases. Cold Spring Harb Perspect Biol 5(8):a011312
113. Nosek J, Tomáška Ľ (2003) Mitochondrial genome diversity: evolution of the molecular architecture and replication strategy. Curr Genet 44(2):73–84
114. Smith DR, Keeling PJ (2013) Gene conversion shapes linear mitochondrial genome architecture. Genome Biol Evol 5 (5):905–912
115. Kolesnikov AA, Gerasimov ES (2012) Diversity of mitochondrial genome organization. Biochemistry 77(13):1424
116. Green BR (2011) Chloroplast genomes of photosynthetic eukaryotes. Plant J 66 (1):34–44
117. Rogalski M, do Nascimento Vieira L, Fraga HP, Guerra MP (2015) Plastid genomics in horticultural species: importance and applications for plant population genetics, evolution, and biotechnology. Front Plant Sci 6:586
118. Janouškovec J, Sobotka R, Lai D, Flegontov P, Koník P, Komenda J et al (2013) Split photosystem protein, linear-mapping topology, and growth of structural complexity in the plastid genome of Chromera velia. Mol Biol Evol 30(11):2447–2462
119. Archibald JM (2007) Nucleomorph genomes: structure, function, origin and evolution. BioEssays 29(4):392–402
120. Funnell BE, Phillips GJ (2004) Plasmid biology. ASM Press, Washington, DC
121. Ravin NV (2011) N15: the linear phage–plasmid. Plasmid 65(2):102–109
122. Shintani M, Sanchez ZK, Kimbara K (2015) Genomics of microbial plasmids: classification and identification based on replication and transfer systems and host taxonomy. Front Microbiol 6:242
123. Burgess DJ (2017) Genetic engineering: CREATE-ing genome-wide designed mutations. Nat Rev Genet 18(2):69
124. Kumar P, Dillon LW, Shibata Y, Jazaeri AA, Jones DR, Dutta A (2017) Normal and cancerous tissues release extrachromosomal circular DNA (eccDNA) into the circulation. Mol Cancer Res 15(9):1197–1205
125. Barbieri M (2016) What is information? Philos Trans R Soc A 374(2063):20150060
126. Gibson DG, Glass JI, Lartigue C, Noskov VN, Chuang RY, Algire MA et al (2010) Creation of a bacterial cell controlled by a chemically synthesized genome. Science 329 (5987):52–56
127. ENCODE Project Consortium (2012) An integrated encyclopedia of DNA elements in

the human genome. Nature 489 (7414):57–74

128. Rands CM, Meader S, Ponting CP, Lunter G (2014) 8.2% of the human genome is constrained: variation in rates of turnover across functional element classes in the human lineage. PLoS Genet 10(7):e1004525
129. Trerotola M, Relli V, Simeone P, Alberti S (2015) Epigenetic inheritance and the missing heritability. Hum Genomics 9(1):17
130. Tang WW, Dietmann S, Irie N, Leitch HG, Floros VI, Bradshaw CR et al (2015) A unique gene regulatory network resets the human germline epigenome for development. Cell 161(6):1453–1467
131. Atkinson TJ, Halfon MS (2014) Regulation of gene expression in the genomic context. Comput Struct Biotechnol J 9(13):1–9
132. Narlikar L, Ovcharenko I (2009) Identifying regulatory elements in eukaryotic genomes. Brief FunctGenomic Proteomic 8 (4):215–230
133. Hershey JW, Sonenberg N, Mathews MB (2012) Principles of translational control: an overview. Cold Spring Harb Perspect Biol 4 (12):a011528
134. Meyer MM (2017) The role of mRNA structure in bacterial translational regulation. Wiley Interdiscip Rev RNA 8(1):e1370
135. Chao JA, Yoon YJ, Singer RH (2012) Imaging translation in single cells using fluorescent microscopy. Cold Spring Harb Perspect Biol 4 (11):a012310
136. Lasko P (2012) mRNA localization and translational control in Drosophila oogenesis. Cold Spring Harb Perspect Biol 4(10): a012294
137. Decker CJ, Parker R (2012) P-bodies and stress granules: possible roles in the control of translation and mRNA degradation. Cold Spring Harb Perspect Biol 4(9):a012286
138. Chekulaeva M, Filipowicz W (2009) Mechanisms of miRNA-mediated post-transcriptional regulation in animal cells. Curr Opin Cell Biol 21(3):452–460
139. Ishino Y, Shinagawa H, Makino K, Amemura M, Nakata A (1987) Nucleotide sequence of the iap gene, responsible for alkaline phosphatase isozyme conversion in Escherichia coli, and identification of the gene product. J Bacteriol 169 (12):5429–5433
140. Makarova KS, Grishin NV, Shabalina SA, Wolf YI, Koonin EV (2006) A putative RNA-interference-based immune system in prokaryotes: computational analysis of the predicted enzymatic machinery, functional analogies with eukaryotic RNAi, and hypothetical mechanisms of action. Biol Direct 1(1):7
141. Hale CR, Zhao P, Olson S, Duff MO, Graveley BR, Wells L et al (2009) RNA-guided RNA cleavage by a CRISPR RNA-Cas protein complex. Cell 139(5):945–956
142. Zhang J, Rouillon C, Kerou M, Reeks J, Brugger K, Graham S et al (2012) Structure and mechanism of the CMR complex for CRISPR-mediated antiviral immunity. Mol cell 45(3):303–313
143. Liu Y, Chen Z, He A, Zhan Y, Li J, Liu L et al (2016) Targeting cellular mRNAs translation by CRISPR-Cas9. Sci Rep 6:29652
144. Waddington CH (1942) The epigenotype. Endeavour 1:18–20
145. Allis CD, Jenuwein T (2016) The molecular hallmarks of epigenetic control. Nat Rev Genet 17:487–500
146. Goodier JL, Kazazian HH (2008) Retrotransposons revisited: the restraint and rehabilitation of parasites. Cell 135(1):23–35
147. Ahmed A (2009) Alternative mechanisms for Tn5 transposition. PLoS Genet 5(8): e1000619
148. Mills RE, Bennett EA, Iskow RC, Devine SE (2007) Which transposable elements are active in the human genome? Trends Genet 23(4):183–191
149. Huda A, Jordan IK (2009) Epigenetic regulation of mammalian genomes by transposable elements. Ann N Y Acad Sci 1178 (1):276–284
150. López-Flores I, Garrido-Ramos MA (2012) The repetitive DNA content of eukaryotic genomes. In: Repetitive DNA, vol 7. Karger Publishers, Basel, pp 1–28
151. Ivics Z, Li MA, Mátés L, Boeke JD, Nagy A, Bradley A, Izsvák Z (2009) Transposon-mediated genome manipulation in vertebrates. Nat Methods 6(6):415–422
152. Chuong EB, Elde NC, Feschotte C (2017) Regulatory activities of transposable elements: from conflicts to benefits. Nat Rev Genet 18(2):71–86
153. Curtis BA, Tanifuji G, Burki F, Gruber A, Irimia M, Maruyama S et al (2012) Algal genomes reveal evolutionary mosaicism and the fate of nucleomorphs. Nature 492 (7427):59–65

154. Howard-Varona C, Hargreaves KR, Abedon ST, Sullivan MB (2017) Lysogeny in nature: mechanisms, impact and ecology of temperate phages. ISME J 11:1511–1520
155. Lindell D, Sullivan MB, Johnson ZI, Tolonen AC, Rohwer F et al (2004) Transfer of photosynthesis genes to and from *Prochlorococcus* viruses. Proc Natl Acad Sci U S A 101:11013–11018
156. Zeidner G, Bielawski JP, Shmoish M, Scanlan DJ, Sabehi G, Béjà O (2005) Potential photosynthesis gene recombination between Prochlorococcus and Synechococcus via viral intermediates. Environ Microbiol 7 (10):1505–1513
157. Sullivan MB, Lindell D, Lee JA, Thompson LR, Bielawski JP, Chisholm SW (2006) Prevalence and evolution of core photosystem II genes in marine cyanobacterial viruses and their hosts. PLoS Biol 4(8):e234
158. Orsi RH, Sun Q, Wiedmann M (2008) Genome-wide analyses reveal lineage specific contributions of positive selection and recombination to the evolution of Listeria monocytogenes. BMC Evol Biol 8(1):233
159. Dunn KA, Bielawski JP, Ward TJ, Urquhart C, Gu H (2009) Reconciling ecological and genomic divergence among lineages of Listeria under an "extended mosaic genome concept". Mol Biol Evol 26 (11):2605–2615
160. Buchrieser C, Rusniok C, Kunst F, Cossart P, Glaser P (2003) Comparison of the genome sequences of Listeria monocytogenes and Listeria innocua: clues for evolution and pathogenicity. Pathog Dis 35(3):207–213
161. Nightingale KK, Windham K, Wiedmann M (2005) Evolution and molecular phylogeny of Listeria monocytogenes isolated from human and animal listeriosis cases and foods. J Bacteriol 187(16):5537–5551
162. Lawrence JG (2002) Gene transfer in bacteria: speciation without species? Theor Popul Biol 61(4):449–460
163. Nesbø CL, Dlutek M, Doolittle WF (2006) Recombination in Thermotoga: implications for species concepts and biogeography. Genetics 172(2):759–769
164. Retchless AC, Lawrence JG (2007) Temporal fragmentation of speciation in bacteria. Science 317(5841):1093–1096
165. Papke RT, Zhaxybayeva O, Feil EJ, Sommerfeld K, Muise D, Doolittle WF (2007) Searching for species in haloarchaea. Proc Natl Acad Sci U SA 104 (35):14092–14097
166. Wang Y, Wang Y, Sheng Y, Huang J, Chen X, AL-Rasheid KA, Gao S (2017) A comparative study of genome organization and epigenetic mechanisms in model ciliates, with an emphasis on Tetrahymena, Paramecium and Oxytricha. Eur J Protistol 61(Pt B):376–387
167. Yerlici VT, Landweber LF (2014) Programmed genome rearrangements in the Ciliate Oxytricha. Microbiol Spectr 2:6. https://doi.org/10.1128/microbiolspec.MDNA3-0025-2014
168. Pilling OA, Rogers AJ, Gulla-Devaney B, Katz LA (2017) Insights into transgenerational epigenetics from studies of ciliates. Eur J Protistol 61(Pt B):366–375
169. Bulic A, Postberg J, Fischer A, Jönsson F, Reuter G, Lipps HJ (2013) A permissive chromatin structure is adopted prior to site-specific DNA demethylation of developmentally expressed genes involved in macronuclear differentiation. Epigenetics Chromatin 6(1):5
170. Bracht JR, Fang W, Goldman AD, Dolzhenko E, Stein EM, Landweber LF (2013) Genomes on the edge: programmed genome instability in ciliates. Cell 152 (3):406–416
171. Forcob S, Bulic A, Jönsson F, Lipps HJ, Postberg J (2014) Differential expression of histone H3 genes and selective association of the variant H3. 7 with a specific sequence class in Stylonychia macronuclear development. Epigenetics Chromatin 7(1):4
172. Blekhman R, Goodrich JK, Huang K, Sun Q, Bukowski R, Bell JT et al (2015) Host genetic variation impacts microbiome composition across human body sites. Genome Biol 16 (1):191
173. Clemente JC, Ursell LK, Parfrey LW, Knight R (2012) The impact of the gut microbiota on human health: an integrative view. Cell 148(6):1258–1270
174. Baquero F, Nombela C (2012) The microbiome as a human organ. Clin Microbiol Infect 18(s4):2–4
175. Sandoval-Motta S, Aldana M, Martínez-Romero E, Frank A (2017) The human microbiome and the missing heritability problem. Front Genet 8:80
176. Gordon H, Moller FT, Andersen V, Harbord M (2015) Heritability in inflammatory bowel disease: from the first twin study to genome-

wide association studies. Inflamm Bowel Dis 21(6):1428

177. Dunn KA, Moore-Connors J, MacIntyre B, Stadnyk A, Thomas NA, Noble A et al (2016) The gut microbiome of pediatric Crohn's disease patients differs from healthy controls in genes that can influence the balance between a healthy and dysregulated immune response. Inflamm Bowel Dis 22(11):2607–2618
178. Dunn KA, Moore-Connors J, MacIntyre B, Stadnyk AW, Thomas NA, Noble A et al (2016) Early changes in microbial community structure are associated with sustained remission after nutritional treatment of pediatric Crohn's disease. Inflamm Bowel Dis 22 (12):2853–2862
179. Bordenstein SR, Theis KR (2015) Host biology in light of the microbiome: ten principles of holobionts and hologenomes. PLoS Biol 13(8):e1002226
180. Gavrilets S (2004) Fitness landscapes and the origin of species (MPB-41), vol 41. Princeton University Press, Princeton, NJ

Chapter 2

Probability, Statistics, and Computational Science

Niko Beerenwinkel and Juliane Siebourg

Abstract

In this chapter, we review basic concepts from probability theory and computational statistics that are fundamental to evolutionary genomics. We provide a very basic introduction to statistical modeling and discuss general principles, including maximum likelihood and Bayesian inference. Markov chains, hidden Markov models, and Bayesian network models are introduced in more detail as they occur frequently and in many variations in genomics applications. In particular, we discuss efficient inference algorithms and methods for learning these models from partially observed data. Several simple examples are given throughout the text, some of which provide the basis for models that are discussed in more detail in subsequent chapters.

Key words Bayesian inference, Bayesian networks, Dynamic programming, EM algorithm, Hidden Markov models, Markov chains, Maximum likelihood, Statistical models

1 Statistical Models

Evolutionary genomics can only be approached with the help of statistical modeling. Stochastic fluctuations are inherent to many biological systems. Specifically, the evolutionary process itself is stochastic, with random mutations and random mating being major sources of variation. In general, stochastic effects play an increasingly important role if the number of molecules, or cells, or individuals of a population is small. Stochastic variation also arises from measurement errors. Biological data is often noisy due to experimental limitations, especially for high-throughput technologies, such as microarrays or next-generation sequencing [1, 2].

Statistical modeling addresses the following questions: What can be generalized from a finite sample obtained from an experiment to the population? What can be learned about the underlying biological mechanisms? How certain can we be about our model predictions?

In the frequentist view of statistics, the observed variability in the data is the result of a fixed true value being perturbed by

Maria Anisimova (ed.), *Evolutionary Genomics: Statistical and Computational Methods*, Methods in Molecular Biology, vol. 1910, https://doi.org/10.1007/978-1-4939-9074-0_2,

random variation, such as, for example, measurement noise. Probabilities are thus interpreted as long-run expected relative frequencies. By contrast, from a Bayesian point of view, probabilities represent our uncertainty about the state of nature. There is no true value, but only the data is real. Our prior belief about an event is updated in light of the data.

Statistical models represent the observed variability or uncertainty by probability distributions [3, 4]. The observed data are regarded as realizations of random variables. The parameters of a statistical model are usually the quantities of interest because they describe the amount and nature of systematic variation in the data. Parameter estimation and model selection are discussed in more detail in the next section. In this section, we first consider discrete, and then continuous random variables and univariate (1-dimensional) before multivariate (n-dimensional) ones. We start by formulating the well-known Hardy–Weinberg principle [5, 6] as a statistical model.

Example 1 (Hardy–Weinberg Model): The Hardy–Weinberg model is a statistical model for the genotypes in a diploid population of infinite size. Let us assume that there are two alleles, denoted A and a, and hence three genotypes, denoted AA, Aa = aA, and aa. Let X be the random variable with state space $\mathcal{X} = \{\text{AA, Aa, aa}\}$ describing the genotype. We parametrize the probability distribution of X by the allele frequency p of A and the allele frequency $q = 1 - p$ of a. The Hardy–Weinberg model is defined by:

$$P(X = \text{AA}) = p^2, \tag{1}$$

$$P(X = \text{Aa}) = 2p(1 - p), \tag{2}$$

$$P(X = \text{aa}) = (1 - p)^2. \tag{3}$$

The parameter space of the model is $\Theta = \{p \in \mathbb{R} \mid 0 \leq p \leq 1\} = [0,1]$, the unit interval. We denote the Hardy–Weinberg model by HW(p) and write $X \sim \text{HW}(p)$ if X follows the distribution (Eqs. 1–3). □

The Hardy–Weinberg distribution $P(X)$ is a discrete probability distribution (or probability mass function) with finite state space: We have $0 \leq P(X = x) \leq 1$ for all $x \in \mathcal{X}$ and $\sum_{x \in \mathcal{X}} P(X = x) = p^2 + 2p(1 - p) + (1 - p)^2 = [p + (1 - p)]^2 = 1$. In general, any statistical model for a discrete random variable with n states defines a subset of the $(n - 1)$-dimensional probability simplex:

$$\Delta_{n-1} = \left\{(p_1, \ldots, p_n) \in [0,1]^n \mid p_1 + \cdots + p_n = 1\right\}. \tag{4}$$

The probability simplex is the set of all possible probability distributions of X, and statistical models can be understood as specific subsets of the simplex [7].

The Hardy–Weinberg distribution is of interest because it arises under the assumption of random mating. A population with major allele frequency p has genotype probabilities given in Eqs. 1–3 after one round of random mating. We find that the new allele frequency:

$$p' = P(\mathrm{AA}) + P(\mathrm{Aa})/2 = p^2 + 2p(1-p)/2 = p, \tag{5}$$

is equal to the one in the previous generation. Thus, genetic variation is preserved under this simple model of sexual reproduction, and the population is at equilibrium after one generation. In other words, Eqs. 1–3 describe the set of all populations at Hardy–Weinberg equilibrium. The parametric representation:

$$\begin{aligned}\{(p_{\mathrm{AA}},p_{\mathrm{Aa}},p_{\mathrm{aa}}) \in \Delta_2 \mid p_{\mathrm{AA}} = p^2,\, p_{\mathrm{Aa}} = 2p(1-p),\\ p_{\mathrm{aa}} = (1-p)^2\},\end{aligned} \tag{6}$$

of this set of distributions is equivalent to the implicit representation as the intersection of the Hardy–Weinberg curve:

$$4\, p_{\mathrm{AA}}\, p_{\mathrm{aa}} - p_{\mathrm{Aa}}^2 = 0 \tag{7}$$

with the probability simplex Δ_2 (Fig. 1).

The simplest discrete random variable is a binary (or Bernoulli) random variable X. The textbook example of a Bernoulli trial is the flipping of a coin. The state space of this random experiment is the set that contains all possible outcomes, namely, whether the coin lands on heads ($X = 0$) or tails ($X = 1$). We write $\mathcal{X} = \{0,1\}$ to denote this state space. The parameter space is the set that contains all possible values of the model parameters. In the coin tossing example, the only parameter is the probability of observing tails, p, and this parameter can take any value between 0 and 1, so we write $\Theta = \{p \mid 0 \le p \le 1\}$ for the parameter space. In general, the event $X = 1$ is often called a "success," and $p = P(X = 1)$ the probability of success.

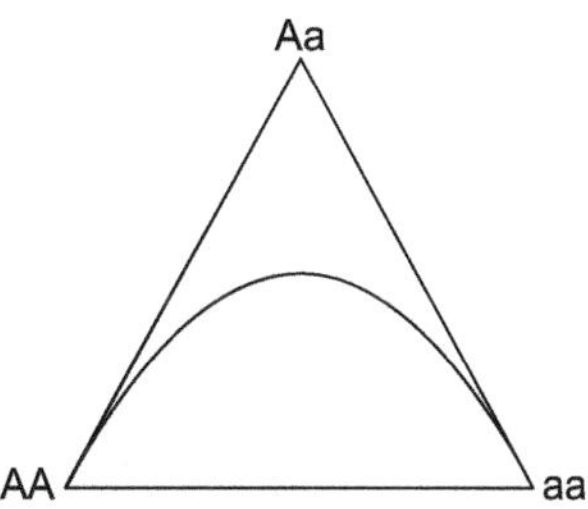

Fig. 1 De Finetti diagram showing the Hardy–Weinberg curve $4\, p_{\mathrm{AA}}\, p_{\mathrm{aa}} - p_{\mathrm{Aa}}^2 = 0$ inside the probability simplex $\Delta_2 = \{(p_{\mathrm{AA}}, p_{\mathrm{Aa}}, p_{\mathrm{aa}}) \mid p_{\mathrm{AA}} + p_{\mathrm{Aa}} + p_{\mathrm{aa}} = 1\}$. Each point in this space represents a population as described by its genotype frequencies. Points on the curve correspond to populations in Hardy–Weinberg equilibrium

Example 2 (Binomial Distribution): Consider n independent Bernoulli trials, each with success probability p. Let X be the random variable counting the number of successes k among the n trials. Then, X has state space $\mathcal{X} = \{0, \ldots, n\}$ and

$$P(X = k) = \binom{n}{k} p^k (1 - p)^{n-k}. \tag{8}$$

This is the binomial distribution, denoted $\mathrm{Binom}(n, p)$. Its parameter space is $\Theta = \mathbb{N} \times [0,1]$. Examples of binomially distributed random variables are the number of "heads" in n successive coin tosses or the number of mutated genes in a group of species. □

Important characteristics of a probability distribution are its expectation (or expected value, or mean) and its variance. They are defined, respectively, as:

$$\mathrm{E}(X) = \sum_{x \in \mathcal{X}} x\, P(X = x), \tag{9}$$

$$\mathrm{Var}(X) = \sum_{x \in \mathcal{X}} [x - \mathrm{E}(X)]^2\, P(X = x). \tag{10}$$

The standard deviation is $\sqrt{\mathrm{Var}(X)}$. For the binomial distribution, $X \sim \mathrm{Binom}(n, p)$, we find $\mathrm{E}(X) = np$ and $\mathrm{Var}(X) = np(1 - p)$.

Example 3 (Poisson Distribution): The Poisson distribution $\mathrm{Pois}(\lambda)$ with parameter $\lambda \geq 0$ is defined as:

$$P(X = k) \;=\; \frac{\lambda^k\, e^{-\lambda}}{k!}, \quad k \in \mathbb{N}. \tag{11}$$

It describes the number X of independent events occurring in a fixed period of time (or space) at average rate λ and independently of the time since (or distance to) the last event. The Poisson distribution has equal expectation and variance, $\mathrm{E}(X) = \mathrm{Var}(X) = \lambda$. □

The Poisson distribution is used frequently as a model for the number of DNA mutations in a gene after a certain time period, where λ is the mutation rate. Both the binomial and the Poisson distribution describe counts of random events. In the limit of large n and fixed product np, the two distributions coincide, $\mathrm{Binom}(n, p) \to \mathrm{Pois}(np)$, for $n \to \infty$.

Example 4 (Shotgun Sequencing): Let us consider a simplified model of the shotgun approach to DNA sequencing. Suppose that n reads of length L have been obtained from a genome of size G. We assume that all reads have the same probability of being sequenced. Then, the probability of hitting a specific base with one read is $p = L/G$, and the average coverage of the sequencing run is $c = np$. Under this model, the number of times X a single base is sequenced is distributed as $\mathrm{Binom}(n, p)$. For large n, we have

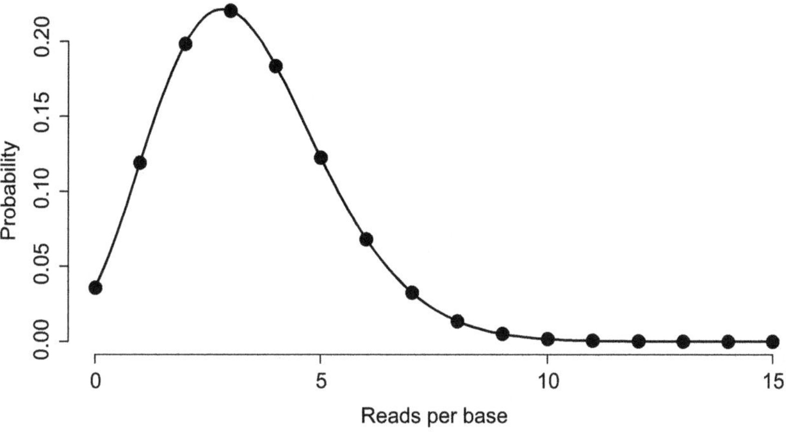

Fig. 2 Coverage distribution of a shotgun sequencing experiment with $n = 10^8$ reads of length $L = 100$ of the human genome of length $G = 3 \cdot 10^9$. The average coverage is $c = np = 3.4$, where $p = L/G$. Dots show the binomial coverage distribution Binom(n, p) and the solid line its approximation by the Poisson distribution Pois(np). Note that the Poisson distribution is also discrete and just shown as a line to distinguish it from the binomial distribution

$$P(X = k) = \binom{n}{k} p^k (1 - p)^{n-k} \approx \frac{c^k \, e^{-c}}{k!}. \tag{12}$$

For example, using next-generation sequencing technology, one might obtain $n = 10^8$ reads of length $L = 100$ bases in a single run. For the human genome of length $G = 3 \cdot 10^9$, we obtain a coverage of $c = 3.4$. The distribution of the number of reads per base pair is shown in Fig. 2. In particular, the fraction of unsequenced positions is $P(X = 0) = e^{-c} = 3.57\%$. □

A continuous random variable X takes values in $\mathcal{X} = \mathbb{R}$ and is defined by a nonnegative function $f(x)$ such that:

$$P(X \in B) = \int_B f(x)dx, \quad \text{for all subsets } B \subseteq \mathbb{R}. \tag{13}$$

The function f is called the probability density function of X. For an interval:

$$P(X \in [a,b]) = P(a \leq X \leq b) = \int_a^b f(x)dx. \tag{14}$$

The cumulative distribution function is

$$F(b) = P(X \leq b) = \int_{-\infty}^b f(x)dx, \quad b \in \mathbb{R}. \tag{15}$$

Thus, the density is the derivative of the cumulative distribution function, $\frac{d}{dx} F(x) = f(x)$.

In analogy to the discrete case, expectation and variance of a continuous random variable are defined, respectively, as:

$$\mathrm{E}(X) = \int_{-\infty}^{\infty} x\, f(x)\, dx, \tag{16}$$

$$\mathrm{Var}(X) = \int_{-\infty}^{\infty} [x - \mathrm{E}(X)]^2 f(x)\, dx. \tag{17}$$

Example 5 (Normal Distribution): The normal (or Gaussian) distribution has the density function:

$$f(x) = (2\pi\sigma^2)^{-1/2} \exp\left[-\frac{(x-\mu)^2}{2\sigma^2}\right]. \tag{18}$$

The parameter space is $\Theta = \{(\mu, \sigma^2) \mid \mu \in \mathbb{R},\ \sigma^2 \in \mathbb{R}_+\}$. A normal random variable $X \sim \mathrm{Norm}(\mu, \sigma^2)$ has mean $\mathrm{E}(X) = \mu$ and variance $\mathrm{Var}(X) = \sigma^2$. Norm(0,1) is called the standard normal distribution. □

The normal distribution is frequently used as a model for measurement noise. For example, $X \sim \mathrm{Norm}(\mu, \sigma^2)$ might describe the hybridization intensity of a sample to a probe on a microarray. Then, μ is the level of expression of the corresponding gene and σ^2 summarizes the experimental noise associated with the microarray experiment. The parameters can be estimated from a finite sample $\{x^{(1)}, \ldots, x^{(N)}\}$, i.e., from N replicate experiments, as the empirical mean and variance, respectively:

$$\overline{x} = \frac{1}{N} \sum_{i=1}^{N} x^{(i)}, \tag{19}$$

$$s^2 = \frac{1}{N-1} \sum_{i=1}^{N} (x^{(i)} - \overline{x})^2. \tag{20}$$

The normal distribution plays a special role in statistics due to the central limit theorem. It asserts that the average $\bar{X}_N = (X^{(1)} + \cdots + X^{(N)})/N$ of N independent (see below) and identically distributed (i.i.d.) random variables $X^{(i)}$ with equal mean μ and variance σ^2 converges in distribution to the standard normal distribution:

$$\sqrt{N}\left(\frac{\bar{X}_N - \mu}{\sigma}\right) \xrightarrow{d} \mathrm{Norm}(0,1), \tag{21}$$

irrespective of the shape of their distribution. As a consequence, many test statistics and estimators are asymptotically normally distributed. For example, the Poisson distribution $\mathrm{Pois}(\lambda)$ is approximately normal $\mathrm{Norm}(\lambda, \lambda)$ for large values of λ.

We often measure multiple quantities at the same time, for example the expression of several genes, and are interested in correlations among the variables. Let X and Y be two random variables with expected values μ_X and μ_Y and variances σ_X^2 and σ_Y^2, respectively. The covariance between X and Y is

$$\mathrm{Cov}(X, Y) = \mathrm{E}[(X - \mu_X)(Y - \mu_Y)] = \mathrm{E}[XY] - \mathrm{E}[X]\mathrm{E}[Y] \tag{22}$$

and the correlation between X and Y is $\rho_{X,Y} = \mathrm{Cov}(X, Y)/(\sigma_X \sigma_Y)$. For observations $(x^{(1)}, y^{(1)}), \ldots, (x^{(N)}, y^{(N)})$, the sample correlation coefficient is

$$r_{x,y} = \frac{\sum_{i=1}^{N} (x^{(i)} - \overline{x})(y^{(i)} - \overline{y})}{(N-1)s_X s_Y}, \tag{23}$$

where s_X and s_Y are the sample standard deviations of X and Y, respectively, defined in Eq. 20.

So far, we have worked with univariate distributions and we now turn to multivariate distributions, i.e., we consider random vectors $X = (X_1, \ldots, X_n)$ such that each X_i is a random variable. For the case of discrete random variables X_i, we first generalize the binomial distribution to random experiments with a finite number of outcomes.

Example 6 (Multinomial Distribution): Let K be the number of possible outcomes of a random experiment and θ_k the probability of outcome k. We consider the random vector $X = (X_1, \ldots, X_K)$ with values in $\mathcal{X} = \mathbb{N}^K$, where X_k counts the number of outcomes of type k. The multinomial distribution $\mathrm{Mult}(n, \theta_1, \ldots, \theta_K)$ is defined as:

$$P(X = x) = \frac{n!}{x_1! \cdots x_K!} \theta_1^{x_1} \cdots \theta_K^{x_K} \tag{24}$$

if $\sum_{k=1}^{K} x_k = n$, and 0 otherwise. The parameter space of the model is $\Theta = \mathbb{N} \times \Delta_{K-1}$. For $K = 2$, we recover the binomial distribution (Eq. 8). Each component X_k of a multinomial vector has expected value $\mathrm{E}(X_k) = n\theta_k$ and $\mathrm{Var}(X_k) = n\theta_k(1 - \theta_k)$. The covariance of two components is $\mathrm{Cov}(X_k, X_l) = -n\theta_k\theta_l$, for $k \neq l$. □

In general, the covariance matrix Σ of a random vector X is defined by:

$$\Sigma_{ij} = \mathrm{Cov}(X_i, X_j) = \mathrm{E}[(X_i - \mu_i)(X_j - \mu_j)], \tag{25}$$

where μ_i is the expected value of X_i. The matrix Σ is also called the variance–covariance matrix because the diagonal terms are the variances $\Sigma_{ii} = \mathrm{Cov}(X_i, X_i) = \mathrm{Var}(X_i)$.

A continuous multivariate random variable X takes values in $\mathcal{X} = \mathbb{R}^n$. It is defined by its cumulative distribution function:

$$F(x) = P(X \leq x), \quad x \in \mathbb{R}^n \tag{26}$$

or, equivalently, by the probability density function:

$$f(x) = \frac{\partial^n}{\partial x_1 \cdots \partial x_n} F(x_1, \ldots, x_n), \quad x \in \mathbb{R}^n. \tag{27}$$

Example 7 (Multivariate Normal Distribution): For $n \geq 1$ and $x \in \mathbb{R}^n$, the multivariate normal (or Gaussian) distribution has density:

$$f(x) = (2\pi)^{-n/2} \det(\Sigma)^{-1/2} \exp\left[-\frac{1}{2}(x-\mu)^t \Sigma^{-1}(x-\mu)\right], \tag{28}$$

with parameter space $\Theta = \{(\mu, \Sigma) \mid \mu = (\mu_1, \ldots, \mu_n) \in \mathbb{R}^n$ and $\Sigma = \left(\sigma_{ij}^2\right) \in \mathbb{R}^{n \times n}\}$, where Σ is the symmetric, positive-definite covariance matrix and μ the expectation. We write $X = (X_1, \ldots, X_n) \sim \text{Norm}(\mu, \Sigma)$ for a random vector with such a distribution. □

We say that two random variables X and Y are independent if $P(X, Y) = P(X)P(Y)$ or, equivalently, if the conditional probability $P(X \mid Y) = P(X, Y)/P(Y)$ is equal to the unconditional probability $P(X)$. If X and Y are independent, denoted $X \perp Y$, then $\text{E}[XY] = \text{E}[X]\text{E}[Y]$ and $\text{Var}(X+Y) = \text{Var}(X) + \text{Var}(Y)$. It follows that independent random variables have covariance zero. However, the converse is only true in specific situations, for example if (X, Y) is multivariate normal, but not in general because correlation captures only linear dependencies.

This limitation can be addressed by using statistical models which allow for a richer dependency structure. Subheading 7 is devoted to Bayesian networks, a family of probabilistic graphical models based on conditional independences. Let X, Y, and Z be three random vectors. Generalizing the notion of statistical independence, we say that X is conditionally independent of Y given Z and write $X \perp Y \mid Z$ if $P(X, Y \mid Z) = P(X \mid Z)P(Y \mid Z)$. Bayes' theorem states that

$$P(Y \mid X) = \frac{P(X \mid Y)P(Y)}{P(X)}, \tag{29}$$

where $P(Y)$ is called the prior probability and $P(Y \mid X)$ the posterior probability. Intuitively, the prior $P(Y)$ encodes our a priori knowledge about Y (i.e., before observing X), and $P(Y \mid X)$ is our updated knowledge about Y a posteriori (i.e., after observing X).

We have $P(X) = \Sigma_Y P(X, Y)$ if Y is discrete, and similarly $P(X) = \int_Y P(X, Y) dY$ if Y is continuous. Here, $P(X)$ is called the marginal and $P(X, Y)$ the joint probability. This summation or integration is known as marginalization (Fig. 3).

Since $P(X) = \Sigma_Y P(X, Y) = \Sigma_Y P(X \mid Y)P(Y)$, Bayes' theorem can also be rewritten as:

$$P(Y \mid X) = \frac{P(X \mid Y)P(Y)}{\sum_{y' \in \mathcal{Y}} P(X \mid y')P(y')}, \tag{30}$$

where $P(y') = P(Y = y')$ and $\mathcal{Y}$ is the state space of Y.

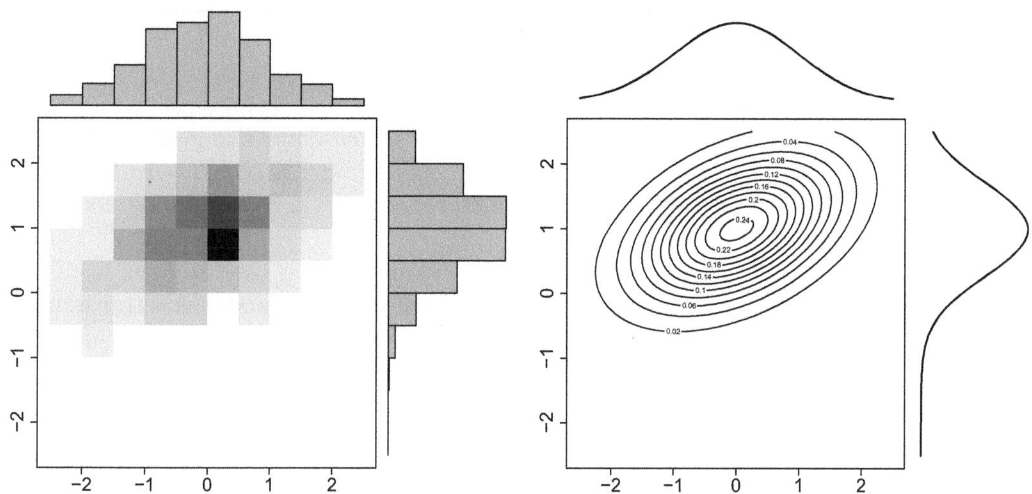

Fig. 3 Marginalization. Left: two-dimensional histogram of a discrete bivariate distribution with the two marginal histograms. Right: contour plot of a two-dimensional Gaussian density with the marginal distributions of each component

Example 8 (Diagnostic Test): We want to evaluate a diagnostic test for a rare genetic disease. The binary random variables D and T indicate disease status ($D = 1$, diseased) and test result ($T = 1$, positive), respectively. Let us assume that the prevalence of the disease is 0.5%, i.e., 0.5% of all people in the population are known to be affected. The test has a false positive rate (probability that somebody is tested positive who does not have the disease) of $P(T = 1 \mid D = 0) = 5\%$ and a true positive rate (probability that somebody is tested positive who has the disease) of $P(T = 1 \mid D = 1) = 90\%$. Then, the posterior probability of a person having the disease given that he or she tested positive is

$$P(D = 1 \mid T = 1) = \frac{P(T = 1 \mid D = 1)P(D = 1)}{P(T = 1 \mid D = 0)P(D = 0) + P(T = 1 \mid D = 1)P(D = 1)} = 0.083, \tag{31}$$

that is, only 8.3% of the positively tested individuals actually have the disease. Thus, our prior belief of the disease status, $P(D)$, has been modified in light of the test result by multiplication with $P(T \mid D)$ to obtain the updated belief $P(D \mid T)$. □

Exercise 9 (Conditional Independence): Let X, Y, and Z be random variables. Using the laws of probability, show that X and Y are conditionally independent given Z (i.e., $X \perp Y \mid Z$) if and only if $P(X \mid Y, Z) = P(X \mid Z)$.

2 Statistical Inference

Statistical models have parameters and a common task is to estimate the model parameters from observed data. The goal is to find the set of parameters with the best model fit. There are two major approaches to parameter estimation: maximum likelihood (ML) and Bayes.

The maximum likelihood approach is based on the likelihood function. Let us consider a fixed statistical model M with parameter space Θ and assume that we have observed realizations $\mathcal{D} = \{x^{(1)}, \ldots, x^{(N)}\}$ of the discrete random variable $X \sim \mathrm{M}(\theta_0)$ for some unknown parameter $\theta_0 \in \Theta$. For the fixed data set $\mathcal{D}$, the likelihood function of the model is

$$L(\theta) = P(\mathcal{D} \mid \theta), \tag{32}$$

where we write $P(\mathcal{D} \mid \theta)$ to emphasize that, here, the probability of the data depends on the model parameter θ. For continuous random variables, the likelihood function is defined similarly in terms of the density function, $L(\theta) = f(\mathcal{D} \mid \theta)$. Maximum likelihood estimation seeks the parameter $\theta \in \Theta$ for which $L(\theta)$ is maximal. Rather than $L(\theta)$, it is often more convenient to maximize $\ell(\theta) = \log L(\theta)$, the log-likelihood function. If the data are i.i.d., then:

$$\ell(\theta) = \sum_{i=1}^{N} \log P(X = x^{(i)} \mid \theta). \tag{33}$$

Example 10 (Likelihood Function of the Binomial Model): Suppose we have observed $k = 7$ successes in a total of $N = 10$ Bernoulli trials. The likelihood function of the binomial model (Eq. 8) is

$$L(p) = p^k \, (1 - p)^{N-k}, \tag{34}$$

where p is the success probability (Fig. 4). To maximize L, we consider the log-likelihood function:

$$\ell(p) = \log L(p) = k\log(p) + (N - k)\log(1 - p) \tag{35}$$

and the likelihood equation $d\ell/dp = 0$. The ML estimate (MLE) is the solution $\hat{p}_{\mathrm{ML}} = k/N = 7/10$. Thus, the MLE of the success probability is just the relative frequency of successes. □

Example 11 (Likelihood Function of the Hardy–Weinberg Model): If we genotype a finite random sample of a population of diploid individuals at a single locus, then the resulting data consists of the numbers of individuals n_{AA}, n_{Aa}, and n_{aa} with the respective genotypes. Assuming Hardy–Weinberg equilibrium (Eqs. 1–3), we want to estimate the allele frequencies p and $q = 1 - p$ of the

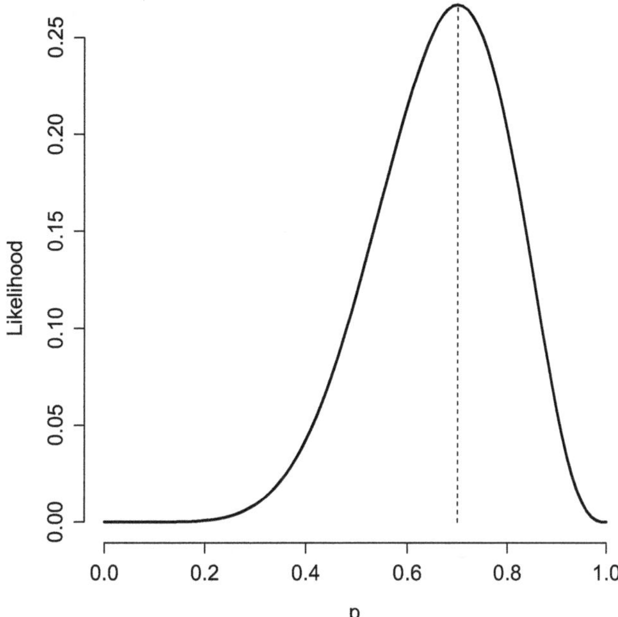

Fig. 4 Likelihood function of the binomial model. The underlying data set consists of $k = 7$ successes out of $N = 10$ Bernoulli trials. The likelihood $L(p) = p^k (1 - p)^{N-k}$ is plotted as a function of the model parameter p, the probability of success (solid line). The MLE is the maximum of this function, $\hat{p}_{\mathrm{ML}} = k/N = 7/10$ (dashed line)

population. The likelihood function of the Hardy–Weinberg model is $L(p) = P(\mathrm{AA})^{n_{\mathrm{AA}}} P(\mathrm{Aa})^{n_{\mathrm{Aa}}} P(\mathrm{aa})^{n_{\mathrm{aa}}}$ and the log-likelihood is

$$\begin{aligned} \ell(p) &= n_{\mathrm{AA}} \log p^2 + n_{\mathrm{Aa}} \log 2p(1-p) + n_{\mathrm{aa}} \log (1-p)^2 \\ &\propto (2n_{\mathrm{AA}} + n_{\mathrm{Aa}}) \log p + (n_{\mathrm{Aa}} + 2n_{\mathrm{aa}}) \log (1-p), \end{aligned} \tag{36}$$

where we have dropped the constant $n_{\mathrm{Aa}} \log 2$. The MLE of $p \in [0, 1]$ can be found by maximizing ℓ. Solving the likelihood equation:

$$\frac{\partial \ell}{\partial p} = \frac{2n_{\mathrm{AA}} + n_{\mathrm{Aa}}}{p} - \frac{n_{\mathrm{Aa}} + 2n_{\mathrm{aa}}}{1-p} = 0 \tag{37}$$

yields the MLE $\hat{p}_{\mathrm{ML}} = (2n_{\mathrm{AA}} + n_{\mathrm{Aa}})/(2N)$, where $N = n_{\mathrm{AA}} + n_{\mathrm{Aa}} + n_{\mathrm{aa}}$ is the total sample size. For example, if we sample $N = 100$ genotypes with $n_{\mathrm{AA}} = 81$, $n_{\mathrm{Aa}} = 18$, and $n_{\mathrm{aa}} = 1$, then we find $\hat{p}_{\mathrm{ML}} = (2 \cdot (81 + 18))/(2 \cdot 100) = 0.9$ for the frequency of the major allele. □

MLEs have many desirable properties. Asymptotically, as the sample size $N \to \infty$, they are normally distributed, unbiased, and have minimal variance. The uncertainty in parameter estimation associated with the sampling variance of the finite data set can be quantified in confidence intervals. There are several ways to construct confidence intervals and statistical tests for MLEs based on the asymptotic behavior of the log-likelihood function $\ell(\theta) = \log L(\theta)$ and its derivatives. For example, the asymptotic normal distribution of the MLE is

$$\hat{\theta}_{\mathrm{ML}} \overset{a}{\sim} \mathrm{Norm}\left(\theta, J(\theta)^{-1}\right), \tag{38}$$

where $I(\theta) = -\partial^2\ell/\partial\theta^2$ is the Fisher information and $J(\theta) = \mathrm{E}[I(\theta)]$ the expected Fisher information. This result gives rise to the Wald confidence intervals:

$$\left[\hat{\theta}_{\mathrm{ML}} \pm z_{1-\alpha/2}\, J(\theta)^{-1}\right], \tag{39}$$

where $z_{1-\alpha/2} = \inf\{x \in \mathbb{R} \mid 1 - \alpha/2 \leq F(x)\}$ is the $(1 - \alpha/2)$ quantile and F the cumulative distribution function of the standard normal distribution. Equation 38 still holds after replacing $J(\theta)$ with the standard error $\mathrm{se}(\hat{\theta}_{\mathrm{ML}}) = [I(\hat{\theta}_{\mathrm{ML}})]^{-\frac{1}{2}}$ or $[J(\hat{\theta}_{\mathrm{ML}})]^{-\frac{1}{2}}$, and it also generalizes to higher dimensions. Other common constructions of confidence intervals include those based on the asymptotic distribution of the score function $S(\theta) = \partial\ell/\partial\theta$ and the log-likelihood ratio $\log(L(\hat{\theta}_{\mathrm{ML}})/L(\theta))$ [8].

We now discuss another more generic approach to quantify parameter uncertainty, not restricted to ML estimation, which is applied frequently in practice due to its simple implementation. Bootstrapping [9] is a resampling method in which independent observations are resampled from the data with replacement. The resulting new data set consists of (some of) the original observations, and under i.i.d. assumptions, the bootstrap replicates have asymptotically the same distribution as the data. Intuitively, by sampling with replacement, one is pretending that the collection of replicates thus obtained is a good proxy for the distribution of data sets that one would have obtained, had we been able to actually replicate the experiment. In this way, the variability of an estimator (or more generally the distribution of any test statistic) can be approximated by evaluating the estimator (or the statistic) on a collection of bootstrap replicates. For example, the distribution of the ML estimator of a model parameter θ can be obtained from the bootstrap samples.

Example 12 (Bootstrap Confidence Interval for the ML Allele Frequency): We use bootstrapping to estimate the distribution of the ML estimator $\hat{p}_{\mathrm{ML}}$ of the Hardy–Weinberg model for the data set $(n_{\mathrm{AA}}, n_{\mathrm{Aa}}, n_{\mathrm{aa}}) = (81, 18, 1)$ of Example 11. For each bootstrap sample, we draw $N = 100$ genotypes with replacement from the original data to obtain random integer vectors of length three summing to 100. The ML estimate is computed for each of a total of B bootstrap samples. The resulting distributions of $\hat{p}_{\mathrm{ML}}$ are shown in Fig. 5, for $B = 100$, 1000, and 10,000. The means of these empirical distributions are 0.899, 0.9004, and 0.9001,

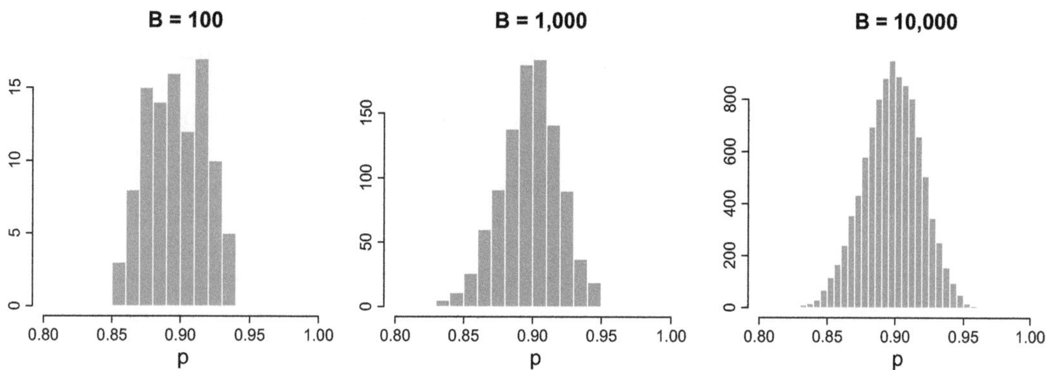

Fig. 5 Bootstrap analysis of the ML allele frequency. The bootstrap distribution of the maximum likelihood estimator $\hat{p}_{ML} = (2n_{AA} + n_{Aa})/(2N)$ of the major allele frequency in the Hardy–Weinberg model is plotted for $B = 100$ (left), $B = 1000$ (center), and $B = 10,000$ (right) bootstrap samples, for the data set $(n_{AA}, n_{Aa}, n_{aa}) = (81, 18, 1)$

respectively, and 95% bootstrap confidence intervals can be derived from the 2.5 and 97.5% quantiles of the distributions. For $B = 100$, 1000, and 10,000, we obtain, respectively, [0.8598, 0.9350], [0.860, 0.940], and [0.855, 0.940]. The basic bootstrap confidence intervals have several limitations, including bias of the bootstrap estimator and skewness of the bootstrap distribution. Other methods exist for constructing confidence intervals from the bootstrap distribution to address some of them [9]. □

The Bayesian approach takes a different point of view and regards the model parameters as random variables [10]. Inference is then concerned with estimating the joint distribution of the parameters θ given the observed data $\mathcal{D}$. By Bayes' theorem (Eq. 30), we have

$$P(\theta \mid \mathcal{D}) = \frac{P(\mathcal{D} \mid \theta)P(\theta)}{P(\mathcal{D})} = \frac{P(\mathcal{D} \mid \theta)P(\theta)}{\int_{\theta \in \Theta} P(\mathcal{D} \mid \theta)P(\theta)\, d\theta}, \tag{40}$$

that is, the posterior probability of the parameters is proportional to the likelihood of the data times the prior probability of the parameters. It follows that, for a uniform prior, the mode of the posterior is equal to the MLE.

From the posterior, credible intervals of parameter estimates can be derived such that the parameter lies in the interval with a certain probability, say 95%. This is in contrast to a 95% confidence interval in the frequentist approach because, there, the parameter is fixed and the interval boundaries are random variables. The meaning of a confidence interval is that 95% of similar intervals would contain the true parameter, if intervals were constructed independently from additional identically distributed data.

The prior $P(\theta)$ encodes our a priori belief in θ before observing the data. It can be used to incorporate domain-specific knowledge into the model, but it may also be uninformative or objective, in which case all observations are equally likely, or nearly so, a priori. However, it can sometimes be difficult to find noninformative priors. In practice, conjugate priors are most often used. A conjugate prior is one that is invariant with respect to the distribution family under multiplication with the likelihood, i.e., the posterior belongs to the same family as the prior. Conjugate priors are mathematically convenient and computationally efficient because the posterior can be calculated analytically for a wide range of statistical models.

Example 13 (Dirichlet Prior): Let $T = (T_1, \ldots, T_K)$ be a continuous random variable with state space Δ_{K-1}. The Dirichlet distribution $\mathrm{Dir}(\alpha)$ with parameters $\alpha \in \mathbb{R}_+^K$ has probability density function:

$$f(\theta_1, \ldots, \theta_K) = \frac{\Gamma\left(\sum_{i=1}^{K} \alpha_i\right)}{\prod_{i=1}^{K} \Gamma(\alpha_i)} \prod_{i=1}^{K} \theta_i^{\alpha_i - 1}, \tag{41}$$

where Γ is the gamma function. The Dirichlet prior is conjugate to the multinomial likelihood: If $T \sim \mathrm{Dir}(\alpha)$ and $(X \mid T = \theta) \sim \mathrm{Mult}(n, \theta_1, \ldots, \theta_K)$, then $(\theta \mid X = x) \sim \mathrm{Dir}(\alpha + x)$. For $K = 2$, this distribution is called the beta distribution. Hence, the beta distribution is the conjugate prior to the binomial likelihood. □

Example 14 (Posterior Probability of Genotype Frequencies): Let us consider the simple genetic system with two loci and two alleles each of Example 1, but without assuming the Hardy–Weinberg model. We regard the observed genotype frequencies $(n_{\mathrm{AA}}, n_{\mathrm{Aa}}, n_{\mathrm{aa}}) = (81, 18, 1)$ as the result of a draw from a multinomial distribution $\mathrm{Mult}(n, \theta_{\mathrm{AA}}, \theta_{\mathrm{Aa}}, \theta_{\mathrm{aa}})$. Assuming a Dirichlet prior $\mathrm{Dir}(\alpha_{\mathrm{AA}}, \alpha_{\mathrm{Aa}}, \alpha_{\mathrm{aa}})$, the posterior genotype probabilities follow the Dirichlet distribution $\mathrm{Dir}(\alpha_{\mathrm{AA}} + n_{\mathrm{AA}}, \alpha_{\mathrm{Aa}} + n_{\mathrm{Aa}}, \alpha_{\mathrm{aa}} + n_{\mathrm{aa}})$. In Fig. 6, the prior $\mathrm{Dir}(10, 10, 10)$ is shown on the left, the multinomial likelihood $P((n_{\mathrm{AA}}, n_{\mathrm{Aa}}, n_{\mathrm{aa}}) = (81, 18, 1) \mid \theta_{\mathrm{AA}}, \theta_{\mathrm{Aa}}, \theta_{\mathrm{aa}})$ in the center, and the resulting posterior $\mathrm{Dir}(10 + 81, 10 + 18, 10 + 1)$ on the right. Note that the MLE is different from the mode of the posterior. As compared to the likelihood, the nonuniform prior has shifted the maximum of the posterior toward the center of the probability simplex. □

We often have two or more competing models and would like to assess which one describes best the given data. For example, we may have observed genotypes from the set {AA, Aa, aa} and want to test whether the Hardy–Weinberg model (Example 1) is a more appropriate description of the genotype data than the multinomial model of the previous Example 14. Intuitively, we might want to select the model that fits the data best, for example, by comparing

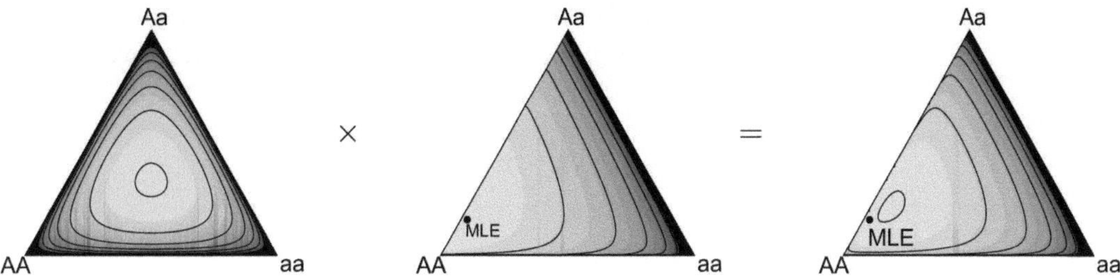

Fig. 6 Dirichlet prior for multinomial likelihood. The Dirichlet prior is conjugate to the multinomial likelihood. Shown are contour lines of the prior Dir(10, 10, 10) on the left, the multinomial likelihood $P((n_{AA}, n_{Aa}, n_{aa}) = (81, 18, 1)|\theta_{AA}, \theta_{Aa}, \theta_{aa})$ in the center, and the resulting posterior Dir(91, 28, 11) on the right. The posterior is the product of prior and likelihood

their likelihoods. However, the Hardy–Weinberg model has only one parameter, namely the allele frequency p, whereas the multinomial model has three parameters subject to the constraint $\theta_{AA} + \theta_{Aa} + \theta_{aa} = 1$. Hence, the number of free parameters is one and two, respectively, for the two models. This difference in the complexity of the models makes a comparison based only on the goodness of fit invalid, because models with more parameters, i.e., higher complexity, can generally provide a better fit. Estimating model complexity and scoring models based on both model complexity and goodness of fit is therefore essential for model comparison and model selection.

The goal of model selection is to find the model that best generalizes to unseen data, rather than just fits the observed data, because we seek the model capable of the most accurate predictions. A model that fits well but generalizes poorly is said to overfit the data. Models that are too complex tend to overfit the data. Model selection can be regarded as finding the right level model complexity for the given data, such that the predictive performance is optimized. This involves defining a criterion of optimality and a procedure for finding the optimal model.

A common frequentist approach to model selection are likelihood ratios. For a data set $\mathcal{D}$, we compare a null model, M_0, to an alternative model, M_1, at given point estimates using the ratio of their likelihoods:

$$\Lambda(\mathcal{D}) = \frac{L(\hat{\theta}_0)}{L(\hat{\theta}_1)} \tag{42}$$

If $\Lambda(\mathcal{D}) < c$, for a defined threshold c, we reject the null model and favor the alternative model. The choice of c should be informed by the distribution of Λ under the null. If the two models are nested, i.e., if M_0 can be obtained from M_1 by specifying a subset of the parameters, then $-2 \log\Lambda$ is approximately χ^2-distributed with degrees of freedom equal to the difference in the number of free parameters between M_1 and M_0.

In the Bayesian framework, it is natural to compare the posterior probabilities of the two models. By Bayes theorem, we have, for $i = 0, 1$:

$$P(\mathrm{M}_i \mid \mathcal{D}) = \frac{P(\mathcal{D} \mid \mathrm{M}_i)P(\mathrm{M}_i)}{P(\mathcal{D})} \tag{43}$$

where:

$$P(\mathcal{D} \mid \mathrm{M}_i) = \int P(\mathcal{D} \mid \theta_i, \mathrm{M}_i)P(\theta_i \mid \mathrm{M}_i)\, d\theta_i \tag{44}$$

is the marginal likelihood. The marginal likelihood accounts for model complexity and for uncertainty in parameter estimates, but is usually analytically intractable and costly to compute. Various approximations of the marginal likelihood exist that give rise to model selection scores, such as the Bayesian information criterion (BIC; *see* Subheading 7) and the Akaike information criterion (AIC) [11].

For Bayesian model comparison, we consider the posterior odds:

$$\frac{P(\mathrm{M}_0 \mid \mathcal{D})}{P(\mathrm{M}_1 \mid \mathcal{D})} = \frac{P(\mathcal{D} \mid \mathrm{M}_0)}{P(\mathcal{D} \mid \mathrm{M}_1)} \frac{P(\mathrm{M}_0)}{P(\mathrm{M}_1)} \tag{45}$$

The ratio of the marginal likelihoods, i.e., the first factor on the right-hand side of Eq. 45, is called the Bayes factor. With equal priors, a Bayes factor larger than 20 is often considered strong support for M_0 over M_1 [12].

Exercise 15 (Poisson Distribution): We wish to model the number of bacterial colonies in a Petri dish and assume that the count data of this experiment follows a Poisson distribution Pois(λ) (Example 3). Derive the log-likelihood function of this model and calculate the MLE of the model parameter λ. Suppose now that the number of bacterial colonies on a Petri dish follows the Poisson distribution with mean $\lambda = 5$. What is the probability of finding exactly three colonies?

3 Hidden Data and the EM Algorithm

We often cannot observe all relevant random variables due to, for example, experimental limitations or study designs. In this case, a statistical model $P(X, Z \mid \theta \in \Theta)$ consists of the observed random variable X and the hidden (or latent) random variable Z, both of which can be multivariate. In this section, we write $X = (X^{(1)}, \ldots, X^{(N)})$ for the random variables describing the N observations and refer to X also as the observed data. The hidden data for this model is $Z = (Z^{(1)}, \ldots, Z^{(N)})$ and the complete data is (X, Z). For convenience, we assume the parameter space Θ to be continuous and the state spaces $\mathcal{X}$ of X and $\mathcal{Z}$ of Z to be discrete.

In the Bayesian framework, one does not distinguish between unknown parameters and hidden data, and it is natural to assess the joint posterior $P(\theta, Z \mid X) \propto P(X \mid \theta, Z)P(\theta, Z)$, which is $P(X, Z \mid \theta)P(\theta)$ if priors are independent, i.e., if $P(\theta, Z) = P(\theta)P(Z)$. Alternatively, if the distribution of the hidden data Z is not of interest, it can be marginalized out. Then, the posterior (Eq. 40) becomes

$$P(\theta \mid X) = \frac{\sum_Z P(X, Z \mid \theta)P(\theta)}{\int_{\theta \in \Theta} \sum_Z P(X, Z \mid \theta)P(\theta)\, d\theta}. \tag{46}$$

In the likelihood framework, it can be more efficient to estimate the hidden data, rather than marginalizing over it. The hidden (or complete-data) log-likelihood is

$$\ell_{\mathrm{hid}}(\theta) = \log P(X, Z \mid \theta) = \sum_{i=1}^{N} \log P(X^{(i)}, Z^{(i)} \mid \theta). \tag{47}$$

For ML parameter estimation, we need to consider the observed log-likelihood:

$$\begin{aligned} \ell_{\mathrm{obs}}(\theta) &= \log P(X \mid \theta) = \log \sum_{Z} P(X, Z \mid \theta) \\ &= \log \sum_{Z^{(1)} \in \mathcal{Z}} \cdots \sum_{Z^{(N)} \in \mathcal{Z}} \prod_{i=1}^{N} P(X^{(i)}, Z^{(i)} \mid \theta). \end{aligned} \tag{48}$$

This likelihood function is usually very difficult to maximize and one has to resort to numerical optimization techniques. Generic local methods, such as gradient descent or Newton's method, can be used, but there is also a more specific local optimization procedure, which avoids computing any derivatives of the likelihood function, called the expectation maximization (EM) algorithm [13].

In order to maximize the likelihood function (Eq. 48), we consider any distribution $q(Z)$ of the hidden data Z and write

$$\ell_{\mathrm{obs}}(\theta) = \log \sum_{Z} q(Z) \frac{P(X, Z \mid \theta)}{q(Z)} = \log \mathrm{E}[P(X, Z \mid \theta)/q(Z)], \tag{49}$$

where the expected value is with respect to $q(Z)$. Jensen's inequality applied to the concave log function asserts that $\log \mathrm{E}[Y] \geq \mathrm{E}[\log Y]$. Hence, the observed log-likelihood is bounded from below by $\mathrm{E}[\log(P(X, Z \mid \theta)/q(Z))]$, or

$$\ell_{\mathrm{obs}}(\theta) \geq \mathrm{E}[\ell_{\mathrm{hid}}(\theta)] + H(q), \tag{50}$$

where $H(q) = -\mathrm{E}[\log q(Z)]$ is the entropy. The idea of the EM algorithm is to maximize this lower bound instead of $\ell_{\mathrm{obs}}(\theta)$ itself. Intuitively, this task is easier because the big sum over the hidden data in Eq. 48 disappears on the right-hand side of Eq. 50 upon taking expectations.

The EM algorithm is an iterative procedure alternating between an E step and an M step. In the E step, the lower bound (Eq. 50) is maximized with respect to the distribution q by setting $q(Z) = P(Z \mid X, \theta^{(t)})$, where $\theta^{(t)}$ is the current estimate of θ, and computing the expected value of the hidden log-likelihood:

$$Q(\theta \mid \theta^{(t)}) = \mathrm{E}_{Z|X,\theta^{(t)}}[\ell_{\mathrm{hid}}(\theta)]. \tag{51}$$

In the M step, Q is maximized with respect to θ to obtain an improved estimate:

$$\theta^{(t+1)} = \arg\max_{\theta} Q(\theta \mid \theta^{(t)}). \tag{52}$$

The sequence $\theta^{(1)}, \theta^{(2)}, \theta^{(3)}, \ldots$ converges to a local maximum of the likelihood surface (Eq. 48). The global maximum and, hence, the MLE is generally not guaranteed to be found with this local optimization method. In practice, the EM algorithm is often run repeatedly with many different starting solutions $\theta^{(1)}$, or with few very reasonable starting solutions obtained from other heuristics or educated guesses.

Example 16 (Naive Bayes): Let us assume that we observe realizations of a discrete random variable $(X_1, \ldots, X_L)$ and we want to cluster observations into K distinct groups. For this purpose, we introduce a hidden random variable Z with state space $\mathcal{Z} = [K] = \{1, \ldots, K\}$ indicating class membership. The joint probability of $(X_1, \ldots, X_L)$ and Z is

$$\begin{aligned} P(X_1, \ldots, X_L, Z) &= P(Z)P(X_1, \ldots, X_L \mid Z) \\ &= P(Z) \prod_{n=1}^{L} P(X_n \mid Z). \end{aligned} \tag{53}$$

The marginalization of this model with respect to the hidden data Z is the unsupervised naive Bayes model. The observed variables X_n are often called features and Z the latent class variable (Fig. 7).

The model parameters are the class prior $P(Z)$, which we assume to be constant and will ignore, and the conditional probabilities $\theta_{n,kx} = P(X_n = x \mid Z = k)$. The complete-data likelihood of observed data $X = (X^{(1)}, \ldots, X^{(N)})$ and hidden data $Z = (Z^{(1)}, \ldots, Z^{(N)})$ is

$$P(X, Z \mid \theta) = \prod_{i=1}^{N} P(X^{(i)}, Z^{(i)} \mid \theta) = \prod_{i=1}^{N} P(Z^{(i)}) \prod_{n=1}^{L} P(X_n^{(i)} \mid Z^{(i)}) \tag{54}$$

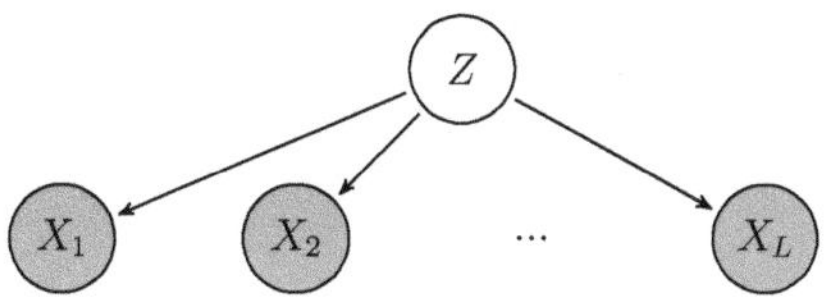

Fig. 7 Graphical representation of the naive Bayes model. Observed features X_n are conditionally independent given the latent class variable Z

$$\propto \prod_{i=1}^{N}\prod_{n=1}^{L}\theta_{n,Z^{(i)}X_n^{(i)}} = \prod_{i=1}^{N}\prod_{n=1}^{L}\prod_{k\in[K]}\prod_{x\in\mathcal{X}}\theta_{n,kx}^{I_{n,kx}(Z^{(i)})}, \tag{55}$$

where $I_{n,kx}(Z^{(i)})$ is equal to one if and only if $Z^{(i)} = k$ and $X_n^{(i)} = x$, and zero otherwise.

To apply the EM algorithm for estimating θ without observing Z, we consider the hidden log-likelihood:

$$\ell_{\text{hid}}(\theta) = \log P(X,Z \mid \theta) = \sum_{i=1}^{N}\sum_{n=1}^{L}\sum_{k\in[K]}\sum_{x\in\mathcal{X}} I_{n,kx}(Z^{(i)})\log\theta_{n,kx}. \tag{56}$$

In the E step, we compute the expected values of $Z^{(i)}$:

$$\begin{aligned}\gamma_{n,kx}^{(i)} = \mathrm{E}_{Z\mid X_n=x,\theta'}\left[Z^{(i)}\right] &= \frac{P(X_n^{(i)} = x \mid Z^{(i)} = k)}{\sum_{k'\in K} P(X_n^{(i)} = x \mid Z^{(i)} = k')} \\ &= \frac{\theta'_{n,kx}}{\sum_{k'\in K}\theta'_{n,k'x}},\end{aligned} \tag{57}$$

where θ' is the current estimate of θ. The expected value $\gamma_{n,kx}^{(i)}$ is sometimes referred to as the responsibility of class k for observation $X_n^{(i)} = x$. The expected hidden log-likelihood can be written in terms of the expected counts $N_{n,kx} = \sum_{i=1}^{N}\gamma_{n,kx}^{(i)}$ as:

$$\mathrm{E}_{Z\mid X,\theta'}[\ell_{\text{hid}}(\theta)] = \sum_{n=1}^{L}\sum_{k\in[K]}\sum_{x\in\mathcal{X}} N_{n,kx}\log\theta_{n,kx}. \tag{58}$$

In the M step, maximization of this sum yields $\hat{\theta}_{n,kx} = N_{n,kx}/\sum_{x'} N_{n,kx'}$. □

4 Markov Chains

A stochastic process $\{X_t, t\in\mathcal{T}\}$ is a collection of random variables with common state space $\mathcal{X}$. The index set $\mathcal{T}$ is usually interpreted as time and X_t is the state of the process at time t. A discrete-time stochastic process $X = (X_1, X_2, X_3, \ldots)$ is called a Markov chain

[14], if $X_{n+1} \perp X_{n-1} \mid X_n$ for all $n \geq 2$ or, equivalently, if each state depends only on its immediate predecessor:

$$P(X_n \mid X_{n-1}, \ldots, X_1) = P(X_n \mid X_{n-1}), \quad \text{for all } n \geq 2. \tag{59}$$

We consider here Markov chains with finite state space $\mathcal{X} = [K] = \{1, \ldots, K\}$ that are homogeneous, i.e., with transition probabilities independent of time:

$$T_{kl} = P(X_{n+1} = l \mid X_n = k), \quad \text{for all } k,l \in [K],\ n \geq 2. \tag{60}$$

The finite-state homogeneous Markov chain is a statistical model denoted MC(Π, T) and defined by the initial state distribution $\Pi \in \Delta_{K-1}$, where $\Pi_k = P(X_1 = k)$, and the stochastic $K \times K$ transition matrix $T = (T_{kl})$.

We can generalize the one-step transition probabilities T_{kl} to:

$$T_{kl}^n = P(X_{n+j} = l \mid X_j = k), \tag{61}$$

the probability of jumping from state k to state l in n time steps. Any $(n + m)$-step transition can be regarded as an n-step transition followed by an m-step transition. Because the intermediate state i is unknown, summing over all possible values yields the decomposition:

$$T_{kl}^{n+m} = \sum_{i=1}^{K} T_{ki}^n T_{il}^m, \quad \text{for all } n,m \geq 1,\ k,l \in [K], \tag{62}$$

known as the Chapman–Kolmogorov equations. In matrix notation, they can be written as $T^{(n+m)} = T^{(n)}T^{(m)}$. It follows that the n-step transition matrix is the n-th matrix power of the one-step transition matrix, $T^{(n)} = T^n$.

A state l of a Markov chain is accessible from state k if $T_{kl}^n > 0$. We say that k and l communicate with each other and write $k \sim l$ if they are accessible from one another. State communication is reflexive ($k \sim k$), symmetric ($k \sim l \Rightarrow l \sim k$), and, by the Chapman–Kolmogorov equations, transitive ($j \sim k \sim l \Rightarrow j \sim l$). Hence, it defines an equivalence relation on the state space. The Markov chain is irreducible if it has a single communication class, i.e., if any state is accessible from any other state.

A state is recurrent if the Markov chain will reenter it with probability one. Otherwise, the state is transient. In finite-state Markov chains, recurrent states are also positive recurrent, i.e., the expected time to return to the state is finite. A state is aperiodic if the process can return to it after any time $n \geq 1$. Recurrence, positive recurrence, and aperiodicity are class properties: if they hold for a state k, then they also hold for all states communicating with k.

A Markov chain is ergodic if it is irreducible, aperiodic, and positive recurrent. An ergodic Markov chain has a unique stationary distribution π given by:

$$\pi_l = \lim_{n\to\infty} T^n_{kl} = \sum_{k=1}^{K} \pi_k T_{kl}, \qquad l\in[K], \quad \sum_{l=1}^{K} \pi_l = 1 \tag{63}$$

independent of the initial distribution Π. In matrix notation, π is the solution of $\pi^t = \pi^t T$.

Example 17 (Two-State Markov Chain): Consider the Markov chain with state space $\{1, 2\}$ and transition probabilities $T_{12} = \alpha > 0$ and $T_{21} = \beta > 0$. Clearly, the chain is ergodic and its stationary distribution π is given by:

$$\begin{pmatrix} \pi_1 & \pi_2 \end{pmatrix} = \begin{pmatrix} \pi_1 & \pi_2 \end{pmatrix} \begin{pmatrix} 1-\alpha & \alpha \\ \beta & 1-\beta \end{pmatrix} \tag{64}$$

or, equivalently, $\alpha\pi_1 = \beta\pi_2$. With $\pi_1 + \pi_2 = 1$, we obtain $\pi^t = (\alpha + \beta)^{-1}(\alpha, \beta)$. □

In Example 17, if $\alpha = 0$, then state 1 is called an absorbing state because once entered it is never left. In evolutionary biology and population genetics, Markov chains are often used to model evolving populations, and the fixation probability of an allele can be computed as the absorption probability in such models.

Example 18 (Wright–Fisher Process): We consider two alleles, A and a, in a diploid population of size N. The total number of A alleles in generation n is described by a Markov chain X_n with state space $\{0, 1, 2, \ldots, 2N\}$. We assume that individuals mate randomly and that maternal and paternal alleles are chosen randomly such that $(X_{n+1} \mid X_n) \sim \text{Binom}(2N, k/(2N))$, where k is the number of A alleles in generation n. The Markov chain has transition probabilities:

$$T_{kl} = \binom{2N}{l} \left(\frac{k}{2N}\right)^l \left(\frac{2N-k}{2N}\right)^{2N-l}. \tag{65}$$

If the initial number of A alleles is $X_1 = k$, then $\text{E}(X_1) = k$. After binomial sampling, $\text{E}(X_2) = 2N(k/(2N)) = k$ and hence $\text{E}(X_n) = k$ for all $n \geq 0$. The Markov chain has the two absorbing states 0 and $2N$, which correspond, respectively, to extinction and fixation of the A allele. To compute the fixation probability h_k of A given k initial copies of it:

$$h_k = \lim_{n\to\infty} P(X_n = 2N \mid X_1 = k), \tag{66}$$

we consider the expected value, which is equal to k, in the limit as $n \to\infty$ to obtain

$$k = \lim_{n \to \infty} \mathrm{E}(X_n) = 0 \cdot (1 - h_k) + 2N \cdot h_k. \tag{67}$$

Thus, the fixation probability is just $h_k = k/(2N)$, the initial relative frequency of the allele. The Wright–Fisher process [15, 16] is a basic stochastic model for random genetic drift, i.e., for the variation in allele frequencies only due to random sampling. □

If we observe data $X = (X^{(1)}, \ldots, X^{(N)})$ from a finite Markov chain MC(Π, T) of length L, then the likelihood is

$$\begin{aligned} L(\Pi, T) &= \prod_{i=1}^{N} P(X^{(i)}) = \prod_{i=1}^{N} P(X_1^{(i)}) \prod_{n=1}^{L-1} P(X_{n+1}^{(i)} \mid X_n^{(i)}) \\ &= \prod_{i=1}^{N} \Pi_{X_1^{(i)}} \prod_{n=1}^{L-1} T_{X_n^{(i)}, X_{n+1}^{(i)}}, \end{aligned} \tag{68}$$

which can be rewritten as:

$$\begin{aligned} L(\Pi, T) &= \prod_{i=1}^{N} \prod_{k \in [K]} \Pi_k^{N_k(X^{(i)})} \prod_{k \in [K]} \prod_{l \in [K]} T_{kl}^{N_{kl}(X^{(i)})} \\ &= \prod_{k \in [K]} \Pi_k^{N_k} \prod_{k \in [K]} \prod_{l \in [K]} T_{kl}^{N_{kl}}. \end{aligned} \tag{69}$$

with $N_{kl}(X^{(i)})$ the number of observed transitions from state k into state l in observation $X^{(i)}$, and $N_{kl} = \sum_{i=1}^{N} N_{kl}(X^{(i)})$ the total number of k-to-l transitions in the data, and similarly $N_k(X^{(i)})$ and N_k the number of times the i-th chain, respectively all chains, started in state k.

Exercise 19 (Markov Chains): Let us consider a simple infectious disease model, where each individual is either healthy (H) or diseased (D). We assume the following two-state Markov chain to describe infection-related disease and recovery via clearance of the pathogen:

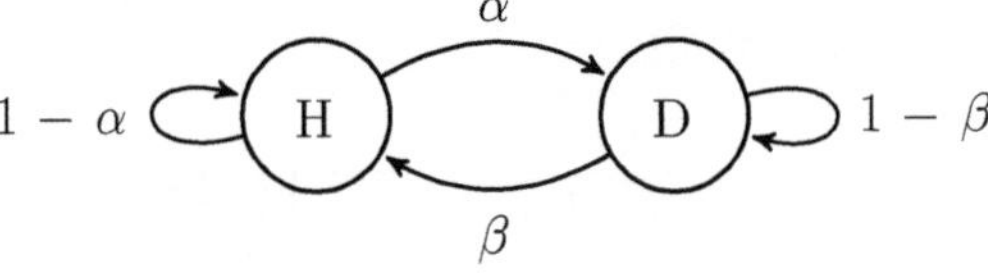

The probability of a healthy individual becoming sick due to infection is $\alpha = 0.6$, and the probability of a diseased individual to clear the infection and recover is $\beta = 0.9$. The initial probabilities for health and disease are $P(\mathrm{H}) = 0.7$ and $P(\mathrm{D}) = 0.3$. Write down the transition matrix T of this Markov chain. What is the probability of observing the disease trajectories DDHHD and HDHDH? Calculate the stationary distribution of the Markov chain.

5 Continuous-Time Markov Chains

A continuous-time stochastic process $\{X(t), t \geq 0\}$ with finite state space $[K]$ is a continuous-time Markov chain if

$$\begin{aligned} P[X(t+s) = l \mid X(s) = k, X(u) = x(u), 0 \leq u < s] \\ = P[X(t+s) = l \mid X(s) = k] \end{aligned} \tag{70}$$

for all $s, t \geq 1$, $k, l, x(u) \in [K]$, $0 \leq u < s$. The chain is homogeneous if Eq. 70 is independent of s. The transition probabilities are then denoted:

$$T_{kl}(t) = P[X(t+s) = l \mid X(s) = k]. \tag{71}$$

It can be shown that the transition matrix $T(t)$ is the matrix exponential of a constant rate matrix R times t:

$$T(t) = \exp(Rt) = \sum_{j=0}^{\infty} \frac{1}{j!}(Rt)^j. \tag{72}$$

Example 20 (Jukes–Cantor Model): Consider a fixed position in a DNA sequence, and let $T_{kl}(t)$ be the probability that, due to mutation, nucleotide k changes to nucleotide l after time t at this position (Fig. 8). The Jukes–Cantor model [17] is the simplest DNA substitution model. It assumes that the transition rates from any nucleotide to any other are equal:

$$R = \begin{pmatrix} -3\alpha & \alpha & \alpha & \alpha \\ \alpha & -3\alpha & \alpha & \alpha \\ \alpha & \alpha & -3\alpha & \alpha \\ \alpha & \alpha & \alpha & -3\alpha \end{pmatrix}. \tag{73}$$

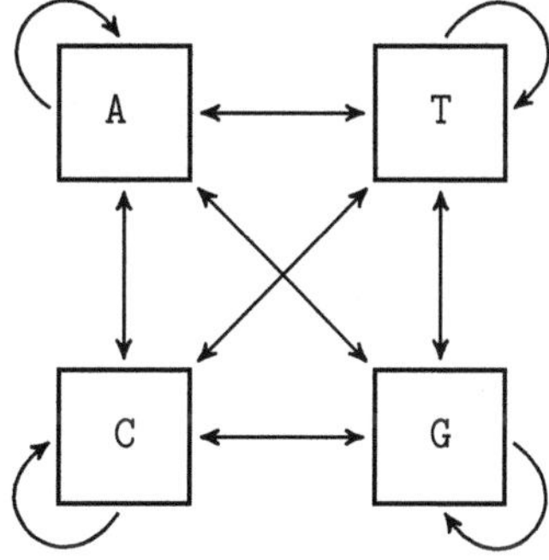

Fig. 8 Nucleotide substitution model. The state space and transitions of a general nucleotide substitution model are shown. For the Jukes–Cantor model (Example 20), all transitions from any nucleotide to any other nucleotide have the same probability $\frac{1}{4}(1 - e^{-4\alpha t})$

The resulting transition matrix $T(t) = \exp(Rt)$ is

$$T(t) = \frac{1}{4}\begin{pmatrix} 1+3e^{-4\alpha t} & 1-e^{-4\alpha t} & 1-e^{-4\alpha t} & 1-e^{-4\alpha t} \\ 1-e^{-4\alpha t} & 1+3e^{-4\alpha t} & 1-e^{-4\alpha t} & 1-e^{-4\alpha t} \\ 1-e^{-4\alpha t} & 1-e^{-4\alpha t} & 1+3e^{-4\alpha t} & 1-e^{-4\alpha t} \\ 1-e^{-4\alpha t} & 1-e^{-4\alpha t} & 1-e^{-4\alpha t} & 1+3e^{-4\alpha t} \end{pmatrix} \tag{74}$$

and the stationary distribution as $t \to \infty$ is uniform, $\pi = (1/4, 1/4, 1/4, 1/4)^t$. □

Example 21 (The Poisson Process): A continuous-time Markov chain $X(t)$ is a counting process, if $X(t)$ represents the total number of events that occur by time t. It is a Poisson process, if in addition $X(0) = 0$, the increments are independent, and in any interval of length t the number of events is Poisson distributed with rate λt:

$$P[X(t+s) - X(s) = k] = P[X(t) = k] = e^{-\lambda t}\frac{(\lambda t)^k}{k!}. \tag{75}$$

The Poisson process is used, for example, to count mutations in a gene. □

Example 22 (Exponential Distribution): The exponential distribution $\mathrm{Exp}(\lambda)$ with parameter $\lambda > 0$ is a common distribution for waiting times. It is defined by the density function:

$$f(x) = \lambda e^{-\lambda x}, \quad \text{for } x \geq 0. \tag{76}$$

If $X \sim \mathrm{Exp}(\lambda)$, then X has expectation $\mathrm{E}(X) = \lambda^{-1}$ and variance $\mathrm{Var}(X) = \lambda^{-2}$. The exponential distribution is memoryless, which means that $P(X > s + t \mid X > t) = P(X > s)$, for all s, $t > 0$. An important consequence of the memoryless property is that the waiting times between successive events are i.i.d. For example, the waiting times τ_n $(n \geq 1)$ between the events of a Poisson process, the sequence of interarrival times, are exponentially distributed, $\tau_n \sim \mathrm{Exp}(\lambda)$, for all $n \geq 1$. □

Exercise 23 (Kimura Model): The Kimura two-parameter model is a DNA substitution model that distinguishes between transitions, i.e., purine-to-purine and pyrimidine-to-pyrimidine substitutions, from transversions, i.e., purine-to-pyrimidine and pyrimidine-to-purine substitutions [18]. It is defined by the rate matrix:

$$R = \begin{pmatrix} -2\beta-\alpha & \beta & \alpha & \beta \\ \beta & -2\beta-\alpha & \beta & \alpha \\ \alpha & \beta & -2\beta-\alpha & \beta \\ \beta & \alpha & \beta & -2\beta-\alpha \end{pmatrix},$$

where α, $\beta \in \mathbb{R}_+$ are the two substitution rates. Assuming that the Markov chain is ergodic, derive its stationary distribution.

6 Hidden Markov Models

A hidden Markov model (HMM) is a statistical model for hidden random variables $Z = (Z_1, \ldots, Z_L)$, which form a homogeneous Markov chain, and observed random variables $X = (X_1, \ldots, X_L)$. Each observed symbol X_n depends on the hidden state Z_n. The HMM is illustrated in Fig. 9. It encodes the following conditional independence statements:

$$Z_{n+1} \perp Z_{n-1} \mid Z_n, \qquad 2 \leq n \leq L-1 \qquad \text{(Markovproperty)} \tag{77}$$

$$X_n \perp X_m \mid Z_n, \qquad 1 \leq m, n \leq L, \; m \neq n \tag{78}$$

The parameters of the HMM consist of the initial state probabilities $\Pi = P(Z_1)$, the transition probabilities $T_{kl} = P(Z_n = l \mid Z_{n-1} = k)$ of the Markov chain, and the emission probabilities $E_{kx} = P(X_n = x \mid Z_n = k)$ of symbols $x \in \mathcal{X}$. The HMM is denoted $\text{HMM}(\Pi, T, E)$. For simplicity, we restrict ourselves here to finite state spaces $\mathcal{Z} = [K]$ of Z and $\mathcal{X}$ of X. The joint probability of (Z, X) factorizes as:

$$\begin{aligned} P(X,Z) &= P(Z_1) \prod_{n=1}^{L-1} P(X_n \mid Z_n) P(Z_{n+1} \mid Z_n) \\ &= \Pi_{Z_1} \prod_{n=1}^{L-1} E_{Z_n, X_n} T_{Z_n, Z_{n+1}}. \end{aligned} \tag{79}$$

The HMM is typically used to model sequence data $x = (x_1, x_2, \ldots, x_L)$ generated by different mechanisms z_n which cannot be observed. Each observation x can be a time series or any other object with a linear dependency structure [19]. In computational biology, the HMM is frequently applied to DNA and protein sequence data, where it accounts for first-order spatial dependencies of nucleotides or amino acids [20].

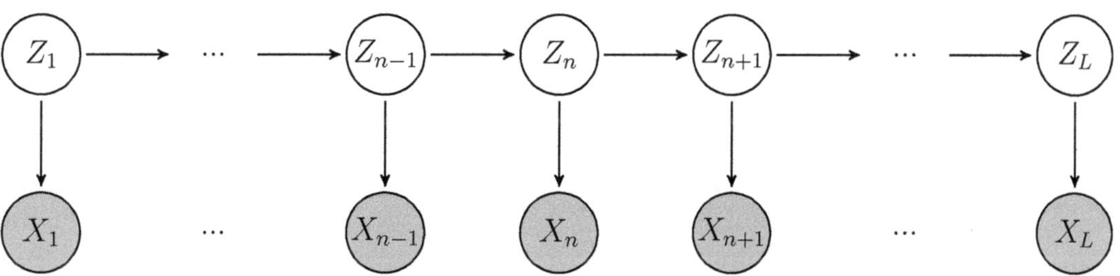

Fig. 9 Hidden Markov model. Shaded nodes represent observed random variables (or symbols) X_n, and clear nodes represent hidden states (or the annotation). Directed edges indicate statistical dependencies which are given, respectively, by transition and emission probabilities among hidden states and between hidden states and observed symbols

Example 24 (CpG Islands): CpG islands are CG-enriched regions in a DNA sequence. They are typically a few hundreds to thousands of base pairs long. We want to use a simple HMM to detect CpG islands in genomic DNA. The hidden states $Z_n \in \mathcal{Z} = \{-, +\}$ indicate whether sequence position n belongs to a CpG island (+) or not (−). The observed sequence is given by the nucleotide at each position, $X_n \in \mathcal{X} = \{\mathtt{A}, \mathtt{C}, \mathtt{G}, \mathtt{T}\}$.

Suppose we observe the sequence $x = (\mathtt{C}, \mathtt{A}, \mathtt{C}, \mathtt{G})$. Then, we can calculate the joint probability of x and any state path z by Eq. 79. For example, if $z = (+, -, -, +)$, then $P(X = x, Z = z) = \Pi_+ \, E_{+,\mathtt{C}} T_{+,-} \, E_{-,\mathtt{A}} T_{-,-} \, E_{-,\mathtt{C}} T_{-,+} \, E_{+,\mathtt{G}}$. □

Typically, one is interested in the hidden state path $z = (z_1, z_2, \ldots, z_L)$ that gave rise to the observation x. For biological sequences, z is often called the annotation of x. In Example 24, the genomic sequence is annotated with CpG islands. For generic parameters, any state path can give rise to a given observed sequence, but with different probabilities. The decoding problem is to find the annotation z^* that maximizes the joint probability:

$$z^* = \underset{z \in \mathcal{Z}}{\operatorname{argmax}} \; P(X = x, Z = z). \tag{80}$$

There are K^L possible state paths such that already for sequences of moderate length, the optimization problem (Eq. 80) cannot be solved by enumerating all paths.

However, there is an efficient algorithm solving (Eq. 80) based on the following factorization along the Markov chain:

$$\begin{aligned}
\max_Z P(X, Z) &= \max_{Z_1, \ldots, Z_L} P(Z_1) \prod_{n=1}^{L-1} P(X_n \mid Z_n) P(Z_{n+1} \mid Z_n) \\
&= \max_{Z_L} P(Z_L \mid Z_{L-1}) P(X_L \mid Z_L) \\
&\quad \Big[\ldots \Big[\max_{Z_2} P(Z_3 \mid Z_2) P(X_2 \mid Z_2) \\
&\quad \Big[\max_{Z_1} P(Z_2 \mid Z_1) P(X_1 \mid Z_1) \cdot P(Z_1) \Big] \Big] \ldots \Big].
\end{aligned} \tag{81}$$

Thus, the maximum over state paths $(Z_1, \ldots, Z_L)$ can be obtained by recursively computing maxima over each Z_n. Each of the L terms in parenthesis defines a probability distribution over K states by maximizing over K values. Hence, the time complexity of the algorithm is $O(LK^2)$, despite the fact that the maximum is over K^L paths. This procedure is known as dynamic programming and it is the workhorse of biological sequence analysis. For HMMs, it is known as the Viterbi algorithm [21].

In order to compute the marginal likelihood $P(X = x)$ of an observed sequence x, we need to sum the joint probability $P(Z = z, X = x)$ over all hidden states $z \in Z$. The length of this sum is K^L, but it can be computed efficiently by the same dynamic programming principle used for the Viterbi algorithm:

$$
\begin{aligned}
\sum_{Z} P(X,Z) &= \sum_{Z_1,\ldots,Z_L} P(Z_1) \prod_{n=1}^{L-1} P(X_n \mid Z_n) P(Z_{n+1} \mid Z_n) \\
&= \sum_{Z_L} P(Z_L \mid Z_{L-1}) P(X_L \mid Z_L) \\
&\quad \Big[\ldots\Big[\sum_{Z_2} P(Z_3 \mid Z_2) P(X_2 \mid Z_2) \\
&\quad \Big[\sum_{Z_1} P(Z_2 \mid Z_1) P(X_1 \mid Z_1) \cdot P(Z_1)\Big]\Big]\ldots\Big].
\end{aligned} \tag{82}
$$

Indeed, this factorization is the same as in Eq. 81 with maxima replaced by sums. The recursive algorithm implementing (Eq. 82) is known as the forward algorithm. In each step, it computes the partial solution $f(n, Z_n) = P(X_1, \ldots, X_n, Z_n)$.

The factorization along the Markov chain can also be done in the other direction starting the recursion from Z_L down to Z_1. The resulting backward algorithm generates the partial solutions $b(n, Z_n) = P(X_{n+1}, \ldots, X_L \mid Z_n)$. From the forward and backward quantities, one can also compute the position-wise posterior state probabilities:

$$
\begin{aligned}
P(Z_n \mid X) &= \frac{P(X, Z_n)}{P(X)} = \frac{P(X_1, \ldots, X_n, Z_n) P(X_{n+1}, \ldots, X_L \mid Z_n)}{P(X)} \\
&= \frac{f(n, Z_n) b(n, Z_n)}{P(X)}.
\end{aligned} \tag{83}
$$

For example, in the CpG island HMM (Example 24), we can compute, for each nucleotide, the probability that it belongs to a CpG island given the entire observed DNA sequence. Selecting the state that maximizes this probability independently at each sequence position is known as posterior decoding. In general, the result will be different from Viterbi decoding.

Example 25 (Pairwise Sequence Alignment): The pair HMM is a statistical model for pairwise alignment of two observed sequences over a fixed alphabet $\mathcal{A}$. For protein sequences, $\mathcal{A}$ is the set of 20 natural amino acids and for DNA sequences, $\mathcal{A}$ consists of the four nucleotides, plus the gap symbol ("-"). At each position of the alignment, a hidden variable $Z_n \in Z = \{\texttt{M}, \texttt{X}, \texttt{Y}\}$ indicates whether there is a (mis-)match (M), an insertion (X), or a deletion (Y) in sequence y relative to sequence x. For example:

```
z= MMMMMMMMMMMMMXXMMMMMMMMMMMMYMMMMYMMMMM
x= CTRPNNNTRKSIRPQIGPGQAFYATGD-IGDI-RQAHC
y= CGRPNNHRIKGLR--IGPGRAFFAMGAIRGGEIRQAHC
```

The emitted symbols are pairs (X_n, Υ_n) of aligned sequence characters with state space $(\mathcal{A} \times \mathcal{A})\backslash\{(-, -)\}$. Thus, a pairwise alignment is a probabilistically generated sequence of pairs of symbols.

The choice of transition and emission probabilities corresponds to fixing a scoring scheme in nonprobabilistic formulations of sequence alignment. For example, the emission probabilities $P[(a, b) \mid \mathrm{M}]$ from a match state encode pairwise amino acid preferences and can be modeled by substitution matrices, such as PAM and BLOSUM [20].

In the pair HMM, computing an optimal alignment between x and y means to find the most probable state path $z^* = \operatorname{argmax}_z P(X = x, \Upsilon = y, Z = z)$, which can be solved using the Viterbi algorithm. Using the forward algorithm, we can also compute efficiently the marginal probability of two sequences being related independent of their alignment, $P(X, \Upsilon) = \Sigma_Z P(X, \Upsilon, Z)$. In general, this probability is more informative than the posterior $P(Z \mid X, \Upsilon)$ of an optimal alignment z^* because many alignments tend to have the same or nearly the same probability such that $P(Z = z^* \mid X, \Upsilon)$ can be very small. Finally, we can also compute the probability of two characters x_n and y_m being aligned by means of posterior decoding. □

Example 26 (Profile HMM): Profile hidden Markov models represent groups of related sequences, such as protein families. They are used for searching homologous sequences and for building multiple sequence alignments. They can be regarded as unrolled versions of the pair HMM. A profile HMM is a statistical model for observed sequences, which are regarded as i.i.d. realizations. It has site-specific emission probabilities $E_n(a) = P(X_n = a)$. In its simplest form allowing only gap-free alignments, the probability of an observation x is just

$$P(X = x) = \prod_{n=1}^{L} E_n(x_i). \tag{84}$$

The matrix $(E_n(a))_{1 \leq n \leq L,\ a \in A}$ is called a position-specific scoring matrix (PSSM).

Profile HMMs can also model indels. Figure 10 shows the hidden state space of such a model. It has match states M_n, which can emit symbols according to the probability tables E_n, insert states I_n, which usually emit symbols in an unspecific manner, and delete states D_n, which do not emit any symbols. The possible transitions between those states allow for modeling alignment gaps of any length.

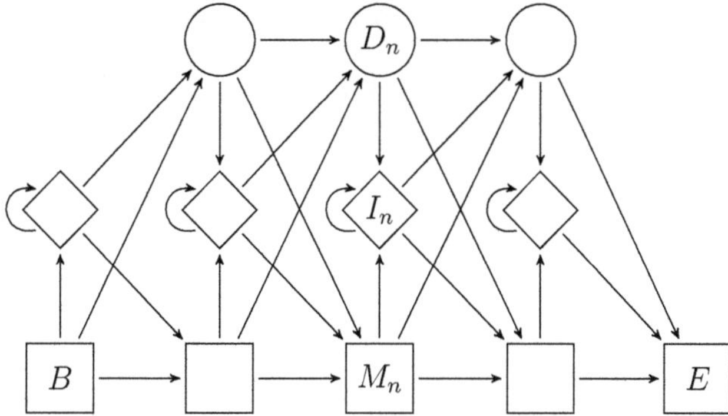

Fig. 10 Profile hidden Markov model. The hidden state space and its transitions are shown for the profile HMM of length $L = 3$. Match states are denoted M_n, insert states I_n, and delete states D_n. B and E denote silent begin and end states, respectively. With match and insert states, probability tables for the emissions of symbols (amino acids or nucleotides, and gaps) are associated

A given profile HMM for a protein family can be used to detect new sequences that belong to the same family. For a query sequence x, we can either consider the most probable alignment of the sequence to the HMM, $P(X = x, Z = z^*)$, or the marginal probability independent of the alignment, $P(X = x) = \Sigma_Z P(X = x, Z)$, to decide about family membership. □

Parameter estimation in HMMs is complicated by the presence of hidden variables. In Subheading 2, the EM algorithm has been introduced for finding a local maximum of the likelihood surface. For HMMs, the EM algorithm is known as the Baum–Welch algorithm [22]. For simplicity, let us ignore the initial state probabilities Π and summarize the parameters of the HMM by $\theta = (T, E)$. For ML estimation, we need to maximize the observed log-likelihood:

$$\begin{aligned} \ell_{\text{obs}}(\theta) &= \log P(X \mid \theta) = \log \sum_Z P(X, Z \mid \theta) \\ &= \log \sum_{Z^{(1)}, \ldots, Z^{(N)}} \prod_{i=1}^{N} P(X^{(i)}, Z^{(i)} \mid \theta), \end{aligned} \tag{85}$$

where $X^{(1)}, \ldots, X^{(N)}$ are the i.i.d. observations. For each observation, we can rewrite the joint probability as:

$$P(X^{(i)}, Z^{(i)} \mid \theta) = \prod_{k \in [K]} \prod_{x \in \mathcal{X}} E_{kx}^{N_{kx}(Z^{(i)})} \cdot \prod_{k \in [K]} \prod_{l \in [K]} T_{kl}^{N_{kl}(Z^{(i)})}, \tag{86}$$

where $N_{kx}(Z^{(i)})$ is the number of x emissions when in state k and $N_{kl}(Z^{(i)})$ the number of k-to-l transitions in state path $Z^{(i)}$ (cf. Eq. 68).

In the E step, the expectation of Eq. 85 is computed with respect to $P(Z \mid X, \theta')$, where θ' is the current best estimate of θ. We use Eq. 86 and denote by N_{kx} and N_{kl} the expected value of $\Sigma_i N_{kx}(Z^{(i)})$ and $\Sigma_i N_{kl}(Z^{(i)})$, respectively, to obtain

$$\begin{aligned} \mathrm{E}[\ell_{hid}(\theta)] &= \sum_{Z} P(Z \mid X, \theta') \log P(X, Z \mid \theta) \\ &= \sum_{Z^{(1)}, \ldots, Z^{(N)}} P(Z \mid X, \theta') \\ &\quad \left[\sum_{k,\ x} N_{kx}(Z^{(i)}) \log E_{kx} + \sum_{k,\ l} N_{kl}(Z^{(i)}) \log T_{kl} \right] \\ &= \sum_{k,\ x} N_{kx} \log E_{kx} + \sum_{k,\ l} N_{kl} \log T_{kl}. \end{aligned} \tag{87}$$

The expected counts N_{kx} and N_{kl} are the sufficient statistics [11] of the HMM, i.e., with respect to the model, they contain all information about the parameters available from the data. The expected counts can be computed using the forward and backward algorithms. In the M step, this expression is maximized with respect to $\theta = (T,\ E)$. We find the MLEs $\hat{T}_{kl} = N_{kl} / \sum_m N_{km}$ and $\hat{E}_{kx} = N_{kx} / \sum_y N_{ky}$.

7 Bayesian Networks

Bayesian networks are a class of probabilistic graphical models which generalize Markov chains and HMMs. The basic idea is to use a graph for encoding conditional independences among random variables (Fig. 11). The graph representation provides not only an intuitive and simple visualization of the model structure, but it is also the basis for designing efficient algorithms for inference and learning in graphical models [23–25].

A Bayesian network (BN) for a set of random variables $X = (X_1, \ldots, X_L)$ consists of a directed acyclic graph (DAG) and local probability distributions (LPDs). The DAG $G = (V, E)$ has vertex set $V = [L]$ and edge set $E \subseteq V \times V$. Each vertex $n \in V$ is identified with the random variable X_n. If there is an edge $X_m \rightarrow X_n$ in G, then X_m is a parent of X_n and X_n is a child of X_m. For each vertex $n \in V$, there is an LPD $P(X_n \mid X_{\mathrm{pa}(n)})$, where $\mathrm{pa}(n)$ is the set of parents of X_n in G. The Bayesian network model is defined as the family of distributions for which the joint probability of X factors into conditional probabilities as:

$$P(X_1, \ldots, X_L) = \prod_{n=1}^{L} P(X_n \mid X_{\mathrm{pa}(n)}). \tag{88}$$

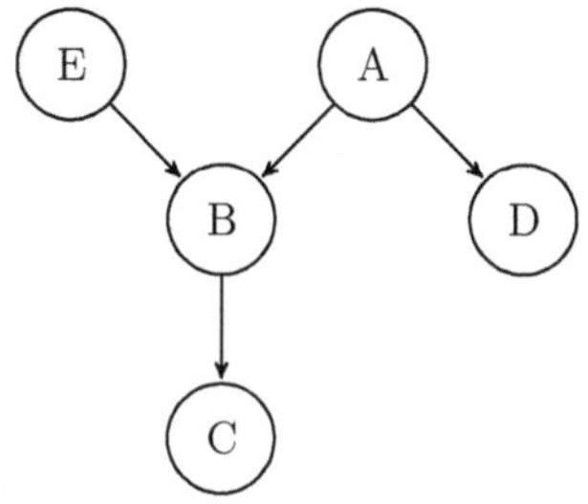

Fig. 11 Example of a Bayesian network. Vertices correspond to random variables and edges represent conditional probabilities. The graph encodes conditional independence statements about the random variables U, V, W, X, Y, and Z. Their joint probability factors according to the graph as $P(U, V, W, X, Y) = P(U)$ $P(Y)P(V \mid U, Y)P(W \mid V)P(X \mid U)$

In this case, we write $X \sim \text{BN}(G, \theta)$, where $\theta = (\theta_1, \ldots, \theta_L)$ denotes the parameters of the LPDs.

For the Bayesian network shown in Fig. 11, we find $P(U, V, W, X, \Upsilon) = P(U)P(\Upsilon)P(V \mid U, \Upsilon)P(W \mid V)P(X \mid U)$. The graph encodes several conditional independence statements about (U, V, W, X, Υ), including, for example, $W \perp \{U, X\} \mid V$.

Example 27 (Markov Chain): A finite Markov chain is a Bayesian network with the DAG $X_1 \to X_2 \to \cdots \to X_L$, denoted C, and joint distribution:

$$P(X_1, \ldots, X_n) = P(X_1)P(X_2 \mid X_1)P(X_3 \mid X_2) \cdots P(X_L \mid X_{L-1}). \tag{89}$$

If $X \sim \text{MC}(\Pi, T)$ is homogeneous, then the LPDs are $\theta_1 = P(X_1) = \Pi$ and $\theta_{n+1} = P(X_{n+1} \mid X_n) = T$ for all $n \in [L-1]$ such that $\text{MC}(\Pi, T) = \text{BN}(C, \theta)$. Similarly, HMMs are Bayesian networks with hidden variables Z and factorized joint distribution given in Eq. 79. □

The meaning of the parameters θ of a Bayesian network depends on the family of distributions that has been chosen for the LPDs. In the general case of a discrete random variable with finite state space, θ_n is a conditional probability table. If each vertex X_n has K possible states, then:

$$\theta_n = \left(P(X_n = a \mid X_{\text{pa}(n)} = b)\right)_{b \in [K]^{pa(n)}, a \in [K]} \tag{90}$$

has $K^{\text{pa}(n)} \times (K-1)$ free parameters. If X_n depends on all other variables, then θ_n has the maximal number of $K^L - 1$ parameters, which is exponential in the number of vertices. If, on the other hand, X_n is independent of all other variables, $\text{pa}(n) = \emptyset$, then θ_n has $(K-1)$ parameters, which is independent of L. For the chain (Example 27) where each vertex has exactly one outgoing and one incoming edge, we find a total of $(K-1) + (L-1)K(K-1)$ free parameters which is of order $O(LK^2)$.

A popular model for continuous random variables X_n is the linear Gaussian model. Here, the LPDs are Gaussian distributions with mean a linear function of the parents:

$$P(X_n \mid X_{\mathrm{pa}(n)}) = \mathrm{Norm}(v_n + w_n^t \cdot X_{\mathrm{pa}(n)},\ \sigma_n^2), \tag{91}$$

with parameters $v_n \in \mathbb{R}$ and $w_i \in \mathbb{R}^{\mathrm{pa}(n)}$ specifying the mean and variance σ_n^2. The number of parameters increases linearly with the number of parents, but only linear relationships can be modeled. All marginal and conditional probabilities of $(X_1, \ldots, X_L)$ are also Gaussians.

Learning a Bayesian network $\mathrm{BN}(G, \theta)$ from data $\mathcal{D}$ can be done in different ways following either the Bayesian or the maximum likelihood approach as introduced in Subheading 2. In general, it involves first finding the optimal network structure:

$$G^* = \underset{G}{\mathrm{argmax}}\ P(G \mid \mathcal{D}), \tag{92}$$

and then estimating the parameters:

$$\theta^* = \underset{\theta}{\mathrm{argmax}}\ P(\theta \mid G^*,\ \mathcal{D}) \tag{93}$$

for the given optimal structure G^*. The first step is a model selection problem as introduced in Subheading 2.

Model selection for Bayesian networks is a particularly hard problem because the number of DAGs increases super-exponentially with the number of vertices rendering exhaustive searches impractical, and the objective function in Eq. 92 is difficult to compute. Recall that the posterior $P(G \mid \mathcal{D})$ is proportional to the product $P(\mathcal{D} \mid G)P(G)$ of marginal likelihood and network prior, and the marginal likelihood:

$$P(\mathcal{D} \mid G) = \int P(\mathcal{D} \mid \theta,\ G)P(\theta \mid G)\ d\theta \tag{94}$$

is usually analytically intractable. Here, $P(\theta \mid G)$ is the prior distribution of parameters given the network topology.

To address this limitation, the marginal likelihood (Eq. 94) can be approximated by a function that is easier to evaluate. A popular choice is the Bayesian information criterion (BIC) [26]:

$$\log P(\mathcal{D} \mid G) \approx \log P(\mathcal{D} \mid \hat{\theta}_{\mathrm{ML}}, G) - \frac{1}{2}\nu \log N, \tag{95}$$

where ν is the number of free parameters of the model and N the size of the data. The BIC approximation can be derived under certain assumptions, including a unimodal likelihood. It replaces computation of the integral (Eq. 94) by evaluating the integrand at the MLE and adding the correction term $-(\nu \log N)/2$, which penalizes models of high complexity.

The model selection problem remains hard even with a tractable scoring function, such as BIC, because of the enormous search space. Local search methods, such as greedy hill climbing or simulated annealing, are often used in practice. They return a local maximum as a point estimate for the best network structure. Results can be improved by running several local searches from different starting topologies.

Often, data are sparse and we will find diffuse posterior distributions of network structures, which might not be represented very well by a single point estimate. In the fully Bayesian approach, we aim at estimating the full posterior $P(G \mid \mathcal{D}) \propto P(\mathcal{D} \mid G)P(G)$. One way to approximate this distribution is to draw a finite number of samples from it. Markov chain Monte Carlo (MCMC) methods generate such a sample by constructing a Markov chain that converges to the target distribution [27].

In the Metropolis–Hastings algorithm [28], we start with a random DAG $G^{(0)}$ and then iteratively generate a new DAG $G^{(n)}$ from the previous one $G^{(n-1)}$ by drawing it from a proposal distribution Q:

$$G^{(n)} \sim Q(G^{(n)} \mid G^{(n-1)}). \tag{96}$$

The new DAG is accepted with acceptance probability:

$$\min\left\{\frac{P(\mathcal{D} \mid G^{(n)})P(G^{(n)})Q(G^{(n-1)} \mid G^{(n)})}{P(\mathcal{D} \mid G^{(n-1)})P(G^{(n-1)})Q(G^{(n)} \mid G^{(n-1)})}, 1\right\} \tag{97}$$

Otherwise, the model is left unchanged and the next sample is drawn. With this acceptance probability, it is guaranteed that the Markov chain is ergodic and converges to the desired distribution. After an initial burn-in phase, samples from the stationary phase of the chain are collected, say $G^{(m)}, \ldots, G^{(N)}$. Any feature f of the network (e.g., the presence of an edge or a subgraph) can be estimated as the expected value:

$$\mathrm{E}(f) = \sum_G f(G)P(G \mid \mathcal{D}) \approx \frac{1}{N}\sum_{n=m}^{N} f(G^{(n)}). \tag{98}$$

A critical point of the Metropolis–Hastings algorithm is the choice of the proposal distribution Q, which encodes the way the network space is explored. Because not all graphs, but only DAGs, are allowed, computing the transition probabilities $Q(G^{(n)} \mid G^{(n-1)})$ is usually the main computational bottleneck.

Parameter estimation, i.e., solving (Eq. 93), can be done along the lines described in Subheading 2 following either the ML or the Bayesian approach. If the model contains hidden random variables, then the EM algorithm (Subheading 3) can be used. However, this approach is feasible only if efficient inference algorithms are available. For hidden Markov models (Subheading 6), the forward and backward algorithms provided an efficient way to compute marginal probabilities and the expected hidden log-likelihood. These algorithms can be generalized to the sum–product algorithm for tree-like graphs and the junction tree algorithm for general DAGs. The computational complexity of the junction tree algorithm is exponential in the size of the largest clique of the so-called moralized graph, which is obtained by dropping edge directions and adding edges between any two vertices that have a common child in the original DAG [11].

Alternatively, if exact inference is computationally too expensive, then approximate inference can be used. For example, Gibbs sampling [29] is an MCMC technique for generating a sample from the joint distribution $P(X_1, \ldots, X_L)$. The idea is to iteratively sample from the conditional probabilities of $P(X_1, \ldots, X_L)$, starting with $X_1^{(n+1)} \sim P(X_1 \mid X_2^{(n)}, \ldots, X_L^{(n)})$ and cycling through all variables in turns:

$$X_j^{(n+1)} \sim P(X_j \mid X_1^{(n+1)}, \ldots, X_{j-1}^{(n+1)}, X_{j+1}^{(n)}, \ldots, X_L^{(n)}) \quad \text{for all } j = 2, \ldots, L. \tag{99}$$

Gibbs sampling can be regarded as a special case of the Metropolis–Hastings algorithm. It is particularly useful, if it is much easier to sample from the conditionals $P(X_k \mid X_{\setminus k})$ than from the joint distribution $P(X_1, \ldots, X_L)$, where $X_{\setminus k}$ denotes all variables X_n except X_k. For graphical models, the conditional probability of each vertex X_k depends only on its Markov blanket $X_{\mathrm{MB}}(k)$, defined as the set of its parents, children, and co-parents (vertices with the same children), $P(X_k \mid X_{\setminus k}) = P(X_k \mid X_{\mathrm{MB}(k)})$.

Example 28 (Phylogenetic Tree Models): A phylogenetic tree model [30] for a set of aligned DNA sequences from different species is a Bayesian network model, where the graph is a tree in which the leaves represent the observed contemporary species and the interior vertices correspond to common extinct ancestors (Fig. 12). The topology (graph structure) S defines the branching order and the branch lengths correspond to (phylogenetic) time. The LPDs are defined by a nucleotide substitution model (Subheading 5).

Let $X^{(i)} \in \{\mathtt{A}, \mathtt{C}, \mathtt{G}, \mathtt{T}\}^L$ denote the i-th column of a multiple sequence alignment of L observed species. We regard the alignment columns as independent observations of the evolutionary process. The character states of the hidden (extinct) ancestors are denoted $Z^{(i)}$. The likelihood of the observed sequence data $X = (X^{(1)}, \ldots, X^{(N)})$ given the tree topology S and the branch lengths t is

$$P(X \mid S, t) = \sum_Z \prod_{i=1}^N P(X^{(i)}, Z^{(i)} \mid S, t), \tag{100}$$

where $P(X^{(i)}, Z^{(i)} \mid S, t)$ factors into conditional probabilities according to the tree structure. This marginal probability can be computed efficiently with an instance of the sum–product algorithm known as the peeling algorithm (or Felsenstein algorithm) [31].

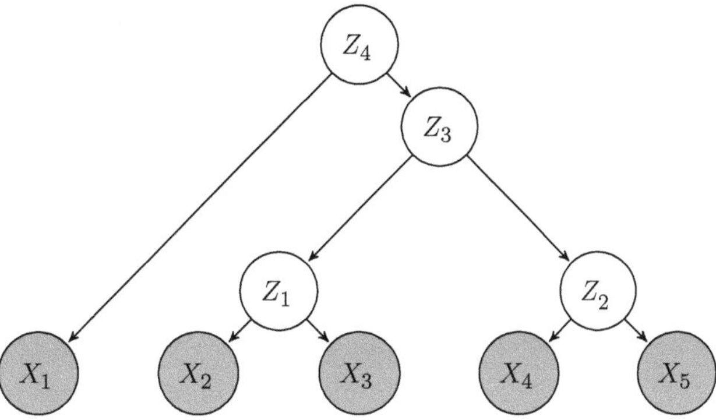

Fig. 12 Phylogenetic tree model. The observed random variables X_i represent contemporary species and the hidden random variables Z_i their unknown common ancestors

For example, in the tree displayed in Fig. 12, each observation X has probability:

$$P(X) = \sum_Z P(X, Z) \tag{101}$$

$$\begin{aligned} = \sum_Z & P(X_1 \mid Z_4) P(X_2 \mid Z_1) P(X_3 \mid Z_1) P(X_4 \mid Z_2) \cdot \\ & \cdot P(X_5 \mid Z_2) P(Z_1 \mid Z_3) P(Z_2 \mid Z_3) P(Z_3 \mid Z_4) P(Z_4) \end{aligned} \tag{102}$$

$$\begin{aligned} = \sum_{Z_4} P(Z_4) P(X_1 \mid Z_4) & \left[\sum_{Z_3} P(Z_3 \mid Z_4) \right. \\ & \left[\sum_{Z_2} P(Z_2 \mid Z_3) P(X_4 \mid Z_2) P(X_5 \mid Z_2) \right] \\ & \left. \cdot \left[\sum_{Z_1} P(Z_1 \mid Z_3) P(X_2 \mid Z_1) P(X_3 \mid Z_1) \right] \right], \end{aligned} \tag{103}$$

where we have omitted the dependency on the branch length t. Several software packages implement ML or Bayesian learning of phylogenetic tree models. □

In the simplest case, we suppose that the observed alignment columns are independent. However, it is more realistic to assume that nucleotide substitution rates vary across sites because of varying selective pressures. For example, there could be differences between coding and noncoding regions, among different regions of a protein (loops, and catalytic sites), or among the three bases of a triplet coding for an amino acid. More sophisticated models can account for this rate heterogeneity. Let us assume site-specific substitution rates r_i such that the local probabilities become

$P(X^{(i)} \mid r_i, t, S)$. To model the distribution of the rates, often a gamma distribution is used.

Example 29 (Gamma Distribution): The gamma distribution $\text{Gamma}(\alpha, \beta)$ is parametrized by a shape parameter α and a rate parameter β. It is defined by the density function:

$$f(x) = \frac{\beta^\alpha}{\Gamma(\alpha)} x^{\alpha-1} e^{-\beta x}, \quad \text{for } x \geq 0. \tag{104}$$

Its expectation is $\text{E}(X) = \alpha/\beta$ and its variance $\text{Var}(X) = \alpha/\beta^2$. The gamma distribution generalizes several other distributions, for example $\text{Gamma}(1, \lambda) = \text{Exp}(\lambda)$ (Example 22). □

Another approach to account for varying mutation rates are phylogenetic hidden Markov models (phylo-HMMs).

Example 30 (Phylo-HMM): Phylo-HMMs [32] combine HMMs and phylogenetic trees into a single Bayesian network model. The idea is to use an HMM along the linear chain of the genomic sequence and, at each position, to condition a phylogenetic tree model on the hidden state (Fig. 13). This architecture allows for modeling different evolutionary histories at different sites of the genome. In particular, the model can account for heterogeneity in the rate of evolution, for example, due to functionally conserved elements, but it also allows for a change in tree topology along the sequence, a situation that can result from recombination [23]. Phylo-HMMs are also used for gene finding. □

Exercise 31 (Inference in Bayesian Networks): Consider the gene network on five genes denoted A, B, C, D, E, with the graph structure displayed below. Gene expression profiles under different conditions C1–C9 have been observed and are summarized in the table below, where a zero indicates that the gene is not expressed and a one that it is expressed.

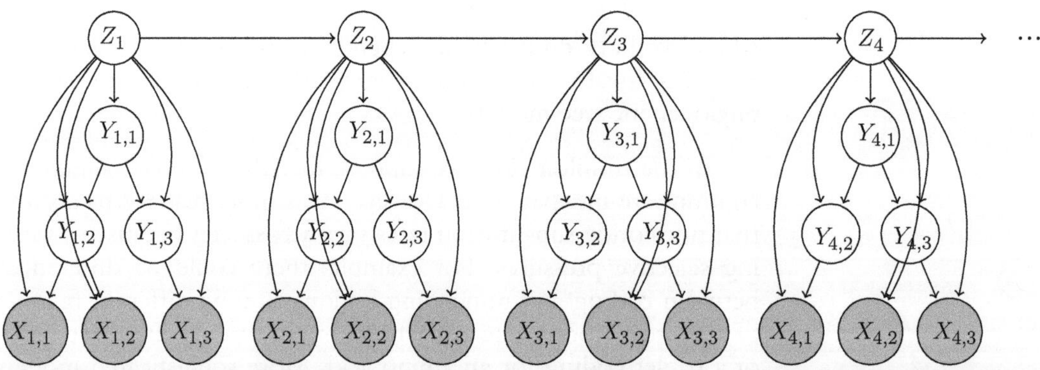

Fig. 13 Phylo-HMM. Shown are the first four positions of a Phylo-HMM. The hidden Markov chain has random variables Z. In the trees, Y denote the hidden common ancestors and X the observed species. Note that the tree topology changes between position 2 and 3

	A	B	C	D	E
C1	0	0	0	0	0
C2	0	0	0	0	1
C3	0	0	0	0	1
C4	1	1	1	1	0
C5	1	0	1	1	0
C6	0	0	0	1	1
C7	1	1	1	1	0
C8	1	0	0	0	1
C9	1	0	0	1	1

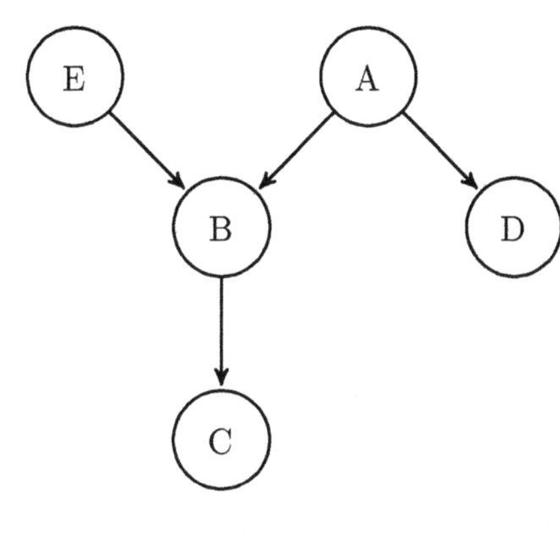

(a) Specify the adjacency matrix of the directed graph.

(b) Determine the local probability distributions for each vertex of the graph. Use conditional counting to determine the conditional probabilities as:

$$P(X_i \mid X_{\text{pa}(i)}) \approx \frac{N(X_i, X_{\text{pa}(i)})}{\sum_k N(X_i = k,\ X_{\text{pa}(i)})},$$

where $N(X_i,\ X_{\text{pa}(i)})$ is the number of joint observations of X_i and its parents.

(c) What is the joined probability of $(X_{\text{A}}, X_{\text{B}}, X_{\text{C}}, X_{\text{D}}, X_{\text{E}})$ for this network?

(d) We now want to determine the most probable explanation for observing a gene C to be active as a result of the influences of its upstream genes A and E. For this, one has to infer the posterior probabilities $P(\text{A} \mid \text{C} = 1)$ and $P(\text{E} \mid \text{C} = 1)$ using Bayes theorem. Here, assume that the probabilities $P(\text{A})$ and $P(\text{E})$ derived from the expression data are suitable prior probabilities. Which constellation is most likely to trigger the expression of C?

References

1. Ewens WJ, Grant GR (2005) Statistical methods in bioinformatics: an introduction, 2nd edn. Springer, Berlin
2. Deonier RC, Tavaré S, Waterman MS (2005) Computational genome analysis: an introduction. Springer, Berlin
3. Davison AC (2009) Statistical models. Cambridge series in statistical and probabilistic mathematics. Cambridge University Press, Cambridge
4. Ross SM (2007) Introduction to probability models. Academic, London

5. Hardy GH (1908) Mendelian proportions in a mixed population. Science 28:49–50
6. Weinberg W (1908) Über den Nachweis der Vererbung beim Menschen. Jahres Wiertt Ver Vaterl Natkd 64:369–382
7. Pachter L, Sturmfels B (2005) Algebraic statistics for computational biology. Cambridge University Press, Cambridge
8. Casella G, Berger RL (2002) Statistical Inference. Thomson Learning, Pacific Grove
9. Efron B, Tibshirani R (1993) An introduction to the bootstrap. Chapman and Hall/CRC, Boca Raton
10. Gelman A, Carlin JB, Stern HS, Rubin DB (2004) Bayesian data analysis, vol 46, 2 edn. Chapman and Hall/CRC, Boca Raton
11. Bishop CM (2006) Pattern recognition and machine leaning. Springer, Berlin
12. Kass R, Raftery A (1995) Bayes factors. J Am Stat Assoc 90:773–795
13. Dempster AP, Laird NM, Rubin DB (1977) Maximum likelihood from incomplete data via the EM algorithm. J R Stat Soc Ser B 39:1–38
14. Norris JR (1998) Markov chains. Cambridge University Press, Cambridge
15. Wright S (1990) Evolution in Mendelian populations. Bull Math Biol 52:241–295
16. Fisher RA (1930) The genetical theory of natural selection. Clarendon Press, Oxford
17. Jukes TH, Cantor CR (1969) Evolution of protein molecules. Mamm Protein Metab 3:21–132
18. Kimura M (1980) A simple method for estimating evolutionary rates of base substitutions through comparative studies of nucleotide sequences. J Mol Evol 16:111–120
19. Rabiner LR (1989) A tutorial on HMM and selected applications in speech recognition. Proc IEEE 77:257–286
20. Durbin R (1998) Biological sequence analysis. Probabilistic models of proteins and nucleic acids. Cambridge University Press, Cambridge
21. Viterbi A (1967) Error bounds for convolutional codes and an asymptotically optimum decoding algorithm. IEEE Trans Inf Theory 13:260–269
22. Baum LE (1972) An inequality and associated maximization technique in statistical estimation for probabilistic functions of Markov processes. Inequalities 3:1–8
23. Husmeier D, Dybowski R, Roberts S (2005) Probabilistic modeling in bioinformatics and medical informatics. Springer, New York
24. Koller D, Friedman N (2009) Probabilistic graphical models: principles and techniques. The MIT Press, Cambridge
25. Jordan MI (1998) Learning in graphical models. Kluwer Academic Publishers, Dordrecht
26. Schwarz G (1978) Estimating the dimension of a model. Ann Stat 6:461–464
27. Neal RM (1993) Probabilistic inference using Markov Chain Monte Carlo methods. Intelligence 62:144
28. Hastings WK (1970) Monte Carlo sampling methods using Markov chains and their applications. Biometrika 57:97–109
29. Geman S, Geman D (1984) Stochastic relaxation, Gibbs distributions, and the Bayesian restoration of images. IEEE Trans Pattern Anal Mach Intell 6:721–741
30. Felsenstein J (2004) Inferring phylogenies. Sinauer Associates, Sunderland
31. Felsenstein J (1981) Evolutionary trees from DNA sequences: a maximum likelihood approach. J Mol Evol 17:368–376
32. Siepel A, Haussler D (2005) Phylogenetic hidden Markov models. Statistical methods in molecular evolution. Springer, New York, pp 325–351

Chapter 3

A Not-So-Long Introduction to Computational Molecular Evolution

Stéphane Aris-Brosou and Nicolas Rodrigue

Abstract

In this chapter, we give a not-so-long and self-contained introduction to computational molecular evolution. In particular, we present the emergence of the use of likelihood-based methods, review the standard DNA substitution models, and introduce how model choice operates. We also present recent developments in inferring absolute divergence times and rates on a phylogeny, before showing how state-of-the-art models take inspiration from diffusion theory to link population genetics, which traditionally focuses at a taxonomic level below that of the species, and molecular evolution. Although this is not a cookbook chapter, we try and point to popular programs and implementations along the way.

Key words Likelihood, Bayes, Model choice, Phylogenetics, Divergence times

1 Introduction

Many books [1–7] and review papers [8–10] have been published in recent years on the topic of computational molecular evolution, so that updating our previous primer on the very same topic [11] may seem redundant. However, the field is continuously undergoing changes, as both models and algorithms become even more sophisticated, efficient, robust, and accurate. This increase in refinement has not been motivated by a desire to complicate existing models, but rather to make an old wish come true: that of having integrated methods that can take unaligned sequences as an input, and simultaneously output the alignment, the tree, and other estimates of interest, in a sound statistical framework justified by sound principles: those of population genetics.

The aim of this chapter is still to provide readers with the essentials of computational molecular evolution, offering a brief overview of recent progress, both in terms of modeling and algorithm development. Some of the details will be left out as they are dealt with by others in this volume. Likewise, the analysis of

Maria Anisimova (ed.), *Evolutionary Genomics: Statistical and Computational Methods*, Methods in Molecular Biology, vol. 1910, https://doi.org/10.1007/978-1-4939-9074-0_3, © The Author(s) 2019

genomic-scale data is briefly touched upon, but the details are left to other chapters.

2 Parsimony and Likelihood

2.1 A Brief Overview of Parsimony

The simplest phylogenetic question pertains to the reconstruction of a rooted tree with three sequences (Fig. 1). The sequences can be made of DNA, RNA, amino acids, or codons, but for the sake of simplicity we focus on DNA throughout this chapter. In the basic example below, based on [12], DNA sequences are assumed to have been sampled from three different species that diverged a "long time ago." In this context, we assume that the data or gene sequences have been aligned (*see* Subheading 6), and that the DNA alignment is:

s_1	ATGACCCCAATACGCAAAACTAACCCCCTAATAAAATTAATTAACCACTCCTTC
s_2	ATGACCCCAATACGGAAAACTAACCCCCAAATAAAATTAATTAACCACTCATTC
s_3	ATGACGCCAATACGCAAAACTAACCGCCTAATAAAATTAATTTACCACTCATTC

The objective is to estimate which of the three fully resolved topologies in Fig. 1 is supported by the data. In order to go further, we recode the data in terms of *site patterns*, which correspond to the patterns observed in each column of our alignment. This recoding implies that columns, or sites, in our alignment evolve according to an identically and independently distributed (iid) process. With this in mind, our alignment can be recoded in the following manner. When all the characters (nucleotides) in a column are identical, the same letter is assigned to each character, for example, x, irrespective of the actual character state. When a substitution occurs in one of the three sequences, we have three corresponding site patterns: xxy, xyx, and yxx, where the order within each site pattern respects the order of the sequences in the alignment, $s_1s_2s_3$.

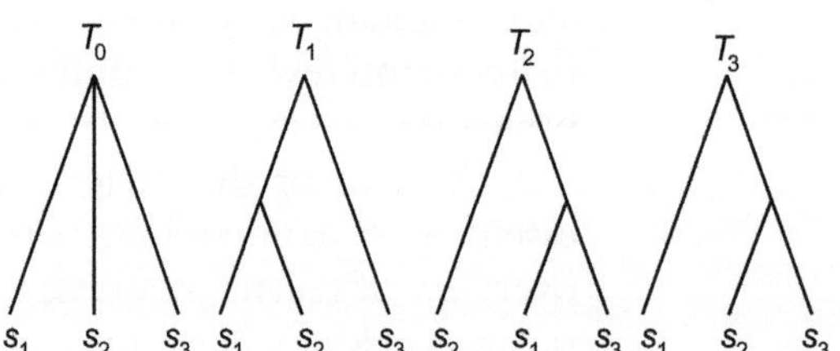

Fig. 1 The simplest phylogenetic problem. With three species, s_1 s_2, and s_3, four rooted trees are possible: T_0, the star tree, and the three resolved topologies T_1–T_3

s_1	`xxxxxxxxxxxxxxxxxxxxxxxxxxxxxxxxxxxxxxxxxxxxxxxxxxyxxx`
s_2	`xxxxxxxxxxxxxxyxxxxxxxxxxxxxyxxxxxxxxxxxxxxxxxxxxxxxxx`
s_3	`xxxxxyxxxxxxxxxxxxxxxxxxxyxxxxxxxxxxxxxxxxyxxxxxxxxxxx`

Table 1
The winning-site strategy

Site pattern	Supported T_i	Count
xxx	T_0	48
xxy	T_1	3
xyx	T_2	2
yxx	T_3	1

The data alignment is reduced to a frequency table of site patterns. In the case of three sequences, only the last three site patterns are informative

The first informative site pattern, xxy, implies that at this particular site, sequences s_1 and s_2 are more similar than any of these to s_3, so that this site pattern supports topology T_1, which groups sequences s_1 and s_2 together (Fig. 1). The most intuitive idea, called the winning-site strategy, is that the topology supported by the data corresponds to the fully resolved topology that has the largest number of site patterns in its favor. In the example shown above, topology T_1 is supported by three columns (with site pattern xxy), topology T_2 by two columns (xyx), and T_3 by one column (yxx; *see* Table 1). This is the intuition behind parsimony, which minimizes the amount of change along a topology. Strictly speaking, unordered parsimony cannot distinguish these three trees as they all require at least one single change. Yet, it can be argued that if tree T_1 is the true tree, site pattern xxy is more likely than any other patterns as xxy requires at least one change along a long branch (the one leading to sequence s_3) while both xyx and yxx require a change along a short branch (see p. 28 *sqq.* in [13]; [12]).

A number of methodological variations exist. A very condensed overview can be found in the books by Durbin [14] or, with more details, Felsenstein [15]. Most computer programs that implement substitution models where sites are iid condense the alignment as an array of site patterns; some, like PAML [16], even output these site patterns.

Note that in obtaining this topology estimate, most of the site columns were discarded from our alignment (all the xxx site patterns, representing 89% of the site in our example above). Most of our data were phylogenetically uninformative (for parsimony). We also failed to take evolutionary time into account, or any process of basic molecular biology, such as the observation that transitions (substitution of a purine [A or G] by a purine, or a pyrimidine by a pyrimidine) are more frequent than transversions (substitution between a purine and a pyrimidine).

2.2 Assessing the Reliability of an Estimate: The Bootstrap

As with any statistical exercise estimating a quantity of interest, we would like to have a confidence interval, taken at a particular level, so that we can gauge the reliability of our estimate. A standard approach to derive confidence intervals is the bootstrap [17], a computational technique that resamples data points with replacement to simulate the distribution of any test statistic under the null hypothesis that is tested. The bootstrap, particularly useful in complicated nonparametric problems where no asymptotic results can be obtained [18], was adapted by Felsenstein to the nonstandard phylogenetic problem [19]. Indeed, the problem is nonstandard in that the object for which we wish to assess accuracy is not a real-valued parameter, but a graph.

The basic idea, clearly explained in [20], consists in resampling columns of the alignment, with replacement, to construct a "synthetic" alignment of the same size as the original alignment. This synthetic or bootstrap replicate is then subjected to the same tree-reconstruction algorithm used on the original data (Fig. 2). This exercise is repeated a large number of times (e.g., $\times\ 10^6$), and the proportion of each original bipartition (internal node) in the set of bootstrapped trees is recorded. In Fig. 2, for instance, the bipartition $s_1s_2|s_3$ is found in two bootstrap trees out of three, so the bootstrap support for this node is 66.7%. In this simple case with three sequences, the bootstrap support for topology T_1 is also 66.7%. This bootstrap proportion for topologies (or for *trees* when branch lengths are taken into account, in a maximum likelihood context, for instance—see below) can be computed very quickly by bootstrapping the sitewise log-likelihood values, instead of the columns of the alignment; this bootstrap is called RELL, for "resampling estimated log-likelihood" [21].

However, this approach is no longer used or cited extensively since 2008 (source: ISI Thompson). One alternative that has gained momentum is the one based on the approximated likelihood ratio test (aLRT) [22], implemented, for instance, in `phyml` [23, 24]. Instead of resampling any quantity (sites or sitewise log-likelihood values), the aLRT tests the null hypothesis that an interior branch length is zero. In spite of being slightly conservative in simulations, the approach is extremely fast and hence highly practical [22].

The meaning of the bootstrap has been a matter of debate for years. As noted before [8] (see also [22]), the bootstrap proportion P can be seen as assessing the correctness of an internal node, and

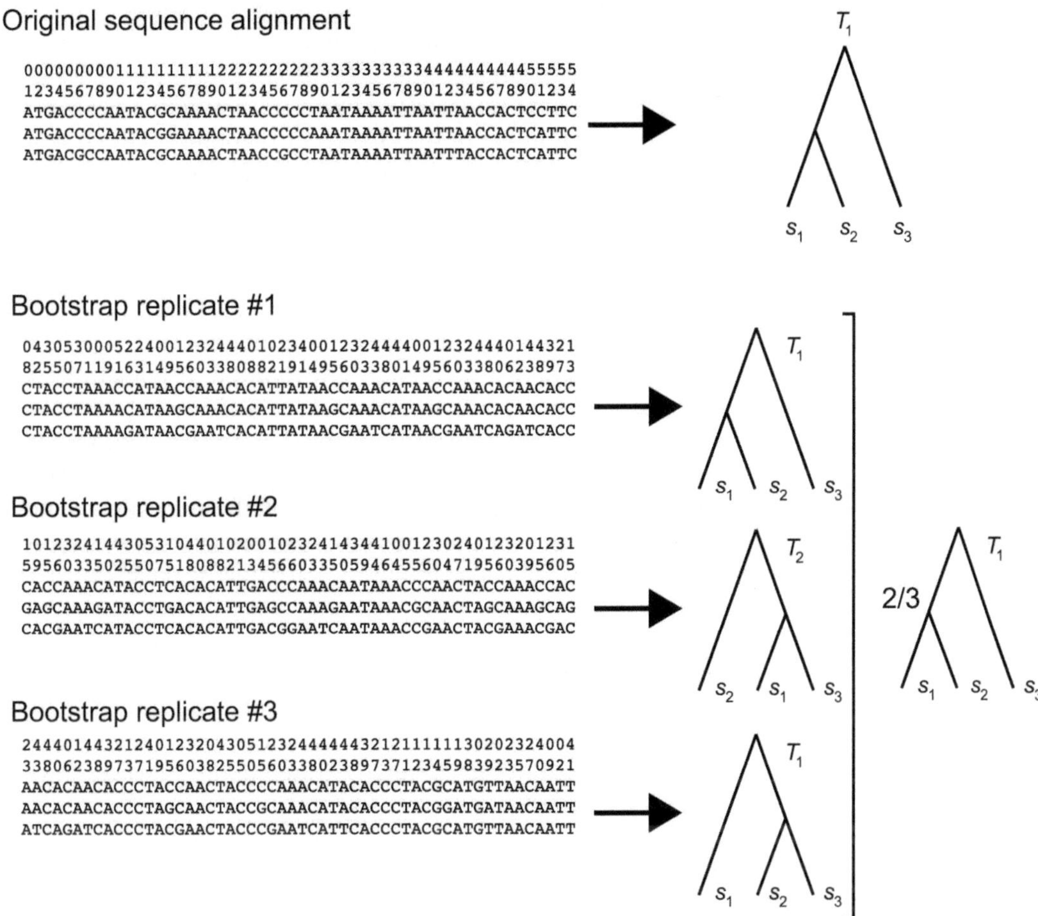

Fig. 2 The (nonparametric) bootstrap. See text for details

failing to do so [25], or $1 - P$ can be interpreted as a conservative probability of falsely supporting monophyly [26]. Since bootstrap proportions are either too liberal or too conservative depending on the actual interpretation of P [27], it is difficult to adjust the threshold below which monophyly can be confidently ruled out [28]. Alternatively, an intuitive geometric argument was proposed to explain the conservativeness of bootstrap probabilities [18] and was further developed into the approximately unbiased or AU test, implemented in CONSEL [29]. In spite of these difficulties, the bootstrap is still widely used—and mandatory in all publications featuring a phylogeny—to assess the confidence one can have in the tree estimated from the data under a particular scheme or model (*see* Subheading 2.9.3 below). Lastly, note that bootstrap support has often been abused [30], as a high value does not necessarily indicate high phylogenetic signal, and can be the result of systematic biases [31] due to the use of the wrong model of evolution, for instance, as detailed below.

2.3 Parsimony and LBA

Now that we have a means of evaluating the support for the different topologies, we can test some of the conditions under which parsimony estimates the correct tree topology. Ideally, a good method should return the correct answer with a probability of one when the number of sites increases to infinity. This desirable statistical property is called consistency. One serious criticism of parsimony is its sensitivity to long branch attraction, or LBA, even in the presence of an infinite amount of data (infinite alignment length) [31]. In other words, parsimony is not statistically consistent.

Different types of model misspecification can lead to LBA, and new ones are continually identified. The topology originally used to demonstrate the artifact is represented in Fig. 3, where two long branches are separated by a shorter one. Felsenstein demonstrated that, under a simple evolutionary process, the artifact or LBA tree is reconstructed. Note that parsimony is not the only phylogenetic method affected by LBA, but because it posits a very simple model of evolution [32–34], parsimony is particularly sensitive to the artifact. In spite of this, one particular journal chose to enforce the use of parsimony, stating that authors should estimate their phylogenies by parsimony but also that, if estimated by some other method, they would need to defend their position "on philosophical grounds" [35]; there is of course no valid scientific justification for taking such a step—derided in the "Twittersphere" as "*#parsimonygate.*"

The LBA artifact has been shown to plague the analysis of numerous data sets, and a number of empirical approaches have been used to detect the artifact [36, 37]. Most recent papers based on multigene analyses (e.g., [38, 39]) now examine carefully the effect of across-site and across-lineage rate variation (in addition to the use of heterogeneous models). For both sites and lineages, the

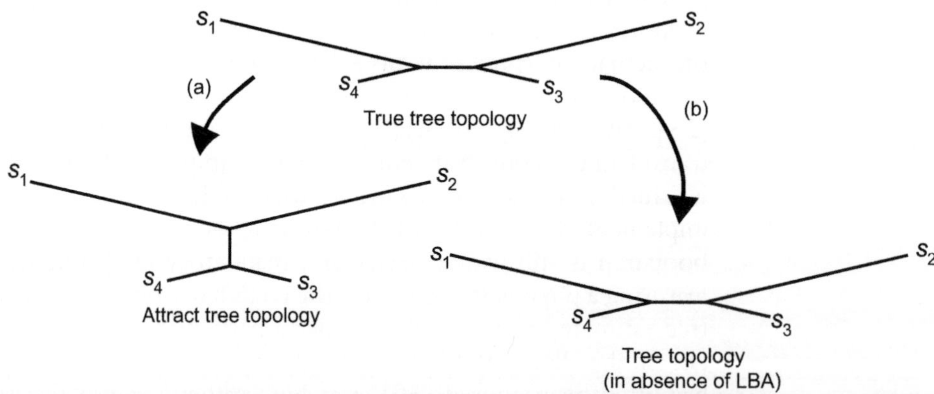

Fig. 3 The long branch attraction artifact. The true tree topology has two long branches separated by a short one. The tree reconstructed under a simple model of evolution (a) is the artifact or LBA tree on the left. The tree reconstructed under the correct model of evolution (b) is the correct tree, on the right

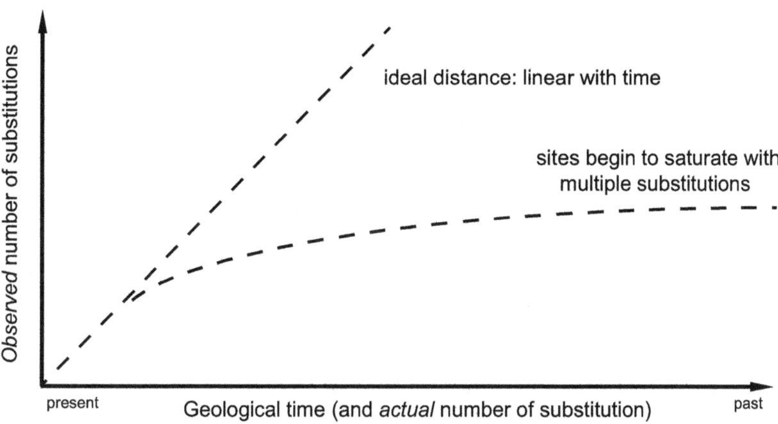

Fig. 4 Saturation of DNA sequences. As time increases, the *observed* number of differences between pairs of sequences reaches a plateau, whereas the *actual* number of substitutions keeps increasing

procedure is the same and consists in successively removing either the sites that evolve the fastest or the taxa that show the longest root-to-tip branch lengths.

2.4 Origin of the Problem

By definition, parsimony minimizes the number of changes along each branch of the tree. When there is only a small number of changes per branch, the method is expected to be accurate. However, when sequences are quite divergent, the parsimony assumption leads to underestimating the actual number of changes (Fig. 4; see also [40]).

Consequently, we would like a tree-reconstruction method that accounts for multiple substitutions. We would also like a method that (1) takes into account less parsimonious as well as most parsimonious state reconstructions (*intervals, tests*), (2) weights changes differently if they occur on branches of different length (*evolutionary time*), and (3) weights different kinds of events (transitions, transversions) differently (*biological realism*). Likelihood methods include such considerations explicitly, as they require modeling the substitution process itself.

2.5 Modeling Molecular Evolution

The basic model of DNA substitution (Fig. 5) is defined on the DNA *state space*, made of the four nucleotides thymine (T), cytosine (C), adenine (A), and guanine (G). Note that T and C are pyrimidines (biochemically, six-membered rings), while A and G are purines (fused five- and six-membered heterocyclic compounds). Depending on these two biochemical categories, two different types of substitutions can happen: transitions within a category, and transversions between categories. Their respective rates are denoted α and β in Fig. 5.

The process we want to model should describe the substitution process of the different nucleotides of a DNA sequence. Again, we

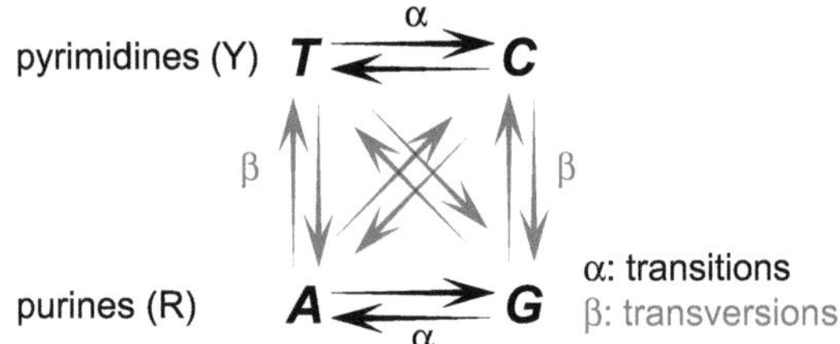

Fig. 5 Molecular evolution 101. Specification of the basic model of DNA substitution

will make the simplifying assumption that sites evolve under a time-homogeneous Markov process and are iid, as above. We can therefore concentrate on one single site for now (e.g., [41]).

At a particular site, we want to describe the change in nucleotide frequency after a short amount of time dt, during which the nucleotide frequency of A, for instance, after dt will change from $f_A(t)$ to $f_A(t + dt)$. According to Fig. 5, $f_A(t + dt)$ will be equal to what we had at time t, $f_A(t)$, minus the quantity of A that "disappeared" by mutation during dt, plus the quantity of A that "appeared" by mutation during dt. Denoting the mutation rate as μ, the quantity of A that "disappeared" by mutation during dt is simply $f_A(t)\mu_A dt$. These mutations away from A generated quantities of T, C, and G, in which we are not interested at the moment since we only want to know what happens to A. There are three different ways to generate A: from either T, C, or G (Fig. 5). Coming from T, mutation will generate $f_T(t)\mu_{T\to A}dt$ of A during dt. Similar expressions exist for C and for G, so that in total, over the three non-A nucleotides, mutation will generate $\Sigma_{i\neq A} f_i(t)\mu_{i\to A}dt$. Mathematically, we can express these ideas as:

$$f_A(t + dt) = f_A(t) - f_A(t)\mu_A dt + \sum_{i\neq A} f_i(t)\mu_{iA} dt \tag{1}$$

Equation 1 describes the change of frequency of A during a short time interval dt. Similar equations can be written for T, C, and G, so that we actually have a system of four equations describing the change in nucleotide frequencies over a short time interval dt:

$$\begin{cases} f_T(t + dt) = f_T(t) - f_T(t)\mu_T dt + \sum_{i\neq T} f_i(t)\mu_{iT} dt \\ f_C(t + dt) = f_C(t) - f_C(t)\mu_C dt + \sum_{i\neq C} f_i(t)\mu_{iC} dt \\ f_A(t + dt) = f_A(t) - f_A(t)\mu_A dt + \sum_{i\neq A} f_i(t)\mu_{iA} dt \\ f_G(t + dt) = f_G(t) - f_G(t)\mu_G dt + \sum_{i\neq G} f_i(t)\mu_{iG} dt \end{cases} \tag{2}$$

which, in matrix notation, can simply be rewritten as:

$$F(t + dt) = F(t) + QF(t)dt \tag{3}$$

with an obvious notation for F, while the *instantaneous rate matrix* Q is

$$Q = \begin{pmatrix} -\mu_T & \mu_{TC} & \mu_{TA} & \mu_{TG} \\ \mu_{CT} & -\mu_C & \mu_{CA} & \mu_{CG} \\ \mu_{AT} & \mu_{AC} & -\mu_A & \mu_{AG} \\ \mu_{GT} & \mu_{GC} & \mu_{GA} & -\mu_G \end{pmatrix} \tag{4}$$

In all the following matrices, we will use the same order for nucleotide: T, C, A, and G, which follows the order in which codon tables are usually written. Recall that μ_{ij} is the mutation rate from nucleotide i to nucleotide j. Note also that the sum of each row is 0.

Let us rearrange the matrix notation from Eq. 3 as:

$$F(t + dt) - F(t) = QF(t)dt \tag{5}$$

and take the variation limit when $dt \to 0$:

$$\frac{dF(t)}{dt} = QF(t) \tag{6}$$

which is a first-order differential equation that can be integrated as:

$$F(t) = e^{Qt}F(0) \tag{7}$$

Very often, this last equation 7 is written as $F(t) = P(t)F(0)$, where $F(0)$ is conveniently taken to be the identity matrix and $P(t) = \{P_{i,j}(t)\} = e^{Qt}$ is the matrix of probabilities of going from state i to j during a finite time duration t. Note that the right-hand side of this equation is a matrix exponentiation, which is not the same as the exponential of all the elements (row and columns) of that matrix. The computation of the term e^{Qt} demands that a spectral decomposition of the matrix Q be realized. This means finding a diagonal matrix D of eigenvalues and a matrix M of (right) eigenvectors so that:

$$P(t) = Me^{Dt}M^{-1} \tag{8}$$

The exponential of the diagonal matrix D is simply the exponential of the diagonal terms.

Except in the simplest models of evolution, finding analytical solutions for the eigenvalues and associated eigenvectors can be tedious. As a result, numerical procedures are employed to solve Eq. 8. Alternatively, a Taylor expansion can be used to *approximate* $P(t)$.

If all entries in Q are positive, any state or nucleotide can be reached from any other in a finite number of steps (all states "communicate") and the base frequencies have a stationary distribution $\pi = (\pi_T, \pi_C, \pi_A, \pi_G)$. This is the steady state reached after an "infinite" amount of time, or long enough for the Markov process to forget its initial state, starting from "random" base frequencies.

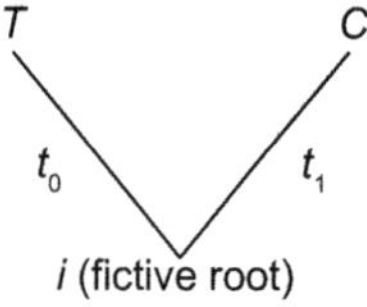

Fig. 6 Likelihood computation on a small tree. See text for details

2.6 Computation on a Tree

Now that we know how to determine the rate of change of nucleotide frequencies during a time interval dt, we can compute the probability of a particular nucleotide change on a tree. The simplest case, though somewhat artificial with only two sequences, is depicted in Fig. 6.

We are looking at a particular nucleotide position, denoted j, for two aligned sequences. The observed nucleotides at this position are T in sequence 1, and C in sequence 2. The branch separating T from C has a total length of $t_0 + t_1$. For the sake of convenience, we set an arbitrary root along this path. The likelihood at site j is then given by the probability of going from the fictive root i to T in t_0, and from i to C in t_1. Any of the four nucleotides can be present at the fictive root. As we do not know which one was there, we sum these probabilities over all possible state, weighted by their prior probabilities, the equilibrium frequencies π_i. In all, we have the likelihood ℓ_j at site j:

$$\ell_j = \sum_{i=\{T,C,A,G\}} \pi_i P_{i,T}(t_0) P_{i,C}(t_1) \tag{9}$$

which is equivalent to the Chapman–Kolmogorov equation [42]. As all the sites are assumed to be iid, the likelihood of an alignment is the product of the site likelihoods in Eq. 9. Because all these sitewise probabilities can be small, and that the product of small numbers can become smaller than what a computer can represent in memory (*underflow*), all computations are done on a logarithmic scale and may include some form of rescaling [43].

Note that this example is somewhat artificial: with only two sequences, we can compute the likelihood directly with $\pi_T P_{T,C}(t_0 + t_1) = \pi_C P_{C,T}(t_0 + t_1)$; the full summation over unknown states as in Eq. 9 is required with three sequences or more. When analyzing a multiple-sequence alignment of S sequences, there will be many nodes in the tree for which the character state is unknown, which means that the summation required will involve many terms. Specifically, the sum will be over 4^{S-3} terms. Fortunately, terms can be factored out of the summation, and a dynamic programing algorithm with a complexity of the order of $\mathcal{O}(4^2 S)$, called the pruning algorithm [44], can be used (*see* [15] for details).

2.7 Substitution Models and Instantaneous Rate Matrices Q

Now that we have almost all the elements to compute the likelihood of a set of parameters, including the tree (i.e., the set of branch lengths *and* the tree topology; *see* Subheading 2.10), the only missing element required to compute the likelihood at each site, as in Eq. 9, for instance, is the specification of the instantaneous rate matrix Q as in Eq. 4. Remember that the $\mu_{i,j}$ represent mutation rates from state (nucleotide) i to j. This matrix is generally rewritten as:

$$Q = \mu \begin{pmatrix} - & r_{TC} & r_{TA} & r_{TG} \\ r_{CT} & - & r_{CA} & r_{CG} \\ r_{AT} & r_{AC} & - & r_{AG} \\ r_{GT} & r_{GC} & r_{GA} & - \end{pmatrix} \tag{10}$$

so that each entry r_{ij} is a rate of change from nucleotide i to nucleotide j. The diagonal entries are left out, indicated by a "−," and are in fact calculated as the negative sum of the off-diagonal entries (as rows sum to 0).

The simplest specification of Q would be that all rates of change are identical, so that Q becomes (leaving out the mutation rate μ and indexing the matrix to indicate the difference):

$$Q_{\mathrm{JC}} = \begin{pmatrix} - & 1 & 1 & 1 \\ 1 & - & 1 & 1 \\ 1 & 1 & - & 1 \\ 1 & 1 & 1 & - \end{pmatrix} \tag{11}$$

which is the model proposed by Jukes and Cantor [45] and often noted "JC" or "JC69". Under the specification of Eq. 11, this model has no free parameter. The process is generally scaled such that the unit of branch lengths can be interpreted as an expected number of substitutions per site.

Of course, this model is extremely simplistic and neglects a fair amount of basic molecular biology. In particular, it overlooks two observations. First, base frequencies are not all equal in actual DNA sequences, but are rather skewed, and second, transitions are more frequent than transversions (*see* Subheading 2.5).

The way to account for this first "biological realism" is as follows. If DNA sequences were made exclusively of *A*s, for instance, that would mean that all mutations are towards the observed base, in this case *A*, whose equilibrium or stationary frequency is π_A. The same reasoning can be used for arbitrary equilibrium frequencies π, so that all relative rates of change in Q become proportional to the vector of equilibrium frequency π of the *target* nucleotide. In other words, the instantaneous rate matrix Q becomes:

$$Q_{\mathrm{F81}} = \begin{pmatrix} - & \pi_C & \pi_A & \pi_G \\ \pi_T & - & \pi_A & \pi_G \\ \pi_T & \pi_C & - & \pi_G \\ \pi_T & \pi_C & \pi_A & - \end{pmatrix} \tag{12}$$

again with the requirement that rows sum to 0. This matrix represents the Felsenstein or F81 model [44]. This model has four parameters (the four base frequencies), but since base frequencies sum to 1, we only have three *free* parameters.

The second "biological realism," accounting for the different rates of transversions and transitions, can be described by saying that transitions occur κ times faster than transversions. From Fig. 5, recall that transitions are mutations from T to C (and vice versa) and from A to G (and vice versa). This translates into:

$$Q_{\mathrm{K80}} = \begin{pmatrix} - & \kappa & 1 & 1 \\ \kappa & - & 1 & 1 \\ 1 & 1 & - & \kappa \\ 1 & 1 & \kappa & - \end{pmatrix} \tag{13}$$

This model is called the Kimura two-parameter model or K80 (or K2P) [46]. The model is alternatively described with the two rates α and β (*see* Fig. 5). In the "κ version" of the model as in Eq. 13, there is only one free parameter.

Of course it is possible to account for both kinds of "biological realism," unequal equilibrium base frequencies and transition bias, all in the same model, whose generator Q becomes:

$$Q_{\mathrm{HKY}} = \begin{pmatrix} - & \pi_C\kappa & \pi_A & \pi_G \\ \pi_T\kappa & - & \pi_A & \pi_G \\ \pi_T & \pi_C & - & \pi_G\kappa \\ \pi_T & \pi_C & \pi_A\kappa & - \end{pmatrix} \tag{14}$$

which corresponds to the Hasegawa–Kishino–Yano or HKY (or HKY85) model [47]. This model has four free parameters: κ and three base frequencies.

The level of "sophistication" goes "up to" the general time-reversible model [48], denoted GTR or REV, which has for generator:

$$Q_{\mathrm{GTR}} = \begin{pmatrix} - & a\pi_C & b\pi_A & c\pi_G \\ a\pi_T & - & d\pi_A & e\pi_G \\ b\pi_T & d\pi_C & - & \pi_G \\ c\pi_T & e\pi_C & \pi_A & - \end{pmatrix} \tag{15}$$

The number of free parameters is now eight (three base frequencies plus five nucleotide propensities). The name is derived from the time-reversibility constraint, which implies that the likelihood is independent of the actual orientation of time.

In fact, there exist only a few "named" additional substitution models [15], most of which are time-reversible models, while a total of 203 models can be derived from GTR [49]. We have focused solely on DNA models in this chapter, but the problem is similar with amino acid or codon models, except that the number of parameters increases quickly. We have also limited ourselves to time-reversible time-homogeneous models, but irreversible non-homogeneous models were developed some time ago [50] and are used, for instance, to root phylogenies [51] or to help alleviate the effects of LBA [39].

2.8 Some Computational Aspects

2.8.1 Optimization of the Likelihood Function

For a given substitution model, how should parameters be estimated, given the (potentially) high dimensionality of the model? Analytical solutions consist in determining when the first derivative of the likelihood function is equal to zero (with a change of sign in the second derivative). However, finding the root of the likelihood function analytically is only possible in the simple case of three sequences of binary characters under the assumption of the molecular clock (*see* Subheading 3.1) [12]. As a result, numerical solutions must be found to maximize the likelihood function.

A number of ideas have been combined to search efficiently for the parameter values that maximize the likelihood function. Most programs will start from a random starting point, for example, $(\theta_1^{(0)}, \theta_2^{(0)})$, denoted by an x in Fig. 7, where we limit ourselves to a two-parameter example. The optimization procedure can follow

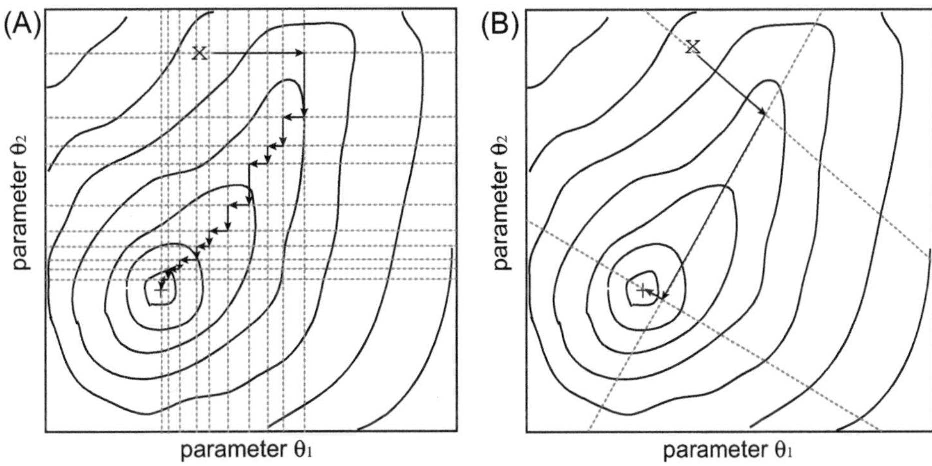

Fig. 7 Two optimization strategies. The likelihood surface of a function with two parameters θ_1 and θ_2 (e.g., two branch lengths) is depicted as a contour plot, whose highest peak is at the + sign. (**a**) Optimization one parameter at a time. (**b**) Optimization of all parameters simultaneously. See text for details

one of the two strategies. In the first one, parameters are optimized one at a time. In Fig. 7a, parameter θ_1 is first optimized to maximize the likelihood function with a line search, which defines a direction along which the other parameter (θ_2) or parameters in the multidimensional case are kept constant. Once $\theta_1^{(1)}$ is found, a new direction is defined to optimize θ_2, and so on so forth until convergence to the maximum of the likelihood function. As shown in Fig. 7a, many iterations can be required, in particular when the parameters θ_1 and θ_2 are correlated. The alternative to optimizing one parameter at a time is to optimize all parameters simultaneously. In this case (Fig. 7b), an initial direction is defined at $(\theta_1^{(0)}, \theta_2^{(0)})$ such that the slope at this point is maximized. The process is repeated until convergence. More technical details can be found in [52]. The simultaneous optimization procedure generally requires fewer steps than optimizing parameters one at a time, but not always. Since the computation of the likelihood function is the most expensive computation of these algorithms, the simultaneous optimization is much more efficient, at least in our toy example.

How general is this result? Simultaneously optimizing parameters of the substitution model, while optimizing branch lengths one at a time, was shown to be more effective on large data sets [43], potentially because of the correlation that exists between some of the parameters entering the Q matrix (*see* Subheading 2.7).

2.8.2 Convergence

Convergence is usually reached either when the increment in the log-likelihood score becomes smaller than an ε value, usually set to a small number such as 10^{-6} (but yet a number larger than the machine ε: the smallest number that a given computer can represent), or when the log-likelihood score has not changed after a predetermined number of iterations. However, none of these stopping rules guarantees that the global maximum of the likelihood function has been found. Therefore, it is generally recommended to run the optimization procedure at least twice, starting from different initial values of the model parameters, and to check that the likelihood score after optimization is the same across the different runs (Fig. 8). If this is not the case, additional runs may be required, and the one with the largest likelihood is chosen for inference (e.g., [53]).

In many instances though, different substitution models will give different tree topologies, and therefore different biological conclusions. One difficulty is therefore to know which model should be used to analyze a particular data set.

2.9 Selection of the Appropriate Substitution Model

One important issue in model selection is about the trade-off between bias and variance [55]: a simple model will fail to capture all the sophistication of the actual substitution process, and will

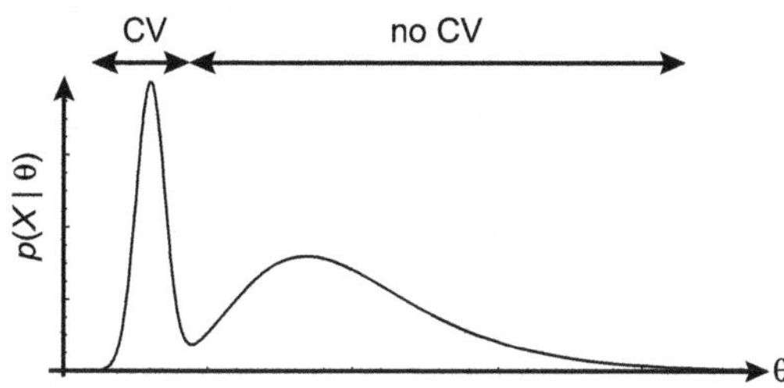

Fig. 8 Likelihood surfaces behaving badly. Schematic of the probability surface of the function $p(X|\theta)$ is plotted as a function of θ. Most line search strategies will converge (CV) to the MLE when the initial value is in the "CV" interval, and fail when it is in the "no CV" interval. Adapted with permission from [54]

therefore be highly biased even if all the parameters can be estimated with tight precision (little variance). Alternatively, a highly parameterized model will "spread" the information available from the data over a large number of parameters, hereby making their estimation difficult (flat likelihood surface; *see* Subheading 2.8.1), with a large variance, in spite of perhaps being a more realistic model with less bias. The objective of most model selection procedure is therefore to find not the *best* model in terms of likelihood score, but the *most appropriate* model, the one that strikes the right balance between bias and variance in terms of number of parameters. However, we argue that optimizing for this bias–variance trade-off works only for statistical procedures, be they, for instance, frequentist (LRT, likelihood ratio test) or Bayesian (BF, Bayes factor). On the other hand, information-theoretic criteria (e.g., AIC, Akaike information criterion) aim at selecting the model that is *approximately* closest to the "true" biological process.

The bias–variance trade-off mainly concerns the comparison of models that are based on the same underlying rationale, for instance, choosing among the 203 models that can be derived from GTR. We may also be interested in comparing models that are based on very different rationales. The likelihood ratio test is suited for assessing the bias–variance trade-off, while Bayesian and information-theoretic approaches, as well as cross-validation (CV), can be used for more general model comparisons. Here we review four approaches to model selection: LRT, BF, AIC, and CV.

2.9.1 The Likelihood Ratio Test

The substitution models presented above have one key property: it is possible to reduce the most sophisticated time-reversible named model (GTR+Γ+I) to any simpler model by imposing some constraints on parameters. As a result, the models are said to be nested, and statistical theory (the Neyman–Pearson lemma) tells us that there is an optimal (most powerful) way of comparing two nested models (a simple null vs. a simple alternative hypothesis) based on the likelihood ratio test or LRT.

The test statistic of the LRT is twice the log-likelihood difference between the most sophisticated model (which by definition is always the one with the highest likelihood—if this is not the case, there is a convergence issue; *see* Subheading 2.8.1) and the simpler model. This test statistic follows asymptotically a χ^2 distribution (under certain regularity conditions), and the degree of freedom of the test is equal to the difference in the number of free parameters between the two models.

The null hypothesis is that the two competing models explain the data equally well. The alternative is that the most sophisticated model explains the data better than the simpler model. If the null hypothesis cannot be rejected at a certain level (type-I error rate), then, based on the argument developed above, the simpler model should be used to analyze the data. Otherwise, if the null hypothesis can be rejected, the more sophisticated model should be used to analyze the data. Note that a test never leads to accepting a null hypothesis; the only outcomes of a test are either *reject* or *fail to reject* a null hypothesis.

Intuitively, we can see the null hypothesis H_0 as stating that a certain parameter θ is equal to θ_0. The maximum likelihood estimate (MLE) is at $\hat{\theta}$, which is our alternative hypothesis H_1, left unspecified. We note the log-likelihood as $\ln p(X|\theta) = \ell(\theta)$, where X represents the data. Under H_0, we have $\theta = \theta_0$, while under H_1 we have $\theta = \hat{\theta}$. The log-likelihood ratio is therefore $\ln LR = \ell(\hat{\theta}) - \ell(\theta_0)$. Under the null H_0, $\ell(\hat{\theta}) = 0$ (by definition). The log-likelihood ratio then reduces to $\ln LR = -\ell(\theta_0)$. We can then take the Taylor expansion of the log-likelihood function ℓ around $\hat{\theta}$, which gives us $\ell \approx \frac{1}{2}(\hat{\theta} - \theta_0)^2 \frac{d^2\ell}{d\theta^2}$ (recall that $\ell(\hat{\theta}) = 0$, so that the first terms of the series "disappear"). Therefore, log-likelihood ratio can be approximated by $-\frac{1}{2}(\hat{\theta} - \theta_0)^2 \frac{d^2\ell}{d\theta^2}$. Recall that Fisher's information is negative reciprocal of the second derivative of the likelihood function, so that:

$$\ln LR \approx \frac{1}{2} \frac{(\hat{\theta} - \theta_0)^2}{var(\theta)} \tag{16}$$

which follows asymptotically half a χ^2 distribution. Hence the usual approximation:

$$2 \ln LR = 2 \times (\ell_1 - \ell_0) \sim \chi^2_k \tag{17}$$

with k being the difference in the number of free parameters between the two models 0 and 1. The important points in this intuitive outline of the proof are that (1) the two hypotheses need to be nested and (2) taking the Taylor expansion around $\hat{\theta}$ requires that the likelihood function be continuous at that point, which implies that ℓ is differentiable left *and* right of $\hat{\theta}$. Therefore, testing points at the boundary of the parameter space cannot be done by approximating the distribution of the test statistic of the LRT by a regular χ^2 distribution, as noted many times in molecular evolution [56–64]. A solution still involves the LRT, but the asymptotic distribution becomes a mixture of χ^2 distributions [65].

An approach that has become popular under the widespread adoption of computer programs such as `ModelTest` [66] and `jModelTest` [67] is the hierarchical LRT (hLRT). This hierarchy goes from the simplest model (JC) to the set of most complex models (+Γ+I), traversing a tree of models. The issue is that there is more than one way to traverse this tree of models, and that depending on which way is adopted, the procedure may end up selecting different models [68, 69].

2.9.2 Information-Theoretic Approaches

Information theory provides us with a number of solutions to circumvent the three limitations of the LRT (nestedness, continuity, and dependency on the order in which models are compared).

The core of the information-based approach is the Kullback–Leibler (KL) distance, or information [70], which measures the distance between an approximating model g and a "true" model f [55]. This distance is computed as:

$$d_{\mathrm{KL}}(f,g) = \int f(x) \ln \frac{f(x)}{g(x|\theta)} dx \tag{18}$$

where θ is a vector of parameters entering the approximating model g and x represents the data. Note that this distance is not symmetric, as typically $d_{\mathrm{KL}}(f, g) \neq d_{\mathrm{KL}}(g, f)$, and that the "true" model f is unknown. The idea is to rewrite $d_{\mathrm{KL}}(f, g)$ in a slightly different form, to make it clear that Eq. 18 is actually a difference between two expectations, both taken with respect to the unknown "truth" f:

$$d_{\mathrm{KL}}(f,g) = E_f[f(x)\ \ln f(x)] - E_f[f(x)\ \ln g(x|\theta)] \tag{19}$$

Equation 19 therefore measures the loss of information incurred by fitting g when the data x actually come from f. As f is unknown, $d_{\mathrm{KL}}(f, g)$ cannot be computed as such.

Two points are key to deriving the criterion proposed by Akaike (*see* [55]). First, we usually want to compare at least two approximating models, g_0 and g_1. We can then measure which one is closest to the "true" process f by taking the difference between their respective Kullback–Leibler distances. In the process, the direct reference to the "true" process cancels out. As a result, the "best" model among g_0 and g_1 is the one that is closest to the "true" process f: it is the model that *minimizes* the distance to f. By setting model parameters to their MLEs, we now deal with *estimated* distances, but these are still with respect to the unknown f.

Second, in the context of a frequentist approach, we would repeat the experiment of sampling data an infinite number of times.

We would then compute the *expected estimated KL distance*, so that model selection can be done on the sole estimated log-likelihood value. Akaike, however, showed that this latter approximation is biased, and must be adjusted by a term that is *approximately equal* to the number of parameters K entering model g (*see* [55]). For "historical reasons" (similarity with asymptotic theory with the normal distribution), the selection criterion is multiplied by 2 to give the well-known definition of the Akaike information criterion or AIC:

$$\mathrm{AIC} = -2 \ln \ell(\hat{\theta}) + 2K \tag{20}$$

Unlike the case of the hLRT, where we were selecting the "most appropriate model" (with respect to the bias–variance trade-off), in the case of AIC we can select the *best* model. This best model is the one that is closest to the "true" unknown model (f), with the smallest relative estimated expected KL distance. The best AIC model therefore minimizes the criterion in Eq. 20.

A small-sample second-order version of AIC exists, where the penalty for extra parameters ($2K$ in Eq. 20) is slightly modified to account for the trade-off between information content in the data and K (*see* [55]). In our experience, we find it advisable to use this small-sample correction irrespective of the actual size of the data, since this correction vanishes in large and informative samples, but corrects for proper model ranking when K becomes very large compared to the amount of information (e.g., in phylogenomics where models are partitioned with respect to hundreds of genes).

The AIC has been shown to tend to favor parameter-rich models [71–75], which has motivated the use and development of alternative approaches in computational molecular evolution. These include, the Bayesian information criterion [76], and the decision theory or DT approach, which is based on ΔAIC weighted by squared branch length differences [71]. Most of these approaches, including the hLRT, have recently been compared in a simulation study that suggests, in agreement with empirical studies [72, 77], that both BIC and DT have the highest accuracy and precision [75].

One particular drawback of these information-theoretic approaches is that they require that every single model of evolution, or at least the most "popular" models (the few named ones), be evaluated. This step can be time-consuming, especially if a full maximum likelihood optimization is performed under each model. A first set of heuristics consists in fixing the tree topology to a tree estimated with a quick distance-based method such as BioNJ [78], and then estimating just the branch lengths and the parameters of the substitution model, as implemented in `jModelTest` [67]. As the optimizations are independent of each other under each substitution model, these computations are typically forked to multiple cores or processors [79]. Further heuristics exist to avoid all these independent optimizations [79], as implemented in `SMS` (*Smart Model Selection* in *PhyML*), which is reported to be cutting runtimes in half without forfeiting accuracy [80].

Note finally that all these approaches are not limited to selecting the most appropriate or the best model of evolution. Disregarding the hLRT, which requires that models be nested (to be able to use the χ^2 approximation; otherwise, *see* [65]), AIC, BIC, etc. allow us to compare non-nested models and, in particular, phylogenetic trees (branch lengths plus topology).

2.9.3 The Bayesian Approach

The Bayesian framework has permitted the development of two main approaches, which are actually two sides of the same coin: one based on finding the model that is the most probable a posteriori, and one based on ranking models and estimating a quantity called the Bayes factor.

In a nutshell, the frequentist approaches developed in the previous sections are based on the likelihood, which is the probability of the data, given the parameters: $p(X|\theta)$. However, this approach may not be the most intuitive, since most practitioners are not interested in knowing the conditional probability of their data, as the data were collected to learn more about the processes that generated them. It can therefore be argued that the Bayesian approach, which considers the probability of the parameters given the data or $p(\theta|X)$, is more intuitive than the frequentist approach. Unlike likelihood, which relies on the function $p(X|\theta)$ and permits point estimation, Bayesian inference is based on the posterior distribution $p(\theta|X)$. This distribution is often summarized by a centrality measure such as its mode, mean, or median. Measures of uncertainty are based on *credibility* intervals, the Bayesian equivalent of *confidence* intervals. Typically, credibility intervals are taken at the 95% cutoff and are called highest posterior densities (HPDs).

The connection between posterior probability and likelihood is made with Bayes' inversion formula, also called Bayes' theorem, by means of a quantity called the prior distribution $p(\theta)$:

$$p(\theta|X) = \frac{p(X|\theta)\ p(\theta)}{p(X)} \tag{21}$$

The prior represents what we think about the process that generated the data, before analyzing the data, and is at the origin of all controversies surrounding Bayesian inference. In practice, priors are more typically chosen based on statistical convenience, and often have nothing to do with our genuine state of knowledge about parameters before observing the available data. We will see in Subheading 3.1 that priors can be used to distinguish between parameters that are confounded in a maximum likelihood analysis (model), so that we argue that the frequentist vs. Bayesian controversy is sterile, and we advocate a more pragmatic approach, that often results in the mixing of both approaches (in their concepts and techniques) [81, 82].

All models have parameters. Subheading 2.7 treats substitution models, which can have eight free parameters in the case of GTR + Γ. Most people are not really interested in these parameters θ or in their estimates $\hat{\theta}$, but have to use them in order to estimate a phylogenetic tree τ. These parameters θ are called *nuisance* parameters because they enter the model but are not the focus of inference. The likelihood solution consists in setting these parameters to their MLE, ignoring the uncertainty with which they can be estimated, while the Bayesian approach will integrate them out, directly accounting for their uncertainty:

$$p(X|\tau) = \int_\Theta p(X|\tau, \theta)p(\theta)\ d\theta \tag{22}$$

One difficulty in Bayesian inference is about the denominator in Eq. 21, as this denominator often has no analytical solution. In spite of being a normalizing constant, $p(X)$ requires integrating out nuisance parameters by means of prior distributions as in Eq. 22. Thus, it is easy to see from Eq. 21 that the posterior distribution of the variable of interest (e.g., τ) can quickly become complicated:

$$p(\tau|X) = \int_\Theta \frac{p(X|\tau, \theta)\ p(\tau)\ p(\theta)}{\sum_T p(X|\tau, \theta)\ p(\tau)\ p(\theta)}\ d\theta \tag{23}$$

where τ and θ are assumed to be independent and the discrete sum is taken over the set T of all possible topologies (*see* Subheading 2.10.1). However, the ratio of posteriors evaluated at two different points will simplify: as the denominator in Eq. 23 is a constant, it will cancel out from the ratio. This simple observation is at the origin of an integration technique for approximating the posterior distribution in Eq. 23: Markov chain Monte Carlo (MCMC) samplers. A very clear introduction can be found in [83].

Building on this, two approaches can be formulated to compare models in a Bayesian framework. The first is to treat the model as a "random variable," and compute its posterior probability. The *best* model is then the one that has the highest posterior probability. This approach is typically implemented in a reversible-jump MCMC (or rjMCMC) sampler (e.g., *see* [49]).

The alternative is to use the Bayesian equivalent of the LRT, the Bayes factor. Rather than comparing two likelihoods, the Bayes factor compares the probability of the data under two models, M_0 and M_1:

$$\mathrm{BF}_{0,1} = \frac{p(X|M_0)}{p(X|M_1)} \tag{24}$$

More specifically, $BF_{0,1}$ evaluates the weight of evidence in favor of model M_0 against model M_1, with $BF_{0,1} > 1$ considered as evidence in favor of M_0. Just as in a frequentist context, where a null hypothesis is significantly rejected at a certain threshold, 5%, 1%, or less depending on different costs or error types, Bayes factors can be evaluated on a specific scale [84]. However, because this scale is just as ad hoc as in a frequentist setting, it might be preferable to use the probability of the data under a particular model $p(X|M_i)$ as a means of ranking models M_i.

The quantity $p(X|M_0)$, which is the denominator in Eq. 23 (where we did not include the dependence on the model in the notation), is called the marginal likelihood. Note that it is also an expectation with respect to a prior probability distribution:

$$p(X|M_0) = \int_{\Theta} p(X|\theta, M_0)\ p(\theta|M_0)\ d\theta \qquad (25)$$

A number of approximations to evaluate Eq. 25 exist and are reviewed in [85] (see also [86, 87]). The simplest one is based on the harmonic mean of the likelihood sampled from the posterior distribution [88], also known as the harmonic mean estimator (HME). The way this estimator is derived demands to understand how integrals can be approximated. Briefly, to compute $I = \int g(\theta)\ p(\theta)\ d\theta$, generate a sample from a distribution $p^{\star}(\theta)$ and calculate the simulation-consistent estimator $I = \sum w_i\ g(\theta)/\sum w_i$ where w_i is the *importance function* $p(\theta)/p^{\star}(\theta)$. Take $g = p(X|\theta)$ and $p^{\star}(\theta) = p(X|\theta)\ p(\theta)/p(X)$, then $\hat{I} = \hat{p}(X|M_0) = \lim_{N\to\infty}\left(\frac{1}{N}\sum\frac{1}{p(X|\theta_i)}\right)^{-1}$ with $\theta \sim p(\theta|X)$ (*see* supplementary information in [89]). As a result, a very simple way to estimate the marginal likelihood and Bayes factors is to take the output of an MCMC sampler and compute the harmonic mean of the likelihood values (not the log-likelihood values) sampled from the posterior distribution.

Because of its simplicity, this estimator is now implemented in most popular programs such as `MrBayes` [90] or `BEAST` [91]. However, it might be considered as the worst estimator possible, because its results are unstable [88, 92] and biased towards the selection of parameter-rich models [86]. An alternative and reliable estimator, based on thermodynamic integration (TI; [86]—also known as path sampling; [93, 94]), is much more demanding in terms of computation. Indeed, it requires running MCMC samplers morphing one model into the other (and vice versa), which can increase computation time by up to an order of magnitude [86]. Improvements of the TI estimator are however available. The stepping-stone (SS) approach builds on importance sampling and TI to speed up the computation while maintaining the accuracy of the standard TI estimator [87, 95].

Moving away from the estimation of marginal likelihoods, an analogue of AIC that can be obtained through the output of an MCMC sampler (AICM) was proposed [96]. In essence, it relies on the asymptotic convergence of the posterior distribution of the log-likelihood on a gamma distribution [97]. As such, it becomes possible to estimate the effective number of parameters as twice the sample variance of posterior distribution of the log-likelihood, which itself can be estimated by a resampling procedure [96]. This gives a very elegant means of estimating AIC, from the posterior simulations. However, although AICM seems to be a more stable measure of model ranking than HME, both TI and SS still seem to outperform this estimator, at least in the case of the comparison of demographic and relaxed molecular clock models [96] (*see* Subheading 3).

2.9.4 Cross-Validation

Cross-validation is another model selection approach, which is extremely versatile in that it can be used to compare any set of models of interest. Besides, the approach is very intuitive. In its simplest form, cross-validation consists in dividing the available data into two sets, one used for "training" and the other one used for "validating." In the training step (TS), the model of interest is fitted to the training data in order to obtain a set of MLEs. These MLEs are then used to compute the likelihood using the validation data (validation step, VS). Because the validation data were not part of the training data, the likelihood values computed during VS can be directly used to compare models, without requiring any explicit correction for model dimensionality.

The robustness of the cross-validation scores can be explored in various ways, such as repeating the above procedure with a switched labeling of training and validation data (hence the expression *cross-validation*). Of course, this simple 2-fold cross-validation could be extended to n-fold cross-validation, where the data are subdivided into n subsets, with $n - 1$ subsets serving for training, and one for validation. Ideally, the procedure is repeated $n - 1$ additional times.

We know of only two examples of its use in phylogenetics, one in the ML framework [98] and one with a Bayesian approach [99]. Given the increasing size of modern data sets, putting aside some of the data for validation is probably not going to dramatically affect the information content of the whole data set. As a result, model selection via cross-validation, which is statistically sound, could become a very popular approach.

2.10 Finding the Best Tree Topology

2.10.1 Counting Trees

Now that we can select a model of evolution (Subheading 2.9) and estimate model parameters (Subheading 2.8) under a particular model (Subheading 2.5), how do we find the optimal tree? The basic example in Subheading 2.1 suggested that we score all possible tree topologies and choose for inference the one that has the highest score. However, a simple counting exercise shows that an exhaustive examination of all possible topologies is not realistic.

Figure 9 shows how to count tree topologies. Starting from the simplest possible unrooted tree, with three taxa, there are three positions where a fourth branch (leading to a fourth taxon) can be added. As a result, there are three possible topologies with four taxa. For each of these, there are four places on the tree where a fifth branch can be added, which leads to a total of $3 \times 5 = 15$ topologies with five taxa. A recursion appears immediately, and it can be shown that the total number of unrooted topologies with n taxa is equal to $1 \times 3 \times \cdots \times 2n - 5$ [100] (*see* [15] for the deeper history), which, as given in [101], is equal to:

$$N_{\text{unrooted}}^{T(n)} = \frac{(2n-5)!}{2^{n-3}(n-3)!} = \frac{2^{n-2}\Gamma(n-\frac{3}{2})}{\sqrt{\pi}} \tag{26}$$

where the Γ function for any real number x is defined as $\Gamma(x) = \int_0^{\infty} t^{x-1}\, e^{-t}\, dt$. An approximation based on Stirling number is also given in [101].

The same exercise can be done for rooted trees (Fig. 10), where the number of possible rooted topologies with n taxa becomes $1 \times 3 \times \cdots \times 2n - 3$, which is

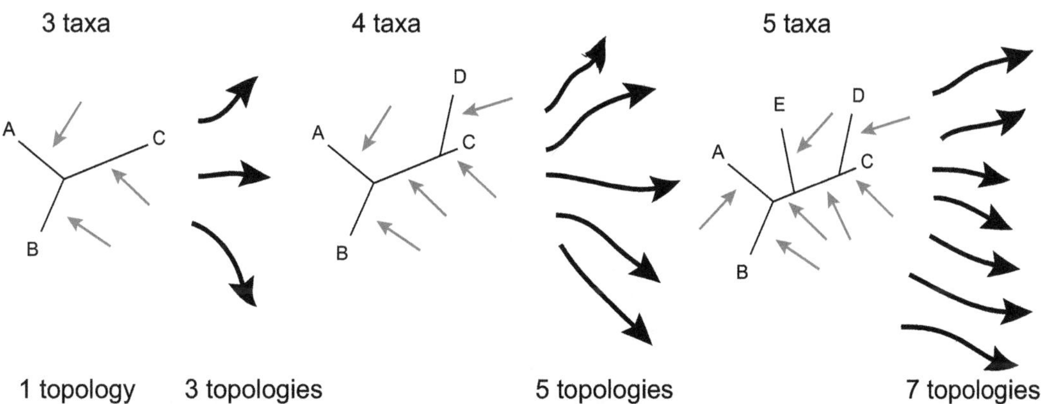

Fig. 9 Procedure to count the number of unrooted topologies. The top line shows the current number of taxa included in the tree below. Gray arrows indicate locations where an additional branch can be grafted to add one taxon. Black arrows show the resulting number of topologies after addition of a branch (taxon). Only one such possible topology is represented at the next step. The bottom line indicates the number of possibilities. These numbers multiply to obtain the total number of trees

Table 2
Counting tree topologies

Number of taxa	Unrooted tree	Rooted trees
3	1	3
4	3	15
5	15	105
6	105	945
10	2,027,025	34,459,425
20	221,643,095,476,699,771,875	8,200,794,532,637,891,559,375

Number of tree topologies are given for the unrooted and rooted cases

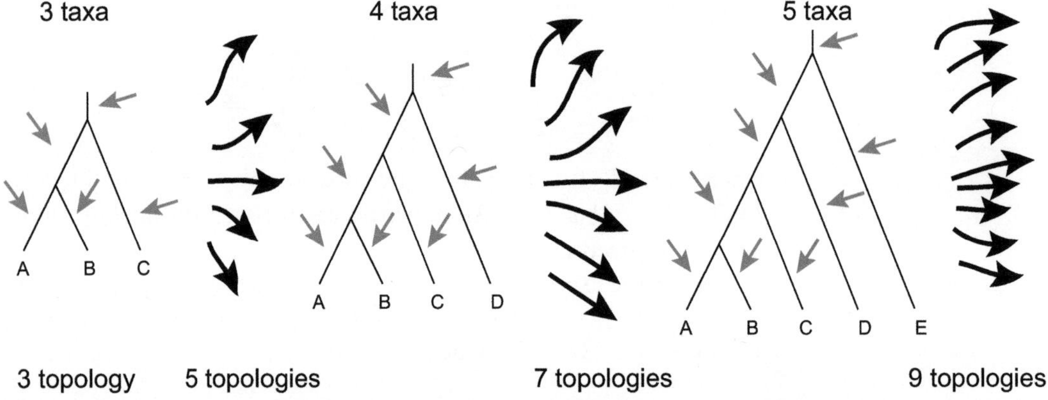

Fig. 10 Procedure to count the number of rooted topologies. *See* Fig. 9 for legend and text for details

$$N_{\text{rooted}}^{T(n)} = \frac{(2n-3)!}{2^{n-2}(n-2)!} = \frac{2^{n-1}\Gamma\left(n-\frac{1}{2}\right)}{\sqrt{\pi}} \tag{27}$$

Note that $N_{\text{unrooted}}^{T(n)} = N_{\text{rooted}}^{T(n-1)}$, as Table 2 clearly suggests.

As a result, the number of possible topologies quickly becomes very large when the number n of sequences increases, even with a very modest n, so that heuristics become necessary to find the best-scoring tree.

2.10.2 Some Heuristics to Find the Best Tree

The simplest approach builds upon the idea presented in Figs. 9 and 10. Stepwise addition, for instance, starts with three sequences drawn at random among the n sequences to be analyzed, and adds sequences one at a time, keeping only the tree that has the highest score at each step (e.g., [52]). However, there is no guarantee that the final tree is the optimal tree [44]. The idea behind branch-and-bound [102], refined in [103], is to have a look-ahead routine that prevents entrapment in suboptimal trees. This routine sets a bound on the trees selected at each round of additions, such that only the trees that have a score at least as good as that of the trees obtained in the next round are kept in the search algorithm. Solutions found by the branch-and-bound algorithm are optimal, but computing time becomes quickly prohibitive with more than 20 sequences.

As a result, most tree-search algorithms will start with a quickly obtained tree, often reconstructed with an algorithm based on pairwise distances such as neighbor-joining [104] or a related approach [78, 105], and then alter the tree randomly until no further improvement is obtained or after a certain number of unsuccessful attempts are reached. Examples of such algorithms include nearest neighbor interchange (NNI), subtree pruning and regrafting (SPR), or tree bisection and reconnection (TBR), see, for instance, [6] for a full description. While the details are of little importance here, the critical point is the extent of topological rearrangement in each case. With, e.g., NNI, each rearrangement can give rise to two topologies. The result is that exploring the topology space is slow, especially in problems with large n. On the other hand, TBR has, among the three methods cited above, the largest number of neighbors. As a result, the topology space is explored quickly, but the optimal tree can be "missed" simply because a dramatic change is attempted, so that the computational cost increases. Alternatively, the chance of finding the optimal tree $\hat{\tau}$ when $\hat{\tau}$ is very different from the current tree is higher when the algorithm can create some dramatic rearrangements. Some programs, such as `PhyML` ver. 3.0, now use a combination of NNI and SPR to address this issue [24]. MCMC samplers that search the tree space implement somewhat similar tree-perturbation algorithms that are either "global," and modify the topology dramatically, or "local" [106] (see also [107] for a correction of the original local moves). As a result, MCMC samplers are affected by the same issues as traditional likelihood methods. Much of the difficulty therefore comes from this kind of trade-off between larger rearrangements that are expected to improve accuracy and the computational burden associated with these extra computations [108].

2.10.3 Cutting Corners with ABC and AI

As some of the above computations can become quite costly (high runtimes, heavy memory footprints, poor scalability with large data sets, etc.), computational workarounds have been and are being explored. One of these resorts to approximate Bayes computing (ABC), which is essentially a likelihood-free approach. First developed in the context of population genetics [109, 110], the driving idea is to bypass the optimization procedures and replace them with simulations in the context of a rejection sampler. In population genetics, the problem could be about a gene tree, which is usually appropriately described by a coalescence tree [111, 112], for which we want to estimate some model parameters. As we are able to simulate trees from such a process, it is possible to place prior distributions on these model parameters, and simulate trees by drawing parameters until the simulated trees "look like" the observed tree. The set of parameters thus drawn approximates the posterior distribution of the corresponding variables. This forms the basis of a naïve rejection sampler, that is quite flexible as it does not even require that a probabilistic model be formulated, but one that can be inefficient, especially if the posterior distribution is far from the prior distribution—which is usually the case. As a result, a number of variations have been described, trying either to correlate sample draws as in MCMC samplers [113] or to resample sequentially from the past [114, 115]. In spite of recent reviews of the computational promises and deliveries of ABC samplers [116–118], the few applications in molecular evolution have been, to date, mostly limited to molecular epidemiology [119–122]. One of the major challenges to estimate a phylogenetic tree from a sequence alignment with ABC is the lack of a proper and efficient simulation strategy: it is possible to simulate trees under various processes (we saw the coalescent above), it is also possible to simulate an alignment from a given (possibly simulated tree), so that in theory one could imagine an ABC algorithm that would use this backward process to estimate phylogenetic trees by comparing a simulated alignment to an "actual" alignment. This, however, would most likely be a very inefficient sampler.

A second area that holds promises is the use of artificial intelligence (AI), and more specifically of machine learning (ML), in molecular evolution. Here again, attempts have been made to using standard ML approaches such as support vector machines [123] to guide the comparison of tree shapes, for instance, [124], which can then be used in epidemiology [121], but estimating a phylogenetic tree has proved more challenging. In one notable exception, an alignment-free distance-based tree-reconstruction method was proposed [125], but its main legacy seems to be in the development of *k*-mers, or unaligned sequences chopped into words of length *k*, to reconstruct phylogenetic trees—in particular in the context of phylogenomics (phylogenetics at a genomics scale) [126, 127]. To the best of our knowledge, nobody has ever tried, yet, to train a neural network or even a deep learning algorithm [128–130] on a database of phylogenetic trees with corresponding alignments such as TreeBASE [131] or PANDIT [132]. As applications of deep learning start emerging in genomics [133] and proteomics [134], it is likely that phylogenetics will come next.

3 Uncovering Processes and Times

3.1 Dating the Tree of Life: Always Deeper?

Similar to the problem of estimating the tree of life, dating the tree of life poses many challenges [135]. Since it was first proposed in 1965 [40], the idea of estimating divergence times has since undergone a dramatic change, and new approaches are regularly proposed. Population geneticists have their own approaches, which are either fully Bayesian [136] or based on approximate Bayesian computation in the coalescent framework [137]. All these approaches make it possible to infer divergence times between recently diverged species, as in the case of humans and chimpanzees, or to date demographic events such as the migrations "out of Africa" of early human populations [138].

In the context of molecular evolution, we are usually interested in estimating deeper divergence times, such as those between species, which are available online, for instance, at www.timetree.org [139], recently revamped and extended to cover close to 100k species [140]. While early "molecular dates" were systematically biased towards ages that are too old [135], we argue here that recent developments in the field have led to more accurate methods and also to a better understanding of methodological limitations.

3.1.1 The Strict Molecular Clock

One quantity that we can estimate when comparing pairs of sequences is the number of differences that exist. This number, estimated as a branch length b, can be corrected for multiple substitutions (*see* Subheading 2.7), but basically remains an expected number of substitutions per site. With "dating" (defined here as the activity of estimating divergence times [141]), we are interested in estimating time t, which relates to the expected numbers of substitutions b according to the following equation:

$$b = \Delta t \times r \quad (28)$$

where Δt is a period of time and r the rate of evolution. In technical terms, times and rates are said to be confounded, because we cannot estimate one without making an assumption about the other.

The molecular clock hypothesis does just this by assuming that rates of evolution are constant in time [40] (see also [142], p. 65). Under this assumption, the estimated tree is ultrametric as in the basic example represented in Fig. 11, which implies that all the tips are level, or equivalently that the distance from root to tip is the same for all branches.

In this example (Fig. 11), the branch length from the fossil-dated node is 0.1 substitutions/site (sub/site), and the fossil was estimated to be present 10 million years ago (MYA). Under the strict molecular clock assumption (equal rates over the whole tree), we can (1) estimate the rate of evolution ($0.1/10 = 0.01$

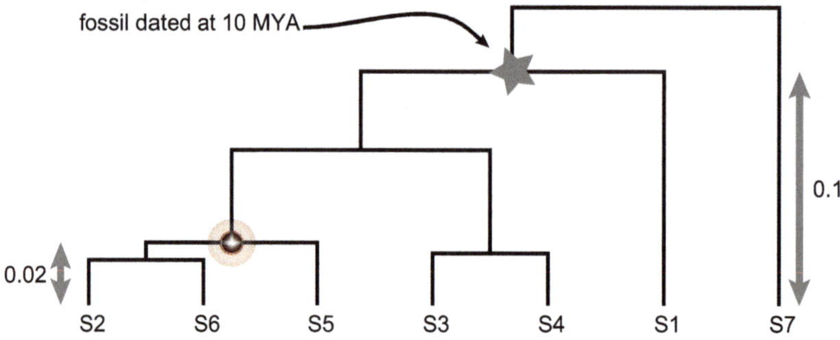

Fig. 11 The strict molecular clock. The tree is ultrametric. The node marked with a star indicates the presence of a fossil, dated in this example to 10 million years ago (MYA). This is the point that we will use to calibrate the clock, that is, to estimate the global rate of evolution. The number of substitutions that accumulated from the marked node to the tips (present) is indicated on the right weights in at 0.1 substitutions/site. The node that is the most recent common ancestor of S2 and S5 is the node of interest. The number of substitutions from this node to the tips is 0.02 substitutions/site

sub/site/my) and (2) date all the other nodes on the tree. For instance, the most recent common ancestor of S2 and S5 is separated from the tips by a branch length of 0.02 sub/site. Its divergence time is therefore $0.02/0.01 = 2$ MYA.

As with any hypothesis, the strict clock can be tested. Tests based on relative rates assess whether two species evolve at the same rate as a third one, used as an outgroup. Originally formulated in a distance-based context [143], likelihood versions have been described [44, 144]. However, because of their low power [145] their use is on the wane. The most powerful test is again the LRT (*see* Subheading 2.9.1). The test proceeds as usual, first calculating the test statistic $2\Delta\ell$ (twice the difference of log-likelihood values). The null hypothesis (strict clock) is nested within the alternative hypothesis (clock not enforced), so that $2\Delta\ell$ follows a χ^2 distribution. The degree of freedom is calculated following Fig. 12. With an alignment of n sequences, we can estimate $n - 1$ divergence times under the null model (disregarding parameters of the substitution model) and we have $2n - 3$ branch lengths under the alternative model. The difference in number of free parameters is therefore $n - 2$, which is our degree of freedom. This version of the test actually assesses whether all tips are at the same distance from the root of the tree [44]. For time-stamped data, serially sampled in time as in the case of viruses, the alternative model incorporates information on tip dates [146].

This linear regression model suggested by the molecular clock hypothesis has often been portrayed as a recipe [147], which gave rise in the late twentieth to early twenty-first century to a veritable cottage industry [148–151], culminating with a paper suggesting that the age of the tree of life might be older than the age of planet Earth [152]. This recipe was put down by two factors: (1) the

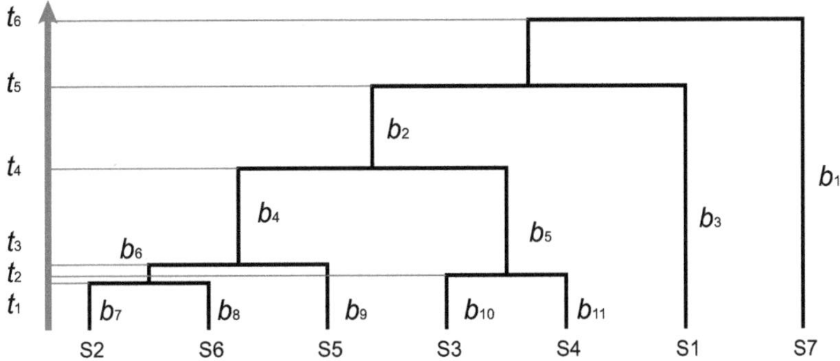

Fig. 12 Testing the strict molecular clock. The divergence times that can be estimated under the strict clock assumption are denoted t_i. The branch lengths that can be estimated without the clock are denoted b_i. In the case depicted, with $n = 7$ sequences, we have $n - 1 = 6$ divergence times and $2n - 3 = 11$ branch lengths

publication of a piece written in a rather unusual style for a scientific paper [153], and (2) new methodological developments. The main points made in [153] are that (1) most of the early dating studies relied on one analysis [149] that used a fossil-based calibration point for the divergence of birds at 310 MYA to estimate a number of molecular dates for vertebrates, and that (2) these molecular dates were then used in subsequent studies as a proxy for calibration points, disregarding their uncertainty. As a result, estimation errors were passed on and amplified from study to study, leading to the nonsensical results in [152].

3.1.2 Local Molecular Clocks

This "debacle" has motivated further theoretical developments in the dating field. The simplest idea is that, if a global clock does not hold for the entire tree, then perhaps groups of related species share the same rate. That is, if a *global* clock does not hold, perhaps the tree can be subdivided into *local* molecular clocks. An initial idea was proposed in the context of quartets of sequences [154] and was later generalized to a tree of any size with any number of local clocks on the tree [155] (constrained by the number of branches on the tree and calibration points). Because of the arbitrariness of such local clocks, methods have been devised to place the clocks on the tree [156] and to estimate the appropriate number of clocks that should be used [157]. A Bayesian approach now estimates all these parameters and their placement in an integrated statistical framework [158].

3.1.3 Correlated Relaxed Clocks

The idea of a correlated relaxed molecular clock goes back to Sanderson [159] (see also [160]), who considered that rates of evolution can change from branch to branch on a tree. By constraining rates of evolution to vary in an autocorrelated manner on a tree, it is possible to devise a method that minimizes the amount of rate change.

The idea of an autocorrelated process governing the evolution of the rates of evolution is attributed to [161] in [159], but could all the same be attributed to Darwin. Thorne et al. developed this idea further in a Bayesian framework [162]. Building upon the basic theory covered in Subheading 2.9.3, the idea is to place prior distributions on the quantities in the right-hand side of Eq. 28. The target distribution is $p(t|X)$. It is proportional to $p(X|t)\ p(t)$ according to Bayes' theorem, but all that we can estimate is

$$p(b|X) = \frac{p(X|b)\ p(b)}{p(X)} = \frac{p(X|r,t)\ p(r,t)}{p(X)} \tag{29}$$

One of the possible ways to expand the joint distribution of rates and times is $p(r, t)$ is $p(r|t)\ p(t)$, which posits a process where rate change depends on the length of time separating two divergences. The "art" is now in choosing prior distributions, conditional on the obvious constraint that rates and times should take positive values. A number of such prior distributions for rates have been proposed and assessed [163] and one of the best-performing model for rates is, in our experience, the log-normal model [162, 164]. The prior on times is either a pure-birth (Yule) model or a birth-and-death process possibly incorporating species sampling effects [165]. If sequences are sampled at the population level, a coalescent process is more appropriate (*see* [112] for an introduction). In this case, the past demography of the sampled sequences can be traced back taking inspiration from spline regression techniques [166, 167] or multiple change-point models [168].

Once these priors are specified, an MCMC sampler will draw from the target distribution in Eq. 29, and marginal distributions for times and rates can easily be obtained. The rationale behind the sampler is represented in Fig. 13. As per Eq. 28, the relationship

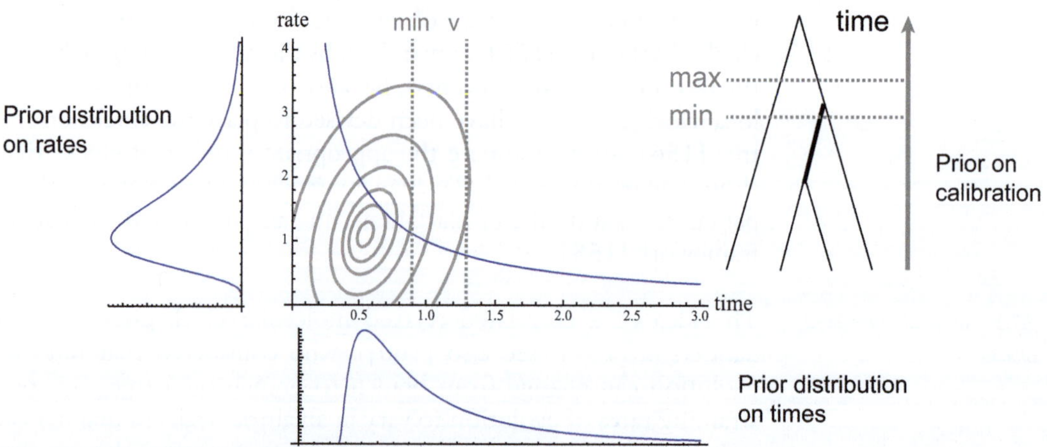

Fig. 13 The relaxed molecular clock. See text for details

between rates and times is the branch of a hyperbolic curve, where the priors on rates and on times define a region of higher posterior probability, symbolized here by a contour plot superimposed on the hyperbolic curve. On top of this, fossil information is incorporated into the analysis as constraints on times. A very influential piece stimulated a discussion about the shape of these prior distributions [153], which was taken up [169], and further developed in [170]. Briefly, fossil information is usually imprecise, as paleontologists can only provide minimum and maximum ages (Fig. 13). Of these two ages, the minimum age is often the most reliable. Under the assumption that the placement of the fossil on the tree is correct, the idea is to place on fossil dates a prior distribution that will be highly skewed towards older (maximum) ages. A "hard bound" can be placed on the minimum age, possibly by shifting this prior distribution by an offset equal to the minimum age, while the tails of the prior distribution will act as "soft bounds," because they do not impose on the tree a strict (or *hard*) constraint. Empirical studies agree, however, that both reliability and precision of fossil calibrations are critical to estimating divergence times [136, 171].

3.1.4 Uncorrelated Relaxed Clocks

Because of the autocorrelation between the rate of each branch and that of its ancestral branch (except for the root, which obviously requires a special treatment), the tree topology is fixed under the autocorrelated models described above. By relaxing this assumption about rate autocorrelation, [172] were able to implement a model that also integrates over topological uncertainty. In spite of the somewhat counter-intuitive nature of the relaxation of the autocorrelated process, as implemented in `BEAST` [91, 173], empirical studies have found this approach to be one of the best-performing (e.g., [157]).

When first published, it was proposed that making use of an uncorrelated relaxed molecular clock could improve phylogenetic inference [172]. The idea was that calibration points and their placement on the tree could act as additional information. However, a simulation study suggests that relaxed molecular clocks might not improve phylogenetic accuracy [174], a result that might be due to the lack of calibration constraints in this particular simulation study.

3.1.5 Some Applications of Relaxed Clock Models

Since the advent of relaxed molecular clocks, two very exciting developments have seen the light of day. The first concerns the inclusion of spatial statistics into dating models [175, 176]. Spatial statistics are not new in population genetics [177] and have been used with success in combination with analyses in computational molecular evolution (e.g., [178]). However, the originality in [176], for instance, is to combine in a single statistical framework molecular data with geographical and environmental information to infer the diffusion of sequences through both space and time. While these preliminary models seem to deal appropriately with natural barriers to gene flow such as coastlines, a more detailed set of constraints on gene flow may further enhance their current predictive power.

The second development coming from relaxed molecular clocks concerns the mapping of ancestral characters onto uncertain phylogenies. This is not a novel topic, as a Bayesian approach was first described in 2004 [179, 180]. The novelty is that we now have the tools to correlate morphological and molecular evolution in terms of their absolute rates and to allow both molecular and morphological rates of evolution to vary in time [181]. Further development will certainly integrate over topological uncertainty. While there has been a heated controversy about the existence of such a correlation in the past [182], all previous studies were using branch length as a proxy for rate of molecular evolution, which is clearly incorrect. We can therefore expect some more accurate results on this topic very soon. More details and examples can be found in recent and extensive reviews [183–185] that further discuss applications to biogeographic studies [186], or extensions to viral [187, 188], as well as other types of genomic [189] and morphological [190] data.

4 Molecular Population Phylogenomics

Population genetics is rich in theory regarding the relative roles of mutation, drift, and selection. Much research in population genomics is now focusing on using this theory to develop statistical procedures to infer past processes based on population-level data, such as those of the 1000-genome project [191], the UK's 10,000 genome project [192], and always more ambitious projects [193]. One limitation of these inference procedures is that they all focus on a thin slice of evolutionary time by studying evolution at the level of populations. If we wish to study longer evolutionary time scales, for example, tens or hundreds of millions of years, we must resort to interspecific data. In such a context, which is becoming intrinsically *phylogenetic*, the most important event is a substitution, that is, a mutation that has been fixed. Yet substitution rates can be defined from several features. In particular, from a population genetics perspective, it is of interest to model both mutational features and selective effects, combining them multiplicatively to specify substitution rates. We review briefly how substitution models that invoke codons as the state space lend themselves naturally to these objectives in a first section below (Subheading 4.1), before explaining the origin (and a shortcoming) of all the approaches developed so far (Subheading 4.2).

4.1 Bridging the Gap Between Population Genetics and Phylogenetics

Assuming a point-mutation process, such that events only change one nucleotide of a codon during a small time interval, Muse and Gaut proposed a codon substitution model with rates specified from the Q_{GTR} nucleotide-level matrix (*see* Subheading 2.7), along with one parameter that modulates synonymous events and another one that modulates nonsynonymous events [194]. In most subsequent formulations, the parameter associated with synonymous events is assumed to be fixed, such that the model only modulates nonsynonymous rates by means of a parameter denoted ω. This parameter has traditionally been interpreted as the nonsynonymous to synonymous rate ratio, and is generally associated with a different formulation of the codon model proposed by Goldman and Yang [195]. More details on codon models can be found in Chapter 4.1 [196]. There continues to be a debate regarding the interpretation of the ω parameter [197, 198]. Regardless of how this issue is settled, it is clear that ω is aimed at capturing the net overall effects of selection, irrespective of the exact nature of these effects.

With the intention to model selective effects themselves, Halpern and Bruno proposed a codon substitution model that combines a nucleotide-level layer, as described above, for controlling mutational features, along with a fixation factor that is proportional to the fixation probability of the mutational event [199]. The fixation factor is in turn specified from an account of amino acid or codon preferences. One objective of the model, then, consists in teasing apart mutation and selection. While [199] proposed their model with site-specific fixation factors, later work has explored simpler specifications, where all sites have the same fixation factor [200]. Other models that aimed at capturing across-site heterogeneities in fixation factors were proposed using nonparametric devices and empirical mixtures [201]. Another core idea behind these approaches is to construct a more appropriate null model against which to test for features of the evolutionary process. This idea has been put into practice for the detection of adaptive evolution in protein-coding genes [202, 203]. Recent developments include sequence-wide fixation factors [9, 197, 204, 205], and we predict that these models will play a role in bridging the gap between molecular evolution at the population and at the species levels.

4.2 Origin of Mutation–Selection Models: The Genic Selection Model

In order to understand a shortcoming of these models, we need to go back to the development of fixation probabilities that took place in the second half of the twentieth century. The basic unit or *quantum* of evolution is a change in allele frequency p. Allele frequencies can be affected by four processes: migration, mutation, selection, and drift. Because of the symmetry between migration and mutation [206], which only differ in their magnitude, these two processes can be treated as one. We are left with three forces:

mutation, selection, and drift. The question is then, what is the fate of an allele under the combined action of these processes? Our development here follows [207] (but *see* [208] for a very clear account).

4.2.1 Fixation Probabilities

Of the three processes affecting allele frequencies, mutation and selection can be seen as directional forces in that their action will shift the distribution of allele frequencies towards a particular point, be it an internal equilibrium, or fixation/loss of an allele. On the other hand, drift is a non-directional process that will increase the variance in allele frequencies across populations, and will therefore spread out the distribution of allele frequencies. This distribution is denoted $\Psi(p, t)$. We also must assume that the magnitude of all three processes, mutation, selection, and drift, is small and of the order of $\frac{1}{2N_e}$, where N_e is the effective population size. To derive the fate of an allele after a certain number of generations, we also need to define $g(p, \varepsilon; dt)$, the probability that allele frequency changes from p to $p + \varepsilon$ during a time interval dt.

In phylogenetics (and population genetics) we are generally interested in predicting the past. The tool making this possible is called the Kolmogorov backward equation, which predicts the frequency of an allele at some time t, given its frequency p_0 at time t_0:

$$\Psi(p, t + dt|p_0) = \int \Psi(p, t|p_0 + \varepsilon)\, g(p_0, \varepsilon; dt)\, d\varepsilon \tag{30}$$

We can take the Taylor expansion of Eq. 30 around p_0, neglect all terms whose order is larger than two ($o(p_0^2)$) and since Ψ is not a function of ε, we obtain:

$$\Psi(p, t + dt|p_0) = \Psi \int g\, d\varepsilon + \frac{\partial \Psi}{\partial p_0} \int \varepsilon g d\varepsilon + \frac{\partial^2 \Psi}{\partial p_0^2} \int \frac{\varepsilon^2}{2} g d\varepsilon \tag{31}$$

This formulation leads to the definition of two terms that represent the directional processes affecting allele frequencies (M) and the non-directional process, or drift (V):

$$\begin{cases} M(p)\, dt &= \int g\, \varepsilon\, d\varepsilon \\ V(p)\, dt &= \int g\, \varepsilon^2 d\varepsilon \end{cases} \tag{32}$$

that we can substitute into Eq. 31. At equilibrium, $\frac{\partial \Psi}{\partial t} = 0$ and, after a bit of calculus, we obtain:

$$\frac{\partial \hat{\Psi}}{\partial p_0} = C\, e^{-\int \frac{2M}{V} dp} \tag{33}$$

Table 3
The standard selection models

Selection coefficients	A_1A_1	A_1A_2	A_2A_2
Genic (positive) selection	$w_1 = 1 + s$	$w_2 = 1 + hs$	$w_3 = 1$
Overdominance	$w_1 = 1$	$w_2 = 1 + s$	$w_3 = 1$

Models are represented for one locus with two alleles, A_1 and A_2, which define three genotypes A_1A_1, A_1A_2, and A_2A_2 of fitness w_1, w_2, and w_3. The selection coefficient is s (positive in this table, but not necessarily so) and the dominance is governed by $h\,(h \in [0, 1])$

for which we need to specify boundary conditions and a model of selection. The boundary conditions are the two absorbing states of the system: (1) once fixed, an allele remains fixed ($\Psi(1, \infty; 1) = 1$) and (2) once lost, an allele remains lost ($\Psi(1, \infty; 0) = 0$). With these two requirements, the probability that the allele frequency is 1 given that it was p_0 in the distant past is the fixation probability:

$$\Psi(1, \infty; p_0) = \frac{\int_0^{p_0} e^{-\int \frac{2M}{V} dp} dp}{\int_0^1 e^{-\int \frac{2M}{V} dp} dp} \tag{34}$$

We therefore only need to compute M and V under a particular model of selection to fully specify the fixation probability of an allele in a mutation–selection-drift system. All that is required now to go further is a selection model.

4.2.2 The Case of Genic Selection

We are now ready to derive an explicit form to $\Psi(1, \infty; p_0)$ in Eq. 34 in the case of the genic selection model (Table 3; [209]). We obtain:

$$\overline{w} = 1 + sp^2 + 2pqhs = 1 + 2phs + sp^2(1 - 2h) \tag{35}$$

which can be approximated by $1 + 2phs$ (the result is exact only when $h = 1/2$). Therefore, $d\overline{w}/dp = 2hs$, and we can calculate the M and V terms to obtain the popular result:

$$\Psi(1, \infty; p_0) = \frac{\int_0^{p_0} e^{-\int \frac{2M}{V} dp} dp}{\int_0^1 e^{-\int \frac{2M}{V} dp} dp} = \frac{e^{-4N_e hs p_0} - 1}{e^{-4N_e hs} - 1} \tag{36}$$

Now, the initial frequency of a mutation in a diploid population of (census) size N is $p_0 = 1/2N$ (following [208]; [207] considered that $p_0 = 1/2N_e$; this debate is beyond the scope of this chapter), which leads to:

$$\Psi\left(1, \infty; \frac{1}{2N}\right) = \frac{e^{-2N_e hs/N} - 1}{e^{-4N_e hs} - 1} \tag{37}$$

If N_e is of the order of N, the numerator of the right-hand side of Eq. 37 becomes approximately $e^{-2hs} - 1$, whose Taylor approximation around $hs = 0$ is simply $-2hs$. We then obtain the result used in [199], and in all the papers that implemented mutation–selection (-drift) models (e.g., [197, 199–201, 204]):

$$\Psi\left(1, \infty; \frac{1}{2N}\right) = \frac{2hs}{1 - e^{-4N_e hs}} \tag{38}$$

Two critical points should be noted here. First, none of the recent codon models [197, 199–202, 204, 210, 211] ever investigated the role of dominance h, as they all consider that the allele under (positive) selection is fully dominant. Second, Table 3 shows that another class of selection models, those based on balancing selection, has never been considered so far. The impact of the selection model on the predictions made by the mutation–selection (-drift) models is currently unknown.

5 High-Performance Computing for Phylogenetics

5.1 Parallelization

Because of the dependency of the likelihood computations on the shape of a particular tree (*see* Subheading 2.6), most phylogenetic computations cannot be parallelized to take advantage of a multiprocessor (or multicore) environment. Nevertheless, two main directions have been explored to speed up computations: first, in computing the likelihood of substitution models that incorporate among-site rate variation and second, in distributing bootstrap replicates to several processors, as both types of computations can be done independently. A third route is explored in Chapter 7.4 [212].

In the first case, among-site rate variation is usually modeled with a Γ distribution [213] that is discretized over a finite (and small) number of categories [214]. The likelihood then takes the form of a weighted sum of likelihood functions, one for each discrete rate category, so that each of these functions can be evaluated independently. The route most commonly used is the plain "embarrassingly parallel" solution, where completely independent computations are farmed out to different processors. Such is the case for bootstrap replicates, for which a version of `PhyML` [24] exists, or in a Bayesian context for independent MCMC samplers [215] (*see* Subheading 2.9.3). The PhyloBayes-MPI package implements distributed likelihood calculations across sites over several compute-cores, allowing for a genuinely parallelized MCMC run [216, 217].

5.2 HPC and Cloud Computing

More recent work has focused on the development of heuristics that make large-scale phylogenetics amenable to high-performance computing (HPC) that are performed on computer clusters.

Because of the algorithmic complexity of resolving phylogenetic trees, an approach based on "algorithmic engineering" was developed [218]. The underlying idea is akin to the training phase in supervised machine learning [123], except that here the target is not the performance of a classifier but that of search heuristics. All of these heuristics reuse parameter estimates, avoid the computation of the full likelihood function for all the bootstrap replicates, or seed the search algorithm for every *n* replicate on the results of previous replicates [218]. For instance, in the "lazy subtree rearrangement" [219], topologies are modified by SPR (*see* Subheading 2.10.2), but instead of recomputing the likelihood on the whole tree, only the branch lengths around the perturbation are re-optimized. This approximation is used to rank candidate topologies, and the actual likelihood is evaluated on the complete tree only for the best candidates. These heuristics now permit the analysis of thousands of sequences in a probabilistic framework [220], but the actual convergence of these algorithms remains difficult to evaluate, especially on very large data sets (e.g., $>10^4$ sequences).

In addition to the reduction of the memory footprint for sparse data matrices [221], an alternative direction to "tweaking likelihood algorithms" has been to take direct advantage of the computing architecture available. One particular effort aims at tapping directly into the computing power of graphics processing units or GPUs, taking advantage of their shared common memory, their highly parallelized architecture, and the comparatively negligible cost of spawning and destroying threads on them. As a result, it is possible to distribute some of the summation entering the pruning algorithm (*see* Subheading 2.6) to different GPUs [222]. The number of programs taking advantage of these developments is widening and includes popular options such as `BEAST` [91] and `MrBayes` [223].

All these fast algorithms can be installed on a local computer cluster, a solution adopted by many research groups since the late 1990s. However, installing a cluster can be demanding and costly because a dedicated room is required with appropriate cooling and power supply (not to mention securing the room, physically). Besides, redundancy requirements, both in terms of power supply and data storage, as well as basic software maintenance and user management, may demand hiring a system administrator. An alternative is to run analyses on a remote HPC server, in the "cloud." Canada, for instance, has a number of such facilities, thanks to national funding bodies (CAC at cac.queensu.ca, SHARCNET at www.sharcnet.ca, or Calcul Quebec at www.calculquebec.ca, just to name a few), and commercial solutions are just a few clicks away (e.g., Amazon Elastic Compute Cloud or EC2). Researchers can obtain access to these HPC solutions according to a number of business models (free, on demand, yearly subscription, etc.) that are associated with a wide spectrum of costs [224]. But in spite of the technical support offered in the price, users usually still have to install their preferred phylogenetic software manually or put a formal request to the team of system administrators managing the HPC facility, all of which is not always convenient.

To make the algorithmic and technological developments described above more accessible, the recent past has seen the emergence of cloud computing [225] dedicated to the phylogenetics community. Examples include the CIPRES Science Gateway (www.phylo.org), or Phylogeny.fr (www.phylogeny.fr, [226]). Many include web portals that do not require that users be well versed in Unix commands, while others may include an application programming interface to cater to the most computer-savvy users. One potential limitation of these services is the bandwidth necessary to transfer large files, and storage requirements—especially in the context of next generation sequencing data. The management of relatively large files will remain a potential issue, unless phylogenetics practitioners are ready to discard these files after analysis, the end product of which is a single tree file a few kilobytes in size, in the same way that people involved in genome projects delete the original image files produced by massively parallel sequencers. Data security or privacy might not be a problem in most applications, except in projects dealing with human subjects or viruses such as HIV that expose the sexual practices of subjects. However, once these various hurdles are out of the way, users could very well imagine running their phylogenetic analyses with millions of sequences from their smartphone while commuting.

6 Conclusions

Although most of the initial applications of likelihood-based methods were motivated by the shortcomings of parsimony, they have now become well accepted as they constitute principled inference approaches that rely on probabilistic logic. Moreover, they allow biologists to evaluate more rigorously the relative importance of different aspects of evolution. The models presented in this chapter have the ability to disentangle rates from times (Subheading 3), or mutation from selection (Subheading 4), while in most cases accounting for the uncertainty about nuisance parameters. But the latest developments described above still make a number of restrictive assumptions (Subheading 4.2), and while many variations in model formulations can be envisaged, they still remain to be explored in practice.

Although some progress has been made in developing integrative approaches (e.g., [176, 181]), throughout this chapter we have assumed that a reliable alignment was available as a starting point. A number of methods exist to co-estimate an alignment and a phylogenetic tree (*see* **Part I** of this book), but the computational requirements and convergence of some of these approaches can be daunting, even on the smallest data sets by today's standards.

This brings us, finally, to the issue of tractability of most of these models in the face of very large data sets. The field of phylogenomics is developing quickly (*see* **Part III**), at a pace that is ever increasing given the output rate of whole genome sequencing projects. Environmental questions are drawing more and more attention, and metagenomes (*see* **Part VI**) will be analyzed in the context of what will soon be called *metaphylogenomics*. Exploring the numerous available and foreseeable substitution models in such contexts will require continued work in computational methodologies. As such, modeling efforts will continue to go hand-in-hand with, and maybe *dependent on*, algorithmic developments [227]. It is also not impossible that in the near future, the use of likelihood-free approach such as ABC or machine learning algorithms in computational molecular evolution be more thoroughly explored.

Acknowledgements

We would like to thank Michelle Brazeau, Eric Chen, Ilya Hekimi, Benoît Pagé, and Wayne Sawtell for their critical reading of a draft of the original chapter, as well as Jonathan Dench and George S. Long for their careful reading of the most recent draft. This work was supported by the Natural Sciences Research Council of Canada (SAB, NR).

References

1. Nei M, Kumar S (2000) Molecular evolution and phylogenetics. Oxford University Press, Oxford
2. Higgs PG, Attwood TK (2005) Bioinformatics and molecular evolution. Blackwell Publishing, Oxford
3. Balding DJ, Bishop MJ, Cannings C (2007) Handbook of statistical genetics, 3rd edn. Wiley, Chichester
4. Salemi M, Vandamme A-M, Lemey P (2009) The phylogenetic handbook: a practical approach to phylogenetic analysis and hypothesis testing, 2nd edn. Cambridge University Press, Cambridge
5. Hall BG (2011) Phylogenetic trees made easy: a how to manual. Sinauer Associates, Sunderland
6. Yang Z (2014) Molecular evolution: a statistical approach. Oxford University Press, Oxford
7. Drummond AJ, Bouckaert RR (2015) Bayesian evolutionary analysis with BEAST. Cambridge University Press, Cambridge
8. Aris-Brosou S, Xia X (2008) Phylogenetic analyses: a toolbox expanding towards Bayesian methods. Int J Plant Genomics 2008:683509
9. Rodrigue N, Philippe H (2010) Mechanistic revisions of phenomenological modeling strategies in molecular evolution. Trends Genet 26:248–252
10. Yang Z, Rannala B (2012) Molecular phylogenetics: principles and practice. Nat Rev Genet 13:303–314
11. Aris-Brosou S, Rodrigue N (2012) The essentials of computational molecular evolution. Methods Mol Biol 855:111–152
12. Yang Z (2000) Complexity of the simplest phylogenetic estimation problem. Proc Biol Sci 267:109–116

13. Sober E (1988) Reconstructing the past: parsimony, evolution, and inference. MIT Press, Cambridge
14. Durbin R, Eddy SR, Krogh A, Mitchison G (1998) Biological sequence analysis: probabilistic models of proteins and nucleic acids. Cambridge University Press, Cambridge
15. Felsenstein J (2004) Inferring phylogenies. Sinauer Associates, Sunderland
16. Yang Z (2007) PAML 4: phylogenetic analysis by maximum likelihood. Mol Biol Evol 24:1586–1591
17. Efron B, Tibshirani R (1993) An introduction to the bootstrap, vol 57. Chapman and Hall, Boca Raton
18. Efron B, Halloran E, Holmes S (1996) Bootstrap confidence levels for phylogenetic trees. Proc Natl Acad Sci USA 93:7085–7090
19. Felsenstein J (1985) Confidence limits on phylogenies: an approach using the bootstrap. Evolution 39:783–791
20. Baldauf SL (2003) Phylogeny for the faint of heart: a tutorial. Trends Genet 19:345–351
21. Hasegawa M, Kishino H (1989) Confidence limits of the maximum-likelihood estimate of the hominoid three from mitochondrial-DNA sequences. Evolution 43:672–677
22. Anisimova M, Gascuel O (2006) Approximate likelihood-ratio test for branches: a fast, accurate, and powerful alternative. Syst Biol 55:539–552
23. Guindon S, Delsuc F, Dufayard J-F, Gascuel O (2009) Estimating maximum likelihood phylogenies with phyml. Methods Mol Biol 537:113–137
24. Guindon S, Dufayard J-F, Lefort V, Anisimova M, Hordijk W, Gascuel O (2010) New algorithms and methods to estimate maximum-likelihood phylogenies: assessing the performance of PhyML 3.0. Syst Biol 59:307–321
25. Hillis DM, Bull JJ (1993) An empirical test of bootstrapping as a method for assessing confidence in phylogenetic analysis. Syst Biol 42:182–192
26. Felsenstein J, Kishino H (1993) Is there something wrong with the bootstrap on phylogenies? A reply to Hillis and Bull. Syst Biol 42:193–200
27. Yang Z, Rannala B (2005) Branch-length prior influences Bayesian posterior probability of phylogeny. Syst Biol 54:455–470
28. Berry V, Gascuel O (1996) On the interpretation of bootstrap trees: appropriate threshold of clade selection and induced gain. Mol Biol Evol 13:999
29. Shimodaira H, Hasegawa M (2001) CONSEL: for assessing the confidence of phylogenetic tree selection. Bioinformatics 17:1246–1247
30. Salichos L, Rokas A (2013) Inferring ancient divergences requires genes with strong phylogenetic signals. Nature 497:327–331
31. Felsenstein J (1978) Cases in which parsimony or compatibility methods will be positively misleading. Syst Zool 27:401–410
32. Tuffley C, Steel M (1997) Links between maximum likelihood and maximum parsimony under a simple model of site substitution. Bull Math Biol 59:581–607
33. Steel M, Penny D (2000) Parsimony, likelihood, and the role of models in molecular phylogenetics. Mol Biol Evol 17:839–850
34. Holder MT, Lewis PO, Swofford DL (2010) The Akaike information criterion will not choose the no common mechanism model. Syst Biol 59:477–485
35. Editors T (2016) Editorial. Cladistics 32:1. https://doi.org/10.1111/cla.12148
36. Philippe H, Zhou Y, Brinkmann H, Rodrigue N, Delsuc F (2005) Heterotachy and long-branch attraction in phylogenetics. BMC Evol Biol 5:50
37. Brinkmann H, van der Giezen M, Zhou Y, de Raucourt GP, Philippe H (2005) An empirical assessment of long-branch attraction artefacts in deep eukaryotic phylogenomics. Syst Biol 54:743–757
38. Hampl V, Hug L, Leigh JW, Dacks JB, Lang BF, Simpson AG, Roger AJ (2009) Phylogenomic analyses support the monophyly of Excavata and resolve relationships among eukaryotic “supergroups”. Proc Natl Acad Sci USA 106:3859–3864
39. Liu H, Aris-Brosou S, Probert I, de Vargas C (2010) A timeline of the environmental genetics of the haptophytes. Mol Biol Evol 27:161–176
40. Zuckerkandl E, Pauling L (1965) Evolutionary divergence and convergence in proteins. In: Bryson V, Vogel HJ (eds) Evolving genes and proteins. Academic, Cambridge, pp 97–166
41. Galtier N, Gascuel O, Jean-Marie A (2005) Markov models in molecular evolution. In: Nielsen R (ed) Statistical methods in molecular evolution. Statistics for biology and health. Springer, New York, pp 3–24
42. Cox DR, Miller HD (1965) The theory of stochastic processes. Chapman and Hall/CRC, Boca Raton
43. Yang Z (2000) Maximum likelihood estimation on large phylogenies and analysis of

adaptive evolution in human influenza virus A. J Mol Evol 51:423–432

44. Felsenstein J (1981) Evolutionary trees from DNA sequences: a maximum likelihood approach. J Mol Evol 17:368–376
45. Jukes JC, Cantor CR (1969) Evolution of protein molecules. In: Munro HN (ed) Mammalian protein metabolism. Academic, New York, pp 21–123
46. Kimura M (1980) A simple method for estimating evolutionary rates of base substitutions through comparative studies of nucleotide sequences. J Mol Evol 16:111–120
47. Hasegawa M, Kishino H, Yano T (1985) Dating of the human-ape splitting by a molecular clock of mitochondrial DNA. J Mol Evol 22:160–174
48. Tavaré S (1986) Some probabilistic and statistical problems in the analysis of DNA sequences. Lect Math Life Sci 17:57–86
49. Huelsenbeck JP, Larget B, Alfaro ME (2004) Bayesian phylogenetic model selection using reversible jump Markov chain Monte Carlo. Mol Biol Evol 21:1123–1133
50. Yang Z, Roberts D (1995) On the use of nucleic acid sequences to infer early branchings in the tree of life. Mol Biol Evol 12:451–458
51. Huelsenbeck JP, Bollback JP, Levine AM (2002) Inferring the root of a phylogenetic tree. Syst Biol 51:32–43
52. Yang Z (2006) Computational molecular evolution. Oxford University Press, Oxford
53. Aris-Brosou S (2005) Determinants of adaptive evolution at the molecular level: the extended complexity hypothesis. Mol Biol Evol 22:200–209
54. Anisimova M, Yang Z (2004) Molecular evolution of the hepatitis delta virus antigen gene: recombination or positive selection? J Mol Evol 59:815–826
55. Burnham KP, Anderson DR (1998) Model selection and inference: a practical information-theoretic approach. Springer, Berlin
56. Anisimova M, Bielawski JP, Yang Z (2001) Accuracy and power of the likelihood ratio test in detecting adaptive molecular evolution. Mol Biol Evol 18:1585–1592
57. Whelan S, Goldman N (2004) Estimating the frequency of events that cause multiple-nucleotide changes. Genetics 167:2027–2043
58. Wong WS, Yang Z, Goldman N, Nielsen R (2004) Accuracy and power of statistical methods for detecting adaptive evolution in protein coding sequences and for identifying positively selected sites. Genetics 168:1041–1051
59. Massingham T, Goldman N (2005) Detecting amino acid sites under positive selection and purifying selection. Genetics 169:1753–1762
60. Zhang J, Nielsen R, Yang Z (2005) Evaluation of an improved branch-site likelihood method for detecting positive selection at the molecular level. Mol Biol Evol 22:2472–2479
61. Anisimova M, Yang Z (2007) Multiple hypothesis testing to detect lineages under positive selection that affects only a few sites. Mol Biol Evol 24:1219–1228
62. Yang Z (2010) A likelihood ratio test of speciation with gene flow using genomic sequence data. Genome Biol Evol 2:200–211
63. Fletcher W, Yang Z (2010) The effect of insertions, deletions, and alignment errors on the branch-site test of positive selection. Mol Biol Evol 27:2257–2267
64. Yang Z, dos Reis M (2011) Statistical properties of the branch-site test of positive selection. Mol Biol Evol 28:1217–1228
65. Self SG, Liang K-Y (1987) Asymptotic properties of maximum likelihood estimators and likelihood ratio tests under nonstandard conditions. J Am Stat Assoc 82:605–610
66. Posada D, Crandall KA (1998) MODELTEST: testing the model of DNA substitution. Bioinformatics 14:817–818
67. Posada D (2008) jModelTest: phylogenetic model averaging. Mol Biol Evol 25:1253–1256
68. Cunningham CW, Zhu H, Hillis DM (1998) Best-fit maximum-likelihood models for phylogenetic inference: empirical tests with known phylogenies. Evolution 52:978–987
69. Pol D (2004) Empirical problems of the hierarchical likelihood ratio test for model selection. Syst Biol 53:949–962
70. Kullback S, Leibler RA (1951) On information and sufficiency. Ann Math Stat 22:79–86
71. Minin V, Abdo Z, Joyce P, Sullivan J (2003) Performance-based selection of likelihood models for phylogeny estimation. Syst Biol 52:674–683
72. Ripplinger J, Sullivan J (2008) Does choice in model selection affect maximum likelihood analysis? Syst Biol 57:76–85
73. Posada D, Crandall KA (2001) Selecting the best-fit model of nucleotide substitution. Syst Biol 50:580–601
74. Abdo Z, Minin VN, Joyce P, Sullivan J (2005) Accounting for uncertainty in the tree

topology has little effect on the decision-theoretic approach to model selection in phylogeny estimation. Mol Biol Evol 22:691–703

75. Luo A, Qiao H, Zhang Y, Shi W, Ho SY, Xu W, Zhang A, Zhu C (2010) Performance of criteria for selecting evolutionary models in phylogenetics: a comprehensive study based on simulated datasets. BMC Evol Biol 10:242
76. Schwarz G (1978) Estimating the dimension of a model. Ann Stat 6:461–464
77. Evans J, Sullivan J (2011) Approximating model probabilities in Bayesian information criterion and decision-theoretic approaches to model selection in phylogenetics. Mol Biol Evol 28:343–349
78. Gascuel O (1997) BIONJ: an improved version of the NJ algorithm based on a simple model of sequence data. Mol Biol Evol 14:685–695
79. Darriba D, Taboada GL, Doallo R, Posada D (2012) jModelTest 2: more models, new heuristics and parallel computing. Nat Methods 9:772–772
80. Lefort V, Longueville J-E, Gascuel O (2017) SMS: smart model selection in PhyML. Mol Biol Evol 34:2422–2424
81. Kleinman CL, Rodrigue N, Bonnard C, Philippe H, Lartillot N (2006) A maximum likelihood framework for protein design. BMC Bioinformatics 7:326
82. Rodrigue N, Philippe H, Lartillot N (2007) Exploring fast computational strategies for probabilistic phylogenetic analysis. Syst Biol 56:711–726
83. Yang Z (2005) Bayesian inference in molecular phylogenetics. In: Gascuel O (ed) Mathematics of evolution and phylogeny. Oxford University Press, Oxford, pp 63–90
84. Jeffreys H (1939) Theory of probability. The International series of monographs on physics. The Clarendon Press, Oxford
85. Kass RE, Raftery AE (1995) Bayes factors. J Am Stat Assoc 90:773–795
86. Lartillot N, Philippe H (2006) Computing Bayes factors using thermodynamic integration. Syst Biol 55:195–207
87. Fan Y, Wu R, Chen MH, Kuo L, Lewis PO (2011) Choosing among partition models in Bayesian phylogenetics. Mol Biol Evol 28:523–32
88. Newton MA, Raftery AE (1994) Approximating Bayesian inference with the weighted likelihood bootstrap. J R Stat Soc B 56:3–48
89. Aris-Brosou S (2003) How Bayes tests of molecular phylogenies compare with frequentist approaches. Bioinformatics 19:618–624
90. Ronquist F, Huelsenbeck JP (2003) MrBayes 3: Bayesian phylogenetic inference under mixed models. Bioinformatics 19:1572–1574
91. Drummond AJ, Rambaut A (2007) BEAST: Bayesian evolutionary analysis by sampling trees. BMC Evol Biol 7:214
92. Raftery AE (1996) Hypothesis testing and model selection. In: Gilks WR, Richardson S, Spiegelhalter DJ (eds) Markov chain Monte Carlo in practice. Chapman & Hall, Boca Raton, pp 163–187
93. Ogata Y (1989) A Monte Carlo method for high dimensional integration. Numer Math 55:137–157
94. Gelman A, Meng X-L (1998) Simulating normalizing constants: from importance sampling to bridge sampling to path sampling. Stat Sci 13:163–185
95. Xie W, Lewis PO, Fan Y, Kuo L, Chen MH (2011) Improving marginal likelihood estimation for Bayesian phylogenetic model selection. Syst Biol 60:150–60
96. Baele G, Lemey P, Bedford T, Rambaut A, Suchard MA, Alekseyenko AV (2012) Improving the accuracy of demographic and molecular clock model comparison while accommodating phylogenetic uncertainty. Mol Biol Evol 29:2157–2167
97. Raftery AE, Newton MA, Satagopan JM, Krivitsky PN (2007) Estimating the integrated likelihood via posterior simulation using the harmonic mean identity. Bayesian Stat 8:1–45
98. Smyth P (2000) Model selection for probabilistic clustering using cross-validated likelihood. Stat Comput 10:63–72
99. Lartillot N, Brinkmann H, Philippe H (2007) Suppression of long-branch attraction artefacts in the animal phylogeny using a site-heterogeneous model. BMC Evol Biol 7 (Suppl 1):S4
100. Cavalli-Sforza LL, Edwards AW (1967) Phylogenetic analysis. Models and estimation procedures. Am J Hum Genet 19:233–257
101. Aris-Brosou S (2003) Least and most powerful phylogenetic tests to elucidate the origin of the seed plants in the presence of conflicting signals under misspecified models. Syst Biol 52:781–793
102. Foulds LR, Penny D, Hendy MD (1979) A general approach to proving the minimality of phylogenetic trees illustrated by an example with a set of 23 vertebrates. J Mol Evol 13:151–166
103. Hendy MD, Penny D (1982) Branch and bound algorithms to determine minimal evolutionary trees. Math Biosci 59:277–290

104. Saitou N, Nei M (1987) The neighbor-joining method: a new method for reconstructing phylogenetic trees. Mol Biol Evol 4:406–425
105. Bruno WJ, Socci ND, Halpern AL (2000) Weighted neighbor joining: a likelihood-based approach to distance-based phylogeny reconstruction. Mol Biol Evol 17:189–197
106. Larget B, Simon D (1999) Markov chain Monte Carlo algorithms for the Bayesian analysis of phylogenetic trees. Mol Biol Evol 16:750
107. Holder MT, Lewis PO, Swofford DL, Larget B (2005) Hastings ratio of the LOCAL proposal used in Bayesian phylogenetics. Syst Biol 54:961–965
108. Whelan S (2007) New approaches to phylogenetic tree search and their application to large numbers of protein alignments. Syst Biol 56:727–740
109. Pritchard JK, Seielstad MT, Perez-Lezaun A, Feldman MW (1999) Population growth of human y chromosomes: a study of Y chromosome microsatellites. Mol Biol Evol 16:1791–1798
110. Beaumont MA, Zhang W, Balding DJ (2002) Approximate Bayesian computation in population genetics. Genetics 162:2025–2035
111. Kingman JFC (1982) The coalescent. Stoch Process Appl 13:235–248
112. Hein J, Schierup MH, Wiuf C (2005) Gene genealogies, variation and evolution: a primer in coalescent theory. Oxford University Press, Oxford
113. Marjoram P, Molitor J, Plagnol V, Tavaré S (2003) Markov chain Monte Carlo without likelihoods. Proc Natl Acad Sci 100:15324–15328
114. Sisson SA, Fan Y, Tanaka MM (2007) Sequential Monte Carlo without likelihoods. Proc Natl Acad Sci 104:1760–1765
115. Toni T, Welch D, Strelkowa N, Ipsen A, Stumpf MP (2009) Approximate Bayesian computation scheme for parameter inference and model selection in dynamical systems. J R Soc Interface 6:187–202
116. Beaumont MA (2010) Approximate Bayesian computation in evolution and ecology. Annu Rev Ecol Evol Syst 41:379–406
117. Sunnåker M, Busetto AG, Numminen E, Corander J, Foll M, Dessimoz C (2013) Approximate Bayesian computation. PLoS Comput Biol 9:e1002803
118. Lintusaari J, Gutmann MU, Dutta R, Kaski S, Corander J (2017) Fundamentals and recent developments in approximate Bayesian computation. Syst Biol 66:e66–e82
119. Ratmann O, Donker G, Meijer A, Fraser C, Koelle K (2012) Phylodynamic inference and model assessment with approximate Bayesian computation: influenza as a case study. PLoS Comput Biol 8:e1002835
120. Zheng Y, Aris-Brosou S (2013) Approximate Bayesian computation algorithms for estimating network model parameters. In: Joint statistical meeting proceedings (2013)—biometrics section, pp 2239–2253
121. Poon AF (2015) Phylodynamic inference with kernel ABC and its application to HIV epidemiology. Mol Biol Evol 32:2483–2495
122. Ibeh N, Aris-Brosou S (2016) Estimation of sub-epidemic dynamics by means of sequential Monte Carlo approximate Bayesian computation: an application to the Swiss HIV cohort study. https://doi.org/10.1101/085993
123. Hastie T, Tibshirani R, Friedman JH (2009) The elements of statistical learning: data mining, inference, and prediction. Springer series in statistics, 2nd edn. Springer, New York
124. Poon AF, Walker LW, Murray H, McCloskey RM, Harrigan PR, Liang RH (2013) Mapping the shapes of phylogenetic trees from human and zoonotic RNA viruses. PLoS One 8:e78122
125. Schwarz RF, Fletcher W, Förster F, Merget B, Wolf M, Schultz J, Markowetz F (2010) Evolutionary distances in the twilight zone—a rational kernel approach. PLoS One 5:e15788
126. Höhl M, Ragan MA (2007) Is multiple-sequence alignment required for accurate inference of phylogeny? Syst Biol 56:206–221
127. Sanderson M, Nicolae M, McMahon M (2017) Homology-aware phylogenomics at gigabase scales. Syst Biol 66:590–603
128. Jordan MI, Mitchell TM (2015) Machine learning: trends, perspectives, and prospects. Science 349:255–260
129. Rusk N (2016) Deep learning. Nat Methods 13:35
130. Esteva A, Kuprel B, Novoa RA, Ko J, Swetter SM, Blau HM, Thrun S (2017) Dermatologist-level classification of skin cancer with deep neural networks. Nature 542:115–118
131. Morell V (1996) TreeBASE: the roots of phylogeny. Science 273:569
132. Whelan S, de Bakker PIW, Quevillon E, Rodriguez N, Goldman N (2006) PANDIT: an evolution-centric database of protein and associated nucleotide domains with inferred trees. Nucleic Acids Res 34:D327–D331

133. Zhou J, Troyanskaya OG (2015) Predicting effects of noncoding variants with deep learning-based sequence model. Nat Methods 12:931–934
134. Tran NH, Zhang X, Xin L, Shan B, Li M (2017) De novo peptide sequencing by deep learning. Proc Natl Acad Sci. https://doi.org/10.1073/pnas.1705691114
135. Benton MJ, Ayala FJ (2003) Dating the tree of life. Science 300:1698–700
136. Rannala B, Yang Z (2007) Inferring speciation times under an episodic molecular clock. Syst Biol 56:453–66
137. Wegmann D, Leuenberger C, Excoffier L (2009) Efficient approximate Bayesian computation coupled with Markov chain Monte Carlo without likelihood. Genetics 182:1207–1218
138. Reich D, Green RE, Kircher M *et al* (2010) Genetic history of an archaic hominin group from Denisova Cave in Siberia. Nature 468:1053–1060
139. Hedges SB, Dudley J, Kumar S (2006) TimeTree: a public knowledge-base of divergence times among organisms. Bioinformatics 22:2971–2972
140. Kumar S, Stecher G, Suleski M, Hedges SB (2017) TimeTree: a resource for timelines, timetrees, and divergence times. Mol Biol Evol 34:1812–1819
141. Welch JJ, Bromham L (2005) Molecular dating when rates vary. Trends Ecol Evol 20:320–327
142. Kimura M (1983) The neutral theory of molecular evolution. Cambridge University Press, Cambridge
143. Sarich VM, Wilson AC (1973) Generation time and genomic evolution in primates. Science 179:1144–1147
144. Muse SV, Weir BS (1992) Testing for equality of evolutionary rates. Genetics 132:269–276
145. Bromham L, Penny D, Rambaut A, Hendy MD (2000) The power of relative rates tests depends on the data. J Mol Evol 50:296–301
146. Rambaut A (2000) Estimating the rate of molecular evolution: incorporating non-contemporaneous sequences into maximum likelihood phylogenies. Bioinformatics 16:395–399
147. Martin AP (2001) Molecular clocks. Encyclopedia of life sciences. Wiley, Hoboken, pp 1–6
148. Wray GA, Levinton JS, Shapiro LH (1996) Molecular evidence for deep Precambrian divergences among Metazoan phyla. Science 274:568–573
149. Kumar S, Hedges SB (1998) A molecular timescale for vertebrate evolution. Nature 392:917–920
150. Wang DY, Kumar S, Hedges SB (1999) Divergence time estimates for the early history of animal phyla and the origin of plants, animals and fungi. Proc Biol Sci 266:163–171
151. Heckman DS, Geiser DM, Eidell BR, Stauffer RL, Kardos NL, Hedges SB (2001) Molecular evidence for the early colonization of land by fungi and plants. Science 293:1129–1133
152. Hedges SB, Chen H, Kumar S, Wang DY, Thompson AS, Watanabe H (2001) A genomic timescale for the origin of eukaryotes. BMC Evol Biol 1:4
153. Graur D, Martin W (2004) Reading the entrails of chickens: molecular timescales of evolution and the illusion of precision. Trends Genet 20:80–86
154. Rambaut A, Bromham L (1998) Estimating divergence dates from molecular sequences. Mol Biol Evol 15:442–448
155. Yoder AD, Yang Z (2000) Estimation of primate speciation dates using local molecular clocks. Mol Biol Evol 17:1081–1090
156. Yang Z (2004) A heuristic rate smoothing procedure for maximum likelihood estimation of species divergence times. Acta Zool Sin 50:645–656
157. Aris-Brosou S (2007) Dating phylogenies with hybrid local molecular clocks. PLoS One 2:e879
158. Drummond AJ, Suchard MA (2010) Bayesian random local clocks, or one rate to rule them all. BMC Biol 8:114
159. Sanderson M (1997) A nonparametric approach to estimating divergence times in the absence of rate constancy. Mol Biol Evol 14:1218
160. Sanderson MJ (2002) Estimating absolute rates of molecular evolution and divergence times: a penalized likelihood approach. Mol Biol Evol 19:101–109
161. Gillespie JH (1991) The causes of molecular evolution. Oxford University Press, Oxford
162. Thorne JL, Kishino H, Painter IS (1998) Estimating the rate of evolution of the rate of molecular evolution. Mol Biol Evol 15:1647–1657
163. Aris-Brosou S, Yang Z (2002) Effects of models of rate evolution on estimation of divergence dates with special reference to the metazoan 18S ribosomal RNA phylogeny. Syst Biol 51:703–714
164. Aris-Brosou S, Yang Z (2003) Bayesian models of episodic evolution support a late

precambrian explosive diversification of the Metazoa. Mol Biol Evol 20:1947–1954

165. Rannala B, Yang Z (1996) Probability distribution of molecular evolutionary trees: a new method of phylogenetic inference. J Mol Evol 43:304–311
166. Pybus OG, Rambaut A, Harvey PH (2000) An integrated framework for the inference of viral population history from reconstructed genealogies. Genetics 155:1429–1437
167. Drummond AJ, Rambaut A, Shapiro B, Pybus OG (2005) Bayesian coalescent inference of past population dynamics from molecular sequences. Mol Biol Evol 22:1185–1192
168. Minin VN, Bloomquist EW, Suchard MA (2008) Smooth skyride through a rough skyline: Bayesian coalescent-based inference of population dynamics. Mol Biol Evol 25:1459–1471
169. Hedges SB, Kumar S (2004) Precision of molecular time estimates. Trends Genet 20:242–247
170. Yang Z, Rannala B (2006) Bayesian estimation of species divergence times under a molecular clock using multiple fossil calibrations with soft bounds. Mol Biol Evol 23:212–226
171. Inoue J, Donoghue PCJ, Yang Z (2010) The impact of the representation of fossil calibrations on Bayesian estimation of species divergence times. Syst Biol 59:74–89
172. Drummond AJ, Ho SYW, Phillips MJ, Rambaut A (2006) Relaxed phylogenetics and dating with confidence. PLoS Biol 4:e88
173. Bouckaert R, Heled J, Kühnert D, Vaughan T, Wu CH, Xie D, Suchard MA, Rambaut A, Drummond AJ (2014) BEAST 2: a software platform for Bayesian evolutionary analysis. PLoS Comput Biol 10:e1003537
174. Wertheim JO, Sanderson MJ, Worobey M, Bjork A (2010) Relaxed molecular clocks, the bias-variance trade-off, and the quality of phylogenetic inference. Syst Biol 59:1–8
175. Lemey P, Rambaut A, Drummond AJ, Suchard MA (2009) Bayesian phylogeography finds its roots. PLoS Comput Biol 5: e1000520
176. Lemey P, Rambaut A, Welch JJ, Suchard MA (2010) Phylogeography takes a relaxed random walk in continuous space and time. Mol Biol Evol 27:1877–1885
177. Guillot G, Santos F, Estoup A (2008) Analysing georeferenced population genetics data with Geneland: a new algorithm to deal with null alleles and a friendly graphical user interface. Bioinformatics 24:1406–1407
178. Nadin-Davis SA, Feng Y, Mousse D, Wandeler AI, Aris-Brosou ST (2010) Spatial and temporal dynamics of rabies virus variants in big brown bat populations across Canada: footprints of an emerging zoonosis. Mol Ecol 19:2120–2136
179. Pagel M, Meade A (2004) A phylogenetic mixture model for detecting pattern-heterogeneity in gene sequence or character-state data. Syst Biol 53:571–581
180. Pagel M, Meade A, Barker D (2004) Bayesian estimation of ancestral character states on phylogenies. Syst Biol 53:673–684
181. Lartillot N, Poujol R (2011) A phylogenetic model for investigating correlated evolution of substitution rates and continuous phenotypic characters. Mol Biol Evol 28:729–744
182. Bromham L, Woolfit M, Lee MS, Rambaut A (2002) Testing the relationship between morphological and molecular rates of change along phylogenies. Evolution 56:1921–1930
183. Ho SYW, Duchêne S (2014) Molecular-clock methods for estimating evolutionary rates and timescales. Mol Ecol 23:5947–5965
184. dos Reis M, Donoghue PCJ, Yang Z (2016) Bayesian molecular clock dating of species divergences in the genomics era. Nat Rev Genet 17:71–80
185. Donoghue PCJ, Yang Z (2016) The evolution of methods for establishing evolutionary timescales. Philos Trans R Soc Lond B Biol Sci. https://doi.org/10.1098/rstb.2016.0020
186. Ho SY, Tong KJ, Foster CS, Ritchie AM, Lo N, Crisp MD (2015) Biogeographic calibrations for the molecular clock. Biol Lett 11:20150194
187. Kühnert D, Wu C-H, Drummond AJ (2011) Phylogenetic and epidemic modeling of rapidly evolving infectious diseases. Infect Genet Evol 11:1825–1141
188. Rieux A, Balloux F (2016) Inferences from tip-calibrated phylogenies: a review and a practical guide. Mol Ecol 25:1911–1924
189. Ho SYW, Chen AXY, Lins LSF, Duchêne DA, Lo N (2016) The genome as an evolutionary timepiece. Genome Biol Evol 8:3006–3010
190. O'Reilly JE, dos Reis M, Donoghue PCJ (2015) Dating tips for divergence-time estimation. Trends Genet 31:637–50
191. 1000 Genomes Project Consortium, Abecasis GR, Altshuler D, Auton A, Brooks LD, Durbin RM, Gibbs RA, Hurles ME, McVean GA (2010) A map of human genome variation from population-scale sequencing. Nature 467:1061–1073

192. UK10K Consortium, Walter K, Min JL, Huang J *et al* (2015) The UK10K project identifies rare variants in health and disease. Nature 526:82–90
193. Ledford H (2016) AstraZeneca launches project to sequence 2 million genomes. Nature 532:427
194. Muse SV, Gaut BS (1994) A likelihood approach for comparing synonymous and nonsynonymous nucleotide substitution rates, with application to the chloroplast genome. Mol Biol Evol 11:715–724
195. Goldman N, Yang Z (1994) A codon-based model of nucleotide substitution for protein-coding DNA sequences. Mol Biol Evol 11:725–736
196. Kosiol C, Anisimova M (2011) Methods for detecting natural selection in protein-coding genes. In: Anisimova M (ed) Evolutionary genomics: statistical and computational methods. Methods in molecular biology series. Humana-Springer, New York
197. Thorne JL, Choi SC, Yu J, Higgs PG, Kishino H (2007) Population genetics without intraspecific data. Mol Biol Evol 24:1667–1677
198. Choi SC, Hobolth A, Robinson DM, Kishino H, Thorne JL (2007) Quantifying the impact of protein tertiary structure on molecular evolution. Mol Biol Evol 24:1769–1782
199. Halpern AL, Bruno WJ (1998) Evolutionary distances for protein-coding sequences: modeling site-specific residue frequencies. Mol Biol Evol 15:910–917
200. Yang Z, Nielsen R (2008) Mutation-selection models of codon substitution and their use to estimate selective strengths on codon usage. Mol Biol Evol 25:568–579
201. Rodrigue N, Philippe H, Lartillot N (2010) Mutation-selection models of coding sequence evolution with site-heterogeneous amino acid fitness profiles. Proc Natl Acad Sci USA 107:4629–4634
202. Rodrigue N, Lartillot N (2017) Detecting adaptation in protein-coding genes using a Bayesian site-heterogeneous mutation-selection codon substitution model. Mol Biol Evol 34:204–214
203. Bloom JD (2017) Identification of positive selection in genes is greatly improved by using experimentally informed site-specific models. Biol Direct 12:1. https://doi.org/10.1186/s13062-016-0172-z
204. Choi SC, Redelings BD, Thorne JL (2008) Basing population genetic inferences and models of molecular evolution upon desired stationary distributions of DNA or protein sequences. Philos Trans R Soc Lond B Biol Sci 363:3931–3939
205. Rodrigue N, Kleinman CL, Philippe H, Lartillot N (2009) Computational methods for evaluating phylogenetic models of coding sequence evolution with dependence between codons. Mol Biol Evol 26:1663–1676
206. Hartl DL, Clark AG (2007) Principles of population genetics, 4th edn. Sinauer Associates, Sunderland
207. Kimura M (1962) On the probability of fixation of mutant genes in a population. Genetics 47:713–719
208. Rice SH (2004) Evolutionary theory: mathematical and conceptual foundations. Sinauer Associates, Sunderland
209. Kimura M (1978) Change of gene frequencies by natural selection under population number regulation. Proc Natl Acad Sci USA 75:1934–1937
210. Tamuri A, dos Reis M, Goldstein R (2012) Estimating the distribution of selection coefficients from phylogenetic data using sitewise mutation-selection models. Genetics 190:1101–1115
211. Rodrigue N (2013) On the statistical interpretation of site-specific variables in phylogeny-based substitution models. Genetics 193:557–564
212. Prins P, Belhachemi D, Möller S, Smant G (2011) Scalable computing in evolutionary genomics. In: Anisimova M (ed) Evolutionary genomics: statistical and computational methods. Methods in molecular biology series. Humana-Springer, New York
213. Yang Z (1993) Maximum-likelihood estimation of phylogeny from DNA sequences when substitution rates differ over sites. Mol Biol Evol 10:1396–1401
214. Yang Z (1994) Maximum likelihood phylogenetic estimation from DNA sequences with variable rates over sites: approximate methods. J Mol Evol 39:306–314
215. Altekar G, Dwarkadas S, Huelsenbeck JP, Ronquist F (2004) Parallel Metropolis coupled Markov chain Monte Carlo for Bayesian phylogenetic inference. Bioinformatics 20:407–415
216. Lartillot N, Rodrigue N, Stubbs D, Richer J (2013) PhyloBayes MPI: phylogenetic reconstruction with infinite mixtures of profiles in a parallel environment. Syst Biol 62:611–615
217. Rodrigue N, Lartillot N (2014) Site-heterogeneous mutation-selection models within the PhyloBayes-MPI package. Bioinformatics 30:1020–1021

218. Stamatakis A, Hoover P, Rougemont J (2008) A rapid bootstrap algorithm for the RAxML Web servers. Syst Biol 57:758–771
219. Stamatakis A, Ludwig T, Meier H (2005) RAxML-III: a fast program for maximum likelihood-based inference of large phylogenetic trees. Bioinformatics 21:456–463
220. Stamatakis A, Göker M, Grimm GW (2010) Maximum likelihood analyses of 3,490 rbcL sequences: scalability of comprehensive inference versus group-specific taxon sampling. Evol Bioinform Online 6:73–90
221. Stamatakis A, Alachiotis N (2010) Time and memory efficient likelihood-based tree searches on phylogenomic alignments with missing data. Bioinformatics 26:i132–i139
222. Suchard MA, Rambaut A (2009) Many-core algorithms for statistical phylogenetics. Bioinformatics 25:1370–1376
223. Ronquist F, Teslenko M, van der Mark P, Ayres DL, Darling A, Höhna S, Larget B, Liu L, Suchard MA, Huelsenbeck JP (2012) MrBayes 3.2: efficient Bayesian phylogenetic inference and model choice across a large model space. Syst Biol 61:539–542
224. Muir P, Li S, Lou S *et al* (2016) The real cost of sequencing: scaling computation to keep pace with data generation. Genome Biol 17:53
225. Schatz MC, Langmead B, Salzberg SL (2010) Cloud computing and the DNA data race. Nat Biotechnol 28:691–693
226. Dereeper A, Guignon V, Blanc G *et al* (2008) Phylogeny.fr: robust phylogenetic analysis for the non-specialist. Nucleic Acids Res 36: W465–W469
227. de Koning AP, Gu W, Pollock DD (2010) Rapid likelihood analysis on large phylogenies using partial sampling of substitution histories. Mol Biol Evol 27:249–265

Part II

Genomic Alignment and Homology Inference

Chapter 4

Whole-Genome Alignment

Colin N. Dewey

Abstract

Whole-genome alignment (WGA) is the prediction of evolutionary relationships at the nucleotide level between two or more genomes. It combines aspects of both colinear sequence alignment and gene orthology prediction and is typically more challenging to address than either of these tasks due to the size and complexity of whole genomes. Despite the difficulty of this problem, numerous methods have been developed for its solution because WGAs are valuable for genome-wide analyses such as phylogenetic inference, genome annotation, and function prediction. In this chapter, we discuss the meaning and significance of WGA and present an overview of the methods that address it. We also examine the problem of evaluating whole-genome aligners and offer a set of methodological challenges that need to be tackled in order to make most effective use of our rapidly growing databases of whole genomes.

Key words Sequence alignment, Whole-genome alignment, Homology map, Toporthology, Genome evolution, Comparative genomics

1 Introduction

When the problem of biological sequence alignment was first described and addressed in the 1970s, sequencing technology was limited to obtaining the sequences of individual proteins or mRNAs or short genomic intervals. As such, classical sequence alignment (as described in Chapter 7 [1]) is typically focused on predicting homologous positions within two or more relatively short and colinear sequences, allowing for the edit events of substitution, insertion, and deletion. Although limited in its scope, this type of alignment remains extremely important today, with gene-sized alignments forming the basis of most evolutionary studies.

Starting in 1995 with the sequencing of the 1.8 Mb-sized genome of the bacterium *H. influenzae* [2], biologists have had access to a different scale of biological sequences, those of whole genomes. DNA sequencing technology has rapidly improved since that time, and as a result, we have seen an explosion in the availability of whole-genome sequences. As of the writing of this chapter, there are 9071 published complete genome sequences (8380

Maria Anisimova (ed.), *Evolutionary Genomics: Statistical and Computational Methods*, Methods in Molecular Biology, vol. 1910, https://doi.org/10.1007/978-1-4939-9074-0_4,

bacterial, 281 archaeal, and 410 eukaryotic), according to the GOLD database [3]. Whole-genome sequencing remains popular, with over 140,000 sequencing projects that are either ongoing or completed.

Along with the ascertainment of these sequences, the problem of whole-genome alignment (WGA) has arisen. As each genome is sequenced, there is interest in aligning it against other available genomes in order to better understand its evolutionary history and, ultimately, the biology of its species. Like classical sequence alignment, WGA is about predicting evolutionarily related sequence positions. However, aligning whole genomes is made more complicated by the fact that genomes undergo large-scale structural changes, such as duplications and rearrangements. In addition, a set of genomes may contain pairs of sequence positions whose evolutionary relationships can be described by any of the three major subclasses of homology: orthology, paralogy, and xenology. As orthologous positions are typically of primary interest, WGA also involves the classification of homologous relationships.

In this chapter, we describe the problem of WGA and the methods that address it. We begin with a thorough definition of the problem and discuss the important downstream applications of WGAs. We then categorize the WGA methods that have been developed and describe the key computational techniques that are used within each category. In addition to describing whole-genome aligners, we also discuss the various approaches that have been used for evaluating the alignments they produce. Lastly, we lay out a number of current methodological challenges for WGA.

2 The Definition and Significance of WGA

2.1 WGA as a Correspondence Between Genomes

In imprecise terms, a WGA is a "correspondence" between genomes. For each segment of a given genome, a WGA tells us where its "corresponding" segments are in other genomes. A segment may be one or more contiguous nucleotide positions within a genome. What does it mean for two genomic segments to "correspond" to each other? In most situations, we consider two segments to be "corresponding" if they are orthologous. Orthologous sequences are those that are evolutionarily related (homologous) and that diverged from their most recent common ancestor (MRCA) due to a speciation event [4]. In contrast, paralogous sequences are homologs that diverged from the MRCA due to a duplication event. Thus, by definition, orthologous sequences are the most closely related pieces of two genomes and, as is more thoroughly discussed later and in Chapter 9 [5], are of primary interest because they are useful for applications such as function prediction and species tree inference. As such, WGA is most commonly taken to be the prediction of orthology between the components of entire genome sequences. When a WGA also predicts paralogy, typically only paralogs whose MRCA is at least as recent as the MRCA of entire set of genomes are considered, as there is extensive ancient homology within extant genomes.

It is important to note that the orthologous relationships between two genomes do not create a one-to-one correspondence. Duplication events that have occurred since the time of the MRCA of the species can result in a genomic segment in one species having multiple orthologous segments in another. This is a particularly important issue when the genome of one lineage has undergone a whole-genome duplication event since the time of the MRCA. In this situation, few segments of the genome of the nonduplicated lineage have a single ortholog in the other genome.

2.2 Toporthology

In many cases, WGAs do not aim to predict all orthologous sequences. Instead, they only predict toporthology (positional orthology), a distinguished subset of orthology [6, 7]. The concept of toporthology captures the notion that not all orthologous relationships are equivalent in terms of the evolutionary history of the genomic context of the orthologs. Figure 1 gives an example scenario in which toporthology helps to distinguish between two orthologous relationships.

The definition of toporthology relies on a classification of duplication events. A duplication event is considered to be "symmetric" if the removal of either copy of the duplicated genomic material (immediately after the event) reverts the genome to its original (preduplication) state. Examples of symmetric duplications are tandem and whole-genome duplications. If only one specific copy can be removed to undo a duplication event, then the event is considered "asymmetric." In the asymmetric case, the removable copy is referred to as the "target," with the other copy referred to as the "source." Retrotransposition and segmental duplication both belong to the asymmetric class.

With this classification of duplication events in hand, we can now define toporthology. Two genomic segments are toporthologous if they are orthologous and neither segment is derived from the target of an asymmetric duplication event since the time of the MRCA of the segments. Thus, two orthologous segments are toporthologous if their evolutionary history (since the MRCA) only involves symmetric duplication events or asymmetric duplications in which their ancestral segment was part of the source copy.

The important property of toporthologs is that, in the absence of rearrangement events, they share the same ancestral genomic context. As the context of a gene or genomic segment has functional consequences, toporthologous sequences are generally expected to be more similar in their function than orthologous sequences that are not toporthologous (atoporthologs) [6]. However, there is no guarantee that toporthologs share a common

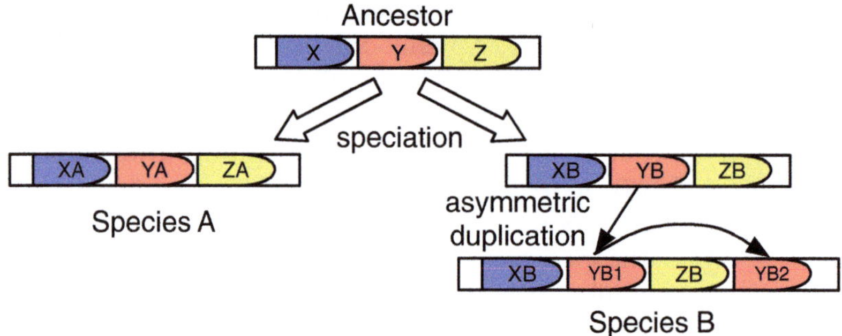

Fig. 1 A hypothetical evolutionary scenario in which the relation of toporthology distinguishes between two ortholog pairs. The bullet-like shapes indicate genomic segments. Both YB1 and YB2 are orthologous to YA. However, only YB1 is toporthologous to YA because YB2 was derived from the target of an asymmetric duplication since the time of the most recent common ancestor, Y, of YB2 and YA

function or that two genomic intervals that have the same function are toporthologs. Thus, a rigorous functional analysis of genomes should consider all classes of homology. Nevertheless, WGAs that focus on toporthology produce a good first approximation to a functional correspondence between genomes.

2.3 Definition and Representation

To be more precise, a WGA is, in general, the prediction of homologous pairs of positions between two or more genome sequences. Often, as we have previously discussed, only orthologous or toporthologous relations are predicted in WGAs. And while alignment is typically focused on homologous relationships *between* sequences, whole-genome comparisons can also include alignments *within* genomes, which represent paralogous sequences.

Note that we define WGA as homology prediction at the level of nucleotides. Although the concept of homology is more commonly used with respect to entire genes or proteins, it is easily used and, in fact, more naturally defined at the level of single nucleotides. Homology of nucleotide positions is established through template-driven nucleotide synthesis, and the definitions of orthology, paralogy, and xenology for nucleotides follow those for genes [7].

While a WGA can be defined as a prediction of homology statements, it is usually represented as a set of nucleotide-level alignment matrices or "blocks," each block made up by segments of the genomes that are both homologous and colinear. Homologous genomic segments are colinear if they have not been broken by a rearrangement event since the time of their MRCA. Since rearrangement events, such as inversions, are common at the scale of entire genomes, WGAs are typically made up of many blocks. In general, a block contains two or more genomic segments, and multiple segments in the same block may belong to the same genome (indicating paralogous sequence). One specific WGA representation, the "threaded blockset" [8], requires that every

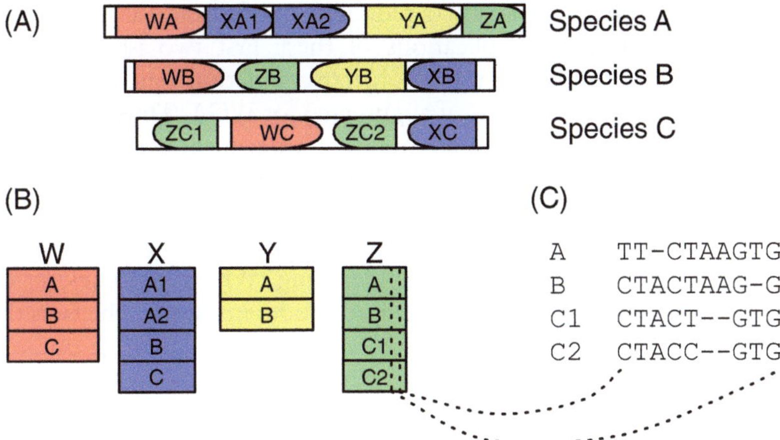

Fig. 2 An example WGA of three genomes represented as a set of alignment blocks. (**a**) The positions of the genomic segments that are in the alignment blocks are shown as shaded bullet-like shapes (the direction of the bullet indicates the orientation of the segment). In this example, not all genomic segments belong to a block (note the unshaded intervals). (**b**) The alignment blocks of the WGA. Note that blocks do not need to contain a segment from all genomes (e.g., block Y) and that some blocks can contain multiple segments from the same genome (e.g., blocks X and Z). (**c**) A slice of alignment block Z, which is a nucleotide-level alignment

position belongs to a block and thus additionally allows a block to contain just a single segment, which would represent a unique genomic sequence. Figure 2 depicts a hypothetical example of a WGA, with some blocks containing both orthologous and paralogous sequences.

As more genomes are added to an alignment or the total evolutionary divergence between them is increased, the blocks in a WGA decrease in size and increase in number. One might imagine that in the limit of an infinite number of genomes or an infinite amount of time, all blocks might have length one (a single column), which makes the concept of an "alignment matrix" irrelevant. However, rearrangements in certain segments of the genome are likely to be highly deleterious to an organism and will thus never be observed. Such segments are referred to as genomic "atoms" [9] and prevent all blocks from becoming single alignment columns.

2.4 Comparison to Other Homology Prediction Tasks

WGA is closely related to classical sequence alignment (the alignment of two or more relatively short and colinear sequences), and most whole-genome aligners rely on classical alignment techniques (e.g., the Needleman–Wunsch [10] and Smith–Waterman [11] pairwise alignment algorithms and heuristics used for multiple alignments) as subroutines. However, there are three key differences between these two classes of alignment. First, and most importantly, classical alignment requires sequences to be colinear, which is often not the case for genome sequences due to rearrangement events. Second, even when restricted to toporthologous relationships, the correspondences between genomes are not one to one, which is also a requirement of classical alignment. Due, in part, to the complications of these first two issues, it is difficult to formulate a useful objective function (such as the sum-of-pairs score for classical alignment) for WGA. Thus, most genome alignment methods are heuristic procedures that lack an explicit objective. A last difference between classical alignment and WGA is the scale of the problem. Classical alignment typically focuses on the alignment of single genes, which are usually on the order of thousands of nucleotides long. Whole genomes, in contrast, are millions to billions of nucleotides in length. The facts that genomes are large and are often neither colinear nor in one-to-one correspondence with other genomes are what make WGA challenging.

Since WGA is often focused on orthologous relationships, it is also related to the "orthology prediction" problem (*see* Chapter 9 [5]). The key difference between the two problems is that orthology prediction is traditionally cast at the level of genes, whereas WGA operates at the level of nucleotides. For most orthology prediction methods, a genome is treated as an unordered set of genes. Whole-genome aligners, on the other hand, consider a genome to be a set of DNA sequences (chromosomes) within which genes are embedded. Thus, a WGA provides orthology predictions for both genes and intergenic regions. Due in part to their treatment of genomes as long nucleotide sequences, current WGA methods rely exclusively on sequence similarity and the ordering of nucleotides in a genome to predict orthology. In contrast, orthology prediction methods often use phylogenetic analyses, which can be more powerful than genome order and sequence similarity information alone. Thus, while the problem of WGA is broader in scope than that of orthology prediction, it is restricted to the analysis of relatively closely related genomes, for which homology of nongenic nucleotides is detectable and gene order is at least partially conserved. Gene-level orthology prediction is more appropriate for distantly related genomes, which may only have detectable homology at the amino acid level and little colinearity.

2.5 Significance

WGAs are powerful because they allow for the analysis of molecular evolution at both large and small scales. At the large scale, one can use such alignments to estimate the frequency and location of rearrangement and duplication events. For example, one might use a WGA between human and mouse to identify colinear orthologous blocks, which are then given to a rearrangement analysis method (e.g., [12]) to determine a most parsimonious set of rearrangement events explaining the current structures of the two genomes. At the small scale, WGAs can be used to examine the rates of substitutions and indels across the entire genome. For example, one might look at alignments of ancestral repeats to estimate the neutral rates of nucleotide evolution. Both small-

and large-scale mutational events identified from WGAs can be used as data for species tree inference. In combination with carefully constructed models of genome evolution at both scales, WGAs also enable the task of ancestral genome reconstruction [13, 14].

Beyond purely evolutionary studies, WGAs are valuable for identifying functional elements within genomes. Each class of functional element within the genome tends to have a unique "evolutionary signature," which can be searched for within WGAs [15]. For example, coding sequences tend to have mutational patterns with a predominance of substitutions at the third positions of codons, which are unlikely to affect the amino acid sequence. This characteristic evolutionary signature of coding sequence has led to the development of comparative gene-finding methods, which often use WGAs (Chapter 6 [16]). Noncoding RNA sequences can also be identified from WGAs but have more complex signatures involving compensatory mutations that maintain base pairing within RNA secondary structures [17]. More generally, one can search for evolutionarily constrained regions within WGAs, which can contain functional elements from a variety of classes [18]. When combined with the knowledge of transcription factor-binding motifs, this approach can be used to identify transcription factor-binding sites with a technique called "phylogenetic footprinting" [19]. The easiest evolutionarily constrained regions to pick out are those of "ultraconserved elements," which maintain high levels of sequence identity across large evolutionary distances and are primarily noncoding components of the genome [20].

WGAs also allow for the transfer of functional information about specific elements from one species to another. As WGAs typically predict orthology and orthologous sequences are likely to have similar functions, WGAs are valuable for function prediction. By aligning at the nucleotide level across the genome, they can aid in function prediction for both genes and nongenic regions, such as those that contain regulatory elements. For example, if we are interested in a specific disease-associated interval in the human genome, we might use an alignment to identify where its mouse orthologs are located. Knowledge of the mouse orthologs would enable us to have a better understanding of the evolutionary history of this genomic region and could lead to genetic manipulation experiments that can only be performed in mice.

3 Methods for WGA

3.1 A Simplistic Approach

It is easier to understand the existing methods for performing WGA by first appreciating the shortcomings of a simplistic approach for comparing whole-genome sequences. One simple approach would be to run BLAST [21], or another similar local alignment tool, between all pairs of genomes. The WGA would then be defined as the union of all significant pairwise local alignments discovered by BLAST. By using a local alignment tool, we avoid the issues of rearrangements and duplications, as sets of local alignments are not constrained to be colinear or in one-to-one correspondence.

While this approach would certainly yield a large set of homology predictions between all pairs of genomes, it has a number of shortcomings. First, by only using a BLAST significance threshold, it makes no distinction between orthology, paralogy, and other refinements of homology. Second, the pairwise alignments that it produces are not guaranteed to be consistent with each other, even though homology, by definition, is a transitive relation. Third, BLAST may miss some homologous sequences that have low similarity but are strongly supported in their relatedness by flanking homologous sequences. BLAST's significance statistics are proven for ungapped sequences and good in practice for sequences with short indels [22], but are not designed for whole-genome comparisons, which often feature large-scale insertions and deletions and heterogeneous substitution rates. Lastly, this approach is overly computationally intensive. For example, it does not take advantage of the fact that homology is a transitive relation, that relationships between sequences are reasonably modeled by a tree, and that homologous sequences between genomes are often found in long colinear segments.

3.2 The Two Major Approaches to WGA

Existing WGA methods attempt to address one or more of the weaknesses of this simple approach. These methods can be loosely classified into two major strategies which we refer to as the "hierarchical" and "local" approaches. The main idea behind the hierarchical approach is to split the WGA problem into a set of global multiple alignment problems. To do this, it first identifies the colinear and homologous (typically orthologous) segments of the genomes. Each set of colinear segments is then given to a specialized genomic global alignment method to produce a nucleotide-level alignment. In contrast, the first step of the "local" approach is to produce a large set of nucleotide-level alignments. Later steps involve the filtering and merging of these alignments to produce sets of pairwise or multiple alignments of homologous (typically orthologous) sequences. Despite their differences, both strategies typically begin with a local alignment step that is similar to the simplistic all-vs.-all alignment of the BLAST approach. A summary of all of the WGA methods described in this chapter and the role they play within one or both approaches is given in Table 1.

Both approaches have advantages and disadvantages. The primary advantage of the hierarchical approach is that it can often be faster and breaks a WGA into a number of independent subproblems that can be solved in parallel. It is faster because the

Table 1
A list of the WGA methods cited in this chapter

Method	Category	Relationships predicted	Pairwise or multiple	References
BLAST	Local alignment	Homology	Pairwise	[21]
BLAT	Local alignment	Homology	Pairwise	[32]
STELLAR	Local alignment	Homology	Pairwise	[33]
LASTZ	Local alignment	Homology	Pairwise	[34]
LAST	Local alignment	Homology	Pairwise	[28]
MUMmer	Local alignment	Orthology	Pairwise	[35]
CHAOS	Local alignment	Homology	Pairwise	[36]
GRIMM-Synteny	Homology mapping	Toporthology	Multiple	[40]
DRIMM-Synteny	Homology mapping	Homology	Multiple	[45]
Mercator	Homology mapping	Toporthology	Multiple	[46]
Enredo	Homology mapping	Homology	Multiple	[47]
OSfinder	Homology mapping	Toporthology	Multiple	[48]
SuperMap	Homology mapping	Homology	Multiple	[49]
Sibelia	Homology mapping	Homology	Multiple	[50]
M-GCAT	Hierarchical WGA	Toporthology	Multiple	[51]
progressiveMauve	Hierarchical WGA	Toporthology	Multiple	[52]
MUGSY	Hierarchical WGA	Toporthology	Multiple	[53]
Cactus	Hierarchical WGA	Homology	Multiple	[54]
MAVID	Global genomic alignment	Colinear homology	Multiple	[60]
LAGAN/Multi-LAGAN	Global genomic alignment	Colinear homology	Pairwise/multiple	[37]
DIALIGN	Global genomic alignment	Colinear homology	Multiple	[36]
SeqAn::T-Coffee	Global genomic alignment	Colinear homology	Multiple	[61]
Pecan	Global genomic alignment	Colinear homology	Multiple	[47]
FSA	Global genomic alignment	Colinear homology	Multiple	[62]
NUCmer/PROmer	Local WGA	Orthology	Pairwise	[35]
MULTIZ/TBA	Local WGA	Homology	Multiple	[8]
AXTCHAIN/CHAINNET	Alignment chaining and filtering	Orthology	Pairwise	[67]

(continued)

Table 1
(continued)

Method	Category	Relationships predicted	Pairwise or multiple	References
PicoInversionMiner	Alignment refinement	Orthology	Pairwise	[68]
Cassis	Alignment refinement	Orthology	Pairwise	[69, 70]
GenAlignRefine	Alignment refinement	Colinear homology	Multiple	[71]
PSAR-Align	Alignment refinement	Colinear homology	Multiple	[73]
Phylo	Alignment refinement	Colinear homology	Multiple	[76, 77]
SLAM	Alignment refinement	Colinear homology	Pairwise	[78]
DOUBLESCAN	Alignment refinement	Colinear homology	Pairwise	[79]
CESAR	Alignment refinement	Colinear homology	Pairwise	[81]
MORPH	Alignment refinement	Colinear homology	Pairwise	[82]
EMMA	Alignment refinement	Colinear homology	Pairwise	[83]
MAFIA	Alignment refinement	Colinear homology	Multiple	[84]
SAPF	Alignment refinement	Colinear homology	Multiple	[85]
REAPR	Alignment refinement	Colinear homology	Multiple	[86]

For each method, the approach it uses or the role it plays within a larger WGA system is given in the "category" column. Each method is labeled as either "pairwise" or "multiple" depending on whether it can be applied to generate multiple alignments. In addition, the primary type of evolutionary relationship predicted by each method is given in the "relationships predicted" column

identification of long colinear and orthologous segments in the genomes can be accurately computed without the need for sensitive nucleotide-level alignments. However, because hierarchical methods do not often use the most sensitive aligners for this step, they tend to miss small rearranged or diverged segments. Thus, the primary advantage of the local method is in its sensitivity to these regions, although "glocal" alignment methods [23], which allow for small rearrangements, can partially ameliorate this weakness of hierarchical methods. Hierarchical methods also run the risk of being overconfident of the colinearity of genomic segments and can thus produce more false-positive aligned positions within sequences predicted to be colinear.

3.3 Local Pairwise Genomic Alignment

Methods for both WGA strategies generally start by finding local alignments between, and perhaps within, the genomes. The Smith–Waterman algorithm is the classical solution to the pairwise local alignment problem, but is generally not used for WGA because it runs in time quadratic in the size of the genomes, which can be large. Instead, most methods adopt a "seed-and-extend" approach for discovering high-scoring local alignments, much like BLAST. This approach first identifies short ungapped matches between the sequences using one of a variety of data structures. It then extends the short matches from both ends using a variant of the Smith–Waterman algorithm, stopping the extension when the score of the alignment drops below a specified threshold. In some cases, nearby and consistent (in terms of order and orientation) local alignments are "chained" together to form larger alignments.

There are a number of techniques used for discovering seeds at the genomic scale for the "seed-and-extend" approach to local alignment. A first distinction between the techniques is whether they find exact or inexact matching seeds. Exact seed discovery is often faster and easier to implement, whereas inexact seeds offer better sensitivity. Seed techniques also vary in whether they use "consecutive" or "spaced" seeds [24]. Consecutive seeds consider matches and mismatches at all positions within a sequence interval, whereas spaced seeds only check for matches at a subset of positions within an interval. The specific subset of positions checked is known as the "seed pattern," and there has been significant work on determining optimal sets of multiple seed patterns (e.g., [25, 26]). It has been shown that carefully chosen spaced seed patterns are superior to consecutive seeds in terms of sensitivity [27]. Lastly, seeds differ in whether their lengths are fixed or adaptive (variable). For WGA, adaptive seeds have been shown to allow for faster local alignment at the same level of sensitivity as fixed seeds [28].

Seed-finding techniques can often be improved by taking advantage of DNA evolutionary models. A generalization of spaced seeds is "subset seeds" [29], which allow subsets of bases to be considered equivalent when determining if there is a match at a given position. Subset seeds are particularly useful for taking into account that transitions are often more common than transversions in genome comparisons. Further taking into account biologically informed substitution patterns is the "translated" seed, which is a match at the amino acid level after translating genomic sequences in all six possible reading frames. Translated seeds enable increased sensitivity in comparisons of more diverged genomes. Lastly, when aligning a genome to a set of genomes for which a multiple WGA has already been constructed, one can take into account the substitution patterns and ancestral sequences inferred from the WGA to devise more sensitive seeds [30, 31].

The choice of seed type is the major determinant of the data structures used for seed discovery. For example, BLAT [32] uses a simple index of all possible k-mers for exact and translated seeds but uses a heuristic of indexing only nonoverlapping k-mers for memory efficiency. STELLAR [33] also uses an index of k-mers but implements an exact algorithm based on filtration for finding all local alignments with an error rate below a given threshold. LASTZ (the successor to BLASTZ [34]), which uses a carefully chosen spaced seed pattern introduced by [24], instead uses a hash table to find both exact and inexact matches. Not to be confused with LASTZ is the more recently developed LAST aligner [28], which uses adaptive seeds with highly configurable patterns that are identified via a suffix array data structure. MUMmer uses a suffix tree to rapidly find all exact consecutive seeds with some minimum length [35]. CHAOS [36], which is a component of the LAGAN-suite of genome alignment tools [37], uses a related structure, a "threaded trie," to find exact and inexact consecutive seeds.

For computational efficiency reasons, the extension step of the seed-and-extend approach typically only allows for ungapped alignments or alignments with short indels. However, genome alignments often feature large indels that are not discovered by extension from a seed. Thus, many local genomic alignment tools use a "chaining" step to link nearby and consistent local alignments discovered by the seed-and-extend strategy. For example, MUMmer includes a module for chaining together nearby exact matches using a variation of the longest increasing subsequence (LIS) problem [38]. CHAOS also uses an LIS-derived algorithm for chaining the inexact consecutive seeds it discovers. Chaining is often followed by more sensitive alignment between chained local alignments. For example, MUMmer runs a variant of Smith–Waterman alignment in between chained matches and LASTZ recursively searches for alignments with more sensitive seeds in between nearby alignments discovered in previous steps.

3.4 The Hierarchical Approach

The hierarchical approach to WGA consists of two steps. First, a high-level homology map between the genomes is constructed. Second, a nucleotide-level alignment is obtained by running a genomic global alignment tool on each homologous and colinear set of genomic segments identified by the homology map. Hierarchical WGA methods vary in the exact techniques used for each step.

The idea behind the hierarchical approach is to separate the problem of identifying rearrangements and duplications from that of obtaining a nucleotide-level alignment. In the absence of rearrangements and duplications, WGA simply reduces to classical sequence alignment although at a much larger scale. Thus, if a WGA problem can be broken into a set of subproblems that do not contain these large-scale events, the numerous methods that have been developed for classical global alignment can be utilized.

The first step of the hierarchical strategy is to construct a homology map between the genomes of interest. A homology map is a collection of sets of genomic intervals, where each set of intervals is required to be homologous and colinear (i.e., free of rearrangements and duplications). Each set represents the sequences that will ultimately form a block within a WGA.

Homology maps generally have the property that each genomic position belongs to at most one set and has all of its homologs contained within that set. For WGA, homology maps are often restricted in the evolutionary relationships that are captured, as only a subset of homologous relationships may be of interest. Typically, only orthologous relationships are captured, forming an "orthology map." When orthology maps are restricted to predicting one-to-one relationships, they are more likely to be representative of toporthology.

The concept of a homology map is closely related to the concepts of "conserved segments" and "syntenic blocks," which generally refer to sets of genomic intervals containing multiple homologous markers (e.g., genes) and featuring conserved orientations and adjacencies of these markers [39, 40]. Unfortunately, these concepts have long been poorly defined, and, as a result, methods for syntenic block identification differ markedly in their output [41]. In addition, methods for identifying syntenic blocks (or closely related concepts) are often focused on identifying sets of genomic intervals that exhibit levels of conservation of marker content or colinearity that exceed what one would expect if markers were randomly shuffled between genomes (e.g., [42–44]). This is in contrast to homology maps, which are concerned with colinear homology, regardless of biological significance. And, in practice, homology maps are intermediate objects in the process of WGA, whereas syntenic block predictions are often of direct interest.

Homology maps are most commonly constructed from local alignments, such as those computed by methods discussed in the previous section. As only a high-level correspondence is desired, these methods are often run in faster but less sensitive configurations. For example, local alignments between just the coding intervals of the genomes can be computed quickly and used for the construction of homology maps that are at least accurate with respect to protein-coding genes.

Although numerous pairwise homology mapping methods exist, in this chapter, we restrict our attention to methods that scale to more than two genomes, as the problem is significantly more challenging in the multiple genome case. Examples of multiple genome homology map methods include GRIMM-Synteny [40], its successor DRIMM-Synteny [45], Mercator [46], Enredo [47], OSfinder [48], SuperMap [49], and Sibelia [50]. The WGA programs M-GCAT [51], progressiveMauve [52], MUGSY [53], and Cactus [54] are integrated hierarchical methods that contain a homology mapping stage.

Many of these methods use graph-based data structures to find a mapping between multiple genomes simultaneously. Kehr et al. [55] characterized the relationships between four commonly used types of graphs: alignment graphs [56], *A-Bruijn* graphs [57, 58], Enredo graphs [47], and Cactus graphs [59]. The most straightforward graph is the alignment graph, which is a mixed graph with vertices representing genomic segments, directed edges representing adjacent segments, and undirected edges representing homologous segments. In an *A-Bruijn* graph, vertices instead represent sets of homologous segments, and directed edges represent adjacencies between pairs of segments (one from each set represented by the connected vertices). Relative to alignment graphs, *A-Bruijn* graphs are more compact and readily reveal the content of each genome. An Enredo graph is very similar to an *A-Bruijn* graph, but has a pair of vertices instead of a single vertex for each set of homologous segments, which captures information regarding the directionality of each segment within a homologous set. Lastly, cactus graphs flip the representation of adjacencies, with vertices corresponding to sets of adjacencies and edges corresponding to sets of homologous segments. Cactus graphs have a natural decomposition that provides advantages for analysis and visualization of WGAs.

Graph-based homology mapping methods generally produce an initial WGA graph using one of the four representations we have discussed and then refine the graph via modifications. Of the homology mapping methods we have listed, GRIMM-Synteny, Mercator, and MUGSY use alignment graphs. DRIMM-Synteny and OSfinder use *A-Bruijn* graphs and Sibelia uses de Bruijn graphs, of which *A-Bruijn* graphs are a generalization. And, as their names suggest, Enredo and Cactus use Enredo and cactus graphs, respectively. These methods use a variety of techniques for graph refinement. For example, MUGSY is unique in its use of flow network algorithms to identify breaks in colinearity. OSfinder uses a novel probabilistic model to determine a maximum likelihood multiple genome orthology map. And Cactus uses a simulated annealing-style algorithm, the *Cactus alignment filter*, to refine an initial cactus graph representing a homology map.

Unlike the graph-based methods that build a map between all genomes simultaneously, the SuperMap and progressiveMauve methods build a multiple genome map by progressively building pairwise maps up a guide tree. The pairwise SuperMap algorithm is essentially a symmetric version of the chaining method used by Shuffle-LAGAN [23], which allows for rearrangements and duplications in its chains of orthologous segments. The progressiveMauve mapping method instead uses a "breakpoint elimination" algorithm to find colinear segments and does not allow for duplications, thus producing output indicative of one-to-one toporthology. This algorithm greedily removes local alignments one by one with the goal of maximizing an objective function that takes into account both the number of breakpoints implied by an alignment and substitution scores.

Once a homology map has been created, any one of a number of genomic global alignment methods can be used to align the orthologous and colinear segments identified by the map. As for our discussion of homology mapping methods, we restrict our attention to global aligners that can handle multiple genomes. Examples of such methods are MAVID [60], MLAGAN [37], DIALIGN [36], SeqAn::T-Coffee [61], PECAN [47], FSA [62], and the *base-level alignment refinement* (*BAR*) algorithm of Cactus [54]. For colinear sequences, the genomic alignment problem is the same as that of classical global alignment but is made more difficult by the fact that the sequences are long (possibly millions of nucleotides in length). Thus, global genomic aligners employ heuristics to speed up the process. By far, the most common heuristic used is to first identify short local alignments, or *anchors*, between the sequences, identify a chain of these anchors, and then perform global alignment between the adjacent chained anchors. This technique is similar to the strategy for hierarchical WGA, but is simpler, due to the fact that rearrangements and duplications do not need to be taken into account. MLAGAN and DIALIGN use the CHAOS local aligner, PECAN and FSA use Exonerate [63], and MAVID and SeqAn::T-Coffee use suffix trees or arrays to find anchors.

In addition to the specific local alignment technique used to speed up the alignment process, global genomic aligners also vary with respect to how they combine local pairwise alignments to build a multiple global alignment. First, MAVID, MLAGAN, SeqAn::T-Coffee, and Pecan all belong to the class of progressive alignment methods, which use a phylogenetic tree to guide their algorithms (*see* Chapter 7 [1]). For the alignment of non-leaf sequences during progressive alignment, MAVID uses maximum likelihood ancestral sequence inference, while MLAGAN, SeqAn:: T-Coffee, and Pecan use a sum-of-pairs objective function. Both SeqAn::T-Coffee and Pecan use a "consistency" technique, which adjusts the score between pairs of positions (or segments) based on the consistency of triplets of pairwise alignments. The nonprogressive methods, DIALIGN, FSA, and BAR, instead put together a multiple alignment by greedily merging consistent local pairwise alignments. While differing in their use of a tree, the FSA, Pecan, and BAR methods take advantage of probabilistic models of sequence alignment and attempt to maximize statistically grounded objective functions, as opposed to the heuristic score-based functions used by the other methods. BAR is unique in its ability to predict breakpoints when aligning groups of sequences that may contain the boundaries of rearrangement events.

Although the hierarchical approach breaks the WGA problem into a large number of subproblems (one per colinear segment set) that can be computed in parallel, it is still a significant computational effort to produce a WGA with this approach, particularly for large eukaryotic genomes. Thus, a number of Web sites host precomputed hierarchical WGAs. Alignments produced by the combination of Pecan with either Enredo or Mercator are hosted at the Ensembl Web site [64]. Similarly, the VISTA Web site [65] hosts

WGAs generated by SuperMap and the LAGAN-suite of genomic aligners. Both sites offer visualizations of the WGAs, which are useful for looking at levels of conservation across genomes.

3.5 The Local Approach

The local approach to WGA bypasses the high-level homology map construction phase of the hierarchical approach and instead begins by identifying a comprehensive set of nucleotide-level pairwise local alignments. The second step of this approach is to combine the pairwise local alignments into a cohesive WGA by filtering out nonorthologous relationships and merging pairwise alignments into multiple alignments. Because there is typically no additional pairwise nucleotide-level alignment performed in the second step, the local alignments generated by the first step are obtained with a more sensitive aligner than that used by hierarchical methods for homology map building. The two primary examples of local WGA methods are MUMmer, a pairwise genome aligner, and MULTIZ/TBA, a multiple genome aligner [8].

MUMmer was one of the first pairwise WGA methods to be developed and was initially targeted at the alignment of prokaryotic-sized genomes. The WGA ability of MUMmer is achieved through a combination of smaller modules that is orchestrated by the NUCmer or PROmer scripts. The first module identifies maximum unique matches (MUMs) between a pair of genomes with a suffix tree data structure. Nearby matches are clustered together, and a high-scoring colinear chain of matches is identified within each cluster. Finally, the matches within the chains are extended with a variant of the Smith–Waterman algorithm, and the resulting extended chains are output as a WGA. The raw WGA output by MUMmer can, in general, include all classes of homologous relationships. However, the chains are typically filtered to leave only those that are highest scoring or that result in a reference position being overlapped by only a single chain. Thus, a filtered WGA from MUMmer is usually representative of orthology.

MULTIZ/TBA, which was instead designed for large eukaryotic genomes, starts by using LASTZ to generate sensitive local pairwise alignments between all pairs of genomes or between a reference genome and all others. MULTIZ is then used to identify local alignment blocks of subsets of genomes that should be combined and to merge these blocks using a banded variant of the Smith–Waterman algorithm. TBA is the program that is used to coordinate this entire process when all pairs of genomes are compared. Thus far, it does not appear that TBA has been used at the whole-genome scale, although MULTIZ is regularly used for reference-based WGAs hosted by the UCSC Genome Browser [66]. For these reference-based WGAs, the ungapped segments of LASTZ alignments are first processed with a chaining program (AXTCHAIN) to establish large colinear alignments between the reference and another genome. In contrast to the output of chaining methods discussed in Subheading 3.3, a chain produced by AXTCHAIN is an ordered set of pairwise local alignments rather than a single long alignment that explicitly aligns between the short local alignments that form the chain. AXTCHAIN chains are typically filtered by the CHAINNET program to retain only the highest-scoring alignment at each position within the reference genome [67]. The remaining alignments, which most likely reflect orthologous relationships, are then combined into multiple alignments with MULTIZ.

3.6 Refining WGAs

Because of the computational complexity of multiple alignment, particularly at the whole-genome scale, methods of both approaches to WGA use heuristics and simplified models to make WGA feasible. For example, most of the methods described in this chapter do not distinguish between different classes of genomic sequence (e.g., genic and intergenic) while constructing nucleotide-level alignments. And many methods disregard small, marginally significant, local alignments for the sake of speed. As a result, at a local level, the results of current WGA methods often leave room for improvement.

To remedy this situation, a number of methods have been developed that may be used to refine WGAs. These methods take as input either a WGA, a single WGA block, or the set of homologous and colinear sequences that make up a WGA block. They can be generally grouped into one of three categories. The first is composed of methods that refine the local structure of a WGA. That is, they redefine the boundaries, or "breakpoints," of the homologous and colinear blocks in the WGA. A secondary category of methods focuses on optimizing individual WGA blocks with respect to an objective function. The last category includes methods that perform alignment while taking into account the structure and evolutionary dynamics of certain classes of genomic elements.

PicoInversionMiner [68] and Cassis [69, 70] are two methods for refining the local structure of a WGA. PicoInversionMiner identifies very small "inplace" inversions between two genomes that are left undetected by an initial WGA. Such inversions are represented by alignments that would typically not have statistically significant scores at the genome level but can be detected via probabilistic models of local sequence evolution. In contrast to PicoInversionMiner, which identifies novel rearrangement events, Cassis refines the coordinates of breakpoints. The refinements produced by Cassis are the result of identifying weak similarities between sequences adjacent to segments of an initial orthology map and extending the boundaries of segments based on these similarities. The BAR algorithm of Cactus, which we have previously discussed in the context of hierarchical WGA, is also an alignment refinement method that identifies breakpoints.

Other methods for refining WGAs focus on improving local colinear multiple alignments with respect to a given objective function. For example, GenAlignRefine [71] attempts to optimize WGA blocks according to the COFFEE objective function [72] using a genetic algorithm. The PSAR-Align method [73] instead realigns blocks to optimize an expected accuracy objective function [74] using pairwise alignment probabilities estimated by the PSAR tool [75] and the sequencing annealing algorithm of the FSA multiple alignment method [62]. Lastly, the Phylo project [76, 77] refines WGAs by "crowd sourcing" the task of optimizing colinear alignment blocks, according to one of a number of objective functions. Phylo casts the multiple alignment problem as a casual game that may be played by "citizen scientists" at the project's website (http://phylo.cs.mcgill.ca/).

Lastly, a number of methods have been developed that can improve the alignments of specific classes of genomic elements, such as gene structures. The primary goal of these methods is generally to improve prediction of genomic elements, but a more accurate alignment often results as a side product. Among the oldest of such methods are comparative gene finders that perform protein-coding gene prediction and pairwise alignment simultaneously. These include SLAM [78] and DOUBLESCAN [79], both of which use pair hidden Markov models [80]. A related method, CESAR [81], was specifically designed for realignment and targets individual coding exons rather than full gene structures. Other methods focus on improving the alignment of noncoding regulatory regions by modeling the evolution of sets of transcription factor-binding sites with known motifs (e.g., MORPH [82], EMMA [83], and MAFIA [84]). Like the comparative gene finders, these methods also use statistical alignment techniques but with models extended to take into the account the conservation of binding sites instead of gene structures. SAPF [85] is also a method aimed at alignment of noncoding regulatory regions but more generally models sequences that are mixtures of "slow" and "fast" evolving elements without knowledge of binding motifs. Lastly, REAPR [86] focuses on the realignment and detection of noncoding RNAs by using alignment models that take into account the conserved secondary structures of such RNAs.

4 Evaluation of WGAs

Just as for small-scale alignment (Chapter 7, [1]), assessing the accuracy of WGAs is hard because we rarely know the true evolutionary history of a set of genome sequences. In fact, the evaluation of WGAs is even harder than that of protein alignments. While protein aligners can be evaluated with "gold standard" benchmarking databases where the truth is established through protein structural information, genome aligners have no benchmarks of real data. In addition, WGAs must be assessed not only for whether they align truly homologous sequences but also for whether they correctly predict orthologous (or toporthologous) relationships. Thus, the evaluation of WGAs is related to that of gene orthology prediction, which is discussed in Chapter 9 [5]. Despite these challenges, a number of creative approaches have been used for determining the accuracy of WGA methods. The approaches generally fall into four categories: (1) simulation, (2) analysis of alignments to annotated regions, (3) comparison with predictions from other methods, and (4) alignment statistics.

Simulated data are appealing for evaluation as we know the entire evolutionary history of the simulated sequences and can thus thoroughly evaluate the accuracy of an alignment. Many of the WGA methods described in this chapter have used simulations for assessing their accuracies [8, 47, 52, 54, 62]. The Alignathon [87], one of the most comprehensive evaluations of WGA methods to date, relied heavily on simulated data sets. This study called attention to one potential pitfall of simulation-based evaluation, which is that the performance of a WGA method may be overestimated when that method was developed or trained with respect to the same simulator used for the assessment.

Simulating the evolution of whole genomes is a challenging task, and it is unclear if the current models used for simulation are close to reality. Such models are highly complex, as they have to account for many different types of evolutionary events, at both the small and large scales. For example, they need to model the random mutations of both single-nucleotide substitutions and megabase-sized inversions. In addition, they also need to model natural selection, which alters the probability of these random mutations becoming fixed within a population. For example, an inversion that cuts an essential gene in half might have a much lower probability of becoming fixed than an inversion with both end points in intergenic regions. Despite these challenging model details, a number of genomic evolution simulators have been developed. Currently, only three simulators model both small-scale events (e.g., substitutions and indels) and large-scale rearrangements and duplications [88–90]. Other simulators focus only on nonrearranging events [8, 91–98] and are thus good for evaluating colinear genomic aligners but not homology mapping methods.

A second class of approaches to evaluating WGAs leverages our knowledge of various classes of elements within the genome. For example, with our understanding that most coding regions are conserved across closely related genomes, the fraction of exons in a genome "covered" by an alignment is an indirect measure of the sensitivity of a WGA [37, 49, 60, 99]. Specificity can also be roughly assessed with coding regions, either by counting the number of coding bases that are aligned to noncoding bases in other genomes [36, 100] or by checking that alignments in coding regions exhibit periodicities in their substitution patterns [99]. A related approach that instead assesses the accuracy of eukaryotic orthology maps is to check if exons from the same gene are mapped in the same order and orientation to other genomes [47]. For the subset of protein-coding and noncoding RNA genes that have curated "gold standard" alignments, the accuracy of a WGA with respect to those genes may be assessed [101]. However, the fact that genic regions are often highly conserved is also a disadvantage of using them for evaluation; the most conserved regions are the easiest to align, and some aligners use exon annotation information or translated matches. Because of these issues, repeat sequences, which are believed to evolve more neutrally, have been used for alignment evaluation [47, 99]. For example, in [99], sensitivity was assessed by alignments of ancestral repetitive elements, and specificity was inferred from the number of alignments to lineage-specific repeat elements (in this study, primate-specific *Alu* repeats).

Another common evaluation technique is to compare whole-genome aligners against other related methods. For example, a WGA produced by one method can be used as the "truth" with which to evaluate the sensitivity and specificity of other WGAs [53]. This technique is useful for judging the similarity of different WGAs but, unfortunately, does not provide much information about accuracy. Another technique is to compare with the results from gene orthology prediction programs [48, 49]. The advantage of this approach is that it provides a more independent test of accuracy, since gene orthology prediction programs generally use different algorithms and information sources to infer orthology. The disadvantages of this approach are that it only provides a gene-level measure of accuracy and does not evaluate alignments of noncoding regions. In addition, since WGA and gene orthology prediction share similar goals, we might expect that future methods will blend techniques from both and thus that this evaluation approach will decrease in usefulness.

A last class of evaluation techniques involves the computation of statistics for WGAs. These statistics can be subdivided into simple descriptive statistics and measures computed via statistical or sampling techniques. One of the most straightforward descriptive statistics of a WGA is the "coverage" or the fraction of the genomes included in an alignment or orthology map block [45, 47, 49, 53, 87]. Generally, the higher the coverage, the more sensitive the WGA is believed to be, although one can easily create high-coverage WGAs with poor sensitivity. As a check of large-scale specificity in mammalian WGAs, the authors of [47] checked the fraction of the X chromosome that was covered by alignments to autosomal chromosomes in other genomes (the assumption being that translocations into and out of the X chromosome are rare in mammals). Some more detailed nucleotide-level statistics of WGAs include the total number of "core" positions [53], which are gap-free alignment columns containing all genomes, and the average level of sequence identity in aligned columns [61].

More sophisticated statistics related to WGA accuracy are computed through the use of statistical or sampling techniques. Just as they are used for BLAST, Karlin and Altschul statistics [102] may be used to assess the significance of local pairwise alignments between genomes. StatSigMA extends these statistics to multiple alignments [103], and StatSigMA-w further extends this technique to detect dubiously aligned regions in WGAs of multiple genomes [104]. Whereas a given local pairwise alignment may be highly significant, the flanks of that alignment may be spurious, and a p-value may be computed assessing the possible "over-alignment" of a flank [105]. Within a multiple alignment, a number of techniques have been developed for estimating the accuracy of the alignment of pairs of residues or entire columns, including simply computing an alignment of reversed sequences [106], computing alignments with bootstrapped guide trees [107], sampling suboptimal multiple alignments [75], and evaluating consistency within a library of alternative alignments [108].

5 Future Challenges

Despite the substantial progress made in WGA methodology development, there are a number of challenges that remain unsolved. First, we are in need of WGA methods that can scale to hundreds or thousands of genomes. Along with ever-improving sequencing technology, we are accumulating whole-genome sequences at an increasing rate. Projects such as the Genome 10K Community of Scientists [109], which aims to collect and sequence the genomes of 10,000 vertebrate species, will further push the WGA problem to new scales. While most WGA algorithms have been made efficient for long genomes, very few are practical for large numbers of genomes. Encouragingly, we are beginning to see methods capable of scaling to thousands of genomes for the simpler task of "core-genome alignment" of highly similar microbial-sized genomes [110]. However, methods scaling to thousands of genomes for the full WGA task or for mammalian-sized genomes do not currently exist. In addition to algorithmic advances, we will also be in need of novel approaches for storing and representing WGAs of thousands of genomes.

Second, advances are needed in the parameterization of WGA methods. Current methods are littered with large numbers of parameters that are often heuristic in nature and not easily determined. In some cases, the default parameters for a WGA method may be markedly suboptimal [111]. One solution to this problem is to adopt probabilistic models, which offer principled approaches to parameter estimation, such as maximum likelihood. In fact, probabilistic models of sequence evolution have already been adopted for the alignment of colinear genomic segments and have been shown to offer improved accuracy [47, 62]. However, we have yet to see a method that integrates probabilistic models of both small- and large-scale changes that is capable of constructing an entire WGA, although the recently introduced "split-alignment" pairwise WGA method is a promising step in this direction [112]. In addition, most WGA alignments use models or scoring schemes that assume homogenous rates of evolution across the genome. This assumption is obviously violated in real data, and new methods will need to be developed that take this into account. Simulated noncoding genomic alignments that represent a heterogeneous mix of evolutionary rates have been developed and should be useful for the development of new WGA methodology [97].

Lastly, more attention must be paid to the fact that a WGA is typically just a single estimate of the evolutionary history of a set of genomes and portions of this estimate may be highly uncertain. Encouragingly, methods for colinear genomic alignment have brought light to this issue at the nucleotide level [62, 113]. However, the issue of uncertainty at the large-scale orthology map level has not been sufficiently studied, perhaps due to the lack of probabilistic models for that level of the WGA problem. In addition, most efforts to address uncertainty in alignments simply assign levels of confidence to the components of a single alignment. It may be more useful to be presented with a set of near-optimal alignments so that alternative evolutionary histories can be examined by downstream analyses [114]. The determination and representation of uncertainty for all scales of a WGA will likely remain a challenging problem as the number of genomes included in alignments increases.

6 Exercises

1. Download the whole-genome aligner MUMmer (http://mummer.sourceforge.net) and FASTA-formatted genome sequences for the species *Helicobacter pylori J99* and *Helicobacter pylori B38* from GenBank (http://www.ncbi.nlm.nih.gov/genbank/, accessions NC000921 and NC012973, respectively). Run the NUCmer or PROmer programs on the two genome sequences. Visualize the resulting alignment with the mummerplot program. How many colinear blocks are there in the alignment? How many inversion events are implied by the alignment?

2. Visit the UCSC Genome Browser (http://genome.ucsc.edu) and browse the human genome version GRCh38/hg38. Search for and view the *CFTR* gene, mutations in which cause the disease cystic fibrosis. Turn on the Net tracks for alignments to

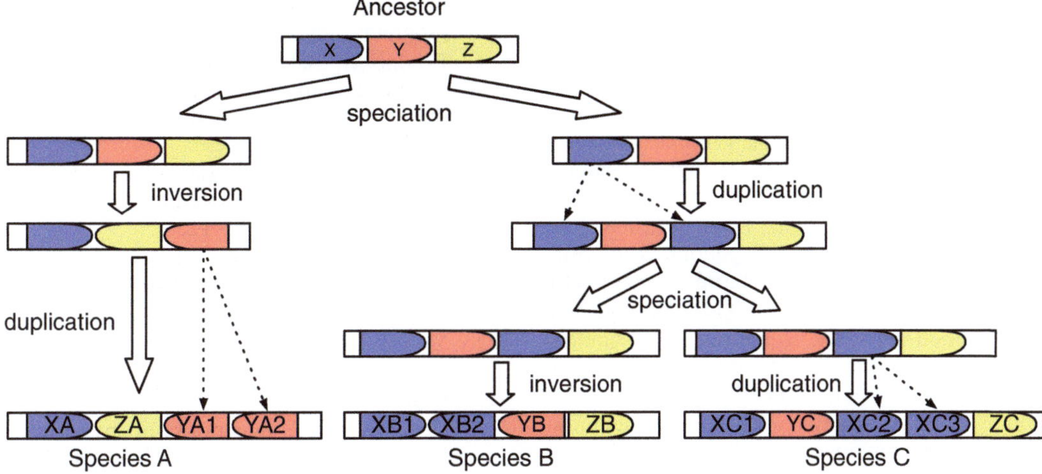

Fig. 3 The evolutionary scenario to be considered for Exercise 3. Each bullet-like shape corresponds to a genomic segment, with the direction of the bullet indicating the orientation of the segment

genomes of non-primate placental mammals by clicking on the "Placental Chain/Net" link (in the "Comparative Genomics" section) and choosing the appropriate configuration. Examine the Mouse Net track in the visualization and note the color of the mouse net alignments. Using the "Chromosome Color Key" (located in between the browser visualization and the track configuration section), identify the chromosome on which the mouse ortholog of *CFTR* is located. Looking at the net alignments for all of the placental mammals, does it appear that *CFTR* has been conserved across this clade?

3. Consider the evolutionary scenario giving rise to the genomes of three species shown in Fig. 3. For each of the relations listed below, give the pairs of genomic segments with that relation.

(a) Orthology

(b) Paralogy

(c) Toporthology

References

1. Löytynoja A (2012) Alignment methods: strategies, challenges, benchmarking, and comparative overview. Methods Mol Biol 855:203–235
2. Fleischmann RD, Adams MD, White O et al (1995) Whole-genome random sequencing and assembly of Haemophilus influenzae Rd. Science 269:496–512
3. Mukherjee S, Stamatis D, Bertsch J et al (2017) Genomes OnLine Database (GOLD) v.6: data updates and feature enhancements. Nucleic Acids Res 45:D446–D456
4. Fitch WM (1970) Distinguishing homologous from analogous proteins. Syst Zool 19:99–113
5. Altenhoff AM, Dessimoz C (2012) Inferring orthology and paralogy. Methods Mol Biol 855:259–279
6. Dewey CN (2011) Positional orthology: putting genomic evolutionary relationships into context. Brief Bioinform 12(5):401–412

7. Dewey CN, Pachter L (2006) Evolution at the nucleotide level: the problem of multiple whole-genome alignment. Hum Mol Genet 15 Spec No 1:R51–R56
8. Blanchette M, Kent WJ, Riemer C et al (2004) Aligning multiple genomic sequences with the threaded blockset aligner. Genome Res 14:708–715
9. Ma J, Ratan A, Raney BJ et al (2008) The infinite sites model of genome evolution. Proc Natl Acad Sci U S A 105:14254–14261
10. Needleman SB, Wunsch CD (1970) A general method applicable to the search for similarities in the amino acid sequence of two proteins. J Mol Biol 48:443–453
11. Smith TF, Waterman MS (1981) Identification of common molecular subsequences. J Mol Biol 147:195–197
12. Tesler G (2002) GRIMM: genome rearrangements web server. Bioinformatics 18:492–493
13. Paten B, Herrero J, Fitzgerald S et al (2008) Genome-wide nucleotide-level mammalian ancestor reconstruction. Genome Res 18:1829–1843
14. Ma J, Zhang L, Suh BB et al (2006) Reconstructing contiguous regions of an ancestral genome. Genome Res 16:1557–1565
15. Stark A, Lin MF, Kheradpour P et al (2007) Discovery of functional elements in 12 Drosophila genomes using evolutionary signatures. Nature 450:219–232
16. Alioto T (2012) Gene prediction. Methods Mol Biol 855:175–201
17. Eddy SR (2002) Computational genomics of noncoding RNA genes. Cell 109:137–140
18. Margulies EH, Blanchette M, NISC Comparative Sequencing Program et al (2003) Identification and characterization of multi-species conserved sequences. Genome Res 13:2507–2518
19. Tagle DA, Koop BF, Goodman M et al (1988) Embryonic epsilon and gamma globin genes of a prosimian primate (Galago crassicaudatus). Nucleotide and amino acid sequences, developmental regulation and phylogenetic footprints. J Mol Biol 203:439–455
20. Bejerano G, Pheasant M, Makunin I et al (2004) Ultraconserved elements in the human genome. Science 304:1321–1325
21. Altschul SF, Gish W, Miller W et al (1990) Basic local alignment search tool. J Mol Biol 215:403–410
22. Altschul SF, Madden TL, Schäffer AA et al (1997) Gapped BLAST and PSI-BLAST: a new generation of protein database search programs. Nucleic Acids Res 25:3389–3402
23. Brudno M, Malde S, Poliakov A et al (2003) Glocal alignment: finding rearrangements during alignment. Bioinformatics 19(Suppl 1):i54–i62
24. Ma B, Tromp J, Li M (2002) PatternHunter: faster and more sensitive homology search. Bioinformatics 18:440–445
25. Sun Y, Buhler J (2005) Designing multiple simultaneous seeds for DNA similarity search. J Comput Biol 12:847–861
26. Xu J, Brown D, Li M et al (2006) Optimizing multiple spaced seeds for homology search. J Comput Biol 13:1355–1368
27. Zhang L (2007) Superiority of spaced seeds for homology search. IEEE/ACM Trans Comput Biol Bioinform 4:496–505
28. Kiełbasa SM, Wan R, Sato K et al (2011) Adaptive seeds tame genomic sequence comparison. Genome Res 21:487–493
29. Kucherov G, Noé L, Roytberg M (2006) A unifying framework for seed sensitivity and its application to subset seeds. J Bioinform Comput Biol 4:553–569
30. Flannick J, Batzoglou S (2005) Using multiple alignments to improve seeded local alignment algorithms. Nucleic Acids Res 33:4563–4577
31. Sun H, Buhler JD (2012) PhyLAT: a phylogenetic local alignment tool. Bioinformatics 28:1336–1344
32. Kent WJ (2002) BLAT–the BLAST-like alignment tool. Genome Res 12:656–664
33. Kehr B, Weese D, Reinert K (2011) STELLAR: fast and exact local alignments. BMC Bioinform 12:S15
34. Schwartz S, Kent WJ, Smit A et al (2003) Human-mouse alignments with BLASTZ. Genome Res 13:103–107
35. Delcher AL, Kasif S, Fleischmann RD et al (1999) Alignment of whole genomes. Nucleic Acids Res 27:2369–2376
36. Brudno M, Chapman M, Göttgens B et al (2003) Fast and sensitive multiple alignment of large genomic sequences. BMC Bioinform 4:66
37. Brudno M, Do CB, Cooper GM et al (2003) LAGAN and Multi-LAGAN: efficient tools for large-scale multiple alignment of genomic DNA. Genome Res 13:721–731
38. Gusfield D (1997) Algorithms on strings, trees, and sequences: computer science and computational biology. Cambridge University Press, Cambridge

39. Nadeau JH, Taylor BA (1984) Lengths of chromosomal segments conserved since divergence of man and mouse. Proc Natl Acad Sci U S A 81:814–818
40. Pevzner P, Tesler G (2003) Genome rearrangements in mammalian evolution: lessons from human and mouse genomes. Genome Res 13:37–45
41. Ghiurcuta CG, Moret BME (2014) Evaluating synteny for improved comparative studies. Bioinformatics 30:i9–i18
42. Wang X, Shi X, Li Z et al (2006) Statistical inference of chromosomal homology based on gene colinearity and applications to Arabidopsis and rice. BMC Bioinform 7:447
43. Proost S, Fostier J, De Witte D et al (2012) i-ADHoRe 3.0–fast and sensitive detection of genomic homology in extremely large data sets. Nucleic Acids Res 40:e11
44. Lucas JMEX, Muffato M, Roest Crollius H (2014) PhylDiag: identifying complex synteny blocks that include tandem duplications using phylogenetic gene trees. BMC Bioinform 15:268
45. Pham SK, Pevzner PA (2010) DRIMM-Synteny: decomposing genomes into evolutionary conserved segments. Bioinformatics 26:2509–2516
46. Dewey CN (2007) Aligning multiple whole genomes with Mercator and MAVID. In: Bergman N (ed) Methods in Molecular Biology, vol 395. Humana, Clifton, NJ, pp 221–236
47. Paten B, Herrero J, Beal K et al (2008) Enredo and Pecan: genome-wide mammalian consistency-based multiple alignment with paralogs. Genome Res 18:1814–1828
48. Hachiya T, Osana Y, Popendorf K et al (2009) Accurate identification of orthologous segments among multiple genomes. Bioinformatics 25:853–860
49. Dubchak I, Poliakov A, Kislyuk A et al (2009) Multiple whole-genome alignments without a reference organism. Genome Res 19:682–689
50. Minkin I, Patel A, Kolmogorov M et al (2013) Sibelia: a scalable and comprehensive synteny block generation tool for closely related microbial genomes. In: Algorithms in bioinformatics, Lecture notes in computer science. Springer, Berlin, pp 215–229
51. Treangen TJ, Messeguer X (2006) M-GCAT: interactively and efficiently constructing large-scale multiple genome comparison frameworks in closely related species. BMC Bioinform 7:433
52. Darling AE, Mau B, Perna NT (2010) progressiveMauve: multiple genome alignment with gene gain, loss and rearrangement. PLoS One 5:e11147
53. Angiuoli SV, Salzberg SL (2011) Mugsy: fast multiple alignment of closely related whole genomes. Bioinformatics 27:334–342
54. Paten B, Earl D, Nguyen N et al (2011) Cactus: algorithms for genome multiple sequence alignment. Genome Res 21:1512–1528
55. Kehr B, Trappe K, Holtgrewe M et al (2014) Genome alignment with graph data structures: a comparison. BMC Bioinform 15:99
56. Kececioglu J (1993) The maximum weight trace problem in multiple sequence alignment. In: Combinatorial pattern matching, Lecture notes in computer science. Springer, Berlin, pp 106–119
57. Pevzner PA, Pevzner PA, Tang H et al (2004) De novo repeat classification and fragment assembly. Genome Res 14:1786–1796
58. Raphael B, Zhi D, Tang H et al (2004) A novel method for multiple alignment of sequences with repeated and shuffled elements. Genome Res 14:2336–2346
59. Paten B, Diekhans M, Earl D et al (2011) Cactus graphs for genome comparisons. J Comput Biol 18:469–481
60. Bray N, Pachter L (2004) MAVID: constrained ancestral alignment of multiple sequences. Genome Res 14:693–699
61. Rausch T, Emde AK, Weese D et al (2008) Segment-based multiple sequence alignment. Bioinformatics 24:i187–i192
62. Bradley RK, Roberts A, Smoot M et al (2009) Fast statistical alignment. PLoS Comput Biol 5:e1000392
63. Slater GSC, Birney E (2005) Automated generation of heuristics for biological sequence comparison. BMC Bioinform 6:31
64. Herrero J, Muffato M, Beal K et al (2016) Ensembl comparative genomics resources. Database (Oxford) 2016:bav096
65. Brudno M, Poliakov A, Minovitsky S et al (2007) Multiple whole genome alignments and novel biomedical applications at the VISTA portal. Nucleic Acids Res 35: W669–W674
66. Casper J, Zweig AS, Villarreal C et al (2018) The UCSC Genome Browser database: 2018 update. Nucleic Acids Res 46:D762–D769
67. Kent WJ, Baertsch R, Hinrichs A et al (2003) Evolution's cauldron: duplication, deletion, and rearrangement in the mouse and human genomes. Proc Natl Acad Sci U S A 100:11484–11489

68. Hou M, Yao P, Antonou A et al (2011) Pico-inplace-inversions between human and chimpanzee. Bioinformatics 27:3266–3275
69. Lemaitre C, Tannier E, Gautier C et al (2008) Precise detection of rearrangement breakpoints in mammalian chromosomes. BMC Bioinform 9:286
70. Baudet C, Lemaitre C, Dias Z et al (2010) Cassis: detection of genomic rearrangement breakpoints. Bioinformatics 26:1897–1898
71. Wang C, Lefkowitz EJ (2005) Genomic multiple sequence alignments: refinement using a genetic algorithm. BMC Bioinform 6:200
72. Notredame C, Holm L, Higgins DG (1998) COFFEE: an objective function for multiple sequence alignments. Bioinformatics 14:407–422
73. Kim J, Ma J (2014) PSAR-align: improving multiple sequence alignment using probabilistic sampling. Bioinformatics 30:1010–1012
74. Schwartz AS, Pachter L (2007) Multiple alignment by sequence annealing. Bioinformatics 23:e24–e29
75. Kim J, Ma J (2011) PSAR: measuring multiple sequence alignment reliability by probabilistic sampling. Nucleic Acids Res 39:6359–6368
76. Kawrykow A, Roumanis G, Kam A et al (2012) Phylo: a citizen science approach for improving multiple sequence alignment. PLoS One 7:e31362
77. Kwak D, Kam A, Becerra D et al (2013) Open-Phylo: a customizable crowd-computing platform for multiple sequence alignment. Genome Biol 14:R116
78. Alexandersson M, Cawley S, Pachter L (2003) SLAM: cross-species gene finding and alignment with a generalized pair hidden Markov model. Genome Res 13:496–502
79. Meyer IM, Durbin R (2002) Comparative ab initio prediction of gene structures using pair HMMs. Bioinformatics 18:1309–1318
80. Durbin R, Eddy S, Korgh A et al (1998) Biological sequence analysis: probabilistic models of proteins and nucleic acids. Cambridge University Press, Cambridge
81. Sharma V, Elghafari A, Hiller M (2016) Coding exon-structure aware realigner (CESAR) utilizes genome alignments for accurate comparative gene annotation. Nucleic Acids Res 44(11):e103
82. Sinha S, He X (2007) MORPH: probabilistic alignment combined with hidden Markov models of cis-regulatory modules. PLoS Comput Biol 3:e216
83. He X, Ling X, Sinha S (2009) Alignment and prediction of cis-regulatory modules based on a probabilistic model of evolution. PLoS Comput Biol 5:e1000299
84. Majoros WH, Ohler U (2010) Modeling the evolution of regulatory elements by simultaneous detection and alignment with phylogenetic pair HMMs. PLoS Comput Biol 6: e1001037
85. Satija R, Pachter L, Hein J (2008) Combining statistical alignment and phylogenetic footprinting to detect regulatory elements. Bioinformatics 24:1236–1242
86. Will S, Yu M, Berger B (2013) Structure-based whole-genome realignment reveals many novel noncoding RNAs. Genome Res 23:1018–1027
87. Earl D, Nguyen N, Hickey G et al (2014) Alignathon: a competitive assessment of whole-genome alignment methods. Genome Res 24:2077–2089
88. Darling ACE, Mau B, Blattner FR et al (2004) Mauve: multiple alignment of conserved genomic sequence with rearrangements. Genome Res 14:1394–1403
89. Edgar RC, Asimenos G, Batzoglou S et al (2011) Evolver: a whole-genome sequence evolution simulator. https://www.drive5.com/evolver
90. Dalquen DA, Anisimova M, Gonnet GH et al (2012) ALF–a simulation framework for genome evolution. Mol Biol Evol 29:1115–1123
91. Stoye J, Evers D, Meyer F (1998) Rose: generating sequence families. Bioinformatics 14:157–163
92. Cartwright RA (2005) DNA assembly with gaps (Dawg): simulating sequence evolution. Bioinformatics 21(Suppl 3):iii31–iii38
93. Pollard DA, Moses AM, Iyer VN et al (2006) Detecting the limits of regulatory element conservation and divergence estimation using pairwise and multiple alignments. BMC Bioinform 7:376
94. Huang W, Nevins JR, Ohler U (2007) Phylogenetic simulation of promoter evolution: estimation and modeling of binding site turnover events and assessment of their impact on alignment tools. Genome Biol 8:R225
95. Varadarajan A, Bradley RK, Holmes IH (2008) Tools for simulating evolution of aligned genomic regions with integrated parameter estimation. Genome Biol 9:R147

96. Fletcher W, Yang Z (2009) INDELible: a flexible simulator of biological sequence evolution. Mol Biol Evol 26:1879–1888
97. Kim J, Sinha S (2010) Towards realistic benchmarks for multiple alignments of non-coding sequences. BMC Bioinform 11:54
98. Arenas M, Posada D (2014) Simulation of genome-wide evolution under heterogeneous substitution models and complex multispecies coalescent histories. Mol Biol Evol 31:1295–1301
99. Margulies EH, Cooper GM, Asimenos G et al (2007) Analyses of deep mammalian sequence alignments and constraint predictions for 1% of the human genome. Genome Res 17:760–774
100. Morgenstern B, Rinner O, Abdeddaïm S et al (2002) Exon discovery by genomic sequence alignment. Bioinformatics 18:777–787
101. Wang AX, Ruzzo WL, Tompa M (2007) How accurately is ncRNA aligned within whole-genome multiple alignments? BMC Bioinform 8:417
102. Karlin S, Altschul SF (1990) Methods for assessing the statistical significance of molecular sequence features by using general scoring schemes. Proc Natl Acad Sci U S A 87:2264–2268
103. Prakash A, Tompa M (2005) Statistics of local multiple alignments. Bioinformatics 21(Suppl 1):i344–i350
104. Prakash A, Tompa M (2007) Measuring the accuracy of genome-size multiple alignments. Genome Biol 8:R124
105. Frith MC, Park Y, Sheetlin SL et al (2008) The whole alignment and nothing but the alignment: the problem of spurious alignment flanks. Nucleic Acids Res 36:5863–5871
106. Landan G, Graur D (2007) Heads or tails: a simple reliability check for multiple sequence alignments. Mol Biol Evol 24:1380–1383
107. Penn O, Privman E, Landan G et al (2010) An alignment confidence score capturing robustness to guide tree uncertainty. Mol Biol Evol 27:1759–1767
108. Chang JM, Di Tommaso P, Notredame C (2014) TCS: a new multiple sequence alignment reliability measure to estimate alignment accuracy and improve phylogenetic tree reconstruction. Mol Biol Evol 31:1625–1637
109. Genome 10K Community of Scientists (2009) Genome 10K: a proposal to obtain whole-genome sequence for 10,000 vertebrate species. J Hered 100:659–674
110. Treangen TJ, Ondov BD, Koren S et al (2014) The Harvest suite for rapid core-genome alignment and visualization of thousands of intraspecific microbial genomes. Genome Biol 15:524
111. Frith MC, Hamada M, Horton P (2010) Parameters for accurate genome alignment. BMC Bioinform 11:80
112. Frith MC, Kawaguchi R (2015) Split-alignment of genomes finds orthologies more accurately. Genome Biol 16:106
113. Lunter G, Rocco A, Mimouni N et al (2008) Uncertainty in homology inferences: assessing and improving genomic sequence alignment. Genome Res 18:298–309
114. Herman JL, Novák Á, Lyngsø R et al (2015) Efficient representation of uncertainty in multiple sequence alignments using directed acyclic graphs. BMC Bioinform 16:108

Chapter 5

Inferring Orthology and Paralogy

Adrian M. Altenhoff, Natasha M. Glover, and Christophe Dessimoz

Abstract

The distinction between orthologs and paralogs, genes that started diverging by speciation versus duplication, is relevant in a wide range of contexts, most notably phylogenetic tree inference and protein function annotation. In this chapter, we provide an overview of the methods used to infer orthology and paralogy. We survey both graph-based approaches (and their various grouping strategies) and tree-based approaches, which solve the more general problem of gene/species tree reconciliation. We discuss conceptual differences among the various orthology inference methods and databases and examine the difficult issue of verifying and benchmarking orthology predictions. Finally, we review typical applications of orthologous genes, groups, and reconciled trees and conclude with thoughts on future methodological developments.

Key words Orthology, Paralogy, Tree reconciliation, Orthology benchmarking

1 Introduction

The study of genetic material almost always starts with identifying, within or across species, *homologous* regions—regions of common ancestry. As we have seen in previous chapters, this can be done at the level of genome segments [1], genes [2], or even down to single residues, in sequence alignments [3]. Here, we focus on genes as evolutionary and functional units. The central premise of this chapter is that it is useful to distinguish between two classes of homologous genes: *orthologs*, which are pairs of genes that started diverging via evolutionary speciation, and *paralogs*, which are pairs of genes that started diverging via gene duplication [4] (Fig. 1, Box 1). Originally, the terms and their definition were proposed by Walter M. Fitch in the context of species phylogeny inference, i.e., the reconstruction of the tree of life. He stated "Phylogenies require orthologous, not paralogous, genes" [4]. Indeed, since orthologs arise by speciation, any set of genes in which every pair is orthologous has by definition the same evolutionary history as the

Adrian M. Altenhoff and Natasha M. Glover are the Joint first authors

Maria Anisimova (ed.), *Evolutionary Genomics: Statistical and Computational Methods*, Methods in Molecular Biology, vol. 1910, https://doi.org/10.1007/978-1-4939-9074-0_5,

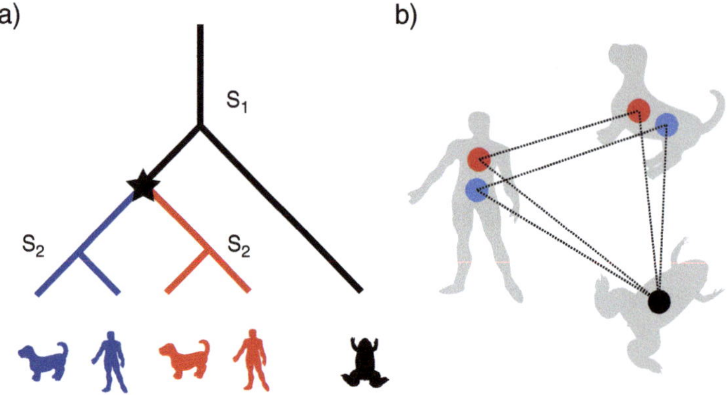

Fig. 1 (**a**) Simple evolutionary scenario of a gene family with two speciation events (S_1 and S_2) and one duplication event (star). The type of events completely and unambiguously define all pairs of orthologs and paralogs: The frog gene is orthologous to all other genes (they coalesce at S_1). The red and blue genes are orthologs between themselves (they coalesce at S_2), but paralogs between each other (they coalesce at star). (**b**) The corresponding orthology graph. The genes are represented here by vertices and orthology relationships by edges. The frog gene forms *one-to-many* orthology with both the human and dog genes, because it is orthologous to more than one sequence in each of these organisms. In such cases, the *bi-directional best-hit* approach only recovers one of the relations (the highest scoring one). Note that in contrary to BBH, the nonsymmetric BeTs approach—simply taking the best genome-wide hit for each gene regardless of reciprocity—would in the situation of a lost blue human gene infer an incorrect orthologous relation between the blue dog and red human gene

underlying species. These days, however, the most frequent motivation for the orthology/paralogy distinction is to study and predict gene function: it is generally believed that orthologs—because they were the same gene in the last common ancestor of the species involved—are likely to have similar biological function. By contrast, paralogs—because they result from duplicated genes that have been retained, at least partly, over the course of evolution—are believed to often differ in function. Consequently, orthologs are of interest to infer function computationally, while paralogs are commonly used to study function innovation.

Box 1: Terminology

Homology is a relation between a pair of genes that share a common ancestor. All pairs of genes in the below figure are homologous to each other.

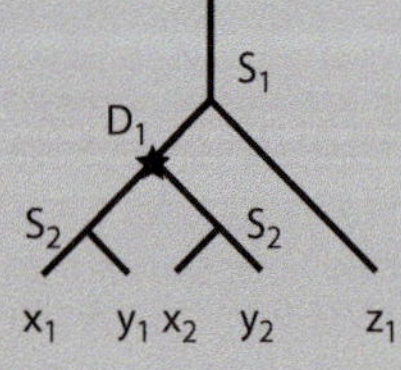

(continued)

Box 1: (continued)

Orthology is a relation defined over a pair of homologous genes, where the two genes have emerged through a speciation event [4]. Example pairs of orthologs are (x_1, y_1) or (x_2, z_1). Orthologs can be further subclassified into one-to-one, one-to-many, many-to-one, and many-to-many orthologs. The qualifiers *one* and *many* indicate for each of the two involved genes whether they underwent an additional duplication after the speciation between the two genomes. Hence, the gene pair (x_1, y_1) is an example of a one-to-one orthologous pair, whereas (x_2, z_1) is a many-to-one ortholog relation.

Paralogy is a relation defined over a pair of homologous genes that have emerged through a gene duplication, e.g., (x_1, x_2) or (x_1, y_2).

In-Paralogy is a relation defined over a triplet. It involves a pair of genes and a speciation event of reference. A gene pair is an in-paralog if they are paralogs and duplicated *after* the speciation event of reference [5]. The pair (x_1, y_2) are in-paralogs with respect to the speciation event S_1.

Out-Paralogy is also a relation defined over a pair of genes and a speciation event of reference. This pair is out-paralogs if the duplication event through which they are related to each other *predates* the speciation event of reference. Hence, the pair (x_1, y_2) are out-paralogs with respect to the speciation event S_2.

Co-orthology is a relation defined over three genes, where two of them are in-paralogs with respect to the speciation event associated to the third gene. The two in-paralogous genes are said to be *co-orthologous* to the third (out-group) gene. Thus, x_1 and y_2 are co-orthologs with respect to z_1.

Homoeology is a specific type of homologous relation in a polyploid species, which thus contain multiple "sub-genomes." This relation describes pairs of genes that originated by speciation and were brought back together in the same genome by allopolyploidization (hybridization) [6]. Thus, in the absence of rearrangement, homoeologs can be thought of as orthologs between sub-genomes.

In this chapter, we first review the main methods used to infer orthology and paralogy, including recent techniques for scaling up algorithms to big data. We then discuss the problem of benchmarking orthology inference. In the last main section, we focus on various applications of orthology and paralogy.

2 Inferring Orthology

Most orthology inference methods can be classified into two major types: graph-based methods and tree-based methods [7]. Methods of the first type rely on graphs with genes (or proteins) as nodes and evolutionary relationships as edges. They infer whether these edges represent orthology or paralogy and build clusters of genes on the basis of the graph. Methods of the second type are based on gene/species tree reconciliation, which is the process of annotating all splits of a given gene tree as duplication or speciation, given the phylogeny of the relevant species. From the reconciled tree, it is trivial to derive all pairs of orthologous and paralogous genes. All pairs of genes which coalesce in a speciation node are orthologs and paralogs if they split at a duplication node. In this section, we present the concepts and methods associated with the two types and discuss the advantages, limitations, and challenges associated with them.

2.1 Graph-Based Methods

Graph-based approaches were originally motivated by the availability of complete genome sequences and the need for efficient methods to detect orthology. They typically run in two phases: a graph construction phase, in which pairs of orthologous genes are inferred (implicitly or explicitly) and connected by edges, and a clustering phase, in which groups of orthologous genes are constructed based on the structure of the graph.

2.1.1 Graph Construction Phase: Orthology Inference

In its most basic form, the graph construction phase identifies orthologous genes by considering pairs of genomes at a time. The main idea is that between any given two genomes, the orthologs tend to be the homologs that diverged least. Why? Because assuming that speciation and duplication are the only types of branching events, the orthologs branched by definition at the latest possible time point—the speciation between the two genomes in question. Therefore, using sequence similarity score as surrogate measure of closeness, the basic approach identifies the corresponding ortholog of each gene through its genome-wide best hit (*BeT*)—the highest scoring match in the other genome [8]. To make the inference symmetric (as orthology is a symmetric relation), it is usually required that BeTs be reciprocal, i.e., that orthology be inferred for a pair of genes g_1 and g_2 if and only if g_2 is the BeT of g_1 and g_1 is the BeT of g_2 [9]. This symmetric variant, referred to as *bi-directional best hit* (*BBH*), has also the merit of being more robust against a possible gene loss in one of the two lineages (Fig. 1).

Inferring orthology from BBH is computationally efficient, because each genome pair can be processed independently and high-scoring alignments can be computed efficiently using dynamic programming [10] or heuristics such as BLAST [11]. Overall, the time complexity scales quadratically in terms of the total number of genes (Box 2). Furthermore, the implementation of this kind of algorithm is simple.

Box 2: Computational Considerations for Scaling to Many Genomes

Time complexity—the amount of time for an algorithm to run as a function of the input—is an important consideration when dealing with big data. This is relevant for inferring orthologs and paralogs due to the massive amounts of sequence data. Thus, it is necessary to consider the time complexity of the inference algorithms, especially when scaling for large and multiple genomes. In computer science, this is commonly denoted in terms of "Big O" notation, which expresses the scaling behavior of the algorithm, up to a constant factor. Below are listed the common time complexities for aspects of some orthology inference algorithms, in order of most efficient to least efficient.

Linear time

- $O(n)$: Optimal algorithm to reconcile rooted, fully resolved gene tree and species tree [12]; Hieranoid algorithm, which recursively merges genomes along the species tree to avoid all-against-all computation [13].

Quadratic time

- $O(n^2)$: The all-against-all stage central to many orthology algorithms scales quadratically, where n is total number of genes.

Cubic time

- $O(n^3)$: The COG database's graph-based clustering merge triplets of homologs which share a common face until no more can be added.

NP-complete

- "Nondeterministic polynomial time," a large class of algorithms for which no solution in polynomial time is known, (e.g. scaling exponentially with respect to the input size), and thus are impractical. NP-complete problems are typically solved approximately, using heuristics. For instance, maximum likelihood gene tree estimation is NP-complete [14].

However, orthology inference by BBH has several limitations, which motivated the development of various improvements (Table 1).

Table 1
Overview of graph-based orthology inference methods and their main properties

Method	In-paralogs	Based on	Grouping strategy	Database	Extra	Available algorithm/DB	References
BBH (best bi-directional hit)	No	BLAST scores	n.a.	–		–/–	[9]
COG	Yes	BLAST scores	Merged adjacent triangles of BeTs	COG/KOG		✓/✓	[8]
EggNOG	Yes	Smith Waterman scores	Hierarchical orthologous groups	EggNOG	Computed at several levels of taxonomic tree	–/✓	[15–17]
Hieranoid	Yes	BLAST scores and HMM profiles	Hierarchical orthologous groups	HieranoiDB		✓/✓	[13, 18]
InParanoid	Yes	BLAST scores	Orthologous groups between pairs of species	InParanoid		✓/✓	[5, 19, 20]
OMA GETHOGS	Yes	ML distance estimates	Hierarchical orthologous groups	OMA Browser	Computed at all levels of the taxonomic tree	✓/✓	[21, 22]
OMA Pairs	Yes	ML distance estimates	Every pair is orthologous	OMA Browser	Detects differential gene loss	✓/✓	[23, 24]
OrthoDB	Yes	Smith Waterman scores	Hierarchical orthologous groups	OrthoDB	Computed at any level of taxonomic tree	✓/✓	[25, 26])
OrthoInspector	Yes	BLAST scores	Only between pairs of species	OrthoInspector		✓/✓	[27, 28]
OrthoMCL	Yes	BLAST scores	MCL clusters	OrthoMCL-DB		✓/✓	[29, 30]
RSD (reciprocal smallest distance)	No	ML distance estimates	Deterministic single-linkage clustering	–		✓/✓	[31–33]

Allowing for More Than One Ortholog

Some genes can have more than one orthologous counterpart in a given genome. This happens whenever a gene undergoes duplication *after* the speciation of the two genomes in question. Since BBH only picks the best hit, it only captures part of the orthologous relations (Fig. 1). The existence of multiple orthologous counterparts is often referred to as *one-to-many* or *many-to-many* orthology, depending whether duplication took place in one or both lineages. To designate the copies resulting from such duplications occurring *after* a speciation of reference, Remm et al. coined the term *in-paralogs* and introduced a method called *InParanoid* that improves upon BBH by potentially identifying all pairs of many-to-many orthologs [5]. In brief, their algorithm identifies all paralogs within a species that are evolutionarily closer (more similar) to each other than to the BBH gene in the other genome. This results in two sets of in-paralogs—one for each species—where all pairwise combinations between the two sets are orthologous relations. Alternatively, it is possible to identify many-to-many orthology by relaxing the notion of "best hit" to "group of best hits." This can be implemented using a score tolerance threshold or a confidence interval around the BBH [23, 34].

Evolutionary Distances

Instead of using sequence similarity as a surrogate for evolutionary distance to identify the closest gene(s), Wall et al. proposed to use direct and proper maximum likelihood estimates of the evolutionary distance between pairs of sequences [31]. This estimate of evolutionary distance is based on the number and type of amino acid substitutions between the two sequences. Indeed, previous studies have shown that the highest scoring alignment is often not the nearest phylogenetic neighbor [35]. Building upon this work, Roth et al. showed how statistical uncertainties in the distance estimation can be incorporated into the inference strategy [36].

Differential Gene Losses

As discussed above, one of the advantages of BBH over BeT is that by virtue of the bi-directional requirement, the former is more robust to gene losses in one of the two lineages. But if gene losses occurred along both lineages, it can happen that a pair of genes mutually closest to one another is in fact paralogs, simply because both their corresponding orthologs were lost—a situation referred to as "differential gene losses." Dessimoz et al. [37] presented a way to detect some of these cases by looking for a third species in which the corresponding orthologs have not been lost and thus can act as *witnesses of non-orthology*.

2.1.2 Clustering Phase: From Pairs to Groups

The graph construction phase yields orthologous relationships between pairs of genes. But this is often not sufficient. Conceptually, information obtained from multiple genes or organisms is often more powerful than that obtained from pairwise comparisons only. In particular, as the use of a third genome as potential witness of non-orthology suggests, a more global view can allow identification and correction of inconsistent/spurious predictions. Practically, it is more intuitive and convenient to work with groups of genes than with a list of gene pairs. Therefore, it is often desirable to cluster orthologous genes into groups.

Tatusov et al. [8] introduced the concept of clusters of orthologous groups (COGs). COGs are computed by using triangles (triplets of genes connected to each other) as seeds and then merging triangles which share a common face, until no more triangle can be added. This clustering can be computed relatively efficient in time $O(n^3)$, where n is the number of genomes analyzed [38]. The stated objective of this clustering procedure is to group genes that have diverged from a single gene in the last common ancestor of the species represented [8]. Practically, they have been found to be useful by many, most notably to categorize prokaryotic genes into broad functional categories.

A different clustering approach was adopted by *OrthoMCL*, another well-established graph-based orthology inference method [29]. There, groups of orthologs are identified by Markov Clustering [39]. In essence, the method consists in simulating a random walk on the orthology graph, where the edges are weighted according to similarity scores. The Markov Clustering process gives rise to probabilities that two genes belong to the same cluster. The graph is then partitioned according to these probabilities and members of each partition form an orthologous group. These groups contain orthologs and "recent" paralogous genes, where the recency of the paralogs can be somewhat controlled through the parameters of the clustering process.

A third grouping strategy consists in building groups by identifying fully connected subgraphs (called "cliques" in graph theory) [23]. This approach has the merits of straightforward interpretation (groups of genes which are all orthologous to one another) and high confidence in terms of orthology within the resulting groups, due to the high consistency required to form a fully connected subgraph. But it has the drawbacks of being hard to compute (clique finding belongs to the NP-complete class of problems, for which no polynomial time algorithm is known; see Box 2) and being excessively conservative for many applications.

As emerges from these various strategies, there is more than one way orthologous groups can be defined, each with different implications in terms of group properties and applications [40]. In fact, there is an inherent trade-off in partitioning the orthology graph into clusters of genes, because orthology is a non-transitive relation: if genes A and B are orthologs and genes B and C are orthologs, genes A and C are not necessarily orthologs, e.g., consider in Fig. 1 the blue human gene, the frog gene, and the red dog gene. Therefore, if groups are defined as sets of genes in which all pairs of genes are orthologs (as with OMA groups), it is not possible to partition A, B, and C into groups capturing all orthologous relations while leaving out all paralogous relations.

2.1.3 Hierarchical Clustering

More inclusive grouping strategies necessarily lead to orthologs and paralogs within the same group. Nevertheless, it can be possible to control the nature of the paralogs included. For instance, as seen above, OrthoMCL attempts at including only "recent" paralogs in its groups. This idea can be specified more precisely by defining groups with respect to a particular speciation event of interest, e.g., the base of the mammals. Such *hierarchical groups* are expected to include orthologs and in-paralogs with respect to the reference speciation—in our example all copies that have descended from a single common ancestor gene in the last mammalian common ancestor. Conceptually, hierarchical orthologous groups can be defined as groups of genes that have descended from a single common ancestral gene within a taxonomic range of interest.

Several resources provide hierarchical clustering of orthologous groups. EggNOG [15] and OrthoDB [25], for example, both implement this concept by applying a COG-like clustering method for various taxonomic ranges. Another example, Hieranoid, produces hierarchical groups by using a guide tree to perform pairwise orthology inferences at each node from the leaves to the root—inferring ancestral genomes at each node in the tree [13, 18]. Similarly, OMA GETHOGs is an approach based on an orthology graph of pairwise orthologous gene relations, where hierarchical orthologous groups are formed starting with the most specific taxonomy and incrementally merges them toward the root [21, 22]. Another method, COCO-CL, identifies hierarchical orthologous groups recursively, using correlations of similarity scores among homologous genes [41] and, interestingly, without relying on a species tree. By capturing part of the gene tree structure in the group hierarchies, these methods try in some way to bridge the gap between graph-based and tree-based orthology inference approaches. We now turn our attention to the latter.

2.2 Tree-Based Methods

At their core, tree-based methods infer orthologs on the basis of gene family trees whose internal nodes are labeled as speciation or duplication nodes. Indeed, once all nodes of the gene tree have been inferred as a speciation or duplication event, it is trivial to establish whether a pair of genes is orthologous or paralogous, based on the type of the branching where they coalesce. Such labeling is traditionally obtained by reconciling gene and species trees. In most cases, gene and species trees have different topologies, due to evolutionary events acting specifically on genes such as duplications, losses, lateral transfers, or incomplete lineage sorting [42]. Goodman et al. [43] pioneered research to resolve these

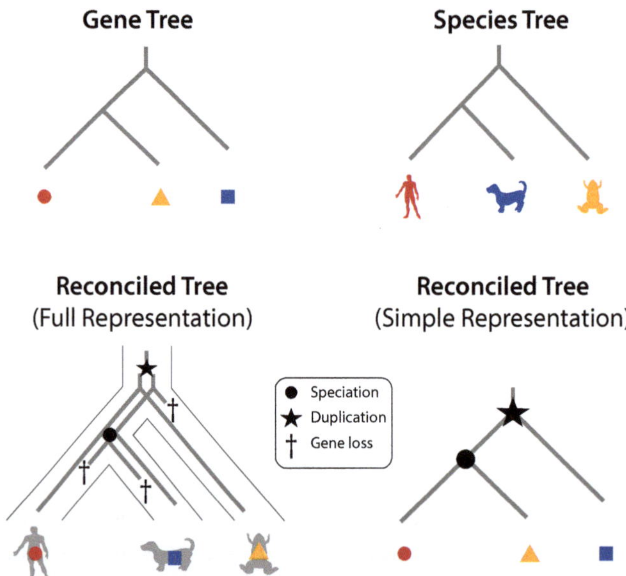

Fig. 2 Schematic example of the gene/species tree reconciliation. The gene tree and species tree are not compatible. Reconciliation methods resolve the incongruence between the two by inferring speciation, duplication, and losses events on the gene tree. The reconciled tree indicates the most parsimonious history of this gene, constrained to the species tree. The simple representation (bottom right) suggests that the human and frog genes are orthologs and that they are both paralogous to the dog gene

incongruences. They showed how the incongruences can be explained in terms of speciation, duplication, and loss events on the gene tree (Fig. 2) and provided an algorithm to infer such events.

Most tree reconciliation methods rely on a parsimony criterion: the most likely reconciliation is the one which requires the least number of gene duplications and losses. This makes it possible to compute reconciliation efficiently and is tenable as long as duplication and loss events are rare compared to speciation events. In their seminal article, Goodman et al. [43] had already devised their reconciliation algorithm under a parsimony strategy. In the subsequent years, the problem was formalized in terms of a map function between the gene and species trees [44], whose computational cost was conjectured [45], and later proved [12, 46] to coincide with the number of gene duplication and losses. These results yielded highly efficient algorithms, either in terms of asymptotic time complexity [12] or in terms of runtimes on typical problem sizes [47]. With these near-optimal solutions, one might think that the tree reconciliation problem has long been solved. As we shall see in the rest of this section, however, the original formulation of the tree reconciliation problem has several limitations in practice, which have stimulated the development of various refinements to overcome them (Table 2).

Table 2
Overview of gene/species tree reconciliation methods and their main properties

Method	Species tree[a]	Rooting[b]	Gene tree uncertainty[c]	Framework[d]	Available Algo/DB	References
BranchClust	Species overlap	Min number of clusters	None	n.a.	–/✓	[48]
DLRSOrthology	Fully resolved	n.a.	n.a.	Probabilistic	✓/–	[49–51]
Ensembl/TreeBeST	Partially resolved	Min dupl + min loss	None	MP	–/✓	[52–54]
HOGENOM	Partially resolved	Min dupl	Multifurcate	MP	✓/✓	[55, 56]
LOFT	Species overlap	Min dupl	None	MP	✓/–	[57]
Orthostrapper	Fully resolved	Min dupl	Bootstrap	MP	✓/–	[58]
PhylomeDB	Species overlap	Outgroup	None	MP	–/✓	[59, 60]
Softparsmap	Partially resolved	Min dupl + min loss	None	MP	✓/–	[61]
Speciation vs. duplication inference (SDI)	Fully resolved	n.a.	None	MP	✓/–	[47]

[a]Required species tree: Fully resolved, multifurcations allowed, computed from species overlap
[b]Approach to root gene tree (n.a. indicates that the initial rooting is assumed to be correct)
[c]Approach taken to handle reconstruction uncertainties of the gene tree (bootstrap, reconcile every bootstrap sample; multifurcate, splits in the gene tree with low support are collapsed)
[d]Used optimization framework (MP, maximum parsimony)

2.2.1 Unresolved Species Tree

A first problem ignored by most early reconciliation algorithms lies in the uncertainty often associated with the species tree, which these methods assume as correct and heavily rely upon.

One way of dealing with the uncertainties is to treat unresolved parts of the species tree as multifurcating nodes (also known as *soft polytomies*). By doing so, the reconciliation algorithm is not forced to choose for a specific type of evolutionary event in ambiguous regions of the tree. This approach is, for instance, implemented in *TreeBeST* [52] and used in the *Ensembl Compara* project [53].

Alternatively, Heijden et al. [57] demonstrated that it is often possible to infer speciation and duplication events on a gene tree without knowledge of the species tree. Their approach, which they call *species overlap*, identifies for a given split the species represented in the two subtrees induced by the split. If at least one species has genes in both subtrees, a duplication event is inferred; else a speciation event is inferred. In fact, this approach is a special case of soft polytomies where all internal nodes have been collapsed. Thus, the only information needed for this approach is a rooted gene tree. Since then, this approach has been adopted in other projects, such as PhylomeDB [59].

2.2.2 Rooting

The classical reconciliation formulation requires both gene and species trees to be rooted. But most models of sequence evolution are time reversible and thus do not allow to infer the rooting of the reconstructed gene tree. One sensible solution is to root a gene tree so that it minimizes the number of duplication events [62]. Thus, this method uses the parsimony principle for both rooting and reconciliation. For cases of multiple optimal rootings, ties can be broken by selecting the tree that minimizes the tree height [63] or by picking the rooting which minimizes the number of gene losses [61].

Another approach is to place the root at the "center of the tree"—also known as "midpoint rooting" [58]. The idea of this method goes back to Farris [64] and is motivated by the concept of a molecular clock. But for most gene families, assuming a constant rate of evolution is inappropriate [65, 66], and thus this approach is not used widely. A newly introduced refinement based on minimizing average deviations among children nodes holds promise of being more robust [67] but still relies on a molecular clock assumption.

For the species tree, the most common and reliable way of rooting trees is by identifying an outgroup species. PhylomeDB uses genes from outgroup species to root gene trees [59]. One main potential problem with this approach is that in many situations, it can be difficult to identify a suitable outgroup. For example, in analysis covering all kingdoms of life, an outgroup species may not be available, or the relevant genes might have been lost

[68]. A suitable out-group needs to be close enough to allow for reliable sequence alignment, yet it must have speciated clearly before any other species separated. Furthermore, ancient duplications can cause outgroup species to carry *in-group* genes. These difficulties make this approach more challenging for automated, large-scale analysis [69].

2.2.3 Gene Tree Uncertainty

Another assumption made in the original tree reconciliation problem is the (topological) correctness of the gene tree. But it has been shown that this assumption is commonly violated, often due to finite sequence lengths, taxon sampling [70, 71], or gene evolution model violations [72]. On the other hand, techniques of expressing uncertainties in gene tree reconstruction via support measures, e.g., bootstrap values, have become well established. Storm and Sonnhammer [58] as well as Zmasek and Eddy [63] independently suggested to extend the bootstrap procedure to reconciliation, thereby reducing the dependency of the reconciliation procedure on any one gene tree while providing a measure of support of the inferred speciation/duplication events. The downsides of using the bootstrap are the high computational costs and interpretation difficulties associated with it [73].

Similarly to how unresolved species tree can be handled, unresolved parts of the gene tree can also be collapsed into multifurcating nodes. For instance, HOGENOM [55] and *Softparsmap* [61] collapse branches with low bootstrap support values.

A third way of tackling this problem consists in simultaneously solving both the gene tree reconstruction and reconciliation problems [74]. They use the parsimony criterion of minimizing the number of duplication events to improve on the gene tree itself. This is achieved by rearranging the local gene tree topology of regions with low bootstrap support such that the number of duplications and losses is further reduced.

2.2.4 Parsimony vs. Likelihood

All the approaches mentioned so far try to minimize the number of gene duplication events. This is generally justified by a parsimony argument, which assumes that gene duplications and losses are rare events. But what if this assumption is frequently violated? Little is known about duplication and loss rates in general [75], but there is strong evidence for historical periods with high gene duplication occurrence rates [76] or gene families specifically prone to massive duplications (e.g., olfactory receptor, opsins, serine/threonine kinases, etc.)

Motivated by this reasoning, Arvestad et al. introduced the idea of a probabilistic model for tree reconciliation [49]. They used a Bayesian approach to estimate the posterior probabilities of a reconciliation between a given gene and species tree using Markov chain Monte Carlo (MCMC) techniques. Arvestad et al. [49]

modeled gene duplication and loss events through a *birth-death process* [77]. In the subsequent years, they refined their method to also model sequence evolution and substitution rates in a unified framework called *gene sequence evolution model with iid rates* (*GSR*) [49, 50].

Perhaps the biggest problem with the probabilistic approach is that it is not clear how well the assumptions of their model (the *birth-death process* with fixed parameters) relate to the true process of gene duplication and gene loss. Doyon et al. [78] compared the maximum parsimony reconciliation trees from 1278 fungi gene families to the probabilistically reconciled trees using gene birth/ death rates fitted from the data. They found that in all but two cases, the maximum parsimony scenario corresponds to the most probable one. This remarkably high level of consistency indicates that in terms of the accuracy of the "best" reconciliation, there is little to gain from using a likelihood approach over the parsimony criterion of minimizing the number of duplication events. But how this result generalizes to other datasets has yet to be investigated.

2.3 Graph-Based vs. Tree-Based: Which Is Better?

Given the two fundamentally different paradigms in orthology inference that we reviewed in this section, one can wonder which is better. Conceptually, tree reconciliation methods have several advantages. In terms of inference, by considering all sequences from all species at the same time, it can also be expected that they can extract more information from the sequences. This in turn should translate into higher statistical power. In terms of their output, reconciled gene trees provide the user more information than pairs or groups of orthologs. For example, the trees display the order of duplication and speciation events, as well as evolutionary distances between these events. In practice, however, these methods have the disadvantage of having much higher computational complexity than their graph-based counterparts. Furthermore, the two approaches are in practice often not that strictly separated. Tree-based methods often start with a graph-based clustering step to identify families of homologous genes. Conversely, several hierarchical grouping algorithms also rely on species trees in their inference.

Thus, it is difficult to make general statements about the relative performance of the two classes of inference methods. One solution that can leverage the unique abilities of both tree-based and graph-based methods is to combine several independent orthology inference methods into one. We discuss this technique in the next section.

3 Meta-methods

In recent years a new class of orthology inference tools has emerged which attempts to make the most out of multiple orthology prediction algorithms—*meta-methods*. These are approaches which combine several individual and distinct methods in order to produce more robust orthology predictions. These meta-methods are able to take advantage of the standardized formats of output which has been a goal of the orthology community [79], as well as the many new and well-established methods out there.

Generally, meta-methods assign a confidence score to a given predicted orthologous relation. In its most basic form, more weight is given to orthologs predicted by the most methods. Some examples include methods which simply take the intersection of several methods, such as GET_HOMOLOGUES [80], COMPARE [81], HCOP [82], and DIOPT [83]. These methods maintain a high level of precision, but since they are based on intersections, they necessarily have a lower recall.

Additionally, post-processing techniques can be used to build upon the base of orthologs found by several methods—thus assigning more sequences as orthologs and improving performance. For example, MOSAIC (Multiple Orthologous Sequence Analysis and Integration by Cluster optimization) [84] uses an iterative graph-based optimization approach that works on ortholog sets predicted by several independent methods. MOSAIC captures orthologs which are missed by some individual methods, producing a 1.6-fold increase in the number of orthologs detected. Another example is the MARIO software, which looks for the intersection of several different orthology methods as seed groups and then progressively adds unassigned proteins to the groups based on HMM profiles [85]. MetaPhOrs' approach integrates phylogenetic and homology information derived from different databases [86]. They demonstrate that the number of independent sources from which an orthology prediction is made, as well as the level of consistency across predictions, can be used as confidence scores.

So far the previously mentioned meta-methods combine independent orthology prediction algorithms and give a higher score based on the more algorithms which predict a given orthologous relation. However, another emerging approach is to use machine learning techniques to recognize patterns among several different orthology inference methods. With this, one can predict previously unknown high-confidence orthologs. WORMHOLE is a tool which uses the information from 17 different orthology prediction methods to train support vector machine classifiers for predicting least diverged orthologs [87]. WORMHOLE was able to strongly re-predict least diverged orthologs in the reference set and also predict previously unclassified orthologous genes.

The type of meta-approach and its associated stringency depends on what the user is going after. For example, if the goal is to get very-high-confidence groups, methods which only combine for the intersection without trying to add more orthologs may be preferable. Studies requiring both high precision and recall may be better suited to use the meta-methods which use post-processing or machine learning to predict orthologs. And as with all methods, it is important to understand which clades the method has been benchmarked in and which orthology tools have been combined. For example, if several methods have the same bias, one will just propagate the bias and end up with a false sense of security because the methods are not independent.

4 Scaling to Many Genomes

In terms of orthology inference, the abundance of genomes now available has resulted in an emphasis on driving down computational processing time via efficient algorithms. When inferring orthology for many genomes, the bottleneck is generally the all-against-all computations—aligning the proteins in every genome against the proteins in every other genome. This is the first step of nearly all graph-based methods. The all-against-all computation has an $O(n^2)$ runtime, meaning it scales quadratically with the number of genomes analyzed (Box 2).

So far, two main techniques for scaling orthology prediction to many genomes have emerged. The first approach is by making the all-against-all comparisons faster. Because comparisons are independent of each other, the most obvious way of doing this is by taking advantage of a high-performance computing cluster, as this is an embarrassingly parallel computing problem. Many methods have implemented this, such as Hieranoid [13], PorthoMCL [88], or OMA [22]. Another way to save time on the all-against-all comparisons is by using very fast algorithms for the homology search. For example, preliminary results of SonicParanoid showed 160–750× speedup of orthology inference compared to InParanoid [89]. Innovations in alignment algorithms with methods such as DIAMOND [90] or MMSeq2 [91] have the potential to greatly reduce the time to do the all-against-all comparisons.

A second approach to efficiently scale up orthology inference to many genomes is by simply avoiding doing the entire all-against-all comparisons. This makes sense, since a significant amount of time is spent comparing unrelated gene pairs. For example, it is possible to avoid aligning many unrelated pairs by exploiting the transitive property of homology. Wittwer et al. [92] did this by first building clusters of homologous sequences with one representative sequence per cluster and subsequently performing the all-against-all within each cluster. Hieranoid avoids unnecessary all-against-all comparisons by using a species tree as a guide, reducing the number of comparisons to $N - 1$ for N genomes, scaling linearly rather than quadratically [18]. Another way to avoid all-by-all comparison is by using a mapping strategy, whereby new proteomes are mapped onto precomputed orthologous groups. This strategy has been successfully implemented with the eggNOG database—each sequence in a new proteome is mapped to a precomputed orthologous cluster based on hidden Markov models. Then, orthology relations and function are transferred to the new sequence from the best matching sequence in the database [93].

5 Benchmarking Orthology

Assessing the quality of orthology predictions is important but difficult. The main challenge is that the precise evolutionary history of entire genomes is largely unknown and thus, predictions can only be validated indirectly, using surrogate measures. To be informative, such measures need to strongly correlate with orthology/paralogy. At the same time, they should be independent from the methods used in the orthology inference process. Concretely, this means that the orthology inference is not based on the surrogate measure and the surrogate measure is not derived from orthology/paralogy.

5.1 Benchmarking Approaches

Several ways of benchmarking orthology inference have been developed in the past years. In the next sections, we go over the main approaches, bringing attention to the advantages and limitations to each.

5.1.1 Functional Conservation

The first surrogate measures proposed revolved around conservation of function [94]. This was motivated by the common belief that orthologs tend to have conserved function, while paralogs tend to have different functions. Indeed, orthologs tend to be more conserved than paralogs in terms of GO annotation similarity [95]. Thus, "for a given evolutionary distance, more accurate orthology inference is likely to be correlated with more functionally similar gene pairs." Hulsen et al. [94] assessed the quality of ortholog predictions in terms of conservation of co-expression levels, domain annotation, and protein-protein interaction partners. Additionally, Altenhoff et al. [96] used similarity of experimentally validated GO annotations as well as Enzyme Commission (EC) numbers as a functional benchmark. Functional benchmarks have an advantage in that many researchers are interested in orthology because they want to find functionally conserved genes, thus making functional tests important for assessing different inference methods. The main limitation of these measures is that it is not so clear how much they correlate with orthology/paralogy. Indeed, it has been argued that the difference in function conservation trends between orthologs and paralogs might be much smaller than commonly assumed and indeed many examples are known of orthologs that have dramatically different functions [97].

5.1.2 Gene Neighborhood Conservation

The fraction of orthologs that have neighboring genes being orthologs themselves is an indicator of consistency and therefore to some extent also of quality of orthology predictions [94]. Although synteny has been used as part of the orthology inference for several algorithms, to date it has not been used as part of large-scale benchmarking efforts. One possible problem is that gene neighborhood can be conserved among paralogs, such as those resulting from whole-genome duplications. Furthermore, some methods use gene neighborhood conservation to help in their inference process, which can bias the assessment done on such measures (principle of independence stated above).

5.1.3 Species Tree Discordance Test

The quality of ortholog predictions can also be assessed based on phylogeny. By definition, the tree relating a set of genes all orthologous to one another only contains speciation splits and has the same topology as the underlying species. We introduced a benchmarking protocol that quantifies how well the predictions from various orthology inference methods agree with undisputed species tree topologies [96, 98]. Thus, the species tree discordance test judges the accuracy of ortholog predictions based on the correctness of the species tree which can be constructed from them. The advantage of this measure is that by virtue of directly ensuing from the definition of orthology, it correlates strongly with it and thus satisfies the first principle. However, the second principle, independence from the inference process, is not satisfied with methods relying on the species tree—typically all reconciliation methods but also most graph-based methods producing hierarchical groups. In such cases, interpretation of the results must be done carefully.

5.1.4 Gold Standard Gene Tree Test

High-quality reference gene trees can also be used to assess orthology inferences. For this, one compares the pairs of orthologs from a given method to pairs of orthologs derived from these expertly curated gene trees [40, 99]. One drawback of this benchmark is that it is limited by the ability to curate the phylogeny—if the evolutionary history of the gene family is ambiguous, the resulting reference tree will unavoidably have mistakes. Another limitation is the small size of most benchmarks of this type. This casts doubts on their generalizability and makes them prone to overfitting.

5.1.5 Subtree Consistency Test

For inference methods based on reconciliation between gene and species trees, Vilella et al. [53] proposed a different phylogeny-based assessment scheme. For any duplication node of the labeled gene tree, a consistency score is computed, which captures the balance of the species found in the two subtrees. Unbalanced nodes correspond to an evolutionary scenario involving extensive gene losses and therefore, under the principle of parsimony, are less likely to be correct. Given that studies to date tend to support the adequacy of the parsimony criterion in the context of gene family dynamics (Subheading 2.2.4), it can be expected that this metric correlates highly with correct orthology/paralogy assignments. However, since virtually all tree-based methods themselves incorporate this very criterion in their objective function (i.e., minimizing the number of gene duplications and losses), the principle of independence is violated, and thus the adequacy of this measure is questionable.

5.1.6 Latent Class Analysis

Chen et al. [100] proposed a purely statistical benchmark based on *latent class analysis* (*LCA*). Given the absence of a definitive answer on whether two given genes are orthologs, the authors argue that by looking at the agreement and disagreement of predictions made by several inference methods on a common dataset, one can estimate the reliability of individual predictors. More precisely, LCA is a statistical technique that computes maximum likelihood estimates of sensitivity and specificity rates for each orthology inference methods, given their predictions and given an error model. This is attractive, because it does not depend on any surrogate measure. However, the results depend on the error model assumed. Thus, we are of the opinion that LCA merely shifts the problem of assessing orthology to the problem of assessing an error model of various orthology inference methods.

5.1.7 Simulated Genomes

Finally, simulated data can be used in benchmarking. By this, the precise evolutionary history of a genome can be validated, in terms of gene duplication, insertion, deletion, and lateral gene transfer [101]. Knowing for certain all aspects of the simulated genomes gives an advantage over assessments based on empirical data, where the true evolutionary history is unknown. On the other hand, how well the simulated data reflect “real” data is debatable.

5.2 Orthology Benchmarking Service

The orthology benchmarking service is a web-based platform for which users can upload their ortholog predictions and run them through a variety of benchmarks. The user must use *quest for orthologs* (QFO) reference proteome set, which is a set of 66 genomes that covers a diverse set of species across all domains [79], to infer pairwise or groups of orthologs. Several phylogenetic and function-based benchmarks are automatically run on the uploaded data, and then summary statistics of the results of each benchmark are reported. The user can compare their method's performance with that of other well-known orthology inference algorithms and choose to make theirs public as well. For each benchmark, a precision-recall curve is reported, allowing for ease of comparison and evaluation of individual inference techniques. Because of the range of benchmarking tests and publicly available methods for comparison, the benchmarking service is useful for both users, who can check which methods work well for their particular problem and for method developers. The orthology benchmarking service can be accessed at http://orthology.benchmarkservice.org.

5.3 Conclusions on Benchmarking

Overall, it becomes apparent that there is no "magic bullet" strategy for orthology benchmarking, as each approach discussed here has its limitations (though some limitations are more serious than others). Nevertheless, comparative studies based on these various benchmarking measures have reported surprisingly consistent findings [40, 94, 96, 98, 100]: these assessments generally observe that there is a trade-off between accuracy and coverage and most common databases are situated on a Pareto frontier. The various assessments concur that the "best" orthology approach is highly dependent on the various possible applications of orthology.

6 Applications

As we have seen so far, there is a large diversity in the methods for orthology inference. The main reason is that, although the methods discussed here all infer orthology as part of their process, many of them have been developed for different reasons and have different ultimate goals. Unfortunately, this is often not mentioned explicitly and tends to be a source of confusion. In this section, we review some of these ultimate goals and discuss which methods and representation of orthology are better suited to address them and why.

As mentioned in the introduction, most interest for orthology is in the context of function prediction and is largely based on the belief that orthologs tend to have conserved function. A conservative approach consists in propagating function between one-to-one orthologs, i.e., pairs of orthologous genes that have not undergone gene duplication since they diverged from one another. Several orthology databases directly provide one-to-one orthology predictions. But even with those that do not, it might still be possible to obtain such predictions, for instance, by selecting hierarchical groups containing at most one sequence in each species or by extracting from reconciled trees' subtrees with no duplication. A more sophisticated approach consists in propagating gene function annotations across genomes on the basis of the full reconciled gene tree. Thomas et al. [102], for instance, proposed a way to assign gene function to uncharacterized proteins using a gene tree and a hidden Markov model (HMM) among gene families. Engelhardt et al.. [103] developed a Bayesian model of function change along reconciled gene trees and showed that their approach significantly improves upon several methods based on pairwise gene function propagation. Ensembl Compara [53] and Panther [102] are two major databases providing reconciled gene trees.

Since Darwin, one traditional question in biology has always been how species are related to each other. As we recall in the introduction of this chapter, Fitch's original motivation for defining orthology was phylogenetic inference. Indeed, the gene tree reconstructed from a set of genes which are all orthologous to each other should by definition be congruent to the species tree. OMA Groups (OMA) have this characteristic and, crucially, are constructed without help of a species tree.

Yet another application associated with orthology are general alignments between genomes, e.g., protein-protein interaction (PPI) network alignments or whole-genome alignments. Finding an optimal PPI network alignment between two genomes on the basis of the network topology alone is a computationally hard problem (i.e., it is an instance of the subgraph isomorphism problem which is NP-complete [104]). Orthology is often used as heuristic to constrain the mapping of the corresponding genes between the two networks and thus to reduce the problem complexity of aligning networks [105]. For whole-genome alignments, people most often use homologous regions and use orthologs as anchor points [106]. These types of application typically rely on ortholog predictions between pairs of genomes, as provided, e.g., by InParanoid [5] or OMA [23].

7 Conclusions and Outlook

The distinction between orthologs and paralogs is at the heart of many comparative genomic studies and applications. The original and generally accepted definition of orthology is based on the evolutionary history of pairs of genes. By contrast, there is a considerable diversity in how groups of orthologs are defined. These differences largely stem from the fact that orthology is a non-transitive relation and therefore, dividing genes into orthologous groups will either miss or wrongly include orthologous relations. This makes it important and worthwhile to identify the type of orthologous group best suited for a given application.

Regarding inference methods, while most approaches can be ordered into two fundamental paradigms—graph-based and tree-based—the difference between the two is shrinking, with graph-based methods increasingly striving to capture more of the evolutionary history. On the other hand, the rapid pace at which new genomes are sequenced limits the applicability of tree-based methods, computationally more demanding.

Benchmarking this large variety of methods remains a hard problem—from a conceptual point as described above but also because of very practical challenges such as heterogeneous data formats, genome versions, or gene identifiers. This has been recognized by the research community and has led to the development of the QFO consortium benchmarking service [96].

Looking forward, we see potential in extending the current model of gene evolution, which is limited to speciation, duplication, and loss events. Indeed, nature is often much more complicated. For instance, lateral gene transfer (LGT) is believed to be a major mode of evolution in prokaryotes. While there has been several attempts at extending tree reconciliation algorithms to detecting LGT [107, 108], the problem is largely unaddressed in typical orthology resources [109]. Another relevant evolutionary process omitted by most methods is whole-genome duplications (WGD). Even though WGD events act jointly on all gene families, with few exceptions [110, 111], most methods consider each gene family independently.

Overall, the orthology/paralogy dichotomy has proved to be useful but also inherently limited. Reducing the whole evolutionary history of homologous genes into binary pairwise relations is bound to be a simplification—and at times an oversimplification. The shift toward hierarchical orthologous groups is thus a promising step toward capturing more features of the evolutionary history of genes. Yet further development will still be needed, as we are nowhere close to grasp the formidable complexity of gene evolution across the full diversity of life.

8 Exercises

Assume the following evolutionary scenario

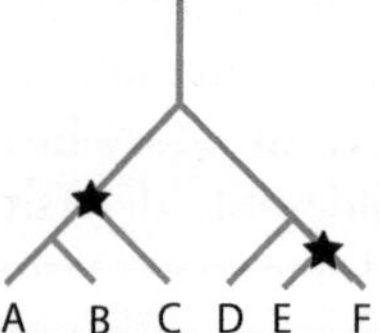

where duplications are depicted as star and all other splits are speciations.

Problem #1: Draw the corresponding orthology graph, where the vertices correspond to the observed genes and the edges indicate orthologous relations between them.

Problem #2: Apply the following two clustering methods on your orthology graph. First, reconstruct all the maximal fully connected subgraphs (cliques) that can be found. Second, reconstruct the COGs. COGs are built by merging triangles of orthologs whenever they share a common face. Remember that in both methods, a gene can only belong to a one cluster.

Acknowledgments

We thank Stefan Zoller for helpful feedback on an earlier version of the manuscript. We gratefully acknowledge support by the Swiss National Science Foundation grant PP00P3_150654 to CD. Adrian M. Altenhoff and Natasha M. Glover contributed equally to this work.

References

1. Dewey CN (2012) Whole-genome alignment. Methods Mol Biol 855:237–257
2. Alioto T (2012) Gene prediction. In: Anisimova M (ed) Evolutionary genomics: statistical and computational methods, vol 1. Humana, Totowa, NJ, pp 175–201
3. Löytynoja A (2012) Alignment methods: strategies, challenges, benchmarking, and comparative overview. In: Anisimova M (ed) Evolutionary genomics: statistical and computational methods, vol 1. Humana, Totowa, NJ, pp 203–235
4. Fitch WM (1970) Distinguishing homologous from analogous proteins. Syst Zool 19:99–113
5. Remm M, Storm CEV, Sonnhammer ELL (2001) Automatic clustering of orthologs and in-paralogs from pairwise species comparisons. J Mol Biol 314:1041–1052
6. Glover NM, Redestig H, Dessimoz C (2016) Homoeologs: what are they and how do we infer them? Trends Plant Sci 21:609–621
7. Kuzniar A, van Ham RCHJ, Pongor S et al (2008) The quest for orthologs: finding the corresponding gene across genomes. Trends Genet 24:539–551
8. Tatusov RL, Koonin EV, Lipman DJ (1997) A genomic perspective on protein families. Science 278:631–637
9. Overbeek R, Fonstein M, D'Souza M et al (1999) The use of gene clusters to infer functional coupling. Proc Natl Acad Sci U S A 96:2896–2901
10. Smith TF, Waterman MS (1981) Identification of common molecular subsequences. J Mol Biol 147:195–197
11. Altschul SF, Madden TL, Schäffer AA et al (1997) Gapped BLAST and PSI-BLAST: a new generation of protein database search programs. Nucleic Acids Res 25:3389–3402
12. Zhang L (1997) On a Mirkin-Muchnik-Smith conjecture for comparing molecular phylogenies. J Comput Biol 4:177–187
13. Schreiber F, Sonnhammer ELL (2013) Hieranoid: hierarchical orthology inference. J Mol Biol 425:2072–2081
14. Chor B, Tuller T (2005) Maximum likelihood of evolutionary trees is hard. In: Proceedings of the 9th annual international conference on research in computational molecular biology. Springer, Berlin, pp 296–310
15. Jensen LJ, Julien P, Kuhn M et al (2008) eggNOG: automated construction and annotation of orthologous groups of genes. Nucleic Acids Res 36:D250–D254
16. Muller J, Szklarczyk D, Julien P et al (2010) eggNOG v2.0: extending the evolutionary genealogy of genes with enhanced non-supervised orthologous groups, species and functional annotations. Nucleic Acids Res 38:D190–D195
17. Huerta-Cepas J, Szklarczyk D, Forslund K et al (2016) eggNOG 4.5: a hierarchical orthology framework with improved functional annotations for eukaryotic, prokaryotic and viral sequences. Nucleic Acids Res 44: D286–D293
18. Kaduk M, Sonnhammer E (2017) Improved orthology inference with Hieranoid 2. Bioinformatics 33:1154–1159
19. Ostlund G, Schmitt T, Forslund K et al (2010) InParanoid 7: new algorithms and tools for eukaryotic orthology analysis. Nucleic Acids Res 38:D196–D203

20. Sonnhammer ELL, Östlund G (2015) InParanoid 8: orthology analysis between 273 proteomes, mostly eukaryotic. Nucleic Acids Res 43:D234–D239
21. Altenhoff AM, Gil M, Gonnet GH et al (2013) Inferring hierarchical orthologous groups from orthologous gene pairs. PLoS One 8:e53786
22. Train C-M, Glover NM, Gonnet GH et al (2017) Orthologous Matrix (OMA) algorithm 2.0: more robust to asymmetric evolutionary rates and more scalable hierarchical orthologous group inference. Bioinformatics 33:i75–i82
23. Dessimoz C, Cannarozzi G, Gil M et al (2005) OMA, a comprehensive, automated project for the identification of orthologs from complete genome data: introduction and first achievements. In: Comparative genomics. Springer, Berlin, pp 61–72
24. Altenhoff AM, Schneider A, Gonnet GH et al (2011) OMA 2011: orthology inference among 1000 complete genomes. Nucleic Acids Res 39:D289–D294
25. Kriventseva EV, Rahman N, Espinosa O et al (2008) OrthoDB: the hierarchical catalog of eukaryotic orthologs. Nucleic Acids Res 36: D271–D275
26. Zdobnov EM, Tegenfeldt F, Kuznetsov D et al (2017) OrthoDB v9.1: cataloging evolutionary and functional annotations for animal, fungal, plant, archaeal, bacterial and viral orthologs. Nucleic Acids Res 45:D744–D749
27. Linard B, Thompson JD, Poch O et al (2011) OrthoInspector: comprehensive orthology analysis and visual exploration. BMC Bioinform 12:11
28. Linard B, Allot A, Schneider R et al (2015) OrthoInspector 2.0: software and database updates. Bioinformatics 31:447–448
29. Li L, Stoeckert CJ Jr, Roos DS (2003) OrthoMCL: identification of ortholog groups for eukaryotic genomes. Genome Res 13:2178–2189
30. Chen F, Mackey AJ, Stoeckert CJ Jr et al (2006) OrthoMCL-DB: querying a comprehensive multi-species collection of ortholog groups. Nucleic Acids Res 34:D363–D368
31. Wall DP, Fraser HB, Hirsh AE (2003) Detecting putative orthologs. Bioinformatics 19:1710–1711
32. DeLuca TF, Wu I-H, Pu J et al (2006) Roundup: a multi-genome repository of orthologs and evolutionary distances. Bioinformatics 22:2044–2046
33. DeLuca TF, Cui J, Jung J-Y et al (2012) Roundup 2.0: enabling comparative genomics for over 1800 genomes. Bioinformatics 28:715–716
34. Fulton DL, Li YY, Laird MR et al (2006) Improving the specificity of high-throughput ortholog prediction. BMC Bioinform 7:270
35. Koski LB, Golding GB (2001) The closest BLAST hit is often not the nearest neighbor. J Mol Evol 52:540–542
36. Roth ACJ, Gonnet GH, Dessimoz C (2008) Algorithm of OMA for large-scale orthology inference. BMC Bioinform 9:518
37. Dessimoz C, Boeckmann B, Roth ACJ et al (2006) Detecting non-orthology in the COGs database and other approaches grouping orthologs using genome-specific best hits. Nucleic Acids Res 34:3309–3316
38. Kristensen DM, Kannan L, Coleman MK et al (2010) A low-polynomial algorithm for assembling clusters of orthologous groups from intergenomic symmetric best matches. Bioinformatics 26:1481–1487
39. Van Dongen SM (2001) Graph clustering by flow simulation. PhD thesis, University of Utrecht
40. Boeckmann B, Robinson-Rechavi M, Xenarios I et al (2011) Conceptual framework and pilot study to benchmark phylogenomic databases based on reference gene trees. Brief Bioinform 12:423–435
41. Jothi R, Zotenko E, Tasneem A et al (2006) COCO-CL: hierarchical clustering of homology relations based on evolutionary correlations. Bioinformatics 22:779–788
42. Nei M (1987) Molecular evolutionary genetics. Columbia University Press, New York
43. Goodman M, Czelusniak J, Moore GW et al (1979) Fitting the gene lineage into its species lineage, a Parsimony strategy illustrated by cladograms constructed from globin sequences. Syst Zool 28:132–163
44. Page RDM (1994) Maps between trees and cladistic analysis of historical associations among genes, organisms, and areas. Syst Biol 43:58–77
45. Mirkin B, Muchnik I, Smith TF (1995) A biologically consistent model for comparing molecular phylogenies. J Comput Biol 2:493–507
46. Eulenstein O (1997) A linear time algorithm for tree mapping. Arbeitspapiere der GMD No. 1046, St
47. Zmasek CM, Eddy SR (2001) A simple algorithm to infer gene duplication and speciation events on a gene tree. Bioinformatics 17:821–828

48. Poptsova MS, Gogarten JP (2007) BranchClust: a phylogenetic algorithm for selecting gene families. BMC Bioinform 8:120
49. Arvestad L, Berglund A-C, Lagergren J et al (2003) Bayesian gene/species tree reconciliation and orthology analysis using MCMC. Bioinformatics 19(Suppl 1):i7–i15
50. Åkerborg Ö, Sennblad B, Arvestad L et al (2009) Simultaneous Bayesian gene tree reconstruction and reconciliation analysis. Proc Natl Acad Sci U S A 106:5714–5719
51. Ullah I, Sjöstrand J, Andersson P et al (2015) Integrating sequence evolution into probabilistic orthology analysis. Syst Biol 64:969–982
52. Li H, Coghlan A, Ruan J et al (2006) TreeFam: a curated database of phylogenetic trees of animal gene families. Nucleic Acids Res 34: D572–D580
53. Vilella AJ, Severin J, Ureta-Vidal A et al (2009) EnsemblCompara GeneTrees: complete, duplication-aware phylogenetic trees in vertebrates. Genome Res 19:327–335
54. Herrero J, Muffato M, Beal K et al (2016) Ensembl comparative genomics resources. Database 2016:bav096
55. Dufayard J-F, Duret L, Penel S et al (2005) Tree pattern matching in phylogenetic trees: automatic search for orthologs or paralogs in homologous gene sequence databases. Bioinformatics 21:2596–2603
56. Penel S, Arigon A-M, Dufayard J-F et al (2009) Databases of homologous gene families for comparative genomics. BMC Bioinform 10(Suppl 6):S3
57. van der Heijden RTJM, Snel B, van Noort V et al (2007) Orthology prediction at scalable resolution by phylogenetic tree analysis. BMC Bioinform 8:83
58. Storm CEV, Sonnhammer ELL (2002) Automated ortholog inference from phylogenetic trees and calculation of orthology reliability. Bioinformatics 18:92–99
59. Huerta-Cepas J, Dopazo H, Dopazo J et al (2007) The human phylome. Genome Biol 8: R109
60. Huerta-Cepas J, Capella-Gutiérrez S, Pryszcz LP et al (2014) PhylomeDB v4: zooming into the plurality of evolutionary histories of a genome. Nucleic Acids Res 42:D897–D902
61. Berglund-Sonnhammer A-C, Steffansson P, Betts MJ et al (2006) Optimal gene trees from sequences and species trees using a soft interpretation of parsimony. J Mol Evol 63:240–250
62. Hallett MT, Lagergren J (2000) New algorithms for the duplication-loss model. In: Proceedings of the fourth annual international conference on computational molecular biology. ACM, New York, NY, pp 138–146
63. Zmasek CM, Eddy SR (2002) RIO: analyzing proteomes by automated phylogenomics using resampled inference of orthologs. BMC Bioinform 3:14
64. Farris JS (1972) Estimating phylogenetic trees from distance matrices. Am Nat 106:645–668
65. Avise JC, Bowen BW, Lamb T et al (1992) Mitochondrial DNA evolution at a turtle's pace: evidence for low genetic variability and reduced microevolutionary rate in the Testudines. Mol Biol Evol 9:457–473
66. Ayala FJ (1999) Molecular clock mirages. Bioessays 21:71–75
67. Tria FDK, Landan G, Dagan T (2017) Phylogenetic rooting using minimal ancestor deviation. Nat Ecol Evol 1:193
68. Huelsenbeck JP, Bollback JP, Levine AM (2002) Inferring the root of a phylogenetic tree. Syst Biol 51:32–43
69. Tarrío R, Rodríguez-Trelles F, Ayala FJ (2000) Tree rooting with outgroups when they differ in their nucleotide composition from the ingroup: the Drosophila saltans and Willistoni groups, a case study. Mol Phylogenet Evol 16:344–349
70. Graybeal A (1998) Is it better to add taxa or characters to a difficult phylogenetic problem? Syst Biol 47(1):9–17
71. Rokas A, Williams BL, King N et al (2003) Genome-scale approaches to resolving incongruence in molecular phylogenies. Nature 425:798–804
72. Yang Z, Goldman N, Friday A (1994) Comparison of models for nucleotide substitution used in maximum-likelihood phylogenetic estimation. Mol Biol Evol 11:316–324
73. Anisimova M, Gascuel O (2006) Approximate likelihood-ratio test for branches: a fast, accurate, and powerful alternative. Syst Biol 55:539–552
74. Durand D, Halldórsson BV, Vernot B (2006) A hybrid micro-macroevolutionary approach to gene tree reconstruction. J Comput Biol 13:320–335
75. Lynch M, Conery JS (2000) The evolutionary fate and consequences of duplicate genes. Science 290:1151–1155
76. Robinson-Rechavi M, Marchand O, Escriva H et al (2001) Euteleost fish genomes are characterized by expansion of gene families. Genome Res 11:781–788

77. Kendall DG (1948) On the generalized "birth-and-death" process. Ann Math Stat 19:1–15
78. Doyon J-P, Hamel S, Chauve C (2012) An efficient method for exploring the space of gene tree/species tree reconciliations in a probabilistic framework. IEEE/ACM Trans Comput Biol Bioinform 9:26–39
79. Gabaldón T, Dessimoz C, Huxley-Jones J et al (2009) Joining forces in the quest for orthologs. Genome Biol 10:403
80. Contreras-Moreira B, Vinuesa P (2013) GET_HOMOLOGUES, a versatile software package for scalable and robust microbial pangenome analysis. Appl Environ Microbiol 79:7696–7701
81. Salgado D, Gimenez G, Coulier F et al (2008) COMPARE, a multi-organism system for cross-species data comparison and transfer of information. Bioinformatics 24:447–449
82. Eyre TA, Wright MW, Lush MJ et al (2007) HCOP: a searchable database of human orthology predictions. Brief Bioinform 8:2–5
83. Hu Y, Flockhart I, Vinayagam A et al (2011) An integrative approach to ortholog prediction for disease-focused and other functional studies. BMC Bioinform 12:357
84. Maher MC, Hernandez RD (2015) Rock, paper, scissors: harnessing complementarity in ortholog detection methods improves comparative genomic inference. G3 5:629–638
85. Pereira C, Denise A, Lespinet O (2014) A meta-approach for improving the prediction and the functional annotation of ortholog groups. BMC Genomics 15(Suppl 6):S16
86. Pryszcz LP, Huerta-Cepas J, Gabaldón T (2011) MetaPhOrs: orthology and paralogy predictions from multiple phylogenetic evidence using a consistency-based confidence score. Nucleic Acids Res 39:e32
87. Sutphin GL, Mahoney JM, Sheppard K et al (2016) WORMHOLE: novel least diverged ortholog prediction through machine learning. PLoS Comput Biol 12:e1005182
88. Tabari E, Su Z (2017) PorthoMCL: parallel orthology prediction using MCL for the realm of massive genome availability. Big Data Anal 2:4
89. Cosentino S, Iwasaki W (2018) SonicParanoid: extremely fast, accurate, and easy orthology inference. Bioinformatics. https://doi.org/10.1093/bioinformatics/bty631
90. Buchfink B, Xie C, Huson DH (2015) Fast and sensitive protein alignment using DIAMOND. Nat Methods 12:59–60
91. Steinegger M, Söding J (2017) MMseqs2 enables sensitive protein sequence searching for the analysis of massive data sets. Nat Biotechnol 35(11):1026–1028
92. Wittwer LD, Piližota I, Altenhoff AM et al (2014) Speeding up all-against-all protein comparisons while maintaining sensitivity by considering subsequence-level homology. PeerJ 2:e607
93. Huerta-Cepas J, Forslund K, Coelho LP et al (2017) Fast genome-wide functional annotation through orthology assignment by eggNOG-Mapper. Mol Biol Evol 34:2115–2122
94. Hulsen T, Huynen MA, de Vlieg J et al (2006) Benchmarking ortholog identification methods using functional genomics data. Genome Biol 7:R31
95. Altenhoff AM, Studer RA, Robinson-Rechavi M et al (2012) Resolving the ortholog conjecture: orthologs tend to be weakly, but significantly, more similar in function than paralogs. PLoS Comput Biol 8:e1002514
96. Altenhoff AM, Boeckmann B, Capella-Gutierrez S et al (2016) Standardized benchmarking in the quest for orthologs. Nat Methods 13:425–430
97. Studer RA, Robinson-Rechavi M (2009) How confident can we be that orthologs are similar, but paralogs differ? Trends Genet 25:210–216
98. Altenhoff AM, Dessimoz C (2009) Phylogenetic and functional assessment of orthologs inference projects and methods. PLoS Comput Biol 5:e1000262
99. Trachana K, Larsson TA, Powell S et al (2011) Orthology prediction methods: a quality assessment using curated protein families. BioEssays 33:769–780
100. Chen F, Mackey AJ, Vermunt JK et al (2007) Assessing performance of orthology detection strategies applied to eukaryotic genomes. PLoS One 2:e383
101. Dalquen DA, Altenhoff AM, Gonnet GH et al (2013) The impact of gene duplication, insertion, deletion, lateral gene transfer and sequencing error on orthology inference: a simulation study. PLoS One 8:e56925
102. Thomas PD, Campbell MJ, Kejariwal A et al (2003) PANTHER: a library of protein families and subfamilies indexed by function. Genome Res 13:2129–2141
103. Engelhardt BE, Jordan MI, Muratore KE et al (2005) Protein molecular function prediction by Bayesian phylogenomics. PLoS Comput Biol 1:e45

104. Cook SA (1971) The complexity of theorem-proving procedures. In: Proceedings of the third annual ACM symposium on theory of computing. ACM, New York, NY, pp 151–158
105. Sharan R, Ideker T (2006) Modeling cellular machinery through biological network comparison. Nat Biotechnol 24:427–433
106. Dewey CN, Pachter L (2006) Evolution at the nucleotide level: the problem of multiple whole-genome alignment. Hum Mol Genet 15 Spec No 1:R51–RR6
107. Górecki P (2004) Reconciliation problems for duplication, loss and horizontal gene transfer. In: Proceedings of the eighth annual international conference on research in computational molecular biology. ACM, New York, NY, pp 316–325
108. Hallett M, Lagergren J, Tofigh A (2004) Simultaneous identification of duplications and lateral transfers. In: Proceedings of the eighth annual international conference on Research in computational molecular biology. ACM, New York, NY, pp 347–356
109. Forslund K, Pereira C, Capella-Gutierrez S et al (2017) Gearing up to handle the mosaic nature of life in the quest for orthologs. Bioinformatics. https://doi.org/10.1093/bioinformatics/btx542
110. Guigó R, Muchnik I, Smith TF (1996) Reconstruction of ancient molecular phylogeny. Mol Phylogenet Evol 6:189–213
111. Bansal MS and Eulenstein O (2008) The multiple gene duplication problem revisited. Bioinformatics 24:i132–i13i138

Chapter 6

Transposable Elements: Classification, Identification, and Their Use As a Tool For Comparative Genomics

Wojciech Makałowski, Valer Gotea, Amit Pande, and Izabela Makałowska

Abstract

Most genomes are populated by hundreds of thousands of sequences originated from mobile elements. On the one hand, these sequences present a real challenge in the process of genome analysis and annotation. On the other hand, they are very interesting biological subjects involved in many cellular processes. Here we present an overview of transposable elements biodiversity, and we discuss different approaches to transposable elements detection and analyses.

Key words Transposable elements, Transposons, Mobile elements, Repetitive elements, Genome analysis, Genome evolution

1 Introduction

Most eukaryotic genomes contain large numbers of repetitive sequences. This phenomenon was described by Waring and Britten a half century ago using reassociation studies [1, 2]. It turned out that most of these repetitive sequences originated in transposable elements (TEs) [3], though the repetitive fraction of a genome varies significantly between different organisms, from 12% in *Caenorhabditis elegans* [4] to 50% in mammals [3], and more than 80% in some plants [5]. With such large contributions to genome sequences, it is not surprising that TEs have a significant influence on the genome organization and evolution. Although much progress has been achieved in understanding the role TEs play in a host genome, we are still far from the comprehensive picture of the delicate evolutionary interplay between a host genome and the invaders. They also pose various challenges to the genomic community, including aspects related to their detection and classification, genome assembly and annotation, genome comparisons, and mapping of genomic variants. They also pose various challenges to the genomic community, including aspects related to their detection and classification, genome assembly and annotation, genome

Maria Anisimova (ed.), *Evolutionary Genomics: Statistical and Computational Methods*, Methods in Molecular Biology, vol. 1910, https://doi.org/10.1007/978-1-4939-9074-0_6, © The Author(s) 2019

comparisons, and mapping of genomic variants. Here we present an overview of TE diversity and discuss major techniques used in their analyses.

2 Discovery of Mobile Elements

Transposable elements were discovered by Barbara McClintock during experiments conducted in 1944 on maize. Since they appeared to influence phenotypic traits, she named them *controlling elements.* However, her discovery was met with less than enthusiastic reception by the genetic community. Her presentation at the 1951 Cold Spring Harbor Symposium was not understood and at least not very well received [6]. She had no better luck with her follow-up publications [7–9] and after several years of frustration decided not to publish on the subject for the next two decades. Not for the first time in the history of science, an unappreciated discovery was brought back to life after some other discovery has been made. In this case it was the discovery of insertion sequences (IS) in bacteria by Szybalski group in the early 1970s [10]. In the original paper they wrote: "Genetic elements were found in higher organisms which appear to be readily transposed from one to another site in the genome. Such elements, identifiable by their controlling functions, were described by McClintock in maize. It is possible that they might be somehow analogous to the presently studied IS insertions" [10]. The importance of McClintock's original work was eventually appreciated by the genetic community with numerous awards, including 14 honorary doctoral degrees and a Nobel Prize in 1983 "for her discovery of mobile genetic elements" (http://nobelprize.org/nobel_prizes/medicine/laureates/1983/).

Coincidently, at the same time as Szybalski "rediscovered" TEs, Susumu Ohno popularized the term *junk DNA* that influenced genomic field for decades [11], although the term itself was used already before [12, 13].[1] Ohno referred to the so-called noncoding sequences or, to be more precise, to any piece of DNA that do not code for a protein, which included all genomic pieces originated in transposons. The unfavorable picture of transposable and transposed elements started to change in early 1990s when some researchers noticed evolutionary value of these elements [14, 15]. With the wheel of fortune turning full circle and advances of genome sciences, TE research is again focused on the role of mobile elements played in the evolution of gene regulation [16–23].

[1] The historical background of the "junk DNA" term was recently discussed by Dan Graur in his excellent blog http://judgestarling.tumblr.com/post/64504735261/the-origin-of-the-term-junk-dna-a-historical

3 Transposons Classification

3.1 Insertion Sequences and Other Bacterial Transposons

The bacterial genome is composed of a core genomic backbone decorated with a variety of multifarious functional elements. These include mobile genetic elements (MGEs) such as *bacteriophages*, *conjugative transposons*, *integrons*, *unit transposons*, *composite transposons*, and *insertion sequences* (IS). Here we elaborate upon the last class of these elements as they are most widely found and described [24].

The ISs were identified during studies of model genetic systems by virtue of their capacity to generate mutations as a result of their translocation [10]. In-depth studies in antibiotic resistance and transmissible plasmids revealed an important role for these mobile elements in formation of resistance genes and promoting gene capture. In particular, it was observed that several different elements were often clustered in "islands" within plasmid genomes and served to promote plasmid integration and excision.

Although these elements sometimes generate beneficial mutations, they may be considered genomic parasites as ISs code only for the enzyme required for their own transposition [24]. While an IS element occupies a chromosomal location, it is inherited along with its host's native genes, so its fitness is closely tied to that of its host. Consequently, ISs causing deleterious mutations that disrupt a genomic mode or function are quickly eliminated from the population. However, intergenically placed ISs have a higher chance to be fixed in the population as they are likely neutral regarding population's fitness [25].

ISs are generally compact (Fig. 1). They usually carry no other functions than those involved in their mobility. These elements contain recombinationally active sequences which define the boundary of the element, together with Tpase, an enzyme, which processes these ends and whose gene usually encompasses the entire length of the element [26]. Majority of ISs exhibit short terminal inverted-repeat sequences (IR) of length 10–40 bp. Several notable exceptions do exist, for example, the IS91, IS110, and IS200/605 families.

The IRs contain two functional domains [27]. One is involved in Tpase binding; the other cleaves and transfers strand-specific reactions resulting in transposition. IS promoters are often positioned partially within the IR sequence upstream of the *Tpase* gene. Binding sites for host-specific proteins are often located within proximity to the terminal IRs and play a role in modulating transposition activity or Tpase expression [28]. A general pattern for the functional organization of Tpases has emerged from the limited numbers analyzed. The N-terminal region contains sequence-specific DNA binding activities of the proteins while the catalytic domain is often localized toward the C-terminal end [28].

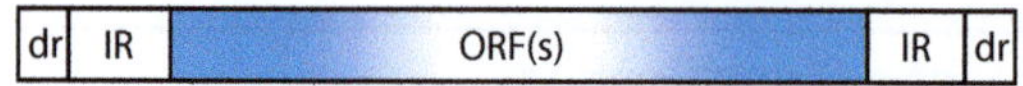

Fig. 1 Schematic representation of insertion sequences (IS). *dr* direct repeats, *IR* inverted repeats, *ORF* open reading frame

Another common feature of ISs is duplication of a target site that results in short direct repeats (DRs) flanking the IS [29]. The length of the direct repeat varies from 2 to 14 base pairs and is a hallmark of a given element. Homologous recombination between two IS elements can result in each having two different DRs [30].

ISs have been classified on the basis of (1) similarities in genetic organization (arrangement of open reading frames); (2) marked identities or similarities in their Tpases (common domains or motifs); (3) similar features of their ends (terminal IRs); and (4) fate of the nucleotide sequence of their target sites (generation of a direct target duplication of determined length). Based on the above rules, ISs are currently classified in 30 families (Table 1) [31].

3.2 Eukaryotic Transposable Elements

The first TE classification system was proposed by Finnegan in 1989 [32] and distinguished two classes of TEs characterized by their transposition intermediate: RNA (class I or retrotransposons) or DNA (class II or DNA transposons). The transposition mechanism of class I is commonly called "copy and paste" and that of class II, "cut and paste." In 2007 Wicker et al. [33] proposed hierarchical classification based on TEs structural characteristics and mode of replication (*see* Table 2 and Fig. 2). Below we present a brief overview of eukaryotic mobile elements that in general follows this classification.

3.2.1 Class I: Mobile Elements

As mentioned above, class I TEs transpose through an RNA intermediary. The RNA intermediate is transcribed from genomic DNA and then reverse-transcribed into DNA by a TE-encoded reverse transcriptase (RT), followed by reintegration into a genome. Each replication cycle produces one new copy, and as a result, class I elements are the major contributors to the repetitive fraction in large genomes. Retrotransposons are divided into five orders: LTR retrotransposons, DIRS-like elements, Penelope-like elements (PLEs), LINEs (*l*ong *in*terspersed *e*lements), and SINEs (*s*hort *in*terspersed *e*lements). This scheme is based on the mechanistic features, organization, and reverse transcriptase phylogeny of these retroelements. Accidentally, the retrotranscriptase coded by an autonomous TE can reverse-transcribe another RNA present in the cell, e.g., mRNA, and produce a retrocopy of it, which in most cases results in a pseudogene.

The LTR retrotransposons are characterized by the presence of *l*ong *t*erminal *r*epeats (LTRs) ranging from several hundred to several thousand base pairs. Both exogenous retroviruses and LTR retrotransposons contain a *gag* gene that encodes a viral

Table 1
Prokaryotic transposable elements as presented in the *IS Finder* database [31]

Family	Typical size range in bp	Direct repeat size in bp	IRs[a]	Number of ORFs
IS*1*	740–4600	0–10	Y	1 or 2
IS*110*	1200–1550	0	Y	1
IS*1182*	1330–1950	0–60	Y	1
IS*1380*	1550–2000	4–5	Y	1
IS1595	700–7900	8	Y	1
IS*1634*	1500–2000	5–6	Y	1
IS*200*/IS*605*	600–2000	0	Y/N	1 or 2
IS*21*	1750–2600	4–8	Y	2
IS*256*	1200–1500	8–9	Y	1
IS*3*	1150–1750	5	Y	2
IS*30*	1000–1700	2–3	Y	1
IS*4*	1150–5400	8–13	Y	1 or more
IS*481*	950–1300	4–15	Y	1
IS*5*	800–1500	2–9	Y	1 or 2
IS*6*	700–900	8	Y	1
IS*607*	1700–2500	0	N	2
IS*630*	1000–1400	2	Y	1 or 2
IS*66*	1350–3000	8–9	Y	1 or more
IS*701*	1400–1550	4	Y	1
IS*91*	1500–2000	0	N	1
IS*982*	1000	3–9	Y	1
IS*As1*	1200–1500	8–10	Y	1
IS*Azo13*	1250–2200	0–4	Y	1
IS*H3*	1225–1500	4–5	Y	1
IS*H6*	1450	8	Y	ISL
IS*Kra4*	1400–2900	0–9	Y	1 or more
IS*L3*	1300–2300	8	Y	ISKra4
IS*Lre2*	1500–2000	9	Y	1
Tn*3*	Over 3000	0	Y	More than 1
IS*NCY*	1300–2400	0–12	Y/N	1 or 2

[a]Presence (Y) or absence (N) of terminal inverted repeats

Table 2
Classification of eukaryotic transposable elements as proposed by Wicker et al. [33]

Class	Order	Superfamily	Phylogenetic distribution
Class I (retrotransposons)	LTR	Copia	Plants, metazoans, fungi
		Gypsy	Plants, metazoans, fungi
		Bel-Pao	Metazoans
		Retrovirus	Metazoans
		ERV	Metazoans
	DIRS	DIRS	Plants, metazoans, fungi
		Ngaro	Metazoans, fungi
		VIPER	Trypanosomes
	PLE	Penelope	Plants, metazoans, fungi
	LINE	R2	Metazoans
		RTE	Metazoans
		Jockey	Metazoans
		L1	Plants, metazoans, fungi
	SINE	tRNA	Plants, metazoans, fungi
		7SL	Plants, metazoans, fungi
		5S	Metazoans
		SVA[a]	Primates
		Retrogenes[a]	Plants, metazoans, fungi
Class II (DNA transposons) Subclass 1	TIR	Tc1-Mariner	Plants, metazoans, fungi
		hAT	Plants, metazoans, fungi
		Mutator	Plants, metazoans, fungi
		Merlin	Metazoans
		Transib	Metazoans, fungi
		P	Plants, metazoans
		PiggyBac	Metazoans
		PIF-harbinger	Plants, metazoans, fungi
		CACTA	Plants, metazoans, fungi
	Crypton	Crypton	Fungi
Class II (DNA transposons) Subclass 2	Helitron	Helitron	Plants, metazoans, fungi
	Maverick	Maverick	Metazoans, fungi

Please note that SVAs and retrogenes are not included in that classification
[a]Not included in the original Wicker classification

particle coat and a *pol* gene that encodes a reverse transcriptase, ribonuclease H, and an integrase, which provide the enzymatic machinery for reverse transcription and integration into the host genome. Reverse transcription occurs within the viral or viral-like particle (GAG) in the cytoplasm, and it is a multistep process [34]. Unlike LTR retrotransposons, exogenous retroviruses contain an *env* gene, which encodes an envelope that facilitates their migration to other cells. Some LTR retrotransposons may contain remnants of an *env* gene, but their insertion capabilities are limited to the originating genome [35]. This would rather suggest that they originated in exogenous retroviruses by losing the *env* gene. However, there is evidence that suggests the contrary, given that

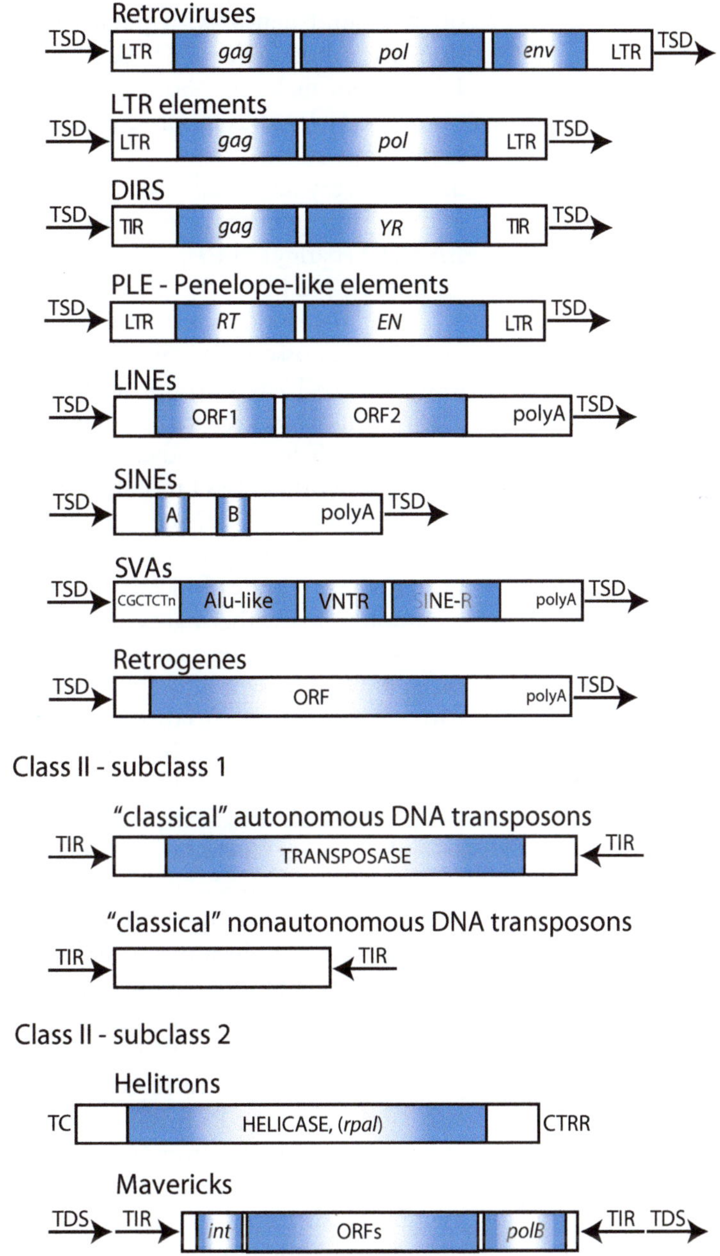

Fig. 2 Structures of eukaryotic mobile elements. See text for detailed discussion

LTR retrotransposons can acquire the *env* gene and become infectious entities [36]. Presently, most of the LTR sequences (85%) in the human genome are found only as isolated LTRs, with the internal sequence being lost most likely due to homologous recombination between flanking LTRs [37]. Interestingly, LTR retrotransposons target their reinsertion to specific genomic sites, often around genes, with putative important functional implications for a host gene [35]. Lander et al. estimated that 450,000 LTR copies make up about 8% of our genome [38]. LTR retrotransposons inhabiting large genomes, such as maize, wheat, or barley, can contain thousands of families. However, despite the diversity, very few families comprise most of the repetitive fraction in these large genomes. Notable examples are Angela (wheat) [39], BARE1 (barley) [40], Opie (maize) [41], and Retrosor6 (sorghum) [42].

The DIRS order clusters structurally diverged group of transposons that possess a tyrosine recombinase (YR) gene instead of an integrase (INT) and do not form target site duplications (TSDs). Their termini resemble either split direct repeats (SDR) or inverted repeats. Such features indicate a different integration mechanism than that of other class I mobile elements. DIRS were discovered in the slime mold (*Dictyostelium discoideum*) genome in the early 1980s [43], and they are present in all major phylogenetic lineages including vertebrates [44]. It has been showed that they are also common in hydrothermal vent organisms [45].

Another order, termed *Penelope*-like elements (PLE), has wide, though patchy distribution from amoebae and fungi to vertebrates with copy number up to thousands per genome [46]. Interestingly, no PLE sequences have been found in mammalian genomes, and apparently they were lost from the genome of *C. elegans* [47]. Although PLEs with an intact ORF have been found in several genomes, including *Ciona* and *Danio*, the only transcriptionally active representative, *Penelope*, is known from *Drosophila virilis*. It causes the hybrid dysgenesis syndrome characterized by simultaneous mobilization of several unrelated TE families in the progeny of dysgenic crosses. It seems that *Penelope* invaded *D. virilis* quite recently, and its invasive potential was demonstrated in *D. melanogaster* [46]. PLEs harbor a single ORF that codes for a protein containing reverse transcriptase (RT) and endonuclease (EN) domains. The PLE RT domain more closely resembles telomerase than the RT from LTRs or LINEs. The EN domain is related to GIY-YIG intron-encoded endonucleases. Some PLE members also have LTR-like sequences, which can be in a direct or an inverse orientation, and have a functional intron [46].

LINEs [48, 49] do not have LTRs; however, they have a poly-A tail at the 3′ end and are flanked by the TSDs. They comprise about 21% of the human genome and among them L1 with about 850,000 copies is the most abundant and best described LINE family. L1 is the only LINE retroposon still active in the human genome [50]. In the human genome, there are two other LINE-like repeats, L2 and L3, distantly related to L1. A contrasting situation has been noticed in the malaria mosquito *Anopheles gambiae*, where around 100 divergent LINE families compose only 3% of its genome [51]. LINEs in plants, e.g., Cin4 in maize and Ta11 in *Arabidopsis thaliana*, seem rare as compared with LTR retrotransposons. A full copy of mammalian L1 is about 6 kb long and contains a PolII promoter and two ORFs. The ORF1 codes for a non-sequence-specific RNA binding protein that contains zinc finger, leucine zipper, and coiled-coil motifs. The ORF1p functions as chaperone for the L1 mRNA [52, 53]. The second ORF encodes an endonuclease, which makes a single-stranded nick in the genomic DNA, and a reverse transcriptase, which uses the nicked DNA to prime reverse transcription of LINE RNA from the 3′ end. Reverse transcription is often unfinished, leaving behind fragmented copies of LINE elements; hence most of the L1-derived repeats are short, with an average size of 900 bp. LINEs are part of the CR1 clade, which has members in various metazoan species, including fruit fly, mosquito, zebrafish, pufferfish, turtle, and chicken [54]. Because they encode their own retrotransposition machinery, LINE elements are regarded as autonomous retrotransposons.

SINEs [48, 49] evolved from RNA genes, such as 7SL and tRNA genes. By definition, they are short, up to 1000 base pair long. They do not encode their own retrotranscription machinery and are considered as nonautonomous elements and in most cases are mobilized by the L1 machinery [55]. The outstanding member of this class from the human genome is the Alu repeat, which contains a cleavage site for the *Alu*I restriction enzyme that gave its name [56]. With over a million copies in the human genome, Alu is probably the most successful transposon in the history of life. Primate-specific Alu and its rodent relative B1 have limited phylogenetic distribution suggesting their relatively recent origins. The mammalian-wide interspersed repeats (MIRs), by contrast, spread before eutherian radiation, and their copies can be found in different mammalian groups including marsupials and monotremes [57]. SVA elements are unique primate elements due to their composite structure. They are named after their main components: SINE, VNTR (a variable number of tandem repeats), and Alu [58]. Usually, they contain the hallmarks of the retroposition, i.e., they are flanked by TSDs and terminated by a poly(A) tail. It seems that SVA elements are nonautonomous retrotransposons mobilized by L1 machinery, and they are thought to be transcribed by RNA polymerase II. SVAs are transpositionally active and are responsible for some human diseases [59]. They originated less than 25 million years ago, and they form the youngest retrotransposon family with about 3000 copies in the human genome [58].

Retro(pseudo)genes are a special group of retroposed sequences, which are products of reverse transcription of a spliced (mature) mRNA. Hence, their characteristic features are an absence of promoter sequence and introns, the presence of flanking direct repeats, and a 3′-end polyadenosine tract [60]. Processed pseudogenes, as sometimes retropseudogenes are called, have been generated in vitro at a low frequency in the human HeLa cells via mRNA from a reporter gene [60]. The source of the reverse transcription machinery in humans and other vertebrates seems to be active L1 elements [61]. However, not all retroposed messages have to end up as pseudogenes. About 20% of mammalian protein-encoding genes lack introns in their ORFs [62]. It is conceivable that many genes lacking introns arose by retroposition. Some genes are known to be retroposed more often than others. For instance, in the human genome there are over 2000 retropseudogenes of ribosomal proteins [63]. A genome-wide study showed that the human genome harbors about 20,000 pseudogenes, 72% of which most likely arose through retroposition [64]. Interestingly, the vast majority (92%) of them are quite recent transpositions that occurred after primate/rodent divergence [64]. Some of the retroposed genes may undergo quite complicated evolutionary paths. An example could be the RNF13B retrogene, which replaced its own parental gene in the mammalian genomes. This retrocopy was duplicated in primates, and the evolution of this primate-specific copy was accompanied by the exaptation of two TEs, Alu and L1, and intron gain via changing a part of coding sequence into an intron leading to the origin of a functional, primate-specific retrogene with two splicing variants [65].

3.2.2 Class II: Mobile Elements

Class II elements move by a conservative cut-and-paste mechanism; the excision of the donor element is followed by its reinsertion elsewhere in the genome. DNA transposons are abundant in bacteria, where they are called insertion sequences (*see* Subheading 3.1), but are present in all phyla. Wicker et al. distinguished two subclasses of DNA transposons based on the number of DNA strands that are cut during transposition [33].

Classical "cut-and-paste" transposons belong to the subclass I, and they are classified as the TIR order. They are characterized by terminal inverted repeats (TIR) and encode a transposase that binds near the inverted repeats and mediates mobility. This process is not usually a replicative one, unless the gap caused by excision is repaired using the sister chromatid. When inserted at a new location, the transposon is flanked by small gaps, which, when filled by host enzymes, cause duplication of the sequence at the target site. The length of these TSDs is characteristic for particular transposons. Nine superfamilies belong to the TIR order, including *Tc1-Mariner*, *Merlin*, *Mutator*, and *PiggyBac*. The second order Crypton consists of a single superfamily of the same name. Originally thought to be limited to fungi [66], now it is clear that they have a wide distribution, including animals and heterokonts [67]. A heterogeneous, small, nonautonomous group of elements MITEs also belong to the TIR order [68], which in some genomes amplified to thousands of copies, e.g., *Stowaway* in the rice genome [69], *Tourist* in most bamboo genomes [70], or *Galluhop* in the chicken genome [71].

Subclass II includes two orders of TEs that, just as those from subclass I, do not form RNA intermediates. However, unlike “classical” DNA transposons, they replicate without double-strand cleavage. Helitrons replicate using a rolling-circle mechanism, and their insertion does not result in the target site duplication [72]. They encode tyrosine recombinase along with some other proteins. Helitrons were first described in plants, but they are also present in other phyla, including fungi and mammals [73, 74]. Mavericks are large transposons that have been found in different eukaryotic lineages excluding plants [75]. They encode various numbers of proteins that include DNA polymerase B and an integrase. Kapitonov and Jurka suggested that their life cycle includes a single-strand excision, followed by extrachromosomal replication and reintegration to a new location [76].

4 Identification of Transposable Elements

With the ever-growing number of sequenced genomes from different branches of the tree of life, there are increasing TE research opportunities. There are several reasons why one would like to analyze TEs and their “offsprings” left in a genome. First of all, they are very interesting biological subjects to study genome structure, gene regulation, or genome evolution. In some cases, they also make genome assembly and annotation quite challenging, especially with the current NGS technology that generates reads shorter than TEs. Nevertheless, TEs should be and are worthy to study. However, it is not a simple task and requires different approaches depending on the level of analysis. We will walk through these different levels starting with raw genome sequences without any annotation and discuss different methods and software used for TE analyses. In principle, we can imagine two scenarios: in the first one, genomic or transcriptome sequences are coming from a species for which there is already some information about the transposon repertoire, for instance, a related genome has been previously characterized or TEs have been studied before. In the second scenario, we have to deal with a completely unknown genome or a genome for which little information exists with regard to TEs. In the former case, one can apply a range of techniques used in comparative genomics or try to search specific libraries of transposons using the “homology search” approach. In the latter, which is basically an approach to identify TEs de novo, first we need to find any repeats in a genome and then attempt characterization and classification of newly identified repetitive sequences. In this approach, we will find *any* repeats, not necessarily transposons. There are many algorithms, and even more software, that can be applied in both approaches.

4.1 De Novo Approaches to Finding Repetitive Elements

There are several steps involved in the de novo characterization of transposons. First, we need to find all the repeats in a genome, then build a consensus of each family of related sequences, and finally classify detected sequences. For the first step, three groups of algorithms exist: the *k*-mer approach, sequence self-comparison, and periodicity analysis.

In the *k*-mer approach, sequences are scanned for overrepresentation of strings of certain length. The idea is that repeats that belong to the same family are compositionally similar and share some oligomers. If the repeats occur many times in a genome, then those oligomers should be overrepresented. However, since repeats and transposons in particular are not perfect copies of a certain sequence, some mismatches must be allowed when oligo frequencies are calculated. The challenge is to determine optimal size of an oligo (*k*-mer) and number of mismatches allowed. Most likely, these parameters should be different for different types of transposons, i.e., low versus high copy number, old versus young transposons, and those from different classes and families. Several programs have been developed based on the *k*-mer idea using a suffix tree data structure including REPuter [77, 78], Vmatch (Kurtz, unpublished; http://www.vmatch.de/), and Repeat-match [79, 80]. Another approach is to use fixed length *k*-mers as seeds and extend those seeds to define repeat's family as it was implemented in ReAS [81], RepeatScout [82], and Tallymer [83]. Another interesting algorithm can be found in the FORRepeats software [84], which uses *factor oracle* data structure [85]. It starts with detection of exact oligomers in the analyzed sequences, followed by finding approximate repeats and their alignment.

The second group of programs developed for de novo detection of repeated sequences is using self-comparison approach. Repeat Pattern Toolkit [86], RECON [87], PILER [88, 89], and BLASTER [90] belong to this group. The idea is to use one of the fast sequence similarity tools, e.g., BLAST [91], followed by clustering search results. The programs differ in the search engine for the initial step, though most are using some of the BLAST algorithms, the clustering method, and heuristics of merging initial hits into a prototype element. For instance, RECON [87], which was developed for the repeat finding in unassembled sequence reads, starts with an all-to-all comparison using WU-BLAST engine. Then, single-linkage clustering is applied to alignment results that is followed by construction of an undirected graph with overlapping. The shortest sequence that contains connected images (aligned subsequences) creates a prototype element. However, this procedure might result in composite elements. To avoid this, all the images are aligned to the prototype element to detect potential illegitimate mergers and split those at every point with a significant number of image ends.

PILER [88, 89] is using a different approach to find initial clusters. Instead of BLAST, it uses PALS (pairwise alignment of long sequences) for the initial alignment. PALS records only hit points and uses banded search of the defined maximum distance to optimize its performance. To further improve performance of the system, PILER uses different heuristics for different types of repeats, i.e., satellites, pseudosatellites, terminal repeats, and interspersed repeats. Finally, a consensus sequence is generated from a multiple sequence alignment of the defined family members.

Dot matrix is a simple method to compare two biological sequences. The graphical output of such an analysis is called a dotplot. Dotplots can be used to detect conserved domains, sequence rearrangements, RNA secondary structure, or repeated sequences. It compares every residue in one sequence to every residue in the other sequence or to every residue of the same sequence in the self-comparison mode. In the latter case, there will be a main diagonal line representing a perfect match and a number of short diagonal lines representing similar regions (red circles in Fig. 3). Interestingly, simple repeats appear as diamond shapes on a main diagonal line or short vertical and horizontal lines outside the main diagonal line (red squares in Fig. 3). The method was introduced to biological analyses almost a half century ago [92, 93]. However, the first easy-to-use software with a graphical interface, DOTTER, was developed much later [94]. The major problem of this approach is the time required for the dotplot calculation, which is of quadratic complexity. This proved to be prohibitive for comparison of the genome-size sequences. One of the solutions to this problem is using a word index for the fast identification of substrings. Gepard implements the suffix array data structure to improve the execution time [95]. It is written in Java, which makes it platform-independent. Gepard enables analyses of sequences at the mega-base level in the matter of seconds, and it takes about an hour to analyze the whole human chromosome I [95]. The example of the dotplot produced by the Gepard is presented in Fig. 3.

4.2 Transposable Elements Determination in NGS Data

With constant improvement of sequencing technology associated with decreasing sequencing cost, the number of new sequenced genomes is exploding. As of January 2019, there are more than 7000 eukaryotic and almost 180,000 prokaryotic genomes publicly available (information retrieved on January 16, 2019, from https://www.ncbi.nlm.nih.gov/genome/browse/). However, this comes with a price; most of the recently sequenced genomes, due to the short read sequencing technology, are available at various levels of

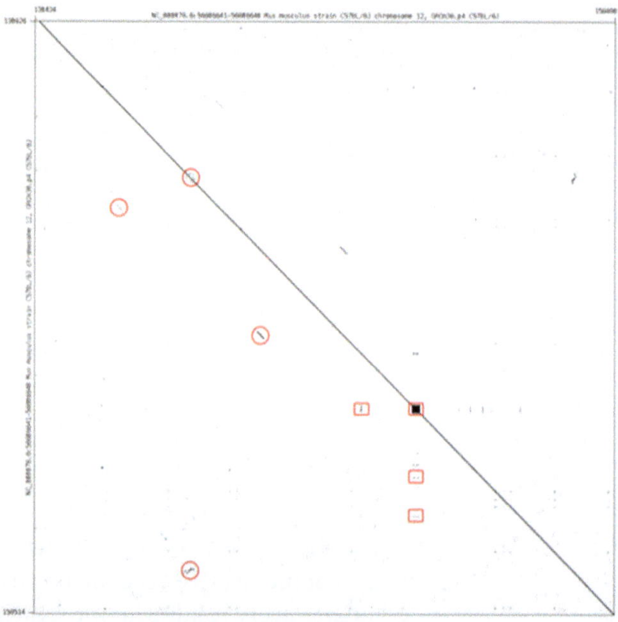

Fig. 3 Graphical output of the Gepard. A 30 kb fragment of mouse chromosome 12 was compared to itself. Similar sequences are represented by diagonal lines if both fragments are located on the same strains or by reverse diagonal lines if the fragments with significant similarity are located on opposite strands. Some of the examples are marked with the red circles. Simple repeats are represented by either diamond shapes on the main diagonal or horizontal and vertical lines. Some of the examples are marked with the red squares

"completeness" or assembly. For most non-model organisms, we are presented with draft assemblies of rather short contigs. Moreover, these genomes usually are not very well annotated, with TEs not being on the annotation priority list. Unfortunately, genome annotation pipelines do not include TE annotation, focusing on protein-coding and RNA-coding genes. To fill the gap, a number of methods have been developed to detect repeats from short reads. Two algorithms dominate in attempts to determine repeats in NGS raw reads: clustering and *k*-mer. Transposome [96] and RepeatExplorer [97] employ the former approach, while RepARK [98], REPdenovo [99], and dnaPipeTE [100] utilize the latter one. Since NGS results in the relatively short reads, assembly of selected sequences into longer contigs representing TEs is required after initial clustering of the raw reads.

4.3 Population-Level Analyses of Transposable Elements

Recent advances in sequencing technology and the sharp decrease in sequencing costs allow genomic studies at population level. Although initially focused on human populations [101–103], recent population studies of other species have been initiated as well [104, 105]. One of the common questions in such studies is

Actual genomic fragment with a TE insertion

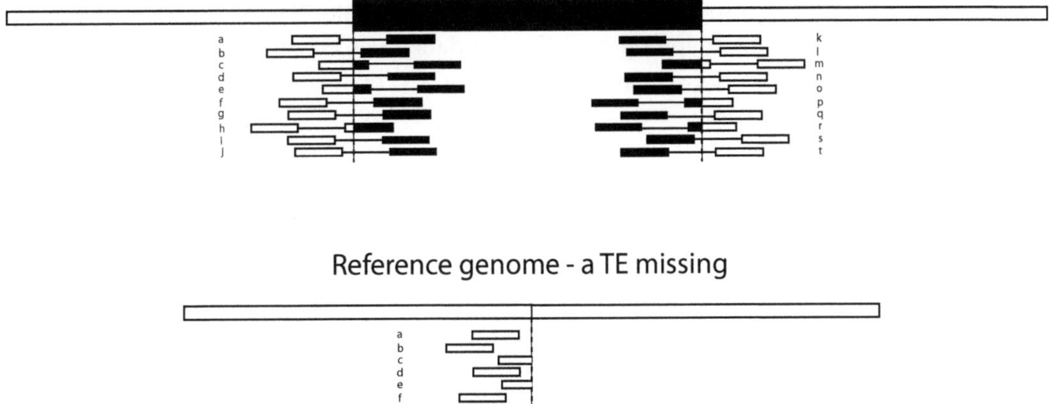

Reference genome - a TE missing

a
b
c
d
e
f
g
h
i
j
k
l
m
n
o
p
q
r
s
t

Fig. 4 Detection of a TE insertion (polymorphic TE) from the NGS data. The upper panel shows real genomic sequence with a TE, which is not present in the reference genome (lower panel). Hypothetical discordant pair-reads (a, b, d, f, g, i, j, k, l, o, q, s, and t) have only one the pairs mapped to the reference genome, while the other would map to a consensus sequence of a TE. The hypothetical split reads (c, e, h, m, p, and r) will have part of the sequence mapped to the reference genome and the other to a TE consensus sequence

how much structural variation (SV) exists in different populations. TE insertions are responsible for about 25% of structural variants in human genomes [106]. In general, any tool designed for detection of SV should work for TE insertion analysis, but specialized software can take advantage of specific expectations related to insertions of TEs. Most of the SV-detection algorithms rely on paired-end reads and are based on discordant read pair mapping and/or split reads mapping (Fig. 4). A discordant pair read is defined as one that is inconsistent with the expected insert size in the library used for sequencing. For example, if the insert size of the library used for sequencing is 300 nt but the reads map to a reference genome within much larger distance or to two different chromosomes, such a pair is considered to be discordant. If, additionally, one of the reads maps to a TE, it might be an indication of a polymorphic TE. Usually some filtering is used to reduce a chance of false positives. These include minimum read number in the cluster mapped to a unique position, quality score of the reads, or consistency in reads orientation. However, the discordant read mapping cannot detect exact insertion position. Therefore another step is required that may include local assembly and split-read mapping.

A split read is defined as a read for which part of it maps uniquely to one position in the genome and the other part to another position. This is, for example, a very common feature of the mapping of RNA-seq data to eukaryotic genomes when reads span two exons. Split reads are being also observed if structural variants exist. In a case of a TE insertion, a part of the read will be mapped to a unique location and the rest to a TE in some other location or may not be mapped at all (Fig. 4).

Different methods for structure variant detection return different results on the same data. Recently published benchmarking demonstrates that TE detection is not an exception [107, 108]. Ewing [107] compared TranspoSeq [109] with two other tools, Tea [110] and TraFIC [111], on the same data sets. Results were not very encouraging as in both comparisons there was a high fraction of insertions detected only by a single program [107]. Similar conclusion was drawn by Rishishwar et al. [108] in a benchmark of larger number of tools including MELT [106], Mobster [112], and RetroSeq [113]. It is clear that different software have different biases, and each one can produce a high number of false positives. It is recommended then to employ several programs for high confidence results. Exhaustive tests run on real and simulated human genome data showed superior performance of MELT [106, 108]. TIPseqHunter is another tool developed to identify transposon insertion sites based on the transpose insertion profiling using next-generation sequencing [114]. It employs machine learning algorithm to ensure high precision and reliability. It is worth to note that all these tools were designed for short read sequencing methods. However, with current development of single-molecule long reads, sequencing technologies such as PacBio and Oxford Nanopore may make these methods irrelevant and obsolete. Long reads should be of superior performance and make TE insertion detection relatively easy with more traditional aligners, such as MegaBLAST [115], BLAT [116], or LAST [117].

4.4 Comparative Genomics of TE Insertions

To understand the general pattern of TE insertions in different genomes and evolutionary dynamics of TE families, a comparative approach is necessary. Although precomputed alignments of different genomes are publicly available, for example, the UCSC Genome Browser includes Multiz alignments of 100 vertebrate genomes [118], not many tools are available for such analyses. One of them is GPAC (genome presence/absence compiler) that creates a table of presence and absence of certain elements based on the precomputed multiple genomes alignment [119] (http://bioinformatics.uni-muenster.de/tools/gpac/index.hbi). The tool is quite generic, but is well suited for the TE comparative analysis (*see* Fig. 5 for an example).

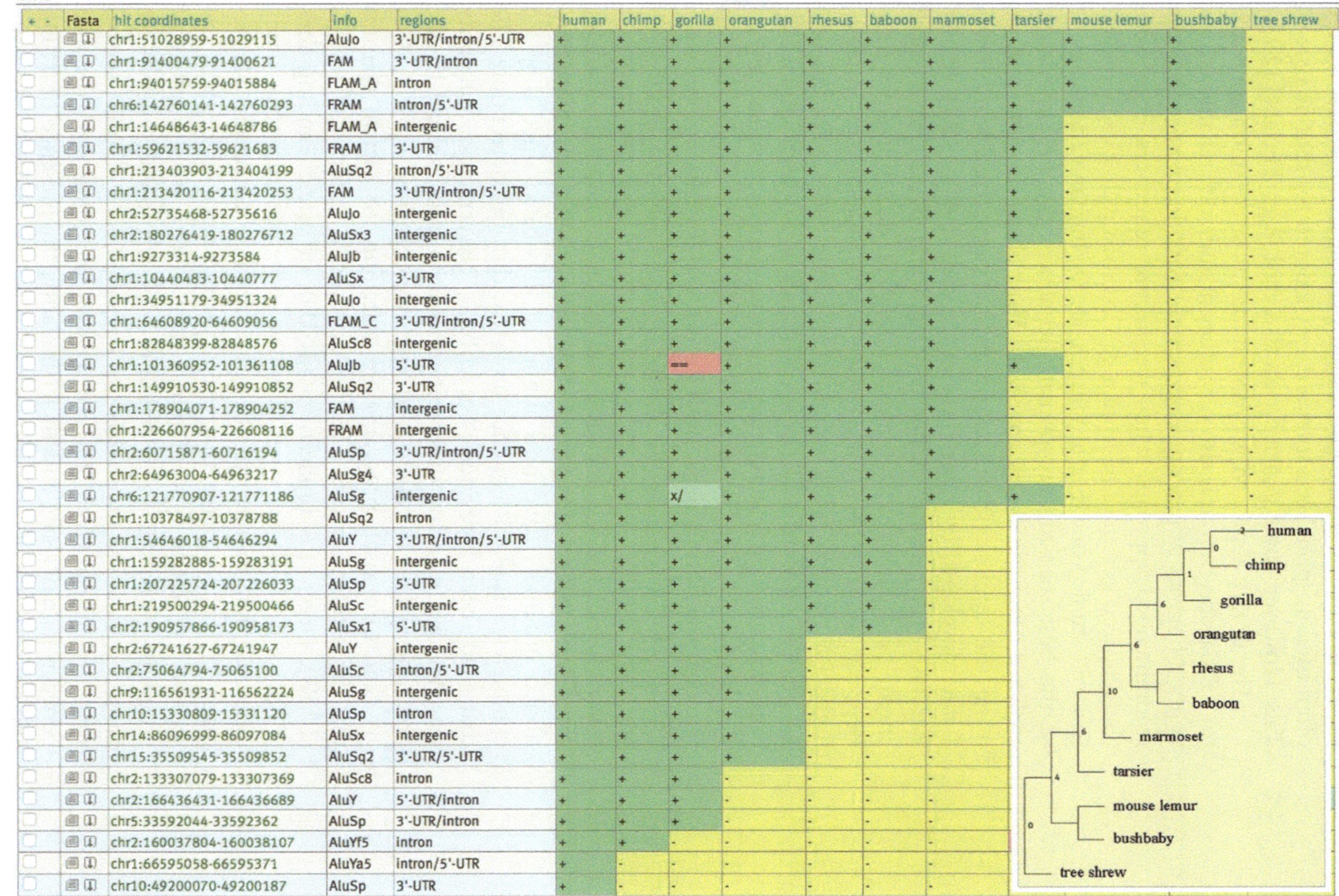

GPAC — Testcase

Fasta	hit coordinates	info	regions	human	chimp	gorilla	orangutan	rhesus	baboon	marmoset	tarsier	mouse lemur	bushbaby	tree shrew
	chr1:51028959-51029115	AluJo	3'-UTR/intron/5'-UTR	+	+	+	+	+	+	+	+	+	+	-
	chr1:91400479-91400621	FAM	3'-UTR/intron	+	+	+	+	+	+	+	+	+	+	-
	chr1:94015759-94015884	FLAM_A	intron	+	+	+	+	+	+	+	+	+	+	-
	chr6:142760141-142760293	FRAM	intron/5'-UTR	+	+	+	+	+	+	+	+	+	+	-
	chr1:14648643-14648786	FLAM_A	intergenic	+	+	+	+	+	+	+	+	-	-	-
	chr1:59621532-59621683	FRAM	3'-UTR	+	+	+	+	+	+	+	+	-	-	-
	chr1:213403903-213404199	AluSq2	intron/5'-UTR	+	+	+	+	+	+	+	+	-	-	-
	chr1:213420116-213420253	FAM	3'-UTR/intron/5'-UTR	+	+	+	+	+	+	+	+	-	-	-
	chr2:52735468-52735616	AluJo	intergenic	+	+	+	+	+	+	+	+	-	-	-
	chr2:180276419-180276712	AluSx3	intergenic	+	+	+	+	+	+	+	+	-	-	-
	chr1:9273314-9273584	AluJb	intergenic	+	+	+	+	+	+	+	-	-	-	-
	chr1:10440483-10440777	AluSx	3'-UTR	+	+	+	+	+	+	+	-	-	-	-
	chr1:34951179-34951324	AluJo	intergenic	+	+	+	+	+	+	+	-	-	-	-
	chr1:64608920-64609056	FLAM_C	3'-UTR/intron/5'-UTR	+	+	+	+	+	+	+	-	-	-	-
	chr1:82848399-82848576	AluSc8	intergenic	+	+	+	+	+	+	+	-	-	-	-
	chr1:101360952-101361108	AluJb	5'-UTR	+	+	==	+	+	+	+	+	-	-	-
	chr1:149910530-149910852	AluSq2	3'-UTR	+	+	+	+	+	+	+	-	-	-	-
	chr1:178904071-178904252	FAM	intergenic	+	+	+	+	+	+	+	-	-	-	-
	chr1:226607954-226608116	FRAM	intergenic	+	+	+	+	+	+	+	-	-	-	-
	chr2:60715871-60716194	AluSp	3'-UTR/intron/5'-UTR	+	+	+	+	+	+	+	-	-	-	-
	chr2:64963004-64963217	AluSg4	3'-UTR	+	+	+	+	+	+	+	-	-	-	-
	chr6:121770907-121771186	AluSg	intergenic	+	+	x/	+	+	+	+	+	-	-	-
	chr1:10378497-10378788	AluSq2	intron	+	+	+	+	+	+	-				
	chr1:54646018-54646294	AluY	3'-UTR/intron/5'-UTR	+	+	+	+	+	+	-				
	chr1:159282885-159283191	AluSg	intergenic	+	+	+	+	+	+	-				
	chr1:207225724-207226033	AluSp	5'-UTR	+	+	+	+	+	+	-				
	chr1:219500294-219500466	AluSc	intergenic	+	+	+	+	+	+	-				
	chr2:190957866-190958173	AluSx1	5'-UTR	+	+	+	+	+	+	-				
	chr2:67241627-67241947	AluY	intergenic	+	+	+	+	-	-	-				
	chr2:75064794-75065100	AluSc	intron/5'-UTR	+	+	+	+	-	-	-				
	chr9:116561931-116562224	AluSg	intergenic	+	+	+	+	-	-	-				
	chr10:15330809-15331120	AluSp	intron	+	+	+	+	-	-	-				
	chr14:86096999-86097084	AluSx	intergenic	+	+	+	+	-	-	-				
	chr15:35509545-35509852	AluSq2	3'-UTR/5'-UTR	+	+	+	+	-	-	-				
	chr2:133307079-133307369	AluSc8	intron	+	+	+	-	-	-	-				
	chr2:166436431-166436689	AluY	5'-UTR/intron	+	+	+	-	-	-	-				
	chr5:33592044-33592362	AluSp	3'-UTR/intron	+	+	+	-	-	-	-				
	chr2:160037804-160038107	AluYf5	intron	+	+	-	-	-	-	-				
	chr1:66595058-66595371	AluYa5	intron/5'-UTR	+	-	-	-	-	-	-				
	chr10:49200070-49200187	AluSp	3'-UTR	+	-	-	-	-	-	-				

Fig. 5 The output table of the GPAC software. Several Alu elements were analyzed for presence/absence in 11 primate species. The human genome was used a reference, and "hit coordinates" refer to that genome along with the information on the annotated elements in the hit region and a type of the region. For each genome, the presence (+) or absence (−) of the hit is presented. x/ denotes that only part of the original insertion (less than 20%) is present in a given genome, and == indicates that more than 80% of the expected sequence is not alignable in a given locus. The optional phylogenetic tree constructed based on the obtained data is shown in the lower right corner

4.5 Classification of Transposable Elements

Once the consensus of a repetitive element has been constructed, it can be subjected to further analyses. There are two major categories of programs dealing with the issue of TE classification: library or similarity-based and signature-based. The latter approach is very often used in specialized software, i.e., tailored for specific type of TEs. However, some general tools also exist, e.g., TEclass [120].

The library approach is probably the most common approach for TE classification. It is also very efficient and quite reliable as long as good libraries of prototype sequences exist. In practice, it is the recommended approach when we analyze sequences from well-characterized genomes or from a genome relatively closely related to a well-studied one. For instance, since the human genome is one of the best studied, any primate sequences can be confidently analyzed using the library approach. Most likely, the first software using the similarity-based approach for repeat classification was *Censor* developed by Jerzy Jurka in the early 1990s [121]. It uses RepBase [122] as a reference collection and BLAST as a search engine [91]. However, the most popular TE detection software is RepeatMasker (RM) (http://www.repeatmasker.org). Interestingly, RM is also using RepBase as a reference collection and AB-BLAST, RM-BLAST, or cross-match as a search engine. In both cases, original search hits are processed by a series of Perl scripts to determine the structure of elements and classify them to one of known TE families. Both Censor and RM also employ user-provided libraries, including “third-party” lineage-specific libraries, e.g., TREP [123]. Over the years, RepeatMasker has become a standard tool for TE analyses, and often its output is used for more biologically oriented studies (see below). The aforementioned programs have one important drawback: since they are completely based on sequence similarity, they can detect only TEs that had been previously described. Nevertheless, similarity searches, like in many other bioinformatics tasks, should be the first approach for the analysis of repetitive elements.

Signature-based programs are searching for certain features that characterize specific TEs, for example, long terminal repeats (LTRs), target site duplications (TSDs), or primer-binding sites (PBSs). Since different types (families) of elements are structurally different, they require specific rules for their detection. Hence, many of the programs that use signature-based algorithms are specific for certain type of transposons. There are a number of programs specialized in detection of LTR transposons, which are based on a similar methodology. They take into account several structural features of LTR retroposons including size, distance between paired LTRs and their similarity, the presence of TSDs, and the presence of replication signals, i.e., the primer-binding site and the polypurine tract (PPTs). Some of the programs check also for ORFs coding for the *gag*, *pol*, and *env* proteins. LTR_STRUC [124] was one of the first programs based on this principle. It uses seed-and-extend strategy to find repeats located within user-defined distance. The candidate regions are extended based on the pairwise alignment to determine cognate LTRs' boundaries. Putative full-length elements are scored based on the presence of TSD, PBS, PPT, and reverse transcriptase ORF. However, because of the heuristics described above, LTR_STRUC is unable to find incomplete LTR transposons and in particular solo LTRs. Another limitation of this program is its Windows-only implementation that significantly prohibits automated large-scale analysis. Several other programs have been developed based on similar principles, e.g., LTR_par [125], find_LTR [126], LTR_FINDER [127], and LTRharvest [128]. Lerat tested performance of these programs [129], and although sensitivity of the methods was acceptable (between 40% and 98%), it was at the expense of specificity, which was very poor. In several cases, the number of falsely assigned transposons exceeded the number of correctly detected ones.

Another group of transposons that have a relatively conserved structure are MITEs and Helitrons. Several specialized programs were developed that take advantage of their specific structure. FINDMITE [130] and MUST [131] are tailored for MITEs, while HelitronFinder [132] and HelSearch [133] were developed for Helitron detection.

A further interesting approach to transposon classification was implemented by Abrusan et al. [120] in the software package called TEclass, which classifies unknown TE consensus sequences into four categories, according to their mechanism of transposition: DNA transposons, LTRs, LINEs, and SINEs. The classification uses support vector machines, random forests, learning vector quantization, and predicts ORFs. Two complete sets of classifiers are built using tetramers and pentamers, which are used in two separate rounds of the classification. The software assumes that the analyzed sequence represents a TE and the classification process is binary, with the following steps: forward versus reverse sequence orientation > DNA versus retrotransposon > LTRs versus nonLTRs (for retroelements) > LINEs versus SINEs (for nonLTR repeats). If the different methods of classification lead to conflicting results, TEclass reports the repeat either as unknown or as the last category where the classification methods agree (http://bioinformatics.uni-muenster.de/tools/teclass/index.hbi).

4.6 Pipelines

Recent years witnessed some attempt to create more complex, global analyses systems. One such a system is REPCLASS [134]. It consists of three classification modules: homology (HOM), structure (STR), and target site duplication (TSD). Each module can be run separately or in the pairwise manner, whereas the final step of the analysis involves integration of the results delivered by each module. There is one interesting novelty in the STR module, namely, implementation of *tRNAscan-SE* [135] to detect tRNA-like secondary structure within the query sequence, one of the signatures of many SINE families. The REPPET is another pipeline for TE sequence analyses. It uses "classical" three-step approach for de novo TE identification: self-alignment, clustering, and consensus sequences generation. However, the pipeline is using a spectrum of different methods at each step, followed by a rigorous TE classification step based on recently proposed classification of TEs [136]. Unfortunately, a complex implementation that makes installation and running the system rather difficult limits usage of the pipeline. The classification step seems to be unreliable as it may annotate lineage-specific TEs in wrong taxonomical lineages (Kouzel and Makalowski, unpublished data).

There are other attempts to create comprehensive systems for "repeatome" analysis. One of them is dnaPipeTE developed for mosquito genomes' analyses [100]. Interestingly, dnaPipeTE works on the raw NGS data, which makes the pipeline well suited for genomes with lower sequencing depth. The raw reads are first subjected to *k*-mer count on the sampled data. The sampling of the data to size less than 0.25× of the genome is required to avoid clustering reads representing unique sequences. The determined repetitive reads are assembled into contigs using Trinity [137]. Although Trinity was originally developed for transcriptome assembly from RNA-seq data, it proves to be very useful for TEs assembly from short reads as it can efficiently determine consensus sequences of closely related transposons. In the next step, dnaPipeTE annotates repeats using RepeatMasker with either built-in or user-defined libraries. This is probably the weakest point of the pipeline as it will not annotate any novel TEs, which have no similar sequences present in the provided libraries. It would be useful to complement this step with model-based or machine learning approaches (*see* Subheading 4.5). After contigs' annotation, copy number of the TEs are estimated using BLAST algorithm [91]. Finally, sequence identity between an individual TE and its consensus sequence is used to determine the relative age of the TEs. The pipeline produces a number of output files including several graphs, i.e., pie chart with the relative proportion of the main repeat classes and graph with the number of base pairs aligned on each TE contig and TE age distribution. Overall, the dnaPipeTE is very efficient, outperforming, according to the authors, RepeatExplorer by severalfold [100].

4.7 Meta-analyses

Most of the software developed are focused on the TE discovery and rarely offer more biological oriented analyses. Consequently, researchers interested in TE biology or using TE insertions as tools for another biological investigations need to utilize other resources. One of them is TinT (transposition in transposition), tool that applies maximum likelihood model of TE insertion probability to estimate relative age of TE families [138] (http://bioinformatics.uni-muenster.de/tools/tint/index.hbi). In the first steps, it takes RepeatMasker output to detect nested retroposons. Then, it generates a data matrix that is used by a probabilistic model to estimate chronology and activity period of analyzed families. The method was applied to resolve the evolutionary history of galliformes [139], marsupials [140], lagomorphs [141], squirrel monkey [142], or elephant shark [143].

Another interesting application that takes advantage of TEs is their use for detecting signatures of positive selection [144], a central goal in the field of evolutionary biology. A typical research scenario for this application would be investigating whether a specific TE fragment exapted into resident genomic features, such as proximal and distal enhancers or exons of spliced transcripts, has undergone accelerated evolution that could be indicative of gain of function events. In short, the test first requires the identification of all genomically interspersed TE fragments that are homolog to the TE segment of interest, which can be done through alignments with a family consensus sequence. Based on multi-species genome alignments, a second step involves identification of lineage-specific substitutions in every single homolog fragment, which are then consolidated into a distribution of lineage-specific substitutions that provides the expectation (null distribution) for a segment evolving largely without specific constraints (neutrally). A significantly higher number of lineage-specific substitutions observed in the TE fragment of interest compared to the null distribution could then be interpreted as a molecular signature of adaptive evolution. However, the possibility of confounding molecular mechanisms, such as GC-biased gene conversion [145–147], needs to be evaluated. We note that building the null distribution based only on data from intergenic regions, where transcription-coupled repair is absent, results in a more liberal estimate of the expected substitutions, which in turn leads to a more conservative estimate of the adaptive evolution. Additionally, building the null distribution requires the detection of many homolog fragments, which limits the applicability of the test to TE families with numerous members in a given genome. Prime examples would be human Alu or murine B1 SINEs. In theory, this test could also be used for detecting signatures of purifying selection by searching for fragments depleted of lineage-specific substitutions. However, the low level or complete lack of lineage-specific substitution is characteristic to many TE fragments, obscuring the effect of potential purifying forces.

5 Concluding Remarks

Annoying junk for some, hidden treasure for others, TEs can hardly be ignored [148]. With their diversity and high copy number in most of the genomes, they are not the easiest biological entities to analyze. Nevertheless, recent years witnessed increased interest in TEs. On the one hand, we observe improvement in computational tools specialized in TE analyses. Table 3 lists some of such tools and

Table 3
Selected resources for transposable elements discovery and analyses

Software	Address
AB-BLAST	http://www.advbiocomp.com/blast.html
ACLAME	http://aclame.ulb.ac.be/
BLASTER suite	http://urgi.versailles.inra.fr/index.php/urgi/Tools/BLASTER
Censor	http://www.girinst.org/censor/download.php
DOTTER	http://sonnhammer.sbc.su.se/Dotter.html
DROPOSON	ftp://biom3.univ-lyon1.fr//pub/drosoposon/
find_ltr	http://darwin.informatics.indiana.edu/cgi-bin/evolution/ltr.pl
FINDMITE	http://jaketu.biochem.vt.edu/dl_software.htm
FORRepeats	http://al.jalix.org/FORRepeats/
Gepard	http://cube.univie.ac.at/gepard
HelitronFinder	http://limei.montclair.edu/HT.html
HelSearch	http://sourceforge.net/project/showfiles.php?group_id=260708
HERVd	http://herv.img.cas.cz/
IRF	http://tandem.bu.edu/irf/irf.download.html
LTR_FINDER	http://tlife.fudan.edu.cn/ltr_finder/
LTR_MINER	http://genomebiology.com/2004/5/10/R79/suppl/s7
LTR_par	http://www.eecs.wsu.edu/~ananth/software.htm
MGEScan-LTR	http://darwin.informatics.indiana.edu/cgi-bin/evolution/daphnia_ltr.pl
MGEScan-nonLTR	http://darwin.informatics.indiana.edu/cgi-bin/evolution/nonltr/nonltr.pl
microTranspoGene	http://transpogene.tau.ac.il/microTranspoGene.html
MITE-Hunter	http://target.iplantcollaborative.org/mite_hunter.html
PILER	http://www.drive5.com/piler/
REannotate	http://www.bioinformatics.org/reannotate/index.html
ReAS	ftp://ftp.genomics.org.cn/pub/ReAS/software/
RECON	http://eddylab.org/software/recon/

(continued)

Table 3
(continued)

Software	Address
RepSeek	http://wwwabi.snv.jussieu.fr/public/RepSeek/
RepeatFinder	http://cbcb.umd.edu/software/RepeatFinder/
RepeatMasker	http://www.repeatmasker.org/
RepeatModeler	http://www.repeatmasker.org/RepeatModeler/
RepeatRunner	http://www.yandell-lab.org/software/repeatrunner.html
Repeat-match	http://mummer.sourceforge.net/
REPET	http://urgi.versailles.inra.fr/index.php/urgi/Tools/REPET
RepMiner	http://repminer.sourceforge.net/index.htm
REPuter	http://bibiserv.techfak.uni-bielefeld.de/reputer/
RetroMap	http://www.burchsite.com/bioi/RetroMapHome.html
SMaRTFinder	http://services.appliedgenomics.org/software/smartfinder/
SoyTEdb	http://www.soytedb.org
Spectral Repeat Finder	http://www.imtech.res.in/raghava/srf/
T-lex	http://petrov.stanford.edu/cgi-bin/Tlex.html
Tallymer	http://www.zbh.uni-hamburg.de/Tallymer/
TARGeT	http://target.iplantcollaborative.org/
TEclass	http://www.bioinformatics.uni-muenster.de/tools/teclass/
TE Displayer	http://labs.csb.utoronto.ca/yang/TE_Displayer/
TE nest	http://www.plantgdb.org/prj/TE_nest/TE_nest.html
TESD	http://pbil.univ-lyon1.fr/software/TESD/
TinT	http://www.bioinformatics.uni-muenster.de/tools/tint/
TIPseqHunter	https://github.com/fenyolab/TIPseqHunter
TRANSPO	http://alggen.lsi.upc.es/recerca/search/transpo/transpo.html
TranspoGene	http://transpogene.tau.ac.il/
Transposon-PSI	http://transposonpsi.sourceforge.net/
TRAP	http://www.coccidia.icb.usp.br/trap/tutorials/
TRF	http://tandem.bu.edu/trf/trf.html
TROLL	http://finder.sourceforge.net/
TSDfinder	http://www.ncbi.nlm.nih.gov/CBBresearch/Landsman/TSDfinder/
WikiPoson	http://www.bioinformatics.org/wikiposon/doku.php
VariationHunter	http://compbio.cs.sfu.ca/software-variation-hunter
Vmatch	http://www.vmatch.de/

the up-to-date list can be found at our web site: http://www.bioinformatics.uni-muenster.de/ScrapYard/. On the other hand, improved tools and new technologies enable biologists to explore new research avenues that might lead to novel, fascinating insights into the biology of mobile elements.

References

1. Waring M, Britten RJ (1966) Nucleotide sequence repetition - a rapidly reassociating fraction of mouse DNA. Science 154 (3750):791–794
2. Britten RJ, Kohne DE (1968) Repeated sequences in DNA. hundreds of thousands of copies of DNA sequences have been incorporated into the genomes of higher organisms. Science 161(841):529–540
3. Makalowski W (2001) The human genome structure and organization. Acta Biochim Pol 48(3):587–598
4. C._elegans_Sequencing_Consortium (1998) Genome sequence of the nematode C. elegans: a platform for investigating biology. Science 282(5396):2012–2018
5. SanMiguel P, Tikhonov A, Jin YK, Motchoulskaia N, Zakharov D, Melake-Berhan A, Springer PS, Edwards KJ, Lee M, Avramova Z, Bennetzen JL (1996) Nested retrotransposons in the intergenic regions of the maize genome. Science 274 (5288):765–768
6. Keller EF (1983) A feeling for the organism: the life and work of Barbara McClintock. W.H. Freeman, San Francisco
7. McClintock B (1950) The origin and behavior of mutable loci in maize. Proc Natl Acad Sci U S A 36(6):344–355
8. McClintock B (1951) Chromosome organization and genic expression. Cold Spring Harb Symp Quant Biol 16:13–47
9. McClintock B (1956) Controlling elements and the gene. Cold Spring Harb Symp Quant Biol 21:197–216
10. Malamy MH, Fiandt M, Szybalski W (1972) Electron microscopy of polar insertions in the lac operon of Escherichia coli. Mol Gen Genet 119(3):207–222
11. Ohno S (1972) So much "junk" DNA in our genome. In: Smith HH (ed) Brookhaven symposia in biology, vol 23. Gordon & Breach, New York, pp 366–370
12. Aronson AI, Bolton ET, Britten RI, Cowie DB, Duerksen JD, McCarthy BJ, McQuillen K, Roberts RB (1960) Biophysics. In: Yearbook 59, vol 59. Carnegie Institution of Washington, Washington, pp 229–279
13. Ehret CF, De Haller G (1963) Origin, development and maturation of organelles and organelle systems of the cell surface in Paramecium. J Ultrastruct Res 23(Suppl 6):1–42
14. Brosius J (1991) Retroposons--seeds of evolution. Science 251(4995):753
15. Makalowski W, Mitchell GA, Labuda D (1994) Alu sequences in the coding regions of mRNA: a source of protein variability. Trends Genet 10(6):188–193
16. Jordan IK, Rogozin IB, Glazko GV, Koonin EV (2003) Origin of a substantial fraction of human regulatory sequences from transposable elements. Trends Genet 19(2):68–72. Pii S0168-9525(02)00006-9
17. Thornburg BG, Gotea V, Makalowski W (2006) Transposable elements as a significant source of transcription regulating signals. Gene 365:104–110. https://doi.org/10.1016/j.gene.2005.09.036. S0378-1119(05)00653-0 [pii]
18. Feschotte C (2008) Transposable elements and the evolution of regulatory networks. Nat Rev Genet 9(5):397–405. https://doi.org/10.1038/nrg2337
19. Mita P, Boeke JD (2016) How retrotransposons shape genome regulation. Curr Opin Genet Dev 37:90–100. https://doi.org/10.1016/j.gde.2016.01.001
20. Chuong EB, Elde NC, Feschotte C (2017) Regulatory activities of transposable elements: from conflicts to benefits. Nat Rev Genet 18 (2):71–86. https://doi.org/10.1038/nrg.2016.139
21. Franke V, Ganesh S, Karlic R, Malik R, Pasulka J, Horvat F, Kuzman M, Fulka H, Cernohorska M, Urbanova J, Svobodova E, Ma J, Suzuki Y, Aoki F, Schultz RM et al

(2017) Long terminal repeats power evolution of genes and gene expression programs in mammalian oocytes and zygotes. Genome Res 27(8):1384–1394. https://doi.org/10.1101/gr.216150.116

22. Wang L, Rishishwar L, Marino-Ramirez L, Jordan IK (2017) Human population-specific gene expression and transcriptional network modification with polymorphic transposable elements. Nucleic Acids Res 45 (5):2318–2328. https://doi.org/10.1093/nar/gkw1286
23. Venuto D, Bourque G (2018) Identifying co-opted transposable elements using comparative epigenomics. Develop Growth Differ 60(1):53–62. https://doi.org/10.1111/dgd.12423
24. Mahillon J, Chandler M (1998) Insertion sequences. Microbiol Mol Biol Rev 62 (3):725–774
25. Wilde C, Escartin F, Kokeguchi S, Latour-Lambert P, Lectard A, Clement JM (2003) Transposases are responsible for the target specificity of IS1397 and ISKpn1 for two different types of palindromic units (PUs), Nucleic Acid Res 31(15):4345–4353
26. Derbyshire KM, Grindley NDF (1996) Cis preference of the IS903 transposase is mediated by a combination of transposase instability and inefficient translation. Mol Microbiol 21(6):1261–1272. https://doi.org/10.1111/j.1365-2958.1996.tb02587.x
27. Ichikawa H, Ikeda K, Amemura J, Ohtsubo E (1990) Two domains in the terminal inverted-repeat sequence of transposon Tn3. Gene 86(1):11–17
28. Maekawa T, Amemura-Maekawa J, Ohtsubo E (1993) DNA binding domains in Tn3 transposase. Mol Gen Genet 236 (2–3):267–274
29. Weinert TA, Schaus NA, Grindley ND (1983) Insertion sequence duplication in transpositional recombination. Science 222 (4625):755–765
30. Turlan C, Chandler M (1995) IS1-mediated intramolecular rearrangements: formation of excised transposon circles and replicative deletions. EMBO J 14(21):5410–5421
31. Siguier P, Gourbeyre E, Varani A, Ton-Hoang B, Chandler M (2015) Everyman's guide to bacterial insertion sequences. Microbiol Spectr 3(2):MDNA3-0030-2014. https://doi.org/10.1128/microbiolspec.MDNA3-0030-2014
32. Finnegan DJ (1989) Eukaryotic transposable elements and genome evolution. Trends Genet 5(4):103–107
33. Wicker T, Sabot F, Hua-Van A, Bennetzen JL, Capy P, Chalhoub B, Flavell A, Leroy P, Morgante M, Panaud O, Paux E, SanMiguel P, Schulman AH (2007) A unified classification system for eukaryotic transposable elements. Nat Rev Genet 8(12):973–982. https://doi.org/10.1038/nrg2165. nrg2165 [pii]
34. Hughes SH (2015) Reverse transcription of retroviruses and LTR retrotransposons. Microbiol Spectr 3(2):MDNA3-0027-2014. https://doi.org/10.1128/microbiolspec.MDNA3-0027-2014
35. Kazazian HH Jr (2004) Mobile elements: drivers of genome evolution. Science 303 (5664):1626–1632. https://doi.org/10.1126/science.1089670. 303/5664/1626 [pii]
36. Malik HS, Henikoff S, Eickbush TH (2000) Poised for contagion: evolutionary origins of the infectious abilities of invertebrate retroviruses. Genome Res 10(9):1307–1318
37. Leib-Mosch C, Haltmeier M, Werner T, Geigl EM, Brack-Werner R, Francke U, Erfle V, Hehlmann R (1993) Genomic distribution and transcription of solitary HERV-K LTRs. Genomics 18(2):261–269. https://doi.org/10.1006/geno.1993.1464
38. Lander ES, Linton LM, Birren B, Nusbaum C, Zody MC, Baldwin J, Devon K, Dewar K, Doyle M, FitzHugh W, Funke R, Gage D, Harris K, Heaford A, Howland J et al (2001) Initial sequencing and analysis of the human genome. Nature 409 (6822):860–921. https://doi.org/10.1038/35057062
39. Wicker T, Stein N, Albar L, Feuillet C, Schlagenhauf E, Keller B (2001) Analysis of a contiguous 211 kb sequence in diploid wheat (Triticum monococcum L.) reveals multiple mechanisms of genome evolution. Plant J 26(3):307–316. tpj1028 [pii]
40. Vicient CM, Kalendar R, Anamthawat-Jonsson K, Schulman AH (1999) Structure, functionality, and evolution of the BARE-1 retrotransposon of barley. Genetica 107 (1–3):53–63
41. SanMiguel P, Gaut BS, Tikhonov A, Nakajima Y, Bennetzen JL (1998) The paleontology of intergene retrotransposons of maize. Nat Genet 20(1):43–45. https://doi.org/10.1038/1695
42. Peterson DG, Schulze SR, Sciara EB, Lee SA, Bowers JE, Nagel A, Jiang N, Tibbitts DC, Wessler SR, Paterson AH (2002) Integration of Cot analysis, DNA cloning, and high-throughput sequencing facilitates genome characterization and gene discovery. Genome

Res 12(5):795–807. https://doi.org/10.1101/gr.226102. Article published online before print in April 2002

43. Zuker C, Lodish HF (1981) Repetitive DNA sequences cotranscribed with developmentally regulated Dictyostelium discoideum mRNAs. Proc Natl Acad Sci U S A 78 (9):5386–5390
44. Goodwin TJ, Poulter RT (2001) The DIRS1 group of retrotransposons. Mol Biol Evol 18 (11):2067–2082
45. Piednoel M, Bonnivard E (2009) DIRS1-like retrotransposons are widely distributed among Decapoda and are particularly present in hydrothermal vent organisms. BMC Evol Biol 9:86. https://doi.org/10.1186/1471-2148-9-86
46. Evgen'ev MB, Arkhipova IR (2005) Penelope-like elements - a new class of retroelements: distribution, function and possible evolutionary significance. Cytogenet Genome Res 110(1–4):510–521. https://doi.org/10.1159/000084984
47. Arkhipova IR (2006) Distribution and phylogeny of Penelope-like elements in eukaryotes. Syst Biol 55(6):875–885. https://doi.org/10.1080/10635150601077683
48. Singer MF (1982) Highly repeated sequences in mammalian genomes. Int Rev Cytol 76:67–112
49. Singer MF (1982) SINEs and LINEs: highly repeated short and long interspersed sequences in mammalian genomes. Cell 28 (3):433–434
50. Mills RE, Bennett EA, Iskow RC, Devine SE (2007) Which transposable elements are active in the human genome? Trends Genet 23(4):183–191. https://doi.org/10.1016/j.tig.2007.02.006
51. Biedler J, Tu Z (2003) Non-LTR retrotransposons in the African malaria mosquito, Anopheles gambiae: unprecedented diversity and evidence of recent activity. Mol Biol Evol 20(11):1811–1825. https://doi.org/10.1093/molbev/msg189. msg189 [pii]
52. Martin SL, Cruceanu M, Branciforte D, Wai-Lun Li P, Kwok SC, Hodges RS, Williams MC (2005) LINE-1 retrotransposition requires the nucleic acid chaperone activity of the ORF1 protein. J Mol Biol 348 (3):549–561. https://doi.org/10.1016/j.jmb.2005.03.003
53. Martin SL (2010) Nucleic acid chaperone properties of ORF1p from the non-LTR retrotransposon, LINE-1. RNA Biol 7 (6):706–711
54. Kapitonov VV, Jurka J (2003) Molecular paleontology of transposable elements in the Drosophila melanogaster genome. Proc Natl Acad Sci U S A 100(11):6569–6574. https://doi.org/10.1073/pnas.0732024100
55. Kajikawa M, Okada N (2002) LINEs mobilize SINEs in the eel through a shared 3′ sequence. Cell 111(3):433–444. S0092867402010413 [pii]
56. Houck CM, Rinehart FP, Schmid CW (1979) A ubiquitous family of repeated DNA sequences in the human genome. J Mol Biol 132(3):289–306
57. Jurka J, Zietkiewicz E, Labuda D (1995) Ubiquitous mammalian-wide interspersed repeats (MIRs) are molecular fossils from the mesozoic era. Nucleic Acids Res 23 (1):170–175
58. Wang H, Xing J, Grover D, Hedges DJ, Han KD, Walker JA, Batzer MA (2005) SVA elements: a hominid-specific retroposon family. J Mol Biol 354(4):994–1007. https://doi.org/10.1016/j.jmb.2005.09.085
59. Ostertag EM, Goodier JL, Zhang Y, Kazazian HH (2003) SVA elements are nonautonomous retrotransposons that cause disease in humans. Am J Hum Genet 73 (6):1444–1451. https://doi.org/10.1086/380207
60. Vanin EF (1985) Processed pseudogenes: characteristics and evolution. Annu Rev Genet 19:253–272
61. Maestre J, Tchenio T, Dhellin O, Heidmann T (1995) mRNA retroposition in human cells: processed pseudogene formation. EMBO J 14:6333–6338
62. Kabza M, Ciomborowska J, Makalowska I (2014) RetrogeneDB--a database of animal retrogenes. Mol Biol Evol 31 (7):1646–1648. https://doi.org/10.1093/molbev/msu139
63. Zhang Z, Harrison P, Gerstein M (2002) Identification and analysis of over 2000 ribosomal protein pseudogenes in the human genome. Genome Res 12:1466–1482
64. Torrents D, Suyama M, Zdobnov E, Bork P (2003) A genome-wide survey of human pseudogenes. Genome Res 13:2559–2567
65. Szcześniak MW, Ciomborowska J, Nowak W, Rogozin IB, Makałowska I (2011) Primate and rodent specific intron gains and the origin of retrogenes with splice variants. Mol Biol Evol 28:33–38

66. Goodwin TJ, Butler MI, Poulter RT (2003) Cryptons: a group of tyrosine-recombinase-encoding DNA transposons from pathogenic fungi. Microbiology 149. (Pt 11:3099–3109
67. Kojima KK, Jurka J (2011) Crypton transposons: identification of new diverse families and ancient domestication events. Mob DNA 2 (1):12. https://doi.org/10.1186/1759-8753-2-12
68. Bureau TE, Wessler SR (1994) Stowaway: a new family of inverted repeat elements associated with the genes of both monocotyledonous and dicotyledonous plants. Plant Cell 6(6):907–916. https://doi.org/10.1105/tpc.6.6.907. 6/6/907 [pii]
69. Feschotte C, Swamy L, Wessler SR (2003) Genome-wide analysis of mariner-like transposable elements in rice reveals complex relationships with stowaway miniature inverted repeat transposable elements (MITEs). Genetics 163(2):747–758
70. Zhou MB, Tao GY, Pi PY, Zhu YH, Bai YH, Meng XW (2016) Genome-wide characterization and evolution analysis of miniature inverted-repeat transposable elements (MITEs) in moso bamboo (Phyllostachys heterocycla). Planta 244(4):775–787. https://doi.org/10.1007/s00425-016-2544-0
71. Wicker T, Robertson JS, Schulze SR, Feltus FA, Magrini V, Morrison JA, Mardis ER, Wilson RK, Peterson DG, Paterson AH, Ivarie R (2005) The repetitive landscape of the chicken genome. Genome Res 15 (1):126–136. https://doi.org/10.1101/gr.2438004
72. Kapitonov VV, Jurka J (2001) Rolling-circle transposons in eukaryotes. Proc Natl Acad Sci U S A 98:8714–8719
73. Hood ME (2005) Repetitive DNA in the automictic fungus Microbotryum violaceum. Genetica 124(1):1–10
74. Pritham EJ, Feschotte C (2007) Massive amplification of rolling-circle transposons in the lineage of the bat *Myotis lucifugus*. Proc Natl Acad Sci U S A 104:1895–1900
75. Pritham EJ, Putliwala T, Feschotte C (2007) Mavericks, a novel class of giant transposable elements widespread in eukaryotes and related to DNA viruses. Gene 390(1–2):3–17. https://doi.org/10.1016/j.gene.2006.08.008. S0378-1119(06)00537-3 [pii]
76. Kapitonov VV, Jurka J (2006) Self-synthesizing DNA transposons in eukaryotes. Proc Natl Acad Sci U S A 103:4540–4545
77. Kurtz S, Schleiermacher C (1999) REPuter: fast computation of maximal repeats in complete genomes. Bioinformatics 15 (5):426–427
78. Kurtz S, Choudhuri JV, Ohlebusch E, Schleiermacher C, Stoye J, Giegerich R (2001) REPuter: the manifold applications of repeat analysis on a genomic scale. Nucleic Acids Res 29(22):4633–4642
79. Delcher AL, Kasif S, Fleischmann RD, Peterson J, White O, Salzberg SL (1999) Alignment of whole genomes. Nucleic Acids Res 27(11):2369–2376
80. Delcher AL, Phillippy A, Carlton J, Salzberg SL (2002) Fast algorithms for large-scale genome alignment and comparison. Nucleic Acids Res 30(11):2478–2483
81. Li RQ, Ye J, Li SG, Wang J, Han YJ, Ye C, Wang J, Yang HM, Yu J, Wong GKS, Wang J (2005) ReAS: recovery of ancestral sequences for transposable elements from the unassembled reads of a whole genome shotgun. PLoS Comput Biol 1(4):313–321. https://doi.org/10.1371/Journal.Pcbi.0010043. Artn E43 [pii]
82. Price AL, Jones NC, Pevzner PA (2005) De novo identification of repeat families in large genomes. Bioinformatics 21:I351–I358. https://doi.org/10.1093/Bioinformatics/Bti1018
83. Kurtz S, Narechania A, Stein JC, Ware D (2008) A new method to compute K-mer frequencies and its application to annotate large repetitive plant genomes. BMC Genomics 9:517. https://doi.org/10.1186/1471-2164-9-517
84. Lefebvre A, Lecroq T, Dauchel H, Alexandre J (2003) FORRepeats: detects repeats on entire chromosomes and between genomes. Bioinformatics 19(3):319–326. https://doi.org/10.1093/Bioinformatics/Btf843
85. Crochemore M, Ilie L, Seid-Hilmi E (2006) Factor oracles. In: Ibarra OH, Yen H-C (eds) Implementation and application of automata. Springer, Berlin, pp 78–89
86. Agrawal P, States D (1994) The Repeat Pattern Toolkit (RPT): analyzing the structure and evolution of the *C. elegans* genome. Proc Int Conf Intell Syst Mol Biol 2:9
87. Bao ZR, Eddy SR (2002) Automated de novo identification of repeat sequence families in sequenced genomes. Genome Res 12 (8):1269–1276. https://doi.org/10.1101/Gr.88502
88. Edgar RC, Myers EW (2005) PILER: identification and classification of genomic repeats. Bioinformatics 21:I152–I158. https://doi.org/10.1093/Bioinformatics/Bti1003

89. Edgar RC (2007) PILER-CR: fast and accurate identification of CRISPR repeats. BMC Bioinform 8:18. https://doi.org/10.1186/1471-2105-8-18
90. Quesneville H, Bergman CM, Andrieu O, Autard D, Nouaud D, Ashburner M, Anxolabehere D (2005) Combined evidence annotation of transposable elements in genome sequences. PLoS Comput Biol 1 (2):166–175. Artn E22. https://doi.org/10.1371/Journal.Pcbi.0010022
91. Altschul SF, Gish W, Miller W, Myers EW, Lipman DJ (1990) Basic local alignment search tool. J Mol Biol 215(3):403–410
92. Fitch WM (1969) Locating gaps in amino acid sequences to optimize the homology between two proteins. Biochem Genet 3 (2):99–108
93. Gibbs AJ, McIntyre GA (1970) The diagram, a method for comparing sequences. Its use with amino acid and nucleotide sequences. Eur J Biochem 16(1):1–11
94. Sonnhammer EL, Durbin R (1995) A dot-matrix program with dynamic threshold control suited for genomic DNA and protein sequence analysis. Gene 167(1–2):GC1–G10
95. Krumsiek J, Arnold R, Rattei T (2007) Gepard: a rapid and sensitive tool for creating dotplots on genome scale. Bioinformatics 23 (8):1026–1028. https://doi.org/10.1093/bioinformatics/btm039
96. Staton SE, Burke JM (2015) Transposome: a toolkit for annotation of transposable element families from unassembled sequence reads. Bioinformatics 31(11):1827–1829. https://doi.org/10.1093/bioinformatics/btv059
97. Novak P, Neumann P, Pech J, Steinhaisl J, Macas J (2013) RepeatExplorer: a Galaxy-based web server for genome-wide characterization of eukaryotic repetitive elements from next-generation sequence reads. Bioinformatics 29(6):792–793. https://doi.org/10.1093/bioinformatics/btt054
98. Koch P, Platzer M, Downie BR (2014) RepARK--de novo creation of repeat libraries from whole-genome NGS reads. Nucleic Acids Res 42(9):e80. https://doi.org/10.1093/nar/gku210
99. Chu C, Nielsen R, Wu Y (2016) REPdenovo: inferring de novo repeat motifs from short sequence reads. PLoS One 11(3):e0150719. https://doi.org/10.1371/journal.pone.0150719
100. Goubert C, Modolo L, Vieira C, ValienteMoro C, Mavingui P, Boulesteix M (2015) De novo assembly and annotation of the Asian tiger mosquito (Aedes albopictus) repeatome with dnaPipeTE from raw genomic reads and comparative analysis with the yellow fever mosquito (Aedes aegypti). Genome Biol Evol 7(4):1192–1205. https://doi.org/10.1093/gbe/evv050
101. Genome of the Netherlands Consortium (2014) Whole-genome sequence variation, population structure and demographic history of the Dutch population. Nat Genet 46 (8):818–825. https://doi.org/10.1038/ng.3021
102. 1000 Genomes Project Consortium, Auton A, Brooks LD, Durbin RM, Garrison EP, Kang HM, Korbel JO, Marchini JL, McCarthy S, McVean GA, Abecasis GR (2015) A global reference for human genetic variation. Nature 526(7571):68–74. https://doi.org/10.1038/nature15393
103. UK10K Consortium, Walter K, Min JL, Huang J, Crooks L, Memari Y, McCarthy S, Perry JR, Xu C, Futema M, Lawson D, Iotchkova V, Schiffels S, Hendricks AE, Danecek P et al (2015) The UK10K project identifies rare variants in health and disease. Nature 526(7571):82–90. https://doi.org/10.1038/nature14962
104. Lack JB, Lange JD, Tang AD, Corbett-Detig RB, Pool JE (2016) A thousand fly genomes: an expanded Drosophila genome nexus. Mol Biol Evol 33(12):3308–3313. https://doi.org/10.1093/molbev/msw195
105. Lynch M, Gutenkunst R, Ackerman M, Spitze K, Ye Z, Maruki T, Jia Z (2017) Population genomics of Daphnia pulex. Genetics 206(1):315–332. https://doi.org/10.1534/genetics.116.190611
106. Gardner EJ, Lam VK, Harris DN, Chuang NT, Scott EC, Pittard WS, Mills RE, Genomes Project Consortium, Devine SE (2017) The Mobile Element Locator Tool (MELT): population-scale mobile element discovery and biology. Genome Res 27 (11):1916–1929. https://doi.org/10.1101/gr.218032.116
107. Ewing AD (2015) Transposable element detection from whole genome sequence data. Mob DNA 6:24. https://doi.org/10.1186/s13100-015-0055-3
108. Rishishwar L, Marino-Ramirez L, Jordan IK (2016) Benchmarking computational tools for polymorphic transposable element detection. Brief Bioinform. https://doi.org/10.1093/bib/bbw072
109. Helman E, Lawrence MS, Stewart C, Sougnez C, Getz G, Meyerson M (2014) Somatic retrotransposition in human cancer revealed by whole-genome and exome

sequencing. Genome Res 24(7):1053–1063. https://doi.org/10.1101/gr.163659.113

110. Lee E, Iskow R, Yang L, Gokcumen O, Haseley P, Luquette LJ III, Lohr JG, Harris CC, Ding L, Wilson RK, Wheeler DA, Gibbs RA, Kucherlapati R, Lee C, Kharchenko PV et al (2012) Landscape of somatic retrotransposition in human cancers. Science 337 (6097):967–971. https://doi.org/10.1126/science.1222077
111. Tubio JMC, Li Y, Ju YS, Martincorena I, Cooke SL, Tojo M, Gundem G, Pipinikas CP, Zamora J, Raine K, Menzies A, Roman-Garcia P, Fullam A, Gerstung M, Shlien A et al (2014) Mobile DNA in cancer. Extensive transduction of nonrepetitive DNA mediated by L1 retrotransposition in cancer genomes. Science 345(6196):1251343. https://doi.org/10.1126/science.1251343
112. Thung DT, de Ligt J, Vissers LE, Steehouwer M, Kroon M, de Vries P, Slagboom EP, Ye K, Veltman JA, Hehir-Kwa JY (2014) Mobster: accurate detection of mobile element insertions in next generation sequencing data. Genome Biol 15(10):488. https://doi.org/10.1186/s13059-014-0488-x
113. Keane TM, Wong K, Adams DJ (2013) RetroSeq: transposable element discovery from next-generation sequencing data. Bioinformatics 29(3):389–390. https://doi.org/10.1093/bioinformatics/bts697
114. Tang Z, Steranka JP, Ma S, Grivainis M, Rodic N, Huang CR, Shih IM, Wang TL, Boeke JD, Fenyo D, Burns KH (2017) Human transposon insertion profiling: analysis, visualization and identification of somatic LINE-1 insertions in ovarian cancer. Proc Natl Acad Sci U S A 114(5):E733–E740. https://doi.org/10.1073/pnas.1619797114
115. Chen Y, Ye W, Zhang Y, Xu Y (2015) High speed BLASTN: an accelerated MegaBLAST search tool. Nucleic Acids Res 43 (16):7762–7768. https://doi.org/10.1093/nar/gkv784
116. Kent WJ (2002) BLAT--the BLAST-like alignment tool. Genome Res 12 (4):656–664. https://doi.org/10.1101/gr.229202. Article published online before March 2002
117. Kielbasa SM, Wan R, Sato K, Horton P, Frith MC (2011) Adaptive seeds tame genomic sequence comparison. Genome Res 21 (3):487–493. https://doi.org/10.1101/gr.113985.110
118. Casper J, Zweig AS, Villarreal C, Tyner C, Speir ML, Rosenbloom KR, Raney BJ, Lee CM, Lee BT, Karolchik D, Hinrichs AS, Haeussler M, Guruvadoo L, Navarro Gonzalez J, Gibson D et al (2018) The UCSC Genome Browser database: 2018 update. Nucleic Acids Res 46(D1): D762–D769. https://doi.org/10.1093/nar/gkx1020
119. Noll A, Grundmann N, Churakov G, Brosius J, Makalowski W, Schmitz J (2015) GPAC-genome presence/absence compiler: a web application to comparatively visualize multiple genome-level changes. Mol Biol Evol 32(1):275–286. https://doi.org/10.1093/molbev/msu276
120. Abrusan G, Grundmann N, DeMester L, Makalowski W (2009) TEclass-a tool for automated classification of unknown eukaryotic transposable elements. Bioinformatics 25 (10):1329–1330. https://doi.org/10.1093/bioinformatics/btp084
121. Jurka J, Klonowski P, Dagman V, Pelton P (1996) Censor - a program for identification and elimination of repetitive elements from DNA sequences. Comput Chem 20 (1):119–121
122. Bao W, Kojima KK, Kohany O (2015) Repbase Update, a database of repetitive elements in eukaryotic genomes. Mob DNA 6:11. https://doi.org/10.1186/s13100-015-0041-9
123. Wicker T, Matthews DE, Keller B (2002) TREP: a database for Triticeae repetitive elements. Trends Plant Sci 7(12):561–562. [pii] S1360-1385(02)02372-5
124. McCarthy EM, McDonald JF (2003) LTR_STRUC: a novel search and identification program for LTR retrotransposons. Bioinformatics 19(3):362–367. https://doi.org/10.1093/Bioinformatics/Btf878
125. Kalyanaraman A, Aluru S (2006) Efficient algorithms and software for detection of full-length LTR retrotransposons. J Bioinforma Comput Biol 4(2):197–216. S021972000600203X [pii]
126. Rho M, Choi JH, Kim S, Lynch M, Tang H (2007) De novo identification of LTR retrotransposons in eukaryotic genomes. BMC Genomics 8:90. https://doi.org/10.1186/1471-2164-8-90. 1471-2164-8-90 [pii]
127. Xu Z, Wang H (2007) LTR_FINDER: an efficient tool for the prediction of full-length LTR retrotransposons. Nucleic Acids Res 35 (Web Server issue):W265–W268. https://doi.org/10.1093/nar/gkm286. gkm286 [pii]
128. Ellinghaus D, Kurtz S, Willhoeft U (2008) LTRharvest, an efficient and flexible software

for de novo detection of LTR retrotransposons. BMC Bioinform 9:18. https://doi.org/10.1186/1471-2105-9-18. 1471-2105-9-18 [pii]

129. Lerat E (2010) Identifying repeats and transposable elements in sequenced genomes: how to find your way through the dense forest of programs. Heredity 104(6):520–533. https://doi.org/10.1038/hdy.2009.165. hdy2009165 [pii]

130. Tu Z (2001) Eight novel families of miniature inverted repeat transposable elements in the African malaria mosquito, Anopheles gambiae. Proc Natl Acad Sci U S A 98 (4):1699–1704. https://doi.org/10.1073/pnas.041593198. 041593198 [pii]

131. Chen Y, Zhou F, Li G, Xu Y (2009) MUST: a system for identification of miniature inverted-repeat transposable elements and applications to Anabaena variabilis and Haloquadratum walsbyi. Gene 436(1–2):1–7. https://doi.org/10.1016/j.gene.2009.01.019. S0378-1119(09)00051-1 [pii]

132. Du C, Caronna J, He L, Dooner HK (2008) Computational prediction and molecular confirmation of Helitron transposons in the maize genome. BMC Genomics 9:51. https://doi.org/10.1186/1471-2164-9-51. 1471-2164-9-51 [pii]

133. Yang L, Bennetzen JL (2009) Structure-based discovery and description of plant and animal Helitrons. Proc Natl Acad Sci U S A 106(31):12832–12837. https://doi.org/10.1073/pnas.0905563106. 0905563106 [pii]

134. Feschotte C, Keswani U, Ranganathan N, Guibotsy ML, Levine D (2009) Exploring repetitive DNA landscapes using REPCLASS, a tool that automates the classification of transposable elements in eukaryotic genomes. Genome Biol Evol 1:205–220. https://doi.org/10.1093/Gbe/Evp023

135. Lowe TM, Eddy SR (1997) tRNAscan-SE: a program for improved detection of transfer RNA genes in genomic sequence. Nucleic Acids Res 25(5):955–964

136. Flutre T, Duprat E, Feuillet C, Quesneville H (2011) Considering transposable element diversification in de novo annotation approaches. PLoS One 6(1):e16526. https://doi.org/10.1371/journal.pone.0016526

137. Grabherr MG, Haas BJ, Yassour M, Levin JZ, Thompson DA, Amit I, Adiconis X, Fan L, Raychowdhury R, Zeng QD, Chen ZH, Mauceli E, Hacohen N, Gnirke A, Rhind N et al (2011) Full-length transcriptome assembly from RNA-Seq data without a reference genome. Nat Biotechnol 29(7):644–U130. https://doi.org/10.1038/nbt.1883

138. Churakov G, Grundmann N, Kuritzin A, Brosius J, Makalowski W, Schmitz J (2010) A novel web-based TinT application and the chronology of the Primate Alu retroposon activity. BMC Evol Biol 10:376. https://doi.org/10.1186/1471-2148-10-376. 1471-2148-10-376 [pii]

139. Kriegs JO, Matzke A, Churakov G, Kuritzin A, Mayr G, Brosius J, Schmitz J (2007) Waves of genomic hitchhikers shed light on the evolution of gamebirds (Aves: Galliformes). BMC Evol Biol 7:190. https://doi.org/10.1186/1471-2148-7-190. 1471-2148-7-190 [pii]

140. Nilsson MA, Churakov G, Sommer M, Tran NV, Zemann A, Brosius J, Schmitz J (2010) Tracking marsupial evolution using archaic genomic retroposon insertions. PLoS Biol 8 (7):e1000436. https://doi.org/10.1371/journal.pbio.1000436

141. Kriegs JO, Zemann A, Churakov G, Matzke A, Ohme M, Zischler H, Brosius J, Kryger U, Schmitz J (2010) Retroposon insertions provide insights into deep lagomorph evolution. Mol Biol Evol 27 (12):2678–2681. https://doi.org/10.1093/molbev/msq162. msq162 [pii]

142. Baker JN, Walker JA, Vanchiere JA, Phillippe KR, St Romain CP, Gonzalez-Quiroga P, Denham MW, Mierl JR, Konkel MK, Batzer MA (2017) Evolution of Alu subfamily structure in the Saimiri lineage of new world monkeys. Genome Biol Evol 9 (9):2365–2376. https://doi.org/10.1093/gbe/evx172

143. Luchetti A, Plazzi F, Mantovani B (2017) Evolution of two short interspersed elements in Callorhinchus milii (Chondrichthyes, Holocephali) and related elements in sharks and the coelacanth. Genome Biol Evol 9(6). https://doi.org/10.1093/gbe/evx094

144. Gotea V, Petrykowska HM, Elnitski L (2013) Bidirectional promoters as important drivers for the emergence of species-specific transcripts. PLoS One 8(2):e57323. https://doi.org/10.1371/journal.pone.0057323

145. Kostka D, Hubisz MJ, Siepel A, Pollard KS (2012) The role of GC-biased gene conversion in shaping the fastest evolving regions of the human genome. Mol Biol Evol 29 (3):1047–1057. https://doi.org/10.1093/molbev/msr279

146. Capra JA, Hubisz MJ, Kostka D, Pollard KS, Siepel A (2013) A model-based analysis of GC-biased gene conversion in the human and chimpanzee genomes. PLoS Genet 9(8):

e1003684. https://doi.org/10.1371/journal.pgen.1003684

147. Gotea V, Elnitski L (2014) Ascertaining regions affected by GC-biased gene conversion through weak-to-strong mutational hotspots. Genomics 103(5–6):349–356. https://doi.org/10.1016/j.ygeno.2014.04.001

148. Makalowski W (2000) Genomic scrap yard: how genomes utilize all that junk. Gene 259 (1–2):61–67. https://doi.org/10.1016/S0378-1119(00)00436-4

Part III

Phylogenomics and Genome Evolution

Chapter 7

Modern Phylogenomics: Building Phylogenetic Trees Using the Multispecies Coalescent Model

Liang Liu, Christian Anderson, Dennis Pearl, and Scott V. Edwards

Abstract

The multispecies coalescent (MSC) model provides a compelling framework for building phylogenetic trees from multilocus DNA sequence data. The pure MSC is best thought of as a special case of so-called "multispecies network coalescent" models, in which gene flow is allowed among branches of the tree, whereas MSC methods assume there is no gene flow between diverging species. Early implementations of the MSC, such as "parsimony" or "democratic vote" approaches to combining information from multiple gene trees, as well as concatenation, in which DNA sequences from multiple gene trees are combined into a single "supergene," were quickly shown to be inconsistent in some regions of tree space, in so far as they converged on the incorrect species tree as more gene trees and sequence data were accumulated. The anomaly zone, a region of tree space in which the most frequent gene tree is different from the species tree, is one such region where many so-called "coalescent" methods are inconsistent. Second-generation implementations of the MSC employed Bayesian or likelihood models; these are consistent in all regions of gene tree space, but Bayesian methods in particular are incapable of handling the large phylogenomic data sets currently available. Two-step methods, such as MP-EST and ASTRAL, in which gene trees are first estimated and then combined to estimate an overarching species tree, are currently popular in part because they can handle large phylogenomic data sets. These methods are consistent in the anomaly zone but can sometimes provide inappropriate measures of tree support or apportion error and signal in the data inappropriately. MP-EST in particular employs a likelihood model which can be conveniently manipulated to perform statistical tests of competing species trees, incorporating the likelihood of the collected gene trees on each species tree in a likelihood ratio test. Such tests provide a useful alternative to the multilocus bootstrap, which only indirectly tests the appropriateness of competing species trees. We illustrate these tests and implementations of the MSC with examples and suggest that MSC methods are a useful class of models effectively using information from multiple loci to build phylogenetic trees.

Key words Introgression, Hybridization, Coalescent, Recombination, Neutrality, Molecular evolution

1 Introduction

The concept of a phylogeny or "species tree," a bifurcating dendrogram graphically depicting the relationships among a group species, is one of the oldest and most powerful icons in all of biology. After Charles Darwin sketched the first species tree

Maria Anisimova (ed.), *Evolutionary Genomics: Statistical and Computational Methods*, Methods in Molecular Biology, vol. 1910, https://doi.org/10.1007/978-1-4939-9074-0_7,

(in *Transmutation of Species*, Notebook B, 1837), he remained fascinated by the image for 22 years, eventually including a species tree as the only figure in *On the Origin of Species* [1]. Though species trees reached their aesthetic apogee with Ernst Haeckel's *Tree of Life* in 1886, the pursuit of ever-more scientifically accurate trees has kept phylogenetics a vibrant discipline for the 150 years since.

Because the direct evolution of species in the past is not observable (not even in the fossil record), relationships among species are often inferred by shared characteristics among extant taxa. Until the 1970s, this effort took place almost exclusively by using morphological characters. Although this approach had many successes, the paucity of characters and the challenges of comparing species with no obvious morphological homologies were persistent problems [2, 3]. When molecular techniques were developed in the late 1960s, it soon became clear that the sheer volume of molecular data that could be collected would represent a vast improvement. When DNA sequences became widely available for a range of species [4], molecular comparisons quickly became *de rigueur* [5–8]. Nonetheless, it was recognized early on that molecular phylogenies had their own suite of problems; the concept that not all gene tree topologies would match the true species tree topology (i.e., would not be speciodendric sensu Rosenberg [9]) was implicit in early empirical allozyme and mitochondrial DNA studies [10, 11]. However, it was generally assumed that the idiosyncratic genealogical history of any one gene, as reconstructed from extant mutations, was an acceptable approximation for the true history of the species given the potentially overwhelming quantity and seductive utility of molecular data [12–15]. Indeed, this assumption is still prevalent in the thinking of those who favor concatenation or supermatrix approaches as a means of combining information from multiple genes that may still differ in their genealogy from each other and from the species tree [16, 17]. In the meantime, the term "phylogeny" frequently became conflated with "gene tree," the entity produced by many of the leading phylogenetic packages of the day. The term "species tree," in use since the late 1970s to emphasize the distinction between lineage histories and gene histories (reviewed in [11, 18]), was only gradually acknowledged, despite the fact that species trees are the rightful heirs to the term "phylogeny" and better encapsulate the true goals of molecular and morphological systematics [19].

1.1 Stopgap Approaches to Gene Tree Heterogeneity

By and large, the ensuing decades of molecular phylogenetics has fulfilled much of its potential, revolutionizing taxonomies and resolving conundrums previously considered intractable. However, as the amount of genetic data per species becomes ever-more voluminous, it has become clear that the conflicts between individual genes with each other and with the overarching species tree, both in topology and branch lengths, can have practical consequences for phylogenetic analysis if not dealt with properly [18–23]. At first, some researchers treated this phenomenon as though it were an information problem: when working with only a few mutations, you were bound to occasionally get unlucky and sequence a gene whose random signal of evolution did not match that of the taxa being studied. The reasoning was surely more and/or longer sequences would fix that problem and cause gene trees to converge [16]. However, as more genes were sequenced, and as the properties of gene lineages within populations were studied in detail [24, 25], the twin realities of gene tree heterogeneity and "incomplete lineage sorting" [11] (ILS) became clear (Figs. 1 and 2). The probability of an event such as incomplete lineage sorting, which if considered alone would lead to inferring the wrong species tree, was worked out theoretically for the four allele/two species case first [26], followed by the three allele/three species case [7, 13] and more general cases [12, 27]. Pamilo and Nei [12] were among those that proposed that the solution was to simply acquire more gene sequences, after which the central tendency of this gene set would point to the correct relationships, a "democratic vote" method, where each gene was allowed to propose its own tree, and the topology with the most "votes" was declared the winner and therefore the true phylogeny. Though generally true for three-species case, it can sometimes produce the wrong topology with four or more species [28]. In fact, we now know that with four or more species, there is an "anomaly zone" for species trees with short branch lengths as measured in coalescence units, in which the addition of more genes for sampled taxa is guaranteed to lead to the wrong species tree topology for the democratic vote method [29, 30]. (Coalescent time units, equivalent to t/Ne where t is the number of generations since divergence and Ne is the effective population size of the lineage, are a convenient unit for discussions of gene tree/species tree heterogeneity. For a clear explanation, see Box 2 of Degnan and Rosenberg [28].) Such anomaly zones may be rare empirically [31], but empirical examples are emerging [32, 33], and even the theoretical possibility remains disconcerting. In addition, because the number of possible tree topologies increases as the double factorial of the number of tips, for species trees with more than four tips, a very large number of genes are required to determine which gene tree is in fact the most frequent. Advanced consensus methods [34] can circumvent some of the problems of the democratic vote by using novel assembly methods, such as rooted triple consensus [35], greedy consensus [36], or supertree methods [37]. However, although such methods suffer from lack of a biological model motivating the method of consensus, approaches such as that proposed by Steel and Rodrigo [38] might help approximate the dynamics of biological models while allowing for faster and more flexible extensions and should be further developed.

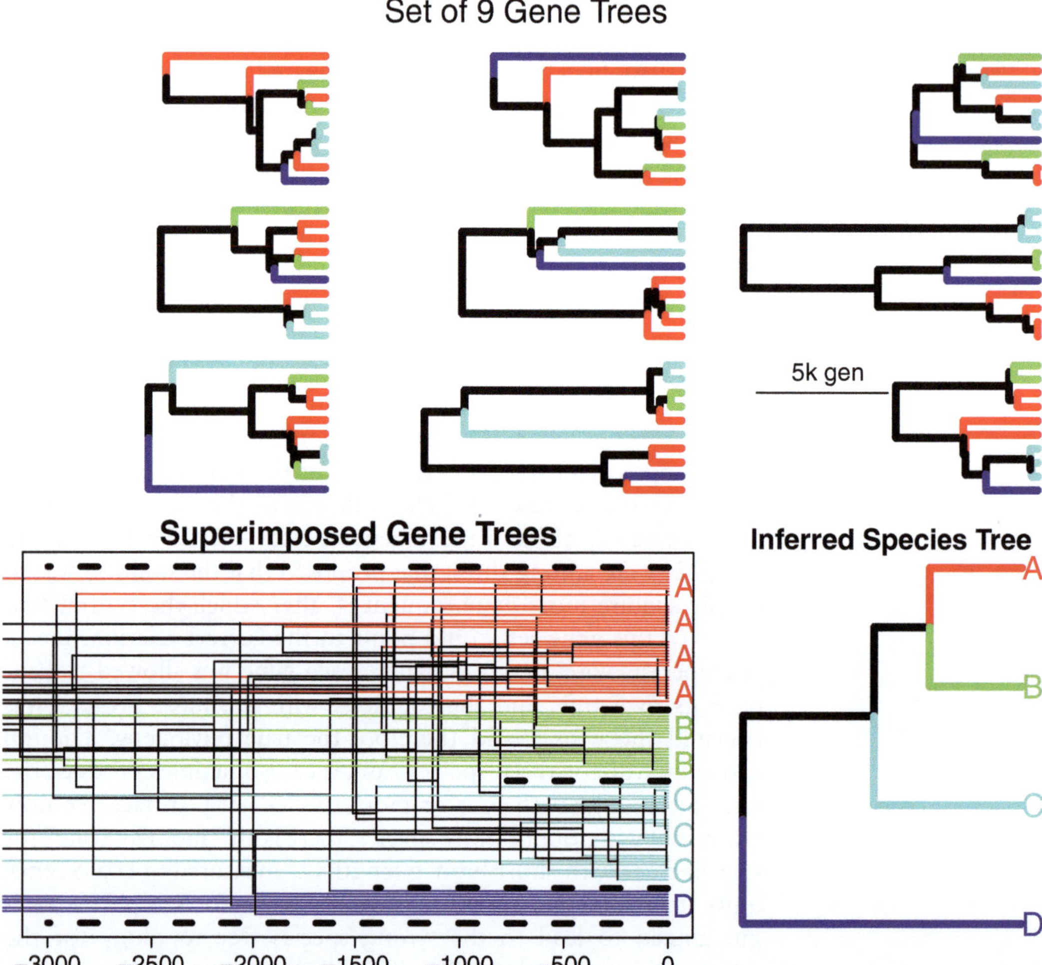

Fig. 1 An example showing the utility of multiple gene trees in producing species tree topologies. (**a**) Nine unlinked loci are simulated (or inferred without error) from a species group with substantial amounts of incomplete lineage sorting. Note that no single gene recovers the correct relationship between clades. Furthermore, despite identical conditions for all nine simulations, no two genes agree on the correct topology, let alone the correct divergence times. (**b**) Superimposing the nine gene trees on top of each other clarifies the relationships. It can be (correctly) inferred that the true tree is perfectly ordered, with (ABC) diverging from D about 1500 generations ago, the (AB)-C split occurring at 800, and A diverging from B about 600 generations ago. Also, the amount of crossbreeding within the recently diverged taxa implies (correctly) that C has the effective smallest population size

The second empirical approach to the problem of conflicting gene trees was to bypass it altogether. Concatenation methods appended the sequence of one gene to that of the next, to create long alignments or supermatrices [39], a technique that in some situations was superior to standard consensus methods in resolving discordance or achieving statistical consistency [40]. But some researchers, including those who questioned the "total-evidence"

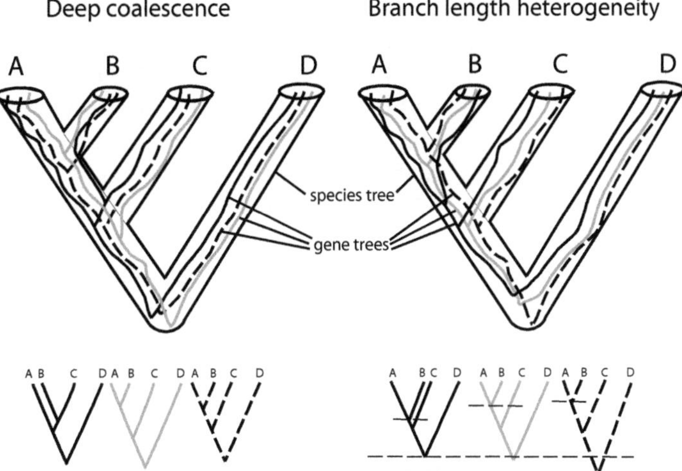

Fig. 2 The relationship between gene trees and species trees. Lines within the species trees indicate gene lineages. Simplified gene trees are shown below each species tree. Whereas gene trees on the left vary due to deep coalescence, gene trees on the right are topologically concordant but vary slightly in branch lengths due to the coalescent. Modified with permission from [19]

approach to systematics (e.g., [41]), advocated against concatenation when, for whatever reason, gene trees appeared to conflict with one another. One problem with the concatenation approach was that it assumed full linkage across the supermatrix, a situation that would obviously not be the case if genes were on different chromosomes. Even when the lineage lengths in a species tree are long in coalescent units, such that gene tree topologies are congruent, the branch lengths of trees of genes on different chromosomes will differ subtly from one another due to the stochasticity of the coalescent process. The early implementations of this method also assumed the same distribution of mutation rates across the sequence, which was clearly not the case if the matrix included coding and noncoding regions. Like democratic vote methods, concatenation of many genes was sometimes defended as sufficient to override the conflicting signal across genes [42, 43], despite widespread acknowledgment that gene tree heterogeneity is ubiquitous and that concatenation can sometimes give the wrong answer, especially although not exclusively in the anomaly zone [44, 45].

Concatenation as a method of combining phylogenomic data still remains popular by default [16, 46], particularly among phylogenetic studies of higher taxa where incomplete lineage sorting is assumed to be rare. However, this logic suffers from two flaws frequently seen in the literature. First, "deep" phylogenetic studies among higher taxa are no more immune to the problems of ILS than are studies among closely related species, because it is the *length* of a given branch, not its *depth* in the tree, that is relevant to probability of gene tree discordance [28]. Detecting such ILS and ruling out gene tree congruence will indeed be more challenging in deep phylogenomic studies, but it should not be assumed that ILS will be less prevalent at deep scales than at shallow scales. Second, current implementations of concatenation represent only one way of species tree construction in which each gene is forced to have the same topology. The real distinction between concatenation and coalescent models is not the presence or absence of ILS but rather the possibility of conditional independence of gene trees as mediated by recombination between genes [47]. Even if all gene trees in an analysis are topologically identical, physically connecting different genes in a single supermatrix does not capture variation in branch lengths that recombination will allow in nature. More effort should be devoted to "supermatrix-like" methods that constrain gene trees to the same topology but allow recombination between genes and conditional independence of branch lengths, since these qualities will influence how signal is accumulated as more genes are added [47]. A final problem with concatenation is that, in a strict sense, concatenation also does not generate species trees, in so far as the method treats all nucleotides as if they were part of a single non-recombining gene, and thus does not distinguish between gene and species trees [19]. In the end, concatenation is best thought of as a special case of more general models of phylogenetic inference that acknowledge gene tree heterogeneity and conditional independence of genes. One such model is the multispecies coalescent model [23, 28, 48]. It is this model that provides the basis for a recent flurry of promising methods that permit efficient and consistent estimation of species trees under a variety of conditions.

2 The Multispecies Coalescent Model

A plausible probabilistic model for analyzing multilocus sequences should involve not only the phylogenetic relationship of species (species tree) but also the genealogical history of each gene (gene tree) and allow different genes to have different histories. Unlike concatenation, such a multispecies coalescent model (MSC) explains the evolutionary history of multilocus sequences through two levels of biological hierarchy, the gene tree and the species tree, rather than just one [23, 49]. Models acknowledging these two levels require an explicit description of how sequences evolve on gene trees, the traditional likelihood equation of Felsenstein [50] and others, as well as how gene trees evolve in the species tree, the likelihood for which was first described by Rannala and Yang [48]. With a few exceptions (described below), the genealogical relationship (gene tree) of neutral alleles can be simply depicted by a coalescence process in which lineages randomly coalesce with each other backward in time. The MSC is a simple application of the single population coalescent model to each branch in a species tree [28]. It holds the standard assumptions found in many neutral coalescent models: no natural selection or gene flow among populations, no recombination within loci but free recombination between loci, random mating and a Wright-Fisher model of inheritance down each branch of the species tree. Despite these seemingly oversimplified assumptions, the pure coalescent model is fundamental in explaining the gene tree-species tree relationship because it forms a baseline for incorporating additional evolutionary forces on top of random drift [28, 49]. More importantly, the pure coalescent model provides an analytic tool to detect the evolutionary forces responsible for the deviation of the observed data (molecular sequences) from those expected from the model.

The coalescent process works, in effect, by randomly choosing ancestors with replacement from the population backward through time for each sequence in the original sample. Eventually, two of these lineages will share a common ancestor, and the lineages are said to "coalesce." The process continues until all lineages coalesce at the most recent common ancestor (MRCA). Multispecies coalescence works the same way but places constraints on how recently the coalescences occur, corresponding to the species' divergence times. Translating this model into computer algorithms for inferring species trees has led to a plethora of models [51–55], some of which first build gene trees by traditional methods and then combine them into a species tree with the highest likelihood or other criteria ("two-step" methods, e.g., [56] or [57]), others of which, particularly Bayesian methods [58–60], simultaneously estimate gene trees and species tree. In general for likelihood or Bayesian approaches, a species tree has been proposed, and the likelihood of each gene tree is evaluated using the MSC, with or without various priors, to evaluate the likelihood of the data (DNA sequences in the case of Bayesian methods or gene trees in the case of likelihood methods like MP-EST [56]) given the species tree or the posterior probability of the species tree. In this way, traditional multispecies coalescent methods are the converse of consensus methods; rather than each locus proposing a potentially divergent species tree, a common species tree is assumed and evaluated, given the sometimes divergent patterns observed among multiple loci.

A number of implementations of this idea have been developed (reviewed by Edwards [19, 54]). Several "two-step" packages are available for moving from independently built gene trees to species trees, including minimization of deep coalescence [61], STEM [62], JIST [63], GLASS [64], STAR, STEAC [65], NJst [66], and ASTRAL [57, 67]. Three methods to date utilize "one-step" Bayesian methods to infer gene trees and the species tree, with the input data being DNA sequences: BEST [58, 68, 69], *BEAST2 [59], and a new model (A00) in the Bayesian Phylogenetics and Phylogeography (bpp) package [70–72]. An additional "one-step" method, SVD Quartets [73], derives species trees directly from aligned, unlinked single-nucleotide polymorphisms using the method of invariants in a coalescent framework. Species tree methods exhibit a number of attractive advantages over concatenation methods in terms of performance. These advantages are not restricted to the anomaly zone, occur across broad regions of tree space, and include less susceptibility to long-branch attraction [74] and missing data [75]. Another attractive aspect of species tree methods and multispecies coalescent models is that they deliver more appropriate estimated levels of confidence that are more evenly spread across genes and appear to be less susceptible to the inflation of posterior probabilities that was early on attributed to Bayesian analyses (e.g., [76, 77]) but may also be due to model misspecification due to concatenation [53]. Bayesian methods are generally agreed to be the most efficient and accurate, capturing all details of the MSC model seamlessly [52]. However, one drawback is that the estimation of larger numbers of parameters (population sizes and divergence times in addition to topologies) can slow computation, may not be relevant in some situations [78], and is generally not possible with the large data sets that are routinely seen today in phylogenomics [59]. Thus far, two-step methods such as ASTRAL, STAR, NJst, and MP-EST have proven the most widely used for large-scale phylogenomic studies, such as the Avian Phylogenomics Project [79] and large-scale phylogenomics of fish [80] and plants [81].

2.1 Sources of Gene Tree/Species Tree Discordance and Violations of the Multispecies Coalescent Model

2.1.1 Population Processes

The "standard" and most common reason why gene trees are not speciodendritic is incomplete lineage sorting, i.e., lineages have not yet been reproductively isolated for long enough for drift to cause complete genetic divergence in the form of reciprocal monophyly of gene trees ([82]; Figs. 1 and 2). This source of gene tree heterogeneity is guaranteed to be ubiquitous, if only because it arises from the finite size populations of all species that have ever come into existence. Almost all the techniques and software packages discussed above are designed to approximate uncertainties in species tree topology arising from this phenomenon.

Delimitation of Species and Diverging Lineages

For recent divergences, the definition of "species" can become problematic for species tree methods [63], and the challenge of delimiting species has, if anything, increased now that the overly conservative strictures of gene tree monophyly as a delimiter of species have been mostly abandoned [82]. This fundamental issue in a phylogenetic study—whether the extent of divergence among lineages warrants species status—has not gone away in the genomic era. However, traditional species tree methods using the MSC need not use "good" species as OTUs; they will work perfectly well on lineages that have recently diverged, so long as they have ceased exchanging genes. The key issue is not whether the OTUs in species tree analyses are in fact species but rather whether they have ceased exchanging genes, which has been shown to compromise traditional MSC methods [83, 84] (see below).

The problem of species delimitation may ultimately be solved by data other than genetics, and today few species concepts use strictly genetic criteria [85]. Some have suggested that the line between a population-level difference and a species-level difference can be drawn empirically and with consistency in well-studied taxa such as birds, using morphological, environmental, and behavioral data simultaneously [86]. Thus, there is some hope that species delimitation can be performed rigorously a priori in many cases. Researchers who opt for delimiting species primarily with molecular data have a wide array of techniques and prior examples available to them, although not all without controversy [71, 87–93]. Recent progress in species delimitation is motivated by the conceptual transition from "biological/reproductive isolation species" to the "lineage species concept," which defines species not in terms of monophyly of gene lineages but as population lineage segments in the species tree [93]. Under that expanded concept, boundaries of species (i.e., lineages in the species tree) can be facilitated by collection and analysis of gene trees in the framework of the multispecies coalescent model [72]. The recent suggestion that coalescent species delimitation methods define only structure but not species [90] was, in our view, already well-established, with confusion stemming largely from the term "species delimitation," as opposed to "delimitation of populations between which gene flow has ceased."

Gene Flow

There are a number of other situations in which the assumptions of the coalescent are violated. MSC models involve a series of isolation events unaccompanied by gene flow. In this regard, they are like the isolation-migration models of phylogeography [94, 95] but without the migration. The assumption of no gene flow naturally restricts their utility, but gene flow of course compromises other methods of phylogenetic inference, including concatenation methods, as well. Additionally, situations in which gene flow yields a prominent molecular signal often are detectable primarily among very closely related species in the realm of phylogeography [96]. If some substantial gene flow continues between species after divergence, then the multispecies coalescent can quickly destabilize, especially for a small number of loci and as the rate of genetic introgression increases (Fig. 6 in [87, 97–99]). We recommend model comparison algorithms like PHRAPL [87] for determining whether a given data set conforms to the assumptions of the MSC.

2.1.2 Molecular Processes

In addition to species delimitation and gene flow, there are at least three mechanisms that generate discordance on the molecular level (Fig. 3). These include horizontal gene transfer (HGT), which can pose a serious risk to phylogenetic analysis; gene duplication, whose risks can be avoided by certain models; and natural selection, which generally poses no direct threat but, depending on its mode of action and consequences for DNA and protein sequences, can be the most challenging of all.

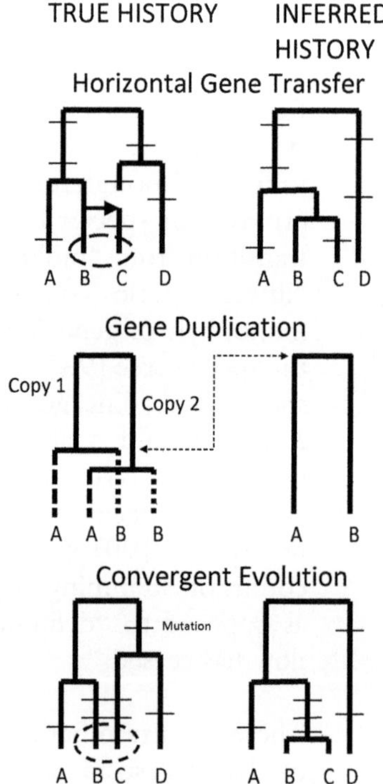

Fig. 3 Three examples of gene histories that depart from the standard multispecies coalescent model. (**a**) A duplication event that precedes a speciation event can lead to incorrect inference of divergence times in the species tree if copy 1 is compared to copy 2. This can be particularly difficult if one of the gene copies has been lost or not sequenced by the researcher. (**b**) Convergent evolution can occur at the molecular level, for example, in certain genes under strong natural selection or highly biased mutational processes. These processes will tend to bring together distantly related taxa in the phylogenetic tree and are likely to be given additional false support by morphological data. (**c**) Horizontal gene transfer causes difficulties in some current species tree methods, because it establishes a spurious lower bound to divergence times. Though rare in eukaryotes, it is by no means unknown and is likely to become a more difficult problem in the future when species trees are based on tens of thousands of loci

Horizontal Gene Transfer

HGT is now known to be so widespread across the Tree of Life, especially in prokaryotes, that some have suggested a web of life may be a more appropriate paradigm for phylogenetic change [100–102]. Growing evidence shows that even eukaryotic genomes contain substantial amounts of "uploaded" genetic material from bacteria, archaea, viruses, and even fellow eukaryotes [103–105]. Even though effective techniques are not yet widely available for detecting HGT in eukaryotes, enough individual cases have been "accidentally" discovered that researchers have given up trying to list them all [103].

The implications of HGT for species tree construction vary depending on the method used. For example, following the standard assumption in coalescent theory that allelic divergences must occur earlier in time than the divergences of species harboring those alleles, some species tree techniques [48, 58], as well as classical approaches (e.g., [13]), assume that the gene tree exhibiting the most recent divergence between taxon A and taxon B establishes a hard upper limit on the divergence time of those species in the species tree. For small sets of genes in taxa where HGT is rare, a researcher would need to be quite unlucky to choose a horizontally transferred gene for analysis. However, as the genomic era advances, it becomes more likely that at least one of the thousands of genes studied will have been transferred horizontally and thus establish a spurious upper bound for clade divergence at the species level. When selective introgression of genes from one species to another is considered, this number of genes coalescing recently between species will increase [106]. Although HGT is clearly a problem for some current methodologies, if transferred genes can first be identified, then they could be extremely useful as genomic markers for monophyletic groups that have inherited such genes and would otherwise be difficult to resolve [107]. However, for other species tree methods that calculate averages of coalescence times, such as STAR [65], HGT events will have less of an impact. Liu et al. [56] examined the effect of HGT on the pseudo-likelihood method MP-EST and predicted that, mathematically, species tree branch lengths may be biased by HGT but that topologies were fairly robust. Davidson et al. [108] found that quartet-based methods, such as ASTRAL-II, were fairly robust to HGT in the presence of ILS. Removal of genes suspected to be transferred via HGT prior to species tree analysis would be warranted; however, some methods to detect such events rely both on having the true species tree already in hand and also on the absence of other mechanisms causing gene tree discordance [109–112]. Recent work aims to incorporate HGT into other mechanisms of gene tree incongruence (reviewed in [113]); how much we need to invest in such synthetic methods will likely depend on the prevalence of HGT in particular taxonomic groups.

Gene Duplication

Gene duplication presents another violation of the basic MSC model (Fig. 3); like HGT, its potential problems are worst when they go unrecognized [49]. Imagine a taxon where a gene of interest duplicated 10 Mya into copy α and copy β; the taxon then split 5 Mya into species 1 and 2. A researcher investigating the daughter species would therefore sequence four orthologous genes, with the potential to compare α1 to β2 and β1 to α2 and thus generate two gene trees where the estimated split time was 10 Mya, rather than 5 Mya. Such a situation will be easily recognized if copy α and β have diverged sufficiently by the time of their duplication, and a number of methods of coalescent analysis have incorporated gene duplication (e.g., [114, 115]; reviewed in [116]). Additionally, failure to recognize the situation may not have drastic consequences for phylogenetic analysis if the paralogs have coalesced very recently or are species-specific, in which case the estimated gene coalescence would be approximately correct no matter which comparison was made. However, if one of the copies has been lost and only one of the remaining copies is sequenced, then the chances of inferring an inappropriately long period of genetic isolation are larger and will increase as the size of the family of paralogs increases. Assessing paralogs in phylogenomic data is a major challenge, particularly in groups like plants and fish, and a growing number of dedicated methods ([117]; assessed in [118]) or filtering protocols [119] for doing so exist. This problem will tend to overestimate gene coalescence times, and some species tree methods depend on minimum isolation times among a large set of genes. These deep coalescences might spuriously increase inferred ancestral population sizes. A systematic search for biases incurred by species tree methods due to gene duplication is needed.

Natural Selection

Natural selection causes yet another violation of the multispecies coalescent model. Selection can cause serious problems in some cases, although in other circumstances it is predicted not to cause problems of phylogenetic analysis [47, 120]. The usual stabilizing selection can be helpful to taxonomists working at high levels because it slows the substitution rate; likewise selective sweeps, directional selection, and genetic surfing [121] tend to clarify phylogenetic relationships by accelerating reciprocal monophyly for genes in rapidly diverging clades. However, challenges to phylogenetic inference are posed by any evolutionary force that may bias the reconstruction of gene trees, including convergent neutral mutations (homoplasy), balancing selection, and selection-driven convergent evolution (e.g., [122]). Balancing selection tends to preserve beneficial alleles at a gene for long periods of time and is probably the most insidious form of selection with respect to accurately reconstructing gene trees and species trees.

2.2 More About Violations and Model Fit of the Multispecies Coalescent Model

Many of the instances of violations of the coalescent model will occur at individual genes and usually will not dominate the signal of the entire suite of genes sampled for phylogenetic analysis. Reid et al. [123] conducted one of the few tests of the fit of the MSC to multilocus phylogenetic data. Although the title of their article suggests that the MSC overall provides a "poor fit" to empirical data, we suggest that their results provide a more hopeful picture. The most important thing is that they investigated the fit of the MSC to individual loci in phylogenetic data sets and were able to identify loci that failed to fit the MSC. They were less successful at identifying the causes of departure from the MSC for individual loci.

More common but still rare are efforts to determine which models of phylogenetic inference, the MSC or concatenation, provide a better fit to empirical phylogenomic data. Edwards et al. [124] and Liu and Pearl [58] both used the Bayesian species tree method BEST [68] to ask using Bayes factors whether the MSC or concatenation fits empirical data sets better. Uniformly, they found that the MSC fit empirical data sets better than concatenation, often by a large margin. However, further work in this area is still needed. Most discussions in the literature have focused on the perceived failings or violations of the MSC by empirical data sets—such as evidence for recombination within loci—even when such failings or assumptions also apply to concatenation [47]. Given that all models are approximations of reality, a better focus would be to ask which model better fits empirical data sets better. The limited research that has been done suggests overwhelmingly that the MSC provides a better fit to empirical data sets than concatenation.

Are there better models for phylogenomics than the MSC? Depending on the data set, almost surely there are (Fig. 4). Several authors working with phylogenomic data sets have suggested that gene flow is detectable, even among lineages that diverged a long time ago (e.g., [129, 130]). The increasing number of reports of hybridization and introgression among phenotypically distinct species suggests that hybridization may be a typical component of speciation and that even phylogenetic models can be improved by incorporating such reticulation (e.g., [47, 106, 131]). The pure MSC is best thought of as a special case of so-called "multispecies network coalescent" models, or MSNC [127, 132–134] (Fig. 4), in which gene flow connects some branches of the species tree. In the end, empiricists will need to decide what level of model fit they are willing to tolerate and which software packages can accommodate the large data sets that are now routine in phylogenomics.

2.2.1 Phylogenetic Outlier Loci

Genes whose phylogenetic signal differs significantly from that of the remainder of data set can be thought of as phylogenetic outliers. These loci are conceptually similar to outliers in population genetics, which have been the focus of many studies (reviewed in

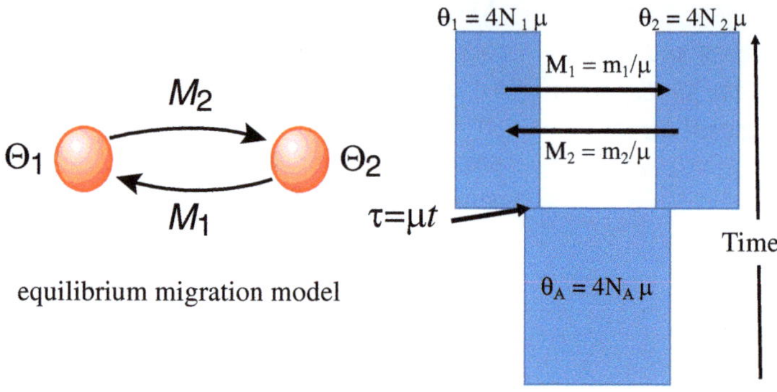

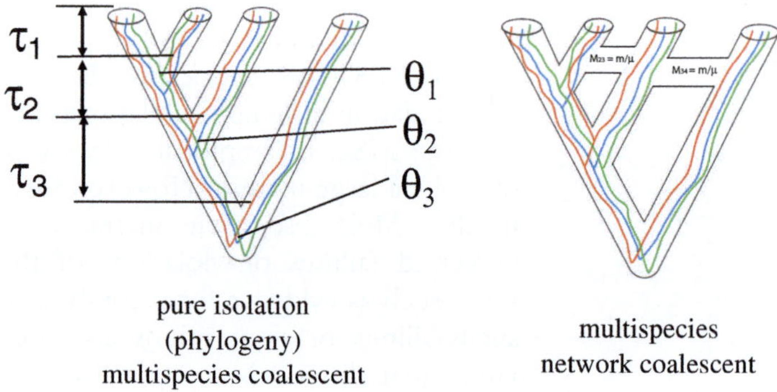

Population size: $\theta = 4N\mu$
Gene flow: $M = m/\mu$
Divergence time: $\tau = \mu t$

Fig. 4 Diversity of phylogeographic models. Species trees estimated by the multispecies coalescent are naturally related to previous phylogeographic models by their shared demographic parameters, usually measured in units of mutation rate or substitutions per site (μ), including genetic diversity or effective population size ($4N\mu$, where $N=$ effective population size; gene flow M/μ, where $M=$ the scaled migration rate; $4Nm$, where m is the number of migrants per generation; and divergence time $\tau = \mu t$, where t is the divergence time in generations). (**a**) Equilibrium migration models as envisioned by early versions of the software MIGRATE [125]. (**b**) Isolation-migration models envisioned by Hey and coworkers [48, 95, 126]. Subscript A indicates ancestral population size. (**c**) Species tree models estimated by the multispecies coalescent [28]. (**d**) Multispecies network coalescent models or phylogenetic network models including divergence and gene flow [127, 128]

[135–137]). However, there has been little work in detecting phylogenomic outliers. Much attention has been paid to particular sites in a data set that differ from the majority and therefore exhibit homoplasy or incongruence with the rest of the data set [76, 138]. The sources of such incongruence are many and can include mutational processes (e.g., gene duplication), HGT, as well homoplasy (e.g., [139, 140]). Incongruence of particular sites, or entire loci, may also be due to technical issues such as contamination, misassembly, mistaken paralogy, annotation mistakes, and alignment errors (e.g., [119]). Here, in an analogy with work in population genetics, we will focus primarily on entire loci that deviate from the expected distribution governed by neutral processes due to natural selection. Understanding the distribution of gene tree topologies expected under the neutral multispecies coalescent [25] is a good starting point for identifying loci that may be targets of natural selection.

2.2.2 Genomic Signals of Phylogenetic Outliers

When faced with a surprising or nonconvergent species tree, one possibility is that an unusual gene tree is to blame. Though techniques for dealing with violations of the coalescent model are in their infancy, researchers do have a few options. Below we list several ideas, some borrowed from classical phylogenetics or from methods used in bioinformatics. It is likely that the several tests constructed to detect phylogenetic outliers in classical phylogenetics can be extended slightly to incorporate the additional variation among genes expected due to the coalescent process. Of course, with larger data sets, at least with some coalescent methods, single anomalous genes may have little effect on the resulting species tree, particularly in species tree methods utilizing summary statistics [65]. However, as pointed out above, species tree methods such as BEST that relies on "hard" boundaries for the species tree by individual genes could be derailed due to the anomalous behavior of even a single gene.

Jackknifing: A straightforward approach to detecting phylogenetic outliers under the multispecies coalescent model is to rerun the analysis n times, where n is the number of loci in the study, leaving one locus out each time. An outlier can then be identified if the analysis that does not include that gene differs from the remaining analyses in which that gene is included. This approach has been applied successfully in fruit flies by Wong et al. [21], who considered their problem resolved when the elimination of one of the ten genes unambiguously resolved a polytomy. There may be other metrics of success that are more robust or sensitive or do not depend as strongly on a priori beliefs about the relationships among taxa. Because some duplications or horizontal transfers may affect only one taxon, whole-tree topology summary statistics are unlikely to be sensitive enough to detect recent events. However, the cophenetic distance of each taxon to its nearest neighbor in the complete species tree could be compared across jackknife results. This procedure will produce a distribution of "typical" distances, and significance can therefore be assigned to highly divergent results. The drawback to such an approach is the computational demand. Species tree analyses on their own can be extremely time consuming to run even once, so jackknifing may prove intractable for studies involving many species and loci (*see* ref. 141).

2.2.3 Simulation Approaches to Detecting Phylogenetic Outliers

Simulating gene trees from a species tree is another method for identifying gene trees that differ from the majority of loci in the data set. Several species tree methods yield estimate of the phylogeny that include branch lengths in coalescent units [56, 57, 70], which are required to simulate gene trees from a species tree. Branch lengths in the estimated species tree can be decomposed into a number of substitutions per site and an estimate of $\theta = 4N\mu$ that are compatible with the original branch length in coalescent units. For example, using any number of algorithms, including maximum likelihood or Bayesian methods, the length of species tree branch lengths in substitutions per site can be approximated by fitting the concatenated alignment of genes to the estimated species tree topology, yielding a tree with the same topology but branch lengths in substitutions per site (μt, where t is the time span of the branch in either generations or years). With these branch lengths in hand, estimates of θ can then be applied to each branch so that the original coalescent units $t/2N \approx \mu t/\theta$ from the species tree are retained. Care needs to be taken to preserve the appropriate ploidy units when simulating gene trees from an estimated species tree. Packages such as MP-EST yield estimates of species tree branch lengths in coalescent units of $4N$ generations, appropriate for diploids, whereas packages such as Phybase [142] simulate gene trees from a species tree in estimates of $2N$ units, appropriate for haploids. Another issue that is important to be aware of is the distinction between gene coalescence times and species tree branch lengths [143, 144]. Whereas species tree branch lengths are estimates of lineage or population branch lengths in the species tree, the DNA sequence alignment that is fitted to the species tree will yield branch lengths reflecting the coalescence time of genes in ancestral species. This discrepancy occurs because gene coalescence times by necessity predate and record a more ancient event than do species divergence times. The discrepancy may represent a small fraction of the branch length if species divergence times are large, but Angelis and dos Reis [143] have suggested that the discrepancy can be quite large even in comparisons of distantly related species, such as exemplars of mammalian orders. There is a great need for methods of molecular dating and combining fossils and DNA data that distinguish between gene coalescence times and speciation times, the latter of which is usually of primary interest.

Once the branch lengths of the species tree are prepared for simulation, gene trees can be simulated using a number of packages (Phybase, [142]; TreeSim, [145]; CoMus, [146]). Even packages traditionally used in phylogeography can be used to simulated gene trees on species trees, given the close relationship between species trees and phylogeographic models like isolation migration [147, 148]. One can then compare the distribution of gene tree topologies and branch lengths observed in one's data set with those simulated under the neutral coalescent model. A common approach is to calculate the distribution of Robinson-Foulds [149] distances among simulated gene trees and compare these to those observed in the original data set. Such approaches have been used to determine if a data set is consistent with the MSC or the percent of the observed gene tree variation that is explained by the MSC. Other statistics, such as the similarity in number of minority gene tree triplets produced by a given species tree at each node, can also be compared to the observed distribution. Song et al. [150] used coalescent simulations using Phybase to propose that the MSC could explain a large (>75%) fraction of the observed gene tree variation in a mammalian data set. Such simulations assume that the gene tree variation observed is biological in origin and not due to errors in reconstruction. They also noted that the near equivalence in frequency of minority triplets in gene trees at various nodes in the mammal tree suggested broad applicability of the neutral coalescent without gene flow or other complicating factors. Still, many papers observe some level of departure of the patterns in the observed data set from those expected under simulation. Usually the source of this departure is unknown. Natural selection or any other force such as HGT or anomalous mutation might be culprits in these cases. Heled et al. [151] proposed a simulation regime that incorporates gene flow between species and thus can be used to test for the effects of migration on gene trees and species tree estimation.

To detect possible phylogenetic outliers, Edwards et al. [152] applied a recently proposed method of detecting gene tree outliers, KDEtrees [153], to a series of phylogenomic data sets. KDEtrees uses the kernel density distribution of gene tree distances to estimate the 95% confidence limits on gene tree topologies in a given data set. Surprisingly, using default parameters, Edwards et al. [152] could not detect a higher-than-expected number of gene tree outliers in any data set, despite the fact that the data sets in several cases contained hundreds of loci. No data set possessed more than the expected 5% of outliers given the test implemented in KDEtrees. Clearly further work is needed to understand the pros and cons of various tests of phylogenetic outliers. For the time being, we can note the robustness of various species tree methods to phylogenetic outliers. One attractive prospect of algorithms for species tree construction that use summary statistics, such as STAR and STEAC, is that these methods are powerful and fast, yet they appear less susceptible to error due to deviations of single genes from neutral expectations. These methods do not utilize all the information in the data and hence can be less efficient than Bayesian or likelihood methods [52], yet they perform well with moderate amounts of gene tree outliers due to processes like HGT.

3 Hypothesis Testing Using the Multispecies Coalescent Model

Hypothesis testing is a cornerstone of phylogenetic analysis but has received little attention in the context of the MSC (*see* ref. 154). Bayesian species tree inference [58, 59, 68–70] provides perhaps the most seamless approach to hypothesis testing. One can relatively easily assess the fit of the collected data to alternative tree topologies and compare the fit using Bayes factors or other approaches. One can also assess the fit of various models of analysis to the collected data [155]. Liu and Pearl [58] and Edwards et al. [124] used Bayes factors to determine whether concatenation or the MSC was a more appropriate model for several data sets; in all cases tested thus far, the MSC provides a far better fit to multilocus data (BF > 10) than does concatenation, in which all gene trees among loci are identical. Further work is needed to apply Bayes factors and likelihood ratio tests to multilocus data.

The bootstrap, introduced to phylogenetics by Felsenstein [156], is the most common statistic applied to phylogenetic trees [157]. In the era of multilocus phylogenetics, the "multilocus bootstrap" of Seo [158] has been recommended as a more suitable approach to assessing confidence limits than the traditional bootstrap. In the traditional bootstrap, sites within a locus, or a series of concatenated loci, are resampled with replacement to create pseudomatrices, which are then subjected to phylogenetic analysis, after which a majority rule consensus tree is usually made. By contrast, in the multilocus bootstrap, sites within loci and the loci themselves are resampled with replacement. In the context of the MSC, resampled pseudomatrices of the same number of loci as the original data set, which may contain duplicates of specific loci due to the random nature of the bootstrap, are then made into gene trees, from which a species tree can be made. The bootstrap and various other measures of branch-specific support [159] have been proposed as a means of assessing confidence in species trees made using the multilocus coalescent. Care should be taken in the comparison of different studies using different measures of support, since not all measures can be directly compared to one another. For example, as pointed out by Liu et al. [160], the measure of posterior support for ASTRAL trees proposed by Sayyari and Mirarab [159] is not the same as traditional bootstrap supports, and we do not yet know how they will scale under different conditions compared to the bootstrap. Edwards [161] summarized knowledge about the use of phylogenomic subsampling, in which data sets of increasing size or signal are analyzed so as to understand the stability and speed of approach to certainty of phylogenetic estimates under the MSC and under concatenation. He found that MSC methods tended to approach phylogenomic certainty more smoothly and monotonically than do concatenation methods, which jump around erratically in their certainty for sometimes conflicting topologies, especially when sampling smaller numbers of genes. Although we cannot simply translate many conclusions from the gene tree era of phylogenetics to the MSC era—for example, contrary to gene tree conclusions, it is not clear for MSC models that more taxa are always better than more loci [74]—many of these discussions about hypothesis testing echo early comparisons of posterior probabilities and bootstrap proportions used in the gene tree era of phylogenetics.

The bootstrap has always provided a means of hypothesis testing that is very indirect with respect to comparing alternative phylogenetic hypotheses. Aside from the tests allowed by Bayesian approaches, there have been few discussions of testing of alternative phylogenetic trees in the era of the multispecies coalescent. In this regard, the pseudo-likelihood model provided by MP-EST [56] provides a convenient framework for hypothesis testing using species trees. This framework is not available in most other species tree methods, including ASTRAL, STAR, and STEAC, since these methods do not employ a likelihood model. MP-EST takes advantage of the likelihood model of Rannala and Yang [48] to assess the fit of a species tree to a collection of gene trees and can thus be used to compare alternative species tree topologies and branch lengths directly.

To conduct a direct comparison of species trees using the likelihood ratio test, we first compare the likelihoods of two trees to find the most probable species tree that can explain the empirical set of gene trees. The likelihood of a set of gene trees given a species tree with branch lengths can be ascertained using functions in Phybase [142]. Let Tree 1 be the null tree and Tree 2 be the alternative tree. The likelihood ratio test statistic is $t = 2$ $(L_{\text{Tree2}} - L_{\text{Tree1}})$, in which L_{Tree1} and L_{Tree2} are the log-likelihoods of the null and alternative hypotheses. The log-likelihood of the null hypothesis can be obtained from the output of the program MP-EST by fitting the branch lengths and topology of Tree 1 to the set of empirical gene trees. Similarly, we can find the log-likelihood of the alternative tree Tree 2 using MP-EST. The null distribution of the test statistic t is approximated by a parametric bootstrap. Specifically, we generate 100 or more bootstrap samples of gene trees under the null tree Tree 1. For each sample of these bootstrapped trees, we calculate the log-likelihoods of the null and alternative trees using the procedure described above. The null distribution of the test statistic t is approximated by the test statistics of the bootstrap samples. If t for the null and alternative species trees is outside the expected distribution of the bootstrap sample statistics, then the result can be considered significant.

We applied this approach to assessing alternative phylogenetic hypotheses to an example from birds (fairy wrens; [162]; Fig. 5). This data set consists of 18 genes and 26 taxa, with loci coming

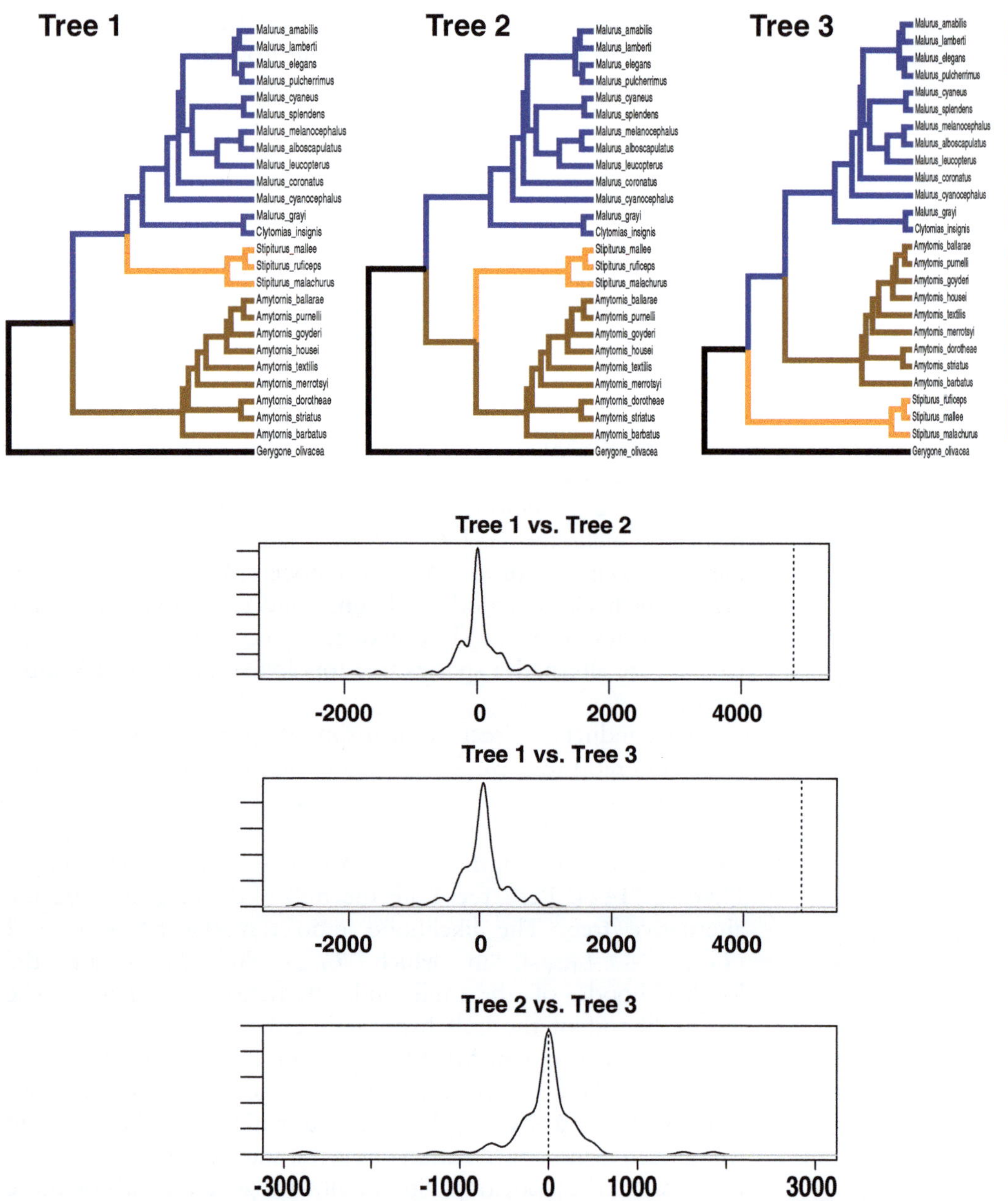

Fig. 5 Example of hypothesis testing of alternative phylogenetic trees under the multispecies coalescent model. Top: alternative phylogenetic hypotheses involving the rearrangement of major groups of Australo-Papuan fairy wrens based on Lee et al. [162]. The three alternative phylogenetic trees are colored to indicate the three major groups whose relationships are being tested. Bottom: results of the likelihood ratio test (LRT) and estimates of confidence limits on the test statistic *t* using parametric bootstrapping. The plots show the distributions of the test statistic *t* resulting from gene trees built from resampled, bootstrapped sequence data. Despite the use of sequence data to generate the bootstrap gene tree distributions, the LRT is only an indirect test of the signal in the sequence data and instead is best thought of as a test of the fit of the estimated gene tree distribution on alternative phylogenies. See main text for further details

from a variety of marker types (exons, introns, anonymous loci). Lee et al. [162] applied a number of MSC approaches to this data set but did not compare alternative trees directly, having only used bootstrap approaches. Here, we consider three-species trees generated from the rearrangement of the three major clades of wrens: the core fairy wrens (*Malurus*), emu-wrens (*Stipitirus*), and grasswrens (*Amytornis*; Fig. 5). Rearranging these major clades results in three alternative rooted species trees. Based on traditional taxonomy and because the gene trees in this data set were highly variable, even among the three major clades, we consider these three alternative hypotheses true alternatives and not "straw men." Rooted maximum likelihood gene trees were built from the alignments of each locus using RaxML [163] and then used as input data for the likelihood ratio test described above. The LRT was applied first to Tree 1 (null) versus Tree 2 and was also applied to Tree 1 versus Tree 3 and Tree 2 versus Tree 3. The results indicate that Tree 1 fits to the empirical gene trees significantly better than does Tree 2 or Tree 3 does ($p < 0.01$), and there is no significant difference between Trees 2 and 3 in their fit to the empirical gene trees ($p = 0.52$). Thus, the LRTs strongly favor Tree 1 over both Tree 2 and Tree 3.

It is important to note that the LRT described above is not a direct test of the phylogenetic signal in the DNA sequence data. Rather, it is a test of the distribution of gene trees inferred from the sequence data and assumes that the gene trees provided as data are without error. It does indirectly test the signal in the sequence data, because if the DNA sequences provide strong and consistent support of the gene trees, then the bootstrapped set of gene trees will be highly similar to one another, and the confidence limits on t will be very tight. By contrast, if the DNA sequence data does not have a strong signal, then the confidence limits on t will be very wide, and it will be difficult to reject alternative species trees. The LRT described here does not involve nested models. If the gene trees are known without error, then the value of t itself can be used to assess significance, assuming a chi-square distribution with 2 degrees of freedom. Further research is needed on methods for comparing and testing alternative species trees in the context of the MSC.

4 Future Directions

Species tree methods are likely to continue to gain ascendancy as the strongest evidence of taxonomic relationship in phylogenetic research. As with any form of evidence, the conclusions of a species tree analysis are fallible, with each method susceptible to biases in the input data. For example, Xi et al. [164] showed that Phyml [165] yields biased gene trees when there is little information in the DNA sequences and can therefore result in biased species trees. This issue is particularly problematic when using MP-EST v. 1.5, which, unlike ASTRAL or MP-EST v. 2.0, does not randomly resolve or appropriately accommodate gene trees with polytomies or 0 or near 0-length branches. This bias may have affected the performance of MP-EST in previous side-by-side comparisons with ASTRAL. In the future, further work should be devoted to discovering and quantifying additional biases in inference of species trees. With the size of phylogenomic data sets increasing, even small biases can be amplified and result in poorly estimated species trees.

Many in the field agree that the most appealing statistical models for species tree inference using the MSC include Bayesian and full-likelihood models [52]. But it is still clear, at least to empiricists, not only that "two-step" methods of species tree inference work quite well in general but also that the large phylogenomic data sets available today prohibit the use of full-likelihood methods. Regardless, we now know that both types of models clearly outperform concatenation across wide swaths of parameter space, especially if one also evaluates the reliability of the confidence limits on the estimate of phylogeny and not only the point estimate of the topology. The major directions for future research in the field of species tree inference therefore include increasing the scalability of computational inference of species trees, further development of frameworks for hypothesis testing using the MSC, developing additional models of divergence with gene flow and network coalescent models (Fig. 4), and improvement in the estimation of gene trees and species trees from SNP data [166]. Linking mutations in species trees and heterogeneous gene trees to diverse phenotypic and ecological data will be another important avenue for the future [167, 168]. We view the MSC, with its application of population genetic models to higher-level systematics, as a key component of the long-term goal of uniting microevolution and macroevolution. Even if it proves incomplete in the long term, the neutral MSC provides a powerful null model for the understanding of genetic diversity across time and space.

5 Practice Problems

1. Consider the following discordant set of gene trees. {Gene 1 = (A:10,(B:8,C:8):2); Gene 2 = (B:9,(A:6,C:6):3); Gene 3 = ((A:4,B:4):4,C:8)}. Assuming that these genes perfectly reflect the time of genetic divergence, and the only cause of discordance is incomplete lineage sorting or deep coalescence, what is the most likely species tree? *Answer: ((A:4,B:4):2,C:6)*
2. Find the data set for 30 noncoding loci from 4 species of Australian grass finches (3 *Poephila*, plus out-group *Taeniopygia*) from Jennings and Edwards [169]. It can be found in the web page for Liang Liu's BEST program: http://faculty.franklin.uga.edu/lliu/content/BEST. Use the Bayesian program

BEST [68] or BPP [70] and the nonparametric method in STAR [65] to estimate the species tree for the four species, using *Taeniopygia* as the out-group. Do you estimate the same topology with both methods? What about the support for the single internal branch? If the support is not the same, what could be causing the difference? *Answer: The BEST or BPP tree should have higher support than the STAR tree, but they both should have the same topology. The STAR tree might have lower support because in the data set about half of the gene trees have a topology differing from the species tree; whereas the full Bayesian model accommodates this variation accurately, nonparametric "two-step" methods interpret this type of gene tree variation as discordance, in conflict with the majority of the gene trees and with the species tree.*

3. For the above data set, make individual gene trees using RaXml [170], and use the likelihood functions and bootstrap capabilities of Phybase [142] to conduct a likelihood ratio test of the two alternative species tree topologies for the four grass finches. Alternatively, you could use the posterior distribution of gene trees generated in BEST to estimate the confidence limits on the test statistic t. Is the tree estimated in question 2 significantly better than alternative trees? *Answer: The LRT indicates that the tree estimated in question 2 is significantly better than alternative trees.*

References

1. Darwin C (1859) On the origin of species, vol Facsimile of 1st Edition. Harvard University Press, Cambridge, p 513
2. Hillis DM (1987) Molecular versus morphological approaches to systematics. Annu Rev Ecol Syst 18:23–42
3. Scotland RW, Olmstead RG, Bennett JR (2003) Phylogeny reconstruction: the role of morphology. Syst Biol 52:539–548
4. Kocher TD, Thomas WK, Meyer A, Edwards SV, Pääbo S, Villablanca FX, Wilson AC (1989) Dynamics of mitochondrial DNA evolution in animals: amplification and sequencing with conserved primers. Proc Natl Acad Sci U S A 86:6196–6200
5. Miyamoto MM, Cracraft J (1991) Phylogeny inference, DNA sequence analysis, and the future of molecular systematics. In: Miyamoto MM, Cracraft J (eds) Phylogenetic analysis of DNA sequences. Oxford University Press, New York, NY, pp 3–17
6. Swofford DL, Olsen GJ, Waddell PJ, Hillis DM (1996) Phylogenetic inference. In: Hillis DM, Moritz C, Mable BK (eds) Molecular systematics. Sinauer, Sunderland, MA
7. Nei M (1987) Molecular evolutionary genetics, vol 512. Columbia University Press, New York
8. Nei M, Kumar S (2000) Molecular evolution and phylogenetics. Oxford University Press, New York
9. Rosenberg NA (2002) The probability of topological concordance of gene trees and species trees. Theor Popul Biol 61:225–247
10. Cavalli-Sforza LL (1964) Population structure and human evolution. Proc R Soc Lond Series B 164:362–379
11. Avise JC, Arnold J, Ball RM, Bermingham E, Lamb T, Neigel JE, Reeb CA, Saunders NC (1987) Intraspecific phylogeography: the mitochondrial DNA bridge between population genetics and systematics. Annu Rev Ecol Syst 18:489–522
12. Pamilo P, Nei M (1988) Relationships between gene trees and species trees. Mol Biol Evol 5:568–583

13. Takahata N (1989) Gene genealogy in three related populations: consistency probability between gene and population trees. Genetics 122:957–966
14. Avise JC (1994) Molecular markers, natural history and evolution. Chapman and Hall, New York
15. Wollenberg K, Avise JC (1998) Sampling properties of genealogical pathways underlying population pedigrees. Evolution 52:957–966
16. Gatesy J, Springer MS (2014) Phylogenetic analysis at deep timescales: unreliable gene trees, bypassed hidden support, and the coalescence/concatalescence conundrum. Mol Phylogenet Evol 80:231–266
17. de Queiroz A, Gatesy J (2007) The supermatrix approach to systematics. Trends Ecol Evol 22:34–41
18. Maddison WP (1997) Gene trees in species trees. Syst Biol 46:523–536
19. Edwards SV (2009) Is a new and general theory of molecular systematics emerging? Evolution 63:1–19
20. Carstens BC, Knowles LL (2007) Estimating species phylogeny from gene-tree probabilities despite incomplete lineage sorting: an example from melanoplus grasshoppers. Syst Biol 56:400–411
21. Wong A, Jensen JD, Pool JE, Aquadro CF (2007) Phylogenetic incongruence in the Drosophila melanogaster species group. Mol Phylogenet Evol 43:1138–1150
22. Knowles LL, Kubatko LS (2010) Estimating species trees: an introduction to concepts and models. In: Knowles LL, Kubatko LS (eds) Estimating species trees: practical and theoretical aspects. Wiley-Blackwell, New York, pp 1–14
23. Liu L, Yu L, Kubatko L, Pearl DK, Edwards SV (2009) Coalescent methods for estimating phylogenetic trees. Mol Phylogenet Evol 53:320–328
24. Neigel JE, Avise JC (1986) Phylogenetic relationships of mitochondrial DNA under various demographic models of speciation. In: Karlin S, Nevo E (eds) Evolutionary processes and theory. Academic, New York, pp 515–534
25. Degnan JH, Salter L (2005) Gene tree distributions under the coalescent process. Evolution 59:24–37
26. Tajima F (1983) Evolutionary relationships of DNA sequences in finite populations. Genetics 105:437–460
27. Mehta RS, Bryant D, Rosenberg NA (2016) The probability of monophyly of a sample of gene lineages on a species tree. Proc Natl Acad Sci U S A 113:8002–8009
28. Degnan JH, Rosenberg NA (2009) Gene tree discordance, phylogenetic inference, and the multispecies coalescent. Trends Ecol Evol 24:332–340
29. Degnan JH, Rosenberg NA (2006) Discordance of species trees with their most likely gene trees. Public Lib Sci Genet 2:762–768
30. Rosenberg NA, Tao R (2008) Discordance of species trees with their most likely gene trees: the case of five taxa. Syst Biol 57:131–140
31. Huang HT, Knowles LL (2009) What is the danger of the anomaly zone for empirical phylogenetics? Syst Biol 58:527–536
32. Sackton TB et al (2018) Convergent regulatory evolution and the origin of flightlessness in palaeognathous birds. bioRxiv. https://doi.org/10.1101/262584
33. Linkem CW, Minin VN, Leaché AD (2016) Detecting the anomaly zone in species trees and evidence for a misleading signal in higher-level skink phylogeny (Squamata: Scincidae). Syst Biol 65:465–477
34. Bryant D (2003) A classification of consensus methods for phylogenetics. In: Janowitz M et al (eds) BioConsensus. American Mathematical Society, Providence, RI, pp 163–183
35. Ewing GB, Ebersberger I, Schmidt HA, von Haeseler A (2008) Rooted triple consensus and anomalous gene trees. BMC Evol Biol 8:118
36. Degnan JH, DeGiorgio M, Bryant D, Rosenberg NA (2009) Properties of consensus methods for inferring species trees from gene trees. Syst Biol 58:35–54
37. Ranwez V, Criscuolo A, Douzery EJ (2010) SuperTriplets: a triplet-based supertree approach to phylogenomics. Bioinformatics 26:i115–i123
38. Steel M, Rodrigo A (2008) Maximum likelihood supertrees. Syst Biol 57:243–250
39. Wiens JJ (2003) Missing data, incomplete taxa, and phylogenetic accuracy. Syst Biol 52:528–538
40. Gadagkar SR, Rosenberg MS, Kumar S (2005) Inferring species phylogenies from multiple genes: concatenated sequence tree versus consensus gene tree. J Exp Zool B Mol Dev Evol 304:64–74
41. Bull JJ, Huelsenbeck JP, Cunningham CW, Swofford DL, Waddell PJ (1993) Partitioning and combining data in phylogenetic analysis. Syst Biol 42:384–397
42. Rokas A, Williams B, King N, Carroll S (2003) Genome-scale approaches to resolving

incongruence in molecular phylogenies. Nature 425:798–804

43. Driskell AC, Ane C, Burleigh JG, McMahon MM, O'Meara BC, Sanderson MJ (2004) Prospects for building the tree of life from large sequence databases. Science 306:1172–1174
44. Rokas A (2006) Genomics. Genomics and the tree of life. Science 313:1897–1899
45. Kubatko LS, Degnan JH (2007) Inconsistency of phylogenetic estimates from concatenated data under coalescence. Syst Biol 56:17–24
46. Wu M, Eisen JA (2008) A simple, fast, and accurate method of phylogenomic inference. Genome Biol 9:R151
47. Edwards SV et al (2016) Implementing and testing the multispecies coalescent model: a valuable paradigm for phylogenomics. Mol Phylogenet Evol 94:447–462
48. Rannala B, Yang Z (2003) Bayes estimation of species divergence times and ancestral population sizes using DNA sequences from multiple loci. Genetics 164:1645–1656
49. Bravo GA et al (2019) Embracing heterogeneity: coalescing the Tree of Life and the future of phylogenomics. PeerJ 7:e6399. https://doi.org/10.7717/peerj.6399
50. Felsenstein J (1981) Evolutionary trees from DNA sequences: a maximum likelihood approach. J Mol Evol 17:368–376
51. Rannala B, Yang ZH (2008) Phylogenetic inference using whole genomes. Annu Rev Genomics Hum Genet 9:217–231
52. Xu B, Yang Z (2016) Challenges in species tree estimation under the multispecies coalescent model. Genetics 204:1353–1368
53. Liu L, Xi Z, Wu S, Davis CC, Edwards SV (2015) Estimating phylogenetic trees from genome-scale data. Ann N Y Acad Sci 1360:36–53
54. Edwards SV (2016) Inferring species trees. In: Kliman R (ed) Encyclopedia of evolutionary biology. Elsevier Inc., New York, pp 236–244
55. Castillo-Ramírez S, Liu L, Pearl D, Edwardsm SV (2010) Bayesian estimation of species trees: a practical guide to optimal sampling and analysis. In: Knowles LL, Kubatko LS (eds) Estimating species trees: practical and theoretical aspects. Wiley-Blackwell, New Jersey, pp 15–33
56. Liu L, Yu L, Edwards S (2010) A maximum pseudo-likelihood approach for estimating species trees under the coalescent model. BMC Evol Biol 10:302
57. Mirarab S, Warnow T (2015) ASTRAL-II: coalescent-based species tree estimation with many hundreds of taxa and thousands of genes. Bioinformatics 31:i44–i52
58. Liu L, Pearl DK (2007) Species trees from gene trees: reconstructing Bayesian posterior distributions of a species phylogeny using estimated gene tree distributions. Syst Biol 56:504–514
59. Ogilvie HA, Bouckaert RR, Drummond AJ (2017) StarBEAST2 brings faster species tree inference and accurate estimates of substitution rates. Mol Biol Evol 34(8):2101–2114
60. Heled J, Drummond AJ (2010) Bayesian inference of species trees from multilocus data. Mol Biol Evol 27:570–580
61. Maddison WP, Knowles LL (2006) Inferring phylogeny despite incomplete lineage sorting. Syst Biol 55:21–30
62. Kubatko LS, Carstens BC, Knowles LL (2009) STEM: species tree estimation using maximum likelihood for gene trees under coalescence. Bioinformatics 25:971–973
63. O'Meara BC (2010) New heuristic methods for joint species delimitation and species tree inference. Syst Biol 59:59–73
64. Mossel E, Roch S (2010) Incomplete lineage sorting: consistent phylogeny estimation from multiple loci. IEEE/ACM Trans Comput Biol Bioinform 7:166–171
65. Liu L, Yu L, Pearl DK, Edwards SV (2009) Estimating species phylogenies using coalescence times among sequences. Syst Biol 58:468–477
66. Liu L, Yu L (2011) Estimating species trees from unrooted gene trees. Syst Biol 60:661–667
67. Mirarab S, Reaz R, Bayzid MS, Zimmermann T, Swenson MS, Warnow T (2014) ASTRAL: genome-scale coalescent-based species tree estimation. Bioinformatics 30:i541–i548
68. Liu L (2008) BEST: Bayesian estimation of species trees under the coalescent model. Bioinformatics 24:2542–2543
69. Liu L, Pearl DK, Brumfield RT, Edwards SV (2008) Estimating species trees using multiple-allele DNA sequence data. Evolution 62:2080–2091
70. Rannala B, Yang Z (2017) Efficient Bayesian species tree inference under the multispecies coalescent. Syst Biol 66:823–842
71. Yang Z (2015) The BPP program for species tree estimation and species delimitation. Curr Zool 61:854–865

72. Yang Z, Rannala B (2010) Bayesian species delimitation using multilocus sequence data. Proc Natl Acad Sci U S A 107:9264–9269
73. Chifman, J Kubatko L (2014) Quartet inference from SNP data under the coalescent model. Bioinformatics 30:3317–3324
74. Liu L, Xi ZX, Davis CC (2015) Coalescent methods are robust to the simultaneous effects of long branches and incomplete lineage sorting. Mol Biol Evol 32:791–805
75. Xi ZX, Liu L, Davis CC (2016) The impact of missing data on species tree estimation. Mol Biol Evol 33:838–860
76. Shen X-X, Hittinger CT, Rokas A (2017) Contentious relationships in phylogenomic studies can be driven by a handful of genes. Nat Ecol Evol 1:0126
77. Suzuki Y, Glazko GV, Nei M (2002) Overcredibility of molecular phylogenies obtained by Bayesian phylogenetics. Proc Natl Acad Sci U S A 99:16138–16143
78. Huang HT, He QI, Kubatko LS, Knowles LL (2010) Sources of error inherent in species-tree estimation: impact of mutational and coalescent effects on accuracy and implications for choosing among different methods. Syst Biol 59:573–583
79. Jarvis ED et al (2014) Whole-genome analyses resolve early branches in the tree of life of modern birds. Science 346:1320–1331
80. Hughes LC et al (2018) Comprehensive phylogeny of ray-finned fishes (Actinopterygii) based on transcriptomic and genomic data. Proc Natl Acad Sci U S A 115:6249–6254
81. Wickett NJ et al (2014) Phylotranscriptomic analysis of the origin and early diversification of land plants. Proc Natl Acad Sci U S A 111: E4859–E4868
82. Avise JC, Ball RMJ (1990) Principles of genealogical concordance in species concepts and biological taxonomy. Oxf Surv Evol Biol 7:45–67
83. Solis-Lemus C, Yang M, Ane C (2016) Inconsistency of species tree methods under gene flow. Syst Biol 65:843–851
84. Stenz NW, Larget B, Baum DA, Ane C (2015) Exploring tree-like and non-tree-like patterns using genome sequences: an example using the inbreeding plant species Arabidopsis thaliana (L.) Heynh. Syst Biol 64:809–823
85. Hudson RR, Coyne JA (2002) Mathematical consequences of the genealogical species concept. Evolution 56:1557–1565
86. Tobias JA, Seddon N, Spottiswoode CN, Pilgrim JD, Fishpool LDC, Collar NJ (2010) Quantitative criteria for species delimitation. Ibis 152:724–746
87. Jackson ND, Carstens BC, Morales AE, O'Meara BC (2017) Species delimitation with gene flow. Syst Biol 66:799–812
88. Leache AD, Zhu T, Rannala B, Yang Z (2018) The spectre of too many species. Syst Biol 66:379
89. Solis-Lemus C, Knowles LL, Ane C (2015) Bayesian species delimitation combining multiple genes and traits in a unified framework. Evolution 69:492–507
90. Sukumaran J, Knowles LL (2017) Multispecies coalescent delimits structure, not species. Proc Natl Acad Sci U S A 114:1607–1612
91. Carstens BC, Pelletier TA, Reid NM, Satler JD (2013) How to fail at species delimitation. Mol Ecol 22:4369–4383
92. Carstens BC, Dewey TA (2010) Species delimitation using a combined coalescent and information-theoretic approach: an example from North American Myotis bats. Syst Biol 59:400–414
93. De Queiroz K (2007) Species concepts and species delimitation. Syst Biol 56:879–886
94. Pinho C, Hey J (2010) Divergence with gene flow: models and data. Annu Rev Ecol Evol Syst 41:215–230
95. Hey J, Nielsen R (2007) Integration within the Felsenstein equation for improved Markov chain Monte Carlo methods in population genetics. Proc Natl Acad Sci U S A 104:2785–2790
96. Carstens BC, Morales AE, Jackson ND, O'Meara BC (2017) Objective choice of phylogeographic models. Mol Phylogenet Evol 116:136–140
97. Wakeley J (2001) The effects of subdivision on the genetic divergence of populations and species. Evolution 54:1092–1101
98. Solís-Lemus C, Yang M, Ané C (2016) Inconsistency of species tree methods under gene flow. Syst Biol 65:843–851
99. Eckert AJ, Carstens BC (2008) Does gene flow destroy phylogenetic signal? The performance of three methods for estimating species phylogenies in the presence of gene flow. Mol Phylogenet Evol 49(3):832–842
100. Doolittle WF, Bapteste E (2007) Pattern pluralism and the Tree of Life hypothesis. Proc Natl Acad Sci U S A 104:2043–2049
101. Boto L (2010) Horizontal gene transfer in evolution: facts and challenges. Proc Biol Sci 277:819–827
102. Soucy SM, Huang J, Gogarten JP (2015) Horizontal gene transfer: building the web of life. Nat Rev Genet 16:472–482

103. Keeling PJ, Palmer JD (2008) Horizontal gene transfer in eukaryotic evolution. Nat Rev Genet 9:605–618
104. Soanes D, Richards TA (2014) Horizontal gene transfer in eukaryotic plant pathogens. Annu Rev Phytopathol 52:583–614
105. Thomas J, Schaack S, Pritham EJ (2010) Pervasive horizontal transfer of rolling-circle transposons among animals. Genome Biol Evol 2:656–664
106. Mallet J, Besansky N, Hahn MW (2015) How reticulated are species? BioEssays 38:140–149
107. Huang J, Gogarten JP (2006) Ancient horizontal gene transfer can benefit phylogenetic reconstruction. Trends Genet 22:361–366
108. Davidson R, Vachaspati P, Mirarab S, Warnow T (2015) Phylogenomic species tree estimation in the presence of incomplete lineage sorting and horizontal gene transfer. BMC Genomics 16(Suppl 10):S1
109. Linz S, Semple C, Stadler T (2010) Analyzing and reconstructing reticulation networks under timing constraints. J Math Biol 61:715–737
110. Linz S, Radtke A, von Haeseler A (2007) A likelihood framework to measure horizontal gene transfer. Mol Biol Evol 24:1312–1319
111. Rasmussen MD, Kellis M (2011) A Bayesian approach for fast and accurate gene tree reconstruction. Mol Biol Evol 28:273–290
112. Rasmussen MD, Kellis M (2007) Accurate gene-tree reconstruction by learning gene- and species-specific substitution rates across multiple complete genomes. Genome Res 17:1932–1942
113. Szöllosi GJ, Tannier E, Daubin V, Boussau B (2015) The inference of gene trees with species trees. Syst Biol 64:e42–e62
114. Sanderson MJ, McMahon MM (2007) Inferring angiosperm phylogeny from EST data with widespread gene duplication. BMC Evol Biol 7(Suppl 1):S3
115. Thomas PD (2010) GIGA: a simple, efficient algorithm for gene tree inference in the genomic age. BMC Bioinform 11:312
116. Boussau B, Szollosi GJ, Duret L, Gouy M, Tannier E, Daubin V (2013) Genome-scale coestimation of species and gene trees. Genome Res 23:323–330
117. Conte MG, Gaillard S, Droc G, Perin C (2008) Phylogenomics of plant genomes: a methodology for genome-wide searches for orthologs in plants. BMC Genomics 9:183
118. Altenhoff AM, Gil M, Gonnet GH, Dessimoz C (2013) Inferring hierarchical orthologous groups from orthologous gene pairs. PLoS One 8:e53786
119. Irisarri I et al (2017) Phylotranscriptomic consolidation of the jawed vertebrate timetree. Nat Ecol Evol 1:1370–1378
120. Edwards SV (2009) Natural selection and phylogenetic analysis. Proc Natl Acad Sci U S A 106:8799–8800
121. Ray N, Excoffier L (2009) Inferring past demography using spatially explicit population genetic models. Hum Biol 81:141–157
122. Castoe TA, de Koning APJ, Kim H-M, Gu W, Noonan BP, Naylor G, Jiang ZJ, Parkinson CL, Pollock DD (2009) Evidence for an ancient adaptive episode of convergent molecular evolution. Proc Natl Acad Sci U S A 106:8986–8991
123. Reid NH, Hird SM, Brown JM, Pelletier TA, McVay JD, Satler JD, Carstens BC (2014) Poor fit to the multispecies coalescent is widely detectable in empirical data. Syst Biol 63:322–333
124. Edwards SV, Liu L, Pearl DK (2007) High-resolution species trees without concatenation. Proc Natl Acad Sci U S A 104:5936–5941
125. Beerli P, Felsenstein J (2001) Maximum likelihood estimation of a migration matrix and effective population sizes in n subpopulations by using a coalescent approach. Proc Natl Acad Sci U S A 98:4563–4568
126. Hey J, Wakeley J (1997) A coalescent estimator of the population recombination rate. Genetics 145:833–846
127. Solis-Lemus C, Bastide P, Ane C (2017) PhyloNetworks: a package for phylogenetic networks. Mol Biol Evol 34:3292–3298
128. Than C, Ruths D, Nakhleh L (2008) PhyloNet: a software package for analyzing and reconstructing reticulate evolutionary relationships. BMC Bioinform 9:322
129. Hallström BM, Janke A (2010) Mammalian evolution may not be strictly bifurcating. Mol Biol Evol 27:2804–2816
130. Kutschera VE, Bidon T, Hailer F, Rodi JL, Fain SR, Janke A (2014) Bears in a forest of gene trees: phylogenetic inference is complicated by incomplete lineage sorting and gene flow. Mol Biol Evol 31:2004–2017
131. Mavárez J, Salazar CA, Bermingham E, Salcedo C, Jiggins CD, Linares M (2006) Speciation by hybridization in Heliconius butterflies. Nature 441:868–871
132. Wen D, Nakhleh L (2017) Coestimating reticulate phylogenies and gene trees from multilocus sequence data. Syst Biol 67 (3):439–457

133. Yu Y, Nakhleh L (2015) A maximum pseudo-likelihood approach for phylogenetic networks. BMC Genomics 16(Suppl 10):S10
134. Stenz NW, Larget B, Baum DA, Ané C (2015) Exploring tree-like and non-tree-like patterns using genome sequences: an example using the inbreeding plant species Arabidopsis thaliana (L.) Heynh. Syst Biol 64(5):809–823
135. Beaumont MA, Balding DJ (2004) Identifying adaptive genetic divergence among populations from genome scans. Mol Ecol 13:969–980
136. Storz JF (2005) Using genome scans of DNA polymorphism to infer adaptive population divergence. Mol Ecol 14:671–688
137. Barrett RDH, Hoekstra HE (2011) Molecular spandrels: tests of adaptation at the genetic level. Nat Rev Genet 12:767–780
138. Swofford DL (1991) When are phylogeny estimates from molecular and morphological data incongruent? In: Miyamoto MM, Cracraft J (eds) Phylogenetic analysis of DNA sequences. Oxford University Press, Oxford, pp 295–333
139. Dufraigne C, Fertil B, Lespinats S, Giron A, Deschavanne P (2005) Detection and characterization of horizontal transfers in prokaryotes using genomic signature. Nucleic Acids Res 33:e6
140. Roettger M, Martin W, Dagan T (2009) A machine-learning approach reveals that alignment properties alone can accurately predict inference of lateral gene transfer from discordant phylogenies. Mol Biol Evol 26:1931–1939
141. Zimmermann T, Mirarab S, Warnow T (2014) BBCA: improving the scalability of *BEAST using random binning. BMC Genomics 15(Suppl 6):S11
142. Liu L, Yu L (2010) Phybase: an R package for species tree analysis. Bioinformatics 26:962–963
143. Angelis K, dos Reis M (2015) The impact of ancestral population size and incomplete lineage sorting on Bayesian estimation of species divergence times. Curr Zool 61:874–885
144. Edwards SV, Beerli P (2000) Perspective: gene divergence, population divergence, and the variance in coalescence time in phylogeographic studies. Evolution 54:1839–1854
145. Stadler T (2011) Simulating trees with a fixed number of extant species. Syst Biol 60:676–684
146. Papadantonakis S, Poirazi P, Pavlidis P (2016) CoMuS: simulating coalescent histories and polymorphic data from multiple species. Mol Ecol Resour 16:1435–1448
147. Anderson CNK, Ramakrishnan U, Chan YL, Hadly EA (2005) Serial SimCoal: a population genetics model for data from multiple populations and points in time. Bioinformatics 21:1733–1734
148. Excoffier L, Foll M (2011) fastsimcoal: a continuous-time coalescent simulator of genomic diversity under arbitrarily complex evolutionary scenarios. Bioinformatics 27:1332–1334
149. Robinson DF, Foulds LR (1981) Comparison of phylogenetic trees. Math Biosci 53:131–147
150. Song S, Liu L, Edwards SV, Wu SY (2012) Resolving conflict in eutherian mammal phylogeny using phylogenomics and the multispecies coalescent model. Proc Natl Acad Sci U S A 109:14942–14947
151. Heled J, Bryant D, Drummond AJ (2013) Simulating gene trees under the multispecies coalescent and time-dependent migration. BMC Evol Biol 13:44
152. Edwards SV, Potter S, Schmitt CJ, Bragg JG, Moritz C (2016) Reticulation, divergence, and the phylogeography–phylogenetics continuum. Proc Natl Acad Sci U S A 113:8025–8032
153. Weyenberg G, Huggins PM, Schardl CL, Howe DK, Yoshida R (2014) KDETREES: non-parametric estimation of phylogenetic tree distributions. Bioinformatics 30:2280–2287
154. Gaither, J Kubatko L (2016) Hypothesis tests for phylogenetic quartets, with applications to coalescent-based species tree inference. J Theor Biol 408:179–186
155. McVay JD, Carstens BC (2013) Phylogenetic model choice: justifying a species tree or concatenation analysis. J Phylogenet Evol Biol 1:114
156. Felsenstein J (1985) Confidence limits on phylogenies: an approach using the bootstrap. Evolution 39:783–791
157. Lemoine F, Domelevo Entfellner JB, Wilkinson E, Correia D, Davila Felipe M, De Oliveira T, Gascuel O (2018) Renewing Felsenstein's phylogenetic bootstrap in the era of big data. Nature 556:452–456
158. Seo TK (2008) Calculating bootstrap probabilities of phylogeny using multilocus sequence data. Mol Biol Evol 25:960–971
159. Sayyari E, Mirarab S (2016) Fast coalescent-based computation of local branch support from quartet frequencies. Mol Biol Evol 33:1654–1668
160. Liu L et al (2017) Reply to Gatesy and Springer: Claims of homology errors and

zombie lineages do not compromise the dating of placental diversification. Proc Natl Acad Sci U S A 114:E9433–E9434

161. Edwards SV (2016) Phylogenomic subsampling: a brief review. Zool Scr 45:63–74
162. Lee JY, Joseph L, Edwards SV (2012) A species tree for the Australo-Papuan Fairy-wrens and Allies (Aves: Maluridae). Syst Biol 61:253–271
163. Stamatakis A (2014) RAxML version 8: a tool for phylogenetic analysis and post-analysis of large phylogenies. Bioinformatics 30:1312–1313
164. Xi Z, Liu L, Davis CC (2015) Genes with minimal phylogenetic information are problematic for coalescent analyses when gene tree estimation is biased. Mol Phylogenet Evol 92:63–71
165. Guindon S, Dufayard JF, Hordijk W, Lefort V, Gascuel O (2009) PhyML: fast and accurate phylogeny reconstruction by maximum likelihood. Infect Genet Evol 9:384–385
166. Leaché AD, Oaks JR (2017) The utility of single nucleotide polymorphism (SNP) data in phylogenetics. Annu Rev Ecol Evol Syst 48:69–84
167. Pease JB, Haak DC, Hahn MW, Moyle LC (2016) Phylogenomics reveals three sources of adaptive variation during a rapid radiation. PLoS Biol 14:e1002379
168. Hahn MW, Nakhleh L (2016) Irrational exuberance for resolved species trees. Evolution 70:7–17
169. Jennings WB, Edwards SV (2005) Speciational history of Australian grass finches (*Poephila*) inferred from 30 gene trees. Evolution 59:2033–2047
170. Stamatakis A (2006) RAxML-VI-HPC: maximum likelihood-based phylogenetic analyses with thousands of taxa and mixed models. Bioinformatics 22:2688–2690

[illegible]

Chapter 8

Genome-Wide Comparative Analysis of Phylogenetic Trees: The Prokaryotic Forest of Life

Pere Puigbò, Yuri I. Wolf, and Eugene V. Koonin

Abstract

Genome-wide comparison of phylogenetic trees is becoming an increasingly common approach in evolutionary genomics, and a variety of approaches for such comparison have been developed. In this article we present several methods for comparative analysis of large numbers of phylogenetic trees. To compare phylogenetic trees taking into account the bootstrap support for each internal branch, the boot-split distance (BSD) method is introduced as an extension of the previously developed split distance (SD) method for tree comparison. The BSD method implements the straightforward idea that comparison of phylogenetic trees can be made more robust by treating tree splits differentially depending on the bootstrap support. Approaches are also introduced for detecting treelike and netlike evolutionary trends in the phylogenetic Forest of Life (FOL), i.e., the entirety of the phylogenetic trees for conserved genes of prokaryotes. The principal method employed for this purpose includes mapping quartets of species onto trees to calculate the support of each quartet topology and so to quantify the tree and net contributions to the distances between species. We describe the applications methods used to analyze the FOL and the results obtained with these methods. These results support the concept of the Tree of Life (TOL) as a central evolutionary trend in the FOL as opposed to the traditional view of the TOL as a "species tree."

Key words Forest of Life, Tree of Life, Phylogenomic methods, Tree comparison, Map of quartets

Abbreviations

BSD	Boot-split distance
CMDS	Classical multidimensional scaling
COG	Clusters of orthologous genes
FOL	Forest of Life
HGT	Horizontal gene transfer
ND	Nodal distance
NUTs	Nearly universal trees
QT	Quartet topology

Electronic supplementary material: The online version of this chapter (https://doi.org/10.1007/978-1-4939-9074-0_8) contains supplementary material, which is available to authorized users.

Maria Anisimova (ed.), *Evolutionary Genomics: Statistical and Computational Methods*, Methods in Molecular Biology, vol. 1910,
https://doi.org/10.1007/978-1-4939-9074-0_8, © The Author(s) 2019

SD	Split distance
TNT	Tree-Net trend
TOL	Tree of Life

1 Introduction

With the advances of genomics, phylogenetics entered a new era that is noted by the availability of extensive collections of phylogenetic trees for thousands of individual genes. Examples of such tree collections are the phylomes that encompass trees for all sufficiently widespread genes in a given genome [1–4] or the "Forest of Life" (FOL) that consists of all trees for widespread genes in a representative set of organisms [5]. It has been known since the early days of phylogenetics that trees built on the same set of species often have different topologies, especially when the set includes distant species, most notably, in prokaryotes [6, 7]. The availability of "forests" consisting of numerous phylogenetic trees exacerbated the problem as an enormous diversity of tree topologies has been revealed. The inconsistency between trees has several major sources: (1) problems with ortholog identification caused primarily by cryptic paralogy; (2) various artifacts of phylogenetic analysis, such as long branch attraction (LBA); (3) horizontal gene transfer (HGT); and (4) other evolutionary processes distorting the vertical, treelike pattern such as incomplete lineage sorting and hybridization [1, 8–10]. In order to obtain robust results in genome-level phylogenetic analysis, for instance, to classify phylogenetic trees into clusters with (partially) congruent topologies or to identify common trends among multiple trees, reliable methods for comparing trees are indispensable.

The number and diversity of tree comparison methods and software have substantially increased in the last few years. The tree comparison methods variously use tree bipartitions, such as partition or symmetric difference metrics [11] and split distance [12]; distance between nodes such as the path length metrics [13], nodal distance [12, 14], and nodal distance for rooted trees [15]; comparison of evolutionary units such as triplets and quartets [16]; subtransfer operations such as subtree transfer distance [17], nearest-neighbor interchanging [18], subtree prune and regraft (SPR) using a rooted reference tree [19], SPR for unrooted trees [20] and tree bisection and reconnection (TBR) [17], and matching pair (MP) distance [21]; (dis)agreement methods such as agreement subtrees [22], disagree [12], corresponding mapping [23], and congruence index [24]; tree reconciliation [25]; and topological and branch lengths methods such as K-tree score [26]. Several algorithms have been proposed to analyze with multi-family trees.

For example, the From Multiple to Single (FMTS) algorithm systematically prunes each gene copy from a multi-family tree to obtain all possible single-gene trees [12] and an algorithm implemented in TreeKO prunes nodes from the input rooted trees in which duplication and speciation events are labeled [27]. Another algorithm employs a variant of the classical Robinson-Foulds method to compare phylogenetic networks [28]. However, to the best of our knowledge, none of the available metrics for tree comparison takes into account the robustness of the branches, a feature that appears important to minimize the impact of artifacts (unreliable parts of a tree) on the outcome of comparative tree analysis. Here, we present the boot-split distance (BSD) method that calculates distances between phylogenetic trees with weighting based on bootstrap values. This method is implemented in the program TOPD/FMTS [12]. In our recent research, we used the BSD method combined with classical multidimensional scaling (CMDS) analysis to explore the main trends in the phylogenetic FOL and to explore the "Tree of Life" (TOL) concept in light of comparative genomics [5, 29].

Since the time (ca 1838) when Darwin drew the famous sketch of an evolutionary tree in his notebook on transmutation of species, with the legend "I think...," the thinking on the "Tree of Life" (TOL) has evolved substantially. The first phylogenetic revolution, brought about by the pioneering work of Zuckerkandl and Pauling [30] and later Woese and coworkers [31], was the establishment of molecular sequences as the principal material for phylogenetic tree construction. The second revolution has been triggered by the advent of comparative genomics when it has been realized that HGT, at least among prokaryotes, was much more common than previously suspected. The first revolution was a triumph of the tree thinking, when a well-resolved TOL started to appear within reach. The second revolution undermines the very foundation of the TOL concept and threatens to destroy it altogether [32–34].

The current views of evolutionary biologists on the TOL span the entire range from acceptance to complete rejection, with a host of moderate positions. The following rough classification may be used to summarize these positions (a) acceptance of the TOL as the dominant trend in evolution: HGT is considered to be rare and overhyped, and most of the observed "transfers" are deemed to be artifacts [35–38]; (b) the TOL is the common history of the (nearly) nontransferable core of genes, surrounded by "vines" of HGT [39–50]; (c) each gene has its own evolutionary history blending HGT and vertical inheritance; a statistical trend might exist in the maze of gene histories, and it could even be treelike [5, 29, 51, 52]; and (d) ubiquity of HGT renders the TOL concept totally obsolete (prokaryotic species and higher taxa do not exist, and microbial "taxonomy" is created by a pattern of biased HGT) [32, 34, 53–58].

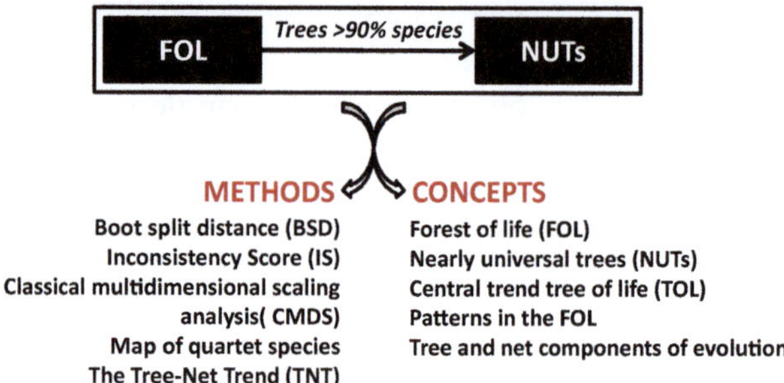

Fig. 1 A schematic of the methods and concepts involved in the FOL analysis

We found that, although different trends and patterns have to be invoked to describe the FOL in its entirety, the main, most robust trend is the "statistical TOL," i.e., the signal of coherent topology that is discernible in a large fraction of the trees in the FOL, in particular, among the nearly universal trees (NUTs) [59, 60].

We further explored the FOL by analysis of species quartets [61]. A quartet is a group of four species which is the minimum evolutionary unit in unrooted phylogenetic trees; each quartet can assume three unrooted tree topologies [16]. We described a quantitative measure of the tree and net signals in evolution that is derived from an analysis of all quartets of species in all trees of the FOL. The results of this analysis indicate that, although diverse routes of netlike evolution jointly dominate the FOL, the pattern of treelike evolution that recapitulates the consensus topology of the NUTs is the single most prominent, coherent trend. Here, we report an extended version of these methodologies introduced to analyze the FOL and its trends, as well as new concepts of prokaryotic evolution under the FOL perspective (Fig. 1).

2 Materials

2.1 The Forest of Life (FOL) and Nearly Universal Trees (NUTs)

We analyzed the set of 6901 phylogenetic trees from [5] that were obtained as follows. Clusters of orthologous genes were obtained from the COG [62] and EggNOG [63] databases from 100 prokaryotic species (59 bacteria and 41 archaea). The species were selected to represent the taxonomic diversity of *Archaea* and *Bacteria* (for the complete list of species, *see* Additional File 1). The BeTs algorithm [62] was used to identify the orthologs with the highest mean similarity to other members of the same cluster ("index orthologs"), so the final clusters contained 100 or fewer genes, with no more than one representative of each species. The sequences in each cluster were aligned using the Muscle program [64] with default parameters and refined using Gblocks [65]. The program Multiphyl [66], which selects the best of 88 amino acid substitution models, was used to reconstruct the maximum likelihood tree of each cluster. The nearly universal trees (NUTs) are defined as trees from COGs that are represented in more than 90% of the species included in the study.

3 Methods

3.1 Boot-Split Distance: A Method to Compare Phylogenetic Trees Taking into Account Bootstrap Support

3.1.1 Boot-Split Distance (BSD)

The BSD method compares trees based on the original split distance (SD) [12] method. Both methods work by collecting all possible binary splits of the two compared trees and calculating the fraction of equal splits, i.e., those splits that are present in both trees (different splits refer to splits that are present in only one of the two trees). Instead of considering all branches as being equal as is the case in SD, the BSD method takes into account the bootstrap values to increase or decrease the SD value proportionally to the robustness of individual internal branches. The BSD value is the average of the BSD in the equal splits (eBSD) and the BSD in the different splits (Eq. 1). Equations 2 and 3 give the formulas to calculate the eBSD and dBSD values, respectively.

$$\mathrm{BSD} = \frac{\mathrm{eBSD} + \mathrm{dBSD}}{2} \tag{1}$$

$$\mathrm{eBSD} = 1 - \left[\frac{e}{a} \cdot M_e\right] \tag{2}$$

$$\mathrm{dBSD} = \frac{d}{a} \cdot M_d \tag{3}$$

Here e is the sum of bootstrap values of equal splits, d is the sum of bootstrap value of different splits, a is the sum of all bootstrap values, M_e is the mean bootstrap value of equal splits, and M_d is the mean bootstrap value of different splits.

The BSD algorithm proceeds in four basic steps to compare pairs of trees (Fig. 2). The first step is to obtain all possible splits from both trees. This procedure implies a binary split of the tree at each internal branch, so that the tree is partitioned into two parts each of which contains at least two species. Then, the common set of leaves between the two trees is obtained, that is, the set of shared species. Only trees with a common leaf set of at least four species can be compared. The third step consists in pruning all splits to the common leaf set of species; at this step, species that are present in only one of the two compared trees are removed from the split list. After this procedure, in partially overlapping trees, the algorithm checks whether each of the splits remains a valid partition, that is, a partition that separates at least two species from the rest of the tree. If a split is not a valid partition, it is removed. Finally, the algorithm calculates the BSD using Eqs. 1–3.

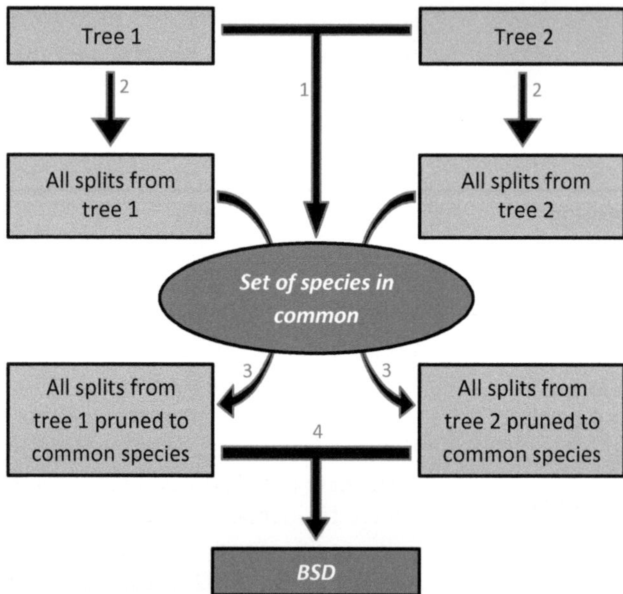

Fig. 2 The main algorithm of the BSD method. The algorithm to calculate the BSD between two trees includes four basic steps: (1) split both trees in all possible partitions, (2) read the common set of species of both trees, (3) prune the splits according with the common leaf set, and (4) calculate the BSD

3.1.2 The BSD Algorithm

There are three possible types of comparisons for trees that do not include paralogs, that is, include one and only one sequence from each of the constituent species (Fig. 3). In the first case, the two trees completely overlap, that is, consist of the same set of species (Fig. 3a). In this case, step 2, the pruning procedure, is not necessary, and the comparison involves only obtaining all possible splits and the calculation of the BSD. In the second case, one of the compared trees is a subset of the other tree (Fig. 3b). In this case, the splits are only pruned and occasionally removed from the bigger tree. In the third case, when the two trees partially overlap or when a tree is a subset of another tree, a pruning procedure is required. In the example shown in Fig. 4, after the pruning procedure (step 3), there is only one remaining split (split: AB|CD) that is repeated several times in both trees. The remaining AB|CD split in Tree 1 is separated by four nodes that have different bootstrap values. In this case, the bootstrap of the remaining split is calculated using Eq. 4, where n is the total number of nodes between the two sides of the split and BS_i is the bootstrap value (adjusted to the 0–1 range) of the node i.

$$\text{Bootstrap} = 1 - \Pi_{i=1}^{n}(1 - \mathrm{BS}_i) \tag{4}$$

The bootstrap value associated with a particular branch of a binary tree is taken as a measure of the probability that the four

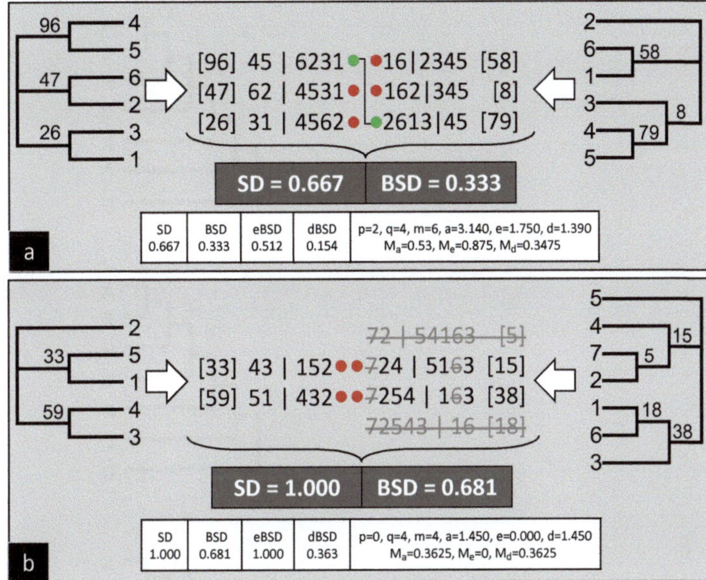

Fig. 3 Examples of the BSD algorithm in single family trees. (**a**) Two trees of the same size. (**b**) Tree 1 is a subtree of the Tree 2. Two trees that partially overlap. *SD* split distance, *BSD* boot-split distance, *eBSD* BSD of equal splits, *dBSD* BSD of different splits, *p* number of equal splits, *q* number of different splits, *m* total number of splits, *a* sum of bootstraps in all splits, *e* sum of bootstraps in equal splits, *d* sum of bootstraps in different splits, M_a mean bootstrap value, M_e mean bootstrap value in equal splits, M_d mean bootstrap value in different splits

subtrees on the opposite ends of this branch are partitioned correctly. To estimate the probability of the correct partitioning of an arbitrary set of four subtrees, the internal branch of the quartet tree is mapped onto each of the internal branches of the original tree. The quartet is considered to be resolved correctly if it is resolved correctly relative to any of these branches. Under the assumption that bootstrap probabilities on individual branches are independent, Eq. 4 is obtained as the estimate of the bootstrap probability for the internal branch of the quartet tree.

3.1.3 Using a Bootstrap Threshold: Pros and Cons

The key question regarding the BSD method is as follows: what is the best approach to phylogenetic tree comparison—using all branches, reliable or not, with the appropriate weighting, or using only branches supported by high bootstrap values? The first option is illustrated in Fig. 3, whereas Fig. 5 shows an example of a tree comparison that employs a bootstrap threshold of 70, i.e., only branches supported by a higher bootstrap are taken into account in the comparison. The second procedure appears reasonable and can be recommended in some cases. However, it is not advisable as a general approach because, when two large trees with varying

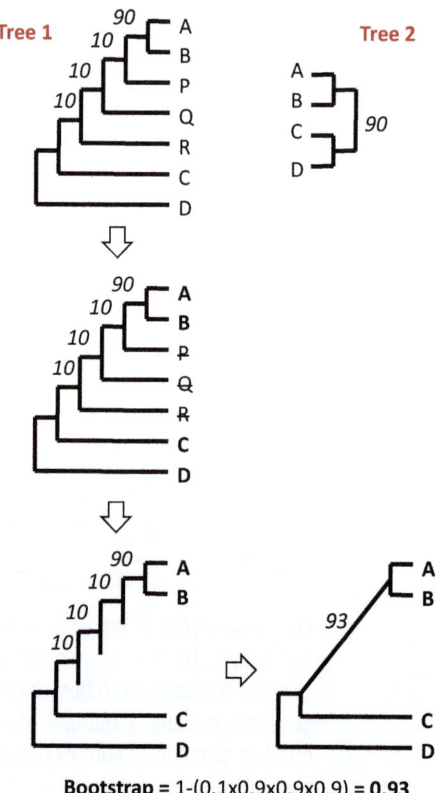

Fig. 4 Calculation of BSD for trees with an unequal numbers of species. The larger tree (1) is pruned prior to the calculation of BSD. The bootstrap value for the only shared internal branch is calculated according to Eq. 4

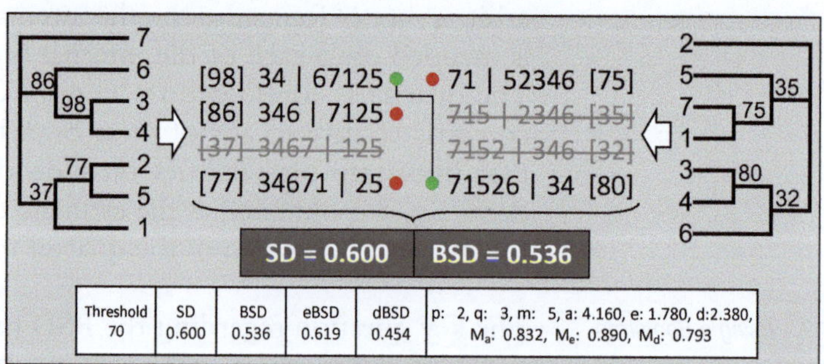

Fig. 5 Example of the BSD algorithm using a bootstrap cutoff. The figure shows the comparison of two phylogenetic trees that takes into account only those branches with bootstrap support greater than 70. *SD* split distance, *BSD* boot-split distance, *eBSD* BSD of equal splits, *dBSD* BSD of different splits, *p* number of equal splits, *q* number of different splits, *m* total number of splits, *a* sum of bootstraps in all splits, *e* sum of bootstraps in equal splits, *d* sum of bootstraps in different splits, M_a mean bootstrap value, M_e mean bootstrap value in equal splits, M_d mean bootstrap value in different splits

bootstrap values are compared, using a strict threshold restricts the comparison to a small subset of robust branches, resulting in an artificially low BSD value. In other words, this procedure artificially inflates the similarity between the two trees by depreciating a large fraction of the branches. In addition, before considering the use of only most supported branches, one should take into account that the BSD method already uses bootstrap values to adjust the distance between trees, so if two trees are topologically similar (low SD) but supported by low bootstrap, the distance value increases (higher BSD), which is one of the advantages of the BSD method (*see* Eqs. 2 and 3).

3.1.4 Testing the BSD Method

The performance of the BSD method was compared with that of the original SD method implemented in the TOPD/FMTS program [12]. Figure 6 shows the correlation of SD and BSD for trees with a number of species from 4 to 15 (a) and from 16 to 100 (b) from a recent large-scale analysis of the FOL [5]. The three-way comparison of SD, BSD, and tree size (number of species) shows a positive correlation between SD and BSD for all tree sizes ($R^2 = 0.8613$ for trees with 4–16 species and $R^2 = 0.7055$ for trees with 16–100 species) (Fig. 6c). However, the SD follows a discrete distribution, which obviously is most conspicuous in the comparisons of small trees (Fig. 6a), whereas, thanks to the use of the bootstrap values, the BSD distribution is continuous (Fig. 7).

Figure 7 shows an example of the comparison (all-against-all) of three trees with six species each that differ in one, two, and three splits, resulting in SD values of 0.33, 0.66, and 1, respectively (Fig. 7a). Also, each tree was compared to itself resulting in a SD of 0. Then, bootstrap values were assigned randomly to the trees in order to compare the trees using the BSD method, and this procedure was repeated 1000 times. The resulting plot (Fig. 7b) shows that, for the comparison of trees with SD of 0 and 1, the BSD values ranged from 0 to 0.5 and from 0.5 to 1, respectively, and in principle, could assume all intermediate values. In the case of the comparisons that differed in one split (SD = 0.33), the BSD value was greater than 0.33 in 75% of the comparison, whereas for the comparisons that differed in two splits (SD = 0.67), 25% of the BSD values were greater than 0.67. Thus, the BSD method for tree comparison offers a better resolution than the SD method, especially, for trees with a small number of species.

Figure 8a shows the results of analysis of six simulated alignments with an increasing level of noise (divergence respect to the initial alignment) in each alignment, i.e., from the alignment 0 (without noise and producing trees with bootstrap values of 100) to alignment 5 with the maximum level of noise. For each alignment, a tree was constructed using the UPGMA method from the web server DendroUPGMA (http://genomes.urv.cat/

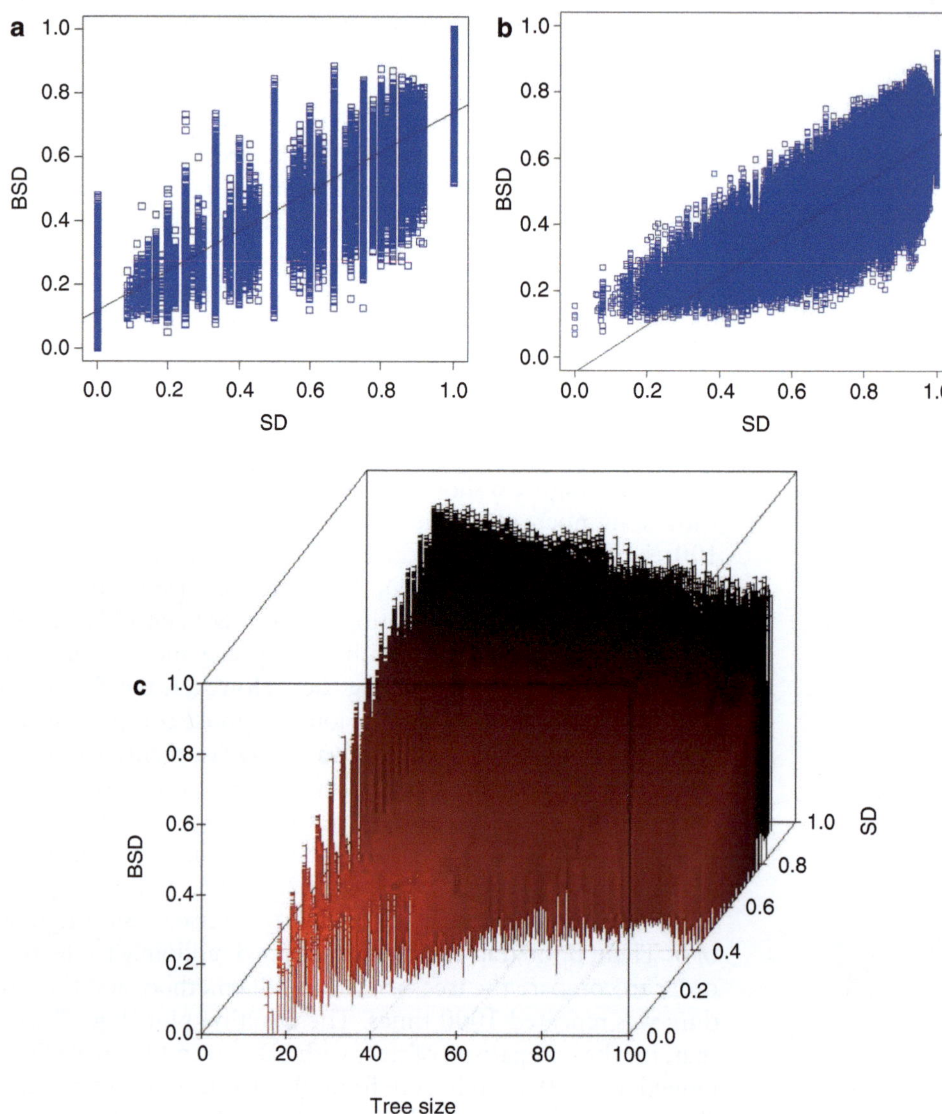

Fig. 6 Correlation of BSD and SD from the all-against-all tree comparisons of 6901 phylogenetic trees. (**a**) Trees containing 4–15 species. (**b**) Trees containing 16–100 species. (**c**) SD, BSD, and tree size for trees containing between 16 and 100 species

UPGMA). Distances were calculated using the Jaccard coefficient, and bootstraps were generated from 100 replicates. The results of the tree comparison (Fig. 8b) using three different methods, namely, nodal distance (ND), SD, and BSD, show that the BSD method presents a continuous distribution resulting in a better resolution of the distances than the other two methods. Indeed, the SD and ND methods fail to discern the similarity between trees after six changes, whereas the BSD method still reports discernible similarity (Fig. 8b). In order to compare the three tree comparison

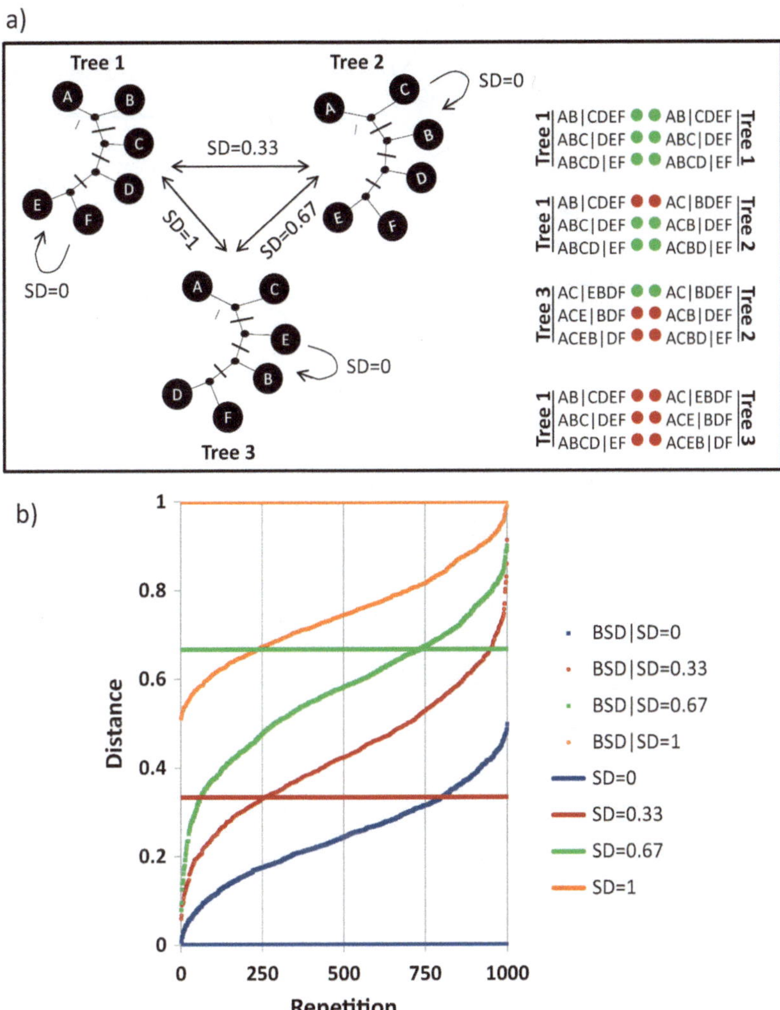

Fig. 7 Comparisons of trees with six taxa. Bootstrap values were assigned randomly in each comparison

methods, the distance reported by each method was normalized to the maximum value in each case, i.e., after 46 changes (maximum number of changes in the simulation), the distance to the initial tree is 1.41, 0.30, and 0.42 for ND, SD, and BSD, respectively. All three distance values indicate that the trees are similar far above the random expectation, supporting the robustness of all methods, but the BSD method presents a better resolution in the tree comparison.

3.1.5 Analysis of Random Trees and the Significance of BSD Results

To assess the significance of the tree comparison by the BSD method, we performed several tree comparisons using random trees containing between 4 and 100 species (Fig. 9). Each test is an all-against-all comparison of 1000 random trees (for complete results *see* Additional File 2). The results from random tree

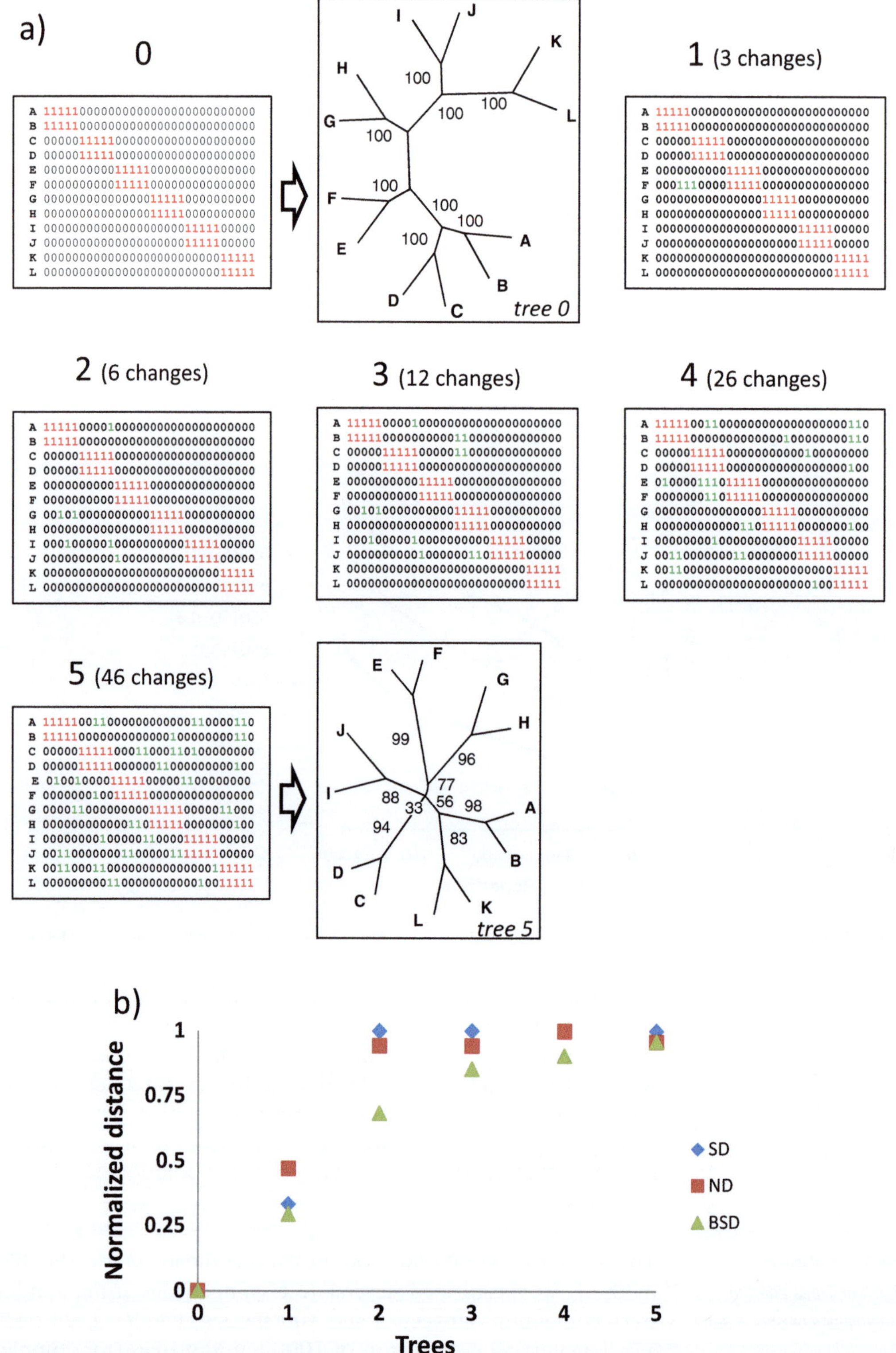

Fig. 8 Comparison of six trees constructed from alignments with increasing noise levels. (**a**) Comparison of trees from six simulated alignments. The UPGMA tree from each alignment was reconstructed with the web

comparison have to be used to determine whether the detected similarities or differences between trees are significantly different from chance [12]. Figure 9 shows that the distance between random trees monotonically increases with the tree size up to a value of approximately 0.75 for BSD and approximately 0.999 for SD. In other words, although BSD is an extension of the SD method, the results obtained by the two methods are not directly comparable. Therefore, to assess whether the similarity between two trees is better than chance, one must consider the method used for the tree comparison (e.g. SD or BSD) and the size of the tree. For example, consider two trees with 15 species each for which the SD method reports a distance of 0.75. This value is far below randomness (Fig. 9), so the conclusion would be that the two trees are nonrandomly similar. However, if the same distance value (0.75) is reported by the BSD method, the conclusion would be the

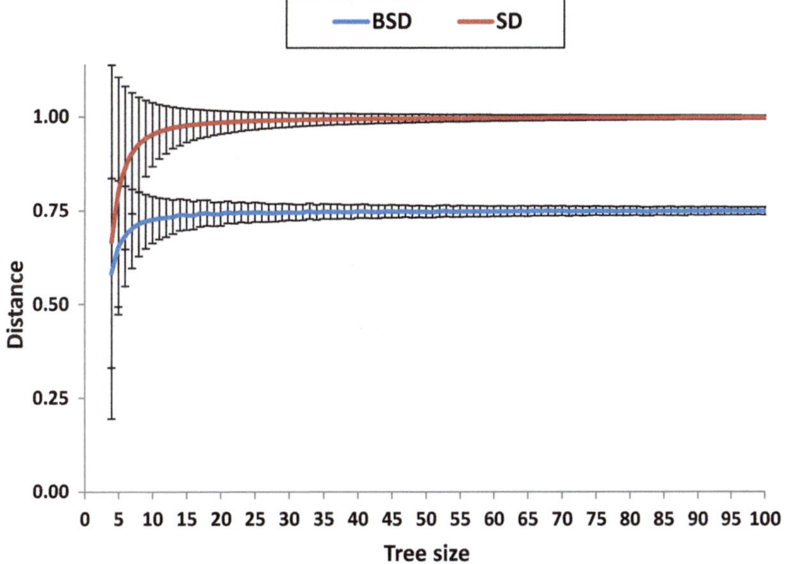

Fig. 9 Random BSD and SD depending on the tree size. Results of the tree comparison of random trees (with different sizes ranging from 4 to 100 species) show that the BSD and SD increase up to 0.75 and 0.999, respectively

Fig. 8 (continued) server DendroUPGMA (http:/genomes.urv.cat/UPGMA) using the Jaccard coefficient as the measure of distance and generating 100 bootstraps replicates. Alignment 0 corresponds to the initial alignment without noise that perfectly separates all branches, resulting in a tree with bootstrap values of 100 for all internal nodes. Alignments 1 to 5 correspond to the derivatives of the initial alignment with increasing noise levels at each step. (**b**) Results of the comparison of each tree [1 to 5] with the initial tree (0). The trees were compared using three methods: split distance (SD), nodal distance (ND), and boot-split distance (BSD). For the purpose of comparison, the results obtained with each of the three methods were normalized to the maximum value in each case

opposite, namely, that the two trees are no more similar than two random trees of 15 species.

Another and probably the most important problem of the comparison of phylogenetic trees is how to interpret the results from a biological perspective. To address this issue, we generated random trees containing from 4 to 100 species and performed 1 to 100 permutations (swap of a pair of branches) in each tree. The resulting tree was then compared with the source tree (Fig. 10a, b). The results show the number of permutations required to obtain a particular BSD value for different tree sizes (number of species). For instance, BSD = 0.3 in the comparison of two trees with

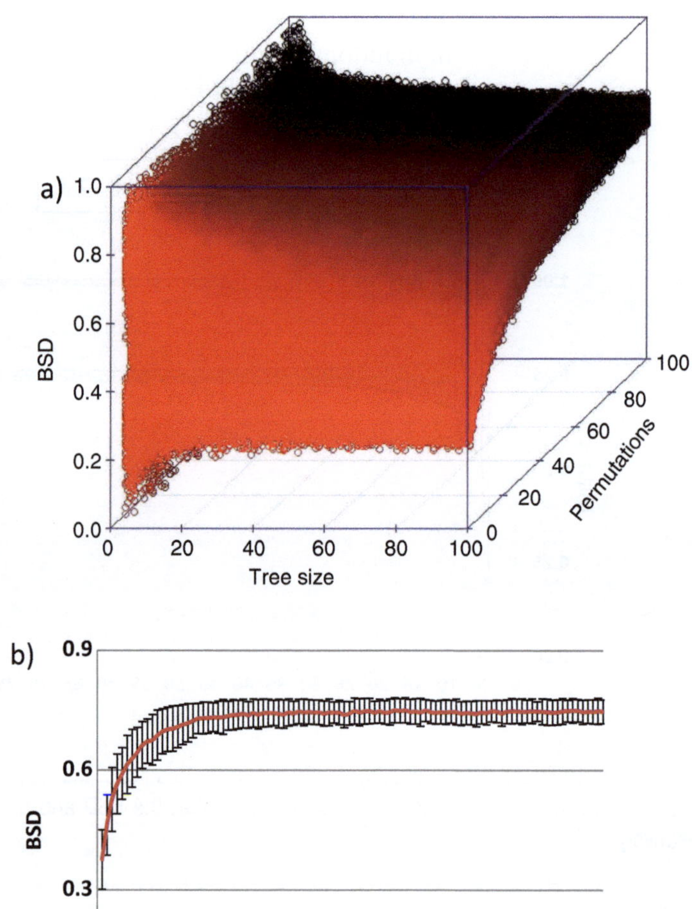

Fig. 10 The number of permutations and the BSD. (**a**) BSD depending on the number of permutations and tree size. (**b**) Mean and standard deviation of the BSD for up to 100 permutations for trees with 20 species

20 species indicates that the two trees are separated by one permutation whereas BSD = 0.6 indicates that the trees are separated by approximately 9 permutations (for the complete listing of equivalences between BSD, SD and the number of permutations, *see* Additional File 3). Considering that each permutation corresponds to an HGT event, the BSD may be construed as the measure of the extent of HGT contributing to the topological difference between the compared trees. Given the discrete distribution of SD values, this measure cannot be used to infer the number of permutations with the same precision as BSD.

3.2 Analysis of Topological Trends in a Set of Phylogenetic Trees

3.2.1 Calculation of the Tree Inconsistency

A key characteristic of the FOL is the degree of the topological (in) consistency between the constituent trees. To quantify this trend, we introduced the inconsistency score (IS), which is the fraction of the times that the splits from a given tree are found in all N trees that comprise the FOL. Thus, the IS may be naturally taken as a measure of how representative of the entire FOL is the topology of the given tree. The IS is calculated using Eqs. 5–7, where N is the total number of trees, X is the number of splits in the given tree, and Y is the number of times the splits from the given tree are found in all trees of the FOL.

$$\mathrm{IS} = \frac{\frac{1}{Y} - \mathrm{IS}_{\min}}{\mathrm{IS}_{\max}} \tag{5}$$

$$\mathrm{IS}_{\min} = \frac{1}{X \cdot N} \tag{6}$$

$$\mathrm{IS}_{\max} = \frac{1}{X} - \mathrm{IS}_{\min} \tag{7}$$

In addition to the calculation of a single value of IS for a given tree by comparing its topology to the topologies of rest of trees in the FOL, IS can be calculated along the depth of the trees, namely, split depth and phylogenetic depth. The split depth was calculated for each unrooted tree according to the number of splits from the tips to the center of the tree. The value of split depth ranged from 1 to 49 ([100 species/2] − 1). The phylogenetic depth was obtained from the branch lengths of a rescaled ultrametric tree, rooted between archaeal and bacterial species, and ranged from 0 to 1. The topology of the ultrametric tree was obtained from the supertree of the 102 NUTs using the CLANN program [67]. The branch lengths from each of the 6901 trees were used to calculate the average distance between each pair of species. The obtained matrix was used to calculate the branch lengths of the supertree of the NUTs. This supertree with branch lengths was then used to construct an ultrametric tree using the program KITSCH from the Phylip package [68] and rescaled to the depth range from 0 to 1. The resulting ultrametric tree was used for the analysis of the dependence of tree inconsistency on phylogenetic depth.

3.2.2 Classical Multidimensional Scaling Analysis

The classical multidimensional scaling (CMDS), also known as principal coordinate analysis, is the multifactorial method best suited to analyze matrices obtained from tree comparison methods like BSD and identify the main trends in a large set of phylogenetic trees. The CMDS embeds n data points implied by a $[n \times n]$ distance matrix into an m-dimensional space ($m < n$) such that, for any $k \in [1, m]$, the embedding into the first k dimensions is the best in terms of preserving the original distances between the points [69, 70]. In our analysis, the data points are distances between trees obtained using the BSD method. The choice of the optimal number of clusters is made using the gap statistics algorithm [71]. The number of clusters for which the value of the gap function for cluster $k + 1$ is not significantly higher than that for cluster k (z-score below 1.96, corresponding to 0.05 significance level) is considered optimal. The CMDS analysis was performed using the K-means function of the R package that implements the K-means algorithm. The CMDS approach has been previously employed by Hillis et al. for phylogenetic tree comparison, with the distances between trees calculated using the Robinson-Foulds distance [72].

3.3 Analysis of Quartets of Species

3.3.1 Definition of Quartets and Mapping Quartets onto Trees

The minimum evolutionary unit in unrooted phylogenetic trees is defined by groups of four species (or quartets), and each quartet may be best represented by the three possible unrooted tree topologies (Fig. 11a). A quartet defined by the set of species A, B, C, and D has three possible unrooted topologies: (1) AB|CD, (2) AC|BD, and (3) AD|BC. To analyze which quartet topology (QT) best represents the relationships among the four species in a quartet, each quartet was compared against the entire set of phylogenetic trees from 100 species (the FOL).

For 100 species, there are 3,921,225 quartets and, accordingly, 11,763,675 topologies (Fig. 11b). A mapping of quartets onto trees is produced using the SD method [12]. A binary version of this method was employed to compare quartets and trees (a quartet is represented in a tree when SD = 0 and not represented when SD > 0). Figure 12a shows an example of quartet mapping onto a set of ten trees. Here q_1 is a resolved quartet, with the topology q_1t_1 supported by eight of the ten trees. By contrast, for q_2, three quartet topologies are equally supported, i.e., the topology of this quartet remains unresolved.

To analyze which of the three possible topologies best represents the almost four million quartets in the FOL, each quartet topology was compared with the entire set of 6901 trees, resulting in a total number of 8.12×10^{10} tree comparisons (Fig. 11b), and the number of trees that support each quartet topology was counted for the entire FOL or for the set of 102 NUTs (Fig. 11b).

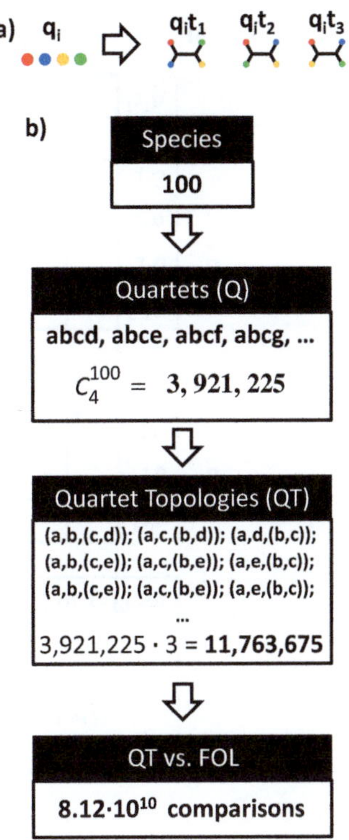

Fig. 11 Quartets and quartet topologies. (**a**) Each quartet (q_i) is defined by a set of four species (different colors denote species) and may be represented by three possible unrooted tree topologies (q_it_i). (**b**) Quartet topologies (QT). In 100 species, the total number of quartets (Q) is 3,921,225. Each quartet may be represented by three distinct QTs, resulting in a total of 11,763,735 QTs. Each QT was mapped onto the FOL, i.e., for each QT, it was determined which of the three topologies is represented in each phylogenetic tree in the FOL (8.12×10^{10} comparisons). Modified from ref. 61

3.3.2 Distance Matrices and Heat Maps

Using the quartet support values for each quartet, a 100×100 between-species distance matrix was calculated as $d_{ij} = 1 - S_{ij}/Q_{ij}$ where d_{ij} is the distance between two species, S_{ij} is the number of trees containing quartets in which the two species are neighbors, and Q_{ij} is the total number of quartets containing the given two species. Then, this distance matrix was used to construct different heat maps using the matrix2png web server ([73], Fig. 12b). In contrast to the BSD method, which is best suited for the analysis of the evolution of individual genes, the distance matrices derived from maps of quartets are used to analyze the evolution of species and to disambiguate treelike evolutionary relationships and "highways" (preferential routes) of HGT.

3.3.3 The Tree-Net Trend (TNT)

The quartet-based between-species distances were used to calculate the Tree-Net Trend (TNT) score. The TNT score is calculated by rescaling each matrix of quartet distances to a 0–1 scale between the

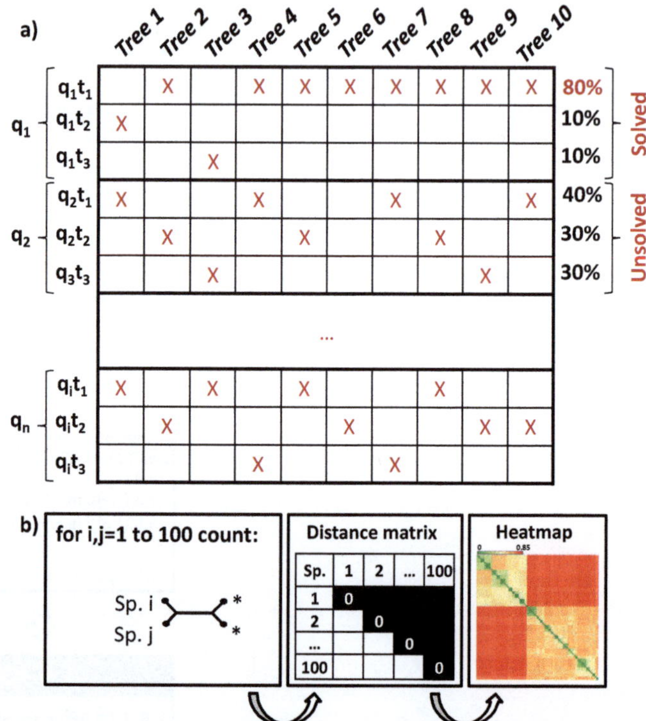

Fig. 12 Mapping quartets. (**a**) Mapping quartets onto a set of ten trees. (**b**) A schematic of the procedure used to reconstruct a species matrix from the map of quartets

supertree-derived matrix (which is taken to represent solely the treelike evolution signal, hence the distance of 0) and the matrix obtained from permuted trees, with distance values around the random expectation of 0.67 (Fig. 13). Two situations may occur in the calculation of the TNT score depending on the relationship between the distance in the supertree matrix (Ds) and the distance in the random matrix (Dr = 0.67). When Ds > Dr (e.g., in comparisons of archaea versus bacteria), $S_{TNT} = (d - Dr)/(Ds - Dr)$, where S_{TNT} is the TNT score and d is the distance between the two compared species in the matrix. When Ds < Dr (in comparisons between closely related species), $S_{TNT} = 1 - ((d - Ds)/(Dr - Ds))$.

4 Phylogenetic Concepts in Light of Pervasive Horizontal Gene Transfer

4.1 Patterns in the Phylogenetic Forest of Life

The reconstruction of the evolutionary trends in the FOL is based on the idea that prokaryotes, effectively, share a common gene pool. This gene pool consists of genes with widely different ranges of phyletic spread, from universal to rare ones only present in a few species [74]. Thus, genes, as the elements of this gene pool, have

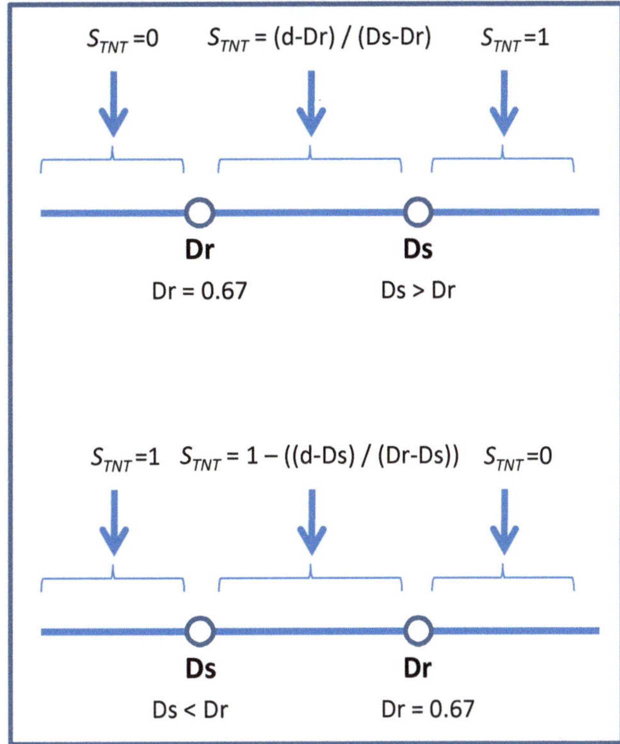

Fig. 13 The Tree-Net Trend (TNT). The figure shows a schematic of the TNT calculation and the rescaling procedure. Modified from ref. 61

their distinct evolutionary histories blending HGT and vertical inheritance (Fig. 14). In principle, the Forest of Life (FOL) encompasses the complete set of phylogenetic trees for all genes from all genomes. However, a comprehensive analysis of the entire FOL is computationally prohibitive (with over 1000 archaeal and bacterial genomes now available and the computational resources accessible to the authors, estimation of the phylogenetic tree for each gene represented in all these genomes would take weeks of computer time) so a representative subset of the trees needs to be selected and analyzed. Previously [5], we defined such a subset by selecting 100 archaeal and bacterial genomes, which are representative of all major prokaryote groups, and building 6901 maximum likelihood (ML) trees for all genes with a sufficient number of homologs and sufficient level of sequence conservation in this set of genomes; for brevity, we refer to this set of trees as the FOL. In this set of almost 7000 trees, only a very small portion of the forest is represented by nearly universal trees (Fig. 14). Furthermore, bacterial and archaeal universal trees are rare as well, as reflected in Fig. 14 by the small peaks around 41 and 59 species, i.e., all archaea and all bacteria, respectively. The dominant pattern in the major part of the FOL is completely different: the FOL is best represented by

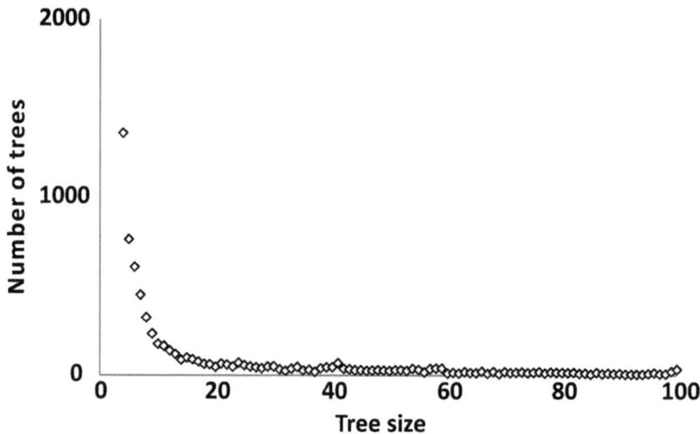

Fig. 14 The Forest of Life (FOL). The distribution of the trees in the FOL by the number of species. Modified from ref. 5

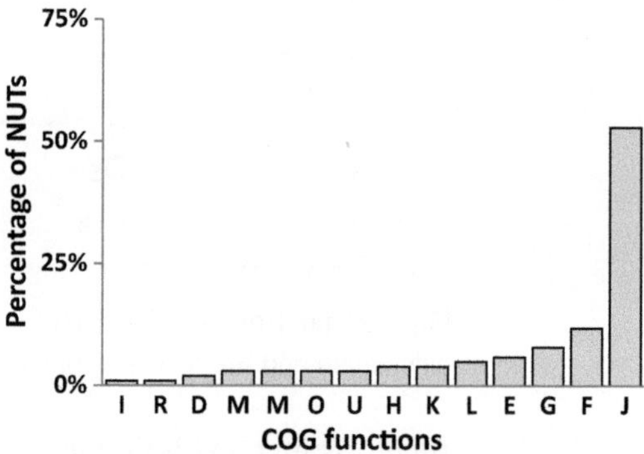

Fig. 15 Distribution of the gene functions among the NUTs. The functional classification of genes was from the COG database [62]

numerous small trees, with about 2/3 of the trees including <20 species (Fig. 14).

4.2 The Nearly Universal Trees (NUTs)

We define the nearly universal trees (NUTs) as trees for those COGs that were represented in more than 90% of the included prokaryotes. This definition yielded 102 NUTs. Not surprisingly, the great majority of the NUTs are genes encoding proteins involved in translation and the core aspects of transcription (Fig. 15). Among the NUTs, only 14 corresponded to COGs that consist of strict 1:1 orthologs (all of them ribosomal proteins), whereas the rest of NUTs included paralogs in some organisms (only the most conserved paralogs were used for tree construction [5]). The 1:1 NUTs were similar to the rest of the NUTs in terms of the connectivity in tree similarity (1-BSD) networks and their positions in the single cluster of NUTs obtained using CMDS.

The 102 NUTs were compared to trees produced by analysis of concatenations of universal proteins [49]. The results showed that most of the NUTs were topologically similar to a tree obtained by the concatenation of 31 universal orthologous genes [5]—in other words, the "Universal Tree of Life" constructed by Ciccarelli et al. [49] was statistically indistinguishable from the NUTs and showed properties of a consensus topology. Not surprisingly, the 1:1 ribosomal protein NUTs were even more similar to the universal tree than the rest of the NUTs, in part because these proteins were used for the construction of the universal tree and, in part, presumably because of the low level of HGT among ribosomal proteins.

4.3 The Tree of Life (TOL) as a Central Trend in the FOL

We analyzed the matrix of all-against-all tree comparisons of the NUTs by embedding them into a 30-dimensional tree space using the CMDS procedure [69, 70]. The gap statistics analysis [71] reveals a lack of significant clustering among the NUTs in the tree space. Thus, all the NUTs seem to belong to one unstructured cloud of points scattered around a single centroid. This organization of the tree space is best compatible with individual trees randomly deviating from a single, dominant topology (which may be denoted the TOL), apparently as a result of random HGT (but in part possibly due to random errors in the tree-construction procedure). Therefore, there is an unequivocal general trend among the NUTs. Although the topologies of the NUTs were, for the most part, not identical, so that the NUTs could be separated by their degree of inconsistency (a proxy for the amount of HGT), the overall high consistency level indicated that the NUTs are scattered in the close vicinity of a consensus tree, with HGT events distributed randomly [5].

Thus, the NUTs present a unique and strong signal of unity that seems to reflect the TOL pattern of evolution. The inconsistency score (IS) among the NUTs ranged from 1.4% to 4.3%, whereas the mean IS value for an equivalent set (102) of randomly generated trees with the same number of species was approximately 80%, indicating that the topologies of the NUTs are highly consistent and nonrandom [5].

To further assess the potential contribution of phylogenetic analysis artifacts to observed inconsistencies between the NUTs, we analyzed these trees with different bootstrap support thresholds (i.e., only splits supported by bootstrap values above the respective threshold value were compared). Particularly low IS levels were detected for splits with high bootstrap support, but the inconsistency was never eliminated completely, suggesting that HGT is a significant contributor to the observed inconsistency among the NUTs (IS ranges from 0.3% to 2.1% and 0.3% to 1.8% for splits with a bootstrap value higher than 70 and 90, respectively) [5].

Analysis of the supernetwork built from the 102 NUTs [5] showed that the incongruence among these trees is mainly concentrated at the deepest levels, with a much greater congruence at shallow phylogenetic depths. The major exception is the unambiguous archaeal-bacterial split that is observed despite the apparent substantial interdomain HGT. Evidence of probable HGT between archaea and bacteria was obtained for approximately 44% of the NUTs (13% from archaea to bacteria, 23% from bacteria to archaea, and 8% in both directions), with the implication that HGT is likely to be even more common between the major branches within the archaeal and bacterial domains [5]. These results are compatible with previous reports on the apparently random distribution of HGT events in the history of highly conserved genes, in particular those encoding proteins involved in translation [75, 76], and on the difficulty of resolving the phylogenetic relationships between the major branches of bacteria [77–79] and archaea [5, 80, 81]. More specifically, archaeal-bacterial HGT has been inferred for 83% of the genes encoding aminoacyl-tRNA synthetases (compared with the overall 44%), essential components of the translation machinery that are known for their horizontal mobility [42, 82]. In contrast, no HGT has been predicted for any of the ribosomal proteins, which belong to an elaborate molecular complex, the ribosome, and hence appear to be non-exchangeable between the two prokaryotic domains [42, 76]. In addition to the aminoacyl-tRNA synthetases, and in agreement with many previous observations ([83] and references therein), evidence of HGT between archaea and bacteria was seen also for the few metabolic enzymes that belonged to the NUTs, including undecaprenyl pyrophosphate synthase, glyceraldehyde-3-phosphate dehydrogenase, nucleoside diphosphate kinase, thymidylate kinase, and others.

4.4 The NUTs Topologies as the Central Trend and Detection Distinct Evolutionary Patterns in the FOL

Using the BSD method, we compared the topologies of the NUTs to those of the rest of the trees in the FOL. Notably, 2615 trees (~38% of the FOL) showed a greater than 50% similarity (P-value <0.05) to at least one of the NUTs, being the mean similarity of the trees to the NUTs approximately 50% (Fig. 16). For a set of 102 randomized trees of the same size as the NUTs, only about 10% of the trees in the FOL showed the same or greater similarity, indicating that the NUTs were strongly and nonrandomly connected to the rest of the FOL.

We then analyzed the structure of the FOL by embedding the 3789 COG trees into a 669-dimensional space using the CMDS procedure [69, 70]. A CMDS clustering of the entire set of 6901 trees in the FOL was beyond the capacity of the R software package used for this analysis; however, the set of COG trees included most of the trees with a large number of species for which the topology comparison is most informative. A gap statistics analysis [69, 70] of K-means clustering of these trees in the tree space revealed distinct clusters of trees in the forest. The FOL is optimally partitioned into seven clusters of trees (the smallest number of clusters for which the gap function did not significantly increase with the increase of the number of clusters) (Fig. 17). Clusters 1, 4, 5, and 6 were enriched

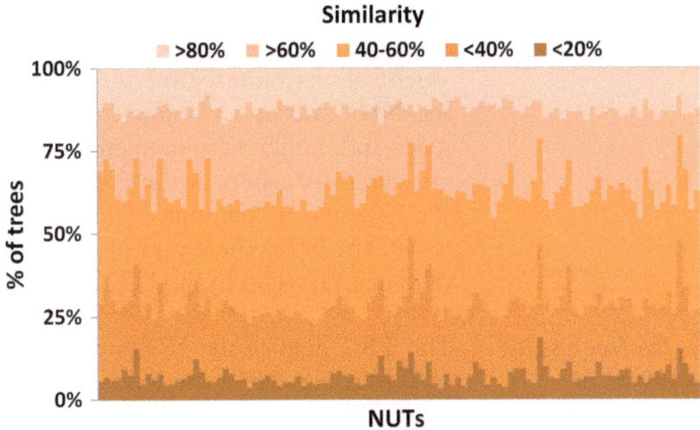

Fig. 16 Topological similarity between the NUTs and the rest of the FOL. Percentage of trees connected to the NUTs at a different percentage of similarity. Modified from ref. 5

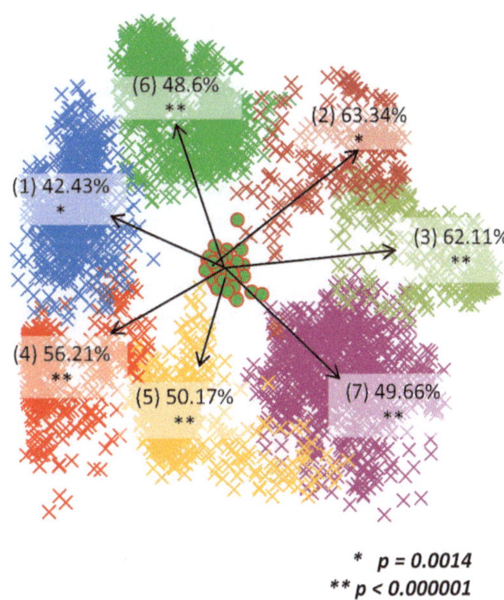

Fig. 17 Clusters and patterns in the FOL. The seven clusters identified in the FOL using the CMDS method and the mean similarity values between the 102 NUTs and all trees from each of the seven clusters are shown. Modified from ref. 5

for bacterial-only trees, all archaeal-only trees belonged to clusters 2 and 3, and cluster 7 consisted entirely of mixed archaeal-bacterial clusters; notably, all the NUTs form a compact group inside cluster 6.

The results of the CMDS clustering (Fig. 17) support the existence of several distinct "attractors" in the FOL. However, we have to emphasize caution in the interpretation of this clustering because trivial separation of the trees by size could be an important contribution. The approaches to the delineation of distinct "groves" within the forest merit further investigation. The most salient observation for the purpose of the present study is that all the NUTs occupy a compact and contiguous region of the tree space and, unlike the complete set of the trees, are not partitioned into distinct clusters by the CMDS procedure. Taken together with the high mean topological similarity between the NUTs and the rest of the FOL, these findings indicate that the NUTs represent a valid central trend in the FOL.

4.5 The Tree and Net Components of Prokaryote Evolution

The TNT map of the NUTs was dominated by the treelike signal (green in Fig. 18a): the mean TNT score for the NUTs was 0.63 (Fig. 19b), so the evolution of the nearly universal genes of

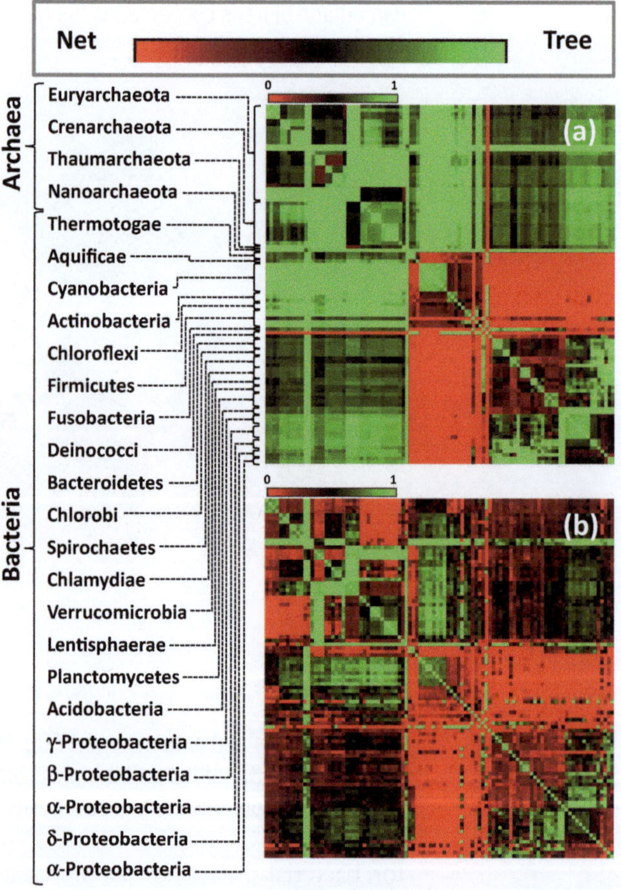

Fig. 18 The Tree-Net Trend (TNT) score heatmaps. (**a**) The 102 NUTs. (**b**) The FOL without the NUTs (6799 trees). The TNT increases from red (low score, close to random, an indication of netlike evolution) to green (high score, close to the supertree topology, an indication of treelike evolution). The species are ordered according to the topology of the supertree of the 102 NUTs. In (**a**), the major groups of archaea and bacteria are denoted. Modified from ref. 61

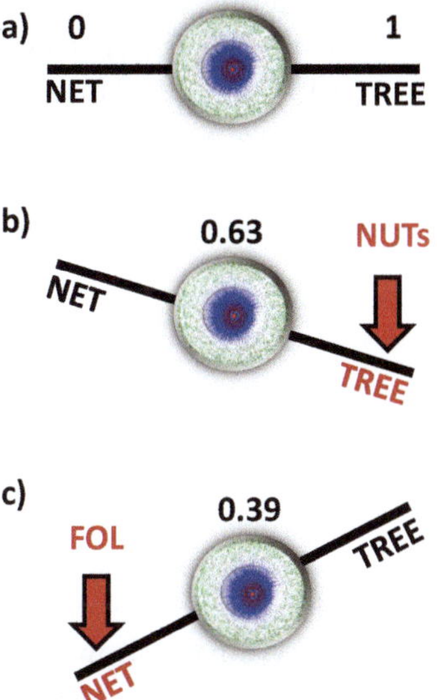

Fig. 19 The Tree-Net Trend in the FOL and in the NUTs. (**a**) A hypothetical equilibrium between the tree and net trends. (**b**) A schematic representation of the tree tendency in the NUTs. (**c**) A schematic representation of the net tendency in the FOL

prokaryotes appears to be almost "two-third treelike" (i.e., reflects the topology of the supertree). The rest of the FOL stood in a stark contrast to the NUTs, being dominated by the netlike evolution, with the mean TNT value of 0.39 (Fig. 19c) (about "60% netlike"). Remarkably, areas of treelike evolution were interspersed with areas of netlike evolution across different parts of the FOL (Fig. 18b). The major netlike areas observed among the NUTs were retained, but additional ones became apparent including Crenarchaeota that showed a pronounced signal of a non-treelike relationship with diverse bacteria as well as some Euryarchaeota (Fig. 18b). The distribution of the tree and net evolutionary signals among different groups of prokaryotes showed a striking split among the NUTs: among the archaea, the tree signal was heavily dominant (mean $TNT_{NUTs\_Archaea} = 0.80 \pm 0.20$), whereas among bacteria the contributions of the tree and net signals were nearly equal (mean $TNT_{NUTs\_Bacteria} = 0.51 \pm 0.38$). Among the rest of the trees in the FOL, archaea also showed a stronger tree signal than bacteria, but the difference was much less pronounced than it was among the NUTs (mean $TNT_{FOL\_Archaea} = 0.47 \pm 0.11$ and mean $TNT_{FOL\_Bacteria} = 0.34 \pm 0.08$). The conclusions on the treelike and netlike components of evolution made here are based on the assumption that the supertree of the NUTs represents the treelike (vertical) signal. We did not perform direct tests of the robustness of these conclusions to the supertree topology. However, observations presented previously [5] suggest that the results are likely to be robust given the coherence of the NUTs topologies as well as the similarity of the supertree topology and the topologies of the individual NUTs to the "Tree of Life" obtained from concatenated sequences of universally conserved ribosomal proteins [49].

5 Conclusions

The analysis of the phylogenetic FOL is a logical strategy for studying the evolution of prokaryotes because each set of orthologous genes presents its own evolutionary history and no single topology may represent the entire forest. Thus, the methods introduced in this article that compare trees without the use of a preconceived representative topology for the entire FOL may be of wide utility in phylogenomics.

We have shown that, although no single topology may represent the entire FOL and several distinct evolutionary trends are detectable, the NUTs contain a strong treelike signal. Although the treelike signal is quantitatively weaker than the sum total of the signals from HGT, it is the most pronounced single pattern in the entire FOL.

Under the FOL perspective, the traditional TOL concept (a single "true" tree topology) is invalidated and should be replaced by a statistical definition. In other words, the TOL only makes sense as a central trend in the phylogenetic forest.

6 Exercises

1. Calculate the split distance (SD) and boot-split distance (BSD) of the following two trees:
   (((A,B)61,C)53,D,E);(((A,C)76,B)38,D,E)
2. Calculate the inconsistency score of the tree X in the "forest of trees" Y.
   X = (((A,B),C),D,E)
   Y = (((A,B),C),D,E); (A,B,(E,D); (((A,C),B),D,E); (A,C,(B, D); (A,B,(C,D); (A,B,(C,E); (A,E,(B,D); (((A,C),D),E,F); (((A,B),D),E,C); (((E,F),A),B,C)

Acknowledgment

The authors' research is supported by the Department of Health and Human Services intramural program (NIH, National Library of Medicine).

References

1. Huerta-Cepas J, Dopazo H, Dopazo J, Gabaldon T (2007) The human phylome. Genome Biol 8:R109
2. Huerta-Cepas J, Bueno A, Dopazo J, Gabaldon T (2008) PhylomeDB: a database for genome-wide collections of gene phylogenies. Nucleic Acids Res 36:D491–D496
3. Frickey T, Lupas AN (2004) PhyloGenie: automated phylome generation and analysis. Nucleic Acids Res 32:5231–5238
4. Sicheritz-Ponten T, Andersson SG (2001) A phylogenomic approach to microbial evolution. Nucleic Acids Res 29:545–552
5. Puigbo P, Wolf YI, Koonin EV (2009) Search for a Tree of Life in the thicket of the phylogenetic forest. J Biol 8:59
6. Felsenstein J (2004) Inferring phylogenies. Sinauer Associates, Sunderland, MA
7. Nei M, Kumar S (2001) Molecular evolution and phylogenetics. Oxford University Press, Oxford
8. Castresana J (2007) Topological variation in single-gene phylogenetic trees. Genome Biol 8:216
9. Soria-Carrasco V, Castresana J (2008) Estimation of phylogenetic inconsistencies in the three domains of life. Mol Biol Evol 25:2319–2329
10. Marcet-Houben M, Gabaldon T (2009) The tree versus the forest: the fungal tree of life and the topological diversity within the yeast phylome. PLoS One 4:e4357
11. Robinson DF, Foulds LR (1981) Comparison of phylogenetic trees. Math Biosci 53:131–147
12. Puigbo P, Garcia-Vallve S, McInerney JO (2007) TOPD/FMTS: a new software to compare phylogenetic trees. Bioinformatics 23:1556–1558
13. Steel MA, Penny D (1993) Distribution of tree comparison metrics - some new results. Syst Biol 42:126–141
14. Bluis J, Shin D-G (2003) Nodal distance algorithm: calculating a phylogenetic tree comparison metric. In: Proceedings of the third IEEE symposium on bioInformatics and bioEngineering, IEEE Computer Society, pp 87–94
15. Cardona G, Llabres M, Rossello F, Valiente G (2009) Nodal distances for rooted phylogenetic trees. J Math Biol 61(2):253–276
16. Estabrook GF, McMorris FR, Meachan A (1985) Comparison of undirected phylogenetic trees based on subtree of four evolutionary units. Syst Zool 34:193–200
17. Allen L, Steel M (2001) Subtree transfer operations and their induced metrics on evolutionary trees. Ann Comb 5:1–15
18. Waterman MS, Steel M (1978) On the similarity of dendrograms. J Theor Biol 73:789–800
19. Beiko RG, Hamilton N (2006) Phylogenetic identification of lateral genetic transfer events. BMC Evol Biol 6:15
20. Hickey G, Dehne F, Rau-Chaplin A, Blouin C (2008) SPR distance computation for unrooted trees. Evol Bioinformatics Online 4:17–27
21. Bogdanowicz D, Giaro K (2017) Comparing phylogenetic trees by matching nodes using the transfer distance between partitions. J Comput Biol 24:422–435
22. Kubicka E, Kubicki G, McMorris FR (1995) An algorithm to find agreement subtrees. J Classif 12:91–99
23. Nye TM, Lio P, Gilks WR (2006) A novel algorithm and web-based tool for comparing two alternative phylogenetic trees. Bioinformatics 22:117–119
24. de Vienne DM, Giraud T, Martin OC (2007) A congruence index for testing topological similarity between trees. Bioinformatics 23:3119–3124
25. Cotton JA, Page RD (2002) Going nuclear: gene family evolution and vertebrate phylogeny reconciled. Proc Biol Sci 269:1555–1561
26. Soria-Carrasco V, Talavera G, Igea J, Castresana J (2007) The K tree score: quantification of differences in the relative branch length and topology of phylogenetic trees. Bioinformatics 23:2954–2956
27. Marcet-Houben M, Gabaldon T (2011) TreeKO: a duplication-aware algorithm for the comparison of phylogenetic trees. Nucleic Acids Res 39:e66
28. Lu B, Zhang L, Leong HW (2017) A program to compute the soft Robinson-Foulds distance between phylogenetic networks. BMC Genomics 18:111
29. Koonin EV, Wolf YI, Puigbo P (2009) The phylogenetic forest and the quest for the elusive tree of life. Cold Spring Harb Symp Quant Biol 74:205–213
30. Zuckerkandl E, Pauling L (1962) Molecular evolution. In: Kasha M, Pullman B (eds) Horizons in biochemistry. Academic, New York, pp 189–225
31. Woese CR (1987) Bacterial evolution. Microbiol Rev 51:221–271

32. Bapteste E, O'Malley MA, Beiko RG, Ereshefsky M, Gogarten JP, Franklin-Hall L et al (2009) Prokaryotic evolution and the tree of life are two different things. Biol Direct 4:34
33. Doolittle WF (2000) Uprooting the tree of life. Sci Am 282:90–95
34. Doolittle WF, Bapteste E (2007) Pattern pluralism and the Tree of Life hypothesis. Proc Natl Acad Sci U S A 104:2043–2049
35. Kurland CG, Canback B, Berg OG (2003) Horizontal gene transfer: a critical view. Proc Natl Acad Sci U S A 100:9658–9662
36. Kurland CG (2005) What tangled web: barriers to rampant horizontal gene transfer. BioEssays 27:741–747
37. Logsdon JM, Faguy DM (1999) Thermotoga heats up lateral gene transfer. Curr Biol 9: R747–R751
38. Genereux DP, Logsdon JM Jr (2003) Much ado about bacteria-to-vertebrate lateral gene transfer. Trends Genet 19:191–195
39. Kunin V, Goldovsky L, Darzentas N, Ouzounis CA (2005) The net of life: reconstructing the microbial phylogenetic network. Genome Res 15:954–959
40. Daubin V, Moran NA, Ochman H (2003) Phylogenetics and the cohesion of bacterial genomes. Science 301:829–832
41. Lerat E, Daubin V, Moran NA (2003) From gene trees to organismal phylogeny in prokaryotes: the case of the gamma-proteobacteria. PLoS Biol 1:E19
42. Woese CR, Olsen GJ, Ibba M, Soll D (2000) Aminoacyl-tRNA synthetases, the genetic code, and the evolutionary process. Microbiol Mol Biol Rev 64:202–236
43. Fitz-Gibbon ST, House CH (1999) Whole genome-based phylogenetic analysis of free-living microorganisms. Nucleic Acids Res 27:4218–4222
44. Hanage WP, Fraser C, Spratt BG (2006) Sequences, sequence clusters and bacterial species. Philos Trans R Soc Lond B Biol Sci 361:1917–1927
45. Eisen JA, Fraser CM (2003) Phylogenomics: intersection of evolution and genomics. Science 300:1706–1707
46. Salzberg SL, White O, Peterson J, Eisen JA (2001) Microbial genes in the human genome: lateral transfer or gene loss? Science 292:1903–1906
47. Galtier N (2007) A model of horizontal gene transfer and the bacterial phylogeny problem. Syst Biol 56:633–642
48. Galtier N, Daubin V (2008) Dealing with incongruence in phylogenomic analyses. Philos Trans R Soc Lond B Biol Sci 363:4023–4029
49. Ciccarelli FD, Doerks T, von Mering C, Creevey CJ, Snel B, Bork P (2006) Toward automatic reconstruction of a highly resolved tree of life. Science 311:1283–1287
50. Choi IG, Kim SH (2007) Global extent of horizontal gene transfer. Proc Natl Acad Sci U S A 104:4489–4494
51. Dagan T, Martin W (2009) Getting a better picture of microbial evolution en route to a network of genomes. Philos Trans R Soc Lond B Biol Sci 364:2187–2196
52. Boucher Y, Douady CJ, Papke RT, Walsh DA, Boudreau ME, Nesbo CL et al (2003) Lateral gene transfer and the origins of prokaryotic groups. Annu Rev Genet 37:283–328
53. Bucknam J, Boucher Y, Bapteste E (2006) Refuting phylogenetic relationships. Biol Direct 1:26
54. Schliep K, Lopez P, Lapointe FJ, Bapteste E (2011) Harvesting evolutionary signals in a forest of prokaryotic gene trees. Mol Biol Evol 28:1393–1405
55. Beiko RG, Doolittle WF, Charlebois RL (2008) The impact of reticulate evolution on genome phylogeny. Syst Biol 57:844–856
56. Doolittle WF, Zhaxybayeva O (2009) On the origin of prokaryotic species. Genome Res 19:744–756
57. Gogarten JP, Townsend JP (2005) Horizontal gene transfer, genome innovation and evolution. Nat Rev Microbiol 3:679–687
58. Gogarten JP, Doolittle WF, Lawrence JG (2002) Prokaryotic evolution in light of gene transfer. Mol Biol Evol 19:2226–2238
59. O'Malley MA, Koonin EV (2011) How stands the Tree of Life a century and a half after The Origin? Biol Direct 6:32
60. Puigbo P, Wolf YI, Koonin EV (2013) Seeing the Tree of Life behind the phylogenetic forest. BMC Biol 11:46
61. Puigbo P, Wolf YI, Koonin EV (2010) The tree and net components of prokaryote evolution. Genome Biol Evol 2:745–756
62. Tatusov RL, Fedorova ND, Jackson JD, Jacobs AR, Kiryutin B, Koonin EV et al (2003) The COG database: an updated version includes eukaryotes. BMC Bioinformatics 4:41
63. Jensen LJ, Julien P, Kuhn M, von Mering C, Muller J, Doerks T et al (2008) eggNOG: automated construction and annotation of orthologous groups of genes. Nucleic Acids Res 36:D250–D254

64. Edgar RC (2004) MUSCLE: multiple sequence alignment with high accuracy and high throughput. Nucleic Acids Res 32:1792–1797
65. Castresana J (2000) Selection of conserved blocks from multiple alignments for their use in phylogenetic analysis. Mol Biol Evol 17:540–552
66. Keane TM, Naughton TJ, McInerney JO (2007) MultiPhyl: a high-throughput phylogenomics webserver using distributed computing. Nucleic Acids Res 35:W33–W37
67. Creevey CJ, McInerney JO (2005) Clann: investigating phylogenetic information through supertree analyses. Bioinformatics 21:390–392
68. Felsenstein J (1996) Inferring phylogenies from protein sequences by parsimony, distance, and likelihood methods. Methods Enzymol 266:418–427
69. Torgerson WS (1958) Theory and methods of scaling. Wiley, New York
70. Gower JC (1966) Some distance properties of latent root and vector methods used in multivariate analysis. Biometrika 53:325–328
71. Tibshirani R, Walther G, Hastie T (2001) Estimating the number of clusters in a data set via the gap statistic. J R Stat Soc B Stat Methodol 63:411–423
72. Hillis DM, Heath TA, St John K (2005) Analysis and visualization of tree space. Syst Biol 54:471–482
73. Pavlidis P, Noble WS (2003) Matrix2png: a utility for visualizing matrix data. Bioinformatics 19:295–296
74. Koonin EV, Wolf YI (2008) Genomics of bacteria and archaea: the emerging dynamic view of the prokaryotic world. Nucleic Acids Res 36:6688–6719
75. Ge F, Wang LS, Kim J (2005) The cobweb of life revealed by genome-scale estimates of horizontal gene transfer. PLoS Biol 3:e316
76. Brochier C, Bapteste E, Moreira D, Philippe H (2002) Eubacterial phylogeny based on translational apparatus proteins. Trends Genet 18:1–5
77. Wolf YI, Rogozin IB, Grishin NV, Koonin EV (2002) Genome trees and the tree of life. Trends Genet 18:472–479
78. Wolf YI, Rogozin IB, Grishin NV, Tatusov RL, Koonin EV (2001) Genome trees constructed using five different approaches suggest new major bacterial clades. BMC Evol Biol 1:8
79. Creevey CJ, Fitzpatrick DA, Philip GK, Kinsella RJ, O'Connell MJ, Pentony MM et al (2004) Does a tree-like phylogeny only exist at the tips in the prokaryotes? Proc Biol Sci 271:2551–2558
80. Brochier-Armanet C, Boussau B, Gribaldo S, Forterre P (2008) Mesophilic Crenarchaeota: proposal for a third archaeal phylum, the Thaumarchaeota. Nat Rev Microbiol 6:245–252
81. Elkins JG, Podar M, Graham DE, Makarova KS, Wolf Y, Randau L et al (2008) A korarchaeal genome reveals new insights into the evolution of the Archaea. Proc Natl Acad Sci U S A 105:8102–8107
82. Wolf YI, Aravind L, Grishin NV, Koonin EV (1999) Evolution of aminoacyl-tRNA synthetases--analysis of unique domain architectures and phylogenetic trees reveals a complex history of horizontal gene transfer events. Genome Res 9:689–710
83. Koonin EV (2003) Comparative genomics, minimal gene-sets and the last universal common ancestor. Nat Rev Microbiol 1:127–136

Chapter 9

The Methodology Behind Network Thinking: Graphs to Analyze Microbial Complexity and Evolution

Andrew K. Watson, Romain Lannes, Jananan S. Pathmanathan, Raphaël Méheust, Slim Karkar, Philippe Colson, Eduardo Corel, Philippe Lopez, and Eric Bapteste

Abstract

In the post genomic era, large and complex molecular datasets from genome and metagenome sequencing projects expand the limits of what is possible for bioinformatic analyses. Network-based methods are increasingly used to complement phylogenetic analysis in studies in molecular evolution, including comparative genomics, classification, and ecological studies. Using network methods, the vertical and horizontal relationships between all genes or genomes, whether they are from cellular chromosomes or mobile genetic elements, can be explored in a single expandable graph. In recent years, development of new methods for the construction and analysis of networks has helped to broaden the availability of these approaches from programmers to a diversity of users. This chapter introduces the different kinds of networks based on sequence similarity that are already available to tackle a wide range of biological questions, including sequence similarity networks, gene-sharing networks and bipartite graphs, and a guide for their construction and analyses.

Key words Sequence similarity network, Evolution, Lateral gene transfer (LGT), Metagenomics, Gene remodeling, Ecology

1 Introduction

An evolutionary biologist is interested in how processes governing evolution have produced the diversity of genes, genomes, organisms, species, and communities that are observed today. For example, a biologist interested in the eukaryotes may wonder what symbiotic partners have contributed to their origins and evolution. Eukaryotic nuclear genomes are chimeric in nature, encoding many genes acquired from their alphaproteobacterial endosymbiont [1–3]. However, in recent years, it has been proposed that the ongoing gain of genes by both microbial [4–6] and multicellular eukaryotes [7, 8] via lateral gene transfer (LGT) has continued to contribute to eukaryotic evolution, though to a lesser extent than

Maria Anisimova (ed.), *Evolutionary Genomics: Statistical and Computational Methods*, Methods in Molecular Biology, vol. 1910, https://doi.org/10.1007/978-1-4939-9074-0_9,

prokaryotes [9]. A biologist interested in prokaryotes may wish to investigate lateral gene transfer to explore the numbers and kinds of genes transferred between bacteria, archaea, and their mobile genetic elements [10–14]. These transfers are important for understanding the accessory genomes of prokaryotes [15–17]. Further, studying gene transfers in real bacterial communities from different environments can help to test the effect of LGT on ecology and evolution of communities [18]. Given the prevalence of introgression [9–11, 19], one interesting question is whether gene transfer has led to the formation of novel fusion genes that combine parts of genes originating from separate domains of life [20]. An ecologist may wish to analyze the distribution of genes and species in the environment [21]. A metagenome analyst may need to overcome an additional challenge exploring the nature of the large proportion of sequences in metagenome datasets that have little or no detectable similarity to characterize sequences and to study the "microbial dark matter" [22].

High-throughput sequencing technologies present new opportunities to investigate these diverse kinds of questions with molecular data; however, they also present challenges in terms of the scale of the analyses. Consequently, a number of network-based methods have recently been developed to expand the toolkit available to molecular biologists [23], and these have already made major contributions to our understanding of molecular evolution. Networks have been used to shed light on the nature of the "microbial dark matter" [24] and used in ecological studies to explore the geographical distribution of organisms or genes [25, 26] or the evolution of different lifestyles [27]. Their suitability for investigating introgressive events has been used to enhance our understanding of the chimeric origin of genes in the eukaryotic proteome [28, 29], the flow of genes between prokaryotes and their mobile genetic elements [30–35], and gene sharing across mobile elements to study the transfer of resistance factors [14, 36]. Networks have also been used to classify highly mosaic viral genomes [37, 38] and identify gene families [39, 40]. These approaches are highly complementary to traditional phylogenetic approaches, highlighted by the development of hybrid approaches and phylogenetic and phylogenomic networks [34, 41–43]. These hybrid networks are beyond the scope of discussion in this chapter but are covered in Chapters 7 and 8.

While the generation and analysis of networks were previously limited to biologists with programming experience, tools have recently been developed to simplify the process and broaden the availability of network analyses of molecular sequence data. This chapter introduces the different kinds of networks that are already available to biologists and a guide to how these networks can be constructed and analyzed for a large range of applications in molecular evolution. More precisely, this chapter will focus on three kinds of network and the types of analyses that are possible using these networks: sequence similarity networks, gene-sharing networks, and multipartite graphs [23].

2 Sequence Similarity Networks (SSNs)

Sequence similarity networks are the bread and butter of network-based molecular sequence analyses, with a huge range of applications in molecular biology. The use of SSNs for molecular sequence analysis first came to the fore in the late 1990s and early 2000s, when SSNs were suggested as a way to analyze the rapid influx of new molecular sequence data due to advances in sequencing technology and reduced cost, as well as to predict gene functions and protein-protein interactions [39, 44–46]. One of the earliest formal and heuristic uses of SSNs was to define the COG groups of homologous families and facilitate prediction of the functions of large numbers of genes based on homology [39, 40]. The need for efficient computation and analyses for large biological databases still pervades; however, more recently SSNs have been increasingly appreciated as useful approaches to describe complex biological systems, including inferring the "social networks" of biological life forms [30], producing maps of genetic diversity [27], detecting distant homologues [47–49], and exploring gene and genome rearrangements [50, 51].

A SSN is a graph in which each node is a sequence and edges connect any two nodes that are similar at the sequence level above a certain threshold (e.g., coverage, percent identity, and *E*-value) as determined by their pairwise alignment (Box 1) (Fig. 1). While the principle behind SSN construction is simple, the expression of similarity data in this structure can enable the use of powerful

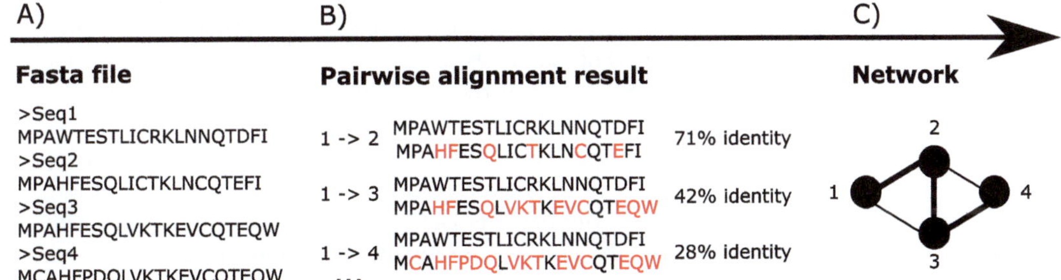

Fig. 1 Constructing a simple sequence similarity network. A set of sequences (protein or DNA) in fasta format (**a**) are aligned in pairs using alignment tools (such as BLAST). These alignments (**b**) are scored with metrics such as the percentage identity between two sequences (the number of identical nucleotides/amino acids displayed above) or the *E*-value of the alignment. In the resulting network (**c**), sequences are represented as nodes. Two sequence nodes are joined with an edge if they can be aligned above a define threshold, with the weight of the edge often based on percentage identity or *E*-value

algorithms for graph analyses to study complex biological phenomena. Construction of a SSN is also frequently the starting point in a diversity of further graph analyses. A SSN can be constructed directly from fasta formatted sequence files using pipelines, such as EGN [52], the updated and faster performing EGN2 (forthcoming), or PANADA [53]. Visualization of networks can be performed with programs such as Cytoscape [54] or Gephi [55], both of which also have a range of internal tools and external plugins for network analysis. While these programs are useful for the visualization and analysis of relatively small networks, it can be difficult to load large and complex networks with a lot of edges (e.g., ≥50,000 edges). In these cases the iGraph library offers an extremely powerful and well-supported implementation of a broad range of commonly used methods for both complex graph generation and analysis in R, Python, and C++ [56]. However, using iGraph requires knowledge of programming in at least one of these languages. An additional package for network analysis in Python is NetworkX [57]. It is our goal here to further generalize network approaches by explaining how evolutionary biologists with less programming knowledge could analyze their data. A list including many of the tools and programs available for SSN generation is available at https://omictools.com.

Box 1: How to Build Your Own Sequence Similarity Network

1. *Dataset assembly*: The first and most important step of SSN construction is the assembly of a dataset of sequences relevant to your biological question, usually in fasta format. This can be used as the initial input for wizards such as EGN or EGN2 [52], which can fully automate the process. The nature of the dataset is highly dependent on the research question, so here we focus on the practicalities of database assembly. To construct the similarity network, all sequences in the dataset are aligned against one another in a similarity search. This similarity search is often the time-limiting step in an analysis, and the total number of searches required is quadratic to the number of sequences in the dataset. For large datasets, it is useful to benchmark the alignment using a subset of the data to estimate the timescale for the alignment. Large datasets can generate huge outputs, not only due to the number of sequences but also the length of their identifier. One way to reduce the output size is to replace each sequence name in the fasta file with a unique integer. The use of integers will reduce disk space use and the memory consumption for any software used to analyze the sequence data.

(continued)

Box 1: (continued)

2. *Similarity search*: To generate a sequence similarity network, all sequences must be aligned against one another in an all-versus-all search, in which the dataset of sequences is searched against a database including the same sequences. For gene networks, the alignment is usually done with a fast pairwise aligner such as BLAST [58, 59] as implemented in EGN [52]. Filters are often used to remove low-complexity sequences from the search, as these can cause artefactual hits (BLAST options --seg yes, -soft-masking true). The BLAST method of alignment will be the focus of future discussion in this chapter; however, alternatives are available including BLAT [60] (also implemented in EGN), SWORD [61], USEARCH [62], and DIAMOND [63]. These alternatives generally include an option to produce a "BLAST" style tabulated output, making them compatible with programs commonly used in network analyses.

   Within alignment tools like BLAST, it is possible to assign thresholds, such as the maximum *E*-value of the alignment. It is not recommended to set minimal thresholds for some parameters (such as % sequence identity) unless required due to memory constraints so that you can generate networks from a single sequence alignment with different thresholds for comparison (e.g., comparison of a 30% similarity threshold to a 90% threshold, where edges will only be drawn between highly similar genes).

   Note: It may be intuitive to use additional CPUs to speed up the alignment process; however, in BLAST it can be more efficient to split the query file and launch multiple searches on separate cores instead of using the BLAST multithreading option. The pairwise alignment step is generally the most time-limiting part of generating a SSN, so benchmarking should be used to establish the optimal settings for the pairwise and/or determine the feasibility of a project given the size of the dataset and the available computational resources.

3. *Filtering similarity search results*: In an all-versus-all similarity search, any given query sequence will have a self-hit in the corresponding database. For example, with sequences A and B, a self-hit is query sequence A matching to sequence A in the database, cases of which must be removed prior to network construction (Fig. 2). When query sequence A in a similarity search is aligned with sequence B in the database, often the reciprocal result is also identified (an alignment between query sequence B and sequence A in the database). These are called reciprocal hits; while the sequences involved

(continued)

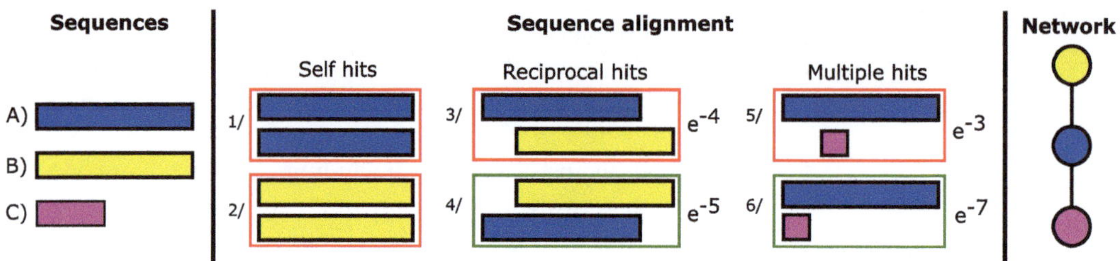

Fig. 2 Filtering sequence similarity results for network construction. In the output of an all-against-all sequence similarity search, there are a number of features that are often filtered out prior to network construction. Self-hits (1/ and 2/), where like sequences are paired in a sequence alignment, are not informative to network construction and are removed (highlighted by the red box surrounding the alignments). In cases where there are reciprocal hits (3/ and 4/) between two sequences, then only the alignment with the highest *E*-value is retained (highlighted with a green box around the retained alignment) to ensure only one edge representing the best possible alignment connects any two nodes in the network. The same is true for cases where a sequence has multiple hits against another sequence, such as when it aligns to another sequence in multiple positions (5/ and 6/)

Box 1: (continued)

are identical, the alignments and scores are not. Retaining both hits would generate two different edges between the same two nodes in a SSN, so generally only the best results from reciprocal hits are retained, based on a score such as the *E*-value (Fig. 2). Finally, a single query sequence may be significantly aligned multiple times in different positions of the same sequence in the database; however, for SSN construction only the best BLAST hit is generally retained (Fig. 2). The selection of the best BLAST hit is again generally often based on the *E*-value. Removing multiple hits against the same sequence allows the generation of an undirected network where a single edge connects two nodes, representing the best possible alignment between these nodes.

4. *Thresholding and network construction*: Constructing a SSN from a BLAST output is conceptually simple; an edge is created between two sequences (nodes) that have been aligned in the sequence similarity search. It is common to apply thresholding criteria such as minimal % ID and/or coverage and/or maximal *E*-value to determine whether an edge is drawn between two sequences in the network (Fig. 1). There are different ways to calculate the % coverage of an alignment. This could be based on the coverage of a single sequence in the alignment, selecting either the query or the database sequence in each alignment or the longest or shortest sequence in each alignment. Alternatively both (mutual coverage) can be used, retaining an alignment

(continued)

Box 1: (continued)

when both values are above a given threshold. Edges above the thresholding criteria can be assigned a weight based on these criteria, producing a weighted sequence similarity network that retains information of the properties of the alignment between two sequences (Fig. 1). It is often useful to construct and compare several SSNs with variable stringencies defining the edges between sequences, for example, to optimize gene family detection within the SSN (discussed below).

2.1 Scalability of Sequence Similarity Network Analysis

As with other computational approaches, the scale of network analysis is limited by the available computational resources. The limiting factor in terms of the size of network it is possible to construct is predominantly governed by the pairwise alignment. All sequences in the dataset need to be aligned against one another in a pairwise manner, meaning the number of alignments is quadratic to the size of the dataset. For example, computing an all-against-all comparison of 1,000,000 sequences requires computation of 10^{12} alignments. BLAST [64] is the standard tool for this step, with a relatively good speed and accuracy for sequence similarity searches; however, the use of BLAST can be a bottleneck for the analysis of large datasets. This is an especially important consideration given the growth in the number of gene and genome sequences available in public databases. Several rapid alignment tools such as BLAT [60], USEARCH [62], Rapsearch [65], and Diamond [63] have been proposed to overcome this issue. For example, Diamond benchmarks suggest that it is almost as accurate as BLAST but is at least three orders of magnitude faster.

A second point to consider from the perspective of scalability is the complexity and size of the graph and the complexity of the algorithms used in their analysis. Algorithms where the number of calculations is linear to the size of the graph can generally be run on huge graphs with sufficient computational resources, for example, finding connected components using the "deep search first" algorithm. Algorithms for community detection (e.g., PageRank [66], Louvain) are also linear and particularly suited for detecting groups of closely related sequences in huge graphs (discussed in Subheading 4). In contrast, computing graph statistics such as the betweenness centrality are not linear to the size of the graph, even using the relatively efficient Brande algorithm for calculation [67], and are therefore more difficult to calculate for huge graphs. This has led to the development of toolkits specifically designed for the analysis of huge graphs (e.g., NetworKit) [68]. A recent book summarizes the challenges of the analysis of huge networks and some of the algorithms that have been developed to face these challenges [69].

2.2 Exploiting Sequence Similarity Networks for Identification of Gene Families

A gene family is usually defined as a group of sequences that are similar at the sequence level, indicative of homology and potentially of shared functions; however, there is no uniform way to define this similarity [70, 71]. One of the early contributions of SSNs in molecular sequence analysis was the construction of the COG database of homologous protein sequences [39, 40]. This study attempted to define gene families based on similarity at the sequence level using the results of sequence similarity searches. Within the results of an all-versus-all BLAST search, groups of at least three proteins encoded by different genomes that were more similar to each other than they were to other proteins found in the same genomes were defined as a likely orthologous gene family. Orthologous gene families are group of genes in different genomes that show sequence similarity, likely as a result of their shared evolutionary history.

The idea of using graphs to identify gene families is now a core part of many graph-based analyses. Members of a gene family aggregate in a sub-network in a SSN. These sub-networks are called connected components (CCs) at these defined thresholds, i.e., clusters of nodes connected by edges either directly or indirectly (via intermediate nodes) (Fig. 3). The size (number of nodes and edges in a CC) and density (the proportion of potential connections between all nodes in a CC that are actually connected by edges in the graph) of CCs will depend on the thresholds used for

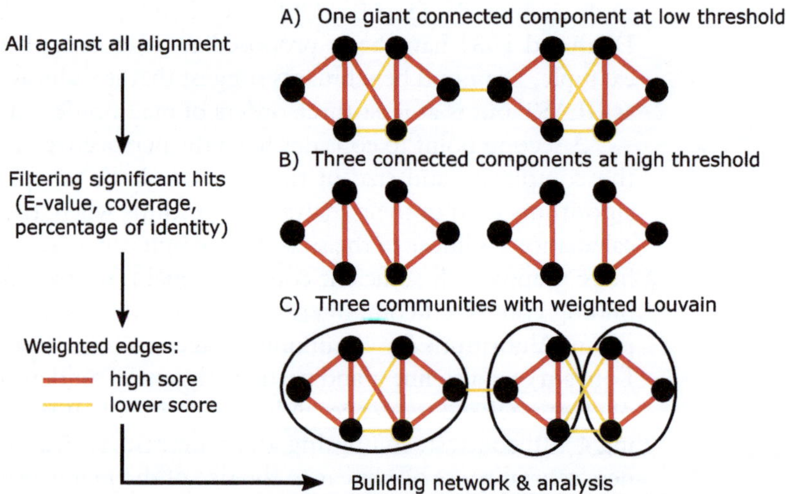

Fig. 3 Louvain community detection in a sequence similarity network. The network is assembled from the results of an all-versus-all alignment, as previously described. Edges can be weighted by *E*-value, percentage of identity, or bitscore. For the purpose of simplification, we consider strong or weak weights rather than actual values. (**a**) A giant connected component at relaxed threshold. (**b**) Three connected components at a more stringent threshold. (**c**) Three communities with Louvain clustering algorithm, taking into account edge weights

constructing the SSN as well as the relationships between sequences in the network. For example, for a given dataset at a given mutual coverage threshold, a threshold of 90% sequence identity will identify a large number of small connected components that only include highly similar genes, while at a threshold of 30% sequence identity, there will be fewer but larger connected components including genes with more variation in sequence similarity. Commonly used thresholds for detecting homologous gene families are an *E*-value $\leq e{-}5$, mutual coverage $\geq$80%, and a percentage of identity $\geq$30% [23].

CCs are often detected in a SSN using the Depth-First Search (DFS) algorithm; however, there are also other approaches for the detection of gene families based on the idea of detecting "communities" [72]. In some cases, a CC can be further separated into communities of sequences that share more similarity to one another than to other sequences in the CC and thus are more highly linked in the SSN (Fig. 3). Communities are commonly identified by using graph clustering algorithms such as Louvain [73], MCL [74], or OMA [75]; however, different clustering algorithms will result in different outputs. The Louvain weighted method is widely used because it is simple to implement and scales very well to large graphs (Figs. 3 and 4) [73]. MCL is a strong deterministic algorithm that has been implemented, for example, in tribeMCL [74] and orthoMCL [76]. A potential drawback of MCL is that it requires user specification of the "inflation index," a parameter which controls cluster granularity (or "tightness"). A high inflation

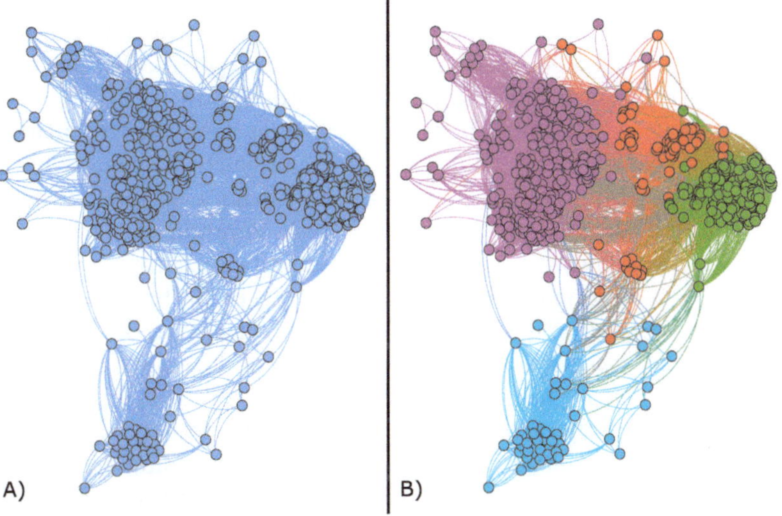

Fig. 4 Giant connected component before and after community detection. (**a**) A single giant connected component from a sequence similarity network. (**b**) The same giant connected component after application of a community detection algorithm. Node colors correspond to the newly assigned communities

index increases the tightness of clustering, producing a larger number of clusters that are smaller on average than those that would be obtained clustering the same dataset using a low inflation index. Selecting an appropriate inflation index is not trivial and requires optimization [74].

A number of the above approaches have been used to compile additional databases of orthology that can act as useful reference datasets. OMA is a program that uses graph-based algorithms and exact Smith-Waterman alignments to identify orthology between genes [77–80]. OMA is also available as a web browser [81] including a database of orthologues that, in 2015, included more than 2000 genomes and more than seven million proteins [75]. SILIX is a software package [82] that aims at building families of homologous sequences by using a transitive linkage algorithm, and HOGENOM [83] is a database that contains families inferred by SILIX for seven million proteins.

In addition to clustering genes into families, valuable information can be extracted from the connected components using network metrics. Highly conserved sequences tend to form CCs where most of the nodes are connected to each other by edges, while sequences from more divergent families will tend to form more sparsely interconnected CCs. This information can be easily assessed for each component using the clustering coefficient. Conserved families will have a clustering coefficient close to 1, even for stringent thresholds. Identifying such conserved families can be useful to produce multiple sequence alignments (MSA) needed for phylogenetic reconstruction, but SSNs have also been demonstrated to unravel relationships between distant homologues by linking distantly related sequences together [24, 29, 48]. In a SSN, two distant sequences A and C which do not share similarity according to BLAST can be linked together due to sequence B which shows similarity to both A and C.

The idea of distant homology has been particularly illuminating regarding chimeric organisms such as eukaryotes which carry homologous genes inherited from a bacterial ancestor and from an archaeal ancestor [29]. A common way to analyze sequence similarity networks is to identify certain "paths" of interest, for example, the shortest possible paths between two nodes. This notion describes the path between two nodes in a connected component that minimizes the sum of the edge weights. Alvarez-Ponce et al. used this approach to explore the topology of connected components in a SSN including the complete proteomes of 14 eukaryotes, 104 prokaryotes (including archaea and bacteria), 2389 viruses, and 1044 plasmids. Eight hundred and ninety-nine CCs contained sequences from all three domains, and of these 208 contained eukaryotic sequences that were not directly similar to one another but only linked to one another via a "eukaryote-archaea-bacteria-eukaryote" shortest path. These are putatively distant homologues in eukaryotes that were present in both the archaeal host of the mitochondrial endosymbiont and in the alpha-proteobacterial endosymbiont, with both copies subsequently retained in eukaryotes and as such strong evidence for the chimeric origin of eukaryotes [29]. This demonstrates the utility of networks in the study of ancient evolutionary relationships including the origin of eukaryotes [28] or rooting the tree of life [84]. Simple path analysis for a network is possible using existing plug-ins within visualization tools such as Cytoscape [54] and Gephi [55].

2.3 Exploiting SSNs to Identify Signatures of "Tinkering" and Gene Fusion

When discussing identification of gene families, we have focused on networks where edges are drawn between protein sequences that show a high enough similarity across their entire length, defined by a high mutual coverage threshold (e.g., 80%). Sequence similarity can also be partial, for example, following gene remodeling or "tinkering" [85] producing new combinations of gene domains via gene fusion and fission events, or through the de novo sequence synthesis of gene extensions, adding to existing sequences. The term "Rosetta Stone sequence" was coined to define the formation of a new fusion protein in a species as the result of the fusion of two proteins that are found separate in another species, with authors originally predicting that these fusions could occur between proteins that physically interact in a common structural complex [86]. One of the earliest applications of sequence similarity searches to identify fusion proteins was an attempt to predict pairs of proteins that may physically interact in an organism based on whether they could be identified as a single "composite" fusion protein in another organism [44]. Beyond predicting protein-protein interactions, this kind of gene remodeling and recycling of existing gene parts has the potential to contribute to the expansion of functional diversity in genomes, creating new and unique combinations of domains and functions [51, 85, 87–91]. Similarity search-based screens have been implemented to identify composite genes and genome rearrangements in a range of prokaryotes [92–94], eukaryotes [87, 95–97], and viruses [98].

Early attempts to identify composite genes were based on the output of sequence similarity searches, but without formalizing the results of search methods into a graph structure. The first attempt to formalize the problem of identifying "composite" genes in networks was the "Neighborhood Correlation" approach, aiming to distinguish genuine multi-domain proteins sharing common ancestry (homologues) from novel multi-domain proteins that share domains due to insertions [99]. The later development of the FusedTriplets and MosaicFinder tools attempted to unify existing graph-based methods for detection of "composite" gene detection [50]. FusedTriplets is a graph-based implementation of the traditional gene-centered method for composite gene identification, originally introduced by Enright et al. [44], with additional cross-

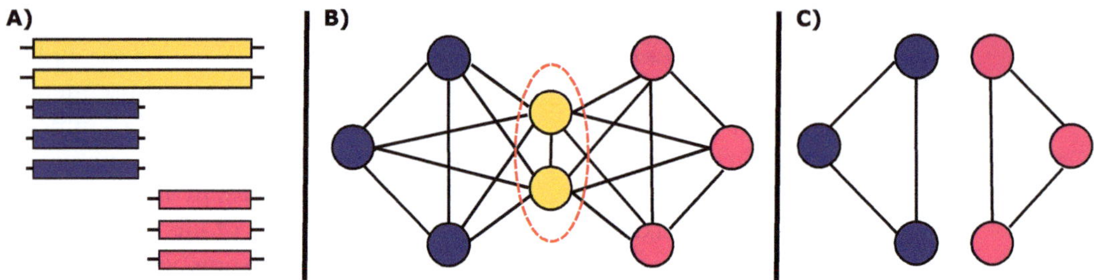

Fig. 5 Composite gene identification using "minimal clique separators." (**a**) A multiple sequence alignment of composite genes (yellow) with two components (blue and magenta). (**b**) The sequence similarity network corresponding to the multiple sequence alignment. The composite genes (yellow) are a minimal clique separator for the network. Their removal (shown in **c**) decomposes the network to the two separate component families

checks on the absence of similarity between the two component genes contributing to a composite gene based on varying thresholds [50, 100]. MosaicFinder is a gene family-centered approach which will only identify highly conserved composite gene families that form "minimal clique separators" (Fig. 5) [50]. This graph topology implies that MosaicFinder may fail to detect divergent (e.g., ancient or fast evolving) composite gene families which will tend to form "quasi-cliques" without perfect separation. CompositeSearch [101] (available at http://www.evol-net.fr/index.php/en/downloads) is a new program designed to overcome this limitation by identifying both conserved and divergent composite gene families (Box 2).

Box 2: How to Identify Composite Genes Using CompositeSearch

1. *BLAST search and filtering*: An all-versus-all BLAST search is carried out as described in Box 1. Filters can be applied on the *E*-value and sequence similarity but should not include a mutual query coverage threshold.
2. *CompositeSearch*: CompositeSearch takes a filtered BLAST output and a list of genes as the initial input. Two search algorithms are implemented: "fastcomposites" detects a list of potential composite genes and "composites" additionally detects potential composite gene families and component gene families. Additional options are included to filter the network based on a number of standard metrics (e.g., *E*-value, sequence similarity, mutual coverage) and set the maximum overlap allowed between different components aligned on the same potential composite gene. The definition of a maximum overlap allows adjustment for the

(continued)

Box 2: (continued)

tendency of BLAST to produce overhanging alignments [100]. The output includes a node, edge, and information file including information on number of nodes, edges, and family connectivity from family detection. Two outputs are included for composite gene detection, a "composites" file with detailed information on each predicted composite gene in fasta format and a "compositesinfo" file, summarizing the data. Similarly, two files provide detailed information on composite gene families and a summary of composite gene families.

3. *Filtering results*: By default, CompositeSearch outputs all possible composite genes in "fast" mode or composite gene families in the full mode. These are given alongside a number of different metrics designed to help to filter families for more confident predictions, including the gene family size, number of composites directly predicted within the gene family, the number of domains, the number of component families, the number of singleton component families (families including only one sequence), the connectivity of the family, and a score based on the overlap between different components mapped to the composite gene.

Recent studies have explored composite gene formation as a source of innovation by "tinkering" [85] during major evolutionary transitions. These can be especially interesting when exploring genome evolution following introgression, raising the possibility of formation of new composite genes using components with different evolutionary origins [20, 51, 102]. For example, the gain of a cyanobacterial endosymbiont at the origin of photosynthetic eukaryotes was accompanied by the transfer of whole cyanobacterial genes to its new host genome, with gene functions related to the role of the plastid [103–105]. Identification of composite genes related to the origin of photosynthetic eukaryotes unraveled novel symbiogenetic composite genes, and unique fusions of genes encoded in the nucleus of photosynthetic eukaryotes that included components derived from the plastid endosymbiont. As with whole genes transferred to the nucleus, several of these components had predicted functions related to the role of the plastid, including redox regulations and light response [51].

2.4 Exploiting SSNs for Ecological Studies

Ecological studies increasingly involve the assembly, analysis, and comparison of large metagenome datasets. In addition to identification of functions and organisms associated with a particular environment, these studies enable the investigation of important hypotheses in microbial ecology at the level of organism or function, such as the often quoted hypothesis that "everything is everywhere, but the environment selects" from Bass Becking: the idea that microbial lineages are limitlessly dispersible in the environment, but the environmental conditions will select for certain lineages and control their distribution rather than any specific geographical separation [21].

Networks are useful for these kinds of ecological studies because existing graph algorithms can be used to investigate the structure of the network. When investigating gene (or gene-sharing networks), it is possible to distinguish nodes by labeling them based on their properties, such as categories for taxonomic or environmental origins (Fig. 6). A simple way to represent this visually is to color nodes based on these properties in Cytoscape or Gephi. A formal way to explore the relationships between node properties is to use network metrics such as conductance [106], modularity [73], and assortativity coefficient (normalized modularity) [107]. Assortativity and conductance are different metrics that attempt to answer the same type of question: do nodes labeled as belonging to a particular category, such as environmental origin, tend to be connected with other nodes labeled as belonging to the same category? More precisely, conductance quantifies whether a given category of nodes shares more edges between themselves than with nodes from different categories. A low conductance approaching zero indicates that nodes of a given category are highly connected to one another, with few connections to nodes from different categories. A higher conductance is indicative that nodes of this category tend to be more sparsely interconnected and share more connections with nodes from different categories. Assortativity is a measure of the preference for a category of nodes in a network to attach to other nodes

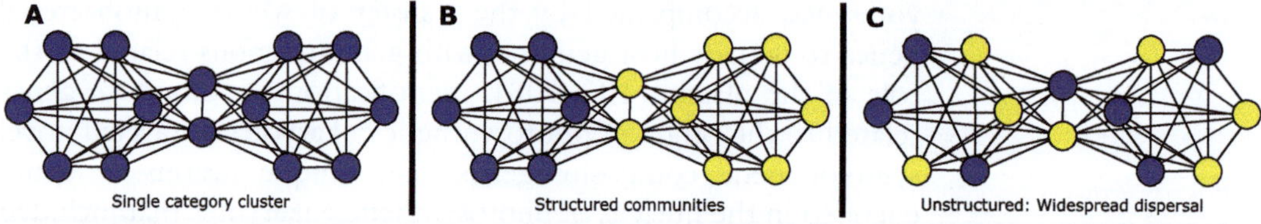

Fig. 6 Exploring distribution of annotations in sequence similarity networks. In this example, nodes within a single connected component are assigned two colors, blue and yellow, corresponding to their having a different categorical annotation (e.g., originating from a different environmental source). Using the example of environmental source, genes in cluster A would all have the same environmental source (blue), indicating an environment-specific cluster of genes. Genes in cluster B are found in two different environmental sources (blue and yellow); however, nodes of the same type are preferentially linked to each other in the network than to genes from different environmental sources. This would result in a positive assortativity coefficient approaching 1 for environment and a low conductance score, suggesting a strong environmental community structure. Genes in cluster C are also found in two different environmental sources; however, there is no clear pattern for the distribution of genes with regard to environment. This network would have an assortativity approaching 0 and a high conductance score

from the same category. Normalized assortativity values range between −1 and 1, where 0 indicates random distribution of categories within the network, 1 indicates that nodes from the same categories tend to be connected to one another in the network, and −1 indicates that nodes from different categories tend to be connected in the network. A detailed description of the algorithms used in these calculations can be found in [108].

2.4.1 Assortativity as a Tool to Study Geographical and Habitat Distributions of Microbes and Genes

Forster et al. used assortativity (among other network statistics, including the previously discussed shortest path analysis) to explore the geographical dispersion patterns of marine ciliates in a network generated from ciliate SSU-rDNA sequences [25]. Sequences were clustered into two different levels of gene family—CCs and Louvain communities (LCs) as previously described. Sequences were assigned categorical labels based on their geographical point of origin (eight locations) or habitat of origin (three habitats), and assortativity was calculated. If sequences, and thus species, are broadly distributed across geographical categories, then assortativity of SSU-rDNA sequences labeled with these geographical categories would be low because similar sequences would be found in different environments. Contrarily, if similar sequences tend to be from the same geographical category, indicative of endemism, then assortativity of sequence geographical origin will be high (Fig. 6). The majority of CCs and LCs showed a positive assortativity for geographical origin, higher than expected by chance, indicative of geographical community structure as opposed to global dispersal of ciliates. Similar approaches were used by Fondi et al. and applied to a collection of environmental metagenome samples to test the "everything is everywhere" hypothesis at the gene pool and functional level. Gene pools were more strongly associated with a particular ecological niche than with specific geographical location, supporting the idea that microbial genes are found everywhere but the environment selects for them [26].

2.4.2 Conductance in the Comparison of Lifestyles and Evolutionary Histories

Conductance is used to explore the clustering of pairs of different node categories in a connected component. In a study by Cheng et al., the proteomes of 84 prokaryote genomes were categorized into four broad redox groups based on their lifestyle, methanogens, obligate anaerobes, facultative anaerobes, and obligate aerobes [27]. For each CC in a pan-proteome sequence similarity network including all 84 genomes, the conductance was calculated for pairs of redox categories and compared to values obtained following random relabelling of the components. The distributions of conductance values for methanogens and for obligate anaerobes groups indicated that the sequences in these groups have features distinct from those in other groups, that anaerobes and aerobes tend to be dissimilar, and that their sequences are more isolated from one another in the SSN than expected by chance.

An additional example of the use of conductance is in exploring the propensity of a gene family to lateral gene transfer. Within a network of archaeal and bacterial genes, CCs showing a low conductance for both archaeal and bacterial sequences indicate that the bacterial and archaeal genes within the corresponding families are structured in two separate and conserved groups (Fig. 6). Structuring gene families into two groups would indicate that there was little or no evidence for lateral gene transfer between archaea and bacteria within this particular gene family. This kind of gene family is rare, with only 86 gene families from 40,584 (0.2%) meeting this criteria [24].

2.5 SSNs in Remote Homologue Identification: Shedding Light on the Microbial Dark Matter

Up to 99% of microbial species are not cultivable and thus have not been studied in isolated culture. Analysis of high-throughput sequencing and metagenomics datasets has shed light on these uncultivable organisms, often referred to as the "microbial dark matter" [109], and in some cases enabled the reconstruction of draft genomes [110–114]. A considerable portion of most metagenome studies have predicted ORFs showing no detectable similarity to any known proteins, termed metaORFans [115]. These can represent 25–85% of the total ORFs identified in metagenomes [22]. Identifying distant homologues of ORFans may help to predict their functions and begin to unravel the microbial dark matter. Recent work by Lopez et al. in 2015 probed the microbial diversity of metagenome datasets from a range of environments including the human gut microbiome, identifying homologues of genes from 86 ancient gene families that are distributed across archaea and bacteria. The majority of these gene families included environmental homologues that were highly divergent from any of their cultured homologues, and many branched deeply with the phylogenetic tree of life, highlighting our limited understanding of diverse elements of the microbial world and hinting at the existence of yet unknown major divisions of life [24] (Fig. 7).

2.6 Exploiting SSNs to Analyze Classifications

Metagenomic and genomic data are providing scientists with a tantalizing amount of sequence data, casting the analysis of the extent of biodiversity as a major research theme in biology [116–120]. In theory, existing organismal and viral classifications are invaluable tools to structure and analyze this biodiversity. However, the way taxonomical classifications are constructed raises questions about their naturalness and their actual application scope [38, 120–128], in particular regarding genetic diversity surveys. There are three major reasons for this. First, organismal and viral diversity is still largely undersampled, which means that existing classifications are incomplete [119, 120]. Therefore, taxonomically unassigned sequences cannot be readily used in class-based genetic diversity surveys, since this dark matter remains outside existing classes. Second, classifications are constructed

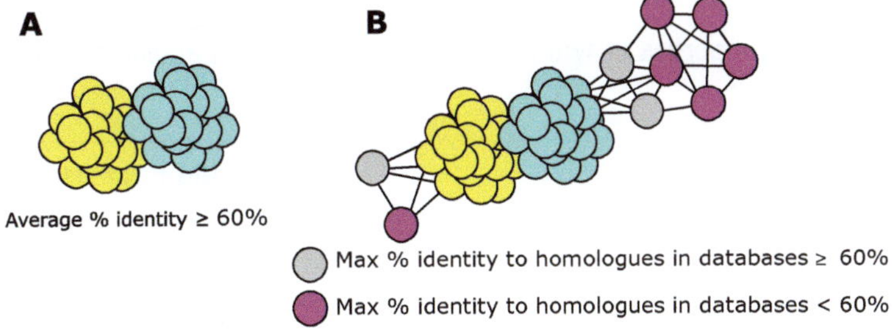

Fig. 7 Remote homologue detection to help characterize the microbial dark matter. (**a**) A hypothetical highly conserved cluster of genes from genomes present in sequence databases, where the average % of identity is high ($\geq$60%). (**b**) The same cluster after addition of divergent environmental sequences to the network. Environmental sequences in gray are more similar to those already identified from genome surveys ($\geq$60% max identity) so are connected directly to the conserved gene cluster in the network. More divergent sequences in pink have $<$60% maximum identity to their homologues in the database. Many of these are only identified as linked to the sequences from the conserved database via intermediate gray nodes. This is the notion of "transitive homology"

using different features (i.e., for viruses, a mix of phylogenetic, morphological, and structural criteria, such as replication properties in cell culture, virion morphology, serology, nucleic acid sequence, host range, pathogenicity, epidemiology, or epizootiology); therefore their classes do not necessarily offer immediate proxies for quantifying genetic diversity per se. Third, evolutionary processes responsible for both genetic and organismal diversity are diverse, and they operate at different tempos and modes in different lineages [49, 123, 129–141]. As a result, genetic diversity within classes and between classes can be heterogeneous, meaning that existing classifications may lack efficiency to discriminate, predict, or compare taxa on genetic bases, potentially hampering diversity studies, a profound practical issue at a time where the analysis of metagenomic sequences is becoming a priority in biology.

Addressing these challenges is notably crucial for viral studies. Recently, the executive committee of the ICTV [142] proposed that network analyses methods that create similarity metrics based on the detection of homologous genes and their genetic divergence constitute a valuable strategy to assist classification of viruses. Consistently, basic network properties and metrics (Table 1) can quantify (1) whether genetic diversity is consistent within and between the classes of existing classifications and (2) describe what classes are the most homogeneous and distinctive in terms of genetic diversity. Three criteria can be used to estimate intra-class genetic heterogeneity (Fig. 8a–c). First, the average edge weights (measured as % of identity, PID) between pairs of sequences from genomes of the

Table 1
Schematic properties of two extreme kinds of taxonomic classes with respect to their genetic diversity

"Ideal" classes	Not ideal classes
Low intra-class genetic diversity (high average PID)	High intra-class genetic diversity (low average PID)
High genetic cohesion (high average CCC)	Low genetic cohesion (low average CCC)
Core components (high maxCore%)	No core components (low maxCore%)
Obvious genetic distinctiveness (high conductance difference with random groups)	Limited genetic distinctiveness (conductance similar to random groups)
Exclusive pangenome (high % of exclusive CC)	No exclusive pangenome (low % of exclusive CC)

The three top properties inform about genetic diversity within classes (intra-class genetic diversity). The last two properties inform about the genetic distinctiveness (core and signature genes) of the classes. Interclass genetic heterogeneity identifies when genetic diversity of a class is not comparable with genetic diversity of another class in the classification. CCC, average proportion of genetic conservation between sequences from the same cluster and from the same taxonomic class; PID, average edge weights (% identity) between two sequences from genomes of the same class

same class provide a trivial measure of intra-class genetic diversity. Second, the average proportion of *C*onserved *C*anonical *C*onnections between sequences from the same connected component and from the same taxonomic class can be exploited (CCC, i.e., in each connected component of the SSN, the total number of edges connecting sequences of a given class i (intra-group edges, denoted E_{ii}) divided by the theoretical maximal number of possible edges between sequences of that class in the connected component (CCC (i) = $2^*E_{ii}/(N_i \times (N_i - 1))$ where N_i is the number of sequences of class i present in the connected component). CCC ranges between 0 and 1. Within a connected component, if all pairs of sequences from the same class are directly connected, CCC equals 1, since all these sequences are more conserved than a given %ID threshold. By contrast, low CCC are observed when sequences from genomes from the same class lack cohesive evolution, for example, when some related sequences evolved so fast that they show less than the minimal similarity required to be directly connected to their homologues in the graph. Third, the genetic consistency of a class can be estimated by (1) identifying what cluster of sequences was present in the largest number of genomes of the class and then (2) by quantifying the proportion (in %) of the class members harboring that most ubiquitous cluster (maxCore%). When maxCore% of a class is <100%, it means that, for this dataset, there is no gene family shared by all members of that class (i.e., no core genes). The SSN structure can also serve to estimate the genetic distinctiveness of each class, i.e., whether sequences from a given class are

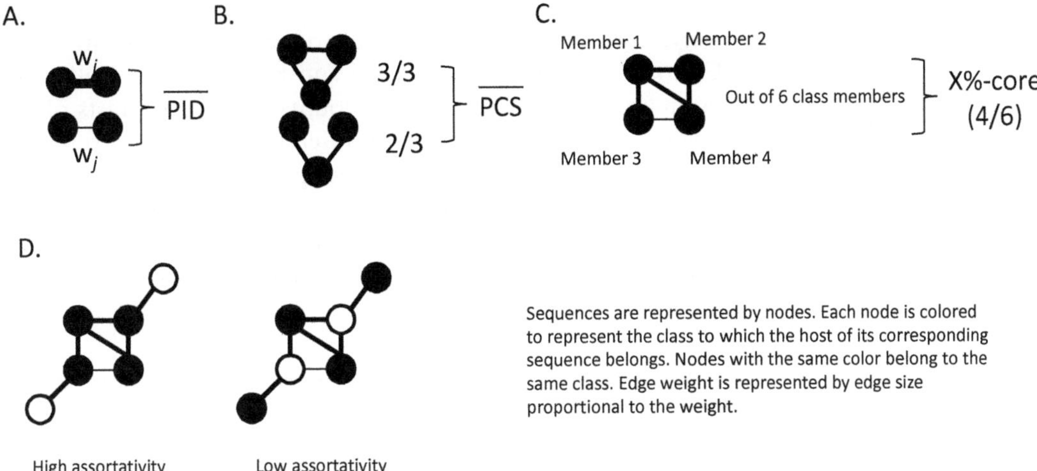

Fig. 8 Intra- and interclasses heterogeneity measurements in weighted similarity networks. Sequences are represented by nodes. Each node is colored to represent the taxonomic class to which its host belongs. Nodes with the same color belong to the same class. Edge weight is represented by edge size proportional to the weight. Subgraphs correspond to clusters of sequences. Direct neighbors have a greater similarity than the threshold set to allow such connections. PID, average edge weights (% identity) between two sequences from genomes of the same class; CCC, average proportion of genetic conservation between sequences from the same cluster and from the same taxonomic class; maxCore%, conductance; and %-exclusive components correspond to the estimates used to assess genetic consistency of classes

more similar to one another than they are to sequences from other classes (Fig. 8d, e). Such sequences could be used as classificatory features to assign members to the class. In a SSN, this property translates to a low ratio of interclass edges over intra-class edges and is measured by conductance (Fig. 8d). Likewise, the proportion of clusters comprised exclusively of sequences from one class, a diagnostic feature of the class, provides an estimate of the class genetic distinctiveness. Genetically highly distinct classes have a high % of such exclusive clusters. Based on these network measures, interclass genetic heterogeneity can simply be diagnosed by contrasting estimates of genetic consistency for all the above measures for each class. There is interclass heterogeneity within a classification when the mean PID, mean CCC, maxCore%, DRC, and % of exclusive components differ between classes.

Such network analyses show that virus classifications face a pragmatic issue: overall genetic distinctiveness allows relatively safe assignments of viral sequences to existing classes; however, genetic diversity of viral taxa of similar ranks differs among the tested classifications. Therefore, virus classifications (especially ICTV classification at the family level) should be used carefully to avoid inaccurate estimates in metagenomic diversity surveys. Classes with broader genetic diversity will tend to be more easily detected in the environment than classes with reduced genetic diversity, since the former will necessarily be associated with more OTUs than the latter. Some alpha- and beta-diversity analyses of environmental data, which rely on counts and on contrasts of the abundance of taxonomic classes in different samples, will thus also be biased. A similar approach could be applied on different types of classified lineages, i.e., to identify what groups of bacteria, archaea, or eukaryotes with comparable taxonomical ranks are the most genetically heterogeneous and what ranks of their classification are the least genetically consistent.

3 Gene-Sharing Networks

Gene-sharing networks are often called "genome networks" as they are best suited for summarizing what genes are shared between different genomes, highlighting routes of gene sharing. The ability to explore gene sharing between all genomes in a network in a simple graph can have useful properties for reflecting microbial social life, inherently inclusive of gene sharing both as a consequence of vertical inheritance and lateral gene transfer (LGT). Bacteriophage and plasmid genomes are typically highly mosaic in nature due to a high level of horizontal gene transfer, making it difficult to classify their genomes [37, 143]. Lima-Mendez et al. proposed the use of gene-sharing networks as a new classification method that tackles this problem of mosaicism by classifying viruses based on their genome's content [37]. Constructing gene-sharing networks using subsets of genes from different functional categories of genes can also be useful in exploring what kinds of genes are being shared by different genomes.

In a gene-sharing network, each genome is represented by a node, and two nodes are connected by an edge when the two corresponding genomes share homologous genes or gene families (Fig. 9). These gene families can be identified from SSNs (of as CCs of LCs) or by alternative methods. In gene-sharing networks, edges can be weighted by the number of genes or gene families shared between the genomes. In this way, gene-sharing networks enable the study of microbial social life, quantitatively displaying the gene families shared between genomes both as a result of vertical transmission and lateral gene transfer.

Gene-sharing networks are useful tools for exploring overall patterns of gene sharing between genomes. Recently, Lord et al. developed BRIDES, a software package that specifically identifies different kinds of patterns in evolving gene-sharing networks after the addition of new genome nodes [144]. However, in gene-sharing networks the kind of gene families that are being shared is often overlooked. To explore how functions are shared between different genomes, gene-sharing networks can be built from genes

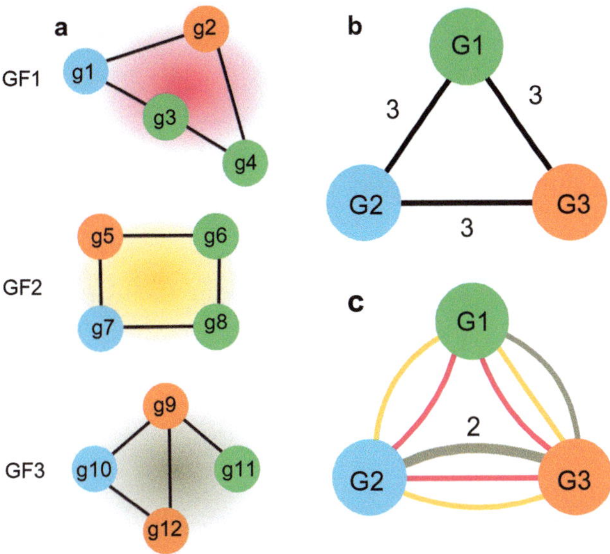

Fig. 9 Translating gene networks to gene-sharing networks. (**a**) Gene network for three gene families. Gene nodes are colored based on their genome of origin. The background color corresponds to the gene family color in part **c**. (**b**) The gene-sharing network corresponding to the gene network in **a**. Edges are weighted on the number of gene families shared by the genomes. (**c**) Multiplex gene-sharing network corresponding to the gene network in **a**. Genomes are connected by multiple edges with colors corresponding to different gene families. These edges are weighted based on the number of genes shared between two genomes for each family

using different subsets of functions (Fig. 10) [29]. An alternative form of the gene-sharing network is the multiplex network. In this network nodes can be linked by edges of different types, for example, each edge representing a different gene family or different functional groups of gene families, thus retaining additional information compared to a simpler gene-sharing network (Fig. 9) [23]. Multiplex networks can be useful for small-scale analyses; however, with large datasets they can rapidly become difficult to interpret and analyze. Importantly, multiplex networks are unimodal projections of bipartite graphs (discussed in the Subheading 14) which can provide greater clarity and have a number of attractive properties for the analysis of larger datasets.

3.1 Classification of Entities Using Gene-Sharing Networks

The possibility of summarizing gene sharing between sets of entities with complex evolutionary histories means that gene-sharing networks can be useful for classifying organisms based on their gene content. Lima-Mendez et al. analyzed bacteriophage genomes to generate two different phage gene-sharing networks that reflect their reticulate evolutionary history [37]. In the first gene-sharing network, phage genomes (nodes) were connected by edges when

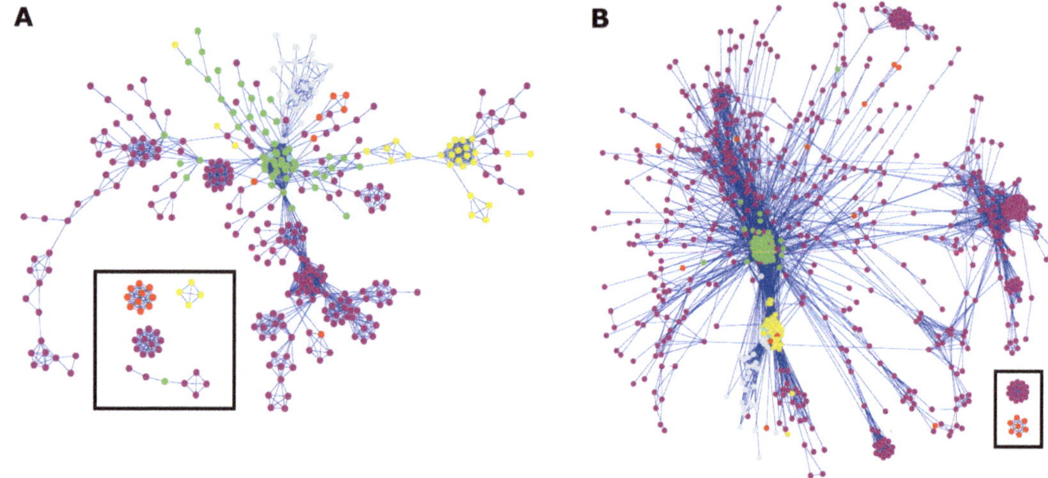

Fig. 10 Functional gene-sharing network reflecting the chimeric nature of eukaryotes. These gene-sharing networks describing how genes in different functional categories are shared between bacteria (green), archaea (yellow), eukaryotes (gray), plasmids (purple), and viruses (red) from a published dataset [29]. In both cases, a giant connected component is shown alongside examples of smaller connected components (**a**) Gene-sharing network for COG category D: cell division control. In this network, sequences of eukaryote origin (gray) cluster with bacterial sequences, reflecting their origin in the alphaproteobacterial endosymbiont that would become the mitochondrion. (**b**) Gene-sharing network for COG category K: transcription machinery. In this network, eukaryote sequence (gray) cluster with archaeal sequences, reflecting the origin of these genes in the archaeal host for the eukaryotic endosymbiont

they shared significant similarity at the sequence level. This gene-sharing network was clustered using the previously discussed MCL algorithm [145], identifying distinct groups of phages with sequence similarity. Following clustering, membership to a particular cluster was reassessed based on shared similarity with viruses in other clusters, reflecting their reticulate evolutionary history, allowing the generation of a matrix assigning a score describing the relative membership of any given viral genome to a particular classification group. In the second approach, Lima-Mendez et al. generated a "module"-based gene-sharing network, where edges are drawn between two phage genomes if they share a "module," in this case defined as a group of genes with similar phylogenetic profiles, enabling the exploration of what kinds of genes are shared between different groups of phages or are "signatures" for a particular group of phage genomes [37].

3.2 Exploring Routes of Gene Sharing in Gene-Sharing Networks

Two network metrics, also useful in the analysis of gene networks, can be used to attempt to identify "hubs" of gene sharing in the context of gene-sharing networks: node "degree" and "betweenness." Both metrics aim to determine the centrality of a node in a network. The degree of a node is simply the number of edges that it is connected to. The betweenness of a node is the frequency at which it is found in all the possible shortest paths between any two nodes in the network. Halary et al. used gene-sharing networks based on DNA sequence similarity to explore gene sharing between prokaryotes and mobile genetic elements [30]. Plasmids were identified as hubs of gene sharing within this pool of genomes, suggesting that they are key vectors for genetic exchange between cellular genomes and a potential DNA reservoir shared by genomes. Phages were more peripheral in the network and mostly linked prokaryotes from the same lineage. Thus, gene-sharing networks provided insights on the evolutionary processes that shape the gene content of prokaryote genomes.

The importance of plasmids in genetic worlds was further highlighted by exploring plasmid gene-sharing networks without inclusion of prokaryote genomes [14, 36]. Connecting 2343 plasmid genomes based on shared gene content in a single graph demonstrated that plasmids tended to cluster based on the phylogenetic class of their corresponding host prokaryote rather than habitat but that more mobile plasmids tended to be more "central" in the graph, indicating that these were hubs of gene sharing. Specifically, routes of gene sharing for gene families including antibiotic resistance markers were identified between actinobacterial plasmids and gammaproteobacterial plasmids, suggesting that Actinobacteria may act as a reservoir for antibiotic resistance genes for Gammaproteobacteria [14].

The finding that plasmids are hubs of gene sharing for prokaryote genomes was supported by analysis of gene sharing in a proteobacterial phylogenomic network including 329 proteobacterial genomes [32]. A phylogenomic network is a type of phylogenetic network that has been constructed from fully sequenced genomes. In this example the phylogenomic network is an alternative to a gene-sharing network, in which genome nodes within a phylogeny are linked by edges if they share genes [34]. This study identified extensive evidence for lateral gene transfer among Proteobacteria, with at least one LGT event inferred in 75% of all gene families. Of these putative LGTs, more were related to plasmid-related genes than phage-related genes, suggesting plasmid conjugation was a more frequent source of gene transfer [32]. Directed graphs exploring directionality of LGT events between 657 prokaryote genomes allowed the polarization of 32,028 putative LGT events finding that frequency of recent events correlates with genome sequence similarity and most LGTs occurring between donor-recipient pairs with <5% difference in GC content, suggesting that there are some barriers to lateral gene transfer between prokaryotes but that these are not insurmountable [31]. Later reconstruction of transduction events linking phage donors and recipients in a phylogenomic network demonstrated that LGT by transduction was generally highest in similar genomes and between clusters of closely related species but that this constraint was occasionally broken, resulting in LGTs over long evolutionary distances [35].

4 Bipartite Graphs

Bipartite graphs are excellent at summarizing what genes are shared between sets of genomes, and as such are ideal for comparative genomics, including for the comparison of genomes reconstructed in metagenomic analyses. The potential to extend this approach to multilevel graphs, adding additional layers of information such as the environment in ecological studies, could provide a powerful summary of gene sharing in relatively complex datasets.

A multilevel network is a network in which edges exclusively connect nodes of different types, i.e., representing different levels of biological organization. Thus, a bipartite graph is a graph with two types of nodes (top and bottom nodes), where edges exclusively connect nodes of different types (Fig. 11) [146]. The types of nodes used can vary widely depending on the biological question, from linking diseases (top nodes) to their associated genes (bottom nodes) in order to explore the association between related disease phenotypes and their genetic causes [147, 148], to exploring the concept of flavor pairings in food based on a graph of ingredients (top nodes) and the flavor compounds they contain (bottom nodes) [149]. For applications in molecular biology, a typical example of a bipartite graph may describe the relationships between genomes (top nodes) and gene families (bottom nodes), with edges between nodes indicating that a genome encodes at least one member of the corresponding gene family (Fig. 11) [23, 33, 38, 150]. This kind of genome to gene family graph is particularly suited for the comparative analysis of the gene content of genomes in microbial communities and for exploring patterns of gene sharing, for example, between distantly related cellular genomes [33] or between cellular genomes and their mobile genetic elements (Corel et al. forthcoming). It is possible to represent all genes shared between a given set of genomes, as a result of both vertical inheritance and horizontal gene transfer, in a single bipartite graph [23].

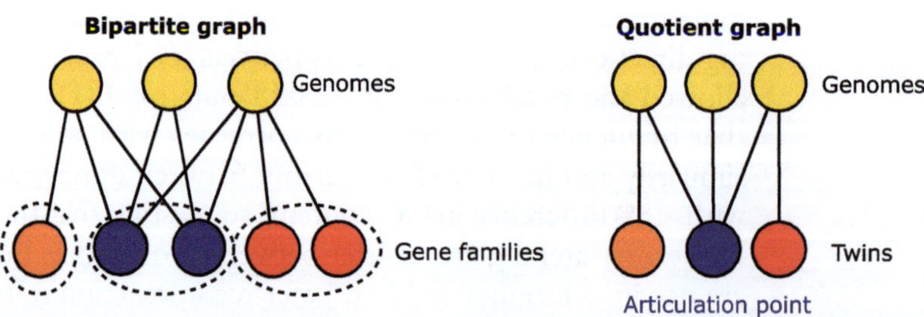

Fig. 11 A bipartite graph and its reduction to a quotient graph: (**a**) An example of a bipartite graph displaying how five gene families are shared between three genomes. (**b**) A reduced form of the bipartite graph in which gene families are combined to "twin" nodes if they share identical taxonomic distributions. A single "articulation point" connects all three genomes

This feature was utilized by Iranzo et al. to explore gene sharing among the entire dsDNA virosphere, a group of entities typified by high rates of molecular evolution and gene transfer [38]. In this case, bipartite modularity was identified in the graph to identify groups of related viral genomes and their shared genes, with the modularity of the graph optimized to Barber's bipartite modularity [151]. A number of additional methods have been developed for detection of module structures within a bipartite graph including for weighted graphs [152]. Two recently developed tools, AcCNET [150] and MultiTwin (forthcoming), have simplified the process of constructing and analyzing multilevel graphs without the need for custom programming (Boxes 3 and 4).

Box 3: Generating Gene-Sharing Networks and Bipartite Graphs

1. *Dataset assembly*: The same rules for dataset assembly as described in SSN generation apply to assembling the dataset for bipartite and gene-sharing graphs. It is especially important to maintain an annotation file that maps gene IDs to their genome of origin.
2. *Definition of gene families*: Gene family identification can be carried out following the construction of sequence similarity networks, as described in Subheading 2. There are a broad range of alternative approaches for construction of gene families that are beyond the scope of discussion in this chapter; however, all of these can also be applied to the generation of gene-sharing and bipartite graphs.
3. *Network construction*: From the definition of gene families, it is possible to construct both gene-sharing networks and bipartite graphs.
   (a) In a gene-sharing network, two genomes are connected by an edge when they encode genes belonging to the same gene family. Generating this kind of network can be automated from BLAST or fasta sequence data using EGN [52].
   (b) In a bipartite graph, there are two types of node, genome nodes and gene family nodes. An edge is drawn between a genome node and a gene family node if that genome encodes a member of the gene family. AcCNET [150] and MultiTwin (forthcoming) tools both include pipelines for generating bipartite graphs from sequence data. MultiTwin can also generate a bipartite graph from two files: a tab-delimited file mapping gene identifiers to their corresponding genome identifier and a tab-delimited file mapping gene identifiers to their corresponding gene family.

Two topological features of bipartite graphs can be used to facilitate studies of gene sharing by an exact decomposition of the bipartite graph: twins and articulation points [23, 153]. A bipartite graph can be reduced to a quotient graph, a reduced variant of the bipartite graph where nodes from the bipartite graph have been combined based on sharing similar properties without the loss of information. For twin nodes ("twins"), this reduction is based on the combination of bottom nodes that have identical neighbors into a single "twin" supernode in the quotient graph (Fig. 11). This is a useful way of reducing the size of large graphs without losing information, but twin nodes also have useful properties for graph interpretation. The genomes supporting a twin node (its neighbors) define a club of genomes that share genes, through common ancestry and/or horizontal transfer, and the number of gene families making up the twin gives a simple description of how many gene families are shared between this club. For example, in any given dataset, any "core" set of gene families encoded by all species in the analysis will be represented by a single twin node. The gene families combined in twin supernodes can be viewed as gene families that are likely to be transmitted together [23]. An articulation point is a node that, when removed, will split the graph into two or more connected components. Within a gene family-genome bipartite graph, articulation points are expected to help to identify "public genetic goods," gene families that are shared by distantly related entities that may confer an advantage independent of genealogy [23, 154], as well as selfish genetic elements such as transposases that also spread across multiple genomes.

Box 4: Considerations for the Construction and Analysis of Bipartite Graphs Using AcCNET and MultiTwin

The default workflow for both ACcNet and MultiTwin takes protein sequence data in fasta format as input and generates a bipartite graph alongside a number of graph summary statistics and outputs for visualization in standard tools (such as Gephi and Cytoscape) but with a number of important differences, including:

- *Graph levels*: Both AcCNET and MultiTwin can generate a bipartite graph using their default workflow; however, MultiTwin can also be used to explore additional graph levels by adding additional node types (e.g., a tripartite graph). Multipartite graphs mean that gene family level annotations can be associated with additional levels of biological information. This may be particularly useful for the comparison of samples in metagenomics studies or time course experiments, allowing gene families to be associated directly with features such as environmental origin or time point.

(continued)

Box 4: (continued)

- *Gene family identification*: AcCNET uses kClust [155] to assemble gene families, a kmer-based method for rapid assembly of clusters of homologous proteins from sequence data. By default, MultiTwin identifies gene families using an all-versus-all BLAST search, followed by identification of connected components at a given threshold, as previously discussed for gene family detection from SSNs. MultiTwin can also be used in a modular way allowing for additional customization, including the use of any custom gene family input in the form of a "community file": a tab-delimited file linking every gene/protein ID to a community identifier, with gene families defined using a clustering method of choice.
- *Edge weighting*: In AcCNET the edge weight is proportional to the inverse of the phylogenetic distance between proteins in a cluster from a given genome to other proteins within the same cluster. In MultiTwin, the default edge weight is based on the number of genes present in a gene family from any given genome.
- *Graph compression*: While both methods can be used to identify "twin" nodes, only MultiTwin generates a quotient graph from these twin nodes and identifies articulation points.

AcCNET is available at: https://sourceforge.net/projects/accnet

MultiTwin is available at: http://www.evol-net.fr/index.php/en/downloads

4.1 Using Bipartite Graphs to Explore Patterns of Gene Sharing Between Diverse Entities

The simplest application of a bipartite graph is the summary of all genes shared between genomes in a single parsable graph, and this feature has been used to explore gene sharing in the dsDNA virome [38], a range of *Escherichia coli* genomes to investigate *the E. coli* pangenome [150] and between a broad range of prokaryotes that include newly discovered organisms [33]. In their analysis of prokaryote genomes, Jaffe et al. used the notion of "twins" to explore patterns of gene sharing between prokaryotes, including Archaea and the recently discovered ultrasmall "Candidate Phyla Radiation" and TM6 bacteria with extremely unusual and reduced genomes. The group found evidence for lateral gene transfer between ultrasmall bacteria and other prokaryotes, consistent with the suggestion that the ultrasmall bacteria may be symbionts [33]. In their exploration of the dsDNA virome, Iranzo et al. used graph module detection, algorithms designed to identify groups of densely connected nodes in a graph, to identify sets of densely connected viral genes and genomes that included viruses with broad host ranges, as well as 14 hallmark viral genes that account for most of the gene sharing between all different viral modules [38].

5 Conclusions

This chapter has offered a brief introduction to the generation of commonly used sequence similarity networks in molecular biology and a guide to how they can be generated and applied to a broad range of studies (Fig. 12). Networks provide a highly scalable framework for the study of an increasingly broad range of applications in molecular biology and evolution and have already contributed to a number of important discoveries in the field. These include exploring patterns of introgression and horizontal transfer across all domains of life and mobile elements, the origin of eukaryotes, the contribution of new genes including novel fusion genes to major evolutionary transitions, shedding light on the "microbial dark matter" in metagenome sequencing datasets and in testing ecological hypotheses about organism and gene distribution and environmental selection. New methods and tools for network analysis are becoming increasingly user-friendly and accessible to biologists without extensive programming experience and enabling network analysis to become a more common part of a biologist toolkit in the analysis of molecular sequence data.

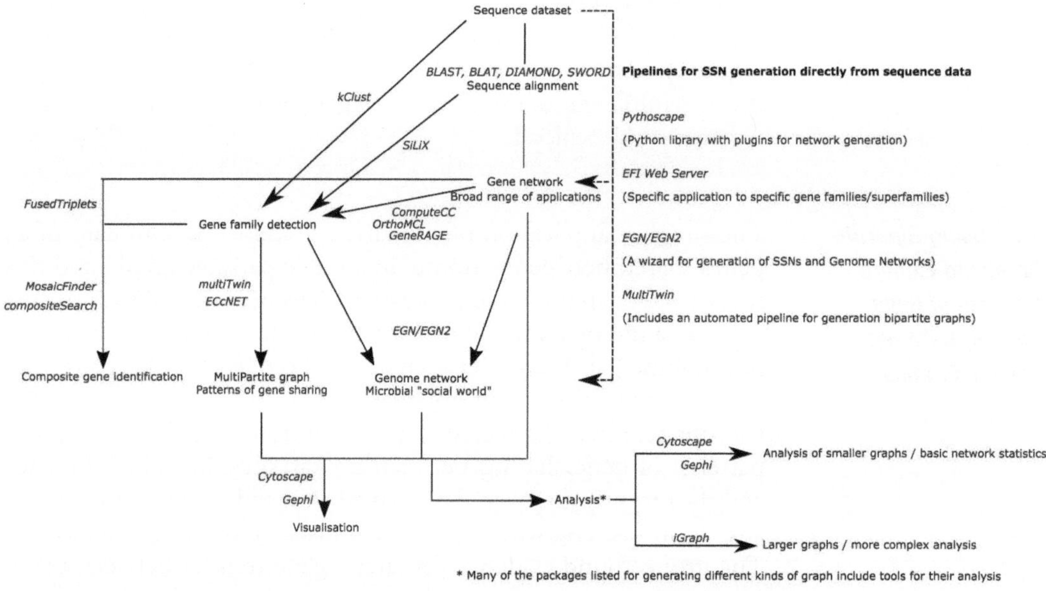

Fig. 12 A workflow highlighting some of the available routes for generation and analysis of SSNs, gene-sharing networks, and bipartite graphs. This workflow highlights just some of the many tools and routes for network construction and analysis

6 Exercises

The exercises use EGN [52] and require access to a local installation of BLAST+ [58] and Perl. The fasta sequence file "example.faa" provided with EGN includes a dataset of protein sequences from Archaea, Bacteria, Eukaryota, and mobile genetic elements, available at http://www.evol-net.fr/index.php/fr/downloads:

1. Perform a manual all-versus-all BLAST using search for a given protein sequence file from the unix terminal (requires local installation of BLAST). The output can be filtered to generate a network:
   (a) Make the blast database using the "*makeblastdb*."
   - Command: "*makeblastdb -dbtype prot -in example.faa –out example*"

   (b) Performing the BLAST search using "blastp," remembering to output data in a tabular format for easy processing.
   - Command: "*blastp -query example.faa -db example -evalue 1e-5 -seg yes -soft_masking true - max_target_-seqs 5000 -outfmt "6 qseqid sseqid evalue pident bitscore qstart qend qlen sstart send slen" -out protein.blastpout*"
2. Generate a SSN using EGN from example.faa (requires local installation of BLAST and download of EGN from http://www.evol-net.fr/index.php/fr/downloads):
   (a) Run EGN from the terminal using "perl egn.1.0.plus.pl" from the programs home directory.

   (b) Follow on-screen prompts sequentially to generate an alignment, filter the output, and generate a gene network with outputs compatible with both Cytoscape and Gephii.
3. Visualize SSN networks:
   (a) In Cytoscape: Import files named "*cc.*.txt*" as a network to visualize that set of connected components.
   - To associate nodes with their annotations, import "*cc*.atr*" as a table.

   (b) In Gephi: Open "*cc*.gxf*" files to import individual connected components from the network into Gephi. Use the "layout" menu to explore different kinds of layouts for the network.

Glossary

Articulation point — A node in a graph whose removal increases the number of connected components of the resulting graph.

Adjacency matrix	A numerical square matrix with row and columns labeled by network nodes, with 1 or 0 in the matrix indicating whether they are connected by an edge in the network.
Assortativity	A measure of the preference for labeled nodes in a network to attach to other nodes with identical labels. This is the Pearson correlation coefficient of the degrees of pairs of linked nodes. Assortativity $= \frac{\text{modularity}}{\text{modularity}_{\text{max}}}$ with modularity defined below and modularity max as the modularity of a perfectly mixed network. $\text{modularity}_{\text{max}} = \frac{1}{2m}\left(2m - \sum_{ij} \frac{k_i k_j}{2m} \delta(c_i - c_j)\right)$.
Betweenness	A centrality measure for a node in a graph. Precisely, this is the proportion of shortest paths between all possible pairs of nodes in a connected component that pass through this node. A betweenness close to 1 is indicative of a highly central gene, whereas close to 0 is more peripheral.
Bipartite graph	A graph with two types of nodes (top and bottom nodes), in which an edge only connects nodes of different types.
Club of genomes	A group of entities that replicated separately but exploit common genetic material that may not trace back to the last common ancestor.
Communities (also called modules)	In graph terminology, a community is defined as a group of nodes that are more connected between themselves than to nodes in the rest of the graph.
Composite gene	A gene that is made up of at least two component parts.
Component genes	Genetic fragments sharing partial similarity to a composite gene.
Conductance	A measure that quantifies whether a given category of nodes shares more edges between themselves than with the rest of the nodes in the graph. A low conductance approaching zero implies that there are few edges shared between this category of nodes and the rest of the graph, while a higher conductance implies more connectivity between that category of nodes and other nodes outside of the category. G a graph, G = {V, E}. With U & G a set of nodes that is assumed to not have more than half the total node. $\bar{U} = G \backslash U$. d(U)

	sum of degree of vertices in U. Conductance $= \frac{\sum_{i \in U, j \in \bar{U}} a_{i,j}}{\min(d(U), d(\bar{U}))}$
Connected component	A subgraph in which any pair of nodes is connected, either directly or indirectly, and that is not connected to the rest of the graph.
Degree	The number of edges connected to a given node.
Endosymbiont	An organism that lives inside another to the mutual benefit of both organisms.
Edge	The link between two nodes in a network.
E-value	The number of alignments in a sequence similarity search expected to be seen by chance searching against a database of a certain size.
Introgression	Descent process through which the genetic material of an entity propagates into different host structures and is replicated within these new host structures.
Lateral gene transfer (LGT; or horizontal gene transfer, HGT)	Movement of genetic material between entities not mediated by vertical descent.
Louvain community	A graph community identified using the Louvain algorithm. Louvain algorithm is based on optimizing modularity.
Network (or graph)	A system of objects (nodes), some pairs of which are linked (edge).
Multipartite graph	Similar to a bipartite graph, but with any number of types of nodes exclusively connected to nodes of other types.
Multiplex graph	A graph where nodes can be connected by edges of different types.
Modularity	The fraction of edges falling within given groups (e.g., communities or functional categories) in a network, minus the fraction of edges that would be expected with a random distribution of edges. With m the total number of vertices, c_i the community of node i, $\delta()$ the Kronecker delta, and k_i the degree of modularity $= \frac{1}{2m}\sum_{ij}\left(A_{ij} - \frac{k_i k_j}{2m}\right)\delta(c_i - c_j)$.
Phylogenomic network	A phylogenetic network constructed from whole genome sequences where genomes are connected based on pairwise relationships including vertical and lateral gene transfer (LGT) events.

Public genetic goods	Common genetic materials shared by clubs of phylogenetically distinct genomes.
Quotient graph	A simplified graph whose nodes represent disjoint subsets of nodes of the original graph; an edge in this new graph connects two such new nodes whenever an edge in the original graph connects at least one element of a new node with at least one from the other.
Supporting genomes	The common set of neighbors that support a "twin" class in a multipartite graph.
Twins	Nodes in a multipartite graph that share identical sets of neighbors.

References

1. Timmis JN, Ayliffe MA, Huang CY, Martin W (2004) Endosymbiotic gene transfer: organelle genomes forge eukaryotic chromosomes. Nat Rev Genet 5:123–135. https://doi.org/10.1038/nrg1271
2. Embley TM, Martin W (2006) Eukaryotic evolution, changes and challenges. Nature 440:623–630. https://doi.org/10.1038/nature04546
3. Williams TA, Foster PG, Cox CJ, Embley TM (2013) An archaeal origin of eukaryotes supports only two primary domains of life. Nature 504:231–236. https://doi.org/10.1038/nature12779
4. Alsmark C, Foster PG, Sicheritz-Ponten T et al (2013) Patterns of prokaryotic lateral gene transfers affecting parasitic microbial eukaryotes. Genome Biol 14:R19. https://doi.org/10.1186/gb-2013-14-2-r19
5. Hirt RP, Alsmark C, Embley TM (2015) Lateral gene transfers and the origins of the eukaryote proteome: a view from microbial parasites. Curr Opin Microbiol 23:155–162. https://doi.org/10.1016/j.mib.2014.11.018
6. Nowack ECM, Price DC, Bhattacharya D et al (2016) Gene transfers from diverse bacteria compensate for reductive genome evolution in the chromatophore of Paulinella chromatophora. Proc Natl Acad Sci U S A 113:12214–12219. https://doi.org/10.1073/pnas.1608016113
7. McCoy JM, Mi S, Lee X et al (2000) Syncytin is a captive retroviral envelope protein involved in human placental morphogenesis. Nature 403:785–789. https://doi.org/10.1038/35001608
8. Kondo N, Nikoh N, Ijichi N et al (2002) Genome fragment of Wolbachia endosymbiont transferred to X chromosome of host insect. Proc Natl Acad Sci U S A 99:14280–14285. https://doi.org/10.1073/pnas.222228199
9. McInerney JO (2017) Horizontal gene transfer is less frequent in eukaryotes than prokaryotes but can be important (retrospective on DOI 10.1002/bies.201300095). BioEssays 39:1700002. https://doi.org/10.1002/bies.201700002
10. Gogarten JP, Doolittle WF, Lawrence JG (2002) Prokaryotic evolution in light of gene transfer. Mol Biol Evol 19:2226–2238
11. Dagan T, Martin W (2007) Ancestral genome sizes specify the minimum rate of lateral gene transfer during prokaryote evolution. Proc Natl Acad Sci U S A 104:870–875. https://doi.org/10.1073/pnas.0606318104
12. Hooper SD, Mavromatis K, Kyrpides NC (2009) Microbial co-habitation and lateral gene transfer: what transposases can tell us. Genome Biol 10:R45. https://doi.org/10.1186/gb-2009-10-4-r45
13. Nelson-Sathi S, Sousa FL, Roettger M et al (2014) Origins of major archaeal clades correspond to gene acquisitions from bacteria. Nature 517:77–80. https://doi.org/10.1038/nature13805
14. Tamminen M, Virta M, Fani R, Fondi M (2012) Large-scale analysis of plasmid relationships through gene-sharing networks. Mol Biol Evol 29:1225–1240. https://doi.org/10.1093/molbev/msr292
15. Lapierre P, Gogarten JP (2009) Estimating the size of the bacterial pan-genome. Trends

Genet 25:107–110. https://doi.org/10.1016/j.tig.2008.12.004

16. Vos M, Hesselman MC, te Beek TA et al (2015) Rates of lateral gene transfer in prokaryotes: high but why? Trends Microbiol 23:598–605. https://doi.org/10.1016/j.tim.2015.07.006
17. McInerney JO, McNally A, O'Connell MJ (2017) Why prokaryotes have pangenomes. Nat Microbiol 2:17040. https://doi.org/10.1038/nmicrobiol.2017.40
18. Niehus R, Mitri S, Fletcher AG, Foster KR (2015) Migration and horizontal gene transfer divide microbial genomes into multiple niches. Nat Commun 6:8924. https://doi.org/10.1038/ncomms9924
19. Hotopp JCD, Clark ME, Oliveira DCSG et al (2007) Widespread lateral gene transfer from intracellular bacteria to multicellular eukaryotes. Science 317:1753–1756. https://doi.org/10.1126/science.1142490
20. Wolf YI, Kondrashov AS, Koonin EV (2000) Interkingdom gene fusions. Genome Biol 1: research0013.1. https://doi.org/10.1186/gb-2000-1-6-research0013
21. Becking LB (1934) Geobiologie of inleiding tot de milieukunde. W.P. Van Stockum & Zoon, Den Haag, The Hague, the Netherlands
22. Lobb B, Kurtz DA, Moreno-Hagelsieb G, Doxey AC (2015) Remote homology and the functions of metagenomic dark matter. Front Genet 6:234. https://doi.org/10.3389/fgene.2015.00234
23. Corel E, Lopez P, Méheust R, Bapteste E (2016) Network-thinking: graphs to analyze microbial complexity and evolution. Trends Microbiol 24:224–237. https://doi.org/10.1016/j.tim.2015.12.003
24. Lopez P, Halary S, Bapteste E (2015) Highly divergent ancient gene families in metagenomic samples are compatible with additional divisions of life. Biol Direct 10:64. https://doi.org/10.1186/s13062-015-0092-3
25. Forster D, Bittner L, Karkar S et al (2015) Testing ecological theories with sequence similarity networks: marine ciliates exhibit similar geographic dispersal patterns as multicellular organisms. BMC Biol 13:16. https://doi.org/10.1186/s12915-015-0125-5
26. Fondi M, Karkman A, Tamminen MV et al (2016) "Every gene is everywhere but the environment selects": global geolocalization of gene sharing in environmental samples through network analysis. Genome Biol Evol 8:1388–1400. https://doi.org/10.1093/gbe/evw077
27. Cheng S, Karkar S, Bapteste E et al (2014) Sequence similarity network reveals the imprints of major diversification events in the evolution of microbial life. Front Ecol Evol 2:72. https://doi.org/10.3389/fevo.2014.00072
28. Thiergart T, Landan G, Schenk M et al (2012) An evolutionary network of genes present in the eukaryote common ancestor polls genomes on eukaryotic and mitochondrial origin. Genome Biol Evol 4:466–485. https://doi.org/10.1093/gbe/evs018
29. Alvarez-Ponce D, Lopez P, Bapteste E, McInerney JO (2013) Gene similarity networks provide tools for understanding eukaryote origins and evolution. Proc Natl Acad Sci U S A 110:E1594–E1603. https://doi.org/10.1073/pnas.1211371110
30. Halary S, Leigh JW, Cheaib B et al (2010) Network analyses structure genetic diversity in independent genetic worlds. Proc Natl Acad Sci U S A 107:127–132. https://doi.org/10.1073/pnas.0908978107
31. Popa O, Hazkani-Covo E, Landan G et al (2011) Directed networks reveal genomic barriers and DNA repair bypasses to lateral gene transfer among prokaryotes. Genome Res 21:599–609. https://doi.org/10.1101/gr.115592.110
32. Kloesges T, Popa O, Martin W, Dagan T (2011) Networks of gene sharing among 329 proteobacterial genomes reveal differences in lateral gene transfer frequency at different phylogenetic depths. Mol Biol Evol 28:1057–1074. https://doi.org/10.1093/molbev/msq297
33. Jaffe AL, Corel E, Pathmanathan J et al (2016) Bipartite graph analyses reveal interdomain LGT involving ultrasmall prokaryotes and their divergent, membrane-related proteins. Environ Microbiol 18:5072–5081. https://doi.org/10.1111/1462-2920.13477
34. Dagan T (2011) Phylogenomic networks. Trends Microbiol 19:483–491. https://doi.org/10.1016/j.tim.2011.07.001
35. Popa O, Landan G, Dagan T (2017) Phylogenomic networks reveal limited phylogenetic range of lateral gene transfer by transduction. ISME J 11:543–554. https://doi.org/10.1038/ismej.2016.116
36. Fondi M, Fani R (2010) The horizontal flow of the plasmid resistome: clues from intergeneric similarity networks. Environ Microbiol 12:3228–3242. https://doi.org/10.1111/j.1462-2920.2010.02295.x

37. Lima-Mendez G, Van Helden J, Toussaint A, Leplae R (2008) Reticulate representation of evolutionary and functional relationships between phage genomes. Mol Biol Evol 25:762–777. https://doi.org/10.1093/molbev/msn023
38. Iranzo J, Krupovic M, Koonin EV (2016) The double-stranded DNA virosphere as a modular hierarchical network of gene sharing. MBio 7:e00978–e00916. https://doi.org/10.1128/mBio.00978-16
39. Tatusov RL, Koonin EV, Lipman DJ (1997) A genomic perspective on protein families. Science 278:631–637
40. Tatusov RL, Galperin MY, Natale DA, Koonin EV (2000) The COG database: a tool for genome-scale analysis of protein functions and evolution. Nucleic Acids Res 28:33–36. https://doi.org/10.1093/nar/28.1.33
41. Huson DH, Scornavacca C (2011) A survey of combinatorial methods for phylogenetic networks. Genome Biol Evol 3:23–35. https://doi.org/10.1093/gbe/evq077
42. Huson DH, Rupp R, Scornavacca C (2011) Phylogenetic networks: concepts, algorithms and applications. Cambridge University Press, New York, NY
43. Nakhleh L (2011) Evolutionary phylogenetic networks: models and issues. In: Problem solving handbook in computational biology and bioinformatics. Springer, New York, pp 125–158
44. Enright AJ, Iliopoulos I, Kyrpides NC, Ouzounis CA (1999) Protein interaction maps for complete genomes based on gene fusion events. Nature 402:86–90. https://doi.org/10.1038/47056
45. Pasternak G, Hochhaus A, Schultheis B, Hehlmann R (1998) Chronic myelogenous leukemia: molecular and cellular aspects. J Cancer Res Clin Oncol 124:643–660
46. Watanabe H, Otsuka J (1995) A comprehensive representation of extensive similarity linkage between large numbers of proteins. Bioinformatics 11:159–166. https://doi.org/10.1093/bioinformatics/11.2.159
47. Park J, Teichmann SA, Hubbard T, Chothia C (1997) Intermediate sequences increase the detection of homology between sequences. J Mol Biol 273:349–354. https://doi.org/10.1006/jmbi.1997.1288
48. Bolten E, Schliep A, Schneckener S et al (2001) Clustering protein sequences--structure prediction by transitive homology. Bioinformatics 17:935–941. https://doi.org/10.1093/bioinformatics/17.10.935
49. Bapteste E, Lopez P, Bouchard F et al (2012) Evolutionary analyses of non-genealogical bonds produced by introgressive descent. Proc Natl Acad Sci U S A 109:18266–18272. https://doi.org/10.1073/pnas.1206541109
50. Jachiet P-A, Pogorelcnik R, Berry A et al (2013) MosaicFinder: identification of fused gene families in sequence similarity networks. Bioinformatics 29:837–844. https://doi.org/10.1093/bioinformatics/btt049
51. Méheust R, Zelzion E, Bhattacharya D et al (2016) Protein networks identify novel symbiogenetic genes resulting from plastid endosymbiosis. Proc Natl Acad Sci U S A 113:3579–3584. https://doi.org/10.1073/pnas.1517551113
52. Halary S, McInerney JO, Lopez P, Bapteste E (2013) EGN: a wizard for construction of gene and genome similarity networks. BMC Evol Biol 13:146. https://doi.org/10.1186/1471-2148-13-146
53. Martin AJM, Walsh I, Di Domenico T et al (2013) PANADA: protein association network annotation, determination and analysis. PLoS One 8:e78383. https://doi.org/10.1371/journal.pone.0078383
54. Shannon P, Markiel A, Ozier O et al (2003) Cytoscape: a software environment for integrated models of biomolecular interaction networks. Genome Res 13:2498–2504. https://doi.org/10.1101/gr.1239303
55. Bastian M, Heymann S, Jacomy M (2009) Gephi: an Open source software for exploring and manipulating networks. Third Int AAAI Conf Weblogs Soc Media. pp 361–362. https://doi.org/10.1136/qshc.2004.010033
56. Csárdi G, Nepusz T (2006) The igraph software package for complex network research. InterJ Complex Syst 1695
57. Hagberg AA, Schult DA, Swart PJ (2008) Exploring network structure, dynamics, and function using NetworkX. In: Varoquaux G, Vaught T, Millman J (eds) Proc. 7th Python Sci. Conf, Pasadena, CA, pp 11–15
58. Camacho C, Coulouris G, Avagyan V et al (2009) BLAST+: architecture and applications. BMC Bioinform 10:421. https://doi.org/10.1186/1471-2105-10-421
59. Altschul SF, Gish W, Miller W et al (1990) Basic local alignment search tool.pdf. J Mol Biol 215:403–410. https://doi.org/10.1016/S0022-2836(05)80360-2
60. Kent WJ (2002) BLAT--the BLAST-like alignment tool. Genome Res 12:656–664.

https://doi.org/10.1101/gr.229202. Article published online before March 2002

61. Vaser R, Pavlović D, Šikić M (2016) SWORD—a highly efficient protein database search. Bioinformatics 32:i680–i684. https://doi.org/10.1093/bioinformatics/btw445
62. Edgar RC (2010) Search and clustering orders of magnitude faster than BLAST. Bioinformatics 26:2460–2461. https://doi.org/10.1093/bioinformatics/btq461
63. Buchfink B, Xie C, Huson DH (2014) Fast and sensitive protein alignment using DIAMOND. Nat Methods 12:59–60. https://doi.org/10.1038/nmeth.3176
64. Altschul SF, Gish W, Miller W et al (1990) Basic local alignment search tool. J Mol Biol 215:403–410. https://doi.org/10.1016/S0022-2836(05)80360-2
65. Ye Y, Choi J-H, Tang H (2011) RAPSearch: a fast protein similarity search tool for short reads. BMC Bioinform 12:159. https://doi.org/10.1186/1471-2105-12-159
66. Page L, Brin S, Motwani R, Winograd T (1998) The PageRank citation ranking: bringing order to the web. Technical Report. Stanford InfoLab
67. Brandes U (2001) A faster algorithm for betweenness centrality*. J Math Sociol 25:163–177. https://doi.org/10.1080/0022250X.2001.9990249
68. Staudt CL, Sazonovs A, Meyerhenke H (2016) NetworKit: a tool suite for large-scale complex network analysis. Network Science 4 (4):508–530. https://doi.org/10.1017/nws.2016.20
69. Teng S-H (2016) Scalable algorithms for data and network analysis. Now Publishers Inc, Hanover, MA
70. Dayhoff MO (1976) The origin and evolution of protein superfamilies. Fed Proc 35:2132–2138
71. Heger A, Holm L (2000) Towards a covering set of protein family profiles. Prog Biophys Mol Biol 73:321–337. https://doi.org/10.1016/S0079-6107(00)00013-4
72. Girvan M, Newman MEJ (2002) Community structure in social and biological networks. Proc Natl Acad Sci U S A 99:7821–7826. https://doi.org/10.1073/pnas.122653799
73. Blondel VD, Guillaume J-L, Lambiotte R, Lefebvre E (2008) Fast unfolding of communities in large networks. J Stat Mech Theory Exp. https://doi.org/10.1088/1742-5468/2008/10/P10008
74. Enright AJ, Van Dongen S, Ouzounis CA (2002) An efficient algorithm for large-scale detection of protein families. Nucleic Acids Res 30:1575–1584. https://doi.org/10.1093/nar/30.7.1575
75. Altenhoff AM, Kunca N, Glover N et al (2015) The OMA orthology database in 2015: function predictions, better plant support, synteny view and other improvements. Nucleic Acids Res 43:D240–D249. https://doi.org/10.1093/nar/gku1158
76. Li L, Stoeckert CJ, Roos DS (2003) OrthoMCL: identification of ortholog groups for eukaryotic genomes. Genome Res 13:2178–2189. https://doi.org/10.1101/gr.1224503
77. Dessimoz C, Cannarozzi G, Gil M et al (2005) OMA, a comprehensive, automated project for the identification of orthologs from complete genome data: introduction and first achievements. Springer, Berlin, pp 61–72
78. Dessimoz C, Boeckmann B, Roth ACJ, Gonnet GH (2006) Detecting non-orthology in the COGs database and other approaches grouping orthologs using genome-specific best hits. Nucleic Acids Res 34:3309–3316. https://doi.org/10.1093/nar/gkl433
79. Roth ACJ, Gonnet GH, Dessimoz C (2008) Algorithm of OMA for large-scale orthology inference. BMC Bioinform 9:518. https://doi.org/10.1186/1471-2105-9-518
80. Altenhoff AM, Gil M, Gonnet GH et al (2013) Inferring hierarchical orthologous groups from orthologous gene pairs. PLoS One 8:e53786. https://doi.org/10.1371/journal.pone.0053786
81. Schneider A, Dessimoz C, Gonnet GH (2007) OMA browser exploring orthologous relations across 352 complete genomes. Bioinformatics 23:2180–2182. https://doi.org/10.1093/bioinformatics/btm295
82. Miele V, Penel S, Duret L (2011) Ultra-fast sequence clustering from similarity networks with SiLiX. BMC Bioinform 12:116. https://doi.org/10.1186/1471-2105-12-116
83. Penel S, Arigon A-M, Dufayard J-F et al (2009) Databases of homologous gene families for comparative genomics. BMC Bioinform 10:S3. https://doi.org/10.1186/1471-2105-10-S6-S3
84. Dagan T, Roettger M, Bryant D, Martin W (2010) Genome networks root the tree of life between prokaryotic domains. Genome Biol Evol 2:379–392. https://doi.org/10.1093/gbe/evq025
85. Jacob F (1977) Evolution and tinkering. Science 196:1161–1166
86. Marcotte EM, Pellegrini M, Ng HL et al (1999) Detecting protein function and

protein-protein interactions from genome sequences. Science 285:751–753

87. Kawai H, Kanegae T, Christensen S et al (2003) Responses of ferns to red light are mediated by an unconventional photoreceptor. Nature 421:287–290. https://doi.org/10.1038/nature01310
88. Kaessmann H (2010) Origins, evolution, and phenotypic impact of new genes. Genome Res 20:1313–1326. https://doi.org/10.1101/gr.101386.109
89. Marsh JA, Teichmann SA (2010) How do proteins gain new domains? Genome Biol 11:126. https://doi.org/10.1186/gb-2010-11-7-126
90. Promponas VJ, Ouzounis CA, Iliopoulos I (2014) Experimental evidence validating the computational inference of functional associations from gene fusion events: a critical survey. Brief Bioinform 15:443–454. https://doi.org/10.1093/bib/bbs072
91. McLysaght A, Guerzoni D (2015) New genes from non-coding sequence: the role of de novo protein-coding genes in eukaryotic evolutionary innovation. Philos Trans R Soc B Biol Sci 370:20140332. https://doi.org/10.1098/rstb.2014.0332
92. Enright AJ, Ouzounis CA (2000) GeneRAGE: a robust algorithm for sequence clustering and domain detection. Bioinformatics 16:451–457. https://doi.org/10.1093/bioinformatics/16.5.451
93. Snel B, Bork P, Huynen M (2000) Genome evolution. Gene fusion versus gene fission. Trends Genet 16:9–11
94. Enright AJ, Ouzounis CA (2001) Functional associations of proteins in entire genomes by means of exhaustive detection of gene fusions. Genome Biol 2:RESEARCH0034
95. Patthy L (2003) Modular assembly of genes and the evolution of new functions. Genetica 118:217–231
96. Nakamura Y, Itoh T, Martin W (2007) Rate and polarity of gene fusion and fission in Oryza sativa and Arabidopsis thaliana. Mol Biol Evol 24:110–121. https://doi.org/10.1093/molbev/msl138
97. Ekman D, Björklund ÅK, Elofsson A (2007) Quantification of the elevated rate of domain rearrangements in metazoa. J Mol Biol 372:1337–1348. https://doi.org/10.1016/j.jmb.2007.06.022
98. Jachiet P-AA, Colson P, Lopez P, Bapteste E (2014) Extensive gene remodeling in the viral world: new evidence for nongradual evolution in the mobilome network. Genome Biol Evol 6:2195–2205. https://doi.org/10.1093/gbe/evu168
99. Song N, Joseph JM, Davis GB et al (2008) Sequence similarity network reveals common ancestry of multidomain proteins. PLoS Comput Biol 4:e1000063. https://doi.org/10.1371/journal.pcbi.1000063
100. Yanai I, Derti A, DeLisi C (2001) Genes linked by fusion events are generally of the same functional category: a systematic analysis of 30 microbial genomes. Proc Natl Acad Sci U S A. https://doi.org/10.1073/pnas.141236298
101. Pathmanathan JS, Lopez P, Lapointe F-J, Bapteste E (2018) CompositeSearch: a generalized network approach for composite gene families detection. Mol Biol Evol 35:252–255. https://doi.org/10.1093/molbev/msx283
102. Dorrell RG, Gile G, McCallum G et al (2017) Chimeric origins of ochrophytes and haptophytes revealed through an ancient plastid proteome. elife. https://doi.org/10.7554/eLife.23717
103. Martin W, Stoebe B, Goremykin V et al (1998) Gene transfer to the nucleus and the evolution of chloroplasts. Nature 393:162–165. https://doi.org/10.1038/30234
104. Martin W, Rujan T, Richly E et al (2002) Evolutionary analysis of Arabidopsis, cyanobacterial, and chloroplast genomes reveals plastid phylogeny and thousands of cyanobacterial genes in the nucleus. Proc Natl Acad Sci U S A 99:12246–12251. https://doi.org/10.1073/pnas.182432999
105. Reyes-Prieto A, Hackett JD, Soares MB et al (2006) Cyanobacterial contribution to algal nuclear genomes is primarily limited to plastid functions. Curr Biol. https://doi.org/10.1016/j.cub.2006.09.063
106. Leskovec J, Lang KJ, Dasgupta A, Mahoney MW (2008) Statistical properties of community structure in large social and information networks. In: Proceeding 17th Int. Conf. World Wide Web - WWW '08. ACM Press, New York, p 695
107. Newman MEJ (2003) Mixing patterns in networks. Phys Rev E 67:26126. https://doi.org/10.1103/PhysRevE.67.026126
108. Newman M (2010) Networks. An introduction. Oxford University Press, Oxford. https://doi.org/10.1093/acprof:oso/9780199206650.001.0001
109. Rappé MS, Giovannoni SJ (2003) The uncultured microbial majority. Annu Rev Microbiol

57:369–394. https://doi.org/10.1146/annurev.micro.57.030502.090759

110. Williams TA, Embley TM (2014) Archaeal? Dark matter? And the origin of eukaryotes. Genome Biol Evol 6:474–481. https://doi.org/10.1093/gbe/evu031
111. Castelle CJJ, Wrighton KCC, Thomas BCC et al (2015) Genomic expansion of domain archaea highlights roles for organisms from new phyla in anaerobic carbon cycling. Curr Biol 25:690–701. https://doi.org/10.1016/j.cub.2015.01.014
112. Brown CT, Hug LA, Thomas BC et al (2015) Unusual biology across a group comprising more than 15% of domain Bacteria. Nature 523:208–211. https://doi.org/10.1038/nature14486
113. Spang A, Saw JH, Jørgensen SL et al (2015) Complex archaea that bridge the gap between prokaryotes and eukaryotes. Nature 521:173–179. https://doi.org/10.1038/nature14447
114. Zaremba-Niedzwiedzka K, Caceres EF, Saw JH et al (2017) Asgard archaea illuminate the origin of eukaryotic cellular complexity. Nature 541:353–358. https://doi.org/10.1038/nature21031
115. Prakash T, Taylor TD (2012) Functional assignment of metagenomic data: challenges and applications. Brief Bioinform 13:711–727. https://doi.org/10.1093/bib/bbs033
116. Hingamp P, Grimsley N, Acinas SG et al (2013) Exploring nucleo-cytoplasmic large DNA viruses in Tara Oceans microbial metagenomes. ISME J 7:1678–1695. https://doi.org/10.1038/ismej.2013.59
117. de Vargas C, Audic S, Henry N et al (2015) Eukaryotic plankton diversity in the sunlit ocean. Science 348:1261605–1261605. https://doi.org/10.1126/science.1261605
118. Sunagawa S, Coelho LP, Chaffron S et al (2015) Structure and function of the global ocean microbiome. Science 348:1261359–1261359. https://doi.org/10.1126/science.1261359
119. Paez-Espino D, Eloe-Fadrosh EA, Pavlopoulos GA et al (2016) Uncovering earth's virome. Nature 536:425–430. https://doi.org/10.1038/nature19094
120. Shi M, Lin XD, Tian JH et al (2016) Redefining the invertebrate RNA virosphere. Nature. https://doi.org/10.1038/nature20167
121. van Regenmortel MH, Mayo MA, Fauquet CM, Maniloff J (2000) Virus nomenclature: consensus versus chaos. Arch Virol 145:2227–2232
122. Gibbs AJ (2000) Virus nomenclature descending into chaos. Arch Virol 145:1505–1507
123. Lawrence JG, Hatfull GF, Hendrix RW (2002) Imbroglios of viral taxonomy: genetic exchange and failings of phenetic approaches. J Bacteriol 184:4891–4905
124. Franklin LR (2007) Bacteria, sex, and systematics. Philos Sci 74:69–95. https://doi.org/10.1086/519476
125. Bapteste E, Boucher Y (2008) Lateral gene transfer challenges principles of microbial systematics. Trends Microbiol 16:200–207. https://doi.org/10.1016/j.tim.2008.02.005
126. Bapteste E, O'Malley MA, Beiko RG et al (2009) Prokaryotic evolution and the tree of life are two different things. Biol Direct 4:34. https://doi.org/10.1186/1745-6150-4-34
127. Andam CP, Williams D, Gogarten JP (2010) Natural taxonomy in light of horizontal gene transfer. Biol Philos 25:589–602. https://doi.org/10.1007/s10539-010-9212-8
128. Koonin EV, Dolja VV (2014) Virus world as an evolutionary network of viruses and capsidless selfish elements. Microbiol Mol Biol Rev 78:278–303. https://doi.org/10.1128/MMBR.00049-13
129. Lederberg J, Tatum EL (1946) Gene recombination in Escherichia coli. Nature 158:558
130. Zinder ND, Lederberg J (1952) Genetic exchange in Salmonella. J Bacteriol 64:679–699
131. Levin BR (1988) Frequency-dependent selection in bacterial populations. Philos Trans R Soc Lond B Biol Sci 319:459–472
132. Rodriguez-Valera F (2004) Environmental genomics, the big picture? FEMS Microbiol Lett 231:153–158
133. Chen I, Christie PJ, Dubnau D (2005) The ins and outs of DNA transfer in bacteria. Science 310:1456–1460. https://doi.org/10.1126/science.1114021
134. Edwards RA, Rohwer F (2005) Viral metagenomics. Nat Rev Microbiol 3:504–510. https://doi.org/10.1038/nrmicro1163
135. Frost LS, Leplae R, Summers AO, Toussaint A (2005) Mobile genetic elements: the agents of open source evolution. Nat Rev Microbiol 3:722–732. https://doi.org/10.1038/nrmicro1235
136. Dagan T, Martin W (2009) Getting a better picture of microbial evolution en route to a network of genomes. Philos Trans R Soc Lond B Biol Sci 364:2187–2196. https://doi.org/10.1098/rstb.2009.0040

137. Kulp A, Kuehn MJ (2010) Biological functions and biogenesis of secreted bacterial outer membrane vesicles. Annu Rev Microbiol 64:163–184. https://doi.org/10.1146/annurev.micro.091208.073413
138. McDaniel LD, Young E, Delaney J et al (2010) High frequency of horizontal gene transfer in the oceans. Science 330:50. https://doi.org/10.1126/science.1192243
139. Dubey GP, Ben-Yehuda S (2011) Intercellular nanotubes mediate bacterial communication. Cell 144:590–600. https://doi.org/10.1016/j.cell.2011.01.015
140. Desnues C, La Scola B, Yutin N et al (2012) Provirophages and transpovirons as the diverse mobilome of giant viruses. Proc Natl Acad Sci U S A 109:18078–18083. https://doi.org/10.1073/pnas.1208835109
141. Kutschera VE, Bidon T, Hailer F et al (2014) Bears in a forest of gene trees: phylogenetic inference is complicated by incomplete lineage sorting and gene flow. Mol Biol Evol 31:2004–2017. https://doi.org/10.1093/molbev/msu186
142. Simmonds P (2014) Methods for virus classification and the challenge of incorporating metagenomic sequence data. J Gen Virol. https://doi.org/10.1099/jgv.0.000016
143. Iranzo J, Koonin EV, Prangishvili D, Krupovic M (2016) Bipartite network analysis of the archaeal virosphere: evolutionary connections between viruses and capsid-less mobile elements. J Virol 90:11043–11055. https://doi.org/10.1128/JVI.01622-16
144. Lord E, Le Cam M, Bapteste É et al (2016) BRIDES: a new fast algorithm and software for characterizing evolving similarity networks using breakthroughs, roadblocks, impasses, detours, equals and shortcuts. PLoS One 11:e0161474. https://doi.org/10.1371/journal.pone.0161474
145. van Dongen SM (2001) Graph clustering by flow simulation. PhD thesis, University of Utrecht
146. Borgatti SP, Everett MG (1997) Network analysis of 2-mode data. Soc Netw 19:243–269. https://doi.org/10.1016/S0378-8733(96)00301-2
147. Goh K-I, Cusick ME, Valle D et al (2007) The human disease network. Proc Natl Acad Sci U S A 104:8685–8690. https://doi.org/10.1073/pnas.0701361104
148. Himmelstein DS, Baranzini SE, Rand V et al (2015) Heterogeneous network edge prediction: a data integration approach to prioritize disease-associated genes. PLoS Comput Biol 11:e1004259. https://doi.org/10.1371/journal.pcbi.1004259
149. Ahn Y-Y, Ahnert SE, Bagrow JP et al (2011) Flavor network and the principles of food pairing. Sci Rep 1:196. https://doi.org/10.1038/srep00196
150. Lanza VF, Baquero F, de la Cruz F, Coque TM (2017) AcCNET (Accessory Genome Constellation Network): comparative genomics software for accessory genome analysis using bipartite networks. Bioinformatics 33:283–285. https://doi.org/10.1093/bioinformatics/btw601
151. Barber MJ (2007) Modularity and community detection in bipartite networks. Phys Rev E 76:66102. https://doi.org/10.1103/PhysRevE.76.066102
152. Beckett SJ (2016) Improved community detection in weighted bipartite networks. R Soc Open Sci 3:140536. https://doi.org/10.1098/rsos.140536
153. Diestel R (2010) Graph theory. Springer, New York
154. McInerney JO, Pisani D, Bapteste E, O'Connell MJ (2011) The public goods hypothesis for the evolution of life on Earth. Biol Direct 6:41. https://doi.org/10.1186/1745-6150-6-41
155. Hauser M, Mayer CE, Söding J (2013) kClust: fast and sensitive clustering of large protein sequence databases. BMC Bioinform 14:248. https://doi.org/10.1186/1471-2105-14-248

Chapter 10

Bayesian Molecular Clock Dating Using Genome-Scale Datasets

Mario dos Reis and Ziheng Yang

Abstract

Bayesian methods for molecular clock dating of species divergences have been greatly developed during the past decade. Advantages of the methods include the use of relaxed-clock models to describe evolutionary rate variation in the branches of a phylogenetic tree and the use of flexible fossil calibration densities to describe the uncertainty in node ages. The advent of next-generation sequencing technologies has led to a flood of genome-scale datasets for organisms belonging to all domains in the tree of life. Thus, a new era has begun where dating the tree of life using genome-scale data is now within reach. In this protocol, we explain how to use the computer program MCMCTree to perform Bayesian inference of divergence times using genome-scale datasets. We use a ten-species primate phylogeny, with a molecular alignment of over three million base pairs, as an exemplar on how to carry out the analysis. We pay particular attention to how to set up the analysis and the priors and how to diagnose the MCMC algorithm used to obtain the posterior estimates of divergence times and evolutionary rates.

Key words Molecular clock, Bayesian analysis, MCMC, Fossil, Phylogeny, Primates, Genome

1 Introduction

The molecular clock hypothesis, which states that the rate of molecular evolution is approximately constant with time, provides a powerful way to estimate the times of divergence of species in a phylogeny. Since its proposal over 50 years ago [1], the molecular clock hypothesis has been used countless times to calibrate molecular phylogenies to geological time, with the ultimate aim of dating the tree of life [2, 3]. Several statistical inference methodologies have been developed for molecular clock dating analyses; however, during the past decade, the Bayesian method has emerged as the method of choice [4, 5], and several Bayesian inference software packages now exist to carry out this type of analysis [6–10].

In this protocol, we will explain how to use the computer program MCMCTree to estimate times of species divergences using genome-scale datasets within the Bayesian inference

Maria Anisimova (ed.), *Evolutionary Genomics: Statistical and Computational Methods*, Methods in Molecular Biology, vol. 1910, https://doi.org/10.1007/978-1-4939-9074-0_10,

framework. Bayesian inference is well suited for divergence time estimation because it allows the natural integration of information from the fossil record (in the form of prior statistical distributions describing the ages of nodes in a phylogeny) with information from molecular sequences to estimate node ages, or geological times of divergence, of a species phylogeny [6, 11]. Another advantage of the Bayesian clock dating method is that relaxed-clock models, which allow for violations of the molecular clock, can be easily implemented as the prior on the evolutionary rates for the branches in the phylogeny [6]. MCMCTree allows analyses to be carried out using two popular relaxed-clock models (the autocorrelated and independent log-normally distributed rates models [12, 13]), as well as under the strict molecular clock. Furthermore, MCMCTree allows the user to build flexible fossil calibrations based on various statistical distributions (such as the uniform, truncated-Cauchy, and skew-*t*, and skew-normal distributions [12, 14, 15]). But perhaps the main advantage of MCMCTree is the implementation of an approximate algorithm to calculate the likelihood [6, 16], which allows the computer analysis of genome-scale datasets to be completed in reasonable amounts of time. The disadvantage of the algorithm is that it only works on fixed tree topologies. Several software packages that perform co-estimation of times and tree topology, but which do not use the approximation, are available [8, 9, 17, 18].

In this protocol, we focus on how to carry out a clock dating analysis with MCMCTree, paying particular attention to diagnosing the MCMC algorithm (the workhorse algorithm within the Bayesian method). Theoretical details of the Bayesian clock dating methods implemented in the program MCMCTree are described in [12–16, 19]. For general introductions to Bayesian statistics and Bayesian molecular clock dating, the reader may consult [20, 21].

2 Software and Data Files

To run the protocol, you will need the MCMCTree and BASEML programs, which are part of the PAML software package for phylogenetic analysis [22]. The source code and compiled versions of the code are freely available from bit.ly/ziheng-paml. All the data files necessary to run the protocol can be obtained from github.com/mariodosreis/divtime. Please create a directory called `divtime` in your computer and download all the data files from the GitHub repository. This protocol was tested with PAML version 4.9e.

You are assumed to have basic knowledge of the command line in Unix or Windows (also known as command prompt, shell, or terminal). Simple tutorials for users of Windows, Mac OS, and Linux are posted at bit.ly/ziheng-software. Install MCMCTree and BASEML in your computer system, and make sure you have the `mcmctree` and `baseml` executables in your system's path (see bit.ly/ziheng-paml for details on how to do this). Finally, it is helpful (but not indispensable) to have knowledge of the R statistical environment (www.r-project.org). R is quite useful to analyze the output of the program, perform convergence diagnostics, and create nice-looking plots. File `R/analysis.R` contains some examples for this tutorial.

In this protocol, we will estimate the divergence times of nine primates and one scandentian (an out-group), using a very long alignment (over three million nucleotides long). This dataset was chosen because it can be analyzed very quickly with MCMCTree and it is thus suitable to illustrate the method. We also provide a dataset of 330 species (276 primates and 4 out-groups) with a shorter alignment, to illustrate time estimation in a taxon-rich dataset (*see* Sect. 5.5 for details).

2.1 Tree and Fossil Calibrations

The phylogenetic tree of the ten species is shown in Fig. 1. The tree encompasses members of all the main primate lineages. The ten species were chosen because they have had their complete genomes sequenced. They are a subset of the 36 mammal species analyzed in [23]. File `data/10s.tree` contains the tree with fossil calibrations in Newick format, which is the format required by MCMCTree. The eight fossil calibrations are shown in Table 1. The calibrations are the same used to estimate primate divergence times in [24]. We discuss fossil calibrations in detail in the "Sampling from the Prior" section. The time unit in the analysis is 100 million years (My). Thus, the calibration B(0.075, 0.10) means the node age is constrained to be between 7.5 and 10 million years ago (Ma).

2.2 Molecular Sequence Data

The molecular data are an alignment of 5614 protein-coding genes from the ten species. All ambiguous codon sites were removed, and thus the alignment contains no missing data. The alignment was separated into two partitions: A partition consisting of all the first and second codon positions (2,253,316 nucleotides long) and a partition of third codon positions (1,126,658 nucleotides long). The alignment is a subset of the larger 36-mammal-species alignment in [23]. *See* also ref. 24. File `10s.phys` in the `data` directory contains the alignment. The alignment is compressed into site patterns (a site pattern is a unique combination of character states in an alignment column) to save disk space.

3 Tutorial

We seek to obtain the posterior distribution (i.e., the estimates) of the divergence times ($\mathbf{t}$) and the molecular evolutionary rates ($\mathbf{r}$, μ, σ^2) for the species in the phylogeny of Fig. 1. Here $\mathbf{t} = (t_{11}, \ldots, t_{19})$ are the nine species divergence times; $\mathbf{r} = (r_{1,12}, \ldots, r_{1,19}, r_{2,12}, \ldots,$

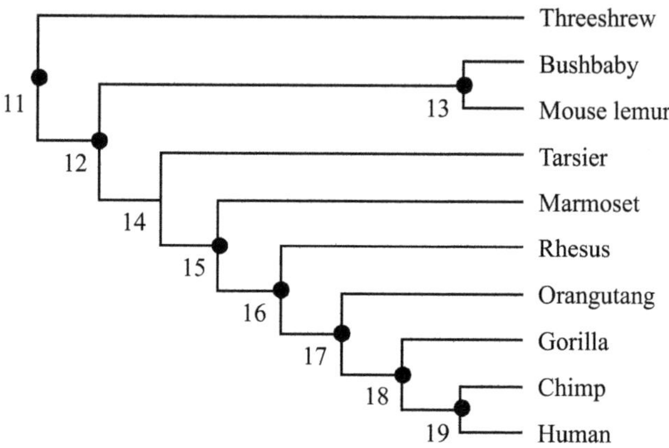

Fig. 1 The tree of ten species. Nodes with fossil calibrations are indicated with black dots (*see* Table 1 for calibration densities). Internal nodes are numbered from 11 to 19 according to the nomenclature used by MCMCTree

Table 1
List of fossil calibrations used in this tutorial

Node[a]	Crown group	MCMCTree calibration[b]
19	Chimp-human	B(0.075, 0.10, 0.01, 0.20)
18	Gorilla-human	B(0.10, 0.132, 0.01, 0.20)
17	Hominidae	B(0.112, 0.28, 0.01, 0.10)
16	Catarrhini	B(0.25, 0.29, 0.01, 0.10)
15	Anthropoidea	ST(0.4754, 0.0632, 0.98, 22.85)
13	Strepsirrhini	B(0.38, 0.58, 0.01, 0.10)
12	Primates	S2N(0.698, 0.65, 0.0365, −3400, 0.650, 0.138, 11409)
11	Euarchonta	G(36, 36.9)

[a]Node numbers as in Fig. 1

[b]B(a, b, p_L, p_U) means the calibration is a uniform distribution between a and b, with probabilities p_L and p_U that the true node age is outside the calibration bounds. ST(location, scale, shape, df) means the calibration is a skew-t distribution. S2N(p, location1, scale1, shape1, location2, scale2, shape2) means the calibration is a p:1 − p mixture of two skew-normal distributions. G(α, β) means the calibration is a gamma distribution with shape α and rate β. See MCMCTree's manual for the full details on fossil calibration formats. The calibrations are from the primate analysis in [24]

$r_{2,19}$) are the $2 \times 8 = 16$ molecular rates, one per branch and partition (i.e., there are eight branches in the tree and two partitions in the molecular data); and $\mu = (\mu_1, \mu_2)$ and $\sigma^2 = (\sigma_1^2, \sigma_2^2)$ are the mean rates and the log-variance of the rates, for each partition. The posterior distribution is

$$f(\mathbf{t}, \mathbf{r}, \mu, \sigma^2 | D) \propto f(\mathbf{t}) f(\mathbf{r} | \mathbf{t}, \mu, \sigma^2) f(\mu) f(\sigma^2) f(D | \mathbf{r}, \mathbf{t}),$$

where $f(\mathbf{t})$ is the prior on times; $f(\mathbf{r}|\mathbf{t}, \mu, \sigma^2)f(\mu)f(\sigma^2)$ is the prior on the branch rates, mean rates, and variances of the log-rates; and $f(D|\mathbf{t}, \mathbf{r})$ is the molecular sequence likelihood. The prior on the times is constructed by combining the birth-death process with the fossil calibration densities (*see* ref. 13 for details). The prior on the rates is constructed under a model of rate evolution, assuming, in this tutorial, that the branch rates are independent draws from a log-normal distribution with mean μ_i and log-variance σ_i^2 [13].

Bayesian phylogenetic inference using MCMC is computationally expensive because of the repeated calculation of the likelihood on a sequence alignment. The time it takes to compute the likelihood is proportional to the number of site patterns in the alignment. Thus, longer alignments take longer to compute. For genome-scale alignments, the computation time is prohibitive.

MCMCTree implements an approximation to the likelihood that speeds computation time substantially, making analysis of genome-scale data feasible. The approximate likelihood method for clock dating was proposed by Thorne et al. [6] and extended within MCMCTree [16]. The method relies on approximating the log-likelihood surface on the branch lengths by its Taylor expansion. Write $\ell(\mathbf{b}_j) = \log f(D|\ \mathbf{b}_j)$ for the log-likelihood as a function of the branch lengths $\mathbf{b}_j = (b_{j,i} = r_{j,i}t_i)$ for the alignment partition j. The Taylor approximation is

$$\ell(\mathbf{b}_j) \approx \ell(\hat{\mathbf{b}}_j) + (\mathbf{b}_j - \hat{\mathbf{b}}_j)^{\mathrm{T}}\mathbf{g}_j + \frac{1}{2}(\mathbf{b}_j - \hat{\mathbf{b}}_j)^{\mathrm{T}}\mathbf{H}_j(\mathbf{b}_j - \hat{\mathbf{b}}_j),$$

where $\hat{\mathbf{b}}_j$ are the maximum likelihood estimates (MLEs) of the branch lengths and $\mathbf{g}_j$ and $\mathbf{H}_j$ are the gradient (vector of first derivatives) and Hessian (matrix of second derivatives) of the log-likelihood surface evaluated at the MLEs for the partition. The approximation can be improved by applying transformations to the branch lengths (*see* ref. 16 for details).

To use the approximation, one first fixes the topology of the phylogeny, and then estimates the branch lengths for each alignment partition on the fixed tree by maximum likelihood. The gradient and Hessian of the log-likelihood are obtained for each partition at the same time as the MLEs of the branch lengths. Note that parameters of the substitution model—such as the transition/transversion ratio, κ, in the HKY model or the α parameter in the discrete gamma model of rate variation among sites—are estimated at this step. Thus, different substitution models will generate different approximations, because they will have different MLEs for the branch lengths, gradient, and Hessian. Note that the time it takes to compute the approximate likelihood depends only on the number of species (which determines the size of $\mathbf{b}$ and $\mathbf{H}$) and not on the alignment length, that is, once $\mathbf{g}$ and $\mathbf{H}$ have been calculated, MCMC sampling on the approximation takes the same time regardless of the length of the original alignment.

3.1 Overview

We will use the approximate likelihood method to speed up the computation of the likelihood on the large genome alignment. The general strategy for the analysis is as follows:

1. *Approximate likelihood calculation*: First, we will calculate the gradient (**g**) and Hessian (**H**) matrix of the branch lengths on the unrooted tree. For this step, we will need to use the MCMCTree and BASEML programs (BASEML will carry out the actual computation of **g** and **H**). The substitution model is chosen at this step.
2. *MCMC sampling from the posterior*: Once **g** and **H** have been calculated and we have decided on our priors, we can use MCMCTree to perform MCMC sampling from the posterior distribution of times and rates. We will then look at the summaries of the posterior (such as posterior mean times and rates and 95% credibility intervals).
3. *Convergence diagnostics*: The MCMC algorithm is a stochastic algorithm that visits regions of the parameter space in proportion to the posterior distribution. Due to its very nature, it is possible that sometimes the MCMC chain is terminated before it has had a chance to explore the parameter space appropriately. The way to guard against this is to run the analysis two or more times and compare the summary statistics from the two (or more) MCMC chains. If the results from different runs are very similar, then convergence to the posterior distribution can be reasonably assumed.
4. *MCMC sampling from the prior*: Finally, we will sample directly from the prior of times and rates. This is particularly important in Bayesian molecular clock dating because in most cases the prior on times may look quite different from the fossil calibration densities specified by the user. Thus, sampling from the prior allows the user to check the soundness of the prior actually used.

Note that in this protocol we assume the user has chosen a suitable sequence alignment and a phylogenetic tree to carry out the analysis. For genome-scale alignments, it is important that the genes chosen among the various species are orthologous and that the alignment has been checked for accuracy. Several chapters in this volume can guide the user in this purpose.

3.2 Calculation of the Gradient and Hessian to Approximate the Likelihood

Go into the `gH` directory, and open the `mcmctree-outBV.ctl` file using your favorite text editor. This control file contains the set of parameters necessary for MCMCTree to carry out the calculations of the gradient and Hessian needed for the approximate likelihood method. Figure 2 shows the contents of the `mcmctree-outBV.ctl` file.

```
  seqfile = ../data/10s.phys
 treefile = ../data/10s.tree

    ndata = 2
  seqtype = 0    * 0: nucleotides; 1:codons; 2:AAs
  usedata = 3    * 0: no data (prior); 1:exact likelihood;
                 * 2: approximate likelihood; 3:out.BV (in.BV)
    clock = 2    * 1: global clock; 2: independent rates; 3: correlated rates

    model = 4    * 0:JC69, 1:K80, 2:F81, 3:F84, 4:HKY85
    alpha = 0.5   * alpha for gamma rates at sites
    ncatG = 5    * No. categories in discrete gamma

cleandata = 0    * remove sites with ambiguity data (1:yes, 0:no)?
```

Fig. 2 The `gH/mcmctree-outBV.ctl` file, with appropriate options to set up calculation of the gradient and Hessian matrix for the approximate likelihood method

```
10

((Bushbaby: 0.029523, Mouse_lemur: 0.019653): 0.006547, (Tarsier: 0.030897, (Marmoset: 0.0

 0.006547  0.029523  0.019653  0.002123  0.030897  0.011754  0.015183  0.003426  0.008716

 -2.114230 -2.618861 21.299836 31.765175 20.801006 -3.019251 -14.909946  8.188538 -3.70464


Hessian

 -2.033e+08  -2.59e+06 -9.717e+06 -4.363e+07  1.799e+06 -5.457e+06  2.055e+06  -1.29e+04
  -2.59e+06  -5.71e+07  2.235e+06  1.475e+06  3.315e+06  1.651e+06  3.436e+06  2.134e+06
 -9.717e+06  2.235e+06 -8.733e+07 -2.954e+06   2.79e+06  7.275e+05  3.371e+06  1.512e+06
 -4.363e+07  1.475e+06 -2.954e+06 -4.622e+08 -5.059e+06 -2.658e+07  3.701e+06 -5.157e+06 -
  1.799e+06  3.315e+06   2.79e+06 -5.059e+06 -5.473e+07  7.951e+05  3.437e+06   2.28e+06
 -5.457e+06  1.651e+06  7.275e+05 -2.658e+07  7.951e+05 -1.403e+08  3.724e+06 -1.163e+07
  2.055e+06  3.436e+06  3.371e+06  3.701e+06  3.437e+06  3.724e+06  -1.25e+08  -1.69e+07
  -1.29e+04  2.134e+06  1.512e+06 -5.157e+06   2.28e+06 -1.163e+07  -1.69e+07 -4.756e+08
  3.483e+06  4.548e+06  4.413e+06 -1.406e+05  4.463e+06  2.246e+06  1.979e+06  1.698e+06
  8.344e+05  2.861e+06  2.023e+06  1.605e+06  2.021e+06 -5.676e+05 -8.424e+05 -1.722e+07 -
  3.625e+06  4.671e+06  4.894e+06  8.939e+05  4.775e+06  2.595e+06  1.699e+06  5.407e+05
  2.701e+06  3.036e+06  2.394e+06  1.777e+06  3.175e+06  6.217e+05 -5.952e+05 -4.592e+06 -
```

Fig. 3 The `gH/out.BV` file produced by BASEML. The first line has the number of species (10), the second line has the tree topology with MLEs of branch lengths, and the MLEs of branch lengths are given again in the third line. The fourth line contains the gradient, **g**, followed by the Hessian, **H**, for partition 1. This file will be renamed `in.BV` and placed into the `mcmc/` directory to carry out MCMC sampling using the approximate likelihood method

The first two items, `seqfile` and `treefile`, indicate the alignment and tree files to be used. The third item, `ndata`, indicates the number of partitions in the sequence file, in this case, two partitions. The fifth item, `usedata`, is very important, as it tells MCMCTree the type of analysis being carried out. The options are

0, to sample from the prior; 1, to sample from the posterior using exact likelihood; 2, to sample from the posterior using approximate likelihood; and 3, to prepare the data for calculation of $\mathbf{g}$ and $\mathbf{H}$. The last is the option we will be using in this step. The next three items, `model`, `alpha`, and `ncatG`, set up the nucleotide substitution model, in this case the HKY + Gamma model [25]. Finally, the `cleandata` option tells MCMCTree whether to remove ambiguous data. Our alignment has no ambiguous sites, so this option has no effect in this case.

Using a terminal, go to the `gH` directory and type

```
$ mcmctree mcmctree-outBV.ctl
```

(Don't type in the $ as this represents the command prompt!) This will start the MCMCTree program. MCMCTree will prepare several `tmp????.*` files and will then call the BASEML program to estimate $\mathbf{g}$ and $\mathbf{H}$. For this step to work correctly, the `baseml` executable must be in your system's path. Once BASEML and MCMCTree have finished, you will notice a file called `out.BV` has been created. Figure 3 shows part of the contents of this file. The first line indicates the number of species (10), followed by the tree with branch lengths estimated under maximum likelihood for the first partition (first and second codon sites). Next, we have the MLEs of the 17 branch lengths (these are the same as in the tree but printed in a different order). Then we have the gradient, $\mathbf{g}_1$, the vector of 17 first derivatives of the likelihood at the branch length MLEs for partition 1. For small datasets, the gradient is usually zero. For large datasets, the likelihood surface is too sharp (i.e., bends downward sharply and it is very narrow at the MLEs), and the gradient is not zero for numerical issues. But this is fine. Next, we have the 17×17 Hessian matrix, $\mathbf{H}_1$, the matrix of second derivatives of the likelihood at the branch length MLEs for partition 1. If you scroll down the file, you will find the second block, with the tree, branch length MLEs, $\mathbf{g}_2$, and $\mathbf{H}_2$ for partition 2 (third codon positions).

3.3 Calculation of the Posterior of Times and Rates

3.3.1 Control File and Priors

Now that we have calculated $\mathbf{g}$ and $\mathbf{H}$, we can proceed to MCMC sampling of the posterior distribution using the approximate likelihood method. Copy the `gH/out.BV` file into the `mcmc` directory, and rename it as `in.BV`. Now go into the `mcmc` directory. There you will find `mcmctree.ctl`, the necessary MCMCTree control file to carry out MCMC sampling from the posterior. Figure 4 shows the contents of the file. The first item, `seed`, is the seed for the random number generator used by the MCMC algorithm. Here it is set to −1, which tells MCMCTree to use the system's clock time as the seed. This is useful, as running the program multiple times will generate different outputs.

```
          seed = -1
       seqfile = ../data/10s.phys
      treefile = ../data/10s.tree
      mcmcfile = mcmc.txt
       outfile = out.txt

         ndata = 2
       seqtype = 0    * 0: nucleotides; 1:codons; 2:AAs
       usedata = 2    * 0: no data (prior); 1:exact likelihood;
                      * 2:approximate likelihood; 3:out.BV (in.BV)
         clock = 2    * 1: global clock; 2: independent rates; 3: correlated rates
       RootAge = '<1.0'  * safe constraint on root age, used if no fossil for root.

         model = 4    * 0:JC69, 1:K80, 2:F81, 3:F84, 4:HKY85
         alpha = 0.5  * alpha for gamma rates at sites
         ncatG = 5    * No. categories in discrete gamma

     cleandata = 0    * remove sites with ambiguity data (1:yes, 0:no)?

       BDparas = 1 1 0   * birth, death, sampling
   kappa_gamma = 6 2     * gamma prior for kappa
   alpha_gamma = 1 1     * gamma prior for alpha

   rgene_gamma = 2 40 1   * gammaDir prior for rate for genes
  sigma2_gamma = 1 10 1   * gammaDir prior for sigma^2     (for clock=2 or 3)

         print = 1   * 0: no mcmc sample; 1: everything except branch rates 2: everything
        burnin = 20000
      sampfreq = 100
       nsample = 20000
```

Fig. 4 The `mcmc/mcmctree.ctl` file necessary to sample from the posterior distribution using the approximate likelihood method

The `mcmcfile` option tells MCMCTree where to save the parameters sampled (divergence times and rates) during the MCMC iterations. Here we will save them to a file named `mcmc.txt`. Once the MCMC sampling has completed, MCMCTree will read the sample from the `mcmc.txt` file and generate a summary of the MCMC output. This summary will be saved to a file called `out.txt` (`outfile` option).

The option `usedata` is set to 2 here, which tells MCMCTree to calculate the likelihood approximately by using the **g** and **H** values saved in the `in.BV` file. Option `clock` sets the clock model. Here we use clock = 2, which assumes rates are identical, independent realizations from a log-normal distribution [7, 26]. Option `RootAge` sets the calibration on the root node of the phylogeny, if none are present in the tree file. In our case, we already have a calibration on the root, so this option has no effect. The next three options, `model`, `alpha`, and `ncatG`, have no effect as the substitution model was chosen during estimation of **g** and **H**.

The following options are very important as they determine the prior used in the analysis. `BDparams` sets the prior on node ages for those nodes without fossil calibrations by using the birth-death process [12]. Here we use 1 1 0, which means node ages are uniformly distributed between present time and the age of the root. Options `kappa_gamma` and `alpha_gamma` set gamma priors for the κ and α parameters in the substitution model. These have no effect as we are using the likelihood approximation. Options `rgene_gamma` and `sigma2_gamma` set the gamma-Dirichlet prior on the mean substitution rate for partitions and for the rate variance parameter, σ^2 [19]. The prior on the mean rate is Gamma(2, 40), which has mean 0.05 substitutions per time 100 My. A symmetric Dirichlet distribution with concentration parameter equal to 1 is used to spread the rate prior across partitions (thus `rgene_gamma = 2 40 1`). *See* ref. 19 for details. The prior on σ^2 is Gamma(1, 10) which has mean 0.1. A Dirichlet is also used to spread the prior across partitions.

The final block of options, `print`, `burnin`, `sampfreq`, and `nsample`, control the length and sampling frequency of the MCMC. We will discard the first 20,000 iterations as the burn-in and then print parameter values to the `mcmc.txt` file every 100 iterations, to a maximum of 20,000 + 1 samples. Thus, our MCMC chain will run for a total of 20,000 + 20,000 × 100 = 2,020,000 iterations.

3.3.2 Running and Summarizing the MCMC

Go into the `mcmc` directory and type

```
$ mcmctree mcmctree.ctl
```

This will start the MCMC sampling. First, MCMCTree will iterate the chain for a set number of iterations, known as the burn-in. During this period, the program will fine-tune the step sizes for proposing parameters in the chain. Once the burn-in is finished, sampling from the posterior will start. Figure 5 shows a screenshot of MCMCTree in action. The leftmost column indicates the progress of the sampling as a percentage of the total (5%, 10% of total iterations, and so on). The next numbers represent the acceptance proportions, which are close to 30% (this is the result of fine-tuning by the program). After the five acceptance proportions, the programs prints a few parameters to the screen and in the last columns the log-likelihood and the time taken.

The above analysis takes about 2 min and 30 s to complete on a 2.2 GHz Intel Core i7 Processor. Once the analysis has finished, you will see that MCMCTree has created several new files in the `mcmc` directory. Rename `mcmc.txt` to `mcmc1.txt` and `out.txt` to `out1.txt`. Now, on the command line, type again

```
$ mcmctree mcmctree.ctl
```

This will run the analysis a second time. The results should be slightly different to the previous run due to the stochastic nature of the algorithm. Once the second run has finished, rename `mcmc.`

```
   0% 0.26 0.39 0.23 0.39 0.28  1.285 1.243 0.588 1.158 0.541 0.321 - 0.192 0.197 -16.9  0:02

(nsteps = 50)
Current Pjump:       0.26200  0.39475  0.23175  0.38650  0.28000  0.27550  0.39200  0.43750
0.40100  0.29725  0.33725  0.27525  0.32275  0.23475  0.23150  0.29875  0.31600  0.27800
0.25300  0.29975  0.29650  0.32575  0.27500  0.61150  0.29850  0.31225  0.35400  0.23200
0.30800  0.28250  0.33050  0.21325  0.22700  0.25900  0.26725  0.26900  0.33150  0.23725
0.31000  0.20700  0.24225  0.61625  0.30675  0.30150  0.32000  0.21975  0.27650  0.22500
0.36650  0.00000
Current finetune:    0.00365  0.00166  0.00586  0.00182  0.00503  0.00697  0.00486  0.00500
0.00835  0.24230  0.21346  0.71942  0.65595  0.01093  0.01230  0.01256  0.00960  0.01492
0.02008  0.02466  0.03547  0.03942  0.04624  0.17077  0.02425  0.04971  0.01513  0.03626
0.03661  0.04475  0.08082  0.00867  0.00949  0.01146  0.00861  0.01133  0.01263  0.02252
0.02728  0.03996  0.03790  0.14736  0.02025  0.04584  0.01209  0.02975  0.02776  0.03389
0.05173  0.00000
New     finetune:    0.00313  0.00232  0.00438  0.00248  0.00465  0.00632  0.00675  0.00806
0.01194  0.23972  0.24532  0.65158  0.71499  0.00829  0.00918  0.01250  0.01020  0.01367
0.01654  0.02463  0.03499  0.04345  0.04183  0.47928  0.02411  0.05210  0.01846  0.02714
0.03776  0.04175  0.09064  0.00592  0.00694  0.00969  0.00755  0.01000  0.01422  0.01728
0.02835  0.02644  0.02976  0.42023  0.02079  0.04611  0.01305  0.02100  0.02527  0.02454
0.06589  0.00000

   5% 0.34 0.30 0.31 0.32 0.28  1.163 0.981 0.622 0.893 0.464 0.295 - 0.129 0.154 -17.0  0:08
  10% 0.35 0.30 0.31 0.32 0.27  1.189 0.943 0.607 0.859 0.457 0.293 - 0.128 0.153 -17.0  0:15
  15% 0.36 0.30 0.30 0.31 0.27  1.156 0.920 0.604 0.837 0.457 0.290 - 0.133 0.160 -17.0  0:22
  20% 0.35 0.30 0.30 0.32 0.26  1.126 0.908 0.600 0.825 0.453 0.290 - 0.137 0.165 -17.0  0:29
  25% 0.36 0.30 0.30 0.31 0.26  1.139 0.912 0.605 0.829 0.458 0.293 - 0.138 0.165 -17.0  0:37
  30% 0.36 0.30 0.30 0.31 0.26  1.153 0.918 0.609 0.834 0.460 0.293 - 0.136 0.163 -17.0  0:43
```

Fig. 5 Screenshot of MCMCTree's output during MCMC sampling of the posterior. Different runs of the program will give slightly different output values

txt to mcmc2.txt and out.txt to out2.txt. If you want to conduct two runs simultaneously, you can create two directories (say r1/ and r2/) and copy the necessary files into them. Then open two terminal windows to start the runs from within each directory.

Using your favorite text editor, open file out1.txt, which contains the summary of the first MCMC run. Scroll to the end of the file (see screenshot, Fig. 6). You will see the time used by the program (in my case 2:32), the posterior means of the parameters sampled, and three phylogenetic trees in Newick format. The first tree simply has internal nodes labelled with a number. This is useful to compare the tree with the posterior means of times at the end of the file. The second tree is the tree with branch lengths in absolute time units. The third tree is like the second by including the 95% credibility intervals (CIs) of the node ages. At the bottom of the file, you have a table with all the divergence times (from t_n11 to t_n19), the mean substitution rates for the two partitions (mu1 and mu2), the rate variation coefficients (sigma2_1 and sigma2_2), and finally the log-likelihood (lnL). The table gives the posterior means, equal-tail CIs, and high-posterior-density CIs. For example, the posterior age of the root (node 11, Fig. 1) is 116.8 Ma (95% CI, 144.2–92.4 Ma) while for the divergence

```
ln Lmax (unconstrained) = -4636133.236961

Time used:  2:26

mean of parameters using all iterations
   1.16785   0.91766   0.60797   0.83447   0.46464   0.29132   0.17725   0.10441   0.08519   0.

Species tree for FigTree.  Branch lengths = posterior mean times; 95% CIs = labels
(1_Tree_shrew, ((2_Bushbaby, 3_Mouse_lemur) 13 , (4_Tarsier, (5_Marmoset, (6_Rhesus, (7_Orangut

(Tree_shrew: 1.167850, ((Bushbaby: 0.607966, Mouse_lemur: 0.607966): 0.309693, (Tarsier: 0.8344

(Tree_shrew: 1.167850, ((Bushbaby: 0.607966, Mouse_lemur: 0.607966) [&95%={0.50317, 0.735468}]:


Posterior mean (95% Equal-tail CI) (95% HPD CI) HPD-CI-width

t_n11          1.1679 (0.9235, 1.4423) (0.9021, 1.4056) 0.5035 (Jnode 18)
t_n12          0.9176 (0.8015, 1.0484) (0.7965, 1.0423) 0.2458 (Jnode 17)
t_n13          0.6080 (0.5032, 0.7355) (0.5019, 0.7337) 0.2318 (Jnode 16)
t_n14          0.8345 (0.7236, 0.9602) (0.7192, 0.9538) 0.2346 (Jnode 15)
t_n15          0.4646 (0.3966, 0.5340) (0.3964, 0.5335) 0.1371 (Jnode 14)
t_n16          0.2913 (0.2526, 0.3380) (0.2499, 0.3333) 0.0833 (Jnode 13)
t_n17          0.1773 (0.1466, 0.2174) (0.1439, 0.2132) 0.0692 (Jnode 12)
t_n18          0.1044 (0.0995, 0.1164) (0.0988, 0.1139) 0.0152 (Jnode 11)
t_n19          0.0852 (0.0758, 0.0981) (0.0746, 0.0958) 0.0212 (Jnode 10)
mu1            0.0269 (0.0221, 0.0334) (0.0217, 0.0328) 0.0111
mu2            0.1110 (0.0898, 0.1396) (0.0877, 0.1364) 0.0488
sigma2_1       0.1370 (0.0607, 0.2833) (0.0484, 0.2511) 0.2027
sigma2_2       0.1634 (0.0755, 0.3201) (0.0625, 0.2883) 0.2258
lnL          -17.0026 (-25.9750, -9.8710) (-24.9110, -9.1170) 15.7940
```

Fig. 6 The end of the `mcmc/out.txt` file produced by MCMCTree at the end of the MCMC sampling of the posterior

between human and chimp (node 19, Fig. 1) is 8.52 Ma (95% CI, 7.58–9.81 Ma).

You will also notice that MCMCTree created a file called `FigTree.tre`. This contains the posterior tree in Nexus format, suitable for plotting in the program FigTree (tree.bio.ed.ac.uk/software/figtree/). Figure 7 shows the posterior tree plotted in FigTree, with the time unit set to 1 My.

3.4 Convergence Diagnostics of the MCMC

Diagnosing convergence of the MCMC chains is extremely important. Several software tools have been written for this purpose. For example, the user-friendly Tracer program (beast.bio.ed.ac.uk/tracer) can be used to read in the `mcmc1.txt` and `mcmc2.txt` files and calculate several convergence statistics. Here we will use R to perform basic convergence tests (check out file `R/analysis.R`).

The first step to assess convergence is to compare the posterior means among the different runs. You can visually inspect the posterior means reported in the `out1.txt` and `out2.txt` files (Fig. 8), although this may be cumbersome. Figure 8a shows a plot, made with R, of posterior times for run 1 vs. those from run 2. You can see that the points fall almost perfectly on the $y = x$ line, indicating that both runs have converged to the same distribution (hopefully the posterior!).

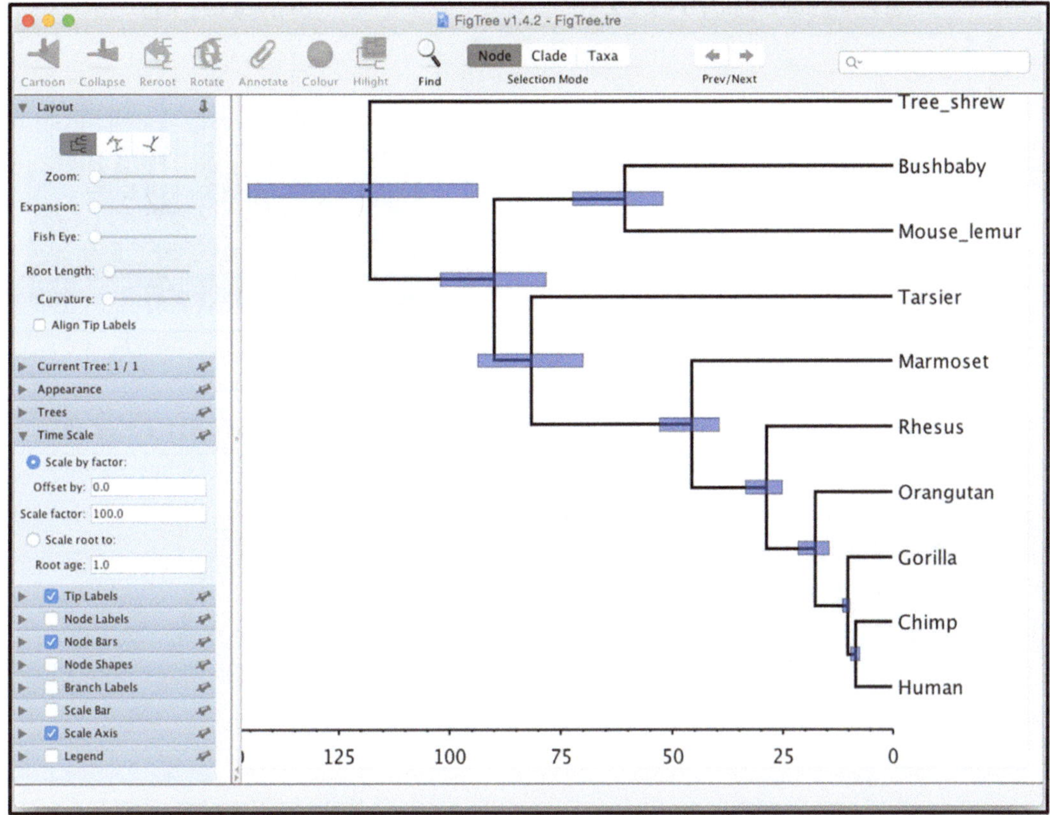

Fig. 7 The dated primate phylogeny with error bars (representing 95% CIs of node ages), drawn with FigTree. The time unit is 1 My

Another useful statistic to be calculated is the effective sample size (ESS). This gives the user an idea about whether an MCMC chain has been run long enough. Tracer calculates ESS automatically for all parameters. Function `coda::effectiveSize` in R will do the same. Figure 9 shows the posterior mean, ESS, posterior variance, and standard error of posterior means calculated with R for run 1 of the MCMC. The longer the ESS, the better. As a rule of thumb, one should seek ESS larger than 1000, although this may not always be practical in phylogenetic analysis. Note in Fig. 9 that some estimates have very low ESSs, while others have substantially higher ESSs. For example, `t_n11` has ESS = 76.1, while `t_n19` has ESS = 1261. Running the analysis again and increasing the total number of iterations (e.g., by increasing `samplefreq` or `nsample`) will lead to higher ESS values for all parameters.

Let v be the posterior variance of a parameter. The standard error of the posterior mean of the parameter is S.E. = $\sqrt{(v/\mathrm{ESS})}$. This is why having large ESS is important: Large ESS leads to small S.E. and better estimates of the posterior mean. For example, for t_n11, the posterior mean is 116.8 Ma, with standard error

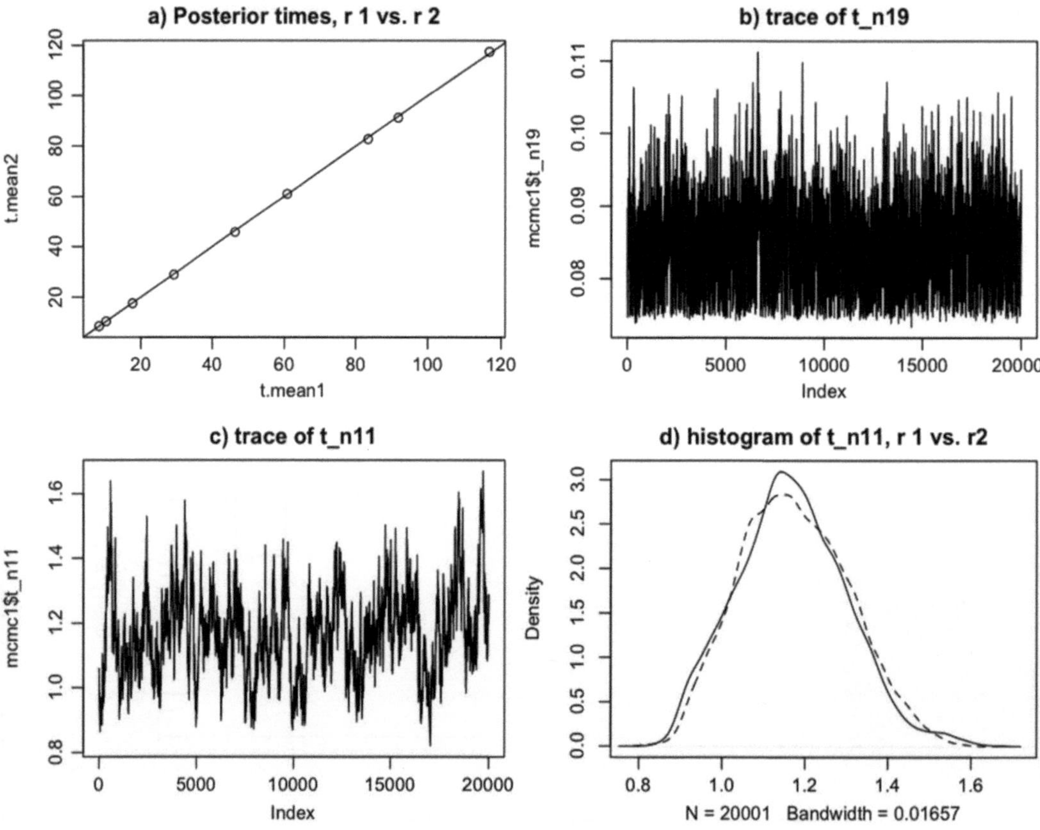

Fig. 8 Convergence diagnostic plots of the MCMC drawn with R (see `R/analysis.R`)

	mean.mcmc	ess.mcmc	var.mcmc	se.mcmc
t_n11	1.16785568	76.14030	1.779905e-02	0.0152894256
t_n12	0.91763459	66.38219	4.085525e-03	0.0078450940
t_n13	0.60801488	151.00623	3.123330e-03	0.0045479066
t_n14	0.83448247	71.93763	3.708967e-03	0.0071803969
t_n15	0.46464686	231.92350	1.211178e-03	0.0022852393
t_n16	0.29131271	353.25425	5.294412e-04	0.0012242361
t_n17	0.17726011	347.03816	3.245757e-04	0.0009670955
t_n18	0.10441651	1035.75332	2.080275e-05	0.0001417203
t_n19	0.08518922	1261.15128	3.363295e-05	0.0001633048
mu1	0.02691074	530.31981	8.464981e-06	0.0001263409
mu2	0.11103179	637.44606	1.577065e-04	0.0004973969
sigma2_1	0.13698819	710.07293	3.298175e-03	0.0021551891
sigma2_2	0.16337732	893.70775	4.046102e-03	0.0021277504
lnL	-17.00256482	20001.00000	1.696800e+01	0.0291265757

Fig. 9 Calculations of posterior mean, ESS, posterior variance, and standard error of the posterior mean in R (see `R/analysis.R`)

1.53 My (Fig. 9). That is, we have estimated the mean accurately to within 2 × 1.53 My = 3.06 My. To reduce the S.E. by half, you need to increase the ESS four times. Note that independent MCMC runs can be combined into a single run. Thus, you may save time by running several MCMC chains in parallel for computationally expensive analyses, although care must be taken to ensure each chain has run long enough to exit the burn-in phase and explore the posterior appropriately.

Trace plots and histograms are useful to spot problems and check convergence. Figure 8b, c shows trace plots for `t_n19` and `t_n11`, respectively. The trace of `t_n19`, which has high ESS, looks like a "hairy caterpillar." Compare it to the trace of `t_n11`, which has low ESS. Visual inspection of a trace plot usually gives a sense of whether the parameter has an adequate ESS without calculating it. Note that both traces are trendless, that is, the traces oscillate around a mean value (the posterior mean). If you see a persistent trend in the trace (such as an increase or a decrease), that most likely means the MCMC did not converge to the posterior and needs a longer burn-in period.

Figure 8d shows the smoothed histograms (calculated using `density` in R) for `t_n11` for the two runs. Notice that the two histograms are slightly different. As the ESS becomes larger, histograms for different runs will converge in shape until becoming indistinguishable. If you see large discrepancies between histograms, that may indicate serious problems with the MCMC, such as lack of convergence due to short burn-in or the MCMC getting stuck in different modes of a multimodal posterior.

3.5 MCMC Sampling from the Prior

Note that fossil calibrations (such as those of Table 1) are represented as statistical distributions of node ages. MCMCTree uses these distributions to construct the prior on times. However, the resulting time prior used by the program may be substantially different from the original fossil calibrations, because the program applies a truncation so that daughter nodes are younger than their ancestors [14, 27]. Thus, it is advisable to calculate the time prior explicitly by running the MCMC with no data so that it can be examined and compared with the fossil calibrations and the posterior.

Go to the `prior` directory and type

```
$ mcmctree mcmctree-pr.ctl
```

This will start the MCMC sampling from the prior. File `mcmctree-pr.ctl` is identical to `mcmc/mcmctree.ctl` except that option usedata has been set to 0. Sampling from the prior is much quicker because the likelihood does not need to be calculated. It takes about 1 min on the Intel Core i7 for MCMCTree to complete the analysis. Rename files `mcmc.txt` and `out.txt` to

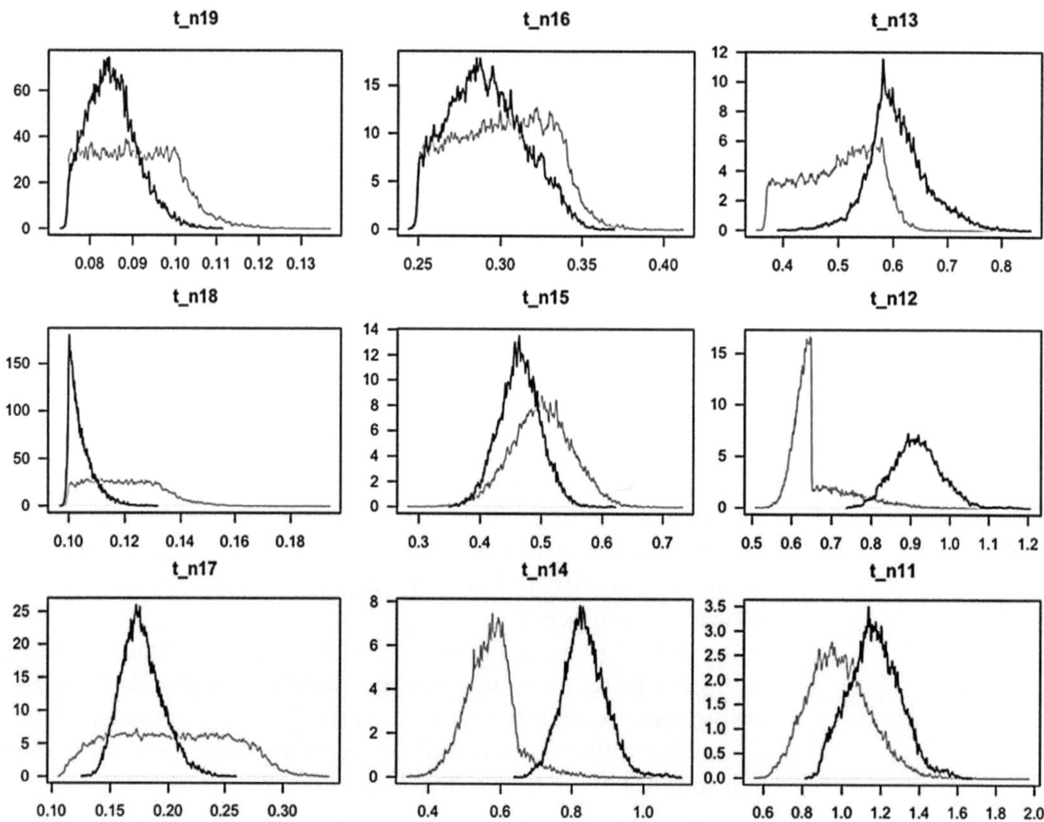

Fig. 10 Prior (gray) and posterior (black) density plots of node ages plotted with R (see `R/analysis.R`)

`mcmc1.txt` and `out1.txt`, and run the analysis again. Rename the new files as appropriate. Check for convergence by calculating the ESS and plotting the traces and histograms.

Figure 10 shows the prior densities of node ages obtained by MCMC sampling (shown in gray) vs. the posterior densities (shown in black). Notice that for four nodes `t_n19`, `t_n18`, `t_n17`, and `t_n16`, the posterior times "agree" with the prior, that is, the posterior density is contained within the prior density. For nodes `t_n15`, `t_n13`, and `t_n11`, there is some conflict between the prior and posterior densities. However, for nodes `t_n14` and `t_n12`, there is substantial conflict between the prior and the posterior. In both cases the molecular data (together with the clock model) suggest the node age is much older than that implied by the calibrations. This highlights the problems in construction of fossil calibrations.

Each fossil calibration represents the paleontologist's best guess about the age of a node. For example, the calibration for the human-chimp ancestor is B(0.075, 0.10, 0.01, 0.20); thus, the calibration is a uniform distribution between 7.5 and 10 million years ago (Ma). The bounds of the calibration are soft, that is, there is a set probability that the bound is violated. In this case the probabilities are 1% for the minimum bound and 20% for the maximum bound. The bound probabilities are asymmetrical because they reflect the nature of the fossil information. Minimum bounds are usually set with confidence because they are based on the age of the oldest fossil member of a clade. For example, the minimum of 7.5 Ma is based on the age of †*Sahelanthropus tchadensis*, recognized as the oldest fossil within the human lineage [28]. On the other hand, establishing maximum bounds is difficult, as absence of fossils for certain clades cannot be interpreted as evidence that the clade in question did not exist during a particular geological time [29]. Our maximum here of 10 Ma represents the paleontologist's informed guess about the likely oldest age of the clade; however, a large probability of 20% is given to allow for the fact that the node age could be older. The conflict between the prior and posterior seen in Fig. 10 evidences this.

Note that when constructing the time prior, the Bayesian dating software must respect the constraints whereby daughter nodes must be younger than their parents. This means that calibration densities are truncated to accommodate the constraint, with the result that the actual prior used on node ages can be substantially different to the calibration density used (*see* Sect. 5.4). Detailed analyses of the interactions between fossil calibrations and the time prior and the effect of truncation are given in [14, 27].

4 General Recommendations for Bayesian Clock Dating

Extensive reviews of best practice in Bayesian clock dating are given elsewhere [4, 20, 21, 30, 31]. Here we give a few brief recommendations.

4.1 Taxon Sampling, Data Partitioning, and Estimation of Tree Topology

In this tutorial we used a small phylogeny to illustrate Bayesian time estimation using approximate likelihood calculation. In practical data analysis, it may be desirable to analyze much larger phylogenies (*see* Sect. 5.5). In large phylogenies, there may be uncertainties in the relationships of some groups. The approximate method discussed here can only be applied to a fixed (known) tree topology. If the uncertainties in the tree are few so that just a handful of tree topologies appear reasonable, the approximate method can be used by analyzing each topology separately [23, 32]. This involves estimating $\mathbf{g}$ and $\mathbf{H}$ for each topology and then running separate MCMC chains on each topology to estimate the times. Several methods to co-estimate divergence times and tree topology are available [8, 9, 17, 18], although they do not implement the approximate likelihood method and are thus unsuitable for the analysis of genome-scale datasets.

We note that partitioning of sites in genomic datasets may have important effects on divergence time estimation. The infinite-sites theory [13, 33] studies the asymptotic behavior of the posterior distribution of times when the amount of molecular data (measured by the number of partitions and the number of sites per partition) increases in a relaxed-clock dating analysis. This theory shows that increasing the number of sites per partition will have minimal effects on time estimation when the sequences per partition are moderately long (>1000 sites, say), but the precision improves when the number of partitions increases, eventually approximating a limit when the number of partitions is infinite. The theory also predicts that very different time estimates may be obtained if the same genomic sequence alignment is analyzed as one partition or as multiple partitions [34]. Furthermore, while more partitions tend to produce more precise time estimates, with narrow CIs, they may not necessarily be more reliable, depending on the correctness of the fossil calibrations and the appropriateness of the partitioning strategies. Unfortunately it is hard to decide on a good partitioning strategy given the genome-scale sequence data, despite efforts to design automatic partitioning strategies for phylogenetic analysis and divergence time estimation [34–36]. Commonly used approaches partition sites in the alignment by codon position or by protein-coding genes of different relative rates [23]. We recommend the use of the infinite-sites plot [14], in which uncertainty in divergence time estimates (measured as the CI width) is plotted against the posterior mean of times. If the scatter points fall on a straight line, information due to the molecular sequence data has reached saturation, and uncertainty in time estimate is predominantly due to uncertainties in fossil calibrations.

4.2 Selection of Fossil Calibrations

Fossil calibrations are one of the most important pieces of information needed to perform divergence time estimation and thus should be chosen after careful consideration of the fossil record, although this may involve some subjectivity [29]. Parham et al. [30] discuss best practice for construction of fossil calibrations. For example, minimum bounds on node ages are normally set to be the age of the oldest fossil member of the crown group. A small probability (say 2.5%) should be set for the probability that the node age violates the minimum bound (e.g., to guard against misidentified or incorrectly dated fossils). Specifying maximum bounds is more difficult, as absence of fossils for a given geological period is not evidence that the clade in question was absent during the period [31]. Current practice is to set the maximum bound to a reasonable value according to the expertise of the paleontologist (*see* ref. 29 for examples), although a large probability (say 10% or even 20%) may be required to guard against badly specified maximum bounds. Calibration densities based on statistical modeling of species diversification, fossil preservation, and discovery are also possible

[15]. In so-called tip-dating approaches, fossil species are included as taxa in the analysis (which may or may not include morphological information for the fossil and extant taxa) [37–39]. Thus, in tip-dating, explicit specification of a fossil calibration density for a node age is not necessary.

4.3 Construction of the Time Prior

The birth-death process with species sampling was used here to construct the time prior for nodes in the phylogeny for which fossil calibrations are not available. Varying the birth (μ), death (λ), and sampling (ρ), parameters can result in substantially different time priors. For example, using $\mu = \lambda = 1$ and $\rho = 0$ leads to a uniform distribution prior on node ages. This diffuse prior appears appropriate for most analyses. Varying the values of μ, λ, and ρ is useful to assess whether the time estimates are robust to the time prior. Parameter configurations can be set up to generate time densities that result in young node ages or in very old node ages (*see* p. 381 in [20] for examples).

4.4 Selection of the Clock Model

In analysis of closely related species (such as the apes), the clock assumption appears to be appropriate for time estimation. A likelihood ratio test can be used to determine whether the strict clock is appropriate for a given dataset [40]. If the clock is rejected, then Bayesian molecular clock dating should proceed using one of the various relaxed-clock models available [7, 13]. In this case, Bayesian model selection may be used to choose the most appropriate relaxed-clock model [41], although the method is computationally expensive and thus only applicable to small datasets. The use of different relaxed-clock models (such as the autocorrelated vs. the independent log-normally distributed rates) may result in substantially different time estimates (*see* ref. 32 for an example). In such cases, repeating the analysis under the different clock models may be desirable.

5 Exercises

5.1 Autocorrelated Rate Model

Modify file `mcmc/mcmctree.ctl` and set `clock = 3`. This activates the autocorrelated log-normal rates model, also known as the geometric Brownian motion rates model [6, 13]. Run the MCMC twice and check for convergence. Compare the posterior times obtained with those obtained under the independent log-normal model (`clock = 2`). Are there any systematic differences in node age estimates between the two analyses? Which clock model produces the most precise (i.e., narrower CIs) divergence time estimates?

5.2 MCMC Sampling with Exact Likelihood Calculation

Modify file `mcmc/mcmctree.ctl` and set `clock = 2` (independent rates), `usedata = 1` (exact likelihood), `burnin = 200`, `sampfreq = 2`, and `nsample = 500`. These last three options will lead to a much shorter MCMC chain, with a total of 1200 iterations. Run the MCMC sampling twice, and check for convergence using the ESS, histograms, and trace plots. How long does it take for the sampling to complete? Can you estimate how long it would take to run the analysis using 2,020,000 iterations, as long as for the approximate method of Sect. 3.3.2? Did the two chains converge despite the low number of iterations?

5.3 Change of Fossil Calibrations

There is some controversy over whether †*Sahelanthropus*, used to set the minimum bound for the human-chimp divergence, is indeed part of the human lineage. The next (younger) fossil in the human lineage is †*Orrorin* which dates to around 6 Ma. Modify file `data/10s.tree` and change the calibration in the human-chimp node to B(0.057, 0.10, 0.01, 0.2). Also change the calibration on the root node to B(0.615, 1.315, 0.01, 0.05). Run the MCMC analysis with the approximate method and again sampling from the prior. Are there any substantial differences in the posterior distributions of times under the new fossil calibrations? Which nodes are affected? How bad is the truncation effect among the calibration densities and the prior?

5.4 Comparing Calibration Densities and Prior Densities

This is a difficult exercise. Use R to plot the prior densities of times sampled using MCMC (the same as in Fig. 10). Now try to work out how to overlay the calibration densities onto the plots. For example, see Fig. 3 in [23] for an idea. First, write functions that calculate the calibration densities. The `dunif` function in R is useful to plot uniform calibrations. Functions `sn::dsn` and `sn::dst` (in the SN package) are useful to plot the skew-t (ST) and skew-normal (SN) distributions. Calibration type S2N (Table 1) is a mixture of two skew-normal distributions [15]. How do the sampled priors compare to the calibration densities? Are there any substantial truncation effects?

5.5 Time Estimation in a Supermatrix of 330 Species

Good taxon sampling is critical to obtaining robust estimates of divergence times for clades. In the `data/` directory, an alignment of the first and second codon positions from mitochondrial protein-coding genes from 330 species (326 primate and 4 out-group species) is provided, `330s.phys`, with corresponding tree topology, `330s.tree`. First, place the fossil calibrations of Table 1 on the appropriate nodes of the species tree. Then obtain the gradient and Hessian matrix for the 330-species alignment using the HKY + G model. Finally, estimate the divergence times on the 330-species phylogeny by using the approximate likelihood method. How does taxon sampling affect node age estimates when comparing the 10-species and 330-species trees? How does uncertainty in node ages in the large tree, which was estimated on a short alignment, compare with the estimates on the small tree, but with a large alignment?

References

1. Zuckerkandl E, Pauling L (1965) Evolutionary divergence and convergence in proteins. In: Bryson V, Vogel HJ (eds) Evolving genes and proteins. Academic, New York, pp 97–166
2. Kumar S (2005) Molecular clocks: four decades of evolution. Nat Rev Genet 6:654–662
3. Bromham L, Penny D (2003) The modern molecular clock. Nat Rev Genet 4:216–224
4. dos Reis M, Donoghue PCJ, Yang Z (2016) Bayesian molecular clock dating of species divergences in the genomics era. Nat Rev Genet 17:71–80
5. Donoghue PCJ, Yang Z (2016) The evolution of methods for establishing evolutionary timescales. Philos Trans R Soc B Biol Sci 371:20160020
6. Thorne JL, Kishino H, Painter IS (1998) Estimating the rate of evolution of the rate of molecular evolution. Mol Biol Evol 15:1647–1657
7. Drummond AJ, Ho SYW, Phillips MJ et al (2006) Relaxed phylogenetics and dating with confidence. PLoS Biol 4:699–710
8. Ronquist F, Teslenko M, Van Der Mark P et al (2012) Mrbayes 3.2: efficient Bayesian phylogenetic inference and model choice across a large model space. Syst Biol 61:539–542
9. Lartillot N, Lepage T, Blanquart S (2009) PhyloBayes 3: a Bayesian software package for phylogenetic reconstruction and molecular dating. Bioinformatics 25:2286–2288
10. Heath TA, Holder MT, Huelsenbeck JP (2012) A Dirichlet process prior for estimating lineage-specific substitution rates. Mol Biol Evol 29:939–955
11. Kishino H, Thorne JL, Bruno WJ (2001) Performance of a divergence time estimation method under a probabilistic model of rate evolution. Mol Biol Evol 18:352–361
12. Yang Z, Rannala B (2006) Bayesian estimation of species divergence times under a molecular clock using multiple fossil calibrations with soft bounds. Mol Biol Evol 23:212–226
13. Rannala B, Yang Z (2007) Inferring speciation times under an episodic molecular clock. Syst Biol 56:453–466
14. Inoue J, Donoghue PCJ, Yang Z (2010) The impact of the representation of fossil calibrations on Bayesian estimation of species divergence times. Syst Biol 59:74–89
15. Wilkinson RD, Steiper ME, Soligo C et al (2011) Dating primate divergences through an integrated analysis of palaeontological and molecular data. Syst Biol 60:16–31
16. Dos Reis M, Yang Z (2011) Approximate likelihood calculation on a phylogeny for Bayesian estimation of divergence times. Mol Biol Evol 8(7):2161–2172
17. Bouckaert R, Heled J, Kühnert D et al (2014) BEAST 2: a software platform for Bayesian evolutionary analysis. PLoS Comput Biol 10 (4):e1003537
18. Höhna S, Landis MJ, Heath TA et al (2016) RevBayes: Bayesian phylogenetic inference using graphical models and an interactive model-specification language. Syst Biol 65:726–736
19. Dos Reis M, Zhu T, Yang Z (2014) The impact of the rate prior on Bayesian estimation of divergence times with multiple loci. Syst Biol 63:555–565
20. Yang Z (2014) Molecular Evolution: A Statistical Approach. Oxford University Press, Oxford
21. Heath TA, Moore BR (2014) Bayesian inference of species divergence times. In: Chen M-H, Kuo L, Lewis PO (eds) Bayesian Phylogenetics: Methods, Algorithms, and Applications. CRC Press, Boca Raton, pp 277–318
22. Yang Z (2007) PAML 4: phylogenetic analysis by maximum likelihood. Mol Biol Evol 24:1586–1591
23. dos Reis M, Inoue J, Hasegawa M et al (2012) Phylogenomic datasets provide both precision and accuracy in estimating the timescale of placental mammal phylogeny. Proc Biol Sci 279:3491–3500
24. dos Reis M, Gunnell G, Barba-Montoya J et al (2018) Using phylogenomic data to explore the effects of relaxed clocks and calibration strategies on divergence time estimation: primates as a test case. Syst Biol 67(4):594–615
25. Yang Z (1996) Among-site rate variation and its impact on phylogenetic analyses. Trends Ecol Evol 11(9):367–372
26. Gillespie JH (1984) The molecular clock may be an episodic clock. Proc Natl Acad Sci U S A 81:8009–8013

27. Warnock RCM, Yang Z, Donoghue PCJ (2012) Exploring uncertainty in the calibration of the molecular clock. Biol Lett 8:156–159
28. Zollikofer CPE, Ponce de León MS, Lieberman DE et al (2005) Virtual cranial reconstruction of Sahelanthropus tchadensis. Nature 434:755–759
29. Benton MJ, Donoghue PCJ (2007) Paleontological evidence to date the tree of life. Mol Biol Evol 24(1):26–53
30. Parham JF, Donoghue PCJ, Bell CJ et al (2012) Best practices for justifying fossil calibrations. Syst Biol 61(2):346–359
31. Ho SYW, Phillips MJ (2009) Accounting for calibration uncertainty in phylogenetic estimation of evolutionary divergence times. Syst Biol 58:367–380
32. Dos Reis M, Thawornwattana Y, Angelis K et al (2015) Uncertainty in the timing of origin of animals and the limits of precision in molecular timescales. Curr Biol 25:2939–2950
33. Zhu T, Reis MD, Yang Z (2014) Characterization of the uncertainty of divergence time estimation under relaxed molecular clock models using multiple loci. Syst Biol 64(2):267–280
34. Angelis K, Alvarez-Carretero S, dos Reis M et al (2018) An evaluation of different partitioning strategies for Bayesian estimation of species divergence times. Syst Biol 67 (1):61–77
35. Lanfear R, Calcott B, Ho SYW et al (2012) PartitionFinder: combined selection of partitioning schemes and substitution models for phylogenetic analyses. Mol Biol Evol 29:1695–1701
36. Duchêne S, Molak M, Ho SYW (2014) ClockstaR: choosing the number of relaxed-clock models in molecular phylogenetic analysis. Bioinformatics 30:1017–1019
37. Heath TA, Huelsenbeck JP, Stadler T (2014) The fossilized birth-death process for coherent calibration of divergence-time estimates. Proc Natl Acad Sci U S A 111:E2957–E2966
38. Ronquist F, Klopfstein S, Vilhelmsen L et al (2012) A total-evidence approach to dating with fossils, applied to the early radiation of the hymenoptera. Syst Biol 61:973–999
39. O'Reilly JE, dos Reis M, Donoghue PCJ (2015) Dating tips for divergence-time estimation. Trends Genet 31(11):637–650
40. Felsenstein J (1981) Evolutionary trees from DNA sequences: a maximum likelihood approach. J Mol Evol 17:368–376
41. Lepage T, Bryant D, Philippe H et al (2007) A general comparison of relaxed molecular clock models. Mol Biol Evol 24:2669–2680

Chapter 11

Genome Evolution in Outcrossing vs. Selfing vs. Asexual Species

Sylvain Glémin, Clémentine M. François, and Nicolas Galtier

Abstract

A major current molecular evolution challenge is to link comparative genomic patterns to species' biology and ecology. Breeding systems are pivotal because they affect many population genetic processes and thus genome evolution. We review theoretical predictions and empirical evidence about molecular evolutionary processes under three distinct breeding systems—outcrossing, selfing, and asexuality. Breeding systems may have a profound impact on genome evolution, including molecular evolutionary rates, base composition, genomic conflict, and possibly genome size. We present and discuss the similarities and differences between the effects of selfing and clonality. In reverse, comparative and population genomic data and approaches help revisiting old questions on the long-term evolution of breeding systems.

Key words Breeding systems, GC-biased gene conversion, Genome evolution, Genomic conflicts, Selection, Transposable elements

1 Introduction

In-depth investigations on genome organization and evolution are increasing and have revealed marked contrasts between species, e.g., evolutionary rates, nucleotide composition, and gene repertoires. However, little is still known on how to link this "genomic diversity" to the diversity of life history traits or ecological forms. Synthesizing previous works in a provocative and exciting book, M. Lynch asserts that variations in fundamental population genetic processes are essential for explaining the diversity of genome architectures while emphasizing the role of the effective population size (N_e) and nonadaptive processes [1]. Life history and ecological traits may influence population genetic parameters, including N_e, making it possible to link species' biology and their genomic organization and evolution (e.g., [2–7])

Among life history traits affecting population genetic processes, breeding systems are pivotal as they determine the way genes are transmitted to the next generation (Fig. 1). Outcrossing,

Maria Anisimova (ed.), *Evolutionary Genomics: Statistical and Computational Methods*, Methods in Molecular Biology, vol. 1910, https://doi.org/10.1007/978-1-4939-9074-0_11, © The Author(s) 2019

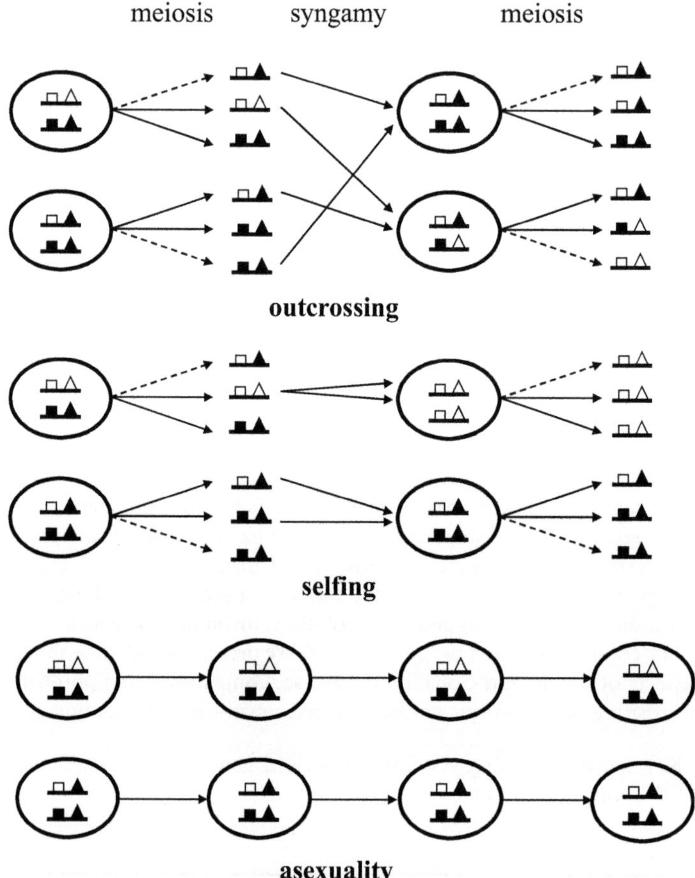

Fig. 1 Reproduction and genotype transmission in outcrossing, selfing, and asexual species. In outcrossers, parental and recombinant (dotted lines) gametes from distinct zygotes are shuffled at generation $n + 1$. In selfers, only gametes produced by a given zygote can mate, which quickly increases homozygosity and reduces the recombination efficacy. Asexuals do not undergo meiosis or syngamy. They reproduce clonally

sexual species (outcrossers) reproduce through the alternation of syngamy (from haploid to diploid) and meiosis (from diploid to haploid), with random mating of gametes from distinct individuals at each generation. Outcrossing is a common breeding system that is predominant in vertebrates, arthropods, and many plants, especially perennials, etc. [8, 9]. Selfing species (selfers) also undergo meiosis, but fertilization only occurs between gametes produced by the same hermaphrodite individual. Consequently, diploid individuals from selfing species are highly homozygous (FIS ~ 1; *see*, for instance, ref. 10)—heterozygosity is divided by two at each generation, and the two gene copies carried by an individual have a high probability of being identical by descent. Selfing is common in various plant families (e.g., *Arabidopsis thaliana*), mollusks, nematodes (e.g., *Caenorhabditis elegans*), and platyhelminthes, among others [8, 9]. Note that many sexual species have intermediate systems in which inbreeding and outbreeding coexist. In organisms with a prolonged haploid phase (such as mosses, ferns, or many algae and fungi), a more extreme form of selfing can occur by taking place during the haploid phase (haploid selfing or intragametophytic selfing), leading instantaneously to genome-wide homozygosity [11]. Clonal asexual species, finally, only reproduce via mitosis, so that daughters are genetically identical to mothers unless a mutation occurs. In diploid asexuals, homologous chromosomes associated in a given zygote do not segregate in distinct gametes—they are co-transmitted to the next generation in the absence of any haploid phase. In contrast to selfing species, individuals from asexual diploid species tend to be highly heterozygous (FIS ~ −1, [12]), since any new mutation will remain at the heterozygote stage forever, unless the same mutation occurs in the homologous chromosome. Clonality is documented in insects (e.g., aphids), crustaceans (e.g., daphnia), mollusks, vertebrates, and angiosperms, among others [13–16]. As for selfing, clonality can also be partial, with sexual reproduction occurring in addition or in alternation with asexual reproduction. In addition to this common form of asexuality, other forms such as automixis imply a modified meiosis in females where unfertilized diploid eggs produce offspring potentially diverse and distinct from their mother, leading to different levels of heterozygosity [13]. This diversity of reproductive systems should be kept in mind, but for clarity we will mainly compare outcrossing, diploid selfing, and clonality.

Through the occurrence, or not, of syngamy, recombination, and segregation, breeding systems affect population genetic parameters (effective population size, recombination rate, efficacy of natural selection; Fig. 2) and thus, potentially, genomic patterns. A large corpus of population genetic theory has been developed to study the causes and consequences of the evolution of breeding systems (Table 1). Thanks to the exponentially growing amount of genomic data, and especially data from closely related species with contrasted breeding systems, it is now possible to test these theoretical predictions. Conversely, genomic data may help in understanding the evolution of breeding systems. Genomes should record the footprints of transitions in breeding systems and help in testing the theory of breeding system evolution in the long run, e.g., the "dead-end hypothesis," which posits that selfers and asexuals are doomed to extinction because of their inefficient selection and low adaptive potential [17, 18]. Since the first edition of this book, several theoretical developments have clarified the population genetics consequences of the different breeding systems, and empirical evidences have been accumulating, partly changing our view of breeding system evolution and consequences, especially for asexual organisms. We first review and update the consequences of

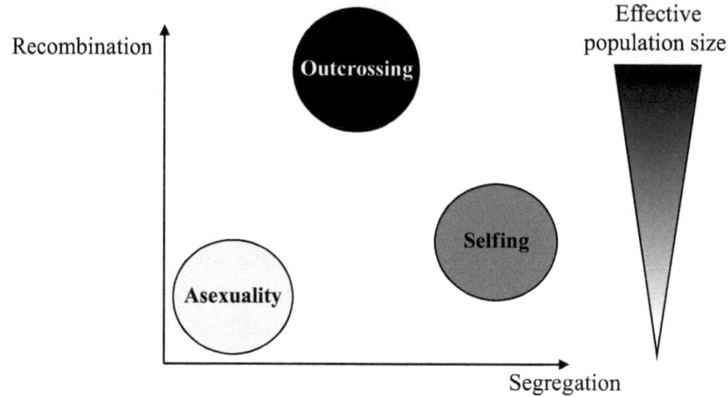

Fig. 2 A schematic representation of the effect of breeding systems on population genetic parameters

Table 1
Summary of the major theoretical predictions regarding breeding systems and evolutionary genomic variables, with outcrossing being taken as reference

	F_{IS}	πS	dN/dS	Codon usage	TE	LD	GC-content
Outcrossing	~0	+	+	+	+	+	+
Selfing	~1	−	++	−	Unclear	++	−
Asexuality	~−1	−	+++	−	Unclear	+++	−

TE transposable element abundance, *LD* linkage disequilibrium

breeding systems on genome evolution and then discuss and re-evaluate how evolutionary genomics shed new light on the old question of breeding system evolution.

2 Contrasted Genomic Consequences of Breeding Systems

2.1 Consequences of Breeding Systems on Population Genetics Parameters

Sex involves an alternation of syngamy and meiosis. In outcrossing sexual species, random mating allows alleles to spread across populations, while segregation and recombination (here in the sense of crossing-over) associated with meiosis generate new genotypic and haplotypic combinations. This strongly contrasts with the case of selfing and asexual species. In such species, alleles cannot spread beyond the lineage they originated from because mating occurs within the same lineage (selfers) or because syngamy is suppressed (asexuals). Recombination, secondly, is not effective in non-outcrossers. In selfers, while physical recombination does occur (r_0), effective recombination (r_e) is reduced because it mainly occurs between homozygous sites, and it completely vanishes under complete selfing: for tight linkage, $r_e = r_0(1 - F_{IS})$, where F_{IS} is the Wright's fixation index [19], whereas for looser linkage, effective recombination is more reduced than predicted by this simple expression [20–22]. In asexuals, physical recombination is suppressed ($r_0 = r_e = 0$). High levels of linkage disequilibrium (nonrandom association of alleles between loci) could therefore be expected in selfers and asexuals. The observed data are mainly consistent with these predictions. In the selfing model species *Arabidopsis thaliana*, LD extends over a few hundreds of kb, while in maize, an outcrosser, LD quickly vanishes beyond a few kb [23]. In a meta-analysis, Glémin et al. [24] also found higher LD levels in selfers than in outcrossers. Beyond pairwise LD, selfing also generates higher-order associations, such as identity disequilibria (the excess probability of being homozygote at several loci, [25]) that alter population genetics functioning compared to outcrossing populations (e.g., [26]).

Theory also predicts that the effective population size, N_e, depends on the breeding system (Fig. 2). First, compared to outcrossers, selfing is expected to directly lower N_e by a factor $1 + F_{IS}$ by reducing the number of independent gametes sampled for reproduction [27]. From a coalescent point of view, selfing reduces coalescent time (again by the same factor $1 + F_{IS}$). Under outcrossing, two gene copies gathered in a same individual either directly coalesce or move apart at the preceding generation. Selfing prolongs the time spent within an individual, hence the probability of coalescing [19, 28]. In diploid asexuals, the picture is less obvious. Since genotypes, not alleles, are sampled, Balloux et al. [12] distinguished between the genotypic and allelic effective size. The genotypic effective size equals N, not $2N$, i.e., the actual population size, similarly to the expectation under complete selfing. On the contrary, the allelic effective size tends toward infinity under complete clonality because genetic diversity within individuals cannot be lost [12]. This corresponds to preventing coalescence as long as gene copies are transmitted clonally [29, 30]. However, very low level of sex (higher than $1/2N$) is sufficient to retrieve standard outcrossing coalescent behavior [29, 30], and as far as natural selection is concerned (see below), the genotypic effective size is what matters [31]. The ecology of selfers and asexuals may also contribute to decreasing N_e as they supposedly experience more severe bottlenecks than outcrossers [32, 33]. On the contrary, higher population subdivision in selfers could contribute to increasing N_e at the species scale. However, Ingvarsson [34] showed that, under most conditions, the extinction/recolonization dynamics is predicted to decrease N_e in selfers, at both the local and metapopulation scale. Finally, because of low or null effective recombination, hitchhiking effects—the indirect effects of selection at a locus on other linked loci—reduce N_e further [35]. Under complete selfing or clonality, because of full genetic linkage, selection at a given locus affects the whole genome. Most forms of selection, and especially directional selection, reduce the number of gene copies contributing to the next generation by removing deleterious alleles to the benefit of advantageous ones. Because of linkage, such a reduction spreads over the rest of the genome, globally reducing the effective population size (sensu lato) in non-outcrossing species. Background selection, the reduction in N_e due to the removal of deleterious mutations at linked loci, can be particularly severe in highly selfing and clonal population, potentially reducing N_e by one order of magnitude or more [22, 36]. And this effect is expected to be stronger in asexuals than in selfers [36]. In the predominantly selfing nematode *C. elegans*, nucleotide diversity has been shown to be reduced genome wide by both background selection [37] and selective sweeps [38], and in a comparative analysis, the effect of linked selection has shown to be more pronounced in selfing than in outcrossing species [39].

As genetic diversity scales positively with $N_e\mu$, where μ is the mutation rate, selfers are expected to be less polymorphic than outcrossers. Asexuals should also exhibit lower genotypic diversity, but the prediction is not clear for allelic diversity (see above). However, because of the lack of recombination, haplotype diversity should be lower for both breeding systems. The effect of selfing on the polymorphism level is well documented, and empirical data mainly agree with the theoretical predictions. Selfing species tend to be more structured, less diverse, and straightforwardly more homozygotes than outcrossers [6, 24, 40, 41]. Much fewer data exist regarding diversity levels in asexuals, but the available datasets confirm that genotypic diversity, at least, is usually low in such species (*see* discussion in ref. 12). At the population level, a recent comparative analysis of sexual and asexual *Aptinothrips rufus* grass thrips confirmed the expected lower nuclear genetic diversity of asexual populations while also evidencing that some asexuals with extensive migration can feature very high mitochondrial genetic diversity [42].

These predictions concerning polymorphism patterns implicitly assumed that mutation rates are the same among species with contrasted breeding systems. However, modifications in breeding systems can also affect various aspects of the species life cycle potentially related to the mutation rate. In asexuals, for instance, loss of spermatogenesis can reduce mutation rates, while loss of the dormant sexual phase can increase them (reviewed in [43]). Mutation rates can also be decreased in non-outcrossers due to the loss of recombination, which can be mutagenic [44, 45]. In selfers, meiosis and physical recombination do occur. However, the specific mutagenic process during meiosis depends on the level of heterozygosity, such as indel-associated mutations (IDAM): heterozygote indels could increase the point mutation rate at nearby nucleotides because of errors during meiosis [46, 47]. Consistent with this prediction, the IDAM process more strongly affects the outcrossing wild rice, *Oryza rufipogon*, than the very recent selfer and weakly

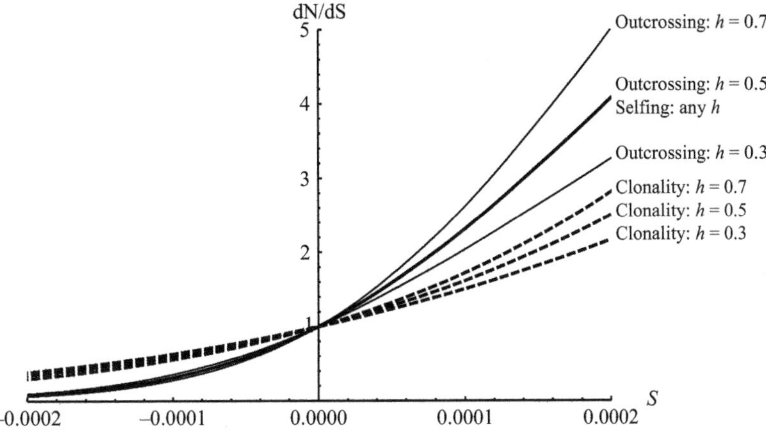

Fig. 3 Substitution rates relative to the neutral case (dN/dS) in outcrossers (thin lines), selfers (bold line), and asexuals (dotted lines) for different mutation dominance levels. The fitness of the resident, heterozygote, and homozygote mutant genotypes are 1, $1 - hs$, and $1 - s$, respectively. For asexuals, it is necessary to consider two substitution rates corresponding to the initial fixation of heterozygotes and the ultimate fixation of complete homozygote mutants from an initially heterozygote population [31]. Population size: $N = 10{,}000$. To highlight the difference between selfers and asexuals due to segregation, demographic and hitchhiking effects reducing N_e in asexuals and selfers are not taken into account

heterozygous domesticated rice, *O. sativa. A. thaliana*, a more ancient and mostly homozygous selfer, is very weakly affected by IDAM [48]. Overall, these processes should globally contribute to lowering mutation rates, and thus polymorphism, in selfing and asexual species.

2.2 Breeding Systems and Selection Efficacy

2.2.1 Drift and Recombination: Parallel Reduction in Selection Efficacy in Selfers and Asexuals?

The effective population size strongly affects the outcome of natural selection. The probability of fixation of a new mutation is a function of the $N_e s$ product, where s is the selection coefficient ([49] and *see* Fig. 3). As N_e is reduced, a higher proportion of mutations behave almost neutrally. Weakly deleterious alleles can thus be fixed, while weakly advantageous ones can be lost. Genetic associations among loci generated by selfing and clonality also induce selective interferences [26, 50]. Because of their reduced effective population size and recombination rate, selection is thus expected to be globally less effective in selfers and asexuals than in outcrossers, which should result in various footprints at the molecular level (Table 1). Assuming that most mutations are deleterious (with possible back compensatory mutations), both the ratio of non-synonymous to synonymous polymorphism, πN/ πS, and the ratio of non-synonymous to synonymous substitutions, dN/dS, are predicted to be higher in selfers and asexuals than in outcrossers. Codon usage should also be less optimized in selfers and asexuals than in outcrossers.

Contrary to polymorphism surveys, few studies have tested these predictions empirically (Table 2). In the few available

Table 2
Summary of studies comparing patterns of molecular evolution between different breeding systems

	Taxonomic group	Groups compared	Dataset	dN/dS	πN/πS	Positive selection	Codon usage	GC-content	Reference
Outcrossing/selfing	Angiosperms	29 selfers/42 outcrossers	Meta-analysis (polymorphism)			+/−		+/−	Glémin et al. [24]
	Angiosperms	10 selfers/18 outcrossers				+			Chen et al. [6]
	Angiosperms (17 phylogenetic datasets)	4–61 selfers or SC/5–125 outcrossers or SI	Chloroplastic genes (matK and rbcL)	+/−[a]					Glémin and Muyle [64]
	Arabidopsis	1 selfer/1 outcrosser	23 nuclear genes + 1 chloroplastic gene	−	−				Wright et al. [61]
	Arabidopsis	1 selfer/1 outcrosser	675/62 nuclear genes			+/−			Foxe et al. [218]
	Arabidopsis/Brassica	1 selfer/2 outcrossers	185 nuclear genes		−				Wright et al. [118]
	Arabidopsis/Capsella	1 selfer/1 outcrosser	257 nuclear genes			+		+	Slotte et al. [83]
	Arabidopsis/Capsella	2 selfers/2 outcrossers	780, 89/120, 257 nuclear genes				+		Qiu et al. [223]
	Capsella	1 selfer/1 outcrosser	Complete genome/transcriptomes			+			Slotte et al. [224], [225]
	Collinsia	1 selfer/1 outcrosser	17 nuclear genes + transcriptomes		−	+			Hazzouri et al. [123]
	Eichornia	3 selfers/1 outcrosser	~8000 nuclear genes		+	+			Ness et al. [221]
	Eichornia paniculata (several populations)	10 selfers/10 outcrossers	Transcriptomes (>16,000 contigs)			+			
	Triticeae	2 selfers/2 outcrossers	52 nuclear genes + 1 chloroplastic gene	−				+/−	Haudry et al. [84]
	Triticeae	9 selfers/10 outcrossers	27 nuclear genes	−					Escobar et al. [63]
	Neurospora	32 homothalics/17 heterothallics	7 nuclear genes	+[b]					Nygren et al. [222]
	Neurospora	1 homothalic/1 heterothallic	>2000 nuclear genes		−				Whittle et al. [226]

	Neurospora	4 homothalic/31 heterothallic	>2700 nuclear genes	+	+/−			Gioti et al. [219]
	Caenorhabditis	2 selfers/4 outcrossers	>1000 nuclear genes	−	+/−			Cutter et al. [62]
	Galba/Physa	2 selfers/2 outcrossers	Transcriptomes (>1000 contigs)	+	+	+	+	Burgarella et al. [66]
Sexuals/ asexuals	*Boechera*	8 asexuals/8 sexuals	Complete genome	+				Lovell et al. [55]
	Oenothera	16 asexuals/16 sexuals	1 nuclear gene (*chiB*)				+	Hersch-Green et al. [85]
	Oenothera	16 asexuals/13 sexuals	Transcriptomes (>4000 contigs)	+	+			Hollister et al. [58]
	Ranunculus	2 sexuals/3 asexuals	Transcriptomes	−				Pellino et al. [57]
	Aphids	4 sexuals/4 asexuals	255 nuclear genes + 10 mitochondrial genes	−				Ollivier et al. [56]
	Campeloma	6 asexuals/12 sexuals	1 mitochondrial gene (*Cytb*)	+				Johnson et al. [53]
	Daphnia	14 asexuals/14 sexuals	Complete mitochondrial genome	+				Paland and Lynch [51]
	Daphnia	11 asexuals/11 sexuals	Complete mitochondrial genome	−				Tucker et al. [60]
	Lineus	2 asexuals/7 sexuals	Transcriptomes (>2800 contigs)		+/−[c]			Ament-Velasquez et al. [59]
	Potamopyrgus	14 asexuals/14 sexuals	Complete mitochondrial genome	+				Neiman et al. [54]
	Rotifers	3 asexuals/2 sexuals	1 nuclear gene (*Hsp 82*)	−				Mark-Welch and Meselson [220]
	Rotifers	3 asexuals/4 sexuals	1 mitochondrial gene (*Cox I*)	+/−	+			Barraclough et al. [217]
	Timema	6 asexuals/7 sexuals	2 nuclear genes + 1 mitochondrial gene	+				Henry et al. [52]

+ and − indicate if theoretical predictions are confirmed or not. Empty cells correspond to nonavailable data

[a]No more significant after controlling for terminal vs. internal branches

[b]Terminal vs. internal branches not controlled

[c]Possible confounding effect of hybrid origin of one asexual lineage

comparative studies, contrasted patterns were observed between selfers and asexuals. Compared to sexual ancestors, recent asexual lineages show a marked increase in the dN/dS ratio in *Daphnia* ([51] but see below), *Timema* stick insects [52], gastropods *Campeloma* [53] and *Potamopyrgus* [54], and the plant *Boechera* [55], in agreement with theoretical predictions (Table 2). However, no significant effect of asexuality on dN/dS was found in four aphid species [56] and in the plant *Ranunculus auricomus* [57]. Bdelloid rotifers, long considered as ancient asexuals (see below), exhibit a higher πN/πS ratio but not a higher dN/dS ratio than comparable sexual groups, suggesting that mildly deleterious mutations can segregate at a higher frequency in asexuals but are eventually removed. A higher πN/πS ratio in asexual lineages than in sexual relatives was reported from transcriptome data in *Oenothera* primroses [58] and *Lineus* nemerteans [59]. Note however that in the latter case, the increased πN/πS is primarily explained by the hybrid nature of the asexual *Lineus pseudolacteus* (Table 2). The recent origin of asexuality through introgression also challenges the interpretation of elevated dN/dS ratio in the mitochondrial genome of asexual lineages of *Daphnia pulex* [51], as less than 1% of mutations on the branches leading to asexual lineages would have arisen after the transition to asexuality [60]. Here, rather than being the direct cause of genomic degradation, asexuality may have evolved in already-degraded lineages.

All predictions are not equally supported by data in selfers. Polymorphism-based measures mostly support reduction in selection efficiency in selfers in various plant species, and this was recently confirmed by a meta-analysis of genome-wide polymorphism data ([6] and *see* Table 2). On the contrary, as far as dN/dS or base composition are compared, most studies, in plants, fungi, and animals, did not find evidence of relaxed selection in selfers (Table 2). A recent origin of selfing is often invoked to explain that effect of selfing is rarely observed in species divergence (e.g., [61, 62–64]), whereas a recent transition to selfing can leave a clear signature of relaxed selection at the polymorphism level [65]. In contrast, in the freshwater snail *Galba truncatula* where selfing is supposed to be old and ancestral to a clade of several species, relaxed selection in the selfing lineage was also observed at the divergence level [66]. The same rationale should apply to asexual species. However, in *Campeloma*, *Potamopyrgus*, *Timema*, and *Boechera*, clonality is also recent, yet the expected patterns are observed at the divergence level. The reduction in N_e could simply be less severe in selfers than in asexuals as predicted by background selection models [36]. Furthermore, complete selfing is hardly ever noted in natural populations; residual outcrossing typically occurs. Among hitchhiking effects, some are very sensitive to the recombination level, such as Muller's ratchet [67], weak Hill-Robertson interferences [50], or hitchhiking of deleterious mutations during selective sweeps [68, 69]. If such mechanisms are the main cause of reduction of N_e in selfers, then even a low recombination rate could be enough to maintain the selection efficacy. This is suggested by genomic patterns across recombination gradients in outcrossing species. In primates, no effect of recombination on the selection efficacy has been detected [70]. In *Drosophila*, Haddrill et al. [71] found little evidence of reduced selection in low recombining regions, except when recombination was fully suppressed, as in Y chromosomes. Differences between selfers and asexuals could thus simply result from different degrees of residual outcrossing. However, as stated above, selfers and asexuals also fundamentally differ as far as segregation is concerned, as we now discuss in more detail.

2.2.2 Segregation: Dealing with Heterozygotes

Selfing affects the selection efficacy by increasing homozygosity and thus exposing recessive alleles to selection. This effect can counteract the effect of reducing N_e. Considering the sole reduction in N_e due to non-independent gamete sampling, selection is less efficient under partial selfing for dominant mutations but more efficient for recessive ones (Fig. 3, and *see* ref. 72). More precisely, Glémin [73] determined the additional reduction in N_e (due to hitchhiking and demographic effects) necessary to overcome the increased selection efficacy due to homozygosity. This additional reduction can be high for recessive mutations. On the contrary, the lack of segregation in asexuals reduces selection efficacy and increases the drift load, as heterozygotes can fix [31]. The effects of selfing and clonality on the fixation probability of codominant, recessive, or dominant mutations are summarized in Fig. 3. Note that segregation may also have indirect effects. When recombination is suppressed, Muller's ratchet is supposed to reduce N_e and contribute to the fixation of weakly deleterious alleles [74]. In selfers, the purging of partially recessive deleterious alleles slows down the ratchet [67], which suggests that the fixation of deleterious alleles at linked loci would be lower in selfers than in asexuals. The same mechanism also contributes to weaker background selection in selfers than in asexuals (see above, [36]). In the extreme case of intra-gametophytic selfing, purging could be even more efficient at removing deleterious alleles [11], as it has been suggested for moss species [75]. Segregation at meiosis could thus partly explain the differences between selfers and asexuals, but more data are clearly needed to confirm this hypothesis.

The two opposite effects of drift and segregation in selfers should also affect adaptive evolution. In outcrossers, new beneficial mutations are more likely to be rapidly lost if recessive, as they are initially present in heterozygotes and masked to selection—a process known as Haldane's sieve [76]. By unmasking these mutations in homozygotes, selfing could help adaptive evolution from recessive mutations [72, 73]. However, this advantage of selfing disappears when adaptation proceeds from pre-existing variation because homozygotes can also be present in outcrossers [77]. Selective interference in selfers also reduces their advantage of not experiencing Haldane's sieve, especially for weakly beneficial mutations [21], and the effect of background should globally reduce the rate of adaptation [73, 77, 78]. Conversely, the lack of segregation in asexuals delays the complete fixation of an advantageous mutation. Once a new advantageous mutation gets fixed in the heterozygotic state, additional lag time until occurrence and fixation of a second mutation is necessary to ensure fixation [79]. Little is known about the dominance levels of new adaptive mutations, but a survey of QTL fixed during the domestication process in several plant species confirmed the absence of Haldane's sieve in selfers compared to outcrossers [80]. This mostly corresponds to strong selection on new mutations or mutations in low initial frequencies in the wild populations. More generally, the effect of selfing on adaptive evolution will depend on the distribution of dominance and selective effects of mutations and the magnitude of genetic drift and linkage.

Few studies have tested for difference in positive selection between selfers and outcrossers. In their survey of sequence polymorphism data in flowering plants, Glémin et al. [24] found, on average, more genes with a signature of positive selection in outcrossers than in selfers assessed by the McDonald-Kreitman test [81]. An extension of this method—where non-synonymous vs. synonymous polymorphism data are used to calibrate the distribution of the deleterious effects of mutations and then attribute the excess non-synonymous divergence observed to positive selection [82]—was applied to one plant [83] and one freshwater snail dataset. In both studies, a large fraction of non-synonymous substitutions was estimated to be adaptive in the outcrossing species (~40% in the plant *Capsella grandiflora* and ~55% in the snail *Physa acuta*), whereas this proportion was not significantly different from zero in the selfer (*Arabidopsis thaliana* and *Galba truncatula*, respectively). Based on methods where the dN/dS ratio is allowed to vary both among branches and sites, a comparative analysis of two outcrossing and two selfing Triticeae species [84] suggested that adaptive substitutions may have specifically occurred in the outcrossing lineages. This would contribute to explaining why selfing lineages did not show a higher dN/dS ratio than outcrossing ones (see above and Table 2). So the data available so far support an increased rate of adaptation in outcrossing species, suggesting that the effects of drift and linkage overwhelm the advantage of avoiding Haldane's sieve. A similar approach was used in *Oenothera* species suggesting also reduced adaptive evolution in clonal compared to sexual lineages [85].

Finally, the classical assumption of a lack of segregation in asexuals must be modulated. First, in some form of asexuality, such as automixis, female meiosis is retained, and diploidy restoration occurs by fusion or duplication of female gametes. Depending on how meiosis is altered, automixis generates a mix of highly heterozygous and highly homozygous regions along chromosomes. The genomes of such species could thus exhibit a gradient of signatures of selfing and diploid clonal evolution [86]. Secondly, mitotic recombination and gene conversion in the germline of asexual lineages can also reduce heterozygosity at a local genomic scale. Mitotic recombination has been well documented in yeast (*see* review in ref. 87) and also occurs in the asexual trypanosome *T. b. gambiense* [88] and in asexual *Daphnia* lineages [60, 89, 90]. If its frequency is of the order or higher than mutation rates, as reported in yeast and *Daphnia*, asexuals would not suffer much from the lack of segregation at meiosis. Especially, during adaptation, the lag time between the appearance of a first beneficial mutation and the final fixation of a mutant homozygote could be strongly reduced [87]. However, such mechanisms of loss of heterozygosity also rapidly expose recessive deleterious alleles in heterozygotes and generate inbreeding-depression-like effects [60].

2.2.3 Selection on Genetic Systems

So far, we have only considered the immediate, mechanistic effects of breeding systems on population genetic parameters. Breeding systems, however, can also affect the evolution of genetic systems themselves, which modulates previous predictions. Theoretical arguments suggested that selfing, even at small rates, greatly increases the parameter range under which recombination is selected for [91–93]. These predictions have been confirmed in a meta-analysis in angiosperms in which outcrossers exhibited lower chiasmata counts per bivalent than species with mixed or selfing mating systems [94]. Higher levels of physical recombination (r_0) could thus help break down LD and reduce hitchhiking effects. This could contribute to explaining why little evidence of long-term genomic degradation has been observed in selfers, compared to asexuals.

Breeding systems may also affect selection on mutation rates. Since the vast majority of mutations are deleterious, mutation rates should tend toward zero, up to physiological costs of further reducing mutation rates being too high (e.g., [95, 96]). Under complete linkage, a modifier remains associated with its "own" mutated genome. Selection should thus favor lower mutation rates in asexuals and selfers (e.g., [95, 96]). However, Lynch recently challenged this view and suggested a lower limit to DNA repair may be set by random drift, not physiological cost [97]. Such a limit should thus be higher in asexuals and selfers. Asexuality is often associated with very efficient DNA repair systems (reviewed in [43]), supporting the view that selection for efficient repair may overwhelm drift in asexual lineages. Alternatively, only groups having high-fidelity repair mechanisms could maintain asexuality in the long run. More formal tests of mutation rate differences between breeding systems are still scarce. The phylogenetic approach revealed no difference in dS, as a proxy of the neutral mutation rate, between *A. thaliana* and *A. lyrata* [61], nor did a mutation accumulation experiment that compared the deleterious genomic mutation rate between *Amsinckia* species with contrasted mating systems [98]. A similar experiment in *Caenorhabditis* showed that the rate of mutational decay was, on average, fourfold greater in gonochoristic outcrossing taxa than in the selfer *C. elegans* [99]. Recent mutation accumulation experiments on *Daphnia pulex* suggested a slightly lower mutation rate in obligate than in facultative asexual genotypes, except for one mutator phenotype which evolved in an asexual subline [90]. Overall, these results do not support Lynch's hypothesis of mutation rates being limited by drift in asexual and selfing species. However, such experiments are still too scarce, and quantifying how mutation rates vary or not with breeding systems is a challenging issue that requires more genomic data.

2.3 Breeding Systems and Genomic Conflicts

Outcrossing species undergo various sorts of genetic conflict. Sexual reproduction directly leads to conflicts within (e.g., for access to mating) and between sexes (e.g., for resource allocations between male and female functions or between offspring). In selfers and asexuals, such conflicts occur because mates are akin or because mating is absent [100, 101]. Outcrossers are also sensitive to epidemic selfish element proliferation and to meiotic drive, because alleles can easily spread over the population through random mating. In contrast, selfers and asexuals should be immune to such genomic conflicts because selection only occurs between selfing or asexual lineages so that selfish elements should be either lost or evolve into commensalists or mutualists [102].

2.3.1 Relaxation of Sexual Conflicts in Selfers and Asexuals

Some genes involved in sexual reproduction are known to evolve rapidly because of recurrent positive selection [103]. Arm races for mating or for resource allocation to offspring are the most likely causes of this accelerated evolution. In selfers and asexuals, selection should be specifically relaxed on these genes, not only because of low recombination and effective size but mainly because the selection pressure per se should be suppressed. According to this prediction, in the outcrosser *C. grandiflora*, 6 out of the 20 genes that show the strongest departure from neutrality are reproductive genes and under positive selection. This contrasts with the selfer *A. thaliana*, for which no reproductive genes are under positive selection [83].

More specifically, two detailed analyses provided direct evidence of relaxed selection associated with sexual conflict reduction. In the predominantly selfer *C. elegans*, some males deposit a copulatory plug that prevents multiple matings. However, other males do not deposit this plug. A single gene (plg-1), which encodes a major structural component of this plug, is responsible for this dimorphic reproductive trait [104]. Loss of the copulatory plug is caused by the insertion of a retrotransposon into an exon of plg-1. This same allele is present in many populations worldwide, suggesting a single origin. The strong reduction in male-male competition following hermaphroditism and selfing evolution explains that no selective force opposes the spread of this loss-of-function allele [104, 105]. In *A. thaliana*, similar relaxed selection has been documented in the MEDEA gene, an imprinted gene directly involved in the male vs. female conflict. MEDEA is expressed before fertilization in the embryo sac and after fertilization in the embryo and the endosperm, a tissue involved in nutrient transfer to the embryo. In *A. lyrata*, an outcrossing relative to *A. thaliana*, MEDEA could be under positive [106] or balancing selection [107], in agreement with permanent conflicting pressures for resource acquisition into embryos between males and females. Conversely, this gene evolved under purifying selection in *A. thaliana*, where the level of conflict is reduced.

Male vs. female diverging interests are also reflected by cyto-nuclear conflicts. When cytoplasmic inheritance is uniparental, as in most species, cytoplasmic male sterility (CMS) alleles favoring transmission via females at the expense of males can spread in hermaphroditic outbreeding species, leaving room for coevolution with nuclear restorers. Maintenance of CMS/non-CMS polymorphism leads to stable gynodioecy [108]. In selfers, CMS mutants also reduce female fitness—because ovules cannot be fertilized—and are thus selected against. In the genus *Silene*, the mitochondrial genome of gynodioecious species exhibits molecular signatures of adaptive and/or balancing selection. This is likely due to cyto-nuclear conflicts as this is not, or is less, observed in hermaphrodites and dioecious [109–111]. Although less studied, cyto-nuclear conflicts are also expected in purely hermaphroditic species. In a recent study in *A. lyrata*, Foxe and Wright [112] found evidence of diversifying selection on members of a nuclear gene family encoding transcriptional regulators of cytoplasmic genes. Some of them show sequence similarity with CMS restorers in rice. Given the putative function of these genes, such selection could be due to ongoing cyto-nuclear coevolution. Interestingly, in *A. thaliana*, these genes do not seem to evolve under similar diversifying selection, as expected in a selfing species where conflicts are reduced.

2.3.2 Biased Gene Conversion as a Meiotic Drive Process: Consequences for Nucleotide Landscape and Protein Evolution

GC-biased gene conversion (gBGC) is a kind of meiotic drive at the base pair scale that can also be strongly influenced by breeding systems. In many species, gene conversion occurring during double-strand break recombination repair is biased toward G and C alleles (reviewed in [113]). This process mimics selection and can rapidly increase the GC content, especially around recombination hotspots [114, 115], and, more broadly, can affect genome-wide nucleotide landscapes. For instance, it is thought to be the main force that shaped the isochore structure of mammals and birds [116]. gBGC has been mostly studied by comparing genomic regions with different rates of (crossing-over) recombination (reviewed in [116]). However, comparing species with contrasted breeding systems offers a broader and unique opportunity to study gBGC. gBGC cannot occur in asexuals because recombination is lacking. Selfing is also expected to reduce the gBGC efficacy because meiotic drive does not occur in homozygotes [117]. To our knowledge, GC content has never been compared between sexual and asexual taxa, but there have been comparisons between outcrossers and selfers.

As expected, no relationship was found between local recombination rates and GC-content in the highly selfing *Arabidopsis thaliana* [117], and Wright et al. [118] suggested that the (weak) differences observed with the outcrossing *A. lyrata* and *Brassica oleracea* could be due to gBGC. Much stronger evidence has been obtained in grasses. Grasses are known to exhibit unusual genomic base composition compared to other plants, being richer and more heterogeneous in GC-content [119], and direct and indirect evidences of gBGC have been accumulating [119, 120–122]. Accordingly, GC-content or equilibrium GC values were found to be higher in outcrossing than in selfing species [24, 84, 120]. Difference in gBGC between outcrossing and selfing lineages has also been found in the plant genus *Collinsia* [123] and in freshwater snails [66], although difference in selection on codon usage cannot be completely ruled out.

gBGC can also affect functional sequence evolution, leaving a spurious signature of positive selection and increasing the mutation load through the fixation of weakly deleterious AT→GC mutations: gBGC would represent a genomic Achilles' heel [124]. Once again, comparing outcrossing and selfing species is useful for detecting interference between gBGC and selection. gBGC is expected to counteract selection in outcrossing species only. The Achilles' heel hypothesis could explain why relaxed selection was not detected in four grass species belonging to the Triticeae tribe [84]. In outcrossing species, but not in selfing ones, dN/dS was found to be significantly higher for genes exhibiting high than low equilibrium GC-content, suggesting that selection efficacy could be reduced because of high substitution rates in favor of GC alleles in these outcrossing grasses. In outcrossing species, gBGC can maintain recessive deleterious mutations for a long time at intermediate frequency, in a similar way to overdominance [125]. This could generate high inbreeding depression in outcrossing species, preventing the transition to selfing. In reverse, recurrent selfing would reduce the load through both purging and the avoidance of gBGC, thus reducing the deleterious effects of inbreeding. Under this scenario, gBGC would reinforce disruptive selection on mating systems. In the long term, gBGC could be a new cost of outcrossing: because of gBGC, not drift, outcrossing species could also accumulate weakly deleterious mutations, to an extent which could be substantial given current estimates of gBGC and deleterious mutation parameters [125]. Whether this gBGC-induced load could be higher than the drift load experienced by selfing species remains highly speculative. Both theoretical works, to refine predictions, and empirical data, to quantify the strength of gBGC and its impact on functional genomic regions, are needed in the future. Grasses are clearly an ideal model for investigating these issues, but comparisons with groups having lower levels of gBGC would also be helpful.

2.3.3 Transposable Elements in Selfers and Asexuals: Purging or Accumulation?

Considering the role of sex in the spread of selfish elements, TEs should be less frequent in selfers and asexuals than in outcrossers because they cannot spread from one genomic background to another through syngamy. However, highly selfing and asexual species derive from sexual outcrossing ancestors, from which they inherit their load of TEs. TE distribution eventually depends on the balance between additional transposition within selfing/clonal lineages on one hand and selection or excision on the other. Following the abandonment of sex, large asexual populations are expected to purge their load of TEs, provided excision occurs, even at very low rates. However, purging can take a very long time, and, without excision, TEs should slowly accumulate, not decline [126]. In small populations, even with excision, a Muller's ratchet-like process drives TE accumulation throughout the genome [126]. Transition from outcrossing to selfing should also rapidly purge TEs, but as for asexuals, in small fully selfing populations, TEs can be retained [127]. Using yeast populations, it was experimentally confirmed that sex increases the spread of TEs [128, 129]. TE numbers were also found to be higher in cyclically sexual than in fully asexual populations of *Daphnia pulex* [130–132] (Table 3), contrary to what was described in the parasitoid wasp *Leptopilina clavipes* and in root knot nematode species (Table 3). It should be noted that several comparative studies on asexual arthropods, nematodes, primroses, and green algae did not evidence any significant effect of breeding system on TE content or evolution (Table 3). At larger evolutionary scales, the putatively ancient asexual bdelloid rotifers strikingly exemplify the fact that

Table 3
Summary of studies comparing transposable element distribution and dynamics between different breeding systems

	Taxonomic group	Groups compared	Age of selfing/ asexuality	TE types	Effect of breeding system	References
Outcrossing/ selfing	*Arabidopsis*	*A. thaliana* (selfer)/ *A. lyrata* (outcrosser)	Recent (0.5–1 M years)	Ac-like DNA TE		
	Arabidopsis	*A. thaliana* (selfer)/ *A. lyrata* (outcrosser)	Recent (0.5–1 M years)	DNA TE, LTR and non-LTR RNA TE (no full genome reference)	No difference in insertion number, purifying selection lower in the selfing species	Lockton and Gaut [140]
	Arabidopsis	*A. thaliana* (selfer)/ *A. lyrata* (outcrosser)	Recent (0.5–1 M years)	DNA TE, LTR and non-LTR RNA TE (full genome reference)	Three times more copies and ten times more specific families in the outcrosser. Recent decrease in TE in number in the selfer	de la Chaux et al. [141]
	Capsella	*C. orientalis* and *C. rubella* (selfers)/ *C. grandiflora* (outcrosser)	Recent (~<1 M years) for *C. orientalis*, very recent (~20,000 years) for *C. rubella*	DNA TE, LTR and non-LTR RNA TE	Slight insertion increase in the recent selfer, strong decrease in the older selfer	Agren et al. [142]
	Solanum	SI and SC species of the *S. lycopersicum* section	Recent (~<1 M years)	*copia*-type RNA TE (ToRTL1, T135, Tnt1)	No effect of MS on TE insertions. Reduced TE sequence diversity in SC lineages. Compatible with a nearly neutral model	Tam et al. [138]
	Caenorhabditis	*C. elegans* (selfer)/ *C. remanei* (outcrosser)	Rather recent (<~4 M years)	Tc1 DNA TE	Purifying selection against TEs less efficient in the selfing species	Dolgin et al. [229]

Sexuals/asexuals	Four asexual angiosperm species	Comparison with sexual plants	Uncertain, maybe between 1 and 10 M years	Ty1/*copia*, Ty3/*gypsy*, and LINE-like RNA TE	Presence of conserved TE in asexuals	Docking et al. [228]
	Oenothera	17 asexual/13 sexual lineages	Unknown	DNA TE, LTR and non-LTR RNA TE	No significant effect	Agren et al. [160]
	Chlamydomonas reinhardtii	Asexual experimental lines	800 asexual generations/100 asexual generations vs. 11 sex events	Two DNA TE (TOC1, *Gulliver*)	No significant effect	Zeyl et al. [235]
	Saccharomyces cerevisiae	Sexual and asexual experimental lines with TE at initial frequency 1%	200–300 asexual generations/8 sex events	Ty3 RNA TE	Higher increase in TE frequency in sexual lines	Zeyl et al. [128]
	Candida albicans	Asexual species, compared with *S. cerevisiae*	Unknown. Rare sex events	LTR RNA TE	More TE families but most of them inactive and lower copy number than in *S. cerevisiae*	Goodwin and Poulter [230]
	Arthropods	Five pairs of asexual/sexual lineages	From very recent (~22 yrs., 10,000–40,000 generations) to old (~10 Myrs)	DNA TE, LTR and non-LTR RNA TE	No difference in any of the five pairs	Bast et al. [226]
	Bdelloid rotifers	Comparison with many other sexual metazoan	Old	LINE-like and *gypsy*-like RNA TE, *Mariner*/TC1-like DNA TE	Absence of RNA TE in asexuals	Arkhipova and Meselson [133]
	Daphnia pulex	Different isolates of the same species	Recent (<200,000 years)	One DNA TE (*Pokey*)	Lower TE insertion in asexuals	Sullender and Crease [130], Valizadeh and Crease [131]
	Daphnia pulex	20 asexuals/20 sexuals isolates	Recent (<200,000 years)	DNA TE, LTR and non-LTR RNA TE	Lower TE insertion but more fixed ones in asexuals. Substantial fraction of TE in asexuals inherited directly from sexuals	Jiang et al. [231]

(continued)

Table 3
(continued)

	Taxonomic group	Groups compared	Age of selfing/asexuality	TE types	Effect of breeding system	References
	Daphnia pulex	Asexual/sexual mutation-accumulation experimental lines	40 asexual generations/at least one sex event	6 DNA TE families	Higher rate of DNA TE loss in cyclical than in obligate parthenogenous lineages	Schaack et al. [233]
	Root-knot nematodes	3 obligate asexuals/1 facultative asexual	Uncertain, maybe between 17 and 40 Myrs	DNA TE, LTR and non-LTR RNA TE	Higher TE content in asexuals	Blanc-Mathieu et al. [157]
	Nematodes	42 species (dioecy, androdioecy, facultative parthenogenesis, strict apomixis)	Uncertain, maybe between 17 and 40 M years	DNA TE, LTR and non-LTR RNA TE	No significant effect of breeding system	Szitenberg et al. [234]
	Leptopilina clavipes	1 sexual/1 asexual (*Wolbachia*-induced) lineages	Recent (<12,000–43,000 generations)	DNA TE, LTR and non-LTR RNA TE	Proliferation of DNA TE and *gypsy*-like RNA TE in asexual lineages	Kraaijeveld et al. [232]

asexuals can purge their load of TEs. Unlike all sexual eukaryotes, they appear to be free of vertically transmitted retrotransposon, while their genome contains DNA transposons, probably acquired via horizontal transfers [133, 134]. Examples of TE accumulation in asexuals are less common, maybe because species are doomed to extinction under this evolutionary scenario [135]. However, the increase in genome size in some apomictic lineages of *Hypericum* species may result from this process [136].

In selfers, the distribution of TEs depends not only on the population size but also on the mode of selection against TEs [127, 137]. Under the "deleterious" model, TE insertions are selected against because they disrupt gene functions. According to the "ectopic exchange" model, TEs are selected against because they generate chromosomal rearrangements through unequal crossing-over between TE at nonhomologous insertion sites. Under the first of these two models, homozygosity resulting from selfing increases the selection efficacy against TEs, while under the second one, under-dominant chromosomal rearrangements are less selected against in selfing than in outcrossing populations [127, 137]. A survey of Ty1-copia-like elements in plants suggests that they are less abundant in self-fertilizing than in outcrossing plants, thus supporting the "deleterious" rather than the "ectopic" exchange model [127]. The distribution of retrotransposons in self-incompatible and self-compatible *Solanum* species also supports the "deleterious" model, even though most insertions are probably neutral [138] (Table 3). In the selfer *Arabidopsis thaliana*, selection efficacy against TEs seems to be reduced compared to its outcrossing sister species *A. lyrata* [139, 140], but comparison of the two complete genomes revealed a higher load of TE in *A. lyrata* and a recent decrease in TE in number in *A. thaliana*, in agreement with the date of transition to selfing [141]. In the *Capsella* genus, while the very recent selfer *C. rubella* possesses a slightly higher number of TEs than the outcrossing *C. grandiflora*, the oldest selfer *C. orientalis* exhibits a significantly reduced load of TE [142] (Table 3). Other selfish elements, such as B chromosomes, are also less frequent in selfers, in support of the view that inbreeding generally prevents selfish element transmission [102].

2.4 Breeding Systems, Ploidy, and Hybridization

Atypical breeding systems are often associated with polyploidy [143], and the reasons for this association are not entirely clear. Polyploid mutants might be more likely to establish as new lineages in selfers and asexuals than in obligate outcrossers if crosses between polyploids and diploids are unfertile or counterselected. This is because at low population frequency a polyploid mutant will experience the disadvantage of mostly mating with diploids—the minority cytotype exclusion principle [144, 145]—unless it reproduces asexually or via selfing. In addition, by doubling gene copy number, polyploidy might alleviate the fitness cost of recessive deleterious mutations being exposed at homozygous state in selfers [146]. Kreiner et al. [147] reported that in Brassicaceae the rate of production of unreduced gametes is higher in asexuals than in outcrossers, suggesting that mating systems can influence not only the establishment rate but also the mutation rate to polyploidy.

Recent genome-wide data analyses have revealed that a number of polyploid selfers or asexuals actually correspond to allopolyploids (e.g., [59, 148–151]), highlighting the possibility that hybridization plays a role in breeding system and ploidy evolution. Hybridization between facultative asexuals might cause immediate transition to obligate asexuality if the two progenitor genomes are so divergent that meiosis is impaired—e.g., due to chromosomal rearrangements, or in case of genetic incompatibilities affecting genes involved in sexual reproduction [16]. Numerous selfing or asexual lineages, either diploid or polyploid, are known to be of hybrid origin (e.g., [13, 152–157]). Hybridization would therefore appear as a potential cause, and polyploidy a potential consequence, of atypical breeding systems [16], but more genome-wide data are obviously needed to draw firm conclusions on these complex relationships.

2.5 Breeding Systems and Genome Size Evolution

As argued above, breeding systems can affect many aspects of genome content and organization. They should also affect the whole genome size. Following Lynch's theory [1], genome size should be higher in selfers and asexuals because of their reduced effective population size, hence reduced ability to get rid of useless, slightly costly sequences. However, the picture is probably more complex. First, because of the recent origin of many selfing and (at least some) asexual lineages, relaxed selection may not have operated longly enough to impact genome size. Second, because of their immunity to selfish element transmission, selfers and asexuals should exhibit lower genome size, especially in groups where TEs are major determinants of genome size. Hence, it is not clear whether genetic drift or resistance to selfish elements (or other processes) is the most important in governing genome size evolution in various breeding systems.

Meta-analyses performed in plants provided equivocal answers. Analysis of the distribution of B chromosomes showed a strong and significant positive association between outcrossing, the occurrence of B chromosomes, and genome size [102, 158]. However, after phylogenetic control, only the association between breeding systems and B chromosomes remains. Whitney et al. [159] simultaneously tested the effect of breeding systems (using outcrossing rate estimates) and genetic drift (using polymorphism data) on genome size in seed plants. Raw data showed a significant effect of both breeding systems and genetic drift, according to theoretical predictions. However, no effect was observed after phylogenetic control, leading the authors to reconsider the hypothesis of a role of nonadaptive processes in genome size evolution. Similarly, phylogenetic comparative analysis of 30 primrose species (*Oenothera*) covering several transitions to asexuality showed no significant relationship between reproductive mode and genome size [160].

Because breeding systems can evolve quickly, more detailed analyses at a short phylogenetic scale are needed to get a clearer picture of their effects on genome size evolution. Moreover, breeding systems are often correlated with other life history traits, such as lifespan, which can make it hard to clarify the causes and consequences of the observed correlations. A detailed analysis of genome size in the *Veronica* genus suggests that selfing, not annuality, is associated with genome size reduction [161]. A comparison of 14 pairs of plant congeneric species with contrasted mating systems also suggested a genome size reduction in selfers [162]. However, this could partly have been due to the four polyploid selfing species of the dataset—polyploidy can lead to haploid genome size reduction because of the loss of redundant DNA following polyploidization. A better understanding can be gained from the comparative analysis of genome composition and organization, not only genome size. In *Caenorhabditis* nematodes, the observed reduction in genome size is not driven by reduction in TEs but by a global loss of all genomic compartments [163]. This pattern contradicts the hypothesis of relaxed selection in selfers against the accumulation of deleterious genomic elements. Alternatively, it could be explained by deletion bias and high genetic drift in selfers. However, in mutation accumulation lines, insertions predominate over deletion in the selfing *C. elegans*, and deletions occurred at the whole gene level instead of being at random among genomic compartments, as predicted under a general deletion bias (*see* discussion in ref. 163). In this genus, Lynch's hypothesis that evolution of genome size should be driven by changes in N_e does not apply. Alternatively, the authors suggested that it is a more direct consequence or even an adaptation to the selfing lifestyle, although the underlying mechanisms still remain unclear.

3 A Genomic View of Breeding System Evolution

Because breeding systems can strongly affect genome structure and evolution, conversely, genomic approaches offer new powerful tools to reconstruct breeding system evolution and to test evolutionary hypotheses, especially concerning long-term evolution.

3.1 Genomic Approaches to Infer Breeding System Evolution

3.1.1 Genomic Characterization of Breeding Systems

Genetic markers have long been used to determine breeding systems and quantify selfing rates or degrees of asexuality. For instance, current selfing rates can be inferred using molecular markers through F_{IS} estimates or preferably—although more time consuming—through progeny analyses [164–166]. Multilocus-based estimates that take identity disequilibrium into account greatly improve the simple F_{IS}-based method that is sensitive to several artifacts such as null alleles ([167], *see* also refs. 168, 169). This method, implemented in the RMES software [167], has proven to give results very similar to progeny-based methods [170]. To take advantage of the information potentially available in sequence data, coalescence-based estimators have also been proposed to infer long-term selfing rates, and they have been implemented more recently in a Bayesian clustering approach in the INSTRUCT software package [171]. However, this approach mostly captures information from recent coalescence events so that such approaches still estimate recent selfing rates [28]. Much more information about long-term selfing rates can be derived from LD patterns [19], but this has not been fully exploited for selfing rate estimators (for instance, LD is not taken into account in INSTRUCT). Similarly, recombination can be inferred using genetic markers or sequence data, and more generally, various methods have been proposed to characterize the degree of clonality in natural populations (for review *see* ref. 172) and recently implemented in the R package RClone [173].

Initially, such methods were applied with few markers, from which only global descriptions of breeding systems were deducible. Thanks to the considerable increase in sequencing facilities, it has become possible to finely characterize temporal and spatial variations in breeding systems. In *A. thaliana*, an analysis of more than 1000 individuals in 77 local stands using more than 400 SNP markers revealed spatial heterogeneity in outcrossing rates. Local "hotspots" of recent outcrossing (up to 15%) were identified, while other stands exhibited complete homozygosity with no detectable outcrossing [174]. Interestingly, at this local scale (from 30 m to 40 km), outcrossing rates have been found to be twofold higher on average in rural than in urban stands; hence, selfing could be associated with higher disturbance in urban stands.

Genomic data may also help characterize breeding systems in species with unknown or ill-characterized life cycles. In yeasts *Saccharomyces cerevisiae* and *S. paradoxus*, the analyses of linkage disequilibrium patterns allowed to quantify the frequency of (rare) sexual reproduction events and the proportion of inbreeding and outcrossing during these events [175, 176]. For instance, in the pico-algae *Ostreococcus*, no sexual form or process has been detected in the lab. However, the occurrence of infrequent recombination (about 1 meiosis for 10 mitoses) inferred from a population genomics approach and the presence of meiosis genes in the genome support the existence of a sexual life cycle [177]. Moreover, a strong negative correlation between chromosome size and GC-content has been observed [178]. In mammals and birds (among others), such a pattern has been interpreted as a long-term effect of gBGC acting on chromosomes with different average recombination rates [116]—small chromosomes having higher recombination rates because of the constraint of at least one chiasmata per chromosome arm. A similar interpretation for *Ostreococcus* is thus appealing. Genomic data also allow to test whether the theoretical signatures of long-term asexuality are observed in putative asexuals. As an example, whole-genome analyses of the trypanosome *T. b. gambiense* demonstrated an independent evolution and divergence of alleles on each homologous chromosome (the "Meselson effect" [179, 180]), which is indicative of strict asexual evolution [88]. In contrast, genomic studies of the putatively ancient asexual bdelloids recently uncovered the occurrence of inter-individual genetic exchanges ([181, 182] *see* below Subheading 3.2.2).

3.1.2 Inferring and Dating Breeding System Transitions

Genomic approaches are also useful for analyzing the dynamics of breeding system evolution. A simple way is to map breeding system evolution on phylogenies, which could provide a raw picture of the frequency and relative timing of breeding system transitions (e.g., [183]). However, these approaches, based on ancestral character reconstruction, are hampered by numerous uncertainties. For instance, in the case of two sister species with contrasting breeding systems, such as *A. thaliana* and *A. lyrata*, it is impossible to know whether *A. thaliana* evolved toward selfing just after divergence (about five million years ago) or only very recently. At a larger phylogenetic scale, inferring rates of transition between characters and ancestral states can be biased if diversification rates differ between characters—this is typically expected with breeding systems for which asexuals and selfers should exhibit higher extinction rates than outcrossers [184].

Thanks to the genomic signatures left by contrasted breeding systems, it is possible to trace back transitions in the past and to date them more precisely. In diploid asexual species, because of the arrest of recombination, the two copies of each gene have diverged independently since the origin of asexuality. After having calibrated the molecular clock, it is thus possible to date this origin from the level of sequence divergence between the two copies. This so-called Meselson effect was observed and quantified in the trypanosome *T. b. gambiense*, suggesting that this species evolved asexually about 10,000 years ago [88]. However, no Meselson effect has been observed in other presumably ancient asexual species such as oribatid mites [185] or darwinulid ostracods [186], while data refute the possibility of cryptic sex. In such cases, it is thus not possible to infer when recombination actually stopped, presumably because of homogenizing processes such as very efficient DNA repair or automixis. Mitotic recombination could also obscure the pattern predicted under this Meselson effect. Of note, when asexuality originates by hybridization (*see* above Subheading 2.4), the last common ancestor of the two copies of a gene dates back to the ancestor of the two parental lineages, which can be much older than the hybridization date, faulting the above-described rationale.

Past transitions from outcrossing to selfing have also been investigated, through either population genomics approaches or the evolutionary analysis of self-incompatibility (SI) genes, which are directly involved in the transition to selfing. Since the evolution of selfing requires the breakdown of SI systems, initially constrained S-locus genes are expected to evolve neutrally after a shift to selfing. In *A. thaliana*, Bechsgaard et al. [187] reasoned that the dN/dS ratio in the selfing lineage should be the average of the neutral dN/dS (i.e., 1) and the outcrossing dN/dS—inferred from sister lineages—weighted by the time spent in the selfing vs. the outcrossing state. They deduced that SRK, one of the major SI genes, became a pseudogene less than 400,000 years ago. SRK, however, is not the only gene involved in SI. Mutations in other genes may have previously disrupted the SI system, thus confusing SRK-based dating. Indeed, coalescence simulations showed that the observed genome-wide pattern of linkage disequilibrium is compatible with the transition to selfing one million years ago [188], suggesting a possible but debated two-step scenario in the evolution of selfing [189, 190]. The persistence of three distinct divergent SRK haplotypes among extant *A. thaliana* individuals also suggests multiple loss of SI [191], but the recent discovery of the co-occurrence of the three haplotypes in Moroccan populations makes possible the evolution of selfing in a single geographic region [192]. In another Brassicaceae, i.e., *Capsella rubella*, analyses of both S-locus and genome-wide genes coupled with coalescence simulations suggested that selfing evolved very recently from the outcrosser *C. grandiflora*, around 50,000 years ago [193, 194] from a potentially large number of founding individuals followed by a strong reduction in N_e [195]. In the tetraploid selfer *Arabidopsis suecica*, which originated as a hybrid between *A. thaliana* and the outcrossing *A. arenosa*, the genomic analysis of the S-locus also revealed the origin of selfing, suggesting an instantaneous loss of SI due to the fixation of nonfunctional alleles from both parents around 16,000 years ago [150].

3.2 Matching Breeding System Evolution Theories with Genomic Data

3.2.1 Testing the Dead-End Hypothesis: Comparison Between Selfing and Asexuality

The expected reduction in N_e in selfers and asexuals may increase the drift load (accumulation of slightly deleterious mutations) and preclude adaptation. Selfing and clonality are thus supposed to be evolutionary dead ends [17, 18]. The twiggy phylogenetic distributions of asexuals [196] and selfers [183] or self-compatible species [197] suggest they are mostly derived recently from outcrossing ancestors (but *see* ref. 198). However, this observation may not be sufficient to support the dead-end hypothesis, and neutral models can also explain this pattern [199–201]. In a comprehensive and epochal phylogenetic study of several Solanaceae genera, Goldberg et al. [202] went further by testing the irreversibility of transitions. Using a phylogenetic method developed for estimating the character effect on speciation and extinction [203, 204], they showed that self-compatible species have both higher speciation and extinction rates—with the resulting net diversification rates being lower—than self-incompatible species. This was the first direct demonstration of the dead-end hypothesis, and additional results have been obtained in *Primula* species [205]. On the contrary, in the *Oenothera* genus, asexuality has been found associated with increased diversification but frequent reversion toward the sexual system, suggesting that the form of asexuality in this group is not an evolutionary dead end [206].

Genomic data also provide an opportunity to investigate the genetic causes of such long-term evolutionary failures. The increased dN/dS ratios reported in asexuals (see above) suggest that deleterious point mutations contribute to the load. However, in *Daphnia* rapid exposure of recessive deleterious alleles through mitotic recombination or gene conversion likely has a much stronger effect on clone persistence than their long-term accumulation under Muller's ratchet [60]. TE could also contribute to the load and to the extinction of asexuals [135], though more data are still needed to unambiguously support this hypothesis (but *see* ref. 136). The pattern in selfers is less clear. While theory globally predicts a reduction in selection efficacy in selfers, models also highlight conditions under which selection can be little affected or even enhanced in selfers [72, 73, 207], especially regarding TE accumulation [127, 137]. Empirical data on both protein and TE evolution have not revealed any strong evidence of long-term accumulation of deleterious mutation in selfers, as compared to outcrossers, whereas polymorphism data mainly support relaxation of selection in selfers (Table 2). This is in agreement with the recent origin of selfing but makes difficult further inference of the underlying causes of higher extinction in selfers as trait-dependent diversification processes alter the relationship between life history traits and rate of molecular evolution [208]. A reduced ability to respond to environmental changes through adaptive evolution could also contribute to long-term extinction in asexuals (but *see* ref. 209) and selfers, especially if standing variation is needed to rescue populations experiencing environmental challenges [77, 210]. Few studies, however, have compared the rate of adaptation in selfers and outcrossers (*see* Table 2). Theoretical predictions regarding this effect, moreover, critically depend on the dominance level of new favorable mutations [72, 73, 77, 210], which are poorly known (but *see* ref. 80).

While several issues remain open, current knowledge suggests that selfers are less prone to extinction than asexuals. The wider distribution of selfing than clonality in plants supports this view [211, 212]. Selfers could go toward extinction more slowly than asexuals, and the causes of their extinction could differ. Since deleterious mutations should accumulate at a slower rate in selfers than in asexuals, as suggested by theory and current data, this process would likely not be sufficient to drive them to extinction. The reduced adaptive potential could be the very cause of their ultimate extinction as initially proposed by Stebbins [18], which could generally occur before sufficient deleterious mutations have accumulated to be detected via molecular measures of divergence. On the contrary, in asexuals, the accumulation of deleterious mutations could be fast enough to leave a molecular signature and contribute to extinction. Alternatively, demographic characteristics associated with uniparental reproduction, such as recurrent bottlenecks, fragmented populations, and extinction/recolonization dynamics, could be sufficient to drive population extension simply because of higher sensitivity to demographic stochasticity (*see* also ref. 213). Genomic degradation would only be the witness of the evolution toward selfing and clonality without being the ultimate cause of their extinctions. These hypotheses need to be further investigated by building more realistic demo-genetic model and by better integrating genomic and ecological approaches.

The literature reviewed above focuses on intrinsic factors that may affect the extinction rate of selfing and asexual species, taken as established lineages, compared to their sexual relatives. Alternatively, Janko et al. [199] suggested that if asexual mutants are produced at a relatively high rate and compete with each other, this would imply a rapid turnover between clonal lineages and a young expected age for extant asexuals, without the need to invoke any fitness effect (*see* also refs. 200, 201). Of note, this model invokes competitive exclusion among clonal lineages, but not between clonal and sexual ones—the ancestral sexual gene pool is assumed to be immune from extinction.

3.2.2 Evading the "Dead End"

The few putatively ancient asexuals known so far seem to escape the mutational load predicted by the dead-end hypothesis and avoid extinction over long evolutionary time scales. For example, fossil evidence and decades of microscopic observations indicate that bdelloid rotifers have apparently persisted for over 40 million years without meiosis, males, or conventional sexual reproduction

[15, 214]. As a matter of fact, the first genome assembly published for these organisms confirmed that their genome structure is incompatible with conventional meiosis [215]. However, two independent studies recently demonstrated that bdelloids could experience genetic exchanges between individuals.

A first article by Debortoli et al. [182] evidenced frequent horizontal exchanges of genetic fragments between individuals of the species *Adineta vaga* (Adinetidae). Such horizontal transfers could be promoted by the peculiar ecology of these rotifers, which experience frequent desiccations damaging their cell and nucleus membranes and thus allowing for the entry of foreign DNA in the cells. In addition, desiccation induces multiple DNA double-strand breaks, facilitating the integration of foreign DNA during repair processes.

Another study by Signorovitch et al. [181] identified a pattern of allele sharing between individuals of the species *Macrotrachela quadricornifera* (Philodinidae) that was incompatible with strict asexual evolution. The authors suggested that bdelloids had evolved an atypical meiotic mechanism similar to what has been described in some species of primroses (*Oenothera*), in which chromosomes organize into a ring during meiosis without requiring homologous chromosome pairing [216]. They advocated that even rare events of such unconventional sex could be enough to generate the observed pattern of allele sharing.

In the absence of conventional meiosis and syngamy, bdelloid rotifers might thus have escaped extinction by maintaining some level of genetic exchanges between individuals, either through horizontal gene transfers or unconventional *Oenothera*-like meiosis. Regardless of the underlying molecular mechanisms, bdelloids should not be considered as "ancient asexual scandals" anymore. These recent results call for a reassessment of the reproductive mode of all supposedly ancient asexuals (*see* Subheading 3.1.1 above). The rise of genomic studies in recent years will greatly contribute to decipher whether putative asexuals evolve as strict asexuals or have developed new alternatives to sex.

4 Conclusion and Prospects

There is a large body of theory on the effects of breeding systems on molecular evolution. However, some of them have not been clearly verified by empirical data, and numerous questions remain. Genomic data have also partly unveiled the complexity of breeding systems, especially in asexual or presumably asexual species. Promising prospects include (1) analysis of the rate and pattern of transition to selfing/asexuality using densely sampled phylogenies with appropriate breeding system distributions combined with genome-wide molecular data, (2) distinguishing between the different forms of selection with a better characterization of the fitness effect of mutations, (3) explicitly accounting for the possible association between breeding system shifts and non-equilibrium demographic dynamics (e.g., bottlenecks in selfers, clone turnover in asexuals). A large theoretical corpus has already been developed, and thanks to the increasing availability of genomic data, qualitative patterns are now rather well described and partly understood. Another challenge in the future is also to make our predictions and tests more quantitative.

5 Questions

1. What population genetic parameters are affected, and how, by selfing and asexuality?
2. What are the potential problems when comparing the dN/dS ratio between selfers and outcrossers or sexuals and asexuals?
3. What is the evolutionary "dead-end hypothesis," and how can we test it using phylogenetic and evolutionary genomic tools?

Acknowledgments

This work was supported by ARCAD, a flagship project of Agropolis Fondation, by an ERC grant (PopPhyl) to N.G. and by the CoGeBi program (grant number ANR-08-GENM-036-01) and a Swiss National Research Found SINERGIA grant.

References

1. Lynch M (2007) The origin of genome architecture, 1st edn. Sinauer, Sunderland, MA
2. Smith SA, Donoghue MJ (2008) Rates of molecular evolution are linked to life history in flowering plants. Science 322(5898):86–89
3. Romiguier J, Gayral P, Ballenghien M, Bernard A, Cahais V, Chenuil A, Chiari Y, Dernat R, Duret L, Faivre N, Loire E, Lourenco JM, Nabholz B, Roux C, Tsagkogeorga G, Weber AA, Weinert LA, Belkhir K, Bierne N, Glemin S, Galtier N (2014) Comparative population genomics in animals uncovers the determinants of genetic diversity. Nature 515(7526):261–263
4. Bromham L, Hua X, Lanfear R, Cowman PF (2015) Exploring the relationships between mutation rates, life history, genome size, environment, and species richness in flowering plants. Am Nat 185(4):507–524
5. Figuet E, Nabholz B, Bonneau M, Mas Carrio E, Nadachowska-Brzyska K, Ellegren H, Galtier N (2016) Life history traits, protein evolution, and the nearly neutral theory in amniotes. Mol Biol Evol 33 (6):1517–1527
6. Chen J, Glemin S, Lascoux M (2017) Genetic diversity and the efficacy of purifying selection across plant and animal species. Mol Biol Evol 34(6):1417–1428
7. Lefebure T, Morvan C, Malard F, Francois C, Konecny-Dupre L, Gueguen L, Weiss-Gayet-M, Seguin-Orlando A, Ermini L, Sarkissian C, Charrier NP, Eme D, Mermillod-Blondin F, Duret L, Vieira C, Orlando L, Douady CJ (2017) Less effective selection leads to larger genomes. Genome Res 27(6):1016–1028
8. Jarne P, Auld JR (2006) Animals mix it up too: the distribution of self-fertilization

among hermaphroditic animals. Evolution 60 (9):1816–1824

9. Vogler DW, Kaliz S (2001) Sex among the flowers: the distribution of plant mating systems. Evolution 55(1):202–204
10. Haldane JBS (1932) The causes of evolution, vol 1, 1st edn. Princeton University Press, Princeton
11. Hedrick PW (1987) Population genetics of intragametophytic selfing. Evolution 41 (1):137–144
12. Balloux F, Lehmann L, de Meeus T (2003) The population genetics of clonal and partially clonal diploids. Genetics 164(4):1635–1644
13. Simon JC, Delmotte F, Rispe C, Crease TJ (2003) Phylogenetic relationships between parthenogens and their sexual relatives: the possible routes to parthenogenesis in animals. Biol J Lin Soc 79:151–163
14. Whitton J, Sears CJ, Baack EJ, Otto SP (2008) The dynamic nature of apomixis in the angiosperms. Int J Plant Sci 169 (1):169–182
15. Schurko AM, Neiman M, Logsdon JM (2009) Signs of sex: what we know and how we know it. Trends Ecol Evol 24(4):208–217
16. Neiman M, Sharbel TF, Schwander T (2014) Genetic causes of transitions from sexual reproduction to asexuality in plants and animals. J Evol Biol 27(7):1346–1359
17. Maynard-Smith J (1978) The evolution of sex. Cambridge University Press, Cambridge
18. Stebbins GL (1957) Self fertilization and population variability in higher plants. Am Nat 91:337–354
19. Nordborg M (2000) Linkage disequilibrium, gene trees and selfing: an ancestral recombination graph with partial self-fertilization. Genetics 154(2):923–929
20. Padhukasahasram B, Marjoram P, Wall JD, Bustamante CD, Nordborg M (2008) Exploring population genetic models with recombination using efficient forward-time simulations. Genetics 178(4):2417–2427
21. Hartfield M, Glemin S (2016) Limits to adaptation in partially selfing species. Genetics 203 (2):959–974
22. Roze D (2016) Background selection in partially selfing populations. Genetics 203 (2):937–957
23. Flint-Garcia SA, Thornsberry JM, Buckler ES IV (2003) Structure of linkage disequilibrium in plants. Annu Rev Plant Biol 54:357–374
24. Glémin S, Bazin E, Charlesworth D (2006) Impact of mating systems on patterns of sequence polymorphism in flowering plants. Proc R Soc Lond B Biol Sci 273 (1604):3011–3019
25. Golding GB, Strobeck C (1980) Linkage disequilibrium in a finite population that is partially selfing. Genetics 94(3):777–789
26. Roze D (2015) Effects of interference between selected loci on the mutation load, inbreeding depression, and heterosis. Genetics 201(2):745–757
27. Pollak E (1987) On the theory of partially inbreeding finite populations. I. Partial selfing. Genetics 117(2):353–360
28. Nordborg M, Donnelly P (1997) The coalescent process with selfing. Genetics 146 (3):1185–1195
29. Ceplitis A (2003) Coalescence times and the Meselson effect in asexual eukaryotes. Genet Res 82(3):183–190
30. Hartfield M, Wright SI, Agrawal AF (2016) Coalescent times and patterns of genetic diversity in species with facultative sex: effects of gene conversion, population structure, and heterogeneity. Genetics 202(1):297–312
31. Haag CR, Roze D (2007) Genetic load in sexual and asexual diploids: segregation, dominance and genetic drift. Genetics 176 (3):1663–1678
32. Schoen DJ, Brown AHD (1991) Intraspecific variation in population gene diversity and effective population size correlates with the mating system in plants. Proc Natl Acad Sci U S A 88:4494–4497
33. Haag CR, Ebert D (2004) A new hypothesis to explain geographic parthenogenesis. Ann Zool Fennici 41:539–544
34. Ingvarsson PK (2002) A metapopulation perspective on genetic diversity and differentiation in partially self-fertilizing plants. Evolution 56(12):2368–2373
35. Gordo I, Charlesworth B (2001) Genetic linkage and molecular evolution. Curr Biol 11(17):R684–R686
36. Agrawal AF, Hartfield M (2016) Coalescence with background and balancing selection in systems with bi- and uniparental reproduction: contrasting partial asexuality and selfing. Genetics 202(1):313–326
37. Thomas CG, Wang W, Jovelin R, Ghosh R, Lomasko T, Trinh Q, Kruglyak L, Stein LD, Cutter AD (2015) Full-genome evolutionary histories of selfing, splitting, and selection in Caenorhabditis. Genome Res 25(5):667–678
38. Andersen EC, Gerke JP, Shapiro JA, Crissman JR, Ghosh R, Bloom JS, Felix MA, Kruglyak L (2012) Chromosome-scale selective sweeps shape Caenorhabditis elegans genomic diversity. Nat Genet 44(3):285–290

39. Coop G (2016) Does linked selection explain the narrow range of genetic diversity across species? bioArxiv. https://doi.org/10.1101/042598
40. Hamrick JL, Godt MJW (1996) Effects of life history traits on genetic diversity in plant species. Philos Trans R Soc Lond B 351 (1345):1291–1298
41. Nybom H (2004) Comparison of different nuclear DNA markers for estimating intraspecific genetic diversity in plants. Mol Ecol 13 (5):1143–1155
42. Fontcuberta Garcia-Cuenca A, Dumas Z, Schwander T (2016) Extreme genetic diversity in asexual grass thrips populations. J Evol Biol 29(5):887–899
43. Normark BB, Judson OP, Moran NA (2003) Genomic signatures of ancient asexual lineages. Biol J Lin Soc 79:69–84
44. Lercher MJ, Hurst LD (2002) Human SNP variability and mutation rate are higher in regions of high recombination. Trends Genet 18(7):337–340
45. Hellmann I, Ebersberger I, Ptak SE, Paabo S, Przeworski M (2003) A neutral explanation for the correlation of diversity with recombination rates in humans. Am J Hum Genet 72 (6):1527–1535
46. Longman-Jacobsen N, Williamson JF, Dawkins RL, Gaudieri S (2003) In polymorphic genomic regions indels cluster with nucleotide polymorphism: quantum genomics. Gene 312:257–261
47. Tian D, Wang Q, Zhang P, Araki H, Yang S, Kreitman M, Nagylaki T, Hudson R, Bergelson J, Chen JQ (2008) Single-nucleotide mutation rate increases close to insertions/deletions in eukaryotes. Nature 455(7209):105–108
48. Hollister JD, Ross-Ibarra J, Gaut BS (2010) Indel-associated mutation rate varies with mating system in flowering plants. Mol Biol Evol 27(2):409–416
49. Kimura M (1962) On the probability of fixation of mutant genes in a population. Genetics 47:713–719
50. Hill WG, Robertson AW (1966) The effect of genetic linkage on the limits to artificial selection. Genet Res 8:269–294
51. Paland S, Lynch M (2006) Transitions to asexuality result in excess amino acid substitutions. Science 311(5763):990–992
52. Henry L, Schwander T, Crespi BJ (2012) Deleterious mutation accumulation in asexual Timema stick insects. Mol Biol Evol 29 (1):401–408
53. Johnson SG, Howard RS (2007) Contrasting patterns of synonymous and nonsynonymous sequence evolution in asexual and sexual freshwater snail lineages. Evolution 61 (11):2728–2735
54. Neiman M, Hehman G, Miller JT, Logsdon JM Jr, Taylor DR (2010) Accelerated mutation accumulation in asexual lineages of a freshwater snail. Mol Biol Evol 27 (4):954–963
55. Lovell JT, Williamson RJ, Wright SI, McKay JK, Sharbel TF (2017) Mutation accumulation in an asexual relative of Arabidopsis. PLoS Genet 13(1):e1006550
56. Ollivier M, Gabaldon T, Poulain J, Gavory F, Leterme N, Gauthier JP, Legeai F, Tagu D, Simon JC, Rispe C (2012) Comparison of gene repertoires and patterns of evolutionary rates in eight aphid species that differ by reproductive mode. Genome Biol Evol 4 (2):155–167
57. Pellino M, Hojsgaard D, Schmutzer T, Scholz U, Horandl E, Vogel H, Sharbel TF (2013) Asexual genome evolution in the apomictic Ranunculus auricomus complex: examining the effects of hybridization and mutation accumulation. Mol Ecol 22 (23):5908–5921
58. Hollister JD, Greiner S, Wang W, Wang J, Zhang Y, Wong GK, Wright SI, Johnson MT (2015) Recurrent loss of sex is associated with accumulation of deleterious mutations in Oenothera. Mol Biol Evol 32(4):896–905
59. Ament-Velasquez SL, Figuet E, Ballenghien M, Zattara EE, Norenburg JL, Fernandez-Alvarez FA, Bierne J, Bierne N, Galtier N (2016) Population genomics of sexual and asexual lineages in fissiparous ribbon worms (Lineus, Nemertea): hybridization, polyploidy and the Meselson effect. Mol Ecol 25(14):3356–3369
60. Tucker AE, Ackerman MS, Eads BD, Xu S, Lynch M (2013) Population-genomic insights into the evolutionary origin and fate of obligately asexual Daphnia pulex. Proc Natl Acad Sci U S A 110(39):15740–15745
61. Wright SI, Lauga B, Charlesworth D (2002) Rates and patterns of molecular evolution in inbred and outbred *Arabidopsis*. Mol Biol Evol 19(9):1407–1420
62. Cutter AD, Wasmuth JD, Washington NL (2008) Patterns of molecular evolution in *Caenorhabditis* preclude ancient origins of selfing. Genetics 178(4):2093–2104
63. Escobar JS, Cenci A, Bolognini J, Haudry A, Laurent S, David J, Glémin S (2010) An integrative test of the dead-end hypothesis of

selfing evolution in Triticeae (poaceae). Evolution 64(10):2855–2872
64. Glémin S, Muyle A (2014) Mating systems and selection efficacy: a test using chloroplastic sequence data in Angiosperms. J Evol Biol 27(7):1386–1399
65. Arunkumar R, Ness RW, Wright SI, Barrett SC (2015) The evolution of selfing is accompanied by reduced efficacy of selection and purging of deleterious mutations. Genetics 199(3):817–829
66. Burgarella C, Gayral P, Ballenghien M, Bernard A, David P, Jarne P, Correa A, Hurtrez-Bousses S, Escobar J, Galtier N, Glémin S (2015) Molecular evolution of freshwater snails with contrasting mating systems. Mol Biol Evol 32(9):2403–2416
67. Charlesworth D, Morgan MT, Charlesworth B (1993) Mutation accumulation in finite outbreeding and inbreeding populations. Genet Res 61:39–56
68. Hartfield M, Glemin S (2014) Hitchhiking of deleterious alleles and the cost of adaptation in partially selfing species. Genetics 196 (1):281–293
69. Hartfield M, Otto SP (2011) Recombination and hitchhiking of deleterious alleles. Evolution 65(9):2421–2434
70. Bullaughey K, Przeworski M, Coop G (2008) No effect of recombination on the efficacy of natural selection in primates. Genome Res 18 (4):544–554
71. Haddrill PR, Halligan DL, Tomaras D, Charlesworth B (2007) Reduced efficacy of selection in regions of the Drosophila genome that lack crossing over. Genome Biol 8(2):R18
72. Charlesworth B (1992) Evolutionary rates in partially self-fertilizing species. Am Nat 140 (1):126–148
73. Glémin S (2007) Mating systems and the efficacy of selection at the molecular level. Genetics 177(2):905–916
74. Charlesworth B, Charlesworth D (1997) Rapid fixation of deleterious alleles can be caused by Muller's ratchet. Genet Res 70 (1):63–73
75. Szovenyi P, Devos N, Weston DJ, Yang X, Hock Z, Shaw JA, Shimizu KK, McDaniel SF, Wagner A (2014) Efficient purging of deleterious mutations in plants with haploid selfing. Genome Biol Evol 6(5):1238–1252
76. Haldane JBS (1937) The effect of variation on fitness. Am Nat 71:337–349
77. Glémin S, Ronfort J (2013) Adaptation and maladaptation in selfing and outcrossing species: new mutations versus standing variation. Evolution 67(1):225–240
78. Kamran-Disfani A, Agrawal AF (2014) Selfing, adaptation and background selection in finite populations. J Evol Biol 27 (7):1360–1371
79. Kirkpatrick M, Jenkins CD (1989) Genetic segregation and the maintenance of sexual reproduction. Nature 339(6222):300–301
80. Ronfort J, Glémin S (2013) Mating system, Haldane's sieve, and the domestication process. Evolution 67(5):1518–1526
81. McDonald JH, Kreitman M (1991) Adaptative protein evolution at the ADH locus in Drosophila. Nature 351:652–654
82. Eyre-Walker A, Keightley PD (2009) Estimating the rate of adaptive molecular evolution in the presence of slightly deleterious mutations and population size change. Mol Biol Evol 26 (9):2097–2108
83. Slotte T, Foxe JP, Hazzouri KM, Wright SI (2010) Genome-wide evidence for efficient positive and purifying selection in *Capsella grandiflora*, a plant species with a large effective population size. Mol Biol Evol 27 (8):1813–1821
84. Haudry A, Cenci A, Guilhaumon C, Paux E, Poirier S, Santoni S, David J, Glémin S (2008) Mating system and recombination affect molecular evolution in four Triticeae species. Genet Res 90(1):97–109
85. Hersch-Green EI, Myburg H, Johnson MT (2012) Adaptive molecular evolution of a defence gene in sexual but not functionally asexual evening primroses. J Evol Biol 25 (8):1576–1586
86. Engelstadter J (2017) Asexual but not clonal: evolutionary processes in automictic populations. Genetics 206(2):993–1009
87. Mandegar MA, Otto SP (2007) Mitotic recombination counteracts the benefits of genetic segregation. Proc Biol Sci 274 (1615):1301–1307
88. Weir W, Capewell P, Foth B, Clucas C, Pountain A, Steketee P, Veitch N, Koffi M, De Meeus T, Kabore J, Camara M, Cooper A, Tait A, Jamonneau V, Bucheton B, Berriman M, MacLeod A (2016) Population genomics reveals the origin and asexual evolution of human infective trypanosomes. Elife 5:e11473
89. Omilian AR, Cristescu ME, Dudycha JL, Lynch M (2006) Ameiotic recombination in asexual lineages of Daphnia. Proc Natl Acad Sci U S A 103(49):18638–18643
90. Keith N, Tucker AE, Jackson CE, Sung W, Lledo JIL, Schrider DR, Schaack S, Dudycha JL, Ackerman M, Younge AJ, Shaw JR, Lynch M (2016) High mutational rates of large-scale

duplication and deletion in Daphnia pulex. Genome Res 26(1):60–69

91. Charlesworth D, Charlesworth B, Strobeck C (1979) Selection for recombination in self-fertilizing species. Genetics 93:237–244
92. Charlesworth D, Charlesworth B, Strobeck C (1977) Effects of selfing on selection for recombination. Genetics 68:213–226
93. Roze D, Lenormand T (2005) Self-fertilization and the evolution of recombination. Genetics 170:841–857
94. Ross-Ibarra J (2007) Genome size and recombination in angiosperms: a second look. J Evol Biol 20(2):800–806
95. Dawson KJ (1998) Evolutionarily stable mutation rates. J Theor Biol 194(1):143–157
96. Kondrashov AS (1995) Modifiers of mutation-selection balance - general-approach and the evolution of mutation-rates. Genet Res 66(1):53–69
97. Lynch M (2010) Evolution of the mutation rate. Trends Genet 26(8):345–352
98. Schoen DJ (2005) Deleterious mutation in related species of the plant genus *Amsinckia* with contrasting mating systems. Evolution 59(11):2370–2377
99. Baer CF, Joyner-Matos J, Ostrow D, Grigaltchik V, Salomon MP, Upadhyay A (2010) Rapid decline in fitness of mutation accumulation lines of gonochoristic (outcrossing) Caenorhabditis nematodes. Evolution 64(11):3242–3253
100. Brandvain Y, Haig D (2005) Divergent mating systems and parental conflict as a barrier to hybridization in flowering plants. Am Nat 166(3):330–338
101. Tazzyman SJ, Abbott JK (2015) Self-fertilization and inbreeding limit the scope for sexually antagonistic polymorphism. J Evol Biol 28(3):723–729
102. Burt A, Trivers R (1998) Selfish DNA and breeding systems in plants. Proc R Soc Lond B 265:141–146
103. Swanson WJ, Vacquier VD (2002) The rapid evolution of reproductive proteins. Nat Rev Genet 3(2):137–144
104. Palopoli MF, Rockman MV, TinMaung A, Ramsay C, Curwen S, Aduna A, Laurita J, Kruglyak L (2008) Molecular basis of the copulatory plug polymorphism in Caenorhabditis elegans. Nature 454 (7207):1019–1022
105. Cutter AD (2008) Reproductive evolution: symptom of a selfing syndrome. Curr Biol 18(22):R1056–R1058
106. Spillane C, Schmid KJ, Laoueille-Duprat S, Pien S, Escobar-Restrepo JM, Baroux C, Gagliardini V, Page DR, Wolfe KH, Grossniklaus U (2007) Positive darwinian selection at the imprinted MEDEA locus in plants. Nature 448(7151):349–352
107. Kawabe A, Fujimoto R, Charlesworth D (2007) High diversity due to balancing selection in the promoter region of the Medea gene in *Arabidopsis lyrata*. Curr Biol 17 (21):1885–1889
108. Budar F, Touzet P, De Paepe R (2003) The nucleo-mitochondrial conflict in cytoplasmic male sterilities revisited. Genetica 117 (1):3–16
109. Houliston GJ, Olson MS (2006) Nonneutral evolution of organelle genes in Silene vulgaris. Genetics 174(4):1983–1994
110. Ingvarsson PK, Taylor DR (2002) Genealogical evidence for epidemics of selfish genes. Proc Natl Acad Sci U S A 99 (17):11265–11269
111. Touzet P, Delph LF (2009) The effect of breeding system on polymorphism in mitochondrial genes of Silene. Genetics 181 (2):631–644
112. Foxe JP, Wright SI (2009) Signature of diversifying selection on members of the pentatricopeptide repeat protein family in Arabidopsis lyrata. Genetics 183(2):663–672, 661SI–668SI
113. Marais G (2003) Biased gene conversion: implications for genome and sex evolution. Trends Genet 19(6):330–338
114. Clement Y, Arndt PF (2013) Meiotic recombination strongly influences GC-content evolution in short regions in the mouse genome. Mol Biol Evol 30(12):2612–2618
115. Glémin S, Arndt PF, Messer PW, Petrov D, Galtier N, Duret L (2015) Quantification of GC-biased gene conversion in the human genome. Genome Res 25(8):1215–1228
116. Duret L, Galtier N (2009) Biased gene conversion and the evolution of mammalian genomic landscapes. Annu Rev Genomics Hum Genet 10:285–311
117. Marais G, Charlesworth B, Wright SI (2004) Recombination and base composition: the case of the highly self-fertilizing plant *Arabidopsis thaliana*. Genome Biol 5(7):R45
118. Wright SI, Iorgovan G, Misra S, Mokhtari M (2007) Neutral evolution of synonymous base composition in the Brassicaceae. J Mol Evol 64(1):136–141
119. Serres-Giardi L, Belkhir K, David J, Glémin S (2012) Patterns and evolution of nucleotide

landscapes in seed plants. Plant Cell 24 (4):1379–1397

120. Muyle A, Serres-Giardi L, Ressayre A, Escobar J, Glémin S (2011) GC-biased gene conversion and selection affect GC content in the *Oryza* genus (rice). Mol Biol Evol 28 (9):2695–2706

121. Rodgers-Melnick E, Vera DL, Bass HW, Buckler ES (2016) Open chromatin reveals the functional maize genome. Proc Natl Acad Sci U S A 113(22):E3177–E3184

122. Li X, Li L, Yan J (2015) Dissecting meiotic recombination based on tetrad analysis by single-microspore sequencing in maize. Nat Commun 6:6648

123. Hazzouri KM, Escobar JS, Ness RW, Killian Newman L, Randle AM, Kalisz S, Wright SI (2013) Comparative population genomics in *Collinsia* sister species reveals evidence for reduced effective population size, relaxed selection, and evolution of biased gene conversion with an ongoing mating system shift. Evolution 67(5):1263–1278

124. Galtier N, Duret L (2007) Adaptation or biased gene conversion? Extending the null hypothesis of molecular evolution. Trends Genet 23(6):273–277

125. Glémin S (2010) Surprising fitness consequences of GC-biased gene conversion: I. Mutation load and inbreeding depression. Genetics 185(3):939–959

126. Dolgin ES, Charlesworth B (2006) The fate of transposable elements in asexual populations. Genetics 174(2):817–827

127. Morgan MT (2001) Transposable element number in mixed mating populations. Genet Res 77(3):261–275

128. Zeyl C, Bell G, Green DM (1996) Sex and the spread of retrotransposon Ty3 in experimental populations of *Saccharomyces cerevisiae*. Genetics 143(4):1567–1577

129. Goddard MR, Greig D, Burt A (2001) Outcrossed sex allows a selfish gene to invade yeast populations. Proc Biol Sci 268 (1485):2537–2542

130. Sullender BW, Crease TJ (2001) The behavior of a *Daphnia pulex* transposable element in cyclically and obligately parthenogenetic populations. J Mol Evol 53(1):63–69

131. Valizadeh P, Crease TJ (2008) The association between breeding system and transposable element dynamics in *Daphnia pulex*. J Mol Evol 66(6):643–654

132. Schaack S, Pritham EJ, Wolf A, Lynch M (2010) DNA transposon dynamics in populations of Daphnia pulex with and without sex. Proc R Soc B Biol Sci 277(1692):2381–2387

133. Arkhipova I, Meselson M (2000) Transposable elements in sexual and ancient asexual taxa. Proc Natl Acad Sci U S A 97 (26):14473–14477

134. Arkhipova IR, Meselson M (2005) Diverse DNA transposons in rotifers of the class Bdelloidea. Proc Natl Acad Sci U S A 102 (33):11781–11786

135. Arkhipova I, Meselson M (2005) Deleterious transposable elements and the extinction of asexuals. BioEssays 27(1):76–85

136. Matzk F, Hammer K, Schubert I (2003) Coevolution of apomixis and genome size within the genus *Hypericum*. Sex Plant Reprod 16:51–58

137. Wright SI, Schoen DJ (1999) Transposon dynamics and the breeding system. Genetica 107(1–3):139–148

138. Tam SM, Causse M, Garchery C, Burck H, Mhiri C, Grandbastien MA (2007) The distribution of copia-type retrotransposons and the evolutionary history of tomato and related wild species. J Evol Biol 20 (3):1056–1072

139. Wright SI, Le QH, Schoen DJ, Bureau TE (2001) Population dynamics of an Ac-like transposable element in self- and cross-pollinating Arabidopsis. Genetics 158 (3):1279–1288

140. Lockton S, Gaut BS (2010) The evolution of transposable elements in natural populations of self-fertilizing Arabidopsis thaliana and its outcrossing relative Arabidopsis lyrata. BMC Evol Biol 10:10

141. de la Chaux N, Tsuchimatsu T, Shimizu KK, Wagner A (2012) The predominantly selfing plant Arabidopsis thaliana experienced a recent reduction in transposable element abundance compared to its outcrossing relative Arabidopsis lyrata. Mob DNA 3(1):2

142. Agren JA, Wang W, Koenig D, Neuffer B, Weigel D, Wright SI (2014) Mating system shifts and transposable element evolution in the plant genus Capsella. BMC Genomics 15:602

143. Otto SP, Whitton J (2000) Polyploid incidence and evolution. Annu Rev Genet 34:401–437

144. Fowler NL, Levin DA (1984) Ecological constraints on the establishment of a novel polyploid in competition with its diploid progenitor. Am Nat 124:703–711

145. Husband BC (2000) Constraints on polyploid evolution: a test of the minority cytotype exclusion principle. Proc Biol Sci 267 (1440):217–223

146. Husband BC (2016) Effect of inbreeding on pollen tube growth in diploid and tetraploid Chamerion angustifolium: do polyploids mask mutational load in pollen? Am J Bot 103(3):532–540
147. Kreiner JM, Kron P, Husband BC (2017) Frequency and maintenance of unreduced gametes in natural plant populations: associations with reproductive mode, life history and genome size. New Phytol 214(2):879–889
148. Xu S, Innes DJ, Lynch M, Cristescu ME (2013) The role of hybridization in the origin and spread of asexuality in Daphnia. Mol Ecol 22(17):4549–4561
149. Douglas GM, Gos G, Steige KA, Salcedo A, Holm K, Josephs EB, Arunkumar R, Agren JA, Hazzouri KM, Wang W, Platts AE, Williamson RJ, Neuffer B, Lascoux M, Slotte T, Wright SI (2015) Hybrid origins and the earliest stages of diploidization in the highly successful recent polyploid *Capsella bursa-pastoris*. Proc Natl Acad Sci U S A 112 (9):2806–2811
150. Novikova PY, Tsuchimatsu T, Simon S, Nizhynska V, Voronin V, Burns R, Fedorenko OM, Holm S, Sall T, Prat E, Marande W, Castric V, Nordborg M (2017) Genome sequencing reveals the origin of the allotetraploid Arabidopsis suecica. Mol Biol Evol 34 (4):957–968
151. Szitenberg A, Salazar-Jaramillo L, Blok VC, Laetsch DR, Joseph S, Williamson VM, Blaxter ML, Lunt DH (2017) Comparative genomics of apomictic root-knot nematodes: hybridization, ploidy, and dynamic genome change. Genome Biol Evol 9(10):2844–2861
152. Svensson O, Smith A, Garcia-Alonso J, van Oosterhout C (2016) Hybridization generates a hopeful monster: a hermaphroditic selfing cichlid. R Soc Open Sci 3(3):150684
153. Beck JB, Alexander PJ, Allphin L, Al-Shehbaz IA, Rushworth C, Bailey CD, Windham MD (2012) Does hybridization drive the transition to asexuality in diploid Boechera? Evolution 66(4):985–995
154. Janko K, Kotusz J, De Gelas K, Slechtova V, Opoldusova Z, Drozd P, Choleva L, Popiolek M, Balaz M (2012) Dynamic formation of asexual diploid and polyploid lineages: multilocus analysis of Cobitis reveals the mechanisms maintaining the diversity of clones. PLoS One 7(9):e45384
155. Lampert KP, Schartl M (2008) The origin and evolution of a unisexual hybrid: Poecilia formosa. Philos Trans R Soc Lond B Biol Sci 363(1505):2901–2909
156. Moon CD, Craven KD, Leuchtmann A, Clement SL, Schardl CL (2004) Prevalence of interspecific hybrids amongst asexual fungal endophytes of grasses. Mol Ecol 13 (6):1455–1467
157. Blanc-Mathieu R, Perfus-Barbeoch L, Aury J-M, Da Rocha M, Gouzy J, Sallet E, Martin-Jimenez C, Bailly-Bechet M, Castagnone-Sereno P, Flot J-F, Kozlowski DK, Cazareth J, Couloux A, Da Silva C, Guy J, Kim-Jo Y-J, Rancurel C, Schiex T, Abad P, Wincker P, Danchin EGJ (2017) Hybridization and polyploidy enable genomic plasticity without sex in the most devastating plant-parasitic nematodes. PLoS Genet 13 (6):e1006777
158. Trivers R, Burt A, Palestis BG (2004) B chromosomes and genome size in flowering plants. Genome 47(1):1–8
159. Whitney KD, Baack EJ, Hamrick JL, Godt MJ, Barringer BC, Bennett MD, Eckert CG, Goodwillie C, Kalisz S, Leitch IJ, Ross-Ibarra J (2010) A role for nonadaptive processes in plant genome size evolution? Evolution 64 (7):2097–2109
160. Agren JA, Greiner S, Johnson MT, Wright SI (2015) No evidence that sex and transposable elements drive genome size variation in evening primroses. Evolution 69(4):1053–1062
161. Albach DC, Greilhuber J (2004) Genome size variation and evolution in *Veronica*. Ann Bot 94(6):897–911
162. Wright S, Ness RW, Foxe JP, Barrett SC (2008) Genomic consequences of outcrossing and selfing in plants. Int J Plant Sci 169 (1):105–118
163. Fierst JL, Willis JH, Thomas CG, Wang W, Reynolds RM, Ahearne TE, Cutter AD, Phillips PC (2015) Reproductive mode and the evolution of genome size and structure in Caenorhabditis nematodes. PLoS Genet 11 (6):e1005323
164. Ritland K (2002) Extensions of models for the estimation of mating systems using n independent loci. Heredity 88(4):221–228
165. Ritland K, Jain S (1981) A model for the estimation of outcrossing rate and gene-frequencies using N independent loci. Heredity 47(1):35–52
166. Koelling VA, Monnahan PJ, Kelly JK (2012) A Bayesian method for the joint estimation of outcrossing rate and inbreeding depression. Heredity (Edinb) 109(6):393–400
167. David P, Pujol B, Viard F, Castella V, Goudet J (2007) Reliable selfing rate estimates from imperfect population genetic data. Mol Ecol 16(12):2474–2487

168. Redelings BD, Kumagai S, Tatarenkov A, Wang L, Sakai AK, Weller SG, Culley TM, Avise JC, Uyenoyama MK (2015) A Bayesian approach to inferring rates of selfing and locus-specific mutation. Genetics 201 (3):1171–1188
169. McClure NS, Whitlock MC (2012) Multilocus estimation of selfing and its heritability. Heredity (Edinb) 109(3):173–179
170. Burkli A, Sieber N, Seppala K, Jokela J (2017) Comparing direct and indirect selfing rate estimates: when are population-structure estimates reliable? Heredity (Edinb) 118 (6):525–533
171. Gao H, Williamson S, Bustamante CD (2007) A Markov chain Monte Carlo approach for joint inference of population structure and inbreeding rates from multilocus genotype data. Genetics 176 (3):1635–1651
172. Halkett F, Simon JC, Balloux F (2005) Tackling the population genetics of clonal and partially clonal organisms. Trends Ecol Evol 20(4):194–201
173. Bailleul D, Stoeckel S, Arnaud-Haond S (2016) RClone: a package to identify Multi-Locus Clonal Lineages and handle clonal data sets in R. Methods Ecol Evol 7:966–970
174. Bomblies K, Yant L, Laitinen RA, Kim ST, Hollister JD, Warthmann N, Fitz J, Weigel D (2010) Local-scale patterns of genetic variability, outcrossing, and spatial structure in natural stands of *Arabidopsis thaliana*. PLoS Genet 6(3):e1000890
175. Tsai IJ, Bensasson D, Burt A, Koufopanou V (2008) Population genomics of the wild yeast Saccharomyces paradoxus: quantifying the life cycle. Proc Natl Acad Sci U S A 105 (12):4957–4962
176. Ruderfer DM, Pratt SC, Seidel HS, Kruglyak L (2006) Population genomic analysis of outcrossing and recombination in yeast. Nat Genet 38(9):1077–1081
177. Grimsley N, Pequin B, Bachy C, Moreau H, Piganeau G (2010) Cryptic sex in the smallest eukaryotic marine green alga. Mol Biol Evol 27(1):47–54
178. Derelle E, Ferraz C, Rombauts S, Rouze P, Worden AZ, Robbens S, Partensky F, Degroeve S, Echeynie S, Cooke R, Saeys Y, Wuyts J, Jabbari K, Bowler C, Panaud O, Piegu B, Ball SG, Ral JP, Bouget FY, Piganeau G, De Baets B, Picard A, Delseny M, Demaille J, Van de Peer Y, Moreau H (2006) Genome analysis of the smallest free-living eukaryote *Ostreococcus tauri* unveils many unique features. Proc Natl Acad Sci U S A 103(31):11647–11652
179. Birky CW (1996) Heterozygosity, heteromorphy, and phylogenetic trees in asexual eukaryotes. Genetics 144(1):427–437
180. Welch DM, Meselson M (2000) Evidence for the evolution of bdelloid rotifers without sexual reproduction or genetic exchange. Science 288(5469):1211–1215
181. Signorovitch A, Hur J, Gladyshev E, Meselson M (2015) Allele sharing and evidence for sexuality in a mitochondrial clade of bdelloid rotifers. Genetics 200(2):581–590
182. Debortoli N, Li X, Eyres I, Fontaneto D, Hespeels B, Tang CQ, Flot JF, Van Doninck K (2016) Genetic exchange among bdelloid rotifers is more likely due to horizontal gene transfer than to meiotic sex. Curr Biol 26 (6):723–732
183. Takebayashi N, Morrell PL (2001) Is self-fertilization an evolutionary dead end? Revisiting an old hypothesis with genetic theories and a macroevolutionary approach. Am J Bot 88(7):1143–1150
184. Goldberg EE, Igic B (2008) On phylogenetic tests of irreversible evolution. Evolution 62 (11):2727–2741
185. Schaefer I, Domes K, Heethoff M, Schneider K, Schon I, Norton RA, Scheu S, Maraun M (2006) No evidence for the 'Meselson effect' in parthenogenetic oribatid mites (Oribatida, Acari). J Evol Biol 19 (1):184–193
186. Schon I, Martens K (2003) No slave to sex. Proc R Soc Lond B 270(1517):827–833
187. Bechsgaard JS, Castric V, Charlesworth D, Vekemans X, Schierup MH (2006) The transition to self-compatibility in *Arabidopsis thaliana* and evolution within S-haplotypes over 10 Myr. Mol Biol Evol 23 (9):1741–1750
188. Tang C, Toomajian C, Sherman-Broyles S, Plagnol V, Guo YL, Hu TT, Clark RM, Nasrallah JB, Weigel D, Nordborg M (2007) The evolution of selfing in *Arabidopsis thaliana*. Science 317(5841):1070–1072
189. Shimizu KK, Tsuchimatsu T (2015) Evolution of selfing: recurrent patterns in molecular adaptation. Annu Rev Ecol Syst 46:593–622
190. Castric V, Billiard S, Vekemans X (2014) Trait transitions in explicit ecological and genomic contexts: plant mating systems as case studies. Adv Exp Med Biol 781:7–36
191. Tsuchimatsu T, Goubet PM, Gallina S, Holl AC, Fobis-Loisy I, Berges H, Marande W, Prat E, Meng D, Long Q, Platzer A, Nordborg M, Vekemans X, Castric V (2017) Patterns of polymorphism at the self-incompatibility locus in 1,083 Arabidopsis

thaliana genomes. Mol Biol Evol 34 (8):1878–1889

192. Durvasula A, Fulgione A, Gutaker RM, Alacakaptan SI, Flood PJ, Neto C, Tsuchimatsu T, Burbano HA, Pico FX, Alonso-Blanco C, Hancock AM (2017) African genomes illuminate the early history and transition to selfing in Arabidopsis thaliana. Proc Natl Acad Sci U S A 114 (20):5213–5218
193. Foxe JP, Slotte T, Stahl EA, Neuffer B, Hurka H, Wright SI (2009) Recent speciation associated with the evolution of selfing in *Capsella*. Proc Natl Acad Sci U S A 106 (13):5241–5245
194. Guo YL, Bechsgaard JS, Slotte T, Neuffer B, Lascoux M, Weigel D, Schierup MH (2009) Recent speciation of *Capsella rubella* from *Capsella grandiflora*, associated with loss of self-incompatibility and an extreme bottleneck. Proc Natl Acad Sci U S A 106 (13):5246–5251
195. Brandvain Y, Slotte T, Hazzouri KM, Wright SI, Coop G (2013) Genomic identification of founding haplotypes reveal the history of the selfing species *Capsella rubella*. PLoS Genet 9 (9):e1003754
196. Judson OP, Normark BB (1996) Ancient asexual scandals. Trends Ecol Evol 11 (2):41–46
197. Igic B, Bohs L, Kohn JR (2006) Ancient polymorphism reveals unidirectional breeding system shifts. Proc Natl Acad Sci U S A 103 (5):1359–1363
198. Ferrer MM, Good-Avila SV (2007) Macrophylogenetic analyses of the gain and loss of self-incompatibility in the Asteraceae. New Phytol 173(2):401–414
199. Janko K, Drozd P, Flegr J, Pannell JR (2008) Clonal turnover versus clonal decay: a null model for observed patterns of asexual longevity, diversity and distribution. Evolution 62(5):1264–1270
200. Janko K (2014) Let us not be unfair to asexuals: their ephemerality may be explained by neutral models without invoking any evolutionary constraints of asexuality. Evolution 68 (2):569–576
201. Schwander T, Crespi BJ (2009) Twigs on the tree of life? Neutral and selective models for integrating macroevolutionary patterns with microevolutionary processes in the analysis of asexuality. Mol Ecol 18(1):28–42
202. Goldberg EE, Kohn JR, Lande R, Robertson KA, Smith SA, Igic B (2010) Species selection maintains self-incompatibility. Science 330 (6003):493–495
203. Fitzjohn RG, Maddison WP, Otto SP (2009) Estimating trait-dependent speciation and extinction rates from incompletely resolved phylogenies. Syst Biol 58(6):595–611
204. Maddison WP, Midford PE, Otto SP (2007) Estimating a binary character's effect on speciation and extinction. Syst Biol 56 (5):701–710
205. de Vos JM, Hughes CE, Schneeweiss GM, Moore BR, Conti E (2014) Heterostyly accelerates diversification via reduced extinction in primroses. Proc Biol Sci 281 (1784):20140075
206. Johnson MT, Fitzjohn RG, Smith SD, Rausher MD, Otto SP (2011) Loss of sexual recombination and segregation is associated with increased diversification in evening primroses. Evolution 65(11):3230–3240
207. Glémin S (2003) How are deleterious mutations purged? Drift versus nonrandom mating. Evolution 57(12):2678–2687
208. Tahir D, Glémin S, Lascoux M, Kaj I (2019) Modeling a trait-dependent diversification process coupled with molecular evolution on a random species tree. J Theor Biol 461:189–203
209. Dalrymple RL, Buswell JM, Moles AT (2015) Asexual plants change just as often and just as fast as do sexual plants when introduced to a new range. Oikos 124(2):196–205
210. Uecker H (2017) Evolutionary rescue in randomly mating, selfing, and clonal populations. Evolution 71(4):845–858
211. Richards AJ (1997) Plant breeding systems, 2nd edn. Chapman & Hall Ltd, London
212. Igic B, Kohn JR (2006) The distribution of plant mating systems: study bias against obligately outcrossing species. Evolution 60 (5):1098–1103
213. Wright SI, Kalisz S, Slotte T (2013) Evolutionary consequences of self-fertilization in plants. Proc Biol Sci 280(1760):20130133
214. Fontaneto D, Barraclough TG (2015) Do species exist in asexuals? Theory and evidence from bdelloid rotifers. Integr Comp Biol 55 (2):253–263
215. Flot JF, Hespeels B, Li X, Noel B, Arkhipova I, Danchin EGJ, Hejnol A, Henrissat B, Koszul R, Aury JM, Barbe V, Barthelemy RM, Bast J, Bazykin GA, Chabrol O, Couloux A, Da Rocha M, Da Silva C, Gladyshev E, Gouret P, Hallatschek O, Hecox-Lea B, Labadie K, Lejeune B, Piskurek O, Poulain J, Rodriguez F, Ryan JF, Vakhrusheva OA, Wajnberg E, Wirth B, Yushenova I, Kellis M, Kondrashov AS, Welch DBM, Pontarotti P,

Weissenbach J, Wincker P, Jaillon O, Van Doninck K (2013) Genomic evidence for ameiotic evolution in the bdelloid rotifer Adineta vaga. Nature 500(7463):453–457

216. Golczyk H, Massouh A, Greiner S (2014) Translocations of chromosome end-segments and facultative heterochromatin promote meiotic ring formation in evening primroses. Plant Cell 26(3):1280–1293
217. Barraclough TG, Fontaneto D, Ricci C, Herniou EA (2007) Evidence for inefficient selection against deleterious mutations in cytochrome oxidase I of asexual bdelloid rotifers. Mol Biol Evol 24:1952–1962
218. Foxe JP, Dar VU, Zheng H, Nordborg M, Gaut BS et al (2008) Selection on amino acid substitutions in Arabidopsis. MBE 25:1375–1383
219. Gioti A, Stajich J, Johannesson H (2013) Neurospora and the dead-end hypothesis: genomic consequences of selfing in the model genus. Evolution 67(12):3600–3616
220. Mark Welch DB, Meselson MS (2001) Rates of nucleotide substitution in sexual and anciently asexual rotifers. Proc Natl Acad Sci USA 98:6720–6724
221. Ness RW, Siol M, Barrett SC (2012) Genomic consequences of transitions from cross- to self-fertilization on the efficacy of selection in three independently derived selfing plants. BMC Genomics 13:611
222. Nygren K, Strandberg R, Wallberg A, Nabholz B, Gustafsson T et al (2011) A comprehensive phylogeny of Neurospora reveals a link between reproductive mode and molecular evolution in fungi. Mol Phylogenet Evol 59:649–663
223. Qiu S, Zeng K, Slotte T, Wright S, Charlesworth D (2011) Reduced efficacy of natural selection on codon usage bias in selfing Arabidopsis and Capsella species. Genome Biol Evol 3:868–880
224. Slotte T, Hazzouri KM, Agren JA, Koenig D, Maumus F et al (2013) The Capsella rubella genome and the genomic consequences of rapid mating system evolution. Nat Genet 45:831–835
225. Brandvain Y, Slotte T, Hazzouri KM, Wright SI, Coop G (2013) Genomic Identification of Founding Haplotypes Reveals the History of the Selfing Species Capsella rubella. PLOS Genetics 9(9): e1003754
226. Whittle CA, Sun Y, Johannesson H (2011) Evolution of synonymous codon usage in Neurospora tetrasperma and Neurospora discreta. Genome Biol Evol 3:332–343
227. Bast J, Schaefer I, Schwander T, Maraun M, Scheu S, Kraaijeveld K (2016) No accumulation of transposable elements in asexual arthropods. Mol Biol Evol 33:697–706
228. Docking TR, Saade FE, Elliott MC, Schoen DJ (2006) Retrotransposon sequence variation in four asexual plant species. J Mol Evol 62:375–387
229. Dolgin ES, Charlesworth B, Cutter AD (2008) Population frequencies of transposable elements in selfing and outcrossing Caenorhabditis nematodes. Genet Res 90:317–329
230. Goodwin TJ, Poulter RT (2008) Multiple LTR-Retrotransposon families in the asexual yeast Candida albicans. Genome Res 10:174–191
231. Jiang X, Tang H, Ye Z, Lynch M (2017) Insertion polymorphisms of mobile genetic elements in sexual and asexual populations of Daphnia pulex. Genome Biol Evol 9:362–374
232. Kraaijeveld K, Zwanenburg B, Hubert B, Vieira C, De Pater S, Alphen V et al (2012) Transposon proliferation in an asexual parasitoid. Mol Ecol 21:3898–3906
233. Schaack S, Choi E, Lynch M, Pritham EJ (2010) DNA transposons and the role of recombination in mutation accumulation in Daphnia pulex. Genome Biol 11:R46
234. Szitenberg A, Cha S, Opperman CH, Bird DM, Blaxter ML, Lunt DH (2016) Genetic drift, not life history or RNAi, determine long-term evolution of transposable elements. Genome Biol Evol 8:2964–2978
235. Zeyl C, Bell G, Da Silva J (1994) Transposon abundance in sexual and asexual populations of Chlamydomonas reinhardtii. Evolution 48:1406–1409

Part IV

Natural Selection and Innovation in Genomic Sequences

Chapter 12

Selection Acting on Genomes

Carolin Kosiol and Maria Anisimova

Abstract

Populations evolve as mutations arise in individual organisms and, through hereditary transmission, may become "fixed" (shared by all individuals) in the population. Most mutations are lethal or have negative fitness consequences for the organism. Others have essentially no effect on organismal fitness and can become fixed through the neutral stochastic process known as random drift. However, mutations may also produce a selective advantage that boosts their chances of reaching fixation. Regions of genomes where new mutations are beneficial, rather than neutral or deleterious, tend to evolve more rapidly due to positive selection. Genes involved in immunity and defense are a well-known example; rapid evolution in these genes presumably occurs because new mutations help organisms to prevail in evolutionary "arms races" with pathogens. In recent years genome-wide scans for selection have enlarged our understanding of the genome evolution of various species. In this chapter, we will focus on methods to detect selection on the genome. In particular, we will discuss probabilistic models and how they have changed with the advent of new genome-wide data now available.

Key words Conserved and accelerated regions, Positive selection scans, Codon models, Selection-mutation models, Polymorphism-aware phylogenetic models

1 Introduction

In the past selection studies mainly focused on the analysis of particular loci such as genes, proteins, or regular elements of interest. With the availability of comparative genomic data, the emphasis has shifted from the study of individual proteins to genome-wide scans for selection.

The search for selection can be performed on different levels comparing homologous nucleotide sequences or protein-coding genes in one or multiple genomes. The evolutionary processes in all these levels can be described by probabilistic models, which set the basis for evaluating selective pressures and selection tests. This book chapter will give an introduction into fundamental properties of the probabilistic models used to detect selection in the Subheading 3 as well as examples of genome-wide scans.

Maria Anisimova (ed.), *Evolutionary Genomics: Statistical and Computational Methods*, Methods in Molecular Biology, vol. 1910, https://doi.org/10.1007/978-1-4939-9074-0_12, © The Author(s) 2019

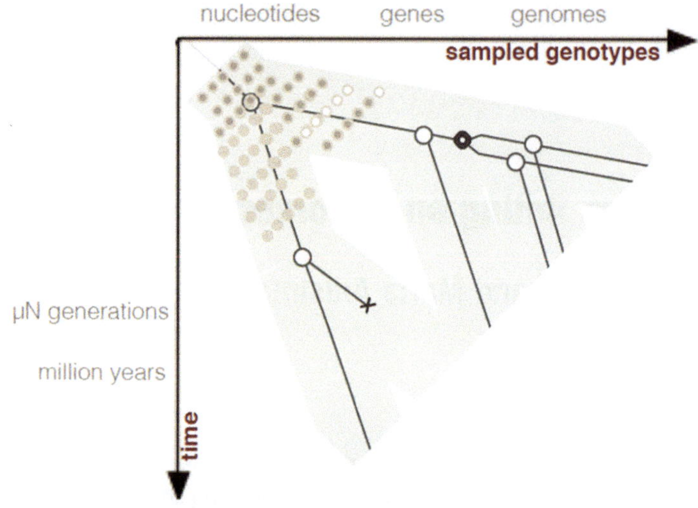

Fig. 1 A diagram illustrating the different data and levels to analyze genomic sequences and the relationship of the various approaches modeling selection

In Fig. 1, we summarize the different data levels and time scales of modeling selection on genomes.

2 Comparative Genome Data

Several whole genome sequence data sets are now available for selection scans. Mammalian genomes are well represented (in particular primates), and insect genomes are becoming more numerous (in particular *Drosophila*). These data can be downloaded as orthologous alignments from the Ensembl [1] and UCSC [2] browsers.

In light of recent advances in DNA sequencing, with so-called next generation sequencing (NGS) technologies that have dramatically reduced the cost and time needed to sequence an organism's entire genome, large-scale (involving many organisms) sequencing projects have been and are currently being undertaken. Just to name a few, genome projects re-sequencing 1000 *D. melanogaster* [3] and 1001 *Arabidopsis* [4] were accomplished, and the 100,000 human genome project [5] is ongoing. These polymorphism data from multiple individuals from several species enable us to detect very recent selection.

Together with the progress in sequencing technologies, algorithmic advances now allow the de novo assembly of genomes from NGS data, including complex mammalian genomes (e.g., giant panda genome [6]). Therefore, not only international consortia but also small groups and individual labs can now envisage to sequence the organisms of their interest. As a consequence platforms for sharing this data have been established. For example, the Genome 10K project aims to assemble a genomic zoo—a collection of DNA sequences representing the genomes of 10,000 vertebrate species, approximately one for every vertebrate genus. All these genomes can be subject to scans for selection, for which we outline methods below.

3 Methods

3.1 Probabilistic Models for Genome Evolution

The statistical modeling of the evolutionary process is of great importance when performing selection studies. When comparing reasonably divergent sequences, counting the raw sequence identity (percentage of sites with observed changes) underestimates the amount of evolution that has occurred because, by chance alone, some sites will have incurred multiple substitutions. In this chapter we discuss maximum likelihood (ML) and Bayesian methods to detect selection based on probabilistic models of character evolution. Such substitution models provide more accurate evolutionary distance estimates by accounting for these unobserved changes and often explicitly model the selection pressures.

One of the primary assumptions made in defining probabilistic substitution models is that future evolution is only dependent on its current state and not on previous (ancestral) states. Statistical processes with this lack of memory are called Markov processes. The assumption itself is reasonable, because during the evolution mutation and natural selection can only act upon the molecules present in an organism and have no knowledge of what came previously. However, some large-scale mutational events, such as recombination [7], gene conversion (e.g., *see* [8, 9]), or horizontal transfer [10] might not satisfy this "memoryless" condition.

To reduce the complexity of evolutionary models, it is often further assumed that each site in a sequence evolves independently from all other sites. There is evidence that the independence of sites assumption is violated. In real proteins, chemical interactions between neighboring sites or the protein structure affects how other sites in the sequence change. Steps have been made toward context-dependent models, where the specific characters at neighboring sites affect the sites evolution (e.g., *see* [11, 12]).

The Markov model asserts that one sequence is derived from another by a series of independent substitutions, each changing one character in the first sequence to another character in the second during the evolution. Thereby we assume independence of evolution at different sites. A continuous-time Markov process is fully defined by its instantaneous rate matrix $Q = \{q_{ij}\}_{i,j=1 \ldots N}$.

The diagonal elements of Q are defined by a mathematical requirement that the rows sum up to zero. For multiple sequence alignments, the substitution process runs in continuous time over a tree representing phylogenetic relations between the sequences. The transition probability matrix $P(t) = \{p_{ij}(t)\} = e^{Qt}$ consists of transition probabilities from residue i to residue j over time t and is found as a solution of the differential equation $dP(t)/dt = P(t)Q$ with $P(0)$ being the identity matrix. In order for tree branches to be measured by the expected number of substitutions per site, the Q-matrix is scaled so that the average substitution rate at equilibrium equals 1.

As a matter of mathematical and computational convenience rather than biological reality, several simplifying assumptions are usually made. Standard substitution models allow any state to change into any other. Such Markov process is called *irreducible* and has a unique *stationary* distribution corresponding to the equilibrium codon frequencies $\pi = \{\pi_i\}$. *Time reversibility* implies that the direction of the change between two states i and j is indistinguishable, so that $\pi_i p_{ij}(t) = \pi_j p_{ji}(t)$. This assumption helps to reduce the number of model parameters and is convenient when calculating the matrix exponential (Q-matrix of a reversible process has only real eigenvectors and eigenvalues [13]). Fully unrestrained Q-matrix for N characters defines an irreversible model with $N(N-1)-1$ free parameters, while for a reversible process this number is $N(N+1)/2-2$.

By comparing how well substitution models explain sequence evolution, and by examining the parameters estimated from data, ML and Bayesian inference can be used to address many biologically important questions. In this section we focus on probabilistic models that are used to detect selection.

3.2 Detecting Regions of Accelerated Genome Evolution

Understanding the forces shaping the evolution of specific lineages is one of the most exciting areas in evolutionary genomics. In particular, regions of accelerated evolution in mammalian and insect species have been studied (e.g., *see* [14]). To eliminate non-functional regions, one strategy is to begin with a search for regions that are conserved through the mammalian history or longer. A likelihood ratio test (LRT) may be used to detect acceleration of rates in a lineage of interest, for example, the human lineage. Such LRT compares the likelihood of the alignment data under two probabilistic models. The null model has a single scale parameter representing shortening (more conserved) and lengthening (less conserved) of all branches of the tree. The alternative model has an additional parameter for the human lineage, which is constraint to be ≥ 1. This extra parameter allows the human branch to be relatively longer (accelerated) than the branches in the rest of the tree.

For example, this approach was used to identify genomic regions that are conserved in most vertebrates but have evolved rapidly in humans. Interestingly, the majority of the human accelerated regions (HARs) were noncoding, and many were located near protein-coding genes with protein functions related to the nervous system [14].

In contrast, the majority of *Drosophila melanogaster* accelerated regions (DMARs) are found in protein-coding regions and primarily result from rapid adaptive change at synonymous sites [15]. This could be because flies have much more compact genomes compared to humans; however, even after considering the genomic content, in *Drosophila* a significant excess of DMARs occur in protein-coding regions. Furthermore, Holloway and colleagues observed a mutational bias from G|C to A|T, and therefore the accelerated divergence in DMARs might be attributed to a shift in codon usage and a fixation of many suboptimal codons.

In a similar manner, amino acid based models search for site- or lineage-specific rate accelerations and residues subject to altered functional constraints. Such sites are likely to be contributing to the change in protein function over time. The advantage of amino acid-based models is that they might be suitable for the analysis of deep divergences of fast-evolving genes, where sequences rapidly saturate over time. Also amino acid methods are not influenced by the effects of codon bias, a topic that is discussed at the end of this chapter. The idea is that adaptive change on the level of amino acid sequences may not necessarily correspond to an adaptive change in protein function but rather to peaks in the protein adaptive landscape reflecting the optimization of the protein function in a particular species to long-term environmental changes. One class of methods for detecting functional divergence searches for a lineage-specific change in the shape parameter of the gamma distribution that is used to model rate heterogeneity (*see* [16–19]). Other methods search for evidence of clade-specific rate shifts at individual sites (*see* [20–26]). For example, Gu [21] proposed a simple stochastic model for estimating the degree of divergence between two pre-specified clusters. The statistical significance was tested using site-specific profiles based on a hidden Markov model, which was used to identify amino acids responsible for these functional differences between two gene clusters. More flexible evolutionary models were incorporated in the maximum likelihood approach applicable to the simultaneous analysis of several gene clusters [27]. This was extended [28] to evaluate site-specific shifts in amino acid properties, in comparison with site-specific rate shifts. Pupko and Galtier [24] used the LRT to compare ML estimates of the replacement rate at an amino acid site in distinct subtrees.

3.3 Codon Models: Site, Branch, and Branch-Site Specificity

3.3.1 Basic Codon Models

In protein-coding sequences, nucleotide sites at different codon positions usually evolve with highly heterogeneous patterns (e.g., [29]). Thus DNA substitution models fail to account for this heterogeneity unless the sequences are partitioned by codon positions for the analysis. But even then, DNA models do not model the structure of genetic code or selection at the protein level. Indeed, one advantage of studying protein-coding sequences at the codon level is the ability to distinguish between nonsynonymous (AA replacing) and synonymous (silent) codon changes. Based on this distinction, the selective pressure on the protein-coding level can be measured by the ratio $\omega = d_N/d_S$ of the nonsynonymous to synonymous substitution rates. The nonsynonymous substitution rate may be higher than the synonymous rate, and thus $\omega > 1$ due to fitness advantages associated with recurrent AA changes in the protein, i.e., positive selection on the protein. In contrast, purifying selection acts to preserve the protein sequence, so that the nonsynonymous substitution rate is lower than the synonymous rate, with $\omega < 1$. Neutrally evolving sequences exhibit similar nonsynonymous and synonymous rates, with $\omega \approx 1$.

First methods that used the ω ratio as a criterion to detect positive selection were based on pairwise estimation of d_N and d_S rates with "counting" methods (e.g., *see* [30]). However, ML estimates of pairwise d_N and d_S based on a codon model were shown to outperform all other approaches [31]. Moreover, a Markov codon model is naturally extended to multiple sequence alignments, unlike the counting methods. This, together with the benefits of the probabilistic framework within which codon models are defined, made codon models very popular in studies of positive selection in protein-coding genes.

The first two codon models were proposed simultaneously in the same issue of Molecular Biology and Evolution [32, 33]. The model of Goldman and Yang [32] included the transition/transversion rate ratio κ, and modeled the selective effect indirectly using a multiplicative factor based on Grantham [34] distances, but was later simplified to estimate the selective pressure explicitly using the ω parameter [35]. The main distinction between the first codon models concerns the way to describe the instantaneous rates with respect to equilibrium frequencies: (1) proportional to the equilibrium frequency of a target codon (as in Goldman and Yang [32]) or (2) proportional to the frequency of a target nucleotide (as in Muse and Gaut [33]).

In 2006, empirical codon models have been estimated (*see* [36, 37]) that summarize substitution patterns from large quantities of protein-coding gene families. In contrast to the parametric codon models that estimate gene-specific parameters (e.g., transition-transversion κ, selective pressure ω, etc.), the empirical codon models do not explicitly consider distinct factors that shape protein evolution. Standard parametric models assume that protein evolution proceeds only by successive single-nucleotide substitutions. However, empirical codon models indicate that model accuracy is significantly improved by incorporating instantaneous doublet and triplet changes. Kosiol et al. [37] also found that the affiliations between codon, the amino acid it encodes, and the physicochemical properties of the amino acid are main driving factors of the process of codon evolution. Neither multiple nucleotide changes nor the strong influence of the genetic code nor amino acid properties form a part of the standard parametric models.

On the other hand, parametric models have been very successful in applications studying biological forces shaping protein evolution of individual genes. Thus combining the advantages of parametric and empirical approaches offers a promising direction. Kosiol, Holmes, and Goldman [37] explored a number of combined codon models that incorporated empirical AA exchangeabilities from ECM while using parameters to study selective pressure, transition/transversion biases, and codon frequencies. Similarly, AA exchangeabilities from (suitable) empirical AA matrices may be used to alter probabilities of nonsynonymous changes, together with traditional parameters ω, κ, and codon frequencies π_j [38]. In 2013, De Maio et al. [39] extended the ECM approach to accommodate site-specific variation of selective pressure and lineage-specific variation. Simulations showed that ECMs allowing for double and triple mutations is more conservative: they reduce the number of false positives and have less power to detect positive selection [39].

3.3.2 Accounting for Variability of Selective Pressures

First codon models assumed constant nonsynonymous and synonymous rates among sites and over time. Although most proteins evolve under purifying selection most of the time, positive selection may drive the evolution in some lineages. During episodes of adaptive evolution, only a small fraction of sites in the protein have the capacity to increase the fitness of the protein via AA replacements. Thus approaches assuming constant selective pressure over time and over sites lack power in detecting genes affected by positive selection. Consequently, various scenarios of variation in selective pressure were incorporated in codon models, making them more powerful at detecting positive selection, and short episodes of adaptive evolution in particular. Evidence of positive selection on a gene can be obtained by a LRT comparing two nested models: a model that does not allow positive selection (constraining $\omega \leq 1$ to represent the null hypothesis) and a model that allows positive selection ($\omega > 1$ is allowed in the alternative hypothesis). Positive selection is detected if a model $\omega > 1$ fits data significantly better compared to the model restricting $\omega \leq 1$ at all sites and lineages. However, the asymptotic null distribution may vary from the standard χ^2 due to boundary problems or if some parameters become not estimable (e.g., *see* [40, 41]).

3.3.3 Case Study: Application of a Genome-Wide Scan of Positive Selection on Six Mammalian Genomes

In 2006, six high-coverage genome assemblies became available for eutherian mammals. The increased phylogenetic depth of this data set permitted Kosiol and colleagues [42] to perform several new lineage- and clade-specific tests using branch-site codon models. Of ~16,500 human genes with high-confidence orthologs in at least two other species, 544 genes showed significant evidence of positive selection using branch-site codon models and standard LRTs.

Interestingly, several pathways were found to be strongly enriched in genes with positive selection, suggesting possible coevolution of interacting genes. A striking example is the complement immunity system, a biochemical cascade responsible for the elimination of pathogens. This system consists of several small proteins found in the blood that cooperate to kill target cells by disrupting their plasma membranes. Of 78 genes associated with this pathway in KEGG (*see* http://www.genome.jp/kegg-bin/show_pathway?map04610 for the complement cascades), nine were under positive selection (FDR < 0.05), and five others had nominal $P < 0.05$. Most of genes under positive selection are inhibitors (DAF, CFH, CFI) and receptors (C5AR1, CR2), but some are part of the membrane attack complex (C7, C9, C8B), which punctures cell membranes to initiate cell lysis. Here we focus on the analysis of these proteins of the membrane attack complex.

First we calculate gene averaged ω value using the basic M0 model [32]. The ML estimates of $\omega < 1$ ($\omega = 0.31$ for C7, $\omega = 0.25$ for C8B, and $\omega = 0.44$ for C9) indicate that most sites in these genes are under purifying selection. However, selection pressure could be variable at different locations of the membrane proteins, and we therefore continue our analysis by applying models that allow for variation in selective pressure across sites.

3.3.4 Selective Variability Among Codons: Site Models

The simplest site models use the general discrete distribution with a pre-specified number of site classes. Each site class i has an independent parameter ω_i estimated by ML together with proportions of sites p_i in each class. Since a large number of site categories require many parameters, three categories are usually used (requiring five independent parameters). To test for positive selection, several pairs of nested site models were defined to represent the null and alternative hypotheses in LRTs. For example, model M1a includes two site classes, one with $\omega_0 < 1$ and another with $\omega_1 = 1$, representing the neutral model of evolution (the null hypothesis). The alternative model M2a extends M1a by adding an extra site class with $\omega_2 \geq 1$ to accommodate sites evolving under positive selection. Significance of the LRT is tested using the χ^2_2-distribution for the M1 vs. M2 comparison. We test the C7 gene for positive selection by the LRT comparing nested models M1a and M2a (Table 1).

Model M2a has two additional parameters compared to model M1a. The resulting LRT statistic is 2(log L2 − log L1) = 2 (−6377.35 − (−6369.67)) = 2 × 7.68 = 15.36. This is much greater than the critical value of the chi-square distribution χ^2 ($df = 2$, at 5%) = 5.99, and we calculate a p-value of $P = 5.0\text{e}{-04}$. However, the M1a vs. M2a comparison for genes C8B and C9 is not significant.

Table 1
Parameter estimates and log-likelihoods for a LRT of positive selection for the complement immunity component C7

M1a (neutral)			
Site class	0	1	
Proportion	$p_0 = 0.69$	$(p_1 = 1 - p_0 = 0.31)$	
ω ratio	$\omega_0 = 0.07$	$(\omega_1 = 1)$	
Log-likelihood L1 = −6377.35			
M2a (selection)			
Site class	0	1	2
Proportion	$p_0 = 0.70$	$p_1 = 0.29$	$(p_2 = 1 - p_0 - p_1 = 0.01)$
ω ratio	$\omega_0 = 0.08$	$(\omega_1 = 1)$	$\omega_2 = 10.89$
Log-likelihood L2 = −6369.67			

The model M2a is the alternative model with a class of sites with $\omega_2 \geq 1$. The null hypothesis M1a is the same model but with $\omega_2 = 1$ fixed

Another LRT can be performed on the basis of the modified model M8 with two site classes: one with sites where the ω ratio is drawn from the beta distribution (with $0 \leq \omega \leq 1$ describing the neutral scenario) and the second, discrete class, with $\omega \geq 1$. Constraining $\omega = 1$ for this second class provides a sufficiently flexible null hypothesis, whereby all evolution can be explained by sites with ω from the beta distribution or from a discrete site class with $\omega = 1$. Significance of the LRT is tested the mixture $\frac{1}{2}\chi_0^2 + \frac{1}{2}\chi_1^2$ for the M8 ($\omega = 1$) vs. M8 comparison. If the LRT for positive selection is found to be significant, specific sites under positive selection may be predicted based on the values of posterior probabilities (PP) to belong to the site class under positive selection (usually PP > 0.95, but *see* [43, 44]). Such posterior probabilities are estimated using the naïve empirical Bayesian approach (NEB, [45]), full hierarchical Bayesian approach ([46]; BEB [44]), or a mid-way approach – the Bayes empirical Bayes (BEB [44]). For a discussion on these approaches, *see* Scheffler and Seoighe [47] and Aris-Brosou [48]. Alternatively, Massingham and Goldman [49] proposed a site-wise likelihood ratio estimation to detect sites under purifying or positive selection.

For the C7 gene, using BEB we identified several amino acids sites to be putatively under selection: residue R at position 223 (PP = 0.94), H at position 239 (PP = 0.93), and N at position 331 (PP = 0.93). Unfortunately, the crystal structures of C7 (as well as C8B and C9) are not known, and we cannot relate the location of amino acids in the protein sequence to relevant 3D data, such as sites of protein-protein interaction or binding sites of the protein. If such structural information were known, it would also be possible to use this biological knowledge in a model that is aware of the position of the different structural elements.

Site models that do not use a priori partitioning of codons (as those described above) are known as random-effect (RE) models. In contrast, fixed-effect (FE) models categorize sites based on a prior knowledge, e.g., according to tertiary structure for single genes, or by gene category for multigene data. Site partitions for FE models can be defined also based on inferred recombination breakpoints, useful for inferences of positive selection from recombining sequences (*see* [50, 51]); although the uncertainty of breakpoint inference is ignored in this way. FE models with each site being a partition should be avoided, as they lead to the "infinitely many parameter trap" (e.g., *see* [52]). Given a biologically meaningful a priori partitioning, FE models are useful to study heterogeneity among partitions. However, a priori information is not always available.

3.3.5 Selective Variability over Time: Branch Models

A simple way to include the variation of the selective pressure over time is by using separate parameters ω for each branch of a phylogeny (known as *free-ratio* model; [35]). Compared with the *one-ratio* model (which assumes constant selection over time), the free-ratio model requires additional $2T - 4$ ω parameters for T species. Figure 2 shows the estimates of the free-ratio model for the C8B gene. Although the ML estimates of ω values on the rodent lineages are visibly higher than on the primate lineages, none of the branches has $\omega > 1$.

Other branch models can be defined by constraining different sets of branches of a tree to have an individual ω. LRTs are used to decide (1) whether selective pressure is significantly different on a pre-specified set of branches and (2) whether these branches are under positive selection.

However, branch models have relatively poor power to detect selection [53] in comparison to branch-site models that are discussed in the next section. Also note that testing of multiple hypotheses on the same data requires a correction, so the overall false-positive rate is kept at the required level (most often 5%). Correction for multiple testing further reduces the power of the method, especially when many hypotheses are tested simultaneously (*see* Subheading 4 later).

3.3.6 Temporal and Spatial Variation of Selective Pressure

Several solutions were proposed to simultaneously account for differences in selective constraints among codons and the episodic nature of molecular evolution at individual sites. One of the first models—model MA [45]—assumes four site classes. Two classes contain sites evolving constantly over time: one under purifying selection with $\omega_0 < 1$; another with $\omega_1 = 1$. The other two site

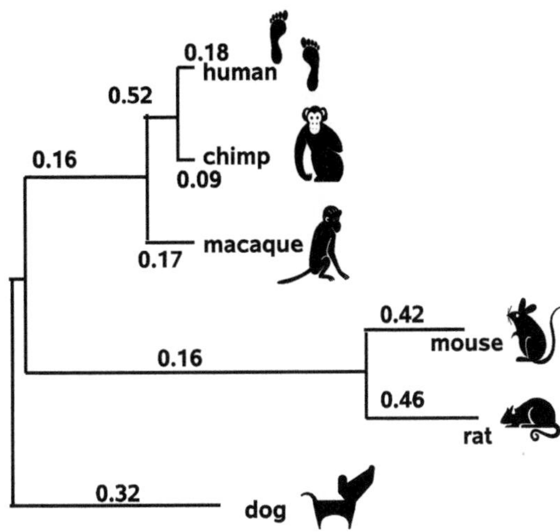

Fig. 2 An estimate of ω for each branch of a six-species phylogeny. Shown is the maximum likelihood estimate for the gene 8B. Each branch is labeled with the corresponding estimate of ω

classes allow selective pressure at a site to change over time on a pre-specified set of branches, known as *the foreground*. The two variable classes are derived from the constant classes so that sites typically evolving with $\omega_0 < 1$ or $\omega_1 = 1$ are allowed to be under positive selection with $\omega_2 \geq 1$ on the foreground. Testing for positive selection on the rodent clade involves a LRT comparing a constrained version of MA (with $\omega_2 = 1$) vs. an unconstrained MA model. Compared to branch models, the branch-site formulation improves the chance of detecting short spills of adaptive pressure in the past even if these occurred at a small fraction of sites.

Returning to our example of gene C8B of the complement pathway, we perform a branch-site LRT for positive selection using the M1a vs. M2a comparison. Thereby we take mouse and the rat lineage, respectively, as foreground branches and all other branches as background branches. Significance of the LRT is tested the mixture $\frac{1}{2}\chi_0^2 + \frac{1}{2}\chi_1^2$ with critical values to be 2.71 at 5%. For the C8B gene, we calculate $2(\log L2 - \log L1) = 2 \times 2.23 = 4.46$ for the mouse lineage and 11.2 for the rate lineage, respectively.

A major drawback of described branch-site models is their reliance on a biologically viable a priori hypothesis. In context of detecting sites and lineages affected by positive selection, one possible solution is to perform multiple branch-site LRTs, each setting a different branch at the foreground [54]. In the example of six species (Fig. 2), a total of nine tests (for an unrooted tree) are necessary in the absence of an a priori hypothesis. Multiple test correction has to be applied to control excessive false inferences. This strategy tends to be conservative but can be sufficiently powerful in detecting episodic instances of adaptation. As with all model-based techniques, precautions are necessary for data with unusual heterogeneity patterns, which may cause deviations from the asymptotic null distribution and thus result in an elevated false-positive rate.

In the case of episodic selection where any combination of branches of a phylogeny can be affected, a Bayesian approach in lieu of the standard LRTs and multiple testing have been suggested. The multiple LRT approach is most concerned with controlling the false-positive rate of selection inference and is less suited to infer the best-fitting selection history. In the hypothetical example (Fig. 2), a total of $2^9 - 1 = 511$ selection histories (excluding the history without selection on any branch) need to be considered. The Bayesian analysis allows a probability distribution over possible selection histories to be computed and therefore permits estimates of prevalence of positive selection on individual branches and clades. Such approach evaluates uncertainty in selection histories using their posterior probabilities and allows robust inference of interesting parameters such as the switching probabilities for gains and losses of positive selection [42].

Other models (e.g., with d_S variation among sites [55]) may be extended to allow changes of selective regimes on different branches. This is achieved by adding further parameters, one per branch, describing the deviation of selective pressure on a branch from the average level on the whole tree under the site model. Such model is parameter-rich and can be used for exploratory purposes on data with long sequences but does not provide a robust way of testing whether $\omega > 1$ on a branch is due to positive selection on a lineage or due to inaccuracy of the ML estimation.

Kosakovsky Pond and Frost [55] suggested detecting lineage-specific variation in selective pressure using the genetic algorithm (GA)—a computational analogue of evolution by natural selection. The GA approach was successfully applied to phylogenetic reconstruction. In the context of detecting lineage-specific positive selection, GA does not require an a priori hypothesis. Instead the algorithm samples regions of the whole hypotheses space according to their "fitness" measured by AIC_C. The branch-model selection with GA may also be adapted to incorporate d_N and d_S among site variation, although this imposes a much heavier computational burden.

In branch and branch-site models, change in selection regime is always associated with nodes of a tree, but the selective pressure remains constant over the length of each branch. Guindon et al. [56] proposed a Markov-modulated model where switches of selection regimes may occur at any site and any time on the phylogeny. In a covarion-like manner, this codon model combines two Markov processes: one governs the codon substitution, while the other specifies rates of switches between selective regimes. These models can be used to study the patterns of the changes in selective pressures over time and across sites, by estimating the relative rates of changes between different selective regimes (purifying, neutral, and positive).

3.3.7 Polymorphism-Aware Phylogenetic Models

Polymorphism-aware phylogenetic models (POMOs, [57, 58]) use polymorphism and divergence data simultaneously to estimate relative mutation rates and scaled selection coefficients. Similar to DNA substitution models, the PoMo approach is based on a continuous-time Markov process to model evolution of hereditary sequences along a species tree. However, not only evolution of a single reference site but rather evolution of a population is considered.

PoMo includes polymorphisms as states of the Markov chain, in addition to the four nucleotide states of classical nucleotide models. Sequence evolution is modeled as a gradual process made by small allele frequency changes. PoMo accounts for ancestral polymorphisms and in particular for ancestral shared polymorphisms and incomplete lineage sorting (when two speciation events are separated by a lapse of time not sufficient for polymorphisms to reach fixation, *see* Maddison and Knowles [59]). The parameters in PoMo do not merely describe substitution rate but are also informative of mutation rates, fixation biases, root nucleotide frequencies, and branch lengths. All these parameters are estimated within a ML framework. De Maio et al. [57] performed a comprehensive study of evolutionary patterns of fourfold-degenerate sites in great apes populations. They show evidence in favor of variation in mutation and fixation rates between genomic regions with different base composition, contributing to the long-standing debate regarding the origin and maintenance of GC content variation (e.g., *see* Eyre-Walker and Hurst [60]). They found that both mutation rates and biased gene conversion vary with GC content. They also found lineage-specific differences, with weaker fixation biases in orangutan species, suggesting a reduced historical effective population size. As PoMo can distinguish between the contributions of mutation and fixation biases, it might also contribute to addressing the problem of disentangling signatures of selection and biased gene conversion (*see* Subheading 4.2).

3.4 Software

The software PHAST (PHylogenetic Analysis with Space/Time models) includes several phylo-HMM-based programs. Two programs in PHAST are particularly interesting in the context of selection studies: PhastCons is a program for conservation scoring and identification of conserved elements (Siepel et al. [61]). PhyloP is designed to compute *p*-values for conservation or acceleration, either lineage-specific or across all branches (Pollard et al. [62]). Recently, the software can also be run through a webportal at http://compgen.cshl.edu/phastweb/.

A variety of codon models to detect selection, including branch-site models and the recent selection-mutation model, are implemented in the CODEML program of PAML [63]. HYPHY is another implementation that includes a large variety of codon models [64]. PoMo has been implemented as part of the IQ-TREE software package (http://www.iqtree.org/) by Schrempf et al. [65].

These programs are primarily developed for maximum likelihood inference on a fixed tree. ML inference of phylogeny under codon models is possible with CodonPhyML, which allows to explicitly account for selection on the protein level [66].

4 Notes/Discussion

With the wider use of codon models to detect selection, some questioned the statistical basis of testing based on branch-site models. In 2004, Zhang found that the original branch-site *test* [67] produced excessive false positives when its assumptions were not met. The modified branch-site test was shown to be more robust to model violations (*see* [43, 68]) and is now commonly used in genome-wide selection scans (e.g., *see* [69]). Recently, however, another simulation study by Nozawa et al. [70] suggested that this modification also showed an excess of false positives. Yang and Dos Reis [52] defended the branch-site test by examining the null distribution and showing that Nozawa and colleagues [70] misinterpreted their simulation results. However, it is clear that even tests with good statistical properties will be affected by data quality and the extent of models violations. Below we list factors that can affect the test and so should be taken into account when analyzing genome-wide data.

4.1 Quality of Multiple Alignments

The impact of the quality of sequence and the alignment is a major concern when performing positive selection scans. For example, in their analysis of 12 genomes Markova-Raina and Petrov [71] found that the results were highly sensitive to the choice of an alignment method. Furthermore, visual analysis indicated that most sites inferred as positively selected are in fact misaligned at the codon level. The rate of false positives ranged ~50% and more depending on the aligner used. Some of these results can be ascribed to the high divergence level of the 12 *Drosophila* species and could be addressed by better filtering of the data. Nevertheless, even in mammals where alignment is easier, problems have been observed.

Bakewell et al. [72] used the branch-site test to analyze ~14,000 genes from the human, chimpanzee, and macaque and detected more genes to be under positive selection on the chimpanzee lineage than on the human lineage (233 vs. 154). The same pattern was also observed by Arbiza et al. [73] and Gibbs et al.

[74]. Mallick et al. [75] re-examined 59 genes detected to be under positive selection on the chimpanzee lineage by Bakewell et al. [72], using more stringent filters to remove less reliable nucleotides and using synteny information to remove misassembled and misaligned regions. They found that with improved data quality, the signal of positive selection disappeared in most of the cases when the branch-site test was applied. It now appears that, as suggested by Mallick et al. [75], the earlier discovery of more frequent positive selection on the chimpanzee lineage than on the human lineage is an artifact of the poorer quality of the chimpanzee genomic sequence. This interpretation is also consistent with a few recent studies analyzing both real and simulated data, which suggest that sequence and alignment errors may cause excessive false positives (*see* [76, 77]). Indeed, most commonly used alignment programs tend to place nonhomologous codons or amino acids into the same column (*see* [78, 79]), generating the wrong impression that multiple nonsynonymous substitutions occurred at the same site and misleading the codon models into detecting positive selection [77]. In 2012, Jordan and Goldman [80] investigated the effect of various multiple alignment and filtering programs on the identification of positive selection. They found that alignment software PRANK [79] and the filter Guidance [81] performed best in simulations. However, it remains very challenging to develop a pipeline to detect positive selection that is robust to errors in the sequences or alignments. Instead we advise to carefully check the alignments of genes that are putatively under selection by any method described here.

4.2 Biased Gene Conversion and Recombination

Mutation rate variation can also cause genomic regions to have different substitution rates without any change in fixation rate. Recent studies of guanine and cytosine (GC)-isochores in the mammalian genome have suggested the importance of another selectively neutral evolutionary process that affects nucleotide evolution. As described in the work of Laurent Duret and others (*see* [82, 83]), biased gene conversion (BGC) is a mechanism caused by the mutagenic effects of recombination combined with the preference in recombination-associated DNA repair toward strong (GC) versus weak (adenine and thymine [AT]) nucleotide pairs at non-Watson-Crick heterozygous sites in heteroduplex DNA during crossover in meiosis. Thus, beginning with random mutations, BGC results in an increased probability of fixation of G and C alleles. In particular, methods looking for accelerated regions in coding DNA but also codon models cannot distinguish positive selection from BGC (*see* [84, 85]). Therefore, the putatively selected genes should be checked for GC content and closeness to recombination hotspots and telomeres.

Most codon models assume a single phylogeny and a constant synonymous rate among sites, implying that rate variation among codons is solely due to the variation of the nonsynonymous rate. Recent studies question whether such assumptions are generally realistic (e.g., *see* [86]) suggesting that failure to account for synonymous rate variation may be one of the reasons why LRTs for positive selection are vulnerable on data with high recombination rates. Some selection scans try to control this problem by checking putatively selected genes for recombination either manually or automated with traditional detection software (e.g., RDP [87]). Also Drummond and Suchard [88] have recently developed a Bayesian approach to detect recombination within a gene.

Another approach is to explicitly consider recombination. For example, Scheffler, Martin, and Seoighe [89] extended codon models with both d_N and d_S site variation and allowed changes of topology at the detected recombination breakpoints. Certainly, fast-evolving pathogens (such as viruses) undergo frequent recombination which often changes either the whole shape of the underlying tree, or only the apparent branch lengths. While the efficiency of the approach depends on the success of inferring recombination breakpoints, the study demonstrated that taking into account alternative topologies achieves a substantial decrease of false-positive inferences of selection while maintaining reasonable power. In principle the correlation structure of a collection of orthologous sequences can be fully described by a network known as an ancestral recombination graph (ARG). However, methods for ARG inferences have not been fast enough for practical use, and for applications on large-scale genomic data, approximations are necessary (Rassmussen et al. [90]).

4.3 Selection on Synonymous Sites

Most selection studies to date focused on detecting selection on the protein, since synonymous changes are often presumed neutral and so unaffected by selective pressures. However, selection on synonymous sites has been documented more than a decade ago. Codon usage bias is known to affect the majority of genes and species. In his seminal work, Akashi [91] demonstrated purifying selection on genes of *Drosophila melanogaster*, where strong codon bias favoring certain (optimal) codons serves to increase the translational accuracy. Pressure to optimize for translational efficiency, robustness, and kinetics leads to synonymous codon bias, which was shown to widely affect mammalian genes [92], as well as genes of fast-evolving pathogens like viruses [93]. The standard approach to study selection on codon usage computes various codon adaptation indexes on full-length protein-coding genes (*see* [94] for review). More recently, methods to study selection on synonymous changes adopted more sophisticated approaches, mainly the following strategies: (1) account for synonymous rate variation within sequences; (2) include codon fitness parameters within a modeling framework that connects population and intraspecific parameters; and (3) allow for selection on synonymous substitutions by introducing the dependency on the rate of protein production and nonsense error rates. Below we elaborate on these approaches.

In the past decade, evidence has accumulated to suggest that codon bias may vary not only between genomes and genes of the same genome but also within genes. Rather than just measuring codon biases in single sequences, a more powerful approach is to model evolution and selection across a set of homologous sequences. Taking the evolutionary perspective into account, Resch et al. [95] conducted a large-scale study of selection on synonymous sites in mammalian genes. They measured selection by comparing the average rate of synonymous substitutions (d_S) to the average substitution rate in the corresponding introns (d_I). While purifying selection was found to affect 28% of genes ($d_S/d_I < 1$), 12% of genes were found to have been affected by positive selection on synonymous sites ($d_S/d_I > 1$). The signal of positive selection correlated with lower predicted mRNA stability compared to genes with negative selection on synonymous sites, suggesting that mRNA destabilization (affecting mRNA levels and translation) could be driving positive selection on synonymous sites.

An increasing number of experimental studies exemplify different scenarios explaining how synonymous mutation may be affected by positive or negative selection. Codon bias to match skews of tRNA abundances may influence translation [96]. Changes at silent sites can disrupt splicing control elements and create new "cryptic" splice sites, as well as mRNA and transcript stability can be affected through preference or avoidance of certain sequence motifs (*see* [92, 97]). Silent changes may affect gene regulation via constraints for efficient binding of miRNA to sense mRNA (e.g., [92, 98]). Selection may act on the choice of synonymous codons near miRNA targets, improving the binding site accessibility, binding efficiency and consequently the function of miRNA itself [99]. Programmed ribosomal frameshifting may be another reason for selection to act on specific codon positions [100]. Speed-dependent protein folding also has been proposed to be a result of selective pressure [101]. According to the co-translational protein folding hypothesis, slower production could cause the protein to take an altered final form (as has been shown in multidrug resistance-1, [102]). Finally, synonymous changes may act to modulate expression by altering mRNA secondary structure, affecting protein abundance [103].

Models of codon evolution currently provide the most powerful approach for studying selection on silent sites. Models with variable synonymous rates (*see* [64, 104]) have been used to evaluate the extent of variability of synonymous rates in a gene and to predict specific sites with most extreme—low or high—synonymous rates (for example *see* [93]). A large-scale study of synonymous rate variation [105] described some intriguing general patterns and showed that the phenomenon is widespread in protein-coding genes. Genes displaying significantly varying synonymous rates increased association with several genetic diseases (especially cancers and diabetes) and were enriched for metabolic pathways. Other studies specifically focusing on human oncogenes revealed that a significant proportion of all cancer driver mutations were synonymous [106]. This suggests that synonymous rates cannot be automatically assumed fitness-neutral. Note that $\omega = d_N/d_S$, an accepted measure of selection on the protein, is not designed to detect selection on synonymous codons, particularly when d_S is assumed constant. Yet, some cautioned that low synonymous rates preserved by purifying selection might erroneously lead to the detection of positive selection on the protein (e.g., Rubinstein et al. [107]). However, the usage of the ω ratio does not rely on the assumption that synonymous sites are neutral (pages 58–59 of Yang [108]; and Section 6.3 of Anisimova and Liberles [109]); rather, it is defined as a ratio of two ratios, comparing the proportions of nonsynonymous and synonymous sites after and before selection has operated on the protein ($\omega = 1$). In general we can assume that the evolutionary forces apply equally to synonymous and nonsynonymous sites. Forces that act differentially on synonymous and nonsynonymous sites should be rare in real data, but they can affect the validity of the ω measure. The only known example of such a natural force is probably synonymous phasing, considered by Xing and Lee [110]. But even in this case, and with a worst case scenario, the estimated effect is very weak. More crucially, an adequate description of mutational processes at the DNA level allows to circumvent biases in the estimation of the ω ratio [106].

Further testing, however, is necessary to decide whether any specific site has been affected by selection on synonymous codon usage. For example, Zhou, Gu, and Wilke [111] suggested distinguishing two types of synonymous substitution rates: the rate of conserving synonymous changes d_{SC} (between "preferred" codons or between "rare" codons) and the rate of non-conserving synonymous changes d_{SN} (between codons from the two different groups "rare" and "preferred"). Silent sites with $d_{SN}/d_{SC} > 1$ may be considered to be under positive selection, and significance can be tested based on a likelihood ratio test. Alternatively, synonymous rates at sites may be compared to the mean substitution rate in the corresponding intron, which can be implemented in a joint codon and DNA model, similar to the approach proposed by Wong and Nielsen [112].

Mutation-selection models include selective and mutational effects separately and allow estimating the fitness of various codon changes (*see* [113–115]). The relative rate of substitution for selected mutations to neutral mutations is given by $\omega = 2\gamma/(1 - e^{-2\gamma})$, where $\gamma = 2Ns$ is the scaled selection coefficient (*see* Exercise 3 for a derivation). Nielsen et al. [114] assumed that all changes between preferred and rare codons have the same fitness (and so the same selection coefficient). They used one selection coefficient for optimal codon usage for each branch of a phylogeny and estimated these jointly with the ω ratio by ML. Using this approach to study ancestral codon usage bias, Nielsen et al. [114] confirmed the reduction in selection for optimal codon usage in *D. melanogaster*. In contrast, Yang and Nielsen [115] estimated individual codon fitness parameters and used them to estimate optimal codon frequencies for a gene across multiple species. LRT is used to test whether the codon bias is due to the mutational bias alone. Nevertheless, one remarkable contribution of the mutation-selection models is the connection they make between the interspecific and population parameters. Exploiting this further should provide insights to how changing demographic factors influence observed intraspecific patterns. Mutation-selection models also allow a new perspective on understanding codon models in the context of fitness landscapes with statistical implications as discussed in Subheading 4.2 of Chapter 13 by Jones, Susko, and Bielawski.

Finally, it is also possible to study selection on synonymous changes by introducing a parametric relationship between fitness and protein production cost. The idea was first described by Gilchrist [116], who assumed that, in addition to mutation and drift, the codon bias evolved under selection to reduce the cost of nonsense errors. Protein production cost can be computed as a ratio of the expected cost to the expected benefit [117]. Kubatko and colleagues [118] have extended a standard codon model to include the difference in protein production due to the usage of different codons (and therefore different elongation probabilities). However, such a model requires position-specific instantaneous rate matrices, and consequently also the probability transition matrices, making the approach computationally very intensive. To circumvent this, a GPU-based implementation was developed and used for phylogeny inference from 104 gene data set from *Saccharomyces cerevisiae*. Based on the standard model selection measure AIC, the new model outperformed the simplest model M0 as well as the mutation-selection model FMutSel of Yang and Nielsen.

5 Exercises

Q1. Amino Acid and Codon Substitution Models

How many parameters need to be estimated in the instantaneous rate matrix Q defining a reversible empirical AA model? How many such parameters are necessary to estimate for a reversible empirical codon model? How many parameters are to be estimated in both cases if a model is nonreversible?

Q2. Positive Selection Scans

1. Go to the UCSC genome browser (http://genome.ucsc.edu). Search for the HAVCR1 (hepatitis A virus cellular receptor 1) in the human genome (assembly GRCh38/hg38) belonging to the mammalian clade. The USCS genome browser tracks provide the summary of previous analysis of coding regions. Switch the "Cons_30_Primates" under "Comparative Genomics" to full and "refresh." Why are only a few bases in the HAVCR1 gene conserved according to the PhastCons track? Click on the "Cons_30_Primates" track to learn more about the conservation scores used.
2. To retrieve the multiple sequence alignments for the HAVCR1 gene, go to "Tools" and "Table Browser" at the top bar of the webpage. This will open a new page. Choose the table "ccdsGene" under the "Genes and Gene Predictions" group and "CCDS" track. Select "CDS FASTA alignment from multiple alignment" option in the output format and "Show nucleotides" to download the aligned coding sequences of the HAVCR1 gene. Alternatively you can retrieve the multiple alignments from Ensembl using BioMart. Here, you have options for more file formats including PHYLIP that is needed for the PAML software.
3. Use the PAML software (http://abacus.gene.ucl.ac.uk/software/paml.html) to test the models for positive selection on any lineage of the mammalian trees by comparing models M1a and M2a with a likelihood ratio test.
4. Use PAML to identify sites under positive selection by using the Bayes Empirical Bayes approach. Do you find the same sites to be under selection as in Fig. 2 of Kosiol et al. [43]?

Q3. Selection-Mutation Models

Selection-mutation rely on a theoretical relationship between the nonsynonymous-synonymous rate ratio ω and the scaled selection coefficient $\gamma = 2Ns$. The probability that a new mutation eventually becomes fixed is

$$\Pr(\text{fixation}) = \left(1 - e^{-2s}\right)/\left(1 - e^{-4Ns}\right) = 2s/\left(1 - e^{-4Ns}\right)$$

if we assume that the selection coefficient s is small and N is large and represents the effective population size, which is constant in time (Kimura and Ohta [119]). Furthermore, assume that synonymous substitutions are neutral and nonsynonymous have equal (and small) selection coefficients. Derive the relationship:

$$\omega = 4s/\left(1 - e^{-4Ns}\right) = 2\gamma/\left(1 - e^{-2\gamma}\right)$$

that combines phylogenetic with population genetic quantities and is crucial for mutation-selection models.

Acknowledgments

C. K. is supported by a grant of the Vienna Science and Technology Fund (WWTF—MA016-061). M. A. receives funding from the Swiss National Science Foundation (grant 31003A_176316).

References

1. Zerbino DR, Achuthan P, Akanni W, Amode MR, Barrell D, Bhai J, Billis K, Cummins C, Gall A, Girón CG, Gil L, Gordon L, Haggerty L, Haskell E, Hourlier T, Izuogu OG, Janacek SH, Juettemann T, To JK, Laird MR, Lavidas I, Liu Z, Loveland JE, Maurel T, McLaren W, Moore B, Mudge J, Murphy DN, Newman V, Nuhn M, Ogeh D, Ong CK, Parker A, Patricio M, Riat HS, Schuilenburg H, Sheppard D, Sparrow H, Taylor K, Thormann A, Vullo A, Walts B, Zadissa A, Frankish A, Hunt SE, Kostadima M, Langridge N, Martin FJ, Muffato M, Perry E, Ruffier M, Staines DM, Trevanion SJ, Aken BL, Cunningham F, Yates A, Flicek P (2018) Ensembl 2018. Nucleic Acids Res 46:D754–D761
2. Casper J, Zweig AS, Villarreal C, Tyner C, Speir ML, Rosenbloom KR, Raney BJ, Lee CM, Lee BT, Karolchik D, Hinrichs AS, Haeussler M, Guruvadoo L, Navarro Gonzalez J, Gibson D, Fiddes IT, Eisenhart C, Diekhans M, Clawson H, Barber GP, Armstrong J, Haussler D, Kuhn RM, Kent WJ (2018) The UCSC Genome Browser database: update 2018. Nucleic Acids Res 46: D762–D769
3. Lack JB, Lange JD, Tang AD, Corbett-Detig RB, Pool JE (2016) A thousand fly genomes: an expanded drosophila genome nexus. Mol Biol Evol 33:3308–3313
4. Weigel D, Mott R (2009) The 1001 Genomes Project for Arabidopsis thaliana. Genome Biol 10:107
5. Turnbull C, Scott RH, Thomas E, Jones L, Murugaesu N, Pretty FB, Halai D, Baple E, Craig C, Hamblin A, Henderson S, Patch C, O'Neill A, Devereaux A, Smith K, Martin AR, Sosinsky A, McDonagh EM, Sultana R, Mueller M, Smedley D, Toms A, Dinh L, Fowler T, Bale M, Hubbard T, Rendon A, Hill S, Caulfield MJ, 100,000 Genomes Project (2018) The 100 000 Genomes Project: bringing whole genome sequencing to the NHS. BMJ 361:k1687
6. Li R, Fan W, Tian G, Zhu H, He L, Cai J, Huang Q, Cai Q, Li B, Bai Y, Zhang Z, Zhang Y, Xuan Z, Wang W, Li J et al (2010) The sequence and de novo assembly of the giant panda genome. Nature 463:311–317
7. Posada D, Crandall KA (2002) The effect of recombination on the accuracy of phylogenetic estimation. J Mol Evol 54:396–402
8. Sawyer S (1989) Statistical tests for detecting gene conversion. Mol Biol Evol 6:526–538
9. Semple C Wolfe KH (1999) Gene duplication and gene conversion in the Caenorhabditis elegans genome. J Mol Evol 48:555–564
10. Doolittle WF (1999) Phylogenetic classification and the universal tree. Science 284:2124–2129
11. Robinson DM, Jones DT, Kishino H, Goldman N, Thorne JL (2003) Protein evolution with dependence among codons due to tertiary structure. Mol Biol Evol 20:1692–1704
12. Choi SC, Holboth A, Robinson DM, Kishino H, Thorne JL (2007) Quantifying the impact of protein tertiary structure on molecular evolution. Mol Biol Evol 24:1769–1782
13. Keilson J (1979) Markov Chain models-rarity and exponentiality. Springer, New York, NY
14. Pollard KS, Salama SR, King B, Kern AD, Dreszer T, Katzman S, Siepel A, Perdersen JS, Berjerano G, Baertsch R, Rosenblum KR, Kent J, Haussler D (2006) Forces shaping the fastest evolving regions in the human genome. PLoS Genet 2(10):e168
15. Holloway AK, Begun DJ, Siepel A, Pollard K (2008) Accelerated sequence divergence of conserved genomic elements in Drosophila melanogaster. Genome Res 18:1592–1601
16. Miyamoto MM, Fitch WM (1995) Testing the covarion hypothesis of molecular evolution. Mol Biol Evol 12:503–513
17. Lockhart PJ, Steel MA, Barbrook AC, Huson DH, Charleston MA, Howe CJ (1998) A covariotide model explains apparent phylogenetic structure of oxygenic photosynthetic lineages. Mol Biol Evol 15:1183–1188
18. Penny D, McComish BJ, Charleston MA, Hendy MD (2001) Mathematical elegance with biochemical realism: the covarion

model of molecular evolution. J Mol Evol 53:711–753

19. Siltberg J, Liberles DA (2002) A simple covarion-based approach to analyse nucleotide substitution rates. J Evol Biol 15:588–594
20. Lichtarge O, Bourne HR, Cohen FE (1996) An evolutionary trace method defines binding surfaces common to protein families. J Mol Evol 257:342–358
21. Gu X (1999) Statistical methods for testing functional divergence after gene duplication. Mol Biol Evol 16:1664–1674
22. Armon A, Graur D, Ben-Tal N (2001) ConSurf: an algorithmic tool for the identification of functional regions in proteins by surface mapping of phylogenetic information. J Mol Biol 307:447–463
23. Gaucher EA, Gu X, Miyamoto MM, Benner SA (2002) Predicting functional divergence in protein evolution by site-specific rate shifts. Trends Biochem Sci 27:315–321
24. Pupko T, Galtier N (2002) A covarion-based method for detecting molecular adaptation: application to the evolution of primate mitochondrial genomes. Proc Biol Sci 269:1313–1316
25. Blouin C, Boucher Y, Roger AJ (2003) Inferring functional constraints and divergence in protein families using 3D mapping of phylogenetic information. Nucleic Acids Res 31:790–797
26. Landau M, Mayrose I, Rosenberg Y, Glaser F, Martz E, Pupko T, Ben-Tal N (2005) ConSurf 2005: the projection of evolutionary conservation scores of residues on protein structures. Nucleic Acids Res 33: W299–W302
27. Gu X (2001) Maximum-likelihood approach for gene family evolution under functional divergence. Mol Biol Evol 18:453–464
28. Gu X (2006) A simple statistical method for estimating type-II (cluster-specific) functional divergence of protein sequences. Mol Biol Evol 23:1937–1945
29. Bofkin L, Goldman N (2007) Variation in evolutionary processes at different codon positions. Mol Biol Evol 24:513–521
30. Hughes AL, Nei M (1988) Pattern of nucleotide substitution at major histocompatibility complex class I loci reveals overdominant selection. Nature 335:167–170
31. Yang Z, Nielsen R (2000) Estimating synonymous and nonsynonymous substitution rates under realistic evolutionary models. Mol Biol Evol 17:32–43
32. Goldman N, Yang Z (1994) A codon-based model of nucleotide substitution for protein-coding DNA sequences. Mol Biol Evol 11:725–736
33. Muse SV, Gaut BS (1994) A likelihood approach for comparing synonymous and nonsynonymous nucleotide substitution rates, with application to the chloroplast genome. Mol Biol Evol 11:715–724
34. Grantham R (1974) Amino acid difference formula to help explain protein evolution. Science 185:862–864
35. Yang Z (1998) Likelihood ratio tests for detecting positive selection and application to primate lysozyme evolution. Mol Biol Evol 15:568–573
36. Schneider A, Cannarozzi GM, Gonnet GH (2005) Empirical codon substitution matrix. BMC Bioinformatics 6:134
37. Kosiol C, Holmes I, Goldman N (2007) An empirical codon model for protein sequence evolution. Mol Biol Evol 24:1464–1479
38. Doron-Faigenboim A, Pupko T (2007) A combined empirical and mechanistic codon model. Mol Biol Evol 24:388–397
39. De Maio N, Holmes I, Schlötterer C, Kosiol C (2013) Estimating empirical hidden Markov models. Mol Biol Evol 30:725–736
40. Whelan S, Goldman N (1999) Distributions of statistics used for the comparison of models of sequence evolution in phylogenetics. Mol Biol Evol 16:1292–1299
41. Anisimova M, Bielawski JP, Yang Z (2001) Accuracy and power of the likelihood ratio test in detecting adaptive molecular evolution. Mol Biol Evol 18:1585–1592
42. Kosiol C, Vinar T, Da Fonseca RR, Hubisz MJ, Bustamante CD, Nielsen R, Siepel A (2008) Patterns of positive selection in six mammalian genomes. PLoS Genet 4: e10000144
43. Anisimova M, Bielawski JP, Yang Z (2002) Accuracy and power of bayes prediction of amino acid sites under positive selection. Mol Biol Evol 19:950–958
44. Yang Z, Wong WS, Nielsen R (2005) Bayes empirical bayes inference of amino acid sites under positive selection. Mol Biol Evol 22:1107–1118
45. Yang Z, Nielsen R, Goldman N, Pedersen AMK (2000) Codon-substitution models for heterogeneous selection pressure at amino acid sites. Genetics 155:431–449
46. Huelsenbeck JP, Dyer KA (2004) Bayesian estimation of positively selected sites. J Mol Evol 58:661–672

47. Scheffler K, Seoighe C (2005) A Bayesian model comparison approach to inferring positive selection. Mol Biol Evol 22:2531–2540
48. Aris-Brosou S, Bielawski JP (2006) Large-scale analyses of synonymous substitution rates can be sensitive to assumptions about the process of mutation. Gene 378:58–64
49. Massingham T, Goldman N (2005) Detecting amino acid sites under positive selection and purifying selection. Genetics 169:1753–1762
50. Kosakovsky Pond SL, Posada D, Gravenor MB, Woelk CH, Frost SD (2006) GARD: a genetic algorithm for recombination detection. Bioinformatics 22:3096–3098
51. Kosakovsky PSL, Posada D, Gravenor MB, Woelk CH, Frost SD (2006) Automated phylogenetic detection of recombination using a genetic algorithm. Mol Biol Evol 23:1891–1901
52. Felsenstein J (2004) Inferring phylogenies. Sinauer Associates, Sunderland, MA
53. Yang Z, Dos Reis M (2011) Statistical properties of the branch-site test of positive selection. Mol Biol Evol 28:1217–1228
54. Anisimova M, Yang Z (2007) Multiple hypothesis testing to detect lineages under positive selection that affects only a few sites. Mol Biol Evol 24:1219–1228
55. Kosakovsky Pond SL, Frost SD (2005) A genetic algorithm approach to detecting lineage-specific variation in selection pressure. Mol Biol Evol 22:478–485
56. Guindon SA, Rodrigo G, Dyer KA, Huelsenbeck JP (2004) Modeling the site-specific variation of selection patterns along lineages. Proc Natl Acad Sci U S A 101:12957–12962
57. De Maio N, Schlötterer C, Kosiol C (2013) Linking great apes genome evolution across time scales using polymorphism-aware phylogenetic models. Mol Biol Evol 30:2249–2262
58. De Maio N, D Schrempf D, Kosiol C (2016) PoMo: an allele frequency-based approach for species tree estimation. Syst Biol 64:1018–1031
59. Maddison W, Knowles L (2006) Inferring phylogeny despite incomplete lineage sorting. Syst Biol 55:21–30
60. Eyre-Walker A, Hurst L (2001) The evolution of isochores. Nat Rev Genet 2:549–555
61. Siepel A, Bejerano G, Pedersen JS, Hinrichs A, Hou M, Rosenbloom K, Clawson H, Spieth J, Hillier LW, Richards S, Weinstock GM, Wilson RK, Gibbs RA, Kent WJ, Miller W, Haussler D (2005) Evolutionarily conserved elements in vertebrate, insect, worm, and yeast genomes. Genome Res 20:1034–1050
62. Pollard KS, Hubisz MJ, Rosenbloom KR, Siepel A (2010) Detection of non-neutral substitution rates on mammalian phylogenies. Genome Res 20:110–121
63. Yang Z (2007) PAML 4: phylogenetic analysis by maximum likelihood. Mol Biol Evol 24:1586–1591
64. Kosakovsky Pond SL, Muse SV (2005) Site-to-site variation of synonymous substitution rates. Mol Biol Evol 22:2375–2385
65. Schrempf D, Minh BQ, De Maio N, von Haeseler A, Kosiol C (2016) Reversible polymorphism-aware phylogenetic models and their application to tree inference. J Theor Biol 407:362–370
66. Gil M, Zanetti MS, Zoller S, Anisimova M (2013) CodonPhyML: fast maximum likelihood phylogeny estimation under codon substitution models. Mol Biol Evol 30:1270–1280
67. Zhang J, Nielsen R, Yang Z (2005) Evaluation of an improved branch-site likelihood method for detecting positive selection at the molecular level. Mol Biol Evol 22:2472–2479
68. Yang Z, Nielsen R (2002) Codon-substitution models for detecting molecular adaptation at individual sites along specific lineages. Mol Biol Evol 19:908–917
69. Vamathevan JJ, Hasan S, Emes RD, Amrine-Madsen H, Rajagopalan D, Topp SD, Kumar V, Word M, Simmons MD, Foord SM, Sanseau P, Yang Z, Holbrook JD (2008) The role of positive selection in determining the molecular cause of species differences in disease. BMC Evol Biol 8:273
70. Nozawa M, Suzuki Y, Nei M (2009) Reliabilities of identifying positive selection by the branch-site and site-prediction methods. Proc Natl Acad Sci U S A 106:6700–6705
71. Markova-Raina P, Petrov D (2011) High sensitivity to aligner and high rate of false positives in the estimates of positive selection in 12 Drosophila genomes. Genome Res 21:863. https://doi.org/10.1101/gr.115949.110
72. Bakewell MA, Shi P, Zhang J (2007) More genes underwent positive selection in chimpanzee than in human evolution. Proc Natl Acad Sci U S A 104:E97
73. Arbiza L, Dopazo J, Dopazo H (2006) Positive selection, relaxation, and acceleration in the evolution of the human and chimp genome. PLoS Comput Biol 2:e38
74. Gibbs RA, Rogers J, Katze MG, Bumgarner R, Weinstock GM, Mardis ER, Remington KA, Strausberg RL, Venter JC, Wilson RK et al (2007) Evolutionary and

biomedical insights from the macaque genome. Science 316:222–234

75. Mallik S, Gnerre S, Muller P, Reich D (2010) The difficulty of avoiding false positives in genome scans for natural selection. Genome Res 19:922–933

76. Schneider A, Souvorov A, Sabath N, Landan G, Gonnet GH (2009) Estimates of positive Darwinian selection are inflated by errors in sequencing, annotation, and alignment. Genome Biol Evol 1:114–118

77. Fletcher W, Yang Z (2010) The effect of insertions, deletions and alignment errors on the branch-site test of positive selection. Mol Biol Evol 27:2257–2267

78. Löytynoja A, Goldman N (2005) An algorithm for progressive multiple alignment of sequences with insertions. Proc Natl Acad Sci U S A 102:10557–10562

79. Löytynoja A, Goldman N (2008) Phylogeny-aware gap placement prevents error in sequence alignment and evolutionary analysis. Science 320:1632–1635

80. Jordan G, Goldman N (2012) The effects of alignment error and alignment filtering on the sitewise detection of positive selection. Mol Biol Evol 29:1125–1139

81. Penn O, Privman E, Landan G, Graur D, Pupko T (2010) An alignment confidence score capturing robustness to guide tree uncertainty. Mol Biol Evol 27:1759–1767

82. Duret L, Semon M, Piganeau G, Mouchiroud D, Galtier N (2002) Vanishing GC-rich isochores in mammalian genomes. Genetics 162:1837–1847

83. Meunier J, Duret L (2004) Recombination drives the evolution of GC content in the human genome. Mol Biol Evol 21:984–990

84. Berglund J, Pollard KS, Webster MT (2009) Hotspots of biased nucleotide substitutions in human genes. PLoS Biol 7:e26

85. Ratnakumar A, Mousset S, Glemin S, Berglund J, Galtier N, Duret L, Webster MT (2010) Detecting positive selection within genomes: the problem of biased gene conversion. Phil Trans R Soc B 365:2571–2580

86. Anisimova M, Nielsen R, Yang Z (2003) Effect of recombination on the accuracy of the likelihood method for detecting positive selection at amino acid sites. Genetics 164:1229–1236

87. Martin DP, Williamson C, Posada D (2005) RDP2: recombination detection and analysis of sequence alignments. Bioinformatics 21:260–262

88. Drummond AJ, Suchard MA (2008) Fully Bayesian tests of neutrality using genealogical summary statistics. BMC Genet 9:68

89. Scheffler K, Martin DP, Seoighe C (2006) Robust inference of positive selection from recombining coding sequences. Bioinformatics 22:2493–2499

90. Rasmussen MD, Hubisz MJ, Gronau I, Siepel A (2014) Genome-wide inference of ancestral recombination graphs. PLoS Genet 10(5): e1004342

91. Akashi H (1994) Synonymous codon usage in Drosophila melanogaster: natural selection and translational accuracy. Genetics 136:927–935

92. Chamary JV, Parmley JL, Hurst LD (2006) Hearing silence: non-neutral evolution at synonymous sites in mammals. Nat Rev Genet 7:98–108

93. Ngandu N, Scheffler K, Moore P, Woodman Z, Martin D, Seoighe C (2009) Extensive purifying selection acting on synonymous sites in HIV-1 Groug M sequences. Virol J 5:160

94. Roth A, Anisimova M, Cannarozzi GM (2012) Measuring codon usage bias. Codon evolution: mechanisms and models. Oxford University Press, New York, NY

95. Resch AM, Carmel L, Marino-Ramirez L, Ogurtsov AY, Shabalina SA, Rogozin IB, Koonin EV (2007) Widespread positive selection in synonymous sites of mammalian genes. Mol Biol Evol 24:1821–1831

96. Cannarozzi GM, Faty M, Schraudolph NN, Roth A, von Rohr P, Gonnet P, Gonnet GH, Barral Y (2010) A role for codons in translational dynamics. Cell 141:355–367

97. Hurst LD, Pál C (2001) Evidence of purifying selection acting on silent sites in BRCA1. Trends Genet 17:62–65

98. Chamary JV, Hurst LD (2005) Biased usage near intron-exon junctions: selection on splicing enhancers, splice site recognition or something else? Trends Genet 21:256–259

99. Gu W, Wang X, Zhai C, Xie X, Zhou T (2012) Selection on synonymous sites for increased accessibility around miRNA binding sites in plants. Mol Biol Evol 29:3037–3044

100. Garcia V, Anisimova M (2018) Accounting for programmed ribosomal frameshifting in the computation of codon usage bias indices. G3 (Bethesda) 8:3173

101. Komar AA (2008) Protein translational rates and protein misfolding: is there any link? In: O'Doherty CB, Byrne AC (eds) Protein

misfolding: new research. Nova Science Publisher Inc, New York, NY

102. Kimichi-Sarfaty C, Oh JM, Kim IW, Sauna ZE, Calcagno AM, Ambudkar SV, Gottesman MM (2007) A silent polymorphism in the MDR1 gene changes substrate specificity. Science 315:525–528

103. Nackley AG, Shabalina SA, Tchivileva IE, Satterfield K, Korchynskyi O, Makarov SS, Maixner W, Diatchenko L (2006) Human catechol-O-methyltransferase haplotypes modulate protein expression by altering mRNA secondary structure. Science 314:1930–1933

104. Mayrose I, Doron-Faigenboim A, Bacharach E, Pupko T (2007) Towards realistic codon models: among site variability and dependency of synonymous and non-synonymous rates. Bioinformatics 23: i319–i327

105. Dimitrieva S, Anisimova M (2014) Unraveling patterns of site-to-site synonymous rates variation and associated gene properties of protein domains and families. PLoS One 9 (7):e102721

106. Martincorena I, Raine KM, Gerstung M, Dawson KJ, Haase K, Van Loo P, Davies H, Stratton MR, Campbell PJ (2017) Cell 171:1029–1041.e21

107. Rubinstein ND, Doron-Faigenboim A, Mayrose I, Pupko T (2011) Evolutionary models accounting for layers of selection in protein-coding genes and their impact on the inference of positive selection. Mol Biol Evol 28:3297–3308

108. Yang Z (2006) Computational molecular evolution. Oxford University Press, New York, NY

109. Anisimova M, Liberles DA (2012) Detecting and understanding natural selection. Codon evolution: mechanisms and models. Oxford University Press, New York, NY

110. Xing Y, Lee C (2006) Alternative splicing and RNA selection pressure--evolutionary consequences for eukaryotic genomes. Nat Rev Genet 7:499–509

111. Zhou T, Gu W, Wilke CO (2010) Detecting positive and purifying selection at synonymous sites in yeast and worm. Mol Biol Evol 27:1912–1922

112. Wong WSW, Nielsen R (2004) Detecting selection in non-coding regions of nucleotide sequences. Genetics 167:949–958

113. Nielsen R, Yang Z (2003) Estimating the distribution of selection coefficients from phylogenetic data with applications to mitochondrial and viral DNA. Mol Biol Evol 20:1231–1239

114. Nielsen R, Bauer DuMont VL, Hubisz MJ, Aquadro CF (2007) Maximum likelihood estimation of ancestral codon usage bias parameters in Drosophila. Mol Biol Evol 24:228–235

115. Yang Z, Nielsen R (2008) Mutation-selection models of codon substitution and their use to estimate selective strengths on codon usage. Mol Biol Evol 25:568–579

116. Gilchrist MA (2007) Combining models of protein translation and population genetics to predict protein production rates from codon usage patterns. Mol Biol Evol 24:2362–2372

117. Gilchrist MA, Shah P, Zaretzki R (2009) Measuring and detecting molecular adaptation in codon usage against nonsense errors during protein translation. Genetics 183:1493–1505

118. Kubatko L, Shah P, Herbei R, Gilchrist MA (2016) A codon model of nucleotide substitution with selection on synonymous codon usage. Mol Phylogenet Evol 94:290–297

119. Kimura M, Ohta T (1969) The average number of generations until fixation of a mutant gene in a finite population. Genetics 61:763–771

Chapter 13

Looking for Darwin in Genomic Sequences: Validity and Success Depends on the Relationship Between Model and Data

Christopher T. Jones, Edward Susko, and Joseph P. Bielawski

Abstract

Codon substitution models (CSMs) are commonly used to infer the history of natural section for a set of protein-coding sequences, often with the explicit goal of detecting the signature of positive Darwinian selection. However, the validity and success of CSMs used in conjunction with the maximum likelihood (ML) framework is sometimes challenged with claims that the approach might too often support false conclusions. In this chapter, we use a case study approach to identify four legitimate statistical difficulties associated with inference of evolutionary events using CSMs. These include: (1) model misspecification, (2) low information content, (3) the confounding of processes, and (4) phenomenological load, or PL. While past criticisms of CSMs can be connected to these issues, the historical critiques were often misdirected, or overstated, because they failed to recognize that the success of any model-based approach depends on the relationship between model and data. Here, we explore this relationship and provide a candid assessment of the limitations of CSMs to extract historical information from extant sequences. To aid in this assessment, we provide a brief overview of: (1) a more realistic way of thinking about the process of codon evolution framed in terms of population genetic parameters, and (2) a novel presentation of the ML statistical framework. We then divide the development of CSMs into two broad phases of scientific activity and show that the latter phase is characterized by increases in model complexity that can sometimes negatively impact inference of evolutionary mechanisms. Such problems are not yet widely appreciated by the users of CSMs. These problems can be avoided by using a model that is appropriate for the data; but, understanding the relationship between the data and a fitted model is a difficult task. We argue that the only way to properly understand that relationship is to perform in silico experiments using a generating process that can mimic the data as closely as possible. The mutation-selection modeling framework (MutSel) is presented as the basis of such a generating process. We contend that if complex CSMs continue to be developed for testing explicit mechanistic hypotheses, then additional analyses such as those described in here (e.g., penalized LRTs and estimation of PL) will need to be applied alongside the more traditional inferential methods.

Key words Codon substitution model, dN/dS, False positives, Maximum likelihood, Mechanistic model, Model misspecification, Mutation-selection model, Parameter confounding, Phenomenological load, Phenomenological model, Positive selection, Reliability, Statistical inference, Site-specific fitness landscape

Maria Anisimova (ed.), *Evolutionary Genomics: Statistical and Computational Methods*, Methods in Molecular Biology, vol. 1910,
https://doi.org/10.1007/978-1-4939-9074-0_13,

1 Introduction

Codon substitution models (CSMs) fitted to an alignment of homologous protein-coding genes are commonly used to make inferences about evolutionary processes at the molecular level (*see* Chapter 10 for examples of different applications of CSMs). Such processes (e.g., mutation and selection) are represented by a vector of parameters θ that can be estimated using maximum likelihood (ML) or Bayesian statistical methods. Here, we focus on ML and for convenience use CSM to indicate a model that is used in conjunction with the ML approach (*see* [21], for an example of the Bayesian approach). Considerable apprehension was expressed about the statistical validity of CSMs during their initial phase of development. In particular were concerns over the risk of falsely inferring that a sequence or codon site evolved by adaptive evolution [11, 22, 23, 46, 60–63, 85]. Many of the studies employed in the critique of CSMs were later shown to be flawed due to statistical errors or incorrect interpretation of results [70, 72, 77, 84]. In their reanalysis of the iconic MHC dataset [24], for example, Suzuki and Nei [61] based their criticism of the ML approach on results that were incorrect due to computational issues [70]. And in simulation studies by Suzuki [60] and Nozawa et al. [46], the branch-site model of Yang and Nielsen [79] was criticized as being too liberal because it falsely inferred positive selection at 32 out of 14,000 simulated sites, despite that this rate (0.0023) was well below the level of significance of the test ($\alpha = 0.05$) [77]. Concerns about the ML approach were eventually mollified by numerous simulation studies showing that the false positive rate is no greater than the specified level of significance of the LRT under a wide range of evolutionary scenarios [2, 3, 29, 31, 37, 70, 77, 82, 85, 86]. The validity and success of the approach is now well established [84], and this has led to the formulation of CSMs of ever-increasing sophistication [31, 41, 48–50, 55, 64, 65].

The most common use of a CSM is to infer whether a given process, such as adaptive evolution somewhere in the gene, the fixation of double and triple mutations, or variations in the synonymous substitution rate, actually occurred when the alignment was generated. Several factors can potentially undermine the reliability of such inferences. These include:

1. **Model misspecification**, which can result in biased parameter estimates;
2. **Low information content**, which can cause parameter estimates to have large sampling errors and can lead to excessive false positive rates;

3. **Confounding**, which can cause patterns in the data generated by one evolutionary process to be attributed to a different process;
4. **Phenomenological load**, which can cause a model parameter to be statistically significant even if the process it represents did not actually occur when the data was generated.

These same factors can impact any model-based effort to make inferences from data generated by complex biological processes, not only to the CSMs described here. The possibility of false inference due to any combination of these factors does not imply that the CSM approach is unreliable in principle. As has been demonstrated by numerous successful applications, CSMs generally extract accurate and useful information provided that the model is well suited for the data at hand [1, 71, 76]. We maintain that the validity of inferences is not a function of the model in and of itself, but is a consequence of the relationship between the model and the data.

Here, we explore this relationship via case studies taken from the historical development of CSMs. Our objective is to be candid about the limitations of CSMs to reliably extract information from an alignment. But, we emphasize that the impact of these limitations (i.e., false positives and confounding) is a consequence of a mismatch between the parameters included in the model and the often limited information contained in the alignment. The case studies are divided into two parts, each corresponding to a distinct phase in the development of CSMs. Phase I is characterized by pioneering efforts to formulate CSMs to account for the most prominent components of variation in an alignment [16, 42]. These include the M-series models that were among the first CSMs to account for variations in selection effects across sites [81], and the branch-site model of Yang and Nielsen [79] (hereafter, YN-BSM) formulated to account for variations in selection effects across both sites and branches. The first pair of case studies exemplifies concerns about the impact of low information content (Case Study A) and model misspecification (Case Study B) on the probability of falsely detecting positive selection in a gene or at a particular codon site. We also include a description of methods recently developed to mitigate the problem of false inference.

Phase II in the historical development is characterized by the general increase in the complexity of CSMs aimed to account for more subtle components of variation in an alignment.[1] Models used to detect temporal changes in site-specific selection effects

[1] The original CSM proposed by Goldman and Yang [16] was in fact quite complex in that it adjusted substitution rates between nonsynonymous codons to account for differences in physicochemical properties using the Grantham matrix [17]. This approach was later abandoned in favor of the simpler formulation now known as M0 [44], e.g., the first M-series model [81].

(e.g., [18, 31, 55]) or "heterotachy" [36] are representative. The movement toward complex parameter-rich models has resulted in a new set of concerns that are not yet widely appreciated. Principal among these is an increase in the possibility of confounding. Two components of the alignment-generating process are confounded if they can produce the same or similar patterns in the data. Such components can be impossible to disentangle without the input of further biological information, and their existence can lead to a statistical pathology that we call phenomenological load (PL). The second pair of case studies illustrates the possibility of false inference due to confounding (Case Study C) and PL (Case Study D). An essential feature of these studies is the use of a much more realistic generating model to produce alignments for the purpose of model evaluation.

Recent discoveries made using the mutation-selection (MutSel; [80]) framework of Halpern and Bruno [19], which is based on a realistic approximation of population dynamics at individual codon sites, have challenged the way we think about the relationship between parameters of traditional CSMs and components of the process of molecular evolution they are meant to summarize (e.g., [25, 26, 56, 57]). Previously, there has been a tendency to think about alignment-generating processes as if they occur in the same way they are modeled by a CSM. This way of thinking can be misleading because mechanisms of protein evolution can differ in important and substantial ways from traditional CSMs. To redress this issue, we begin this chapter with a brief overview of the conceptual foundations of MutSel as a more realistic way of thinking about the actual process of molecular evolution. This material is followed by a novel presentation of the ML statistical framework intended to illustrate potential limitations in what can reasonably be inferred when a CSM is fitted to data.

2 Conceptual Foundations

2.1 How Should We Think About the Alignment-Generating Process?

A codon substitution model represents an attempt to explain the way a target protein-coding gene changed over time by a combination of mutation, selection (purifying as well as adaptive), and drift. Adaptive evolution occurs at each site within a protein in response to a hierarchy of effects, including, but not limited to, changes in the network of the protein's interactions, changes in the functional properties of that network, and changes in both the cellular and organismal environment over time. The result of the complex interplay between these effects is typically viewed through the narrow lens of an alignment of homologous sequences X obtained from extant species, possibly accompanied by a tree topology τ (for our purposes, it is always assumed that τ is known). The information contained in X is evidently insufficient to resolve all of the effects of the true generating process, which would in any case be difficult or even impossible to parameterize with any accuracy. It is therefore necessary to base the formulation of a CSM on a number of simplifying assumptions. The usual assumptions include that:

1. Sites evolved independently;
2. Each site evolved via a homogenous substitution process over the tree (formally, by a Markov process governed by a substitution rate matrix Q);
3. The selection regime at a site is determined by Q_j drawn from a small set of possible substitution rate matrices $\{Q_1, \ldots, Q_k\}$;
4. All sites share a common vector of stationary frequencies and evolved via a common mutation process.

The elements q_{ij} of a substitution rate matrix Q are typically defined for codons $i \neq j$ as follows [44]:

$$q_{ij} = \begin{cases} 0 & \text{if } i \text{ and } j \text{ differ by more than one nucleotide} \\ \pi_j & \text{for synonymous transversions} \\ \kappa\pi_j & \text{for synonymous transitions} \\ \omega\pi_j & \text{for nonsynonymous transversions} \\ \omega\kappa\pi_j & \text{for nonsynonymous transitions} \end{cases} \tag{1}$$

where κ is the transition bias and π_i is the stationary frequency of the ith codon, both assumed to be the same for all codon sites. The ratio $\omega = dN/dS$ of the nonsynonymous substitution rate dN to the synonymous substitution rate dS (both adjusted for "opportunity"[2]) quantifies the stringency of selection at the site, with values closer to zero corresponding to sites that are more strongly conserved. We follow standard notation and use $\hat{\omega}$ to represent the maximum likelihood estimate (MLE) of ω obtained by fitting Eq. 1 to an alignment.

Equation 1 provides the building block for most CSMs, yet it is unsuitable as a means to think about the substitution process at a site. For instance, the rate ratio in Eq. 1 is assumed to be the same for all nonsynonymous pairs of codons. If interpreted mechanistically, this is tantamount to the assumption that the amino acid occupying a site has fitness f and all other amino acids have fitness $f + df$, and that, with each substitution, the newly fixed amino acid changes its fitness to f and the previous occupant changes it fitness

---

[2] Single-nucleotide (SN) mutations that are nonsynonymous occur more frequently than those that are synonymous due to idiosyncrasies in the genetic code. This is accounted for in the formulation of dN and dS, so that dN can be interpreted as the proportion of nonsynonymous SN mutations that are fixed. Likewise, dS is the proportion of synonymous SN mutations that are fixed. *See* Jones et al. [25] for a discussion of various interpretations of dN/dS.

to $f + df$. Such a narrow view of the substitution process, akin to frequency-dependent selection [6, 25], is conceptually misleading for the majority of proteins. To be clear, CSMs are undoubtedly a valuable tool to make inferences about the evolution of a protein (e.g., [8, 52, 71, 76]); our point is that they do not necessarily provide the best way to *think* about the process.

The way we think about the substitution process should not be limited to unrealistic assumptions used to formulate a tractable CSM. It is more informative to conceptualize evolution at a codon site using the traditional metaphor of a fitness landscape upon which greater height represents greater fitness as depicted in Fig. 1. If sites are assumed to evolve independently, a **site-specific fitness landscape** can be defined for the hth site by a vector of fitness coefficients $\mathbf{f}^h$ and its implied vector of equilibrium codon frequencies $\boldsymbol{\pi}^h$. Combined with a model for the mutation process, $\boldsymbol{\pi}^h$ determines the evolutionary dynamics at the site, or the way it "moves" over its landscape (more formally, the way mutation and fixation events occur at a codon site in a population over time). This provides a way to think about evolution at a codon site in terms of three possible dynamic regimes: **shifting balance**, under which the site moves episodically away from the peak of its fitness landscape (i.e., the fittest amino acid) via drift and back again by positive selection (Fig. 1a); **adaptive evolution**, under which a change in the landscape is followed by movement of the site toward its new fitness peak (Fig. 1b); and **neutral** or **nearly neutral evolution**, under which drift dominates and the site is free to move over a relatively flat landscape limited primarily by biases in the mutation process. This way of thinking about the alignment-generating process is encapsulated by the MutSel framework [6, 7, 25]. The precise relationship between the MutSel framework and the three dynamic regimes will be presented in Case Study C.

2.2 What Is the Objective of Model Building?

CSMs have become increasingly complex with the addition of more free parameters since the introduction of the M-series models in Yang et al. [81]. The *prima facie* objective of this trend is to produce models that provide better mechanistic explanations of the data. The assumption is that this will lead to more accurate inferences about evolutionary processes, particularly as the volume of genetic data increases [35]. However, the significance of a new model parameter is assessed by a comparison of site-pattern distributions without reference to mechanism. Combined with the possibility of confounding, this feature of the ML framework means that the objective of improving model fit does not necessarily coincide with the objective of providing a better representation of the mechanisms of the true generating process.

Given any CSM with parameters θ_{M}, it is possible to compute a vector $\boldsymbol{P}$ that assigns a probability to each of the 61^N possible site patterns for an N-taxon alignment (i.e., a multinomial distribution

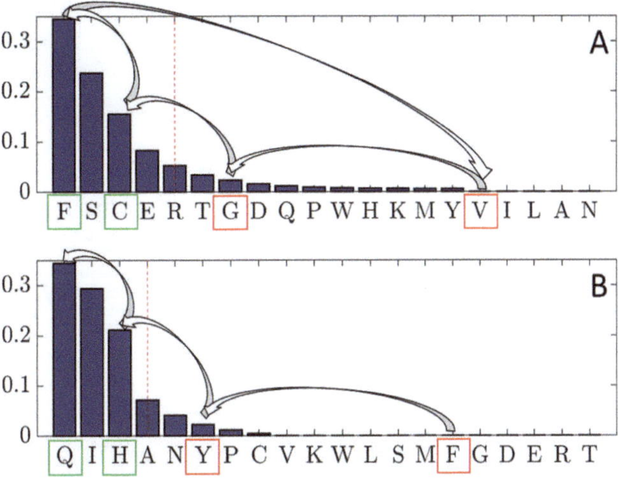

Fig. 1 It can be useful to think of the substitution process at a site as movement on a site-specific fitness landscape. The horizontal axis in each figure shows the amino acids at a hypothetical site in order of their stationary frequencies indicated by the height of the bars. Frequency is a function of mutation and selection, but can be construed as a proxy for fitness. The site-specific *dN*/*dS* ratio [25] is a function of the amino acid that occupies the site, and can be <1 (left of the red dashed line) or >1 (right of the dashed red line). (**a**) Suppose phenylalanine (F, TTT) is the fittest amino acid. The site-specific *dN*/*dS* ratio is much less than one when occupied by F because any nonsynonymous mutation will always be to an amino acid that is less fit. Nevertheless, it is possible for an amino acid such as valine (V, GTT) to be fixed on occasion, provided that selection is not too stringent. When this happens, *dN*/*dS* at the site is temporarily elevated to a value greater than one as positive selection moves the site back to F by a series of replacement substitutions, e.g., V (GTT) $\rightarrow$ G (GGT) $\rightarrow$ C (TGT) $\rightarrow$ F (TTT). We call the episodic recurrence of this process **shifting balance** on a static fitness landscape. Shifting balance on a landscape for which all frequencies are approximately equal corresponds to **nearly neutral** evolution (not depicted), when *dN*/*dS* is always ≈ 1. (**b**) Now, consider what happens following a change in one or more external factors that impact the functional significance of the site. The relative fitnesses of the amino acids might change from that depicted in **a** to that in **b** for instance, where glutamine (Q) is fittest. If at the time of the change the site is occupied by F (as is most likely), then *dN*/*dS* would be temporarily elevated as positive selection moves the site toward its new peak at Q, e.g., F (TTT) $\rightarrow$ Y (TAT) $\rightarrow$ H (CAT) $\rightarrow$ Q (CAA). This process of **adaptive evolution** is followed by a return to shifting balance once the site is occupied by Q

for 61^N categories). We refer to $\boldsymbol{P} = P_{\mathrm{M}}(\theta_{\mathrm{M}})$ as the site-pattern distribution for that model. Figure 2 depicts the space of all possible site-pattern distributions for an N-taxon alignment. Each ellipse represents the family of distributions $\{P_{\mathrm{M}}(\theta_{\mathrm{M}})|\theta_{\mathrm{M}} \in \Omega_{\mathrm{M}}\}$, where Ω_{M} is the vector space of all possible values of θ_{M}. For example, $\{P_{\mathrm{M0}}(\theta_{\mathrm{M0}})|\theta_{\mathrm{M0}} \in \Omega_{\mathrm{M0}}\}$ is the family of distributions that can be

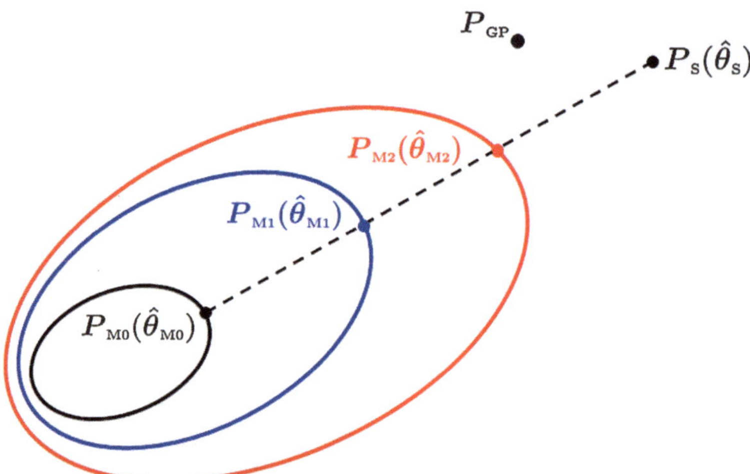

Fig. 2 The $(61^N - 1)$-dimensional simplex containing all possible site-pattern distributions for an N-taxon alignment is depicted. The innermost ellipse represents the subspace $\{P_{M0}(\theta_{M0})|\theta_{M0} \in \Omega_{M0}\}$ that is the family of distributions that can be specified using M0, the simplest of CSMs. This is nested in the family of distributions that can be specified using M1 (blue ellipse), a hypothetical model that has the same parameters as M0 plus some extra parameters. Similarly, M1 is nested in M2 (red ellipse). Whereas models are represented by subspaces of distributions, the true generating process is represented by a single point P_{GP}, the location of which is unknown. The empirical site-pattern distribution $P_S(\hat{\theta}_S)$ corresponds to the saturated model fitted to the alignment; with large samples, $P_S(\hat{\theta}_S) \approx P_{GP}$. For any other model M, the member $P_M(\hat{\theta}_M) \in \{P_M(\theta_M) \mid \theta_M \in \Omega_M\}$ most consistent with X is the one that minimizes deviance, which is twice the difference between the maximum log-likelihood of the data under the saturated model and the maximum log-likelihood of the data under M

specified using M0, the simplest CSM that assumes a common substitution rate matrix Q for all sites and branches. This is nested inside $\{P_{M1}(\theta_{M1})|\theta_{M1} \in \Omega_{M1}\}$, where M1 is a hypothetical model that is the same as M0 but for a few extra parameters. Likewise, M1 is nested in M2. The location of the site-pattern distribution for the true generating process is represented by P_{PG}. Its location is fixed but unknown. It is therefore not possible to assess the distance between it and any other distribution. Instead, comparisons are made using the site-pattern distribution inferred under the saturated model.

Whereas a CSM $\{P_M(\theta_M)|\theta_M \in \Omega_M\}$ can be thought of as a family of multinomial distributions for the 61^N possible site patterns, the fitted saturated model $P_S(\hat{\theta}_S)$ is the unique distribution defined by the MLE $\hat{\theta}_S = (y_1/n, \ldots, y_m/n)^T$, where $y_i > 0$ is the observed frequency of the ith site pattern, m is the number of unique site patterns, and n is the number of codon sites. In other words, the fitted saturated model is the empirical site-pattern distribution for a given alignment. Because it takes none of the mechanisms of mutation or selection into account, ignores the phylogenetic relationships between sequences, and excludes the possibility of site patterns that were not actually observed (i.e., $y_i/n = 0$ for site patterns i not observed in X), $P_S(\hat{\theta}_S)$ can be construed as the maximally phenomenological explanation of the observed alignment. An alignment is always more likely under the saturated model than it is under any other CSM. $P_S(\hat{\theta}_S)$ therefore provides a natural benchmark for model improvement.

For any alignment, the MLE over the family of distributions $\{P_M(\theta_M)|\theta_M \in \Omega_M\}$ is represented by a fixed point $P_M(\hat{\theta}_M)$ in Fig. 2. $P_M(\hat{\theta}_M)$ is the distribution that minimizes the statistical deviance between $P_M(\theta_M)$ and $P_S(\hat{\theta}_S)$. Deviance is defined as twice the difference between the maximum log-likelihood (LL) of the data under the saturated model and the maximum log-likelihood of the data under M:

$$D(\hat{\theta}_M, \hat{\theta}_S) = 2\{\ell(\hat{\theta}_S \mid X) - \ell(\hat{\theta}_M \mid X)\} \tag{2}$$

A key feature of deviance is that it always decreases as more parameters are added to the model, corresponding to an increase in the probability of the data under that model. For example, suppose $\{P_{M2}(\theta_{M2})|\theta_{M2} \in \Omega_{M2}\}$ is the same family of distributions as $\{P_{M1}(\theta_{M1})|\theta_{M1} \in \Omega_{M1}\}$ but for the inclusion of one additional parameter ψ, so that $\theta_{M2} = (\theta_{M1}, \psi)$. The improvement in the probability of the data under $P_{M2}(\hat{\theta}_{M2})$ over its probability under $P_{M1}(\hat{\theta}_{M1})$ is assessed by the size of the reduction in deviance induced by ψ:

$$\begin{aligned} \Delta D(\hat{\theta}_{M1}, \hat{\theta}_{M2}) &= D(\hat{\theta}_{M1}, \hat{\theta}_S) - D(\hat{\theta}_{M2}, \hat{\theta}_S) \\ &= 2\{\ell(\hat{\theta}_{M2} \mid X) - \ell(\hat{\theta}_{M1} \mid X)\} \end{aligned} \tag{3}$$

Equation 3 is just the familiar log-likelihood ratio (LLR) used to compare nested models under the maximum likelihood framework.

Given this measure of model improvement, the de facto objective of model building is not to provide a mechanistic explanation of the data that more accurately represents the true generating process, but only to move closer to the site-pattern distribution of the fitted saturated model. Real alignments are limited in size, so there will always be some distance between $P_S(\hat{\theta}_S)$ and P_{GP} due to sampling error (as represented in Fig. 2). But even with an infinite number of codon sites, when $P_S(\hat{\theta}_S)$ converges to P_{GP}, the criterion of minimizing deviance does not inevitably lead to a better explanation of the data because of the possibility of confounding. Two processes are said to be confounded if they can produce similar patterns in the data. Hence, if ψ represents a process E that did not actually occur when the data was generated, and if E is confounded with another process that did occur, the LLR in Eq. 3 can still be significant. Under this scenario, the addition of ψ to M1 would engender movement toward $P_S(\hat{\theta}_S)$ and P_{GP}, but the new model M2 would also provide a worse mechanistic explanation of the data because it would falsely indicate that E occurred. The possibility of confounding and its impact on inference is demonstrated in Case Study D.

3 Phase I: Pioneering CSMs

The first effort to detect positive selection at the molecular level [24] relied on heuristic counting methods [43]. Phase I of CSM development followed with the introduction of formal statistical approaches based on ML [16, 42]. The first CSMs were used to infer whether the estimate $\hat{\omega}$ of a single nonsynonymous to synonymous substitution rate ratio averaged over all sites and branches was significantly greater than one. Such CSMs were found to have low power due to the pervasiveness of synonymous substitutions at most sites within a typical gene [76]. An early attempt to increase the statistical power to infer positive selection was the CSM designed to detect $\hat{\omega} > 1$ on specific branches [78]. Models accounting for variations in ω across sites were subsequently developed, the most prominent of which are the M-series models [78, 81]. These were accompanied by methods to identify individual sites under positive selection. The quest for power culminated in the development of models that account for variations in the rate ratio across both sites and branches. The appearance of various branch-site models (e.g., [4, 10, 79, 86]) marks the end of Phase I of CSM development.

Two case studies are employed in this section to illustrate some of the inferential challenges associated with Phase I models. We use Case Study A to examine the impact of low information content on the inference of positive selection at individual codon sites. The subject of this study is the M1a vs M2a model contrast applied to the *tax* gene of the human T-cell lymphotropic virus type I (HTLV-I; [63, 82]). We use Case Study B to illustrate how model misspecification (i.e., differences between the fitted model and the generating process) can lead to false inferences. The subject of this study is the Yang–Nielsen branch-site model (YN-BSM; [79]) applied to simulated data.

3.1 Case Study A: Low Information Content

To study the impact of low information content on inference, we use a pair of nested M-series models known as M1a and M2a [70, 82]. Under M1a, sites are partitioned into two rate-ratio categories, $0 < \omega_0 < 1$ and $\omega_1 = 1$ in proportions p_0 and $p_1 = 1 - p_0$. M2a includes an additional category for the proportion of sites $p_2 = 1 - p_0 - p_2$ that evolved under positive selection with

$\omega_2 > 1$. The use of multiple categories permits two levels of inference. The first is an omnibus likelihood ratio test (LRT) for evidence of positive selection somewhere in the gene, which is conducted by contrasting a pair of nested models. For example, the contrast of M1a vs M2a is made by computing the distance $\text{LLR} = \Delta D(\hat{\theta}_{\text{M1a}}, \hat{\theta}_{\text{M2a}})$ between the two models and comparing the result to the limiting distribution of the LLR under the null model. In this case, the limiting distribution of LLR is often taken to be χ_2^2 [75], which would be correct under regular likelihood theory because the models differ by two parameters. The second level of inference is used to identify individual sites that underwent positive selection. This is conducted only if positive selection is inferred by the omnibus test (e.g., if LLR > 5.99 for the M1a vs M2a contrast at the 5% level of significance). Let c_0, c_1, and c_2 represent the event that a given site pattern x falls into the stringent ($0 < \hat{\omega}_0 < 1$), neutral ($\hat{\omega}_1 = 1$), or positive ($\hat{\omega}_2 > 1$) selection category, respectively. Applying Bayes' rule:

$$\Pr(c_2 \mid x, \hat{\theta}_{\text{M2a}}) = \frac{\Pr(x \mid c_2, \hat{\theta}_{\text{M2a}})\hat{p}_2}{\sum_{k=0}^{2} \Pr(x \mid c_k, \hat{\theta}_{\text{M2a}})\hat{p}_k} \tag{4}$$

Sites with a sufficiently high posterior probability (e.g., $\Pr(c_2 \mid x, \hat{\theta}_{\text{M2a}}) > 0.95$) are inferred to have undergone positive selection. Equation 4 is representative of the naive empirical Bayes (NEB) approach under which MLEs ($\hat{\theta}_{\text{M2a}}$) are used to compute posterior probabilities.

The NEB approach ignores potential errors in parameter estimates that can lead to false inference of positive selection at a site (i.e., a false positive). The resulting false positive rate can be especially high for alignments with low information content. An example setting with low information content arises when there are a substantial number of invariant sites, since these provide little information about the substitution process. The issue of low information content is well illustrated by the extreme case of the *tax* gene, HTLV-I [63]. The alignment consists of 20 sequences with 181 codon sites, 158 of which are invariant. The 23 variable sites have only one substitution each: 2 are synonymous and 21 are nonsynonymous. The high ratio of nonsynonymous-to-synonymous substitutions suggests that the gene underwent positive selection. This hypothesis was supported by analytic results: the LLR for the M1a vs M2a contrast was 6.96 corresponding to a p-value of approximately 0.03 [82]. The omnibus test therefore supported the conclusion that the gene underwent positive selection. However, the MLE for p_2 under M2a was $\hat{p}_2 = 1$. Using this value in Eq. 4 gives $\Pr(c_2 \mid x, \hat{\theta}_{\text{M2a}}) = 1$ for all sites, including the 158 invariable sites. Such an unreasonable result can occur under NEB because, despite the possibility of large sampling errors in

MLEs due to low information, $\hat{\theta}_{M2a}$ is treated as a known value in Eq. 4.

Bayes empirical Bayes (BEB; [82]), a partial Bayesian approach under which rate ratios and their corresponding proportions are assigned discrete prior distributions (cf. [21]), was proposed as an alternative to NEB. Numerical integration over the assumed priors tends to provide better estimates of posterior probabilities, particularly in cases where information content is low. Using BEB in the analysis of the *tax* gene, for example, the posterior probability was $0.91 < \Pr(c_2 \mid x,\hat{\theta}_{M2a}) < 0.93$ for the 21 sites with a single nonsynonymous change and $0.55 < \Pr(c_2 \mid x,\hat{\theta}_{M2a}) < 0.61$ for the remaining sites [82]. Hence, the BEB approach mitigated the problem of low information content, as the posterior probability of positive selection at invariant sites was reduced. An alternative to BEB is called smoothed bootstrap aggregation (SBA) [38]. SBA entails drawing site patterns from X with replacement (i.e., bootstrap) to generate a set of alignments $\{X_1, \ldots, X_m\}$ with similar information content as X. The MLEs $\{\hat{\theta}_i\}_{i=1}^{m}$ for the vector of model parameters θ are then estimated by fitting the CSM to each $X_i \in \{X_1, \ldots, X_m\}$. A kernel smoother is applied to these values to reduce sampling errors. The mean value of the resulting smoothed $\{\hat{\theta}_i\}_{i=1}^{m}$ is then used in Eq. 4 in place of the MLE for θ obtained from the original alignment to estimate posterior probabilities. This approach was shown to balance power and accuracy at least as well as BEB. But, SBA has the advantage that it can accommodate the uncertainty of all parameter estimates (not just those of the ω distribution, as in BEB) and is much easier to implement. When SBA was applied to the *tax* gene, the posterior probabilities for positive selection were further reduced: $0.87 < \Pr(c_2 \mid x,\hat{\theta}_{M2a}) < 0.89$ for the 21 sites with a single nonsynonymous change, and $0.55 < \Pr(c_2 \mid x,\hat{\theta}_{M2a}) < 0.60$ for the remaining sites [38].

The problem of low information content was fairly obvious in the case of the *tax* gene, as 158 of the 181 codon sites within that dataset were invariant. However, it can sometimes be unclear whether there is enough variation in an alignment to ensure reliable inferences. It would be useful to have a method to determine whether a given data set might be problematic. An MLE $\hat{\theta}$ will always converge to a normal distribution centered at the true parameter value θ with variance proportional to $1/n$ as the sample size n (a proxy for information content) gets larger, provided that the CSM satisfies certain "regularity" conditions (a set of technical conditions that must hold to guarantee that MLEs will converge in distribution to a normal, and that the LLR for any pair of nested models will converge to its expected chi-squared distribution). This expectation makes it possible to assess whether an alignment is sufficiently informative to obtain the benefits of regularity. The first step is to generate a set of bootstrap alignments $\{X_1, \ldots, X_m\}$. The CSM can then be fitted to these to produce a sample distribution $\{\hat{\theta}_i\}_{i=1}^{m}$ for the MLE of any model parameter θ. If the alignment is sufficiently informative with respect θ, then a histogram of $\{\hat{\theta}_i\}_{i=1}^{m}$ should be approximately normal in distribution. Serious departures from normality (e.g., a bimodal distribution) indicate unstable MLEs, which are a sign of insufficient information or an irregular modeling scenario. Mingrone et al. [38] recommend using this technique with real data as a means of gaining insight into potential difficulties of parameter estimation using a given CSM.

3.1.1 Irregularity and Penalized Likelihood

Issues associated with low information content can be made worse by violations of certain regularity conditions. For example, M2a is the same as M1a but for two extra parameters, p_2 and ω_2. Usual likelihood theory would therefore predict that the limiting distribution of the LLR is χ_2^2. However, this result is valid only if the regularity conditions hold. Among these conditions is that the null model is not obtained by placing parameters of the alternate model on the boundary of parameter space. Since M1a is the same as M2a but with $p_2 = 0$, this condition is violated. The same can be said for many nested pairs of Phase I CSMs, such as M7 vs M8 [81] or M1 vs branch-site Model A [79]. Although the theoretical limiting distribution of the LLR under some irregular conditions has been determined by Self and Liang [54], those results do not include cases where one of the model parameters is unidentifiable under the null [2]. Since M1a is M2a with $p_2 = 0$, the likelihood under M1a is the same for any value of ω_2. This makes ω_2 unidentifiable under the null. The limiting distribution for the M1a vs M2a contrast is therefore unknown [74].

A penalized likelihood ratio test (PLRT; [39]) has been proposed to mitigate problems associated with unidentifiable parameters. Under this method, the likelihood function for the alternate model (e.g., M2a) is modified so that values of p_2 closer to zero are penalized. This has the effect of drawing the MLE for p_2 away from the boundary, and can interpreted as a way to "regularize" the model. PLRT seems to be more useful in cases where the analysis of a real alignment produces a small value of $\hat{p}_2$ accompanied by an unrealistically large value of $\hat{\omega}_2$. This can happen because $\hat{\omega}_2$ is influenced by fewer and fewer site patterns as $\hat{p}_2$ approaches zero, and is therefore subject to larger and larger sampling errors. In addition, $\hat{\omega}_2$ and $\hat{p}_2$ tend to be negatively correlated, which further contributes to the large sampling errors. For example, Mingrone et al. [39] found that M2a fitted to a 5-taxon alignment with 198 codon sites without penalization gave $(\hat{p}_2, \hat{\omega}_2) = (0.01, 34.70)$. These MLEs, if taken at face value, suggest that a small number of sites in the gene underwent positive selection. However, such a large rate ratio is difficult to believe given that its estimate is consistent with only approximately 2 codon sites (e.g., an estimated 1% of the 198 sites or ≈2 sites). Using the PLRT, the MLEs were $(\hat{p}_2,\hat{\omega}_2) = (0.09,1.00)$. These suggest that selection pressure was nearly neutral at a significant proportion of sites in the gene. In this case, the rate ratio is consistent with 9% of the 198 sites or ≈18 sites and is therefore less likely to be an artifact of sampling error. We expect this approach to be useful in a wide variety of evolutionary applications that rely on mixture models to make inferences (e.g., [13, 34, 47, 66]).

Other approaches for dealing with low information content in the data for an individual gene include the empirical Bayes approach of Kosiol et al. [33] and the parametric bootstrapping methods of Gibbs [14]. Both methods exploit the additional information content available from other genes. Kosiol et al. [33] adopted an empirical Bayes approach, where ω values varied over edges and genes according to a distribution. Because empirical posterior distributions are used, the approach is more akin to detecting sites under positive selection (e.g., using NEB) than formal testing. By contrast, Gibbs [14] adopted a test-based approach and utilized parametric bootstrapping [15] to approximate the distribution of the likelihood ratio statistic using data from other genes to obtain parameter sets to use in the bootstrap. Whereas this approach can attenuate issues associated with low information content, it can also be computationally expensive, especially when applied to large alignments.

3.2 Case Study B: Model Misspecification

The mechanisms that give rise to the diversity of site patterns in a set of homologous genes are highly complex and not fully understood. CSMs are therefore necessarily simplified representations of the true generating process, and are in this sense misspecified. The extent to which misspecification might cause an omnibus LRT to falsely detect positive selection was of primary concern during Phase I of model development. We use a particular form of the YN-BSM called Model A [79] to illustrate this issue. In its original form, the omnibus LRT assumes a null under which a proportion p_0 of sites evolved under stringent selection with $\omega_0 = 0$ and the remaining sites evolved under a neutral regime with $\omega_1 = 1$ on all branches of the tree (i.e., model M1 in [44]). This is contrasted with Model A, which is the same as M1 except that it assumes that some stringent sites and some neutral sites evolved under positive selection with $\omega_2 > 1$ on a prespecified branch called the foreground branch. The omnibus test contrasting M1 with Model A was therefore designed to detect a subset of sites that evolved adaptively on the same branch of the tree.

During this period of model development, the standard method to test the impact of misspecification on the reliability of

Table 1
Rate ratios (ω) for regimes X and Z taken from Zhang [85]

Sites	1–20	21–40	41–60	61–80	81–100	101–120	121–140	141–160	161–180	181–200
ω regime X	1.00	1.00	0.80	0.80	0.50	0.50	0.20	0.20	0.00	0.00
ω regime Z	1.00	1.00	1.00	1.00	1.00	1.00	1.00	1.00	1.00	1.00

an omnibus LRT was to generate alignments in silico using a more complex version of the CSM to be tested as the generating model. This usually involved adding more variability in ω across sites and/or branches than assumed by the fitted CSM while leaving all other aspects of the generating model the same. In Zhang [85], for example, alignments were generated using site-specific rate matrices, as in Eq. 1, with rate ratios ω specified by predetermined selection regimes, two of which are shown in Table 1. In one simulation, 200 alignments were generated using regime Z on a single foreground branch and regime X on all of the remaining branches of a 10 or 16 taxon tree. The gene therefore underwent a mixture of stringent selection and neutral evolution over most of the tree (regime X), but with complete relaxation of selection pressure on the foreground branch (regime Z). Positive selection did not occur at any of the sites. Nevertheless, the M1 vs Model A contrast inferred positive selection in 20–55% of the alignments, depending on the location of the foreground branch. Such a high rate of false positives was attributed to the mismatch between the process used to generate the data compared to the process assumed by the null model M1 [85].

The branch-site model was subsequently modified to allow $0 < \omega_0 < 1$ instead of $\omega_0 = 0$ (Modified Model A in [86]). Furthermore, the new null model is specified under the assumption that some proportion p_0 of sites (the stringent sites) evolved under stringent selection with $0 < \omega_0 < 1$ everywhere in the tree except on the foreground branch, where those same sites evolved neutrally with $\omega_2 = 1$. All other sites in the alignment (the neutral sites) are assumed to have evolved neutrally with $\omega_1 = 1$ everywhere in the tree. This is contrasted with the Modified Model A, which assumes that some of the stringent sites and some of the neutral sites evolved under positive selection with $\omega_2 > 1$ on the foreground. Hence, unlike the original omnibus test that contrasts M1 with Model A, the new test contrasts Modified Model A with $\omega_2 = 1$ against Modified Model A with $\omega_2 > 1$. These changes to the YN-BSM were shown to mitigate the problem of false inference. For example, using the same generating model with regimes X and Z, the modified omnibus test falsely inferred positive selection in only 1–7.5% of the alignments, consistent with the 5% level of significance of the test [86].

This case study demonstrates how problems associated with model misspecification were traditionally identified, and how they could be completely corrected through relatively minor changes to the model. However, the generating methods employed by studies such as Zhang [85] and Zhang et al. [86], although sophisticated for their time, produced alignments that were highly unrealistic compared to real data. For example, it was recently shown that a substantial proportion of variation in many real alignments might be due to selection effects associated with shifting balance over static site-specific fitness landscapes [25, 26]. This process results in random changes in site-specific rate ratios, or heterotachy, that cannot be replicated using traditional CSMs as the generating model. While the mitigation of statistical pathologies due to low information content (e.g., using BEB or SBA) or model misspecification (e.g., by altering the null and alternative hypotheses or the use of penalized likelihood) were critical advancements during Phase I of CSM development, other statistical pathologies went unrecognized due to reliance on unrealistic simulation methods. This issue is taken up in the next section.

4 Phase II: Advanced CSMs

A typical protein-coding gene evolves adaptively only episodically [59]. The evidence of adaptive evolution of this type can be very difficult to detect. For example, it is assumed under the YN-BSM that a random subset of sites switched from a stringent or neutral selection regime to positive selection together on the same set of foreground branches. The power to detect a signal of this kind can be very low when the proportion of sites that switched together is small [77]. Perhaps encouraged by the reliability of Phase I models demonstrated by extensive simulation studies [2, 3, 29, 31, 37, 70, 77, 82, 85, 86], combined with experimental validation of results obtained from their application to real data [1, 71, 76], investigators began to formulate increasingly complex and parameter-rich CSMs [31, 41, 48, 50, 55, 64, 65]. The hope was that carefully selected increases in model complexity would yield greater power to detect subtle signatures of positive selection overlooked by Phase I models. The introduction of such CSMs marks the beginning of Phase II of their historical development.

Phase II models fall into three broad categories:

1. The first consists of Phase I CSMs modified to account for more variability in selection effects across sites and branches than previously assumed, with the aim of increasing the power to detect subtle signatures of positive selection (e.g., the branch-site random effects likelihood model, BSREL; [31]).

2. The second category includes Phase I CSMs modified to contain parameters for mechanistic processes not directly associated with selection effects. Many such models have been motivated by a particular interest in the added mechanism (e.g., the fixation of double and triple mutations; [26, 40, 83]), or by the notion that increasing the mechanistic content of a CSM can only improve inferences about selection effects (e.g., by accounting for variations in the synonymous substitution rate; [30, 51]).
3. The third category of models abandons the traditional formulation of Eq. 1 in favor of a substitution process expressed in terms of explicit population genetic parameters, such as population size and selection coefficients [45, 48–50, 64, 65].

An example of the first category of models is BSREL, which accounts for variations in selection effects across sites and over branches by assuming a different rate-ratio distribution $\{(\omega_i^b, p_i^b) : i = 1, \ldots, k_b)\}$ for each branch b of a tree [31]. BSREL was later found to be more complex than necessary, so an adaptive version was formulated to allow the number of components k_b on a given branch to adjust to the apparent complexity of selection effects on that branch (aBSREL; [55]). A further reduction in model complexity led to the formulation of the test known as BUSTED (for branch-site unrestricted statistical test for episodic diversification; [41]), which we use to illustrate the problem of confounding in Case Study C. An example of the second category of models is the addition of parameters for the rate of double and triple mutations to traditional CSMs, the most sophisticated version of which is RaMoSSwDT (for Random Mixture of Static and Switching sites with fixation of Double and Triple mutations; [26]). This model is used in Case Study D to illustrate the problem of phenomenological load.

Models in the third category are the most ambitious CSMs currently in use, and are far more challenging to fit to real alignments than traditional models. One of the most impressive examples of their application is the site-wise mutation-selection model (swMutSel; [64, 65]) fitted to a concatenated alignment of 12 mitochondrial genes (3598 codon sites) from 244 mammalian species. Based on the mutation-selection framework of Halpern and Bruno [19], swMutSel estimates a vector of selection coefficients for each site in an alignment. This and similar models (e.g., [48–50]) appear to be reliable [58], but require a very large number of taxa (e.g., hundreds). Phase II models of this category are therefore impractical for the majority of empirical datasets. Here, we utilize MutSel as an effective means to generate realistic alignments with plausible levels of variation in selection effects across sites and over time rather than as a tool of inference.

4.1 Case Study C: Confounding

By expressing the codon substitution process in terms of explicit population genetic parameters, the MutSel framework facilitates the investigation of complex evolutionary dynamics, such as shifting balance on a fixed fitness landscape or adaptation to a change in selective constraints (i.e., a peak shift; [6, 25]) that are missing from alignments generated using traditional methods. Specifically, by assigning a different vector of fitness coefficients for the 20 amino acids to each site, MutSel can generate more variation in rate ratio across sites and over time than has been realized in the past simulation studies (e.g., Table 1). In this way, MutSel provides the basis of a generating model that can be adjusted to produce alignments that closely mimic real data [26]. MutSel therefore serves to connect demonstrably plausible evolutionary dynamics to the pathology we refer to as confounding.

Under MutSel, the dynamic regime at the *h*th codon site (e.g., shifting balance, neutral, nearly neutral, or adaptive evolution) is uniquely specified by a vector of fitness coefficients $\mathbf{f}^h = f_1^h, \ldots, f_m^h$. It is generally assumed that mutation to any of the three stop codons is lethal, so $m = 61$ for nuclear genes and $m = 60$ for mitochondrial genes. And, although it is not a requirement, it is typical to assume that the f_j^h are constant across synonymous codons [25, 57]. Given $\mathbf{f}^h$, the elements of a site-specific instantaneous rate matrix A^h can be defined as follows for all $i \neq j$ (cf. Eq. 1):

$$A_{ij}^h \propto \begin{cases} \mu_{ij} & \text{if } s_{ij}^h = 0 \\ \mu_{ij} \dfrac{s_{ij}^h}{1 - \exp\left(-s_{ij}^h\right)} & \text{otherwise} \end{cases} \quad (5)$$

where μ_{ij} is the rate at which codon i mutates to codon j and $s_{ij}^h = 2N_e(f_j^h - f_i^h)$ is the scaled selection coefficient for a population of haploids with effective population size N_e. The probability that the new mutant j is fixed is approximated by $s_{ij}^h/\{1 - \exp(-s_{ij}^h)\}$ [9, 28].

The rate matrix A^h defines the dynamic regime for the site as illustrated in Fig. 3. The bar plot shows codon frequencies $\boldsymbol{\pi}^h = \pi_1^h, \ldots, \pi_m^h$ sorted in descending order. A site spends most of its time occupied by codons to the left or near the "peak" of its landscape. The codon-specific rate ratio for the site (dN_i^h/dS_i^h for codon i) is low near the peak (red line plot in Fig. 3) since mutations away from the peak are seldom fixed. However, if selection is not too stringent, the site will occasionally drift to the right into the "tail" of its landscape. When this occurs, the codon-specific rate

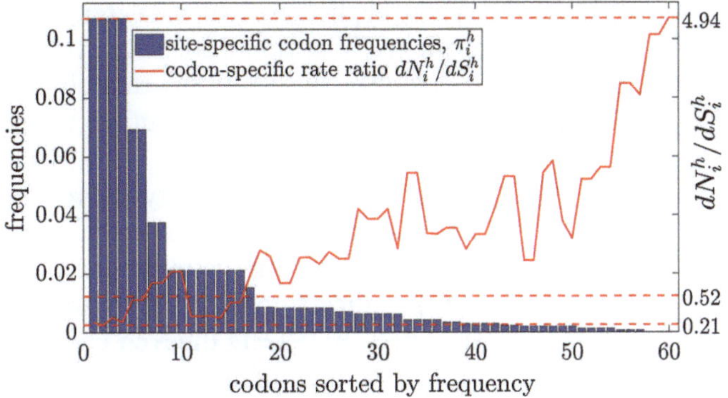

Fig. 3 Fitness coefficients for the 20 amino acids were drawn from a normal distribution centered at zero and with standard deviation $\sigma = 0.001$. Bars show the resulting stationary frequencies (a proxy for fitness) sorted from largest to smallest. They compose a metaphorical site-specific landscape over which the site is imagined to move. The solid red line shows the codon-specific rate ratio dN_i^h/dS_i^h for the sorted codons. This varies depending on the codon currently occupying the site, and can be greater than one following a chance substitution into the tail (to the right) of the landscape. In this case, the codon-specific rate ratio for the site ranged from 0.21 to 4.94 with a temporally averaged site-specific rate ratio of $dN^h/dS^h = 0.52$

ratio will be elevated for a time until a combination of drift and positive selection moves the site back to its peak. This dynamic between selection and drift is reminiscent of Wright's shifting balance. It implies that, when a population is evolving on a fixed fitness landscape (i.e., with no adaptive evolution), its gene sequences can nevertheless contain signatures of temporal changes in site-specific rate ratios (heterotachy), and that these might include evidence of transient elevation to values greater than one (i.e., positive selection). Such signatures of positive selection due to shifting balance can be detected by Phase II CSMs [25].

For example, BUSTED [41] was developed as an omnibus test for episodic adaptive evolution. The underlying CSM was formulated to account for variations in the intensity of selection over both sites and time modeled as a random effect. This is in contrast to the YN-BSM, which treats temporal changes in rate ratio as a fixed effect that occurs on a prespecified foreground branch (although the sites under positive selection are still a random effect). We therefore refer to the CSM underlying BUSTED as the random effects branch-site model (RE-BSM) to serve as a reminder of this important distinction. Under RE-BSM, the rate ratio at each site and branch combination is assumed to be an independent draw from the distribution $\{(\omega_0, p_0), (\omega_1, p_1), (\omega_2, p_2)\}$. In this way, the model accounts for variations in selection effects both across sites and over time. BUSTED contrasts the null hypothesis that $\omega_0 \leq \omega_1 \leq \omega_2 = 1$ with the alternative that $\omega_0 \leq \omega_1 \leq 1 \leq \omega_2$.

When applied to real data, rejection of the null is interpreted as evidence of episodic adaptive evolution.

Unlike the YN-BSM that aims to detect a subset of sites that underwent adaptive evolution together on the same foreground branches (i.e., coherently), BUSTED was designed to detect heterotachy similar to the type predicted by the mutation-selection framework: shifting balance on a static fitness landscape. Jones et al. [25] recently demonstrated that plausible levels of shifting balance can produce signatures of episodic positive selection that can be detected. BUSTED inferred episodic positive selection in as many as 40% of alignments generated using the MutSel framework. Significantly, BUSTED was correct to identify episodic positive selection in these trials. Even though the generating process assumed fixed site-specific landscapes (so there was no episodic adaptive evolution), and the long-run average rate ratio at each site was necessarily less than one [57], positive selection nevertheless did sometimes occur by shifting balance. This illustrates the general problem of confounding. Two processes are said to be confounded if they can produce the same or similar patterns in the data. In this case, episodic adaptive evolution (i.e., the evolutionary response to changes in site-specific landscapes) and shifting balance (i.e., evolution on a static fitness landscape) are confounded because they can both produce rate-ratio distributions that indicate episodic positive selection. The possibility of confounding underlines the fact that there are limitations in what can be inferred about evolutionary processes based on an alignment alone.

4.2 Case Study D: Phenomenological Load

Phenomenological load (PL) is a statistical pathology related to both model misspecification (Case Study B) and confounding (Case Study C) that was not recognized during Phase I of CSM development. When a model parameter that represents a process that played no role in the generation of an alignment (i.e., a misspecified process) nevertheless absorbs a significant amount of variation, its MLE is said to carry PL [26]. This is more likely to occur when the misspecified process is confounded with one or more other processes that did play a role in the generation of the data, and when a substantial proportion of the total variation in the data is unaccommodated by the null model [26]. PL increases the probability that a hypothesis test designed to detect the misspecified process will be statistically significant (as indicated by a large LLR) and can therefore lead to the incorrect conclusion that the misspecified process occurred. Critically, Jones et al. [26] showed that PL was only detected when model contrasts were fitted to data generated with realistic evolutionary dynamics using the MutSel model framework.

To illustrate the impact of PL, we consider the case of CSMs modified to detect the fixation of codons following simultaneous double and triple (DT) nucleotide mutations. The majority of

CSMs currently in use assume that codons evolve by a series of single-nucleotide substitutions, with the probability for DT changes set to zero. However, recent model-based analyses have uncovered evidence for DT mutations [32, 68, 83]. Early estimates of the percentage of fixed mutations that are DT were perhaps unrealistically high. Kosiol et al. [32], for example, estimated a value close to 25% in an analysis of over 7000 protein families from the Pandit database [69]. Alternatively, when estimates were derived from a more realistic site-wise mutation-selection model, DT changes comprised less than 1% of all fixed mutations [64]. More recent studies suggest modest rates of between 1% and 3% [5, 20, 27, 53]. Whatever the true rate, several authors have argued that it would be beneficial to introduce a few extra parameters into a standard CSM to account for DT mutations (e.g., [40, 83]). The problem with this suggestion is that episodic fixation of DT mutations can produce signatures of heterotachy consistent with shifting balance.

Recall the comparison of M1, a CSM containing parameters represented by the vector θ_1, and M2, the same model but for the inclusion of one additional parameter ψ, so that $\theta_2 = (\theta_1, \psi)$. The parameter ψ will reduce the deviance of M2 compared to M1 by some proportion of the baseline deviance between the simplest CSM (M0) and the saturated model $P_S(\hat{\theta}_S)$. We call this the percent reduction in deviance (PRD) attributed to $\hat{\psi}$:

$$\mathrm{PRD}(\hat{\psi}) = \frac{\Delta D(\hat{\theta}_{\mathrm{M1}}, \hat{\theta}_{\mathrm{M2}})}{\Delta D(\hat{\theta}_{\mathrm{M0}}, \hat{\theta}_{\mathrm{S}})} \tag{6}$$

Suppose M1 and M2 were fitted to an alignment and that the $\mathrm{LLR} = \Delta D(\hat{\theta}_{\mathrm{M1}}, \hat{\theta}_{\mathrm{M2}})$ was found to be statistically significant. This would lead an analyst to attribute the $\mathrm{PRD}(\hat{\psi})$ to real signal for the process ψ was meant to represent, possibly combined with some PL and noise. Now, consider the case in which the process represented by ψ did not actually occur (i.e., it was not a component of the true generating process). Under this scenario, $\mathrm{PRD}(\hat{\psi})$ would contain no signal, but would be entirely due to PL plus noise. When this is known to be the case, we set $\mathrm{PRD}(\hat{\psi}) = \mathrm{PL}(\hat{\psi})$. As illustrated below, $\mathrm{PL}(\hat{\psi})$ can be large enough to result in rejection of the null, and therefore lead to a false conclusion about the data generating process.

We illustrate PL by contrasting the model RaMoSS with a companion model RaMoSSwDT that accounts for the fixation of DT mutations via two rate parameters, α (the double mutation rate) and β (the triple mutation rate) [26]. RaMoSS combines the standard M-series model M3 with the covarion-like model CLM3 (cf., [12, 18]). Specifically, RaMoSS mixes (with proportion p_{M3}) one model with two rate-ratio categories $\omega_0 < \omega_1$ that are constant over the entire tree with a second model (with proportion

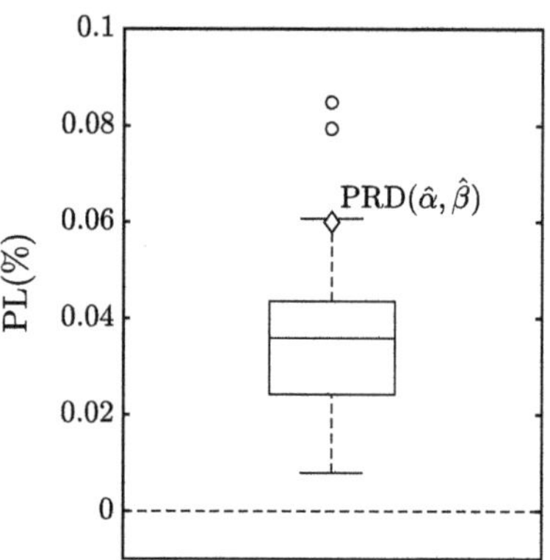

Fig. 4 The box plot depicts the distribution of the phenomenological load (PL) carried by $(\hat{\alpha}, \hat{\beta})$ produced by fitting the RaMoSS vs RaMoSSwDT contrast to 50 alignments generated under MutSel-mmtDNA: the circles represent outliers of this distribution. The diamond is the percent reduction in deviance for the same parameters estimated by fitting RaMoSS vs RaMoSSwDT to the real mtDNA alignment

$p_{CLM3} = 1 - p_{M3}$) under which sites switch randomly in time between $\omega'_0 < \omega'_1$ at an average rate of δ switches per unit branch length. Fifty alignments were simulated to mimic a real alignment of 12 concatenated H-strand mitochondrial DNA sequences (3331 codon sites) from 20 mammalian species as distributed in the PAML package [73]. The generating model, MutSel-mmtDNA [26], was based on the mutation-selection framework and produced alignments with single-nucleotide mutations only. Since DT mutations are not fixed under MutSel-mmtDNA, the PRD carried by $(\hat{\alpha}, \hat{\beta})$ in each trial can be equated to PL (plus noise). The resulting distribution of PL$(\hat{\alpha}, \hat{\beta})$ is shown as a boxplot in Fig. 4.

Although DT mutations were not fixed when the data was generated, shifting balance on a static landscape can produce similar site patterns as a process that includes rare fixation of DT mutations (site patterns exhibiting both synonymous and nonsynonymous substitutions; [26]).[3] DT and shifting balance are therefore confounded. And since shifting balance tends to occur at a substantial proportion (approximately 20%) of sites when an alignment is generated under MutSel-mmtDNA, DT mutations were falsely inferred by the LRT in 48 of 50 trials at the 5% level of significance (assuming LLR $\approx \chi^2_2$ for the two extra parameters α and β in RaMoSSwDT compared to RaMoSS). The PRD $(\hat{\alpha}, \hat{\beta})$ when RaMoSS vs RaMoSSwDT was fitted to the real mmtDNA is

[3] It has previously been noted that the rapid fixation of compensatory mutations following substitution to an unstable base pair (e.g., AT→GT→GC) can also produce site patterns that suggest fixation of DT mutations [74, p. 46].

shown as a diamond in the same plot. Although $(\hat{\alpha},\hat{\beta})$ estimated from the real mmtDNA were found to be highly significant (LLR = 84, p-value << 0.001), the PRD$(\hat{\alpha},\hat{\beta})$ was found to be just under the 95th percentile of PL$(\hat{\alpha},\hat{\beta})$ (PRD = 0.060% compared to the 95th percentile of PL = 0.061). The evidence for DT mutations in the real data is therefore only marginal, and it is reasonable to suspect that its PRD$(\hat{\alpha},\hat{\beta})$, if not entirely the result of PL, is at least partially caused by PL.

5 Discussion

CSMs have been subjected to a certain degree of censure, particularly during Phase I of their development [11, 22, 23, 46, 60–63, 85]. We maintain that it is not the model in and of itself, or the maximum likelihood framework it is based on, that gives rise to statistical pathologies, but the relationship between model and data. This principle was illustrated by our analysis of the history of CSM development, which we divided into two phases. Phase I was characterized by the formulation of models to account for differences in selection effects across sites and over time that comprise the major component of variation in an alignment. Starting with M0, such models represent large steps toward the fitted saturated model in Fig. 2, and also provide a better representation of the true generating process. The main criticism of Phase I models was the possibility of falsely inferring positive selection in a gene or at an individual codon site [62, 63, 85]. But, the most compelling empirical case of false positives was shown to be the result of inappropriate application of a complex model to a sparse alignment [63]. Methods for identifying (bootstrap) and dealing with (BEB, SBA, and PLRT) low information content were illustrated in Case Study A.

The other big concern that arose during Phase I development was the possibility of pathologies associated with model misspecification. The method used to identify such problems was to fit a model to alignments generated under a scenario contrived to be challenging, as illustrated in Case Study B. There, the omnibus test based on Model A of the YN-BSM was shown to result in an excess of false positives when fitted to alignments simulated using the implausible but difficult "XZ" generating scenario (e.g., with complete relaxation of selection pressure at all sites on one branch of the tree; Table 1). Subsequent modifications to the test reduced the false positive rate to acceptable values. Hence, Case Study B underlines the importance of the model–data relationship. However, it is not clear whether a model adjusted to suit an unrealistic data-generating process is necessarily more reliable when fitted to a real alignment. This difficulty highlights the need to find ways, for the purpose of model testing and adjustment, to generate alignments that mimic real data as closely as possible.

Confidence in the CSM approach, combined with the exponential increase in the volume of genetic data and the growth of computational power, spurred the formulation CSMs of ever-increasing complexity during Phase II. The main issue with these models, which has not been widely appreciated, is confounding. Two processes are confounded if they can produce the same or similar patterns in the data. It is not possible to identify such processes when viewed through the narrow lens of an alignment (i.e., site patterns) alone. This was illustrated by Case Study C, where shifting balance on a static landscape was shown to be confounded with episodic adaptive evolution [7, 25]. Confounding can lead to what we call phenomenological load, as demonstrated in Case Study D. In that analysis, the parameters (α, β) were assigned a specific mechanistic interpretation, the rate at which double and triple mutations arise. It was shown that (α, β) can absorb variations in the data caused by shifting balance; hence, the MLEs $(\hat{\alpha}, \hat{\beta})$ resulted in a significant reduction in deviance in 48/50 trials (Fig. 4), and therefore improved the fit of the model to the data. However, the absence of DT mutations in the generating process invalidated the intended interpretation of $(\hat{\alpha}, \hat{\beta})$. This result underlines that a better fit does not imply a better mechanistic representation of the true generating process.

It is natural to assume that a better mechanistic representation of the true generating process can be achieved by adding parameters to our models to account for more of the processes believed to occur. The problem with this assumption is that the metric of model improvement under ML (reduction in deviance) is independent of mechanism. A parameter assigned a specific mechanistic interpretation is consequently vulnerable to confounding with other processes that can produce the same distribution of site patterns. As CSMs become more complex, it seems likely that the opportunity for confounding will only increase. It would therefore be desirable to assess each new model parameter for this possibility using something like the method shown in Fig. 4 whenever possible. The idea is to generate alignments using MutSel or some other plausible generating process in such a way as to mimic the real data as closely as possible, but with the new parameter set to its null value. To provide a second example, consider the test for changes in selection intensity in one clade compared to the remainder of the tree known as RELAX [67]. Under this model, it is assumed that each site evolved under a rate ratio randomly drawn from $\boldsymbol{\omega}_R = \{\omega_1, \ldots, \omega_k\}$ on a set of prespecified reference branches, and from a modified set of rate ratios $\boldsymbol{\omega}_T = \{\omega_1^m, \ldots, \omega_k^m\}$ on test branches, where m is an exponent. A value $0 < m < 1$ moves the rate ratios in $\boldsymbol{\omega}_T$ closer to one compared to their corresponding values in $\boldsymbol{\omega}_R$, consistent with relaxation of selection pressure at all sites on the test branches. Relaxation is indicated when the contrast of the null hypothesis that $m = 1$ versus the alternative that $m < 1$ is statistically significant. The distribution of PL($\hat{m}$) can be estimated from alignments generated with $m = 1$. The PRD($\hat{m}$) estimated from the real data can then be compared to this to assess the impact of PL (cf. Fig. 4). This approach is predicated on the existence of a generating model that could have plausibly produced the site patterns in the real data. Jones et al. [26] present a variety of methods for assessing the realism of a simulated alignment, although further development of such methods is warranted. Software based on MutSel is currently available for generating data that mimic large alignments of 100-plus taxa (Pyvolve; [56]). Other methods have been developed to mimic smaller alignments of certain types of genes (e.g., MutSel-mmtDNA; [25]). It is only by the use of these or other realistic simulation methods that the relationship between a given model and an alignment can be properly understood.

References

1. Anisimova M, Kosiol C (2009) Investigating protein-coding sequence evolution with probabilistic codon substitution models. Mol Biol Evol 26:255–271
2. Anisimova M, Bielawski JP, Yang ZH (2001) Accuracy and power of the likelihood ratio test in detecting adaptive molecular evolution. Mol Biol Evol 18:1585–1592
3. Anisimova M, Bielawski JP, Yang ZH (2002) Accuracy and power of Bayes prediction of amino acid sites under positive selection. Mol Biol Evol 19:950–958
4. Bielawski JP, Yang ZH (2004) A maximum likelihood method for detecting functional divergence at individual codon sites, with application to gene family evolution. J Mol Evol 59:121–132
5. De Maio N, Holmes I, Schlötterer C, Kosiol C (2013) Estimating empirical codon hidden Markov models. Mol Biol Evol 30:725–736
6. dos Reis M (2013). http://arxiv:1311.6682v1. Last accessed 26 Nov 2013
7. dos Reis M (2015) How to calculate the non-synonymous to synonymous rate ratio protein-coding genes under the Fisher-Wright mutation-selection framework. Biol Lett 11:1–4.
8. Field SF, Bulina MY, Kelmanson IV, Bielawski JP, Matz MV (2006) Adaptive evolution of multicolored fluorescent proteins in reef-building corals. J Mol Evol 62:332–339
9. Fisher R (1930) The distribution of gene ratios for rare mutations. Proc R Soc Edinb 50:205–220
10. Forsberg R, Christiansen FB (2003) A codon-based model of host-specific selection in parasites, with an application to the influenza a virus. Mol Biol Evol 20:1252–1259
11. Friedman R, Hughes AL (2007) Likelihood-ratio tests for positive selection of human and mouse duplicate genes reveal nonconservative and anomalous properties of widely used methods. Mol Phylogenet Evol 542:388–393
12. Galtier N (2001) Maximum-likelihood phylogenetic analysis under a covarion-like model. Mol Biol Evol 18:866–873
13. Gaston D, Susko E, Roger AJ (2011) A phylogenetic mixture model for the identification of functionally divergent protein residues. Bioinformatics 27:2655–2663
14. Gibbs RA (2007) Evolutionary and biomedical insights from the Rhesus macaque genome. Science 316:222–234
15. Goldman N (1993) Statistical tests of models of DNA substitution. J Mol Evol 36:182–198
16. Goldman N, Yang ZH (1994) Codon-based model of nucleotide substitution for protein-coding DNA-sequences. Mol Biol Evol 11:725–736
17. Grantham R (1974) Amino acid difference formula to help explain protein evolution. Science 862–864

18. Guindon S, Rodrigo AG, Dyer KA, Huelsenbeck JP (2004) Modeling the site-specific variation of selection patterns along lineages. Proc Natl Acad Sci USA 101:12957–12962
19. Halpern AL, Bruno WJ (1998) Evolutionary distances for protein-coding sequences: modeling site-specific residue frequencies. Mol Biol Evol 15:910–917
20. Harris K, Nielsen R (2014) Error-prone polymerase activity causes multinucleotide mutations in humans. Genome Res 9:1445–1554
21. Huelsenbeck JP, Dyer KA (2004) Bayesian estimation of positively selected sites. J Mol Evol 58:661–672
22. Hughes AL (2007) Looking for Darwin in all the wrong places: the misguided quest for positive selection at the nucleotide sequence level. Heredity 99:364–373
23. Hughes AL, Friedman R (2008) Codon-based tests of positive selection, branch lengths, and the evolution of mammalian immune system genes. Immunogenetics 60:495–506
24. Hughes AL, Nei M (1988) Pattern of nucleotide substitution at major histocompatibility complex class-1 loci reveals overdominant selection. Nature 335:167–170
25. Jones CT, Youssef N, Susko E, Bielawski JP (2017) Shifting balance on a static mutation-selection landscape: a novel scenario of positive selection. Mol Biol Evol 34:391–407
26. Jones CT, Youssef N, Susko E, Bielawski JP (2018) Phenomenological load on model parameters can lead to false biological conclusions. Mol Biol Evol 35:1473–1488
27. Keightley P, Trivedi U, Thomson M, Oliver F, Kumar S, Blaxter M (2009) Analysis of the genome sequences of three Drosophila melanogaster spontaneous mutation accumulation lines. Genet Res 19:1195–1201
28. Kimura M (1962) On the probability of fixation of mutant genes in a population. Genetics 47:713–719
29. Kosakovsky Pond SL, Frost SDW (2005) Not so different after all: a comparison of methods for detecting amino acid sites under selection. Mol Biol Evol 22:1208–1222
30. Kosakovsky Pond SL, Muse SV (2007) Site-to-site variations of synonymous substitution rates. Mol Biol Evol 22:2375–2385
31. Kosakovsky Pond SL, Murrell B, Fourment M, Frost SDW, Delport W, Scheffler K (2011) A random effects branch-site model for detecting episodic diversifying selection. Mol Biol Evol 28:3033–3043
32. Kosiol C, Holmes I, Goldman N (2007) An empirical codon model for protein sequence evolution. Mol Biol Evol 24:1464–1479
33. Kosiol C, Vinař T, daFonseca RR, Hubisz MJ, Bustamante CD, Nielsen R, Siepel A (2008) Patterns of positive selection in six mammalian genomes. PLoS Genet 4:1–17
34. Lartillot N, Philippe H (2004) A Bayesian mixture model for across-site heterogeneities in the amino-acid replacement process. Mol Biol Evol 21:1095–1109
35. Liberles DA, Teufel AI, Liu L, Stadler T (2013) On the need for mechanistic models in computational genomics and metagenomics. Genome Biol Evol 5:2008–2018
36. Lopez P, Casane D, Phillipe H (2002) Heterotachy, and important process of protein evolution. Mol Biol Evol 19:1–7
37. Lu A, Guindon S (2013) Performance of standard and stochastic branch-site models for detecting positive selection among coding sequences. Mol Biol Evol 31:484–495
38. Mingrone J, Susko E, Bielwaski JP (2016) Smoothed bootstrap aggregation for assessing selection pressure at amino acid sites. Mol Biol Evol 33:2976–2989
39. Mingrone J, Susko E, Bielwaski JP (2018) Modified likelihood ratio tests for positive selection (submitted). Bioinformatics, Advance Access https://doi.org/10.1093/bioinformatics/bty1019
40. Miyazawa S (2011) Advantages of a mechanistic codon substitution model for evolutionary analysis of protein-coding sequences. PLoS ONE 6:20
41. Murrell B, Weaver S, Smith MD, Wertheim JO, Murrell S, Aylward A, Eren K, Pollner T, Martin DP, Smith DM, Scheffler K, Pond SLK (2015) Gene-wide identification of episodic selection. Mol Biol Evol 32:1365–1371
42. Muse SV, Gaut BS (1994) A likelihood approach for comparing synonymous and nonsynonymous nucleotide substitution rates, with applications to the chloroplast genome. Mol Biol Evol 11:715–724
43. Nei M, Gojobori T (1986) Simple methods for estimating the numbers of synonymous and nonsynonymous nucleotide substitutions. Mol Biol Evol 3:418–426
44. Nielsen R, Yang ZH (1998) Likelihood models for detecting positively selected amino acid sites and applications to the HIV-1 envelope gene. Genetics 148:929–936
45. Nielsen R, Yang Z (2003) Estimating the distribution of selection coefficients from phylogenetic data with applications to mitochondrial and viral DNA. Mol Biol Evol 20:1231–1239
46. Nozawa M, Suzuki Y, Nei M (2009) Reliabilities of identifying positive selection by the branch-site and the site-prediction methods. Proc Natl Acad Sci USA 106:6700–6705

47. Pagel M, Meade A (2004) A phylogenetic mixture model for detecting pattern-heterogeneity in gene sequence or character-state data. Syst Biol 53:571–581
48. Rodrigue N, Lartillot N (2014) Site-heterogeneous mutation-selection models with the PhyloBayes-MPI package. Bioinformatics 30:1020–1021
49. Rodrigue N, Lartillot N (2016) Detection of adaptation in protein-coding genes using a Bayesian site-heterogeneous mutation-selection codon substitution model. Mol Biol Evol 34:204–214
50. Rodrigue N, Philippe H, Lartillot N (2010) Mutation-selection models of coding sequence evolution with site-heterogeneous amino acid fitness profiles. Proc Natl Acad Sci USA 107:4629–4634
51. Rubinstein ND, Doron-Faigenboim A, Mayrose I, Pupko T (2011) Evolutionary model accounting for layers of selection in protein-coding genes and their impact on the inference of positive selection. Mol Biol Evol 28:3297–3308
52. Sawyer SL, Emerman M, Malik HS (2007) Discordant evolution of the adjacent antiretroviral genes trim22 and trim5 in mammals. PLoS Pathog 3:e197
53. Schrider D, Hourmozdi J, Hahn M (2014) Pervasive multinucleotide mutational events in eukaryotes. Curr Biol 21:1051–1054
54. Self SG, Liang KY (1987) Asymptotic properties of maximum likelihood estimators and likelihood ratio test under nonstandard conditions. J Am Stat Assoc 82:605–610
55. Smith MD, Wertheim JO, Weaver S, Murrell B, Scheffler K, Pond SLK (2015) Less is more: an adaptive branch-site random effects model for efficient detection of episodic diversifying selection. Mol Biol Evol 32:1342–1353
56. Spielman S, Wilke CO (2015) Pyvolve: a flexible Python module for simulating sequences along phylogenies. PLoS ONE 10:1–7
57. Spielman S, Wilke CO (2015) The relationship between dN/dS and scaled selection coefficients. Mol Biol Evol 34:1097–1108
58. Spielman S, Wilke CO (2016) Extensively parameterized mutation-selection models reliably capture site-specific selective constraints. Mol Biol Evol 33:2990–3001
59. Struder RA, Robinson-Rechavi M (2009) Evidence for an episodic model of protein sequence evolution. Biochem Soc Trans 37:783–786
60. Suzuki Y (2008) False-positive results obtained from the branch-site test of positive selection. Genes Genet Syst 83:331–338
61. Suzuki Y, Nei M (2001) Reliabilities of parsimony-based and likelihood-based methods for detecting positive selection at single amino acid sites. Mol Biol Evol 18:2179–2185
62. Suzuki Y, Nei M (2002) Simulation study of the reliability and robustness of the statistical methods for detecting positive selection at single amino acid sites. Mol Biol Evol 19:1865–1869
63. Suzuki Y, Nei M (2004) False-positive selection identified by ML-based methods: examples from the Sig1 gene of the diatom Thalassiosira weissflogii and the tax gene of the human T-cell lymphotropic virus. Mol Biol Evol 21:914–921
64. Tamuri AU, dos Reis M, Goldstein RA (2012) Estimating the distribution of selection coefficients from phylogenetic data using sitewise mutation-selection models. Genetics 190:1101–1115
65. Tamuri AU, Goldman N, dos Reis M (2014) A penalized-likelihood method to estimate the distribution of selection coefficients from phylogenetic data. Genetics 197:257–271
66. Wang H, Li K, Susko E, Rodger AJ (2008) A class frequency mixture model that adjusts for site-specific amino acid frequencies and improves inference of protein phylogeny. BMC Evol Biol 8:1–13
67. Wertheim JO, Murrell B, Smith MD, Pond SLK, Scheffler K (2014) Relax: detecting relaxed selection in a phylogenetic framework. Mol Biol Evol 32:820–832
68. Whelan S, Goldman N (2004) Estimating the frequency of events that cause multiple-nucleotide changes. Genetics 167:2027–2043
69. Whelan S, de Bakker PIW, Quevillon E, Rodriguez N, Goldman N (2006) Pandit: an evolution-centric database of protein and associated nucleotide domains with inferred trees. Nucleic Acids Res 34(Database issue): D327–D331
70. Wong WSW, Yang ZH, Goldman N, Nielsen R (2004) Accuracy and power of statistical methods for detecting adaptive evolution in protein coding sequences and for identifying positively selected sites. Genetics 168:1041–1051
71. Yang ZH (2005) The power of phylogenetic comparison in revealing protein function. Proc Natl Acad Sci USA 102:3179–3180
72. Yang ZH (2006) On the varied pattern of evolution in 2 fungal genomes: a critique of Hughes and Friedman. Mol Biol Evol 23:2279–2282
73. Yang ZH (2007) PAML4: phylogenetic analysis by maximum likelihood. Mol Biol Evol 24:1586–1591

74. Yang ZH (2014) Molecular evolution: a statistical approach. Oxford University Press, Oxford
75. Yang ZH (2017) PAML: phylogenetic analysis by maximum likelihood. http://abacus.gene.ucl.ac.uk/software/pamlDOC.pdf
76. Yang ZH, Bielawski JP (2000) Statistical methods for detecting molecular adaptation. Trends Ecol Evol 15:496–503
77. Yang ZH, dos Reis M (2011) Statistical properties of the branch-site test of positive selection. Mol Biol Evol 28:1217–1228
78. Yang ZH, Nielsen R (1998) Synonymous and nonsynonymous rate variation in nuclear genes of mammals. J Mol Evol 46: 409–418
79. Yang ZH, Nielsen R (2002) Codon-substitution models for detecting molecular adaptation at individual sites along specific lineages. Mol Biol Evol 19:908–917
80. Yang ZH, Nielsen R (2007) Mutation-selection models of codon substitution and their use to estimate selective strengths on codon usage. Mol Biol Evol 25:568–579
81. Yang ZH, Nielsen R, Goldman N, Pedersen AMK (2000) Codon-substitution models for heterogeneous selection pressure at amino acid sites. Genetics 155:431–449
82. Yang ZH, Wong SWS, Nielsen R (2005) Bayes empirical Bayes inference of amino acid sites under positive selection. Mol Biol Evol 22:1107–1118
83. Zaheri M, Dib L, Salamin N. (2014) A generalized mechanistic codon model. Mol Biol Evol 31:2528–2541
84. Zhai W, Nielsen R, Goldman N, Yang ZH (2012) Looking for Darwin in genomic sequences – validity and success of statistical methods. Mol Biol Evol 20:2889–2893
85. Zhang J (2004) Frequent false detection of positive selection by the likelihood method with branch-site models. Mol Biol Evol 21:1332–1339
86. Zhang J, Nielsen R, Yang ZH (2005) Evaluation of an improved branch-site likelihood method for detecting positive selection at the molecular level. Mol Biol Evol 22:2472–2479

Chapter 14

Evolution of Viral Genomes: Interplay Between Selection, Recombination, and Other Forces

Stephanie J. Spielman, Steven Weaver, Stephen D. Shank, Brittany Rife Magalis, Michael Li, and Sergei L. Kosakovsky Pond

Abstract

Natural selection is a fundamental force shaping organismal evolution, as it both maintains function and enables adaptation and innovation. Viruses, with their typically short and largely coding genomes, experience strong and diverse selective forces, sometimes acting on timescales that can be directly measured. These selection pressures emerge from an antagonistic interplay between rapidly changing fitness requirements (immune and antiviral responses from hosts, transmission between hosts, or colonization of new host species) and functional imperatives (the ability to infect hosts or host cells and replicate within hosts). Indeed, computational methods to quantify these evolutionary forces using molecular sequence data were initially, dating back to the 1980s, applied to the study of viral pathogens. This preference largely emerged because the strong selective forces are easiest to detect in viruses, and, of course, viruses have clear biomedical relevance. Recent commoditization of affordable high-throughput sequencing has made it possible to generate truly massive genomic data sets, on which powerful and accurate methods can yield a very detailed depiction of when, where, and (sometimes) how viral pathogens respond to various selective forces.

Here, we present recent statistical developments and state-of-the-art methods to identify and characterize these selection pressures from protein-coding sequence alignments and phylogenies. Methods described here can reveal critical information about various evolutionary regimes, including whole-gene selection, lineage-specific selection, and site-specific selection acting upon viral genomes, while accounting for confounding biological processes, such as recombination and variation in mutation rates.

Key words Virus evolution, Molecular evolution, Recombination, Positive selection, Relaxed selection, Phylogenetics, Codon models

1 Introduction

Natural selection is a powerful evolutionary force that shapes genomes of all living organisms. A variety of computational approaches have been developed to measure the strength and direction of selection directly from genomic data. Given an alignment of homologous gene sequences, the strength of natural selection acting on a given gene or genes can be measured in a phylogenetic

Maria Anisimova (ed.), *Evolutionary Genomics: Statistical and Computational Methods*, Methods in Molecular Biology, vol. 1910, https://doi.org/10.1007/978-1-4939-9074-0_14,

context using *codon models* [1, 3]. A typical analysis on viral genomes, for example, might be performed for a single gene represented by isolates from different individuals (e.g., sequences from many HIV-1-infected hosts) or from different hosts (e.g., primate lentiviruses).

In the context of codon models, selection is typically measured using dN/dS (also referred to as ω, or Ka/Ks), which represents the ratio of the non-synonymous evolutionary rate (dN) to the synonymous evolutionary rate (dS). The synonymous evolutionary rate is used to provide a baseline rate of neutral evolution because the average selective effect of a synonymous substitution is assumed to be negligible compared to the effect of a non-synonymous substitution. [1] The selective regime can be deduced by establishing, with a degree of statistical confidence, that dN/dS differs from unity, i.e., the neutral expectation where $dN/dS = 1$. Diversifying, balancing, or (sometimes) directional selection yields $dN/dS > 1$, whereas purifying selection effects $dN/dS < 1$. Comparative methods for selection detection **estimate** dN/dS, or dS and dN separately, at sites and/or branches and perform a statistical test to establish on which side of the neutral expectation the inferences fall. As with any statistical procedure applied to finite data, each inference can be a false positive or a false negative, although methods typically take care to control the rates of both.

While the question *"Is this gene under selection?"* is an obvious one, the nearly universally applicable answer to this question is "yes.'. That is because a functional gene is (or has been) subject to some form of selection, e.g., negative selection to maintain essential features. On the other extreme is the question that has an immediate biological significance: *"Is changing a leucine to an arginine at position 209 in gene X along a specific branch in the phylogeny adaptive?"*. Without additional information, such as a carefully experimentally measured fitness impact of introducing said substitution, current comparative sequence approaches cannot answer this question. Indeed, such a scenario presents a sample size of one, which cannot be statistically meaningful.

In this chapter, we present a collection of statistical methods, each of which is designed to carefully address a biological question somewhere on the spectrum between the two extremes: sufficiently specific to be interesting, yet general enough to be answerable based only on the evolutionary history of homologous sequences. We will not discuss the technical details of codon substitution methods here (for details, please see one of the excellent available reviews Anisimova and Kosiol [1], Delport et al. [3], Yang [44], or the primary methods papers including Goldman and Yang [8],

[1] We note that there are a variety of well-documented situations where synonymous substitutions can have strong effects on fitness [11, 30].

Kosakovsky Pond and Muse [13], Muse and Gaut [27], Nielsen and Yang [29]). Instead, we present each method operationally (*"How and when does one use this method?"*), by addressing the following points:

1. What biological question is the method designed to answer?
2. What are the recommended applications?
3. What is the statistical procedure and statistical test used to establish significance for this method?
4. How should one interpret positive and negative test results?
5. Rules of thumb for when this method is likely to work well, and when it is not.

We conclude by discussing how inferences can accommodate potentially confounding biological processes, including intragenic recombination and mutation rate variation. It is critical to model these processes, both in their own right and because ignoring their effects could bias selection inference tools and yield misleading results.

2 Materials

The data used throughout the following tutorials and exercises are available at https://github.com/veg/evogenomics_hyphy. A "README" file in the top directory of this repository provides a detailed description of all contents. Importantly, all datasets used here reside in the `datasets` directory. Please refer to http://www.hyphy.org for instructions on downloading and installing HyPhy to your system. All exercises have been validated using version `2.3.4`. Throughout, we will use the `hyphymp` executable (MP = multiprocessor). For all analyses, you will need the following information:

(a) the **full path** to all files being analyzed (alignment and tree), e.g., `/home/user/data/alignment.fna`,

(b) the genetic code (in almost all cases, universal), and

(c) level of statistical significance; suggestions are given below.

All methods will produce a final file of results in `JSON` (JavaScript Object Notation) format, a highly extensible format that is simple, relatively compact, and both machine- and human-readable. `JSON` output files can be visually and interactively examined within our new web application, `hyphy-vision`, accessible at vision.hyphy.org.

All methods employ the general time reversible (GTR) nucleotide model for initial branch length optimization and correcting nucleotide substitution biases, followed by fitting a Muse–Gaut model (with general time reversible nucleotide biases) to obtain preliminary dN/dS estimates (*see* Kosakovsky Pond and Frost [12] for a detailed model description) for selection inference. Codon frequencies are estimated using the CF3x4 procedure [15]. In our view, the historical rationale for using simpler evolutionary models (e.g., K80, F81, or HKY85), namely, computational cost, to fit nucleotide data is no longer relevant.

Finally, we recommend different P-value thresholds depending on the given analysis method. As site-level methods (FEL, SLAC, and MEME) tend to be conservative on biological data, we recommend significance as $P \leq 0.1$ (or posterior probability ≥ 0.9 for FUBAR). By contrast, we recommend significance as $P \leq 0.05$ for alignment-wide methods—BUSTED, RELAX, and aBSREL.

3 Methods

3.1 How to Run a Selection Analysis

There is a uniform workflow to run any of the described methods, either locally (on one's own computer and/or a high-performance computing environment) in HyPhy or using the Datamonkey web-service, available at www.datamonkey.org. The version of HyPhy that supports all of the analyses is a command-line program, i.e., it must be run from a terminal prompt (similar to most other bioinformatics packages) in Linux or Mac OS X. It is also possible to run the program in Windows, with an appropriate POSIX emulation environment (e.g., MinGW) installed.

To execute a selection analysis locally, the following steps will need to be taken.

1. Prepare your coding sequence alignment. In general, any duplicate sequences should be removed before analysis. Most importantly, it is imperative that the sequence alignment be in the correct reading frame, meaning that alignment must be performed with codon structure in mind. A common approach to ensure this criterion is met is to generate the alignment using translated amino-acid data and then back-translate to the original nucleotide sequences.
2. Prepare a phylogenetic tree from the multiple sequence alignment. Note that certain analyses may require a labeled phylogenetic tree, as indicated within each subsequent tutorial. Keep in mind that for most selection analyses, a tree topology is a nuisance parameter. Hence, while it is advisable to use good practices when inferring trees, minor errors in tree inference tend to have minor effects on gene- and site-level inference. A notable exception occurs when lineage-specific selection is investigated; in this case, ensuring high-quality tree topologies is important.

3. An essential and *strongly* recommended step before analyzing data for selection is to screen sequences for recombination. If recombinant sequences are naively analyzed without an appropriate phylogenetic correction, inference results are likely to be biased (Posada et al. [33]) (see the section on **Screening sequences for recombination** later in this chapter).
4. Prepare your data (alignment and phylogeny) for input to HyPhy. There are three ways to provide a dataset for HyPhy analysis, each of which will trigger a different analysis prompt at runtime:
   - Two separate files containing the alignment and phylogeny, respectively. In this circumstance, HyPhy issues two successive prompts: the first for the file containing the alignment, and the second for the file containing the tree.
   - A single file containing an alignment in one of the formats supported by `HyPhy` (FASTA, MEGA, and PHYLIP), with a Newick-formatted phylogeny included at the bottom of this file. In this circumstance, HyPhy issues two successive prompts: the first for the file containing the alignment, and the second asking whether to accept the tree found in the file (provide the affirmative response, e.g., "y," to accept it).
   - A NEXUS file containing both the alignment and phylogeny. In this circumstance, HyPhy automatically accepts the provided phylogeny and therefore only issues a single prompt for the file containing the alignment. This is also the format that can be used to specify partitioned data, which is necessary to account for recombination.
5. Execute the appropriate method in HyPhy, selecting options suitable for the specific analysis.

Each method will provide live on-the-screen progress updates and, when finished, a text summary of the analysis. The output is generated in Markdown,[2] which can either be read directly as text or formatted using one of many Markdown viewers.

When an analysis is finished, HyPhy will write a JSON file with numerous details about the analysis to disk. By convention, this file will be placed in the same directory as the input alignment file, with the added `<method>.json` extension, e.g., `flu_ha.nex.BUSTED.json` for an input alignment named `flu_ha.nex` analyzed by the method BUSTED. All results contained in this JSON file can be explored visually within a web browser using a web application from the `hyphy-vision` suite of tools, accessible at vision.hyphy.org. Since JSON files can be easily accessed by

[2] With the exception of GARD.

scripting and data-analysis languages, these are also well-suited for incorporation into pipelines.[3]

When run through www.datamonkey.org, this entire workflow is automated: one simply uploads an alignment, selects options for the analysis, and waits for the job to finish. Once the job has completed, the results will be displayed in an interactive application within the web browser. Note that Datamonkey will automatically remove duplicate sequences before executing any analysis.

3.2 BUSTED

What Biological Question Is the Method Designed to Answer?:
Is there evidence that some sites in the alignment have been subject to positive diversifying selection, either pervasive (throughout the evolutionary tree) or episodic (only on some lineages)? In other words, BUSTED asks whether a given gene has been subject to positive, diversifying selection at any site, at any time [26]. If a priori information about lineages of interest is available (e.g., due to migration, change in the environment, etc.), then BUSTED can be restricted to test for selection only on a subset of tree lineages, potentially boosting power.

Recommended Applications

1. **Annotating** a collection of alignments with a binary attribute: Has this alignment been subject to positive diversifying selection (yes/no)? [34].
2. Testing **small- or low-divergence alignments** (i.e., $\leq\sim 10$ sequences) for evidence of positive diversifying selection, where neither branch- nor site-level methods have sufficient power to detect weak, but present, signal.

Statistical Test Procedure:
Each (branch, site) pair evolves with $\omega_1 \leq \omega_2 \leq 1$, or $\omega_3 \geq 1$, with the ratio chosen independently of other (branch, site) pairs with probability p_1, p_2, p_3 (normalized to sum to 1). The three-rate ω distribution is estimated jointly from the entire alignment, i.e., rates are shared by all (branch,site) combinations. Therefore, BUSTED is technically a "branch-site" model [16], although it is not intended to detect individual sites which drive signal of selection.

[3] Note that the method GARD does not provide markdown output or a JSON, and output is in a different format. This may be updated in a future HyPhy release.

The test for episodic diversifying selection is performed by comparing the full model versus the nested null model, where ω_3 is constrained to 1. Statistical significance is obtained by the likelihood ratio test, assuming the χ_2^2 asymptotic distribution of the likelihood ratio statistic under the null model.

When only some of the branches are chosen for testing, and the remainder are designated as the background, two independent three-rate ω distributions are fitted: one for the test branches, and one for the background branches. Testing for selection is carried out by constraining the distribution on the test branches as described above.

Example Analysis:
To begin, we will perform a BUSTED analysis using a dataset of primate-specific KSR2, kinase suppressor of RAS2, genes from Enard et al. [5]. This gene has been implicated as a so-called 'virus-interacting protein,' and previous work has suggested it has experienced adaptation in mammalian lineages due to selective pressures exerted by viruses [5]. We will test all lineages for positive selection (rather than specifying a subset of "test" branches), thereby asking the question: "Has KSR2 been subject to diversifying selection at some time during evolution in primates?"

To run BUSTED, open a terminal session and enter `HYPHYMP` from the command line to launch the HyPhy analysis menu. Enter `1` (Selection Analyses) and then `5` to reach the BUSTED analysis menu, and supply values for the following prompts:

1. **Choose genetic code.** This option tells HyPhy which translation table to use for codon-level analyses. Enter `1` to use the Universal genetic code.
2. **Select a coding sequence alignment file.** Provide the full path to the dataset of interest: `/path/to/data/ksr2.fna`.
3. **A tree was found in the data file...Would you like to use it (y/n)?** Enter "y" to use the tree.
4. **Choose the set of branches to test for selection.** Enter `1` to test all branches for selection.

BUSTED will now run to completion, printing status indicators to screen while it runs. For an example of how this output will look when rendered into HTML (or similarly, PDF), see this link: http://bit.ly/2vsRZrh.

Listing 1 Partial BUSTED screen output:

```
### Branches to test for selection in the BUSTED analysis
* Selected 15 branches to test in the BUSTED analysis: 'HUM, PAN, Node6, GOR,
Node5, PON, Node4, GIB, Node3, MAC, BAB, Node12, Node2, MAR, BUS'

### Obtaining branch lengths and nucleotide substitution biases under the
nucleotide GTR model
* Log ( L ) = -5768.01, AIC - c = 11582.06 (23 estimated parameters)

### Obtaining the global omega estimate based on relative GTR branch lengths
and nucleotide substitution biases
* Log ( L ) = -5342.48, AIC - c = 10745.17 (30 estimated parameters)
* non - synonymous / synonymous rate ratio for *test* = 0.0342

### Improving branch lengths , nucleotide substitution biases, and global dN/dS
ratios under a full codon model
* Log ( L ) = -5333.46, AIC - c = 10727.13 (30 estimated parameters)
* non - synonymous / synonymous rate ratio for *test* = 0.0307

### Performing the full ( dN / dS > 1 allowed) branch-site model fit
* Log ( L ) = -5319.67, AIC - c = 10707.62 (34 estimated parameters)
* For * test * branches , the following rate distribution for branch-site
combinations was inferred

|        Selection mode         |     dN/dS    |Proportion, %|      Notes      |
|-------------------------------|--------------|-------------|-----------------|
|      Negative selection       |    0.024     |   99.151    |                 |
|      Negative selection       |    0.085     |    0.812    |                 |
|    Diversifying selection     |   118.143    |    0.037    |                 |

### Performing the constrained (dN/dS > 1 not allowed) model fit
* Log ( L ) = -5326.18, AIC - c = 10718.63 (33 estimated parameters)
* For * test * branches under the null (no dN/dS > 1 model), the following
rate distribution for branch-site combinations was inferred

|      Selection mode      |    dN/dS     |Proportion, %|        Notes         |
|--------------------------|--------------|-------------|----------------------|
|     Negative selection   |   0.000      |   10.598    |                      |
|     Negative selection   |   0.000      |   86.086    | Collapsed rate class |
|      Neutral evolution   |   1.000      |    3.316    |                      |

----
## Branch - site unrestricted statistical test of episodic diversification
[BUSTED]
Likelihood ratio test for episodic diversifying positive selection, **p =
0.0015**.
```

Interpreting Results:
The results printed to the terminal indicate a highly significant result (P = 0.0015) in the test for whole-gene selection. Analysis with BUSTED therefore provides robust evidence that KSR2 experienced episodic positive selection in the primates. Because we performed the original BUSTED analysis on the entire tree (i.e., without a specified set of test branches), we do not know from this result along which lineages KSR2 was subject to positive selection. We can conclude only that a non-zero proportion of sites on some lineage(s) in the primate tree experienced diversifying selection pressure.

The output additionally provided information about the specific BUSTED model fits to the test data, including the inferred ω distributions and corresponding weights. The BUSTED alternative model (shown under the output header `Performing the full ( dN/dS >1 allowed) branch-site model fit`) found that a very small proportion (only ~0.037%) of sites evolved under a very large ω of over 100 (118.143). Importantly, neither of these estimates is precise because they were derived from a small subset of the data. As such, all the BUSTED tests establish the fact that the proportion of sites along test lineages (here, the entire phylogeny) with $\omega > 1$ is non-zero. For example, if BUSTED had inferred a rate category of $\omega = 10$ on a different gene, it would *not* be correct to claim that this gene evolves under weaker selection than does KSR2. A formal statistical test would have to be carried out to establish such a claim.

Conversely, had the result not been statistically significant, we would not be able to reject the null hypothesis that no positive selection had occurred in KSR2. Importantly, however, a negative finding *would not* unequivocally rule out the presence of positive selection. This outcome could be due to a lack of statistical power wherein the provided data did not contain a sufficiently strong selection.

BUSTED's fixed a priori assumption of model complexity (a three-rate ω distribution) may lead to over-parameterized (or under-parameterized) models. For example, in the constrained model for KSR2, two of the three rate classes have the same value of $\omega(0.0)$, implying that one of them is unnecessary. HyPhy will report this to the screen as a diagnostic message `Collapsed rate class`, but there is no corrective action that needs to be taken. These messages simply point to *low-complexity* data.

We will additionally take this opportunity to showcase the visual power of our accompanying web browser, HyPhy-Vision. Figure 1 displays the rendering of the output `ksr2.fna.BUSTED.json` as it appears in HyPhy-Vision. On this site, users can interactively view and explore inference results, view figures and charts, and perform other tasks.

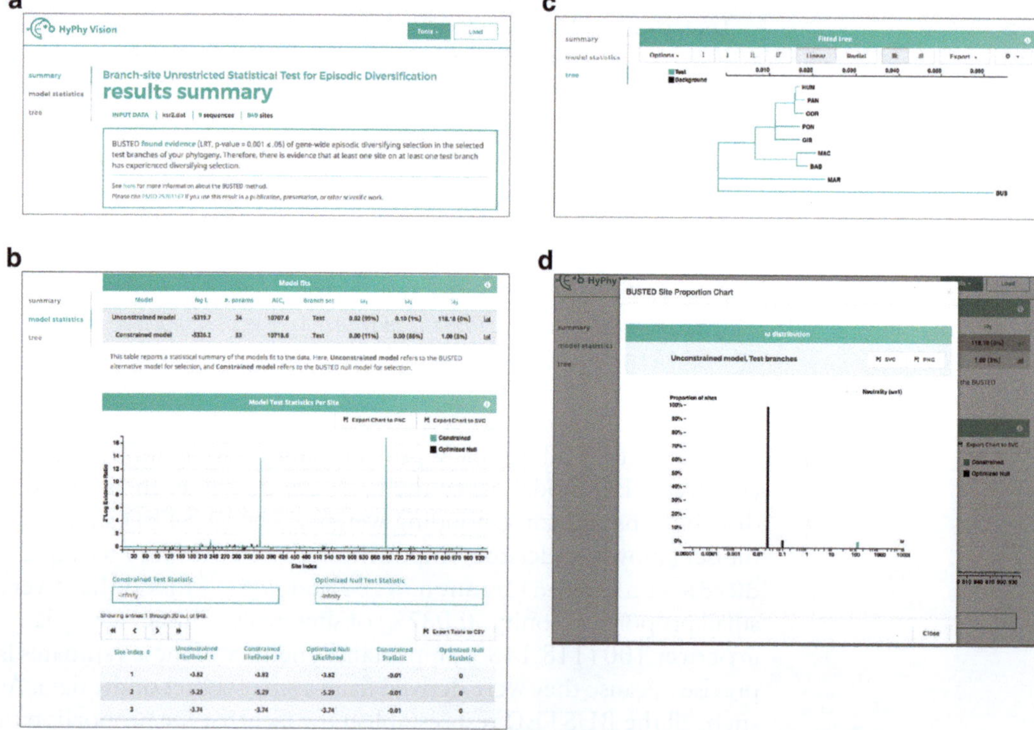

Fig. 1 Example analysis visualization in HyPhy-Vision of BUSTED results. (**a**) The **summary** section provides a brief overview of the analysis performed, including information about the inputted data (which can be downloaded via the linked file name) and primary results from the hypothesis test performed. (**b**) The **model statistics** section provides information about models fitted to the data. In BUSTED, this section additionally includes an interactive display of site evidence ratios, which can be interpreted as a *descriptive* measure for which sites may have contributed to the selection signal. (**c**) The **tree** section displays the phylogeny as fitted under all inferred models and data partitions, if specified. Tree views can be toggled under the *Options* drop-down menu. (**d**) Graphical views of each model's inferred ω distribution can be viewed when clicking on a given row's plot icon in the **Model fits** table seen in (**b**)

Rules of Thumb for BUSTED Use

1. Best applied to small- or medium-sized datasets (e.g., up to 100 sequences). Larger datasets will take longer to run and may not be well described by a fixed complexity model.
2. If one suspects that only a small subset of lineages is subject to selection, e.g., because the phenotype, environment, or fitness changed along those branches, designating those a priori as the test set will significantly boost power.
3. In simulation studies, BUSTED performs best when a sufficient proportion (5–10%) of branch site combinations is subject to positive diversifying selection, and the effect size (ω value) is reasonably large (e.g., ≥ 3).

3.3 RELAX

What Biological Question Is the Method Designed to Answer?:
*Is there evidence that the strength of selection has been relaxed (or conversely intensified) on a specified group of lineages (*Test*) relative to a set of reference lineages (*Reference*)? We note that the RELAX framework can perform this specific hypothesis test as well as fit a suite of descriptive models which address, for example, overall rate differences between test and reference branches or lineage-specific inferences of selection relaxation. We focus our attention here on RELAX's hypothesis testing abilities. More information about descriptive analyses is available on hyphy.org as well as in RELAX's primary publication [43]. Importantly, RELAX* is not *designed to detect diversifying selection specifically.*

Recommended Applications

1. Testing for a **systematic shift** (relaxation/intensification) in the distribution of selection pressure associated with major biological transitions such as hosting switching in viruses [6] or lifestyle evolution in bacteria (i.e., transition from free-living to endosymbiotic lifestyle [43]).
2. **Comparing selective regimes** between two subsets of branches in the tree, e.g., to investigate selective differences among transmission routes in HIV-1 [42].

Statistical Test Procedure:
Given a tree with at least two sets of branches, one of which is designated as Test, *and the other as* Reference, *the core version of RELAX compares two nested models, which follow the same general framework as BUSTED. Each (branch, site) combination is drawn independently from a 3-rate* ω *distribution. The evolutionary rates for* Test *branches are functions of those for* Reference *branches. Specifically,* $\omega_{\text{Test}} = \omega_{\text{Reference}}^{K}$, *where* K *is the relaxation or intensification parameter. The alternative model infers* K *from the data, and the null model sets* K = *1. Statistical significance is obtained by the likelihood ratio test, assuming the* χ_1^2 *asymptotic distribution under the null model. A significant result of* K > *1 indicates that selection strength has been* intensified *along the test branches, and a significant result of* K < *1 indicates that selection strength has been* relaxed *along the test branches. In other words, for* K < *1 the* Test ω *values shrink toward neutrality (*ω = *1) relative to* Reference, *and for* K > *1 they move away from neutrality.*

If some branches in the tree belong to neither the *T*est or the *R*eference set, they are allocated to a group with its own (*Unclassified*) distribution of ω, which is uncoupled from the testing procedure.

Example Analysis:
We will perform a RELAX analysis using a dataset of Influenza A PB2 subunit sequences from Tamuri et al. [41]. The PB2 subunit, which is part of influenza's RNA polymerase complex, has emerged as a critical determinant of influenza infectivity and, as a consequence, host range [9, 18]. The dataset we examine here contains sequences from both avian host and human host strains.[4] *Previous studies have shown that this host switch is correlated with significant shifts in selection pressures and preferred amino acids at key sites in PB2 [36, 40, 41]. We now re-analyze this dataset using RELAX to ask a different but related question: "Was the shift from avian to human hosts associated with a relaxation of selection pressures in Influenza A PB2?"*

RELAX requires an a priori specification of test and reference lineages, although not all lineages in a tree need to be classified. As such, you must label your test (and reference, if desired) branches in the input phylogeny. We provide an online widget to assist with tree labeling at http://phylotree.hyphy.org. The dataset we have provided for this analysis already has a labeled phylogeny, with the human host lineages labeled as "test."

To run RELAX, open a terminal session and enter `HYPHYMP` from the command line to launch the HyPhy analysis menu. Enter `1` (Selection Analyses) and then `7` to reach the RELAX analysis menu, and supply values for the following prompts:

1. **Choose genetic code.** Enter `1` to use the Universal genetic code.
2. **Select a coding sequence alignment file.** Provide the full path to the dataset of interest: `/path/to/data/pb2.fna`.
3. **A tree was found in the data file...Would you like to use it (y/n)?** Enter "y" to use the tree.
4. **Choose the set of branches to test for selection.** This option asks you to specify the *label* inside your tree used to specify the test lineages. You can either select all unlabeled branches, or HyPhy will show all labels it found in the tree you provided.

[4] The original dataset in Tamuri et al. [41] contained 401 sequences. For the purposes of this chapter, we analyze a subset of this alignment with only 35 sequences (20 from avian and 15 from human hosts), thereby achieving a tractable runtime on a personal machine.

Enter 1 to select the branches labeled as "test" as the test set in RELAX analysis. Note that when multiple labels are present in your tree, HyPhy will issue an additional prompt to choose the set of *Reference* branches, in the event that some branches should remain *Unclassified.*

5. **Analysis type.** This option asks you to specify the scope of RELAX analysis. Selecting "Minimal" will run the RELAX hypothesis test, and selecting "All" will run hypothesis testing and fit two additional descriptive models, described earlier. Here, we will perform only hypothesis testing to determine whether the data shows evidence for a relaxation or intensification of selection intensity between the test and reference lineages. Enter the option 2 to run the "Minimal" analysis.

RELAX will now run to completion, printing status indicators to screen while it runs.

Listing 2 Partial RELAX screen output:

```
### Obtaining branch lengths and nucleotide substitution biases under the
nucleotide GTR model
* Log ( L ) = -16755.26, AIC - c = 33660.66 (75 estimated parameters)

### Obtaining the global omega estimate based on relative GTR branch lengths
and nucleotide substitution biases
* Log ( L ) = -14410.97, AIC - c = 28988.46 (83 estimated parameters)
* non - synonymous / synonymous rate ratio for *Reference* = 0.0401
* non - synonymous / synonymous rate ratio for *Test* = 0.0604

### Improving branch lengths , nucleotide substitution biases, and global dN/dS
ratios under a full codon model
* Log ( L ) = -14354.67, AIC - c = 28875.86 (83 estimated parameters)
* non - synonymous / synonymous rate ratio for *Reference* = 0.0358
* non - synonymous / synonymous rate ratio for *Test* = 0.0609

### Fitting the alternative model to test K != 1
* Log ( L ) = -14337.22, AIC - c = 28849.02 (87 estimated parameters)
* Relaxation / intensification parameter (K) = 0.73
* The following rate distribution was inferred for **test** branches

| Selection mode | dN/dS |Proportion, %| Notes |
|-------------------------------|-------------|------------|------------------|
| Negative selection | 0.031 | 94.752 | |
| Negative selection | 0.086 | 2.951 | |
| Diversifying selection | 1.406 | 2.297 | |
```

```
*  The  following  rate  distribution was inferred for **reference** branches

|          Selection mode          |    dN/dS     |Proportion, %|         Notes      |
|--------------------------------- |--------------|-------------|--------------------|
|         Negative selection       |    0.009     |   94.752    |                    |
|         Negative selection       |    0.035     |    2.951    |                    |
|       Diversifying selection     |    1.591     |    2.297    |                    |

###  Fitting  the  null  ( K  :=  1)  model
*  Log ( L )  =  -14342.33,  AIC - c  =  28857.22 (86 estimated parameters)
*  The  following  rate  distribution for test/reference branches was inferred

|          Selection mode          |    dN/dS     |Proportion, %|       Notes        |
|--------------------------------- |--------------|-------------|--------------------|
|         Negative selection       |    0.010     |   94.149    |                    |
|         Negative selection       |    0.021     |    3.391    |                    |
|       Diversifying selection     |    1.735     |    2.460    |                    |

----
##  Test  for  relaxation  ( or  intensification) of selection [RELAX]
Likelihood  ratio  test  ** p  =    0.0014**.
> Evidence  for  * relaxation  of  selection* among **test** branches _relative_ to
the **reference** branches at P<=0.05
----
```

Interpreting Results:
On this data, RELAX has inferred a relaxation parameter K = *0.73 with a highly significant* P = *0.0014. Therefore, there is evidence to reject the null hypothesis that selection pressure has not been shifted in the test (here, human host) lineages. We instead have strong evidence that selection has been* relaxed *(because the inferred* K < *1) in the human host lineages. In other words, selection in the test branches has generally moved towards neutrality (*$\omega = 1$*) compared to the reference branches. This finding is consistent with the evolutionary changes that typically occur during a virus host-switching event, wherein selection stringency will be reduced to facilitate viral adaptation.*

Keep in mind that RELAX defines relaxation (or intensification) in a fairly restrictive fashion. In other words, all selective regimes (i.e., all ω rates), both negative and positive, must weaken or strengthen. Therefore, certain relaxation scenarios, for example, when only positive selection is relaxed but negative selection is maintained, may result in a non-significant RELAX test even though selection has changed.

Rules of Thumb for RELAX Use

1. Always provide a labeled phylogeny indicating which branches to include in the "test" lineages. You can additionally label "reference" lineages if you wish to keep some branches as unclassified. It is convenient to use the `phylotree.js` online widget at http://phylotree.hyphy.org/ to label branches before analysis.

3.4 aBSREL

It is often of interest to determine whether a specific lineage or lineage(s) have been subject to selection. Such analyses have historically been performed using the so-called branch or branch-site class of models, which allow evolutionary rates to vary across branches or across sites **and** branches [16, 45, 46]. Early versions of branch-site models allowed users to compare selection pressure on a pre-selected branch sets of "foreground" branches to a pre-selected set of "background" branches, on which positive selection was disallowed [45, 46]. (Note that this approach is similar to how BUSTED performs gene-wide selection inference [26].) Later efforts demonstrated that disallowing positive selection on background branches could lead to highly elevated false positive rates and advocated a strategy wherein any branch, regardless of data partition, could evolve at any rate [16]. This strategy has been described as the BS-REL model in HyPhy [16]. However, in BS-REL, each branch was constrained to have three rate categories, an assumption with little justification.

Since then, we have developed a greatly improved branch-site model called aBSREL ("adaptive branch-site random effects likelihood"). Rather than assuming that each branch should be fit with three rate classes, aBSREL infers, using small-sample Akaike Information Criterion correction (AICc), the optimal number of rate categories per branch. In this manner, computational complexity and the number of parameters are greatly reduced, leading to a tractable runtime for larger datasets that could not otherwise be studied with earlier branch-site models.

What Biological Question Is the Method Designed to Answer?:
Like classical branch-site models, aBSREL asks whether some proportion of sites is subject to positive selection along specific branches or lineages of a phylogeny.

Recommended Applications

1. **Exploratory testing** for evidence of lineage-specific positive diversifying selection in small- to medium-sized alignments (up to 100 sequences).

2. **Targeted testing** of branches selected a priori for positive diversifying selection. This includes alignments with prohibitive runtimes under older branch-site models (up to ~1000 sequences) [37].

Statistical Test Procedure:
aBSREL uses the information-theoretic criterion AIC_c *to automatically determine the complexity of the evolutionary process at every branch [37]. As a heuristic optimization, aBSREL will always examine branches in order from longest to shortest, because longer branches tend to be the ones requiring more complex models. In this adaptive model, one rate class is allowed to assume any value of* $\omega > 1$, *whereas for any other inferred rate class is constrained as* $\omega \leq 1$. *In the null model, all* ω *categories are constrained as* $\omega \leq 1$. *For any branch inferred to have sufficient rate variation (i.e., more than one rate category) where one rate category is described by* $\omega > 1$, *aBSREL will proceed to fit a null model to this branch. In other words, if the maximum-inferred* $\omega \leq 1$ *on a branch, the null model will have the same exact fit as the alternative model, and the resulting* P*-value is 1. The test for lineage-specific diversifying selection is performed by comparing the full model versus the nested null model, and statistical significance is obtained by the likelihood ratio test. Significance is evaluated using a mixture of* $50\%\chi_0^2$, $20\%\chi_1^2$, *and* $30\%\chi_2^2$ *distributions (proportions determined via simulations Smith et al. [37]). Finally, aBSREL will correct all* P*-values obtained from individual tests for multiple comparisons using the Bonferroni–Holm procedure to control family-wise false-positive rates (i.e., the probability of generating one or more false positives, when all null hypotheses are correct).*

One can either select a specific set of branches in order to test a specific a priori hypothesis or one can perform an exploratory analysis across the entire phylogeny by testing all branches for selection. The former approach may have substantially more power to detect selection, especially if only a few branches in a large tree are chosen, due to the decreased volume of multiple testing. However, the approach does carry the risk of failing to identify branches subject to positive selection that have not been included in the test set.

Example Analysis:
Here, we will demonstrate aBSREL use and interpretation using a dataset of HIV-1 env *sequences collected from an epidemiologically linked donor–recipient transmission pair [7]. This dataset can be found in the provided file* `hiv1_transmission.fna`.

To run aBSREL, open a terminal session and enter `HYPHYMP` from the command line to launch the HyPhy analysis menu. Enter `1` (Selection Analyses) and then `6` to reach the aBSREL analysis menu, and supply values for the following prompts:

1. **Choose genetic code.** This option tells HyPhy which translation table to use for codon-level analyses. Enter `1` to use the Universal genetic code.
2. **Select a coding sequence alignment file.** Provide the full path to the dataset of interest: `/path/to/hiv1_transmission.fna`.
3. **A tree was found in the data file...Would you like to use it (y/n)?** Enter "y" to use the included tree.
4. **Choose the set of branches to test for selection.** You can now select on which branches aBSREL should conduct a formal hypothesis test for positive selection. Enter `1` to test all branches for selection.

aBSREL will now run to completion, printing status indicators to screen while it runs (some output abbreviated).

Listing 3 Partial aBSREL screen output:

```
### Obtaining branch lengths and nucleotide substitution biases under the nucleotide GTR model
* Log ( L ) = -5524.50, AIC - c = 11153.08 (52 estimated parameters)

### Fitting the baseline model with a single dN/dS class per branch, and no site-to-site variation.
* Log ( L ) = -5402.40, AIC - c = 11009.72 (102 estimated parameters)
* Branch - level non - synonymous / synonymous rate ratio distribution has median 0.66, and 95% of the weight in 0.00--5.41

### Determining the optimal number of rate classes per branch using a step up procedure
```

```
| Branch     |Length|Rates|   Max. dN/dS     |   Log(L)   |  AIC-c   |Best AIC-c so far|
|-----------|-------|-----|----------------|------------|---------|----------------|
| 0564 _22   | 0.01 | 2 |  1.96 (52.27%) |  -5402.41 | 11013.78 |    11009.72    |
| 0564 _7    | 0.01 | 2 |  0.74 ( 5.19%) |  -5402.40 | 11013.76 |    11009.72    |
| Separator  | 0.01 | 2 | 197.32 ( 3.95%) |  -5397.53 | 11004.02 |    11004.02    |
| Separator  | 0.01 | 3 | 180.22 ( 4.08%) |  -5397.53 | 11008.06 |    11004.02    |
| 0564 _4    | 0.01 | 2 |  29.79 ( 2.15%) |  -5394.37 | 11001.74 |    11001.74    |
| 0564 _4    | 0.01 | 3 |  29.78 ( 2.15%) |  -5394.37 | 11005.78 |    11001.74    |
| 0564 _3    | 0.01 | 2 | 126.86 ( 3.14%) |  -5388.59 | 10994.22 |    10994.22    |
| 0564 _3    | 0.01 | 3 | 135.96 ( 3.05%) |  -5388.59 | 10998.25 |    10994.22    |
| 0564 _9    | 0.01 | 2 |  10.01 ( 8.61%) |  -5388.37 | 10997.82 |    10994.22    |
 ...
| Node53     | 0.00 | 2 |  1.00 (100.00%) |  -5371.63 | 10976.46 |    10971.76    |
| 0557 _6    | 0.00 | 2 | 27.66 (100.00%) |  -5371.32 | 10975.83 |    10971.76    |
| 0557 _21   | 0.00 | 2 |   0.25 ( 1.96%) |  -5371.30 | 10975.80 |    10971.76    |
| 0557 _7    | 0.00 | 2 |   0.25 ( 1.96%) |  -5371.30 | 10975.80 |    10971.76    |

### Rate class analyses summary
* 38 branches with **1** rate classes
* 6 branches with **2** rate classes

### Improving parameter estimates of the adaptive rate class model
* Log ( L ) = -5370.66, AIC - c = 10970.49 (114 estimated parameters)

### Testing selected branches for selection

|     Branch       |  Rates   |      Max. dN/dS      | Test LRT | Uncorrected p-value |
|------------------|----------|----------------------|----------|---------------------|
|     0564 _22     |    1     |   1.22 (100.00%)     |   0.11   |      0.43015        |
|     0564 _7      |    1     |   0.61 (100.00%)     |   0.00   |      1.00000        |
|    Separator     |    2     |  197.72 ( 3.95%)     |  14.13   |      0.00029        |
|     0564 _4      |    2     |   28.89 ( 2.15%)     |   4.81   |      0.03281        |
|     0564 _3      |    2     |  127.66 ( 3.14%)     |  14.06   |      0.00030        |
|     0564 _9      |    1     |   0.72 (100.00%)     |   0.00   |      1.00000        |
|     0564 _1      |    1     |   1.07 (100.00%)     |   0.01   |      0.48208        |
...
|     0557 _21     |    1     |   1.00 (100.00%)     |   0.00   |      1.00000        |
|     0557 _7      |    1     |   1.00 (100.00%)     |   0.00   |      1.00000        |
----
### Adaptive branch site random effects likelihood test
Likelihood ratio test for episodic diversifying positive selection at Holm-
Bonferroni corrected _p = 0.0500_ found **3** branches under selection among **44**
tested.

* Node35 , p - value = 0.00018
* Separator , p - value = 0.01251
* 0564 _3 , p - value = 0.01266
```

Interpreting Results:
The first printed markdown table (`"Determining the optimal number of rate classes per branch using a step up procedure"`) *summarizes the model selection process. For example, when two* ω *rates were assigned to branch* Separator, *this improved the* AIC_c *score of the fit (compared to the single-rate model) from 11, 009.72 to 11, 004.02. However, allocating three* ω *rates to the same branch worsens the score to 11, 008.06. Therefore the aBSREL model will use two* ω *rates at the branch.*

The second printed markdown table (`"Testing selected branches for selection"`) shows the results of tests for episodic selection on individual branches. At branch *0564_4*, for example, the tested model includes two ω rates, with the positive selection class taking on value 28.89 (2.15% proportion of the mixture). Constraining this rate to range between 0 and 1 yields the likelihood ratio test statistic of 4.81, which maps to a P-value (before multiple test correction) of 0.03281.

Finally, aBSREL reports three branches under episodic diversifying selection pressure. Further examination of results using HyPhy-Vision shows that these branches are found (a) along the transmission event from donor to recipient, and (b) within a highly diverged clade in the donor (Fig. 2). The first finding is consistent with an expected increase in evolutionary rate when a virus infects a new host and encounters novel host immunity, and the second finding is consistent with intrahost adaptive dynamics of the donor's long-term HIV infection. Importantly, a close examination of the markdown-output table under the header `"Testing selected branches for selection"` reveals several nodes with uncorrected P-values whose significance was lost upon applying the Bonferroni–Holm correction, e.g., `0564_4` whose uncorrected $P = 0.03281$. This result illustrates the potential loss of power incurred by this aBSREL exploratory analysis.

Rules of Thumb for aBSREL Use

1. A priori identification of branches to test for selection will generally increase power to detect selection on those branches. That said, to maintain statistical robustness, we *strongly discourage* performing multiple separate tests for selection on different branch sets. Such an approach will necessarily introduce false positives. In such a case, we recommend performing an exploratory analysis wherein all branches are considered.
2. Exploratory analyses of very large datasets are unlikely to yield many significant results, because correcting for multiple testing will reduce power as the number of branches grows, while the

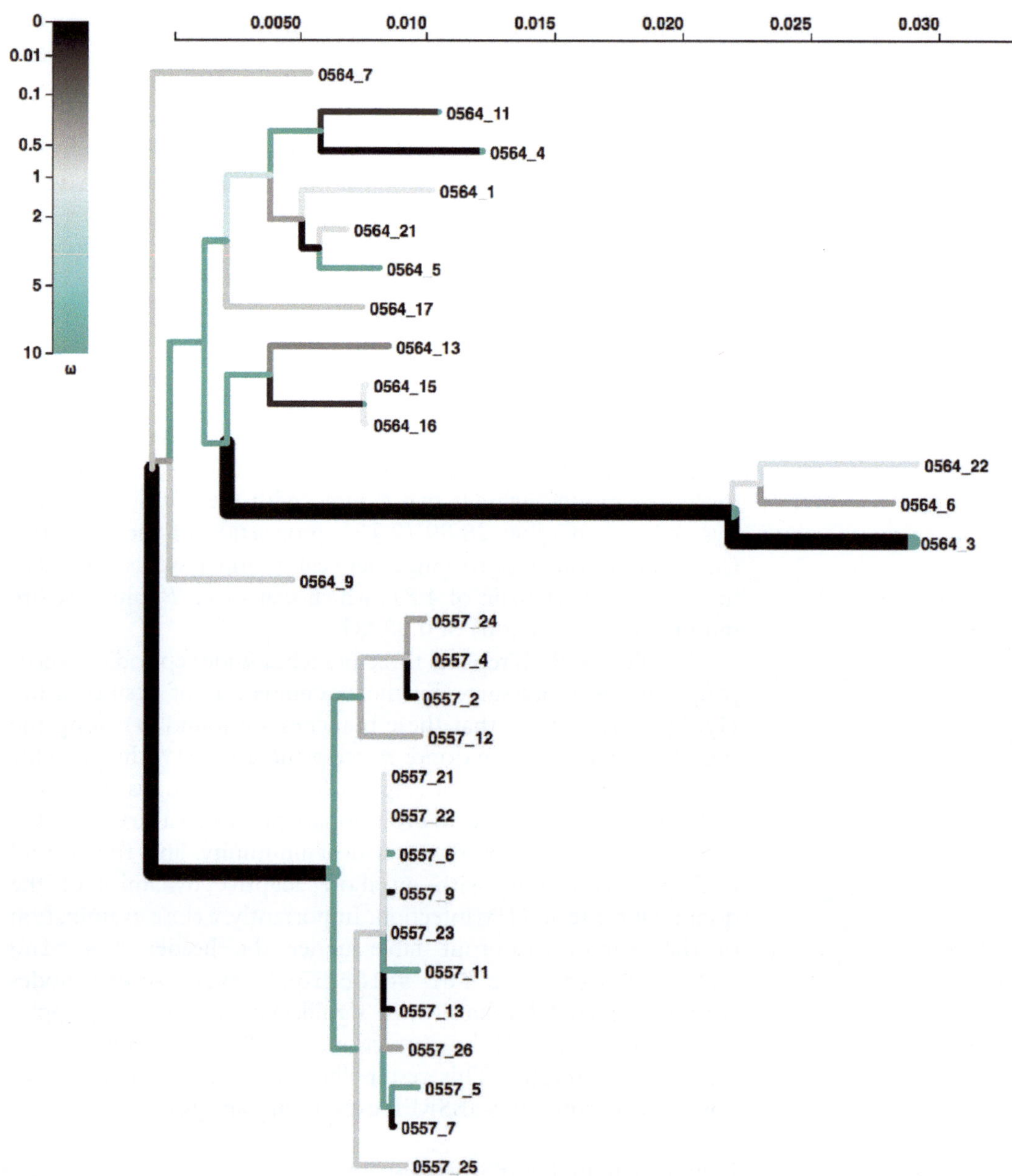

Fig. 2 `HyPhy-Vision` tree viewer depicting the fitted aBSREL Adaptive model to HIV-1 data. Branches are colored by their inferred ω distribution, as indicated in the legend. Lineages identified as positive selection at $P < 0.05$ after correction for multiple testing are shown with thick branches, with color distributions representing the relative values and proportions of inferred ω categories. Note that taxon labels beginning with "0554" represent HIV-1 sequences derived from the donor patient, and labels beginning with "0557" represent HIV-1 sequences derived from the recipient patient

amount of statistical signal does not increase for larger datasets. One option is to thin out large phylogenies (before performing any testing), retaining major clades and lineages of interest.

3.5 Site-Level Selection: MEME, FEL, SLAC, and FUBAR

What Biological Question Is the Method Designed to Answer?:
The methods FEL, SLAC, and FUBAR address the question: Which site(s) in a gene are subject to pervasive, *i.e., consistently across the entire phylogeny, diversifying selection? MEME addresses a more general question: Which site(s) in a gene are subject to pervasive or* episodic, *i.e., only on a single lineage or subset of lineages, diversifying selection?*

Recommended Applications

1. **MEME** is the sole method in HyPhy for detecting selection at individual sites that considers both pervasive and episodic selection. MEME is therefore our recommended method if maximum power is desired.
2. The phenomenon of pervasive selection is generally most prevalent in pathogen evolution and any biological system influenced by evolutionary arms race dynamics (or balancing selection), including adaptive immune escape by viruses. As such, FEL, SLAC, and FUBAR are ideally suited to identify sites under positive selection which represent candidate sites subject to strong selective pressures across the entire phylogeny. Each of these methods has a particular use case as well:
   - **FEL** is our recommended method for analyzing small-to-medium size datasets when one wishes only to study pervasive selection at individual sites.
   - **FUBAR** is our recommended method for detecting pervasive selection at individual sites on large (> 500 sequences) datasets for which other methods have prohibitive runtimes, unless you have access to a computer cluster.
   - **SLAC** provides legacy functionality as a counting-based method adapted for phylogenetic applications. In general, this method will be the least statistically robust.

Statistical Test Procedure:
Each method presented here employs a distinct algorithmic approach to inferring selection. FEL uses maximum likelihood to fit a codon model to each site, thereby estimating a value for dN *and* dS *at each site. FEL tests for selection with the likelihood ratio test using the χ^2_1 distribution, asking whether the* dN *estimate is significantly greater than the inferred* dS *estimate.*

(continued)

SLAC represents the most basic inference method and is an extension of the Suzuki–Gojobori counting-based method [39] for phylogenetically related sequences (as opposed to sequence pairs). SLAC uses maximum likelihood to infer ancestral characters for each site across the phylogeny and then directly counts the number of synonymous and non-synonymous changes which have occurred at each site over evolutionary time. SLAC then tests for selection by testing whether or not there are too many or too few non-synonymous changes compared to what is expected under neutrality. The neutral expectation is derived based on the phylogeny-wide estimated numbers of synonymous and non-synonymous nucleotide sites at a given codon. The statistical test employs the binomial distribution to compute significance, e.g., how likely is it to observe 13 non-synonymous and 1 synonymous substitutions at a site, if the expected synonymous to non-synonymous substitution count ratio under neutrality is 1:4?

MEME employs a mixed-effects maximum likelihood approach. For each site, MEME infers two ω *rate classes and corresponding weights representing the probability that the site evolves under each rate class at a given branch. To this end, MEME infers a single* α *(*dS*) parameter and two separate* β *(*dN*) parameters,* β_- *and* β_+*. The* ω *rates per site, therefore, consist of* β_+/α *and* β_-/α*. MEME uses this framework to fit a null and alternative model each, both models enforcing the constraint* $\beta_- \leq \alpha$*. The null model disallows positive selection by enforcing the constraint* $\beta_+ \leq \alpha$*, whereas the alternative model places no constraint on* β_+*. MEME uses the likelihood ratio test to compare between null and alternative model fits, with significance assessed using the mixture of* $33\%\chi_0^2$, $30\%\chi_1^2$, *and* $37\%\chi_2^2$.

FUBAR takes a Bayesian approach to selection inference and is a particular case of statistical models developed in the context of document classification (latent Dirichlet allocation). The key innovation to FUBAR's approach is its use of an a priori specified grid of dN *and* dS *values (typically 20 × 20), spanning the range of negative, neutral, and positive selection regimes, whose likelihoods can be pre-computed and used throughout analysis (rather than having to re-compute likelihoods during optimization as traditional random-effects approaches do [12, 29]). This approach, combined with other algorithmic advances, speeds computation time by at least an order of magnitude compared to FEL, while yielding comparable statistical performance. FUBAR estimates every model parameter except the proportion of sites allocated to each grid point using simple (and fast) nucleotide models. The proportions are estimated using an MCMC procedure, and non-neutral evolution at each site is inferred using a straightforward naive*

(continued)

empirical Bayes approach [29]. Sites are called positively or negatively selected if the corresponding posterior probabilities are sufficiently high.

Note that FEL and SLAC report both positively and negatively selected sites, but MEME and FUBAR report only sites under positive selection.

Example Analysis:
We will demonstrate the use and interpretation of site-level methods using data from influenza strain H3N2 (the "Hong Kong flu"), the primary circulating strain of seasonal influenza since the late 1960s. We specifically will assess selection on the H3 hemagglutinin, the influenza surface protein which is responsible for host cell binding. Hemagglutinin experiences rapid evolution triggered by host immune escape, and previous studies have identified numerous signatures of positive diversifying selection in H3 sequences with a particular concentration around the host-binding domain [28].

We base analyses here on an alignment from Meyer and Wilke [22] of H3 sequences sampled over time since the 1991–1992 influenza season. We removed all partial and strongly outlying sequences (i.e., those with excessive divergence) from the original dataset before proceeding, yielding 2555 sequences to comprise our "full" H3 dataset. We further subsetted this alignment to two smaller alignments with comparable numbers of taxa but spanning different evolutionary time frames: The first smaller alignment ("trunk") contains 163 sequences sampled along the influenza H3 trunk, whereas the second smaller alignment ("shallow") contains 121 sequences sampled from a single clade (Fig. 3). Therefore, while these two smaller datasets contain a comparable number of sequences, the trunk dataset spans a much longer time frame and contains substantially more sequence divergence relative to the shallow dataset. Indeed, the trunk dataset has a total tree length (sum of branch lengths, in units substitutions/site/unit time) of 0.43, whereas the shallow dataset had a total tree length of 0.12, meaning that the trunk dataset contains nearly four times the amount of sequence divergence seen in the shallow dataset. We have compiled results for all three datasets analyzed with all four methods (Table 1). We now describe, using the trunk dataset as an example, how to run each of these analyses in HyPhy.

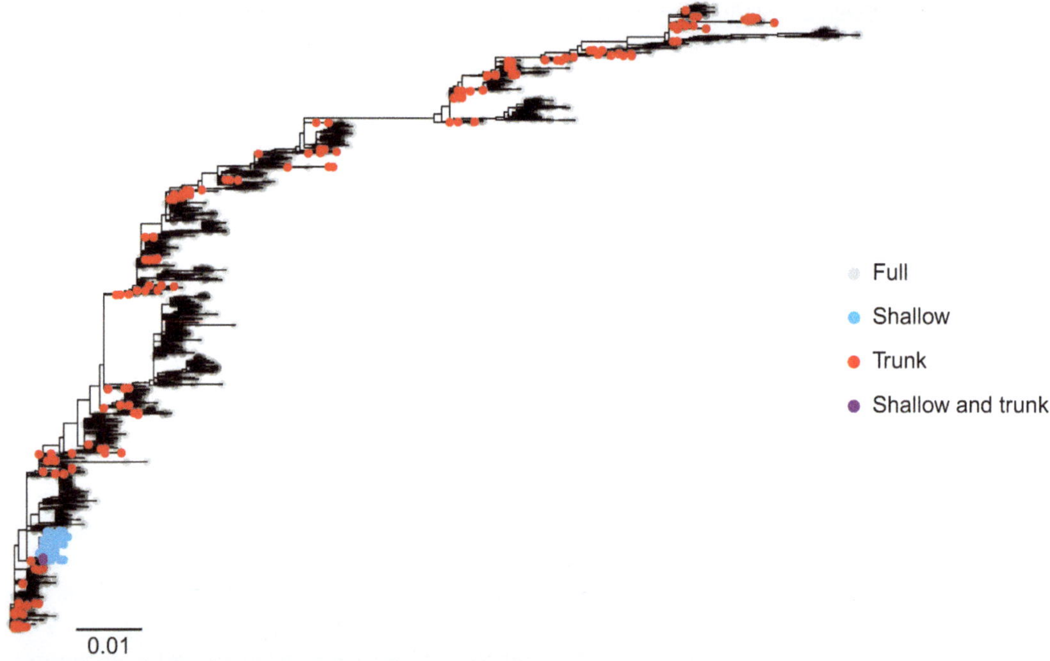

Fig. 3 Phylogeny of H3 hemagglutinin sequences analyzed here. Tip colors indicate those selected for each dataset

Table 1
Sites identified as positively selected across the H3 datasets analyzed here

Dataset	Method	Sites under selection at $P \leq 0.1$*
Full H3	MEME	(16) **19, 47, 61, 69, 110**, 151, **154, 156, 173, 208, 236, 241, 277, 278, 292, 538**
Full H3	FEL	(15) **19, 47, 61, 69, 110, 154, 156, 173, 236, 237, 241, 277, 278, 292, 538**
Full H3	SLAC	(19) **19, 47, 61, 69, 110**, 137, **154, 156**, 158, **173**, 189, **208, 236, 237, 241, 277, 278, 292**, 505, 546
Full H3	FUBAR	(13) **47, 61, 69, 110, 154, 160, 173, 208, 236, 237, 241, 278, 538**
Shallow H3	MEME	(2) **49**, 320
Shallow H3	FEL	(2) **49, 241**
Shallow H3	SLAC	*None*
Shallow H3	FUBAR	(3) **19, 49, 241**
Trunk H3	MEME	(6) **64, 154**, 171, **208, 242**, 402
Trunk H3	FEL	(3) **64, 154, 208**
Trunk H3	SLAC	(2) **154, 208**
Trunk H3	FUBAR	(6) **61, 64, 69, 154, 208, 242**

Bold sites are those identified by multiple methods for a given dataset. ***Bold italicized*** sites are those identified in more than one dataset, generally by more than one method. Numbers in parentheses give the total number of positively selected sites identified with the given method and dataset

* For FUBAR, significance is assessed as posterior probability ≥ 0.9

FEL: Launch HyPhy from the command line, and enter options `1` (Selection Analyses) and then `2` to reach the FEL analysis menu, and supply values for the following prompts:

1. **Choose genetic code**. Enter `1` to use the Universal genetic code.
2. **Select a coding sequence alignment file**. Provide the full path to the dataset of interest: `/path/to/data/h3_trunk.fna`.
3. **A tree was found in the data file...Would you like to use it (y/n)?**. Enter "y" to use the tree.
4. **Choose the set of branches to test for selection**. This option allows you to specify which branches along which site-level inference should be performed. Enter `1` to test all branches for selection.
5. **Use synonymous rate variation?**. This option asks you to specify whether the *dS* parameter in the codon model should be allowed to vary across sites ("Yes") or be fixed to 1 at all sites ("No"). Enter `1` to use a model with synonymous rate variation.
6. **Select the *P*-value used to perform the test at (permissible range = [0,1], default value = 0.1)**. Provide the default threshold of `0.1`.

FEL will now run to completion and print status indicators to the screen, including results for any site found to be under selection (either positive or negative). Abbreviated results are shown below.

Listing 4 Partial FEL screen output:

```
### Obtaining branch lengths and nucleotide rates under the GTR model
* Log ( L ) = -7506.06

### Obtaining the global omega estimate based on relative GTR branch lengths
and nucleotide substitution biases
* Log ( L ) = -7302.10
* non - synonymous / synonymous rate ratio for *test* = 0.2923

### Improving branch lengths , nucleotide substitution biases, and global dN/dS
ratios under a full codon model
* Log ( L ) = -7289.65
* non - synonymous / synonymous rate ratio = 0.2598

### For partition 1 these sites are significant at p <=0.1

| Codon | Partition | alpha | beta | LRT | Selection detected? |
|:----------:|:-----------:|:---------:|:-------:|:----------:|:-------------------:|
...
```

```
|    146    |      1     |  3.818  | 0.000 |    7.336   |   Neg. p = 0.0068  |
|    152    |      1     |  1.968  | 0.000 |    3.634   |   Neg. p = 0.0566  |
|    154    |      1     |  0.000  | 3.912 |    4.652   |   Pos. p = 0.0310  |
|    159    |      1     |  4.413  | 0.716 |    2.972   |   Neg. p = 0.0847  |
|    164    |      1     |  2.082  | 0.000 |    2.713   |   Neg. p = 0.0995  |
|    176    |      1     |  1.659  | 0.000 |    2.986   |   Neg. p = 0.0840  |
|    177    |      1     |  6.393  | 0.000 |    8.421   |   Neg. p = 0.0037  |
|    181    |      1     |  1.928  | 0.000 |    3.286   |   Neg. p = 0.0699  |
|    190    |      1     |  2.085  | 0.000 |    2.715   |   Neg. p = 0.0994  |
|    201    |      1     |  1.645  | 0.000 |    3.370   |   Neg. p = 0.0664  |
|    208    |      1     |  0.000  | 3.625 |    4.668   |   Pos. p = 0.0307  |
...

  ###  **  Found  _3_  sites  under  pervasive positive diversifying and _115_
sites under negative selection at p <= 0.1**
```

Inference details for codons with significant likelihood ratio tests for positive or negative selection are reported to the screen.

Codon	The codon where non-neutral evolution has been detected.
Partition	Allows one to keep track which subset of the alignment a particular site belongs to. This is important for recombination-corrected partition analyses.
alpha	Site-specific synonymous substitution rate
beta	Site-specific non-synonymous substitution rate
LRT	Site-specific likelihood ratio test statistic for non-neutral evolution (alpha $\neq$ beta)
Selection detected?	Selection classification (positive or negative) and the corresponding *P*-value

Note that the "Codon" and "Partition" columns are common to all site-specific analyses.

MEME and SLAC: SLAC and MEME follow identical menu prompts as FEL, with the exception that only FEL will prompt for synonymous rate variation. Instead, SLAC has a different prompt for Step 5: **Select the number of samples used to assess ancestral reconstruction uncertainty**. If this number is positive, then HyPhy will draw samples from the distribution of ancestral states and use them to measure whether or not inference is sensitive to ancestral inference uncertainty. When you encounter this option, provide the default value of `100` (or `0` to forego sampling). MEME does not emit any additional prompts.

Listing 5 Partial SLAC screen output:

```
...
###   For   partition   1   these   sites are significant at p <=0.1

|    Codon  | Partition|      S     |     N     |     dS    |      dN     |Selection detected?|
|:--------: |:-------: |:--------: |:------:  |:-------: |:-------: |:----------------:  |
  ...
|    146    |     1    |   3.000   | 0.000    | 3.000     |   0.000    |    Neg. p = 0.037    |
|    154    |     1    |   0.000   | 8.000    | 0.000     |   4.000    |    Pos. p = 0.039    |
|    177    |     1    |   3.000   | 0.000    | 4.038     |   0.000    |    Neg. p = 0.020    |
|    208    |     1    |   0.000   | 6.000    | 0.000     |   2.994    |    Pos. p = 0.089    |

...
###   Ancestor   sampling   analysis

> Generating   100   ancestral   sequence samples to obtain confidence intervals

Resampling   results   for   partition 1

|Codon|Part.|S[median,IQR] |N[median,IQR] |dS[median, IQR]|dN[median, IQR]|p-value[median, IQR]|
|:---:|:---:|:---------:|-----------:|---------:|----------:|---------------------:|
...
|146 |1|3.00 [3.00-3.00]|0.00 [0.00-0.00]|3.00 [3.00-3.00] |0.00 [0.00-0.00] |0.04 [0.04-0.04].|
|154 |1 |0.00 [0.00-0.00]|8.00 [8.00-8.00]| 0.00 [0.00-0.00]|4.00 [4.00-4.00]|0.04 [0.04-0.04]  |
|177|1|3.00 [3.00-4.00]| 0.00 [0.00-0.00]| 4.04 [4.04-5.38]|0.00 [0.00-0.00]|0.02 [0.01-0.02]  |
|208|1| 0.00 [0.00-0.00]| 6.00 [6.00-6.00]| 0.00 [0.00-0.00]|2.99 [2.99-2.99]|0.09 [0.09-0.09]  |

  ...
```

SLAC reports several key quantities for codons with significant *P*-values for positive or negative selection to the screen.

S The number of synonymous substitutions inferred at this site

NS The number of non-synonymous substitutions inferred at this site

dS Estimated site-specific synonymous rate

dN Estimated site-specific non-synonymous rate

Selection detected? Selection classification (positive or negative) and the corresponding *P*-value for the binomial test

If the user elected to perform ancestral resampling, another table is reported, showing how much these quantities are affected by ancestral state reconstruction uncertainty. For example, at codon 177, some ancestral reconstructions yielded 3 synonymous substitutions, whereas others yielded 4; however, this was not sufficient to move the *P*-value on different sides of the threshold.

Listing 6 Partial MEME screen output:

```
...
| Codon |Partition| alpha |beta+ | p+ |LRT |Episodic selection detected?|#branches|
|:------:|:-------: |:----:|:----: |:---: |:---:|:----------------------: |:--------: |
| 64 | 1 | 0.000 |14.717|0.204|3.512| Yes, p = 0.0816 | 5 |
| 154 | 1 | 0.000 |35.302|0.145|5.334| Yes, p = 0.0317 | 8 |
| 171 | 1 | 0.000 |45.005|0.017|5.753| Yes, p = 0.0256 | 1 |
| 208 | 1 | 0.000 |59.749|0.089|5.554| Yes, p = 0.0283 | 6 |
| 242 | 1 | 1.839 |34.114|0.216|4.273| Yes, p = 0.0549 | 7 |
| 402 | 1 | 0.000 |10.476|0.091|3.493| Yes, p = 0.0824 | 2 |

### ** Found _6_ sites under episodic diversifying positive selection at p
<= 0.1**
```

MEME prints information only about codons subject to positive selection, since MEME does not directly test for negative selection.

alpha	Site-specific synonymous substitution rate
beta+	Site-specific non-synonymous substitution rate for the positive selection category
p+	Site-specific weight (~ proportion of branches) assigned for the positive selection category
LRT	Site-specific likelihood ratio test statistic for episodic diversifying selection (beta+ > 1 and p+ > 0)
Episodic selection detected?	Selection classification (yes) and the corresponding *P*-value
# branches	An exploratory estimate of the number of individual branches which have sufficient empirical Bayes support for positive selection; since MEME pools signal from multiple branches, there may be overall evidence for selection, without necessarily implicating any individual branches.

FUBAR: To run FUBAR, launch HyPhy from the command line, and enter options `1` (Selection Analyses) and then `4` to reach the FUBAR analysis menu, and supply values for the following prompts[5]:

[5] Note that for all prompts with default values, simply pressing `enter` will choose this default.

1. **Choose genetic code**. Enter `1` to use the Universal genetic code.
2. **Select a coding sequence alignment file**. Provide the full path to the dataset of interest: `/path/to/data/h3_trunk.fna`.
3. **A tree was found in the data file...Would you like to use it (y/n)?**. Enter "y" to use the tree.
4. **Number of grid points per dimension.** This option controls how **fine** the FUBAR analysis is by setting the range of possible dN and dS values that can be inferred, along an $N \times N$ grid. We will use the default value of `20` (leading to a 20×20 grid of dN/dS ratios). FUBAR will now pre-compute likelihoods for each value in the grid.
5. **Number of MCMC chains to run**. This option determines the number of Markov Chain Monte Carlo chains to run during Bayesian inference of evolutionary rates. Enter the default value of `5` to run 5 chains.
6. **The length of each chain**. This option controls for how long each MCMC chain should be run. Enter the default value of `2000000` to run each chain for two million generations (thus obtaining two million samples).
7. **Use this many samples as burn-in**. This option determines how many initial samples drawn from the MCMC chain should be discarded as burn-in, as is standard in Bayesian analyses. Enter the default value of `1000000`, leading to a final value of one-million draws per chain.
8. **How many samples should be drawn from each chain**. This option determines the final number of samples to draw from the full set of one-million draws per chain. Enter the default value of `100`.
9. **The concentration parameter of the Dirichlet prior**. This option controls the shape of the Dirichlet prior distribution. Enter the default value of `0.5`.

Listing 7 Partial FUBAR screen output:

```
...
###   Tabulating   site - level   results
|   Codon  |Partition| alpha |  beta  |  N.eff   |Posterior prob for positive selection|
|:-------:|:-------:|:----:|:----:  |:-------:|:----------------------------------:|
|    61    |    1     | 0.753 |  4.365|   64.549 |         Pos. posterior = 0.9262        |
|    64    |    1     | 0.753 |  3.920|   77.106 |         Pos. posterior = 0.9095        |
|    69    |    1     | 0.730 |  4.447|   64.182 |         Pos. posterior = 0.9325        |
|   154    |    1     | 0.637 |  6.595|   53.312 |         Pos. posterior = 0.9826        |
```

```
|  208    |    1    | 0.622 |  5.908|   55.794 |        Pos. posterior = 0.9731        |
|  242    |    1    | 2.215 | 12.055| 1489.879 |        Pos. posterior = 0.9131        |
----
##   FUBAR   inferred   6   sites   subject to diversifying positive selection at
posterior probability >= 0.9
Of  these ,   0.36  are  expected  to be false positives (95% confidence interval
of 0-2 )
```

Like other site analyses, FUBAR will print a number of inferences about each individual site detected to be under pervasive positive selection

alpha	The posterior estimate of the synonymous substitution rate at a site
beta	The posterior estimate of the non-synonymous substitution rate at a site
N.eff	An estimate of the effective sample size for inferring positive selection at this site; smaller values (e.g., < 20) imply that the MCMC procedure may have failed to sample the parameter space well, and longer chains (or more chains) might be warranted
Posterior prob for positive selection	The estimated posterior probability for pervasive diversifying selection ($dN/dS > 1$).

Interpreting Results:
Sites identified as positively selected by each method, across all three datasets, are given in Table 1. In general, we expect MEME to be the most comprehensive and robust of all site-level methods because it uniquely considers both pervasive and episodic selection [24]. In addition, power studies have shown that FUBAR is expected to outperform FEL and SLAC under most circumstances [25]. Finally, we expect that SLAC will be the least robust method due to its reliance on a relatively naive counting-based approach [12].

These expectations are generally borne out in the results obtained here in our brief study of H3 selection. For the full H3 dataset of 2555 sequences, MEME identified 16 sites, and FEL identified 15 sites under positive selection. All sites were identical except for the following: MEME uniquely identified sites 151 and 208, and FEL uniquely identified with 237. Interestingly, site 208 was additionally identified as positively selected by all methods on the trunk H3 dataset. Combined, these results demonstrate

MEME's ability to identify sites subject to both pervasive and episodic selection, as site 208 appears to be under pervasive selection only along the H3 trunk. Because FEL uses a less stringent test statistic distribution (χ_1^2) to call significance, occasionally sites subject to pervasive selection near the significance thresholds may be detected by FEL but missed by MEME (e.g., site 237, with FEL reporting $P = 0.08$ and MEME reporting $P = 0.105$).

FUBAR identified two fewer selected sites in the full H3 alignment compared to FEL (which is a directly comparable test), missing sites 19 (posterior 0.83), 277 (posterior 0.59), and 292 (posterior 0.89) relative to FEL, but adding site 160 (FEL $P = 0.8$).

In addition to differences across methods, we expect to see some important differences for sites inferred across the full, shallow, and trunk H3 datasets. Because the trunk and full H3 datasets span similar time frames, we expect sites returned for these two datasets to have the most overlap. In addition, sites found to be under selection in the shallow lineage may not be detected across the full H3 phylogeny, as selection may have been fleeting, weak, or constrained to the specific shallow clade examined here. For example, site 49 was specifically selected in the shallow H3 lineage alone, as indicated by three of the four methods. In contrast, sites 19 and 241 were found to be selected in both the shallow and the full H3 datasets, but this signal was not apparent when the trunk lineage was examined independently, perhaps because these sites experience only transient changes that do not propagate along the trunk.

What are some potential reasons for seeing discrepancies in inferences across H3 datasets? The site 154, for example, is positively selected in both the full H3 phylogeny and the trunk H3 lineage, but not the shallow H3 lineage. This result suggests that site 154 may have experienced pervasive selection throughout H3 evolution, but its signal in the shallow clade alone was either too weak to detect or selection was attenuated in the shallow clade. In addition, sites which appeared only in the shallow clade analyses may have experienced lineage-specific selection where the signal was too weak to detect when the entire phylogeny was considered.

Furthermore, while MEME, FEL, and FUBAR were able to detect selected sites in the shallow H3 lineage, SLAC did not identify any such sites. This is because SLAC requires a large number of substitutions, which are unlikely to have occurred in the shallow sample, to achieve significance. Overall, we emphasize that in many cases different site-level methods will **not** identify exactly the same set of sites under selection, although, as the H3 example shows, the agreement between is typically good.

Rules of Thumb for Site-Level Detection of Selection

1. Small datasets, i.e., ≤ 10 sequences (especially when coupled with low divergence), are unlikely to yield any sites under selection. Consider using gene-wide methods like BUSTED or aBSREL to look for selection in these cases.
2. On large datasets (e.g., > 500 sequences), all methods tend to give similar results (but see the MEME exception below), hence the default method of choice is FUBAR, since its run time is dramatically shorter than FEL or MEME, and its statistical performance is better than SLAC.
3. MEME tends to be the most sensitive method, because it is the only one designed to detect episodic selection. Indeed, sometimes SLAC, FEL, or FUBAR may all call a site subject to episodic positive selection site negatively selected, if a burst of selection is followed by strong conservation. MEME is often able to tease the two processes apart and correctly call such sites positively selected. Hence, MEME should be the preferred method, unless computationally prohibitive.
4. We cannot universally recommend running all the available methods on a given dataset and then aggregating the results, as done in Table 1, for several reasons. Firstly, while it may be tempting to use agreement between all methods as a hedge against false positives, i.e., calling a site selected only if all the methods agreed on it, reduces the power of the analysis to that of the least sensitive method. Secondly, while comparing the sites on which methods disagree can potentially reveal critical information (e.g., a site detected by MEME but not FUBAR may be under strong episodic selection), considerable effort and diligence must be put into disentangling meaningful biological differences from statistical artifacts. Thirdly, statistical strategy must be informed before the analysis commences by deciding which is more important to optimize: does one care more about specificity (reducing false positives) or sensitivity (reducing false negatives)? For example, if little is known about a gene, it may be advisable to generate the most inclusive list of sites that could be subject to selection for subsequent testing using other approaches; in this case, the most sensitive method or the union of all methods may be appropriate.
5. We strongly recommend against performing multiple testing or false discovery rate correction on individual site results. Firstly, methods are calibrated to not generate excessive false positives on strictly neutral data. In most genes, most sites will be under relatively strong negative selection, making the statistical testing procedure conservative. Secondly, multiple testing

corrections will nearly always yield no significant results on small to moderate sized datasets. Thirdly, some key assumptions of methods for correcting false discovery rates are not applicable for site-level testing. For example, a typical collection of results from site-level testing will contain very few, if any, true sites with *P*-values supporting neutrality ($dN/dS = 1$).

3.6 Screening Sequences for Recombination

A critical aspect of sequence analysis we have not yet covered is the detection of and correction for intragenic *recombination* in an alignment of homologous sequences. Because recombination is such a key biological process in many viral pathogens, we strongly advocate screening an alignment for recombination before proceeding with additional analyses, unless there is a sound biological reason to discount (i.e., intragenic recombination Influenza A is negligibly rare). Indeed, because recombination causes different regions of an alignment to be related by different phylogenies, its presence can heavily influence selection detection and other downstream applications.

There are many computational approaches to finding evidence of recombination in a sequence alignment [32], however at their core, many such methods look for evidence of phylogenetic incongruence. Here, we demonstrate one such method, GARD (genetic algorithms for recombination detection) that we have found to perform very well among a wide range of approaches on simulated data [14]. Note that at this time, GARD will not produce a JSON file as output but instead several text files containing inference information, as well as a final *partitioned alignment* for downstream use if recombination was detected.

3.7 GARD

What Biological Question Is the Method Designed to Answer?:
Have sequences in the given alignment undergone recombination, and if so what are the recombination breakpoints and segment-specific phylogenies?

Recommended Applications:

GARD is geared towards mapping the breakpoints and detecting segments of the alignment which can be adequately described by a single tree topology. Therefore, alignments, particularly alignments of viral sequences, should be screened for the presence of recombination before performing any selection inference. The NEXUS output from GARD can be directly used as input for most downstream selection detection analyses.

Statistical Test Procedure:
GARD employs a genetic algorithm to find a solution to a complex optimization problem by mimicking processes of biological evolution (mutation, recombination, and selection) in a population of competing solutions. In this application of genetic algorithms, we are evolving a population of "chromosomes" that specify different numbers and locations of recombination breakpoints in the alignment with the objective of detecting topological incongruence, i.e., support for different phylogenies by separate regions of the alignment. The "fitness" of each chromosome is determined by using maximum likelihood methods to evaluate a separate phylogeny for each non-recombinant fragment defined by the breakpoints (e.g., to the left and to the right of a breakpoint in Fig. 4), and computing a goodness of fit (AIC_c) for each such model. The genetic algorithm searches for the number and placement of breakpoints yielding the best AIC_c and also reports confidence values for inferred breakpoint locations based on the contribution of each considered model weighted by how well the model fit the data. For computational expedience, the current implementation of GARD infers topologies for each segment using neighbor joining [37] based on the TN93 pairwise distance estimator [41] and then fits a user-specified nucleotide evolutionary model using maximum likelihood to obtain AIC_c scores.

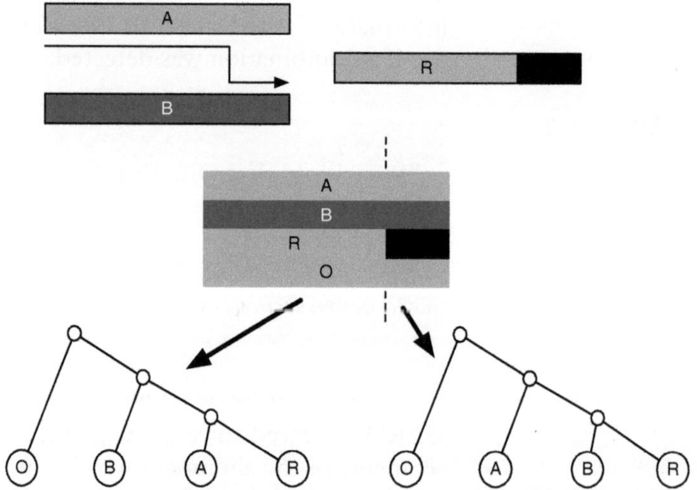

Fig. 4 Phylogenetic incongruence caused by the presence of a recombinant sequence in an alignment. Sequence R is a product of homologous recombination between sequences A and B. Phylogenies reconstructed from sequences A, B, R and an outgroup sequence (O) will differ based on which part of the alignment is being considered. To the left of the breakpoint, R clusters with A, whereas to the right of the breakpoint R clusters with B

Example Analysis 1: *We will demonstrate the use of GARD, as well as its benefits for downstream analysis, using a dataset consisting of 13 glycoprotein sequences from Cache Valley Fever virus (*`cvf.fna`*). We will first use GARD to detect recombination in this dataset, and then we will process both the GARD-informed data and the original alignment (with no recombination assumed) with FEL to see how the presence of recombination may confound selection inference.*

Importantly, GARD specifically requires the use of HyPhy's MPI-enabled executable, HYPHYMPI. To run GARD from the command line, you will need an operating system with a MPI headers and libraries installed so that this executable can be compiled. Here, we will describe how to use GARD from the command line, but we emphasize that GARD is fully implemented and available on www.datamonkey.org and takes the same input options described here.

To run GARD, open a terminal session and start HYPHYMPI in the appropriate MPI environment (e.g., MPIRUN *in OpenMPI) from the command line to launch the HyPhy analysis menu. Enter* `12` *(Recombination) and then* `1` *to reach the GARD analysis menu, and supply values for the following prompts:*

1. **Nucleotide file to screen**: Provide the full path to the dataset of interest: `/path/to/data/cvf.fna`.
2. **Please enter a 6-character model designation (e.g., 010010 defines HKY85)**. This option controls which nucleotide substitution model is to be used for analysis, using PAUP notational shorthand. The six-character shorthand allows the user to specify the entire spectrum from F81 (000000) to GTR (012345), which we recommend as default option. Provide the value `012345` for this prompt.
3. **Rate variation options**. This option determines how site-to-site rate variation should be modeled. The option `None` will discount site-to-site rate variation, allowing the analysis to run several times faster than other options but also creating the risk of mistaking rate heterogeneity for recombination. As such, we can only recommend this option for extremely small alignments (i.e., 3–5 sequences). The option `General Discrete` (the default) models rate variation using an N bin general discrete distribution, and option `Beta-Gamma` models rate variation using an adaptively discretized distribution, a more flexible version of the standard Gamma+4 model. Enter option `2` to select the General Discrete model.
4. **How many distribution bins [2–32]?**. If rate variation was selected in the previous step, this option allows the user to decide how many different rate classes should be included in the model. We recommend using 3 rate classes by default, as both General Discrete and Beta-Gamma distributions are flexible enough to reliably capture rate variability in the majority of alignments with only a few rate classes. Therefore, enter the value `3`.

5. **Save results to**. For this option, provide a full path to the output file to which you would like GARD to write results. The supplied file name will ultimately contain an HTML-formatted summary of the analysis. HyPhy will generate several other files with names obtained by appending suffixes (as in `<file name>_suffix`) to the main result file. In particular, the `_finalout` file stores the original alignment in NEXUS format with inferred non-recombinant sections of the alignment saved in the `ASSUMPTIONS` block and trees inferred for each partition in the `TREES` block. This NEXUS file can be input into many recombination-aware analyses in HyPhy and other programs that can read NEXUS. The `_ga_details` file contains two lines of information about each model examined by the genetic algorithm: its AICc score and the location of breakpoints in the model. Finally, the `_ga_splits` file stores information about the location of breakpoints and trees inferred for each alignment region under the best model found by the GA.

   GARD will now run to completion, printing status indicators to screen while it runs:

Listing 8 Partial GARD output:

```
Fitting a baseline nucleotide model...
Done with single partition analysis. Log(L) = -5921.9511901113, c-AIC = 11914.85153276497
Starting the GA ...

GENERATION 2 with 1 breakpoints (~0% converged)
Breakpoints    c-AIC  Delta c-AIC [BP    1]
          0 11914.85
          1 11804.56      110.291     1393
GA has considered          92/        328 (92 over all runs) unique models
Total run time             0 hrs 0 mins 2 seconds
Throughput                 46.00 models/second
Allocated time remaining 999 hrs 59 mins 58 seconds (approx. 165599908 more models.)
...
GENERATION 52 with 4 breakpoints (~100% converged)
Breakpoints    c-AIC  Delta c-AIC [BP    1] [BP    2] [BP    3] [BP    4]
          0 11914.85
          1 11804.56      110.291     1445
          2 11783.92       20.638      617     1490
          3 11778.94        4.978      587      962     1475
          4 11778.94        0.000      587      962     1475
GA has considered          268/    473490550 (1356 over all runs) unique models
Total run time             0 hrs 4 mins 2 seconds
Throughput                  5.60 models/second
Allocated time remaining 999 hrs 55 mins 58 seconds (approx. 20170544.82644628 more models.)
Performing the final optimization...
```

Interpreting Results:
GARD found evidence of recombination in this dataset with three breakpoints, yielding a 135.9 point AIC_c improvement over the model without recombination. Among all models with three breakpoints in the Cache Valley Virus glycoprotein alignment, the best model places them at nucleotides 587, 962, and 1475. Importantly, if GARD had reported that the best model had 0 breakpoints, we could conclude that no evidence of recombination had been found. Note that because genetic algorithms are stochastic, there is no guarantee that replicate runs will converge to exactly the same quantitative results. When there is a strong signal of recombination breakpoints in the data, however, the qualitative results (number and general location of breakpoints) should be fairly robust.

Example Analysis 2: *The NEXUS file that GARD produced is a* partitioned dataset, *wherein different groups of sites are described by different trees. Most HyPhy selection analyses discussed here,*[6] *including MEME, FUBAR, FEL, SLAC, and BUSTED, are able to analyze partitioned data. To demonstrate the importance of screening for recombination, we will now compare results for a FEL analysis performed on the original alignment of 13 Cache Valley Virus glycoproteins, as well as on the GARD-inferred partitioned alignment. All steps here were carried out as described earlier in this chapter.*

Interpreting Results:
FEL inference on the GARD-processed partitioned Cache Valley Virus data does not detect sites under selection at $P \leq 0.1$. *By contrast, FEL inference on the unpartitioned Cache Valley Virus data (i.e., not pre-screened for recombination) detects three positively selected sites at* $P \leq 0.1$ *(212, 516, and 558 at* $P = 0.08$, $P = 0.03$, *and* $P = 0.09$, *respectively). From these results, we can clearly tell that not screening or recombination has the potential for adverse consequence including an increased false positive rate as seen here. As such, we strongly encourage users to screen alignments for recombination if such processes are suspected before proceeding to selection detection.*

3.8 Accounting for Synonymous Rate Variation

A critical genomic process that one must consider when detecting selection is the phenomenon of *synonymous rate variation*, wherein the rate of synonymous codon evolution (represented by *dS* in the

[6] Note that neither aBSREL nor RELAX accepts partitioned data because they require a consistent phylogeny to define branch sets.

context of codon models and representing mutation rate) varies across species, genes, and even intragenic positions. In particular, intragenic synonymous rate variation has been identified across domains of life [11, 20, 30] and can arise from a variety of evolutionary processes, including selection on mRNA secondary structure [2], gene expression [4], GC-biased gene conversion [10], and other neutral mutation processes. For example, even the genomic context of a given nucleotide can influence its mutation rate; indeed, experimental work has shown that GC-neighboring sites can feature up to a 75-fold increase in mutation rate [20, 38]. In addition, the synonymous rate at certain sites may be elevated due to the mutational vulnerability of the non-template DNA strand during transcription [20]. These processes must be accounted for in order to ensure an appropriate baseline *dS* is used when testing for selection.

We demonstrate the importance of considering synonymous rate variation for selection inference using a dataset of 10 mammalian CD2 genes, which code for a specific T-cell surface adhesion molecule [21]. We use FEL to detect selection in this dataset under two specifications: with synonymous rate variation ("yes" in prompt 4 in the FEL analysis menu), and without synonymous rate variation ("no" in prompt 4 in the FEL analysis menu).

Interpreting Results:
At P $\leq$ *0.1, analysis of CD2* with *synonymous rate variation revealed a total of 14 sites under positive selection. By contrast, CD2 analysis with FEL without* dS *variation only detected four sites under positive selection (Fig. 5). Similarly, analysis with* dS *variation revealed 27 sites under purifying selection, but analysis without* dS *variation revealed only 15 sites under purifying selection. Most importantly, all sites detected when* dS *was fixed to 1 were a subset of the sites identified by the model with* dS *variation (Fig. 5). Together, these results demonstrate that ignoring* dS *variation can induce both an increased false negative rate regarding positive selection detection and an overall decrease in power to detect any selective regime. We acknowledge that it is possible that the opposite conclusion might be true, namely, that additional sites identified by FEL with* dS *variation might instead be false positives. However, in our experience, this is much less frequently the case [12].*

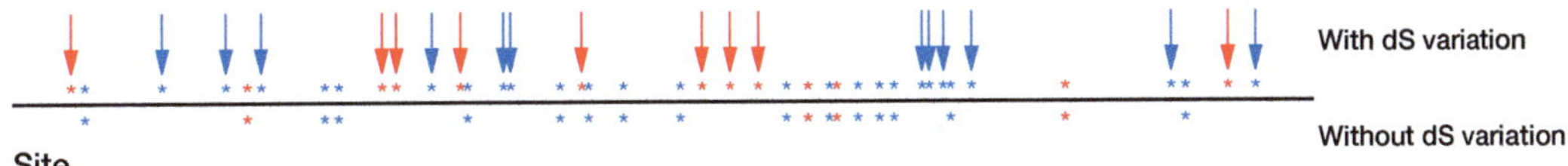

Fig. 5 Sites identified as positively (red) and negatively (blue) selected in CD2 at $P \leq 0.1$ by FEL run with (above the line) and without *dS* variation (below the line). Sites with arrows represent those identified as selected by FEL with *dS* variation that were *not* identified by FEL when *dS* variation was ignored

4 Tips

Here we provide some helpful notes on HyPhy usage.

- An actively maintained board for usage questions and filing bug reports is available at `https://github.com/veg/hyphy/issues`.
- Each HyPhy analysis described here will export a JSON file. This file can either be uploaded to **HyPhy-Vision** for visual examination, or it can be easily parsed using a standard scripting language using standard packages, for example, the `json` package in Python or the `jsonLite` package in R. All fields used in these output files are defined in http://hyphy.org.
- Mac OS(X) users may need to install a new set of compilers (i.e., gcc-6) that are compatible with openMP in order to have full functionality from the HYPHYMP executable, as is described on the HyPhy website.

5 Exercises

1. Earlier, we performed a BUSTED analysis without designating a specific subset of test lineages. For this exercise, we will analyze the HIV-1 transmission dataset with BUSTED in two different ways: testing all branches, and testing only recipient-derived HIV-1 sequences. The input data for this exercise, with an appropriately labeled phylogeny, is available in `exercises/hiv1_transmission_exercise1.fna`. For select branches labeled `All` or `test` as the test lineages.
    - Is there evidence (compare model fits using the small sample AIC) that `test` branches have a different selective regime than the rest of the tree?
    - The entire dataset should provide evidence for episodic diversification, but the recipient only analysis should return a negative result. What does this mean biologically, i.e., where does the selection signal come from?

2. Investigate the effect of recombination of site-specific inference of episodic selection using MEME. Run MEME on `exercises/cvf.fna` (single partition data, i.e., assuming no recombination), and then on the same dataset screened for recombination using GARD `exercises/cvf_gard.nexus`, testing for selection on `all` branches, with `P=0.1`. Compare the list of sites detected to be under selection by the two analyses.
   - Which analysis generated more positive results?
   - Do you think these results are true or false positives? How does this compare to the FEL analysis we described in the text?
   - Compare site-wise estimates of substitution rates (e.g., α) between the two analyses. Is there a discernible bias introduced by not accounting for recombination?
3. When analyzing intraspecies or intrahost data, dN/dS estimates may be inflated due to the fact that not all observed sequence variation are due to substitutions, but some are simply mutations that have not yet been filtered by selection [17, 23, 31, 35]. In other words, dN/dS may be elevated by intraspecies/intrahost polymorphism that should not necessarily be attributed to positive selection. One simple approach to mitigating this undesirable effect is to restrict site-specific analyses to `Internal` branches only. Internal branches are less likely to contain spurious polymorphic variants because they encompass at least one process on which selection can act (i.e., transmission and/or multiple rounds of replication). Apply MEME and FEL to an intrahost sample of HIV-1 sequences, found in `exercises/JS1774.nex`, from an infected individual analyzed in Lorenzo-Redondo et al. [19] first choosing to test `All` branches, and next choosing `Internal` branches.
4. Compare the lists of selected sites between All/Internal analyses. How different are they?
5. Use RELAX to formally test whether or not selective regimes (dN/dS distributions) are different between terminal and internal branches in `exercises/JS1774.nex`.

References

1. Anisimova M, Kosiol C (2009) Investigating protein-coding sequence evolution with probabilistic codon substitution models. Mol Biol Evol 26:255–271
2. Chamary JV, Hurst LD (2005) Evidence for selection on synonymous mutations affecting stability of mRNA secondary structure in mammals. Genome Biol 6:R75
3. Delport W, Scheffler K, Seoighe C (2009) Models of coding sequence evolution. Brief Bioinform 10(1):97–109
4. Drummond DA, Wilke CO (2008) Mistranslation-induced protein misfolding as a dominant constraint on coding-sequence evolution. Cell 134:341–352

5. Enard D, Cai L, Gwennap C, Petrov DA (2016) Viruses are a dominant driver of protein adaptation in mammal. eLife 5:e12469
6. Forni D, Cagliani R, Clerici M, Sironi M (2017) Molecular evolution of human coronavirus genomes. Trends Microbiol 25(1):35–48. ISSN 0966-842X
7. Frost SDW, Liu Y, Kosakovsky Pond SL, Chappey C, Wrin T, Petropoulos CJ, Little SJ, Richman DD (2005) Characterization of Human Immunodeficiency Virus Type 1 (HIV-1) envelope variation and neutralizing antibody responses during transmission of HIV-1 Subtype B. J Virol 79:6523–6527
8. Goldman N, Yang Z (1994) A codon-based model of nucleotide substitution for protein-coding DNA sequences. Mol Biol Evol 11:725–736
9. Graef KM, Vreede FT, Lau Y-F, McCall AW, Carr SM, Subbarao K, Fodor E (2010) The PB2 subunit of the Influenza virus RNA Polymerase affects virulence by interacting with the mitochondrial antiviral signaling protein and inhibiting expression of beta interferon. J Virol 84:8433–8445
10. Harrison RJ, Charlesworth B (2011) Biased gene conversion affects patterns of codon usage and amino acid usage in the Saccharomyces sensu stricto group of yeasts. Mol Biol Evol 28:117–129
11. Hershberg R, Petrov D (2008) Selection on codon bias. Annu Rev Genet 42:287–299
12. Kosakovsky Pond SL, Frost SWD (2005) Not so different after all: a comparison of methods for detecting amino acid sites under selection. Mol Biol Evol 22:1208–1222
13. Kosakovsky Pond SL, Muse SV (2005) Site-to-site variation of synonymous substitution rates. Mol Biol Evol 22:2375–2385
14. Kosakovsky Pond SL, Posada D, Gravenor MB, Woelk CH, Frost SDW (2006) Automated phylogenetic detection of recombination using a genetic algorithm. Mol Biol Evol 23 (10):1891–901
15. Kosakovsky Pond SL, Delport W, Muse SV, Scheffler K (2010) Correcting the bias of empirical frequency parameter estimators in codon models. PLoS One 5:e11230
16. Kosakovsky Pond SL, Murrell B, Fourment M, Frost SD, Delport W, Scheffler K (2011) A random effects branch-site model for detecting episodic diversifying selection. Mol Biol Evol 28:3033–3043
17. Kryazhimskiy S, Plotkin JB (2008) The population genetics of *dN/dS*. PLoS Genet 4: e1000304
18. Labadie K, Dos Santos Afonso E, Rameix-Welti M-A, van der Werf S, Naffakh N (2007) Host-range determinants on the PB2 protein of influenza A viruses control the interaction between the viral polymerase and nucleoprotein in human cells. Virology 362:271–282
19. Lorenzo-Redondo R, Fryer HR, Bedford T, Kim E-Y, Archer J, Pond SLK, Chung Y-S, Penugonda S, Chipman JG, Fletcher CV, Schacker TW, Malim MH, Rambaut A, Haase AT, McLean AR, Wolinsky SM (2016) Persistent HIV-1 replication maintains the tissue reservoir during therapy. Nature 530 (7588):51–56
20. Lynch M, Ackerman MS, Gout J-F, Long H, Sung W, Thomas WK, Foster PL (2016) Genetic drift, selection and the evolution of the mutation rate. Nature 17(11):704–714
21. Lynn DJ, Freeman AR, Murray C, Bradley DG (2005) A genomics approach to the detection of positive selection in cattle: adaptive evolution of the t-cell and natural killer cell-surface protein cd2. Genetics 170(3):1189–1196
22. Meyer AG, Wilke CO (2015) Geometric constraints dominate the antigenic evolution of Influenza H3N2 Hemagglutinin. PLoS Pathog 11:e1004940
23. Mugal CF, Wolf JBW, Kaj I (2014) Why time matters: codon evolution and the temporal dynamics of *dN/dS*. Mol Biol Evol 31:212–231
24. Murrell B, Wertheim JO, Moola S, Weighill T, Scheffler K, Kosakovsky Pond SL (2012) Detecting individual sites subject to episodic diversifying selection. PLoS Genet 8(7): e1002764
25. Murrell B, Moola S, Mabona A, Weighill T, Scheward D, Kosakovsky Pond SL, Scheffler K (2013) FUBAR: a fast, unconstrained Bayesian AppRoximation for inferring selection. Mol Biol Evol 30:1196–1205
26. Murrell B, Weaver S, Smith MD, Wertheim JO, Murrell S, Aylward A, Eren K, Pollner T, Martin DP, Smith DM, Scheffler K, Kosakovsky Pond SL (2015) Gene-wide identification of episodic selection. Mol Biol Evol 32:1365–1371
27. Muse SV, Gaut BS (1994) A likelihood approach for comparing synonymous and non-synonymous nucleotide substitution rates, with application to the chloroplast genome. Mol Biol Evol 11:715–724
28. Nelson MI, Holmes EC (2007) The evolution of epidemic influenza. Nat Rev Genet 8 (3):196–205
29. Nielsen R, Yang Z (1998) Likelihood models for detecting positive selected amino acid sites

and applications to the HIV-1 envelope gene. Genetics 148:929–936

30. Plotkin JB, Kudla G (2011) Synonymous but not the same: the causes and consequences of codon bias. Nature Rev Genet 12:32–42
31. Pond SLK, Frost SDW, Grossman Z, Gravenor MB, Richman DD, Brown AJL (2006) Adaptation to different human populations by HIV-1 revealed by codon-based analyses. PLoS Comput Biol 2(6):e62
32. Posada D, Crandall KA (2001) Evaluation of methods for detecting recombination from DNA sequences: computer simulations. Proc Natl Acad Sci U S A 98(24):13757–13762
33. Posada D, Crandall KA, Holmes EC (2002) Recombination in evolutionary genomics. Annu Rev Genet 36:75–97
34. Price SA (2015) Comparative genomics of amphibian-like Ranaviruses, nucleocytoplasmic large DNA viruses of Poikilotherms. Evol Biol Online 11:71–82
35. Rocha EPC, Maynard Smith J, Hurst LD, Holden MTG, Cooper JE, Smith NH, Feil EJ (2006) Comparisons of *dN/dS* are time dependent for closely related bacterial genomes. J Theor Biol 239:226–235
36. Rodrigue N (2013) On the statistical interpretation of site-specific variables in phylogeny-based substitution models. Genetics 193:557–564
37. Smith MD, Wertheim JO, Weaver S, Murrell B, Scheffler K, Kosakovsky Pond SL (2015) Less is more: an adaptive branch-site random effects model for efficient detection of episodic diversifying selection. Mol Biol Evol 32:1342–1353
38. Sung W, Ackerman MS, Gout J-F, Miller SF, Williams E, Foster PL, Lynch M (2015) Asymmetric context-dependent mutation patterns revealed through mutation–accumulation experiments. Mol Biol Evol 32(7):1672–1683
39. Suzuki Y, Gojobori T (1999) A method for detecting positive selection at single amino acid sites. Mol Biol Evol 16:1315–1328
40. Tamuri AU, dos Reis M, Hay AJ, Goldstein RA (2009) Identifying changes in selective constraints: host shifts in influenza. PLoS Comput Biol 5(11):e1000564
41. Tamuri AU, dos Reis M, Goldstein RA (2012) Estimating the distribution of selection coefficients from phylogenetic data using sitewise mutation–selection models. Genetics 190:1101–1115
42. Tully DC, Ogilvie CB, Batorsky RE, Bean DJ, Power KA, Ghebremichael M, Bedard HE, Gladden AD, Seese AM, Amero MA, Lane K, McGrath G, Bazner SB, Tinsley J, Lennon NJ, Henn MR, Brumme ZL, Norris PJ, Rosenberg ES, Mayer KH, Jessen H, Kosakovsky Pond SL, Walker BD, Altfeld M, Carlson JM, Allen TM (2016) Differences in the selection bottleneck between modes of sexual transmission influence the genetic composition of the HIV-1 founder virus. PLoS Pathog 12(5):e1005619
43. Wertheim JO, Murrell B, Smith MD, Kosakovsky Pond SL, Scheffler K (2015) RELAX: detecting relaxed selection in a phylogenetic framework. Mol Biol Evol 32(3):820–832
44. Yang Z (2006) Computational molecular evolution. Oxford University Press, Oxford
45. Yang Z, Nielsen R (2002) Codon-substitution models for detecting molecular adaptation at individual sites along specific lineages. Mol Biol Evol 19:908–917
46. Zhang J, Nielsen R, Yang Z (2005) Evaluation of an improved branch-site likelihood method for detecting positive selection at the molecular level. Mol Biol Evol 22:2472–2479

Chapter 15

Evolution of Protein Domain Architectures

Sofia K. Forslund, Mateusz Kaduk, and Erik L. L. Sonnhammer

Abstract

This chapter reviews current research on how protein domain architectures evolve. We begin by summarizing work on the phylogenetic distribution of proteins, as this will directly impact which domain architectures can be formed in different species. Studies relating domain family size to occurrence have shown that they generally follow power law distributions, both within genomes and larger evolutionary groups. These findings were subsequently extended to multi-domain architectures. Genome evolution models that have been suggested to explain the shape of these distributions are reviewed, as well as evidence for selective pressure to expand certain domain families more than others. Each domain has an intrinsic combinatorial propensity, and the effects of this have been studied using measures of domain versatility or promiscuity. Next, we study the principles of protein domain architecture evolution and how these have been inferred from distributions of extant domain arrangements. Following this, we review inferences of ancestral domain architecture and the conclusions concerning domain architecture evolution mechanisms that can be drawn from these. Finally, we examine whether all known cases of a given domain architecture can be assumed to have a single common origin (monophyly) or have evolved convergently (polyphyly). We end by a discussion of some available tools for computational analysis or exploitation of protein domain architectures and their evolution.

Key words Protein domain, Protein domain architecture, Superfamily, Monophyly, Polyphyly, Convergent evolution, Domain evolution, Kingdoms of life, Domain co-occurrence network, Node degree distribution, Power law, Parsimony

1 Introduction

1.1 Overview

By studying the domain architectures of proteins, we can understand their evolution as a modular phenomenon, with high-level events enabling significant changes to take place in a time span much shorter than required by point mutations only. This research field has become possible only now in the -omics era of science, as both identifying many domain families in the first place and acquiring enough data to chart their evolutionary distribution require access to many completely sequenced genomes. Likewise, the conclusions drawn generally consider properties averaged for entire

Maria Anisimova (ed.), *Evolutionary Genomics: Statistical and Computational Methods*, Methods in Molecular Biology, vol. 1910, https://doi.org/10.1007/978-1-4939-9074-0_15,

species or organism groups or entire classes of proteins, rather than properties of single genes.

We will begin by introducing the basic concepts of domains and domain architectures, as well as the biological mechanisms by which these architectures can change. The remainder of the chapter is an attempt at answering, from the recent literature, the question of which forces shape domain architecture evolution and in what direction. The underlying issue concerns whether it is fundamentally a random process or whether it is primarily a consequence of selective constraints. We end by outlining some available software tools and resources for analysis of domain architectures and their evolution.

1.2 Protein Domains

Protein domains are high-level parts of proteins that either occur alone or together with partner domains on the same protein chain. Most domains correspond to tertiary structure elements and are able to fold independently. All domains exhibit evolutionary conservation, and many either perform specific functions or contribute in a specific way to the function of their proteins. The word domain strictly refers to a distinct region of a specific protein, an instance of a domain family. However, domain and domain family are often used interchangeably in the literature.

1.3 Domain Databases

By identifying recurring elements in experimentally determined protein 3D structures, the various domain families in structural domain databases such as SCOP [1] and CATH [2] were gathered. New 3D structures allow assignment to these classes from semiautomated inspection. The SUPERFAMILY [3] database assigns SCOP domains to all protein sequences by matching them to hidden Markov models (HMMs) that were derived from SCOP superfamilies, i.e., proteins whose evolutionary relationship is evidenced structurally. The Gene3D [4] database is similarly constructed but based on domain families from CATH.

This approach resembles the methodology used in pure sequence-based domain databases such as Pfam [5]. In these databases, conserved regions are identified from sequence analysis and background knowledge, to make multiple sequence alignments. From these, HMMs are built that are used to search new sequences for the presence of the domain represented by each HMM. All such instances are stored in the database. The HMM framework ensures stability across releases and high quality of alignments and domain family memberships. The stability allows annotation to be stored along with the HMMs and alignments. The InterPro database [6] is a meta-database of domains combining the assignments from several different source databases, including Pfam. The Conserved Domain Database (CDD) is a similar meta-database that also contains additional domains curated by the NCBI [7]. SMART [8] is a manually curated resource focusing primarily on signaling and extracellular domains. ProDom [9] is a comprehensive domain database automatically generated from sequences in UniProt [10]. Likewise, ADDA [11] is automatically generated by clustering subsequences of proteins from the major sequence databases, though it has not been updated for some time. Genome3D [12] is a recent consensus database which brings together several domain prediction tools as well as the SCOP and CATH databases for describing representative domain arrangements in a series of trusted, well-annotated genomes.

Since the domain definitions from different databases only partially overlap, results from analyses often cannot be directly compared. In practice, however, choice of database appears to have little effect on the main trends reported by the studies described here.

1.4 Domain Architectures

The terms "domain architecture" or "domain arrangement" generally refer to the domains in a protein and their order, reported in N- to C-terminal direction along the amino acid chain. Another recurring term is domain combinations. This refers to pairs of domains co-occurring in proteins, either anywhere in the protein (the "bag-of-domains" model) or specifically pairs of domains being adjacent on an amino acid chain, in a specific N- to C-terminal order [13]. The latter concept is expanded to triplets of domains, which are subsequences of three consecutive domains, with the N- and C-termini used as "dummy" domains. A domain X occurring on its own in a protein thus produces the triplet N-X-C [14].

1.5 Mechanisms for Domain Architecture Change

Most mutations are point mutations: substitutions, insertions, or deletions of single nucleotides. While conceivably enough of these might create a new domain from an old one or noncoding sequence or remove a domain from a protein, in practice we are interested in mechanisms whereby the domain architecture of a protein changes instantly or nearly so (but see below for an overview of recent work on the origin of new domains). Figure 1 shows some examples of ways in which domain architectures may mutate. In general, adding or removing domains requires genetic recombination events. These can occur either through errors made by systems for repairing DNA damage such as homologous [16, 17] or nonhomologous (illegitimate) [18, 19] recombination or through the action of mobile genetic elements such as DNA transposons [20] or retrotransposons [21, 22]. Recombination can cause loss or duplication of parts of genes, entire genes or much longer chromosomal regions.

In organisms that have introns, exon shuffling [23, 24] refers to the integration of an exon from one gene into another, for instance, through chromosomal crossover, gene conversion, or mobile genetic elements. Exons could also be moved around by being

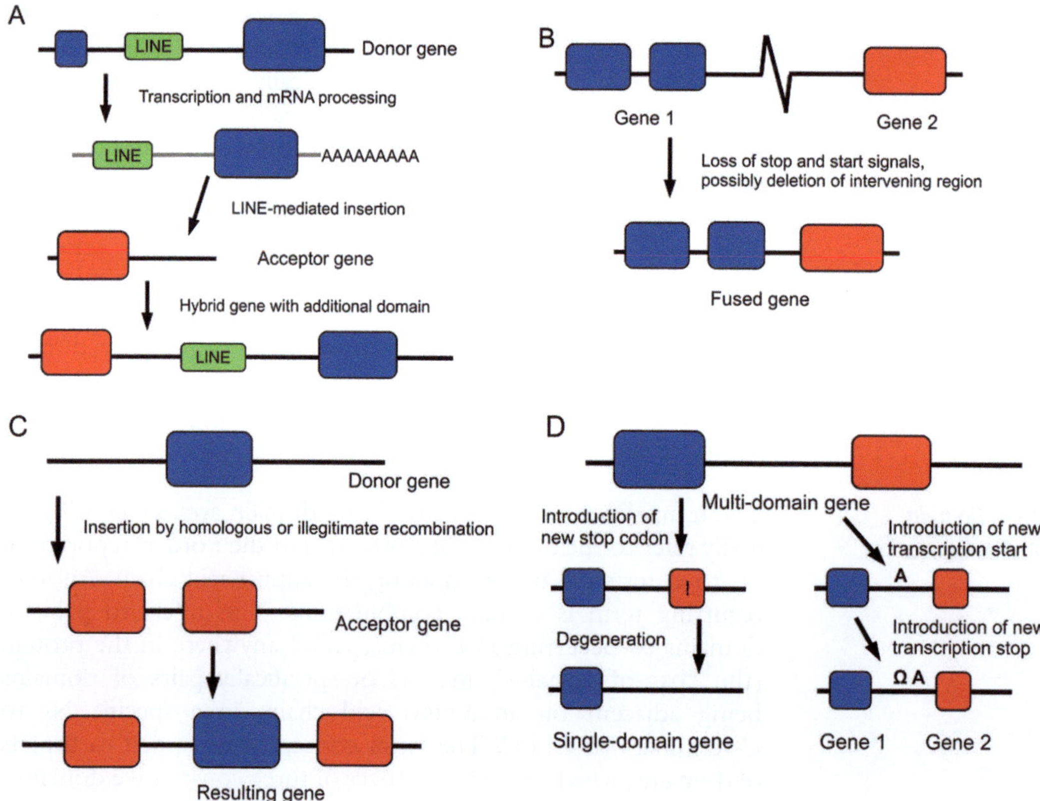

Fig. 1 Examples of mutations that can change domain architectures. Adapted from Buljan et al. [25]. (**a**) Gene fusion by a mobile element. LINE refers to a Long Interspersed Nuclear repeat Element, a retrotransposon. The reverse transcriptase encoded within the LINE causes its mRNA to be reverse-transcribed into DNA and integrated into the genome, making the domain-encoding blue exon from the donor gene integrate along with it in the acceptor gene. (**b**) Gene fusion by loss of a stop signal or deletion of much of the intergenic region. Genes 1 and 2 are joined together into a single, longer gene. (**c**) Domain insertion through recombination. The blue domain from the donor gene is inserted within the acceptor gene by either homologous or illegitimate recombination. (**d**) Right: Gene fission by introduction of transcription stop (the letter Ω) and start (the letter A). Left: Domain loss by introduction of a stop codon (exclamation mark) with subsequent degeneration of the now untranslated domain

brought along by mobile genetic elements such as retrotransposons [24, 25].

Two adjacent genes can be fused into one if the first one loses its transcription stop signals. Point mutations can cause a gene to lose a terminal domain by introducing a new stop codon, after which the "lost" domain slowly degrades through point mutations as it is no longer under selective pressure [26]. Alternatively, a multi-domain gene might be split into two genes if both a start and a stop signal are introduced between the domains. Novel domains could arise, for instance, through exonization, whereby an intronic or intergenic region becomes an exon, after which subsequent mutations would fine-tune its folding and functional properties [25, 27].

Recent literature (*see*, e.g., [28]) has discussed the possibility of de novo domain creation through a variety of mutational mechanisms, with some support for this occurring more often than previously thought [29, 30]. The majority of such new domains arise as novel genes from noncoding sequence but may subsequently recombine to join with older domains. Furthermore, young domains in vertebrates tend more often to occur at the N-terminal of a protein and tend to experience higher relative rates of non-synonymous substitution than older domains, which may reflect the nature of the mechanisms through which novel domains arise. Moore, Bornberg-Bauer et al. explore the relative prevalence of domain loss, duplication, and de novo origination in arthropods [31] and plants [32], suggesting such novel domains most frequently are associated with environmental adaptations.

2 Distribution of the Sizes of Domain Families

Domain architectures are fundamentally the realizations of how domains combine to form multi-domain proteins with complex functions. Understanding how these combinations come to be requires first that we understand how common the constituent domains of those architectures are and whether there are selective pressures determining their abundances. Because of this, the body of work concerning the sizes and species distributions of domain families becomes important to us.

Comprehensive studies of the distributions and evolution of protein domains and domain architectures are possible as genome sequencing technologies have made many entire proteomes available for bioinformatic analysis. Initial work [33–35] focused on the number of copies that a protein family, either single domain or multi-domain, has in a species. Most conclusions from these early studies appear to hold true for domains, for supra-domains (see below) and for domain architectures [36–38]. In particular, these all exhibit a ***dominance of the population by a selected few*** [35], i.e., a small number of domain families are present in a majority of the proteins in a genome, whereas most domain families are found only in a small number of proteins.

Looking at the frequency N of families of size X (defined as the number of members in the genome), in the earliest studies, this frequency was modeled as the power law

$$N = cX^{-a}$$

where a is an exponent parameter. The power law is a special case of the generalized Pareto distribution (GPD) [39]:

$$N = c(i + X)^{-a}$$

Power law distributions arise in a vast variety of contexts: from human income distributions, connectivity of internet routers, word usage in languages, and many other situations ([34, 35, 40, 41], *see* also [42], for a conflicting view). Luscombe et al. [35] described a number of other genomic properties that also follow power law distributions, such as the occurrence of DNA "words," pseudogenes, and levels of gene expression. These distributions fit much better than the alternative they usually are contrasted against, an exponential decay distribution. The most important difference between exponential and power law distributions in this context concerns the fact that the latter has a "fat tail," that is, while most domain families occur only a few times in each proteome, most domains in the proteome still belong to one of a small number of families.

Later work ([39, 43], *see* also [44]) demonstrated that proteome-wide domain occurrence data fit the general GPD better than the power law but that it also asymptotically fits a power law as $X \gg i$. The deviation from strict power law behavior depends on proteome size in a kingdom-dependent manner [43]. Regardless, it is mostly appropriate to treat the domain family size distribution as approximately (and asymptotically) power law-like, and later studies typically assume this.

The power law, but not the GPD, is scale-free in the sense of fulfilling the condition

$$f(ax) = g(a)f(x)$$

where $f(x)$ and $g(x)$ are some functions of a variable x and where a is a scaling parameter, that is, studying the data at a different scale will not change the shape of function. This property has been extensively studied in the literature and is connected to other attributes, notably when it occurs in network degree distributions (i.e., frequency distributions of edges per node). Here it has been associated with properties such as the presence of a few central and critical hubs (nodes with many edges to other nodes), the similarity between parts and the whole (as in a fractal), and the growth process called preferential attachment, under which nodes are more likely to gain new links the more links they already have. However, the same power law distribution may be generated from many different network topologies with different patterns of connectivity. In particular, they may differ in the extent that hubs are connected to each other [42]. It is possible to extend the analysis by taking into account the distribution of degree pairs along network edges, but this is normally not done.

What kind of evolutionary mechanisms give rise to this kind of distribution of gene or domain family sizes within genomes? In one model by Huynen and van Nimwegen [33], every gene within a gene family will be more or less likely to duplicate, depending on the utility of the function of that gene family within the particular lineage of organisms studied, and they showed that such a model matches the observed power laws. While they claimed that any model that explains the data must take into account family-specific probabilities of duplication fixation, Yanai and coworkers [45] proposed a simpler model using uniform duplication probability for all genes in the genome and also reported a good fit with data.

Later, more complex birth-death [43] and birth-death-and-innovation (BDIM) [29, 34, 39, 46] models were introduced to explain the observed distributions, and from investigating which model parameter ranges allow this fit, the authors were able to draw several far-ranging conclusions. First, the asymptotic power law behavior requires that the rates of domain gain and loss are asymptotically equal. Karev et al. [39] interpreted this as support for a punctuated equilibrium-type model of genome evolution, where domain family size distributions remain relatively stable for long periods of time but may go through stages of rapid evolution, representing a shift between different BDIM evolutionary models and significant changes in genome complexity. Like Huynen and van Nimwegen [33], they concluded that the likelihood of fixated domain duplications or losses in a genome directly depend on family size. The family will however only grow as long as new copies can find new functional niches and contribute to a net benefit for survival, i.e., as long as selection favors it.

Aside from Huynen and van Nimwegen's, none of the models discussed depend very strongly on family-specific selection to explain the abundances of individual gene families, nor do they exclude such selection. Some domains may be highly useful to their host organism's lifestyle, such as cell-cell connectivity domains to an organism beginning to develop multicellularity. Expansion of these domain families might therefore become more likely in some lineages than in others. To what extent these factors actually affect the size of domain families remains to be fully explored. Karev et al. [39] suggested that the rates of domain-level change events themselves—domain duplication and loss rates, as well as the rate of influx of novel domains from other species or de novo creation—must be evolutionarily adapted, as only some such parameters allow the observed distributions to be stable. Van Nimwegen [47] investigated how the number of genes increases in specific functional categories as total genome size increases. He found that the relationship matches a power law, with different coefficients for each functional class remaining valid over many bacterial lineages. Ranea et al. found similar results. Also, Ranea et al. [48] showed that, for domain superfamilies inferred to be present in the last universal common ancestor (LUCA), domains associated with metabolism have significantly higher abundance than those associated with translation, further supporting a connection between the function of a domain family and how likely it is to expand.

Extending the analysis to multi-domain architectures, Apic et al. [37] showed that the frequency distribution of multi-domain family sizes follows a power law curve similar to that reported for individual domain families. It therefore seems likely that the basic underlying mechanisms should be similar in both cases, i.e., that duplication of genes, and thus their domain architectures, is the most important type of event affecting the evolution of domain architectures.

Have the trends described above stood the test of time as more genomes have been sequenced and more domain families have been identified? We considered the 1943 UniProt proteomes covered by version 30.0 of Pfam, plotted the frequency Υ of domain families that have precisely X members as a function of X, and fit a power law curve to this. Figure 2a shows the resulting plots for three representative species, one complex eukaryote (*Homo sapiens*),

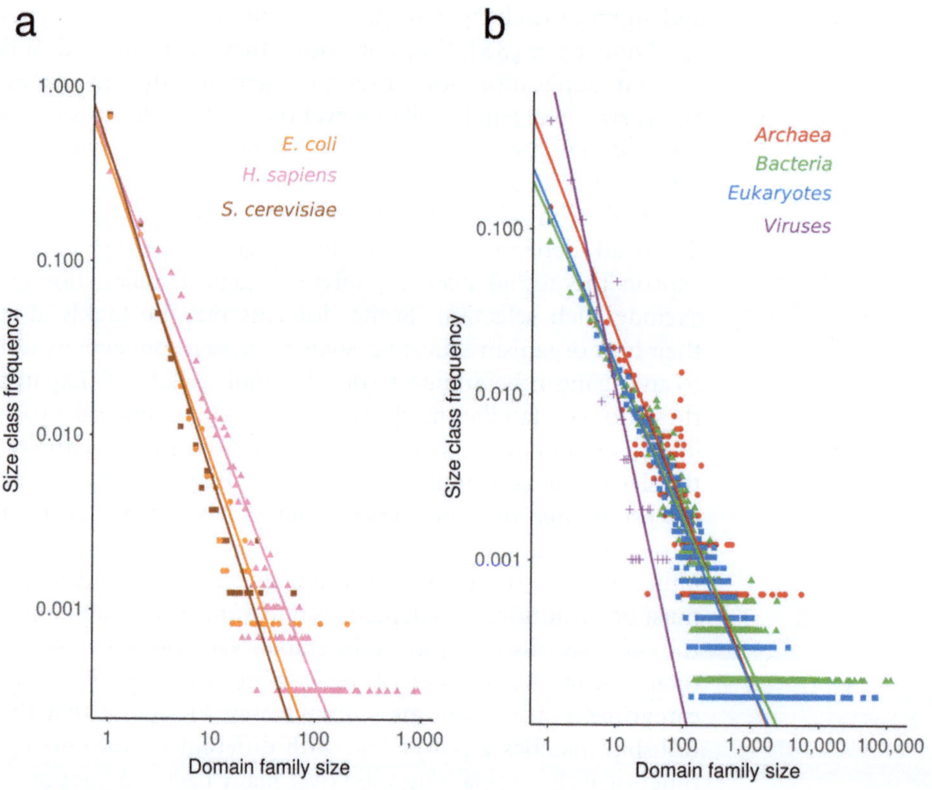

Fig. 2 (**a**) Distribution of domain family sizes in three selected species. Power law distributions were fitted to these curves such that for frequency f of families of size X, $f = cX^a$. For *S. cerevisiae*, $a = -1.9$, for *E. coli*, $a = -1.7$, and for *H. sapiens*, $a = -1.5$. (**b**) Distribution of domain family sizes across the three kingdoms. Power law distributions were fitted to these curves such that for frequency f of families of size X, $f = cX^a$. For bacteria, $a = -0.9$, for archaea, $a = -1.1$, for eukaryotes, $a = -0.8$, and for viruses, $a = -1.9$

one simple eukaryote (*Saccharomyces cerevisiae*), and one prokaryote (*Escherichia coli*). Figure 2b shows the corresponding plots for all domains in all complete eukaryotic, bacterial, and archaeal proteomes. The power law curve fits decently well, with slopes becoming less steep for the more complex organisms, whose distributions have relatively more large families. The power law-like behavior suggests that complex organisms with large proteomes were formed by heavily duplicating domains from relatively few families. Figures 3a, b show equivalent plots, not for single domains but for entire multi-domain architectures. The curve shapes and the relationship between both species and organism groups are similar, indicating that the evolution of these distributions have been similar.

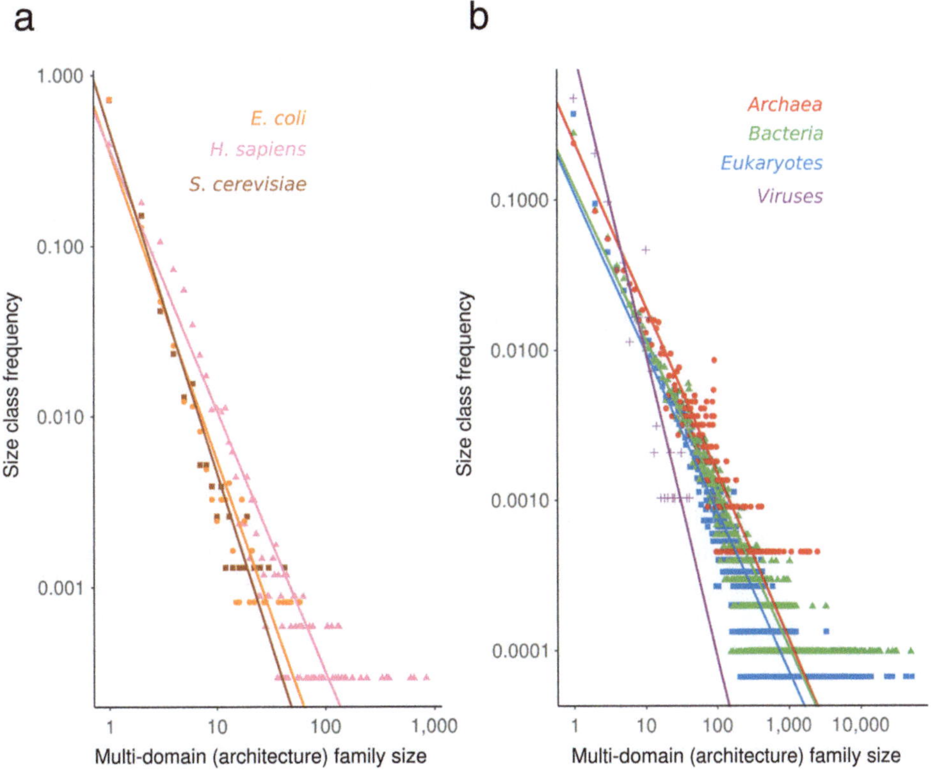

Fig. 3 (**a**) Distribution of multi-domain (architecture) family sizes in three selected species. Power law distributions were fitted to these curves such that for frequency f of families of size X, $f = cX^a$. For *S. cerevisiae*, $a = -2.0$, for *E. coli*, $a = -1.8$, and for *H. sapiens*, $a = -1.5$. (**b**) Distribution of multi-domain (architecture) family sizes across the three kingdoms. Power law distributions were fitted to these curves such that for frequency f of families of size X, $f = cX^a$. For bacteria, $a = -1.0$, for archaea, $a = -1.1$, for eukaryotes, $a = -1.1$, and for viruses, $a = -2.0$

3 Kingdom and Age Distribution of Domain Families and Architectures

How old are specific domain families or domain architectures? With knowledge of which organism groups they are found in, it is possible to draw conclusions about their age and whether lineage-specific selective pressures have determined their kingdom-specific abundances. Domain families and their combinations have arisen throughout evolutionary history, presumably by new combinations of pre-existing elements that may have diverged beyond recognition or by processes such as exonization. We can estimate the age of a domain family by finding the largest clade of organisms within which it is found, excluding organisms with only xenologs, i.e., horizontally transferred genes [14]. The age of this lineage's root is the likely age of the family. The same holds true for domain combinations and entire domain architectures. This methodology allows us to determine how changing conditions at different points in evolutionary history, or in different lineages, have affected the evolution of domain architectures.

Apic et al. [36] analyzed the distribution of SCOP domains across 40 genomes from archaea, bacteria, and eukaryotes. They found that a majority of domain families are common to all three kingdoms of life and thus likely to be ancient. Kuznetsov et al. [43] performed a similar analysis using InterPro domains and found that only about one fourth of all such domains were present in all three kingdoms, but a majority was present in more than one of them. Lateral gene transfer or annotation errors can cause a domain family to be found in one or a few species in a kingdom without actually belonging to that kingdom. To counteract this, one can require that a family must be present in at least a reasonable fraction of the species within a kingdom for it to be considered anciently present there. For instance, using Gene3D assignments of CATH domains to 114 complete genomes, mainly bacterial, Ranea et al. [48] isolated protein superfamily domains that were present in at least 90% of all the genomes and at least 70% of the archaeal and eukaryotic genomes, respectively. Under these stringent cutoffs for considering a domain to be present in a kingdom, 140 domains, 15% of the CATH families found in at least one prokaryote genome, were inferred to be ancient. Chothia and Gough [49] performed a similar study on 663 SCOP superfamily domains evaluated at many different thresholds and found that while 516 (78%) superfamilies were common to all three kingdoms at a threshold of 10% of species in each kingdom, only 156 (24%) superfamilies were common to all three kingdoms at a threshold of 90%. They also showed that for prokaryotes, a majority of domain instances (i.e., not domain families but actual domain copies) belong to common superfamilies at all thresholds below 90%.

Extending to domain combinations, Apic et al. [36] reported that a majority of SCOP domain pairs are unique to each kingdom but also that more kingdom-specific domain combinations than expected were composed only of domain families shared between all three kingdoms. This would imply a scenario where the independent evolution of the three kingdoms mainly involved creating novel combinations of domains that existed already in their common ancestor.

Several studies have reported interesting findings on domain architecture evolution in lineages closer to ourselves: in metazoa and vertebrates. Ekman et al. [50] claimed that new metazoa-specific domains and multi-domain architectures have arisen roughly once every 0.1–1 million years in this lineage. According to their results, most metazoa-specific multi-domain architectures are a combination of ancient and metazoa-specific domains. The latter category are however mostly found as novel single-domain proteins. Much of the novel metazoan multi-domain architectures involve domains that are versatile (see below) and exon-bordering (allowing for their insertion through exon shuffling). The novel domain combinations in metazoa are enriched for proteins associated with functions required for multicellularity—regulation, signaling, and functions involved in newer biological systems such as immune response or development of the nervous system, as previously noted by Patthy [23]. They also showed support for exon shuffling as an important mechanism in the evolution of metazoan domain architectures. Itoh et al. [51] added that animal evolution differs significantly from other eukaryotic groups in that lineage-specific domains played a greater part in creating new domain combinations. Nasir et al. [52] analyzed the age and taxonomic distribution of domains drawing on species phylogenies reconstructed from domain repertoires, concluding among other things that most widespread domains are relatively old and suggesting high numbers of both domain gain and loss in the evolution of the three organismal superkingdoms. Bacterial and archaeal genes have tended to gain or lose domains encoding aspects of metabolic capacity, whereas those of eukaryotes—including multicellular ones—have gained domains enabling more elaborate extracellular processes such as immunity and regulatory capacities.

In the most recent datasets, what is the distribution of domains and domain combinations across the three kingdoms of life? Looking at the set of UniProt proteomes represented in version 30.0 of Pfam, the distribution of domains across the three kingdoms are as displayed in the Venn diagram of Fig. 4a. Figure 4b, c show the equivalent distributions of immediate neighbors and triplets of domains, respectively, and Fig. 4d the distribution of multi-domain architectures across kingdoms. The numbers are somewhat biased toward bacteria as 56% of the UniProt proteomes are from this kingdom. However, with this high coverage of all kingdoms

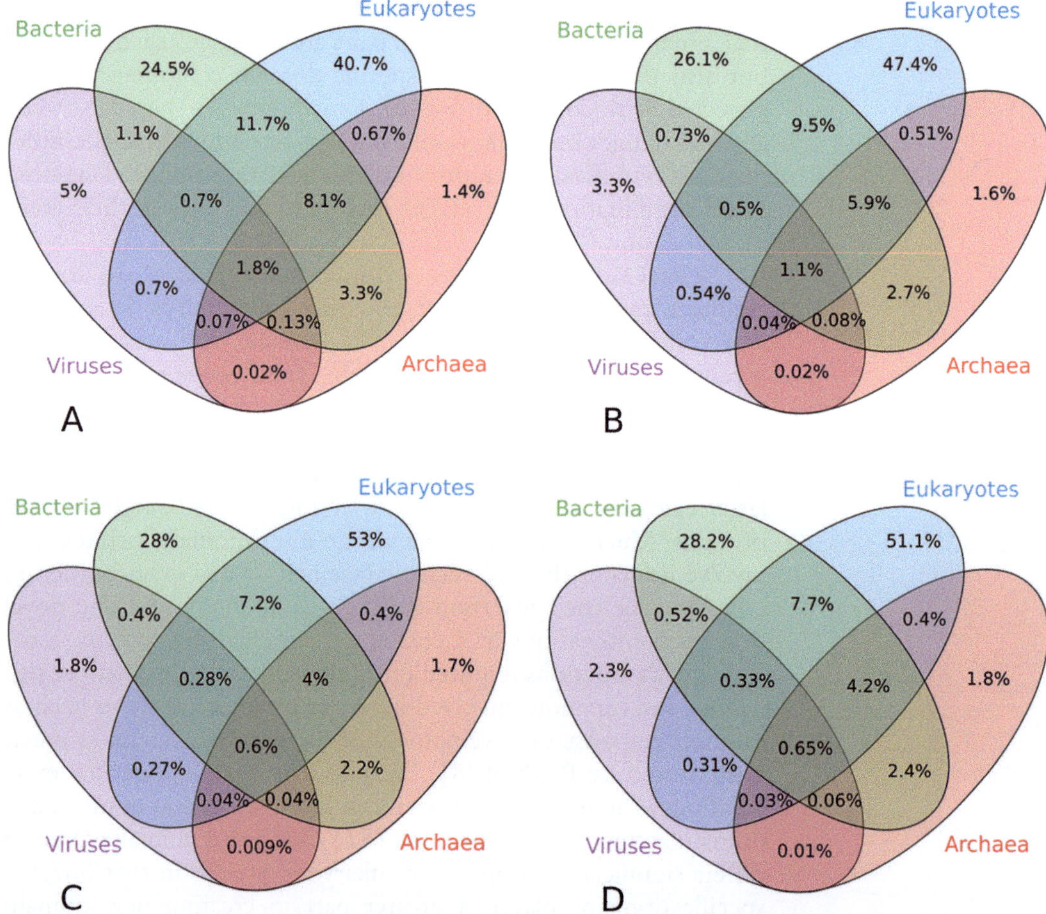

Fig. 4 (**a**) Kingdom distribution of unique domains. Values are given as percentages of the total, 10,330 domains. (**b**) Kingdom distribution of unique domain pairs. Values are given as percentages of the total, 31,287 domain pairs. (**c**) Kingdom distribution of unique domain triplets. Values are given as percentages of the total, 33,662 domain triplets. (**d**) Kingdom distribution of unique multi-domain architectures. Values are given as percentages of the total, 23,238 multi-domain architectures

(506 eukaryotic, 94 archaeal, and 1090 bacterial proteomes, as well as 253 viral entities), the results should be robust in this respect. Compared to most previous reports, we see a striking difference in that a much smaller portion of domains are shared between all kingdoms. There are some potential artifacts which could affect this analysis. If lateral gene transfer is very widespread, we may overestimate the number of families present in all three kingdoms. Moreover, there are cases where separate Pfam families are actually distant homologs of each other, which could lead to underestimation of the number of ancient families. To counteract this, we make use of Pfam clans, considering domains in the same clan to be equivalent. While not all distant homologies have yet been

registered in the clan system, performing the analysis on the clan level reduces the risk of such underestimation.

Our finding that 10% of all Pfam-A domains are present in all three main kingdoms is strikingly lower than in the earlier works and is even lower than reported by Ranea et al. [48], who used very stringent cutoffs. However, a direct comparison of statistics for Pfam domains/clans and CATH superfamilies is difficult. The decrease in ancient families that we observe may be a consequence of the massive increase in sequenced genomes and/or that the recent growth of Pfam has added relatively more kingdom-specific domains. We further found that only 1.5% of all domains or domain combinations are unique to archaea, suggesting that known representatives of this lineage have undergone very little independent evolution and/or that most archaeal gene families have been horizontally transferred to other kingdoms. The trend when going from domain via domain combinations to whole architectures is clear—the more complex patterns are less shared between the kingdoms. In other words, each kingdom has used a common core of domains to construct its own unique combinations of multi-domain architectures.

4 Domain Co-occurrence Networks

A multi-domain architecture connects individual domains with each other. There are several ways to derive these connections and quantify the level of co-occurrence. The simplest method is to consider all domains on the same amino acid chain to be connected, but we can also limit the set of co-occurrences we consider to, e.g., immediate neighbor pairs or triplets. Regardless of which method is used, the result is a domain co-occurrence network, where nodes represent domains and where edges represent the existence of proteins in which members of these families co-occur. Figure 5 shows an example of such a network and the set of domain architectures which defines it. This type of explicit network representation is explored in several studies, notably by Itoh et al. [51], Przytycka et al. [53], and Kummerfeld and Teichmann [13]. It is advantageous as it allows the introduction of powerful analysis tools developed within the engineering sciences for use with artificial network structures such as the World Wide Web. The patterns of co-occurrences that we observe should be a direct consequence of the constraints and conditions under which domain architectures evolve, and because of this, the study of these patterns becomes relevant for understanding such factors.

The frequency distribution of node degrees in the domain co-occurrence network has been fitted to a power law [36] and a more general GPD as well [40]. The closer this approximation holds, the more the network will have the scale-free property.

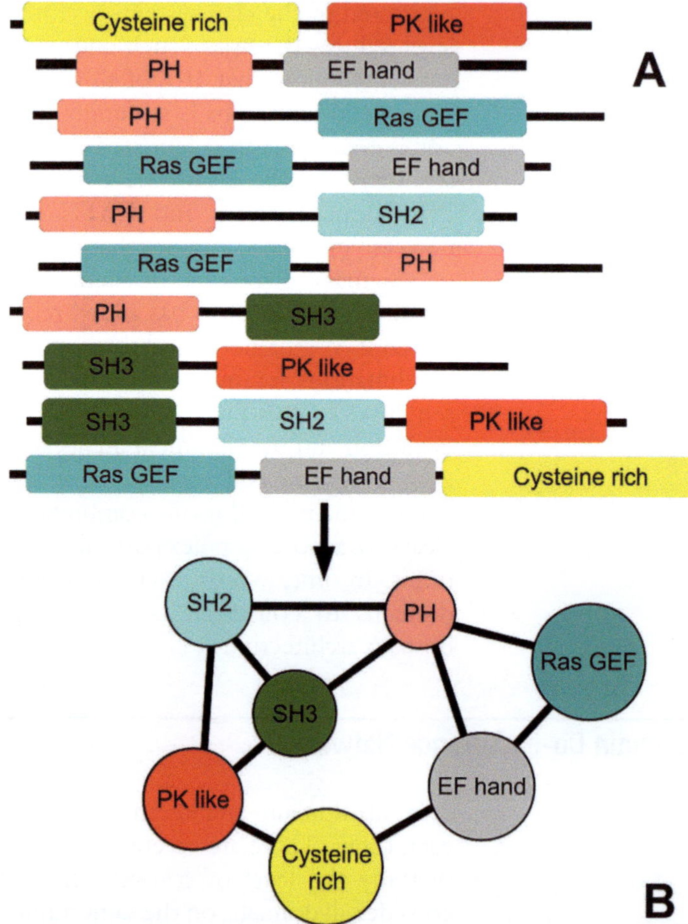

Fig. 5 Example of protein domain co-occurrence network, adapted from Kummerfeld and Teichmann [13]. (**a**) Sample set of domain architectures. The lines represent proteins and the boxes their domains in N- to C-terminal order. (**b**) Resulting domain co-occurrence (neighbor) network. Nodes correspond to domains and are linked by an edge if at least one domain exists where the two domains are found adjacent to each other along the amino acid chain

This property can be thought of as a hierarchy in the network, where the more centrally connected nodes link to more peripheral nodes with the same relative frequency at each level. In the context of domains, this means that a small number of domains co-occur with a high number of other domains, whereas most domains only have a few neighbors—usually some of the highly connected hubs. The most highly connected domains are referred to as promiscuous [54], mobile, or versatile [14, 55, 56]. Many such hub domains are involved in intracellular or extracellular signaling, protein-protein interactions and catalysis, and transcription regulation. In general, these are domains that encode a generic function, e.g., phosphorylation, which is reused in many contexts by additional domains that

Table 1
The 20 most densely connected hubs with regard to immediate domain neighbors, according to Pfam 30.0

Identifier	Name	Number of different immediate neighbors
CL0023	P-loop containing nucleoside triphosphate hydrolase superfamily	415
CL0063	FAD/NAD(P)-binding Rossmann fold superfamily	390
CL0123	Helix-turn-helix clan	358
CL0016	Protein kinase superfamily	192
CL0159	Ig-like fold superfamily (E-set)	148
CL0020	Tetratricopeptide repeat superfamily	146
CL0028	Alpha/beta-hydrolase fold	140
CL0172	Thioredoxin-like	136
CL0036	Common phosphate-binding site TIM barrel superfamily	136
CL0219	Ribonuclease H-like superfamily	127
CL0058	Tim barrel glycosyl hydrolase superfamily	120
CL0257	N-acetyltransferase-like	115
CL0167	Zinc beta-ribbon	114
CL0072	Ubiquitin superfamily	112
CL0125	Peptidase clan CA	106
CL0186	Beta propeller clan	105
CL0021	OB fold	101
CL0192	Family A G protein-coupled receptor-like superfamily	101
CL0015	Major facilitator superfamily	97
CL0220	EF-hand-like superfamily	95

confer substrate specificity or localization. Table 1 shows the domains (or clans) with the highest numbers of immediate neighbors in Pfam 30.0.

One way of evolving a domain co-occurrence network that follows a power law is by "preferential attachment" [53, 57]. This means that new edges (corresponding to proteins where two domains co-occur) are added with a probability that is higher the more edges these nodes (domains) already have, resulting in a power law distribution.

Apic et al. [37] considered a null model for random domain combination, in which a proteome contains domain combinations with a probability based on the relative abundances of the domains only. They showed that this model does not hold and that far fewer domain combinations than expected under it are actually seen. If most domain duplication events are gene duplication events that do not change domain architecture—or at the very least do not disrupt domain pairs—then this finding is not unexpected, nor does it require or exclude any particular selective pressure to keep these domains together in proteins. There is growing support for the idea that separate instances of a given domain architecture in general descend from a single ancestor with that architecture [58], with polyphyletic evolution of domain architectures occurring only in a small fraction of cases [53, 59, 60].

Itoh et al. [51] performed reconstruction of ancestral domain architectures using maximum parsimony, as described in the next section. This allowed them to study the properties of the ancestral domain co-occurrence network and thus explore how network connectivity has altered over evolutionary time. Among other things, they found increased connectivity in animals, particularly of animal-specific domains, and suggest that this phenomenon explains the high connectivity for eukaryotes reported by Wuchty [40]. For non-animal eukaryotes, they reported a correlation between connectivity and age, such that older domains had relatively higher connectivity, with domains preceding the divergence of eukaryotes and prokaryotes being the most highly connected, followed by early eukaryotic domains. In other words, early eukaryotic evolution saw the emergence of some key hub proteins, while the most prominent eukaryotic hubs emerged in the animal lineage. Parikesit et al. [61] studied the functional annotation of co-occurring domains in eukaryotes, concluding that while these may have different associated functional descriptors, these descriptors usually tend to fall within the same overall category within the gene ontology. Co-occurring domains thus tend to contribute to the same overall process type rather than have very widely divergent functional annotations. Hsu et al. [62] constructed a network linking domain architectures (i.e., each node is a multi-domain architecture, as opposed to in a regular domain co-occurrence network) where parsimonious reconstruction suggests evolution of one from the other, identifying "highly evolvable" architectures as hubs in this network. Proteins with such architectures were reported to be more widespread, less often essential, more often duplicated, and more often associated with gene functions involved in specific adaptation of organisms.

What is the degree distribution of current domain co-occurrence networks? We again used the domain architectures from all complete proteomes in version 30.0 of Pfam and considered the network of immediate neighbor relationships, i.e., nodes (domains) have an edge between them if there is a protein where they are adjacent. Each domain was assigned a degree as its number

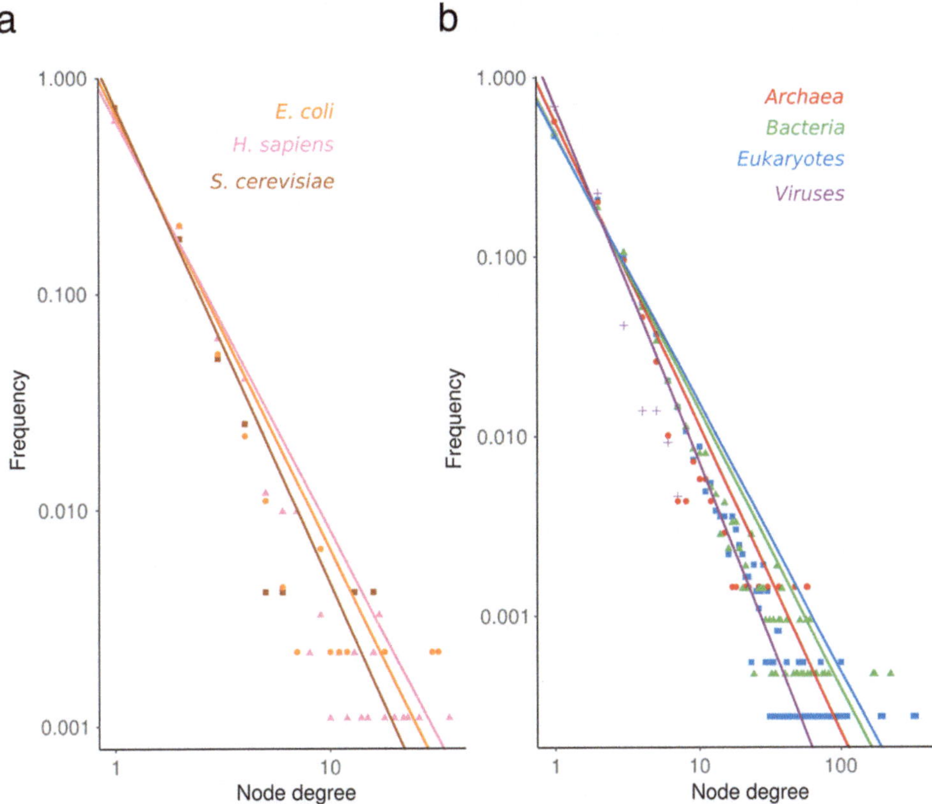

Fig. 6 (**a**) Distribution of domain co-occurrence network node degrees in three selected species. Power law distributions were fitted to these curves such that for frequency f of families of size X, $f = cX^a$. For *S. cerevisiae*, $a = -2.2$, for *E. coli*, $a = -2.0$, and for *H. sapiens*, $a = -1.9$. (**b**) Distribution of domain co-occurrence network node degrees across the three kingdoms. This corresponds to a network where two domains are connected if any species within the kingdom has a protein where these domains are immediately adjacent. Power law distributions were fitted to these curves such that for frequency f of families of size X, $f = cX^a$. For bacteria, $a = -1.6$, for archaea, $a = -1.7$, for eukaryotes, $a = -1.5$, and for viruses $a = -2.0$

of links to other domains. We then counted the frequency with which each degree occurs in the co-occurrence network. Figure 6a shows this relationship for the set of domain architectures found in the same species as for Figs. 2a, and 6b shows the equivalent plots for the three kingdoms as found among the complete proteomes in Pfam. Regressions to a power law have been added to the plots. The presence of a power law-like behavior of this type implies that few domains have very many immediate neighbors, while most domains have few immediate neighbors. Note that the observed degrees in our dataset were strongly reduced by removing all sequences with a stretch longer than 50 amino acids lacking domain annotation.

5 Supra-domains and Conserved Domain Order

As we have seen, whole multi-domain architectures or shorter stretches of adjacent domains are often repeated in many proteins. These only cover a small fraction of all possible domain combinations. Are the observed combinations somehow special? We would expect selective pressure to retain some domain combinations but not others, since only some domains have functions that would synergize together in one protein. Often, co-occurring domains require each other structurally or functionally, for instance, in transcription factors where the DNA-binding domain provides substrate specificity, whereas the trans-activating domain recruits other components of the transcriptional machinery [63]. Vogel et al. [38] identified series of domains co-occurring as a fixed unit with conserved N- to C-terminal order but flanked by different domain architectures and termed them supra-domains. By investigating their statistical overrepresentation relative to the frequency of the individual domains in the set of nonredundant domain architectures (where "nonredundant" is crucial, as otherwise, e.g., whole-gene duplication would bias the results), they identified a number of such supra-domains. Many ancient domain combinations (shared by all three kingdoms) appear to be such selectively preserved supra-domains.

How conserved is the order of domains in multi-domain architectures? In a recent study, Kummerfeld and Teichmann [13] built a domain co-occurrence network with directed edges, allowing it to represent the order in which two domains are found in proteins. As in other studies, the distribution of node degrees fits a power law well. Most domain pairs were only found in one orientation. This does not seem required for functional reasons, as flexible linker regions should allow the necessary interface to form also in the reversed case [58], but may rather be an indication that most domain combinations are monophyletic. Weiner and Bornberg-Bauer [64] analyzed the evolutionary mechanisms underlying a number of reversed domain order cases and concluded that independent fusion/fission is the most frequent scenario. Although domain reversals occur in only a few proteins, it actually happens more often than was expected from randomizing a co-occurrence network [13]. That study also observed that the domain co-occurrence network is more clustered than expected by a random model and that these clusters are also functionally more coherent than would be expected by chance.

6 Domain Mobility, Promiscuity, or Versatility

While some protein domains co-occur with a variety of other domains, some are always seen alone or in a single architecture in all proteomes where they are found. A natural explanation is that some domains are more likely to end up in a variety of architectural contexts than others due to some intrinsic property they possess. Is such domain versatility or promiscuity a persistent feature of a given domain, and does it correlate with certain functional or biological properties of the domain?

Several ways of measuring domain versatility have been suggested. One measure, NCO [40], counts the number of other domains found in any architectures where the domain of interest is found. Another measure, NN [37], instead counts the number of distinct other domains that a domain is found adjacent to. Yet another measure, NTRP [65], counts the number of distinct triplets of consecutive domains where the domain of interest is found in the middle. All of these measures can be expected to be higher for common domains than for rare domains, i.e., variations in domain abundance (the number of proteins a domain is found in) can hide the intrinsic versatility of domains. Therefore, three different studies [14, 55, 66] formulated *relative domain versatility* indices that aim to measure versatility independently of abundance. It is worth noting that most studies have considered only immediately adjacent domain neighbors in these analyses, a restriction based on the assumption that those are more likely to interact functionally than domains far apart on a common amino acid chain. More recent work [67] introduced a *network versatility* metric which can classify domains as being central or peripheral with regard to the large-scale structure of their bigram network (i.e., the network-linking domains found adjacent in proteins), observing how peripheral such domains exhibit relatively higher primary sequence conservation suggestive of adaptation to more specific functions, whereas the core domains may be more multifunctional.

The first relative versatility study was presented by Vogel et al. [66], who used as their domain dataset the SUPERFAMILY database applied to 14 eukaryotic, 14 bacterial, and 14 archaeal proteomes. They modeled the number of unique immediate neighbor domains as a power law function of domain abundance, performed a regression on this data, and used the resulting power law exponent as a relative versatility measure. Basu et al. [55] used Pfam and SMART [8] domains and measured relative domain versatility for 28 eukaryotes as the immediate neighbor pair frequency normalized by domain frequency. They then defined promiscuous domains as a class according to a bimodality in the distribution of the raw numbers of unique domain immediate neighbor pairs.

Weiner et al. [14] used Pfam domains for 10,746 species in all kingdoms and took as their relative versatility measure the logarithmic regression coefficient for each domain family across genomes, meaning that it is not defined within single proteomes.

To what extent is high versatility an intrinsic property of a certain domain? Vogel et al. [66] only examined large groups of domains together and therefore did not address this question for single domains. Basu et al. [55] and Weiner et al. [14] instead analyzed each domain separately and concluded that there are strong variations in relative versatility at this level. Their results are very different in detail, however, reflected by the fact that only one domain family (PF00004, AAA ATPase family) is shared between the ten most versatile domains reported in the two studies. As they used fairly similar domain datasets, it would appear that the results strongly depend on the definition of relative versatility. Another potential reason for the different results is that Basu's list was based on eukaryotes only, while Weiner's analysis was heavily biased toward prokaryotes. Furthermore, the top ten list in Basu et al. [55] and their follow-up paper [56] only overlap by four domains, yet the main difference is that in the latter study all 28 eukaryotes were considered, while the former study was limited to the subset of 20 animal, plant, and fungal species. The choice of species thus seems pivotal for the results when using this method. They also used different methods for calculating the average value of relative versatility across many species, which may influence the results.

Does domain versatility vary between different functional classes of domains? Vogel et al. [66] found no difference in relative versatility between broad functional or process categories or between SCOP structural classes. In contrast to this, Basu et al. [55] reported that high versatility was associated with certain functional categories in eukaryotes. However, no test for the statistical significance of these results was performed. Weiner et al. [14] also noted some general trends but found no significant enrichment of gene ontology terms in versatile domains. This does not necessarily mean that no such correlation exists, but more research is required to convincingly demonstrate its strength and its nature. More recently, Cromar et al. [68] analyzed domain architectures in eukaryotic extracellular matrix proteomes, noting that these structures are organized around a set of versatile domains under the weighted bigram metric of Basu et al. [55].

Another important question is to what extent domain versatility varies across evolutionary lineages. Vogel et al. [66] reported no large differences in average versatility for domains in different kingdoms. The versatility measure of Basu et al. [55] can be applied within individual genomes, which means that according to this measure domains may be versatile in one organism group but not in another, as well as gain or lose versatility across evolutionary time. They found that more domains were highly versatile in animals than in other eukaryotes. Modeling versatility as a binary property defined for domains in extant species, they further used a maximum parsimony approach to study the persistence of versatility for each domain across evolutionary time and concluded that both gain and loss of versatility are common during evolution. Inferring ancestral domain architectures, Cohen-Gihon et al. [69] report an increase in versatility in many domains during eukaryotic evolution, in particular around the divergence of Bilateria. Weiner at al. [14] divided domains into age categories based on distribution across the tree of life and reported that the versatility index is not dependent on age, i.e., domains have equal chances of becoming versatile at different times in evolution. This is consistent with the observation by Basu et al. [55] that versatility is a fast-evolving and varying property. When measuring versatility as a regression within different organism groups, Weiner et al. [14] found slightly lower versatility in eukaryotes, which is in conflict with the findings of Basu et al. [55]. Again, this underscores the strong dependence of the method and dataset on the results.

Further properties reported to correlate with domain versatility include sequence length, where Weiner et al. [14] found that longer domains are significantly more versatile within the framework of their study, while at the same time, shorter domains are more abundant and hence may have more domain neighbors in absolute numbers. Basu et al. [55] further reported that more versatile domains have more structural interactions than other domains. To determine which of these reported correlations that genuinely reflect universal biological trends, further comprehensive studies are needed using more data and uniform procedures. This would hopefully allow the results from the studies described here to be validated and any conflicts between them to be resolved.

Basu et al. [55] further analyzed the phylogenetic spread of all immediate domain neighbor pairs ("bigrams") containing domains classified as promiscuous. The main observation this yielded was that although most such combinations occurred in only a few species, most promiscuous domains are part of at least one combination that is found in a majority of species. They interpreted this as implying the existence of a reservoir of evolutionarily stable domain combinations from which lineage-specific recombination may draw promiscuous domains to form unique architectures. Later work by Hsu et al. [70] analyzed the domain co-occurrence networks centered on each domain family, classifying such subnetworks as being either mostly starlike, taillike, or tetragon-like, with promiscuous domains forming cores of starlike architecture networks in this representation.

7 Principles of Domain Architecture Evolution

What mutation events can generate new domain architectures, and what is their relative predominance? The question can be approached by comparing protein domain architectures of extant proteins. This is based on the likely realistic assumption that most current domain architectures evolved from ancestral domain architectures that can still be found unchanged in other proteins. Because of this, in pairs of most similar extant domain architectures, one can assume that one of them is ancestral. This agrees well with results indicating that most groups of proteins with identical domain architectures are monophyletic. By comparing the most similar proteins, several studies have attempted to chart the relative frequencies of different architecture-changing mutations.

Björklund et al. [71] used this particular approach and came to several conclusions. First, changes to domain architecture are much more common by the N- and C-termini than internally in the architecture. This is consistent with several mechanisms for architecture changes such as introduction of new start or stop codons or mergers with adjacent genes, and similar results have been found in several other studies [15, 25, 26]. Furthermore, insertions or deletions of domains ("indels") are more common than substitutions of domains, and the events in question mostly concern just single domains, except in cases with repeats expanding with many domains in a row [72]. In a later study, the same group made use of phylogenetic information as well, allowing them to infer directionality of domain indels [50]. They then found that domain insertions are significantly more common than domain deletions.

Weiner et al. [26] performed a similar analysis on domain loss and found compatible results—most changes occur at the termini (*see* also discussion in [28]). Moreover, they demonstrated that terminal domain loss seldom involves losing only part of a domain, or rather, that such partial losses quickly progress into loss of the entire domain. However, it is important to ensure such observations are not confounded by cases where errors in gene boundary recognition make domain detection less accurate [73].

There is some support [23, 74, 75] for exon shuffling to have played an important part in domain evolution, and there are a number of domains that match intron borders well, for example, structural domains in extracellular matrix proteins. While it may not be a universal mechanism, exon shuffling is suggested to have been particularly important for vertebrate evolution [23].

Recognizing the potential role of gene duplications in domain architecture evolution, Grassi et al. [76] analyzed domain architecture shifts following either whole-genome duplication (WGD) or smaller-scale gene duplication events in yeast. Surviving WGD duplicates had retained ancestral architecture in ca 95% of cases, with approximately the same chance of architecture change in WGD as under local duplication. Genes retained over time from either type of duplication were enriched for a core of commonly occurring domains but with a subset of rarer domains additionally enriched in retained WGD duplicates compared to locally duplicated genes. The former category more often was associated with housekeeping-type gene functions, whereas the latter more often involved adaptive functions. Functional change was generally larger than architectural change following duplication. Zhang et al. [77] similarly studied domain architecture evolution in plants, noting that lineage-specific architecture expansions largely can be explained from differential retention of genes following successive whole-genome duplications. Another form of domain duplication particularly relevant in plants is amplification of the numbers of domain repeats in proteins, discussed, e.g., by Sharma and Pandey [78].

8 Inferring Ancestral Domain Architectures

The above analyses, based on pairwise comparison of extant protein domain architectures, cannot tally ancestral evolutionarily events nearer the root of the tree of life. With ancestral architectures, one can directly determine which domain architecture changes have taken place during evolution and precisely chart how mechanisms of domain architecture evolution operate, as well as gauge their relative frequency. A drawback is that since we can only infer ancestral domain architectures from extant proteins, the result will depend somewhat on our assumptions about evolutionary mechanisms. On the upside, it should be possible to test how well different assumptions fit the observed modern-day protein domain architecture patterns.

Attempts at such reconstructions have been made using parsimony. Given a gene tree and the domain architectures at the leaves, dynamic programming can be used in order to find the assignment of architectures to internal nodes that require the smallest number of domain-level mutation events. This simple model can be elaborated by weighting loss and gain differently or by requiring that a domain or an architecture can only be gained at most once in a tree (Dollo parsimony) [79].

An early study of Snel et al. [80] considered 252 gene trees across 17 fully sequenced species and used parsimony to minimize the number of gene fission and fusion events occurring along the species tree. Their main conclusion, that gene fusions are more common than gene fissions, was subsequently supported by a larger study by Kummerfeld and Teichmann [81], where fusions were found to be about four times as common as fissions in a most parsimonious reconstruction. Fong et al. [82] followed a similar procedure on yet more data and concluded that fusion was 5.6 times as likely as fission.

Buljan and Bateman [15] performed a similar maximum parsimony reconstruction of ancestral domain architectures. They too observed that domain architecture changes primarily take place at the protein termini, and the authors suggested that this might largely occur because terminal changes to the architecture are less likely to disturb overall protein structure. Moreover, they concluded from reconciliation of gene and species trees that domain architecture changes were more common following gene duplications than following speciation but that these cases did not differ with respect to the relative likelihood of domain losses or gains.

Recently, Buljan et al. [25] presented a new ancestral domain architecture reconstruction study which assumed that gain of a domain should take place only once in each gene tree, i.e., Dollo parsimony [79]. Their results also support gene fusion as a major mechanism for domain architecture change. The fusion is generally preceded by a duplication of either of the fused genes. Intronic recombination and insertion of exons are observed but relatively rarely. They also found support for de novo creation of disordered segments by exonization of previously noncoding regions. More recently still a method for domain architecture history reconstruction using a network construct called a *plexus* was described [83]. Yang and Bourne [84] further described another parsimony-based reconstruction approach, as did Wu et al. [85], reporting that histories of signaling and development proteins are enriched for gene fusion/fission events. Stolzer et al. [86] present another method for domain architecture history inference, made available through the Notung software.

9 Polyphyletic Domain Architecture Evolution

There appears to be a "grammar" for how protein domains are allowed to be combined. If nature continuously explores all possible domain combinations, one would expect that the allowed combinations would be created multiple times throughout evolution. Such independent creation of the same domain architecture can be called convergent or polyphyletic evolution, whereas a single original creation event for all extant examples on an architecture would be called divergent or monophyletic evolution. This is relevant for several reasons, not least because it determines whether or not we can expect two proteins with identical domain architectures to have the same history along their entire length.

A graph theoretical approach to answer this question was taken by Przytycka et al. [53], who analyzed the set of all proteins containing a given superfamily domain. The domain architectures of these proteins define a domain co-occurrence network, where edges connect two domains both found in a protein, regardless of sequential arrangement. The proteins of such a set can also be placed in an evolutionary tree, and the evolution of all multi-domain architectures containing the reference domain can be expressed in terms of insertions and deletions of other domains along this tree to form the extant domain architectures. The question, then, is whether or not all leaf nodes sharing some domain arrangement (up to and including an entire architecture) stem from a single ancestral node possessing this combination of domains. For monophyly to be true for all architectures containing the reference domain, the same companion domain cannot have been inserted in more than one place along the tree describing the evolution of the reference domain. By application of graph theory and Dollo parsimony [79], they showed that monophyly is only possible if the domain co-occurrence network defined by all proteins containing the reference domain is chordal, i.e., it contains no cycles longer than three edges.

Przytycka et al. [53] then evaluated this criterion for all superfamily domains in a large-scale dataset. For domains where the co-occurrence network contained fewer than 20 nodes (domains), the chordal property and hence the possibility of complete monophyly of all domain combinations and domain architectures containing that domain held. By comparing actual domain co-occurrence networks with a preferential attachment null model, they showed that far more architectures are potentially monophyletic than would be expected under a pure preferential attachment process. This finding is analogous to the observation by Apic et al. [37] that most domain combinations are duplicated more frequently (or reshuffled less) than expected by chance. In other words, gene duplication is much more frequent than domain recombination [66]. However, for many domains that co-occurred with more than 20 other different domains, particularly for domains previously reported as promiscuous, the chordal property was violated, meaning that multiple independent insertions of the same domain, relative to the reference domain phylogeny, must be assumed.

A more direct approach is to do complete ancestral domain architecture reconstruction of protein lineages and to search for concrete cases that agree with polyphyletic architecture evolution. There are two conceptually different methodologies for this type of analysis. Either one only considers architecture changes between nodes of a species tree, or one considers any node in a reconstructed gene tree. The advantage of using a species tree is that one avoids the inherent uncertainty of gene trees, but on the other hand, only events that take place between examined species can be observed.

Gough [59] applied the former species-tree-based methodology to SUPERFAMILY domain architectures and concluded that polyphyletic evolution is rare, occurring in 0.4–4% of architectures.

The value depends on methodological details, with the lower bound considered more reliable.

The latter gene-tree-based methodology was applied by Forslund et al. [60] to the Pfam database. Ancestral domain architectures were reconstructed through maximum parsimony of single-domain phylogenies which were overlaid for multi-domain proteins. This strategy yielded a higher figure, ranging between 6% and 12% of architectures depending on dataset and whether or not incompletely annotated proteins were removed. The two different approaches thus give very different results. The detection of polyphyletic evolution is in both frameworks dependent on the data that is used—its quality, coverage, filtering procedures, etc. The studies used different datasets which makes it hard to compare. However, given that their domain annotations are more or less comparable, the major difference ought to be the ability of the gene-tree method to detect polyphyly at any point during evolution, even within a single species. It should be noted that domain annotation is by no means complete—only a little less than half of all residues are assigned to a domain [5]—and this is clearly a limiting factor for detecting architecture polyphyly. The numbers may thus be adjusted considerably upwards when domain annotation reaches higher coverage. A later study by Zmasek and Godzik [87] reports much higher rates (25–75%) still of polyphyletic evolution of eukaryotic multi-domain architectures, arguing that previous datasets were too small to have the power to reveal this.

Future work will be required to provide more reliable estimates of how common polyphyletic evolution of domain architectures is. Any estimate will depend on the studied protein lineage, the versatility of the domains, and methodological factors. A comprehensive and systematic study using more complex phylogenetic methods than the fairly ad hoc parsimony approach, as well as effective ways to avoid overestimating the frequency of polyphyletic evolution due to incorrect domain assignments or hidden homology between different domain families, may be the way to go. At this point all that can be said is that polyphyletic evolution of domain architectures definitely does happen, but relatively rarely, and that it is more frequent for complex architectures and versatile domains. A detailed case study was made recently of netrin domain-containing proteins, where polyphyletic evolution in metazoa seems well-supported [88]; these authors further suggest the term *merology* for such polyphyletic evolution. A series of papers by Nagy and Patthy et al. [73, 89, 90] further elaborates on challenges faced within this line of research; they report strong confounding influence of gene prediction errors. They further propose the term *epaktology* for gene similarity resulting from the independent acquisition of two proteins by the same additional domain. The authors suggest such cases inflate both estimates of terminal domain changes and estimates of gene fusion-driven changes in domain architecture. Beyond such changes, whether correctly inferred or not, the authors describe internal domain shuffling as an important mechanism for how domain architecture evolution has occurred.

10 Conclusions

As access to genomic data and to increasing amounts of compute power has grown during the last decade-and-a-half, so has our knowledge of the overall patterns of domain architecture evolution. Still, no study is better than its underlying assumptions, and differences in the representation of data and hypotheses mean that results often cannot be directly compared. Overall, however, the current state of the field appears to support some broad conclusions.

Domain and multi-domain family sizes, as well as numbers of co-occurring domains, all approximately follow power laws, which implies a scale-free hierarchy. This property is associated with many biological systems in a variety of ways. In this context, it appears to reflect how a relatively small number of highly versatile components have been reused again and again in novel combinations to create a large part of the domain and domain architecture repertoire of organisms. Gene duplication is the most important factor to generate multi-domain architectures, and as it outweighs domain recombination, only a small fraction of all possible domain combinations is actually observed. This is probably further modulated by family-specific selective pressure, though more work is required to demonstrate to what extent. Most of the time, all proteins with the same architecture or domain combination stem from a single ancestor where it first arose, but there remains a fraction of cases, particularly with domains that have very many combination partners, where this does not hold.

Most changes to domain architectures occur following a gene duplication and involve the addition of a single domain to either protein terminus. The main exceptions to this occur in repeat regions. Exon shuffling played an important part in animals by introducing a great variety of novel multi-domain architectures, reusing ancient domains as well as domains introduced in the animal lineage.

In this chapter, we have reexamined with the most up-to-date datasets many of the analyses done previously on less data and found that the earlier conclusions still hold true. Even though we are at the brink of amassing enormously much more genome and proteome data thanks to the new generation of sequencing technology, there is no reason to believe that this will alter the fundamental observations we can make today on domain architecture evolution. However, it will permit a more fine-grained analysis, and also there will be a greater chance to find rare events, such as independent creation of domain architectures. Furthermore, careful application of more complex models of evolution with and without selection pressure may allow us to determine more closely to what extent the process of domain architecture evolution was shaped by selective constraints.

11 Materials and Methods

Updated statistics were generated from the data in Pfam 30.0. All UniProt proteins in the SwissPfam set for Pfam 30.0 were included. These span 1090 bacteria, 506 eukaryotes, and 94 archaea. All Pfam-A domains regardless of type were included. However, as stretches of repeat domains are highly variable, consecutive subsequences of the same domain were collapsed into a single pseudodomain, if it was classified as type Motif or Repeat, as in several previous works [50, 60, 66, 82].

Domains were ordered within each protein based on their sequence start position. In the few cases of domains being inserted within other domains, this was represented as the outer domain followed by the nested domain, resulting in a linear sequence of domain identifiers. As long regions without domain assignments are likely to represent the presence of as-yet uncharacterized domains, we excluded any protein with unassigned regions longer than 50 amino acids (more than 95% of Pfam-A domains are longer than this). This approach is similar to that taken in previous works [59, 60, 71]. Other studies [50, 72] have instead performed additional, more sensitive domain assignment steps, such as clustering the unassigned regions to identify unknown domains within them.

Pfam domains are sometimes organized in clans, where clanmates are considered homologous. A transition from a domain to another of the same clan is thus less likely to be a result of domain swapping of any kind and more likely to be a result of sequence divergence from the same ancestor. Because of this, we replaced all Pfam domains that are clan members with the corresponding clan.

The statistics and plots were generated using a set of Perl and R scripts, which are available upon request. Power law regressions were done using the R nls function. For reasons of scale, the regression for a power law relation such as

$$N = cX^{-a}$$

was performed on the equivalent relationship

$$\log(X) = (1/a)(\log(c) - \log(N))$$

for the parameters a and c, with the exception of the data for Fig. 6, where instead the relationship

$$\log(N) = \log(c) - a\log(X)$$

was used. Moreover, because species or organism group datasets were of very different size, raw counts of domains were converted to frequencies before the regression was performed.

12 Online Domain Database Resources

For further studies or research into this field, the first and most important stop will be the domain databases. Table 2 presents a selection of domain databases in current use.

Table 2
A selection of protein domain databases

Database	URL	Notes	Reference
ADDA	http://ekhidna.biocenter.helsinki.fi/sqgraph/pairsdb	Automatic clustering of protein domain sequences	[11]
CATH	http://www.cathdb.info	Based solely on experimentally determined 3D structures	[2]
CDD	http://www.ncbi.nlm.nih.gov/Structure/cdd/cdd.shtml	Meta-database joining together domain assignments from many different sources, as well as some unique domains	[7]
Gene3D	http://gene3d.biochem.ucl.ac.uk	Bioinformatic assignment of sequences to CATH domains using hidden Markov models	[4]
InterPro	http://www.ebi.ac.uk/interpro	Meta-database joining together domain assignments from many different sources	[6]
Pfam	http://pfam.sanger.ac.uk	Domain families are defined from manually curated multiple alignments and represented using hidden Markov models	[5]
ProDom	http://prodom.prabi.fr	Automatically derived domain families from proteins in UniProt	[9]
SCOP	http://scop.mrc-lmb.cam.ac.uk	Based solely on experimentally determined 3D structures	[1]
SMART	http://smart.embl-heidelberg.de	Domain families are defined from manually curated multiple alignments and represented using hidden Markov models	[8]
SUPERFAMILY	http://supfam.cs.bris.ac.uk	Bioinformatic assignment of sequences to SCOP domains using hidden Markov models trained on the sequences of domains in SCOP	[3]
Genome3D	http://genome3d.eu/	Meta-database joining together domain assignments from many different sources, operating on the architecture level for a set of selected genomes	[12]

13 Domain Architecture Analysis Software

Several software tools have been described and made available that allow for analysis and visualization of domain architectures and their evolution. A selection of such tools is shown in Table 3.

A few of these tools allow domain architecture evolution analysis by visualizing each protein's domain architecture along a protein sequence tree. An example is the web tool TreeDom [96] which, given a protein domain family and an anchor sequence, fetches the family from Pfam and builds a tree with the nearest neighbors of the anchor sequence. An example output from TreeDom is shown in Fig. 7, in which a nonredundant set of representative proteomes were queried. Here one can see that while the NUDIX domain of the anchor sequence tends to co-occur with two other domains (zf-NADH-PPase and NUDIX-like), it also has recombined with many other domains over the course of evolution.

Other tools allow different types of analyses, for instance, searching for similar domain architectures or showing taxonomic distributions. Some of the protein domain databases listed in Table 2 include variants of such analyses, while external tools typically offer more specialized functionality. For example, the Pfam website allows searching for domain content, while the java tool PfamAlyzer allows searching Pfam for particular domain architecture patterns specified with a given domain order and spacing [94].

The RAMPAGE/RADS tools [95] make use of domain assignments for rapid homology searching. DoMosaics [92] is a software

Table 3
A selection of online software applying protein domain architecture evolution analysis

Tool	URL	Description	Reference
CDART	https://www.ncbi.nlm.nih.gov/Structure/lexington/lexington.cgi	Searches for proteins with similar domain architecture	[91]
DoMosaics	http://www.domosaics.net/	Visualizes domain evolution using trees	[92]
FACT	http://fact.cibiv.univie.ac.at/	Searches for functionally equivalent proteins by scoring domain architecture similarities	[93]
PfamAlyzer	http://pfam.xfam.org/search	Searches Pfam for proteins with specific domain architecture patterns	[94]
RADS/RAMPAGE	http://rads.uni-muenster.de/	Homology searching by aligning multiple domains instead of residues	[95]
TreeDom	http://treedom.sbc.su.se/	Graphical web tool for analyzing domain architecture evolution using Pfam	[96]

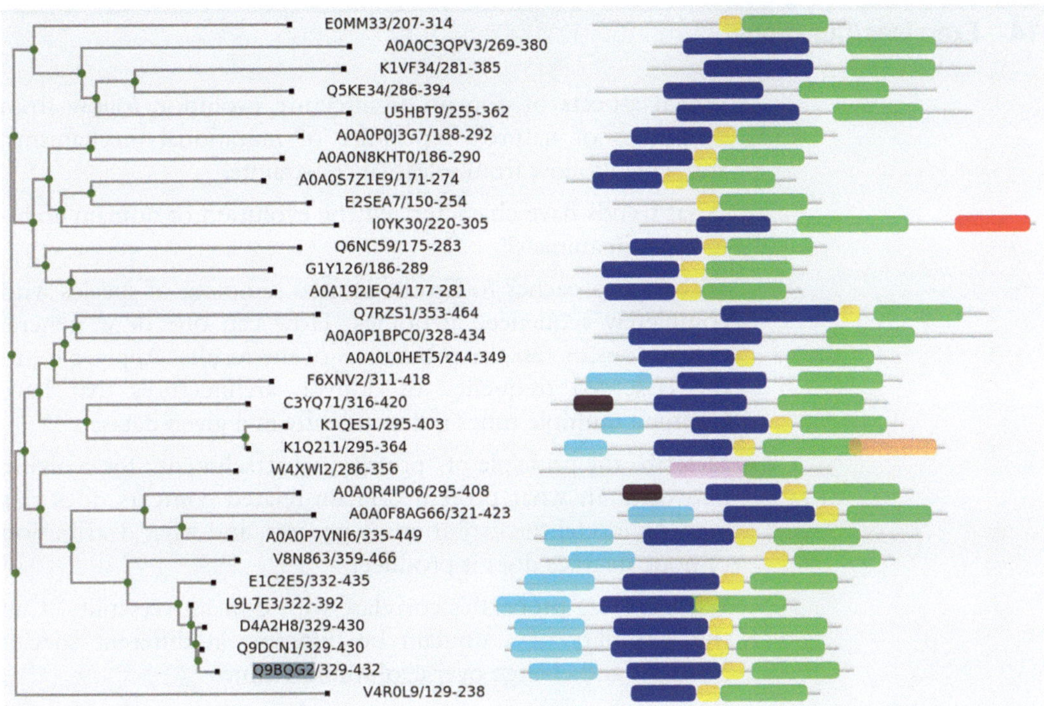

Fig. 7 TreeDom output using as query the NUDIX domain (PF00293), the human NUDT12 (Q9BQG2) protein, 30 closest sequences, and RP15 (representative proteomes at 15% co-membership). The domains are green, NUDIX; blue, NUDIX-like (PF09296); yellow, zf-NADH-PPase (PF09297); red, Ocnus (PF05005); cyan, Ank_2 (PF12796); black, Ank_5 (PF13857); orange, Prefoldin (PF02996); and pink, Fibrinogen_C (PF00147)

tool that can act as a wrapper for domain annotation tools, allowing detailed visualization and analysis of domain architectures, as does DomArch [97]. The DAAC algorithm [98] explicitly transfers functional annotation to query sequences based on domain architectural similarity to annotated homologs, as does FACT [93]. In the same vein, similarity measures between architectures are available using the WDAC [99] tool and in ADASS [100]. Domain architecture similarity is used for orthology detection in the porthoDom software [68]. The DOGMA tool makes use of domain content data to assess completeness of a proteome or transcriptome [101].

14 Exercises/Questions

- Which aspects of domain architecture evolution follow from properties of nature's repertoire of mutational mechanisms, and which follow from selective constraints?
- What trends have characterized the evolution of domain architectures in animals?
- Discuss approaches to handle limited sampling of species with completely sequenced genomes. How can one draw general conclusions or test the robustness of the results? Apply, e.g., to the observed frequency of domain architectures that have emerged multiple times independently in a given dataset.
- Describe the principle of "preferential attachment" for evolving networks. In what protein domain-related contexts does this seem to model the evolutionary process, and what distribution of node degrees does it produce?
- What protein properties correlate with domain versatility? Can the versatility of a domain be different in different species (groups) and change over evolutionary time?
- What protein domain-related properties differ between prokaryotes and eukaryotes?

References

1. Chandonia J-M, Fox NK, Brenner SE (2017) SCOPe: manual curation and artifact removal in the structural classification of proteins – extended database. Comput Res Mol Biol 429(3):348–355
2. Dawson NL, Lewis TE, Das S, Lees JG, Lee D, Ashford P et al (2017) CATH: an expanded resource to predict protein function through structure and sequence. Nucleic Acids Res 45(D1):D289–D295
3. Wilson D, Pethica R, Zhou Y, Talbot C, Vogel C, Madera M et al (2009) SUPERFAMILY—sophisticated comparative genomics, data mining, visualization and phylogeny. Nucleic Acids Res 37(suppl_1): D380–D386
4. Lam SD, Dawson NL, Das S, Sillitoe I, Ashford P, Lee D et al (2016) Gene3D: expanding the utility of domain assignments. Nucleic Acids Res 44(D1):D404–D409
5. Finn RD, Mistry J, Tate J, Coggill P, Heger A, Pollington JE et al (2010) The Pfam protein families database. Nucleic Acids Res 38 (suppl_1):D211–D222
6. Finn RD, Attwood TK, Babbitt PC, Bateman A, Bork P, Bridge AJ et al (2017) InterPro in 2017—beyond protein family and domain annotations. Nucleic Acids Res 45 (D1):D190–D199
7. Marchler-Bauer A, Derbyshire MK, Gonzales NR, Lu S, Chitsaz F, Geer LY et al (2015) CDD: NCBI's conserved domain database. Nucleic Acids Res 43(D1):D222–D226
8. Letunic I, Doerks T, Bork P (2015) SMART: recent updates, new developments and status in 2015. Nucleic Acids Res 43(D1): D257–D260
9. Bru C, Courcelle E, Carrère S, Beausse Y, Dalmar S, Kahn D (2005) The ProDom database of protein domain families: more emphasis on 3D. Nucleic Acids Res 33(suppl_1): D212–D215
10. UniProt Consortium (2017) UniProt: the universal protein knowledgebase. Nucleic Acids Res 45(D1):D158–D169
11. Heger A, Wilton CA, Sivakumar A, Holm L (2005) ADDA: a domain database with global coverage of the protein universe. Nucleic Acids Res 33(suppl 1):D188–D191
12. Lewis TE, Sillitoe I, Andreeva A, Blundell TL, Buchan DWA, Chothia C et al (2015) Genome3D: exploiting structure to help users

understand their sequences. Nucleic Acids Res 43(D1):D382–D386

13. Kummerfeld SK, Teichmann SA (2009) Protein domain organisation: adding order. BMC Bioinformatics 10(1):39
14. Weiner J, Moore AD, Bornberg-Bauer E (2008) Just how versatile are domains? BMC Evol Biol 8(1):285
15. Buljan M, Bateman A (2009) The evolution of protein domain families. Biochem Soc Trans 37(4):751
16. Orozco-Mosqueda M d C, Altamirano-Hernandez J, Farias-Rodriguez R, Valencia-Cantero E, Santoyo G (2009) Homologous recombination and dynamics of rhizobial genomes. Res Microbiol 160(10):733–741
17. Heyer W-D, Ehmsen KT, Liu J (2010) Regulation of homologous recombination in eukaryotes. Annu Rev Genet 44:113–139
18. Brissett NC, Doherty AJ (2009) Repairing DNA double-strand breaks by the prokaryotic non-homologous end-joining pathway. Biochem Soc Trans 37(3):539
19. van Rijk A, Bloemendal H (2003) Molecular mechanisms of exon shuffling: illegitimate recombination. Genetica 118(2):245–249
20. Feschotte C, Pritham EJ (2007) DNA transposons and the evolution of eukaryotic genomes. Annu Rev Genet 41:331–368
21. Cordaux R, Batzer MA (2009) The impact of retrotransposons on human genome evolution. Nat Rev Genet 10(10):691–703
22. Gogvadze E, Buzdin A (2009) Retroelements and their impact on genome evolution and functioning. Cell Mol Life Sci 66(23):3727
23. Patthy L (2003) Modular assembly of genes and the evolution of new functions. In: Long M (ed) Origin and evolution of new gene functions. Springer, Dordrecht, pp 217–231
24. Liu M, Grigoriev A (2004) Protein domains correlate strongly with exons in multiple eukaryotic genomes – evidence of exon shuffling? Trends Genet 20(9):399–403
25. Buljan M, Frankish A, Bateman A (2010) Quantifying the mechanisms of domain gain in animal proteins. Genome Biol 11(7):R74
26. Weiner J, Beaussart F, Bornberg-Bauer E (2006) Domain deletions and substitutions in the modular protein evolution. FEBS J 273(9):2037–2047
27. Schmidt EE, Davies CJ (2007) The origins of polypeptide domains. Bioessays 29 (3):262–270
28. Bornberg-Bauer E, Huylmans A-K, Sikosek T (2010) How do new proteins arise? Nucl Acids Seq Topol 20(3):390–396
29. Demuth JP, Hahn MW (2009) The life and death of gene families. Bioessays 31(1):29–39
30. Toll-Riera M, Albà MM (2013) Emergence of novel domains in proteins. BMC Evol Biol 13 (1):47
31. Moore AD, Bornberg-Bauer E (2012) The dynamics and evolutionary potential of domain loss and emergence. Mol Biol Evol 29(2):787–796
32. Kersting AR, Bornberg-Bauer E, Moore AD, Grath S (2012) Dynamics and adaptive benefits of protein domain emergence and arrangements during plant genome evolution. Genome Biol Evol 4(3):316–329
33. Huynen MA, van Nimwegen E (1998) The frequency distribution of gene family sizes in complete genomes. Mol Biol Evol 15 (5):583–589
34. Qian J, Luscombe NM, Gerstein M (2001) Protein family and fold occurrence in genomes: power-law behaviour and evolutionary model1lEdited by J. Thornton. J Mol Biol 313(4):673–681
35. Luscombe NM, Qian J, Zhang Z, Johnson T, Gerstein M (2002) The dominance of the population by a selected few: power-law behaviour applies to a wide variety of genomic properties. Genome Biol 3(8):research0040.1
36. Apic G, Gough J, Teichmann SA (2001) Domain combinations in archaeal, eubacterial and eukaryotic proteomes1lEdited by G. von Heijne. J Mol Biol 310(2):311–325
37. Apic G, Huber W, Teichmann SA (2003) Multi-domain protein families and domain pairs: comparison with known structures and a random model of domain recombination. J Struct Funct Genomics 4(2–3):67–78
38. Vogel C, Berzuini C, Bashton M, Gough J, Teichmann SA (2004) Supra-domains: evolutionary units larger than single protein domains. J Mol Biol 336(3):809–823
39. Karev GP, Wolf YI, Rzhetsky AY, Berezovskaya FS, Koonin EV (2002) Birth and death of protein domains: a simple model of evolution explains power law behavior. BMC Evol Biol 2:18–18
40. Wuchty S (2001) Scale-free behavior in protein domain networks. Mol Biol Evol 18 (9):1694–1702
41. Rzhetsky A, Gomez SM (2001) Birth of scale-free molecular networks and the number of distinct DNA and protein domains per genome. Bioinformatics (Oxford, England) 17(10):988–996
42. Li L, Alderson D, Doyle JC, Willinger W (2005) Towards a theory of scale-free graphs:

definition, properties, and implications. Internet Math 2(4):431–523

43. Kuznetsov VA, Pickalov VV, Senko OV, Lnott GD (2002) Analysis of the evolving proteomes: predictions of the number of protein domains in nature and the number of genes in eukaryotic organisms. J Biol Syst 10 (04):381–407
44. Koonin EV, Wolf YI, Karev GP (2002) The structure of the protein universe and genome evolution. Nature 420(6912):218–223
45. Yanai I, Camacho CJ, DeLisi C (2000) Predictions of gene family distributions in microbial genomes: evolution by gene duplication and modification. Phys Rev Lett 85 (12):2641–2644
46. Eirin-Lopez JM, Rebordinos L, Rooney AP, Rozas J (2012) The birth-and-death evolution of multigene families revisited. Genome Dyn 7:170–196
47. van Nimwegen E (2003) Scaling laws in the functional content of genomes. Trends Genet 19(9):479–484
48. Ranea JAG, Sillero A, Thornton JM, Orengo CA (2006) Protein superfamily evolution and the last universal common ancestor (LUCA). J Mol Evol 63(4):513–525
49. Chothia C, Gough J (2009) Genomic and structural aspects of protein evolution. Biochem J 419(1):15
50. Ekman D, Björklund ÅK, Elofsson A (2007) Quantification of the elevated rate of domain rearrangements in metazoa. J Mol Biol 372 (5):1337–1348
51. Itoh M, Nacher JC, Kuma K, Goto S, Kanehisa M (2007) Evolutionary history and functional implications of protein domains and their combinations in eukaryotes. Genome Biol 8(6):R121
52. Nasir A, Kim KM, Caetano-Anollés G (2014) Global patterns of protein domain gain and loss in superkingdoms. PLoS Comput Biol 10 (1):e1003452
53. Przytycka T, Davis G, Song N, Durand D (2005) Graph theoretical insights into evolution of multidomain proteins. In: Miyano S, Mesirov J, Kasif S, Istrail S, Pevzner PA, Waterman M (eds) Res. Comput. Mol. Biol. 9th Annu. Int. Conf. RECOMB 2005 Camb. MA USA May 14-18 2005 Proc. Springer, Berlin, pp 311–325
54. Marcotte EM, Pellegrini M, Ng H-L, Rice DW, Yeates TO, Eisenberg D (1999) Detecting protein function and protein-protein interactions from genome sequences. Science 285(5428):751
55. Basu MK, Carmel L, Rogozin IB, Koonin EV (2008) Evolution of protein domain promiscuity in eukaryotes. Genome Res 18 (3):449–461
56. Basu MK, Poliakov E, Rogozin IB (2009) Domain mobility in proteins: functional and evolutionary implications. Brief Bioinform 10 (3):205–216
57. Barabási A-L, Albert R (1999) Emergence of scaling in random networks. Science 286 (5439):509
58. Bashton M, Chothia C (2002) The geometry of domain combination in proteins11Edited by J. Thornton. J Mol Biol 315(4):927–939
59. Gough J (2005) Convergent evolution of domain architectures (is rare). Bioinformatics 21(8):1464–1471
60. Forslund K, Henricson A, Hollich V, Sonnhammer ELL (2008) Domain tree-based analysis of protein architecture evolution. Mol Biol Evol 25(2):254–264
61. Parikesit AA, Stadler PF, Prohaska SJ (2017) Large-scale evolutionary patterns of protein domain distributions in eukaryotes. BioRxiv
62. Hsu C-H, Chiang AWT, Hwang M-J, Liao B-Y (2016) Proteins with highly evolvable domain architectures are nonessential but highly retained. Mol Biol Evol 33 (5):1219–1230
63. Brivanlou AH, Darnell JE (2002) Signal transduction and the control of gene expression. Science 295(5556):813
64. Weiner J III, Bornberg-Bauer E (2006) Evolution of circular permutations in multidomain proteins. Mol Biol Evol 23(4):734–743
65. Tordai H, Nagy A, Farkas K, Bányai L, Patthy L (2005) Modules, multidomain proteins and organismic complexity. FEBS J 272 (19):5064–5078
66. Vogel C, Teichmann SA, Pereira-Leal J (2005) The relationship between domain duplication and recombination. J Mol Biol 346(1):355–365
67. Xie X, Jin J, Mao Y (2011) Evolutionary versatility of eukaryotic protein domains revealed by their bigram networks. BMC Evol Biol 11 (1):242
68. Bitard-Feildel T, Kemena C, Greenwood JM, Bornberg-Bauer E (2015) Domain similarity based orthology detection. BMC Bioinformatics 16(1):154
69. Cohen-Gihon I, Fong JH, Sharan R, Nussinov R, Przytycka TM, Panchenko AR (2011) Evolution of domain promiscuity in eukaryotic genomes-a perspective from the

inferred ancestral domain architectures. Mol Biosyst 7(3):784–792

70. Hsu C-H, Chen C-K, Hwang M-J (2013) The architectural design of networks of protein domain architectures. Biol Lett 9 (4):20130268
71. Björklund ÅK, Ekman D, Light S, Frey-Skött J, Elofsson A (2005) Domain rearrangements in protein evolution. J Mol Biol 353 (4):911–923
72. Björklund ÅK, Ekman D, Elofsson A (2006) Expansion of protein domain repeats. PLoS Comput Biol 2(8):e114
73. Nagy A, Szlama G, Szarka E, Trexler M, Banyai L, Patthy L (2011) Reassessing domain architecture evolution of metazoan proteins: major impact of gene prediction errors. Genes. 2(3):449–501
74. Doolittle RF, Bork P (1993) Evolutionarily mobile modules in proteins. Sci Am 269 (4):50–56
75. Moore AD, Björklund ÅK, Ekman D, Bornberg-Bauer E, Elofsson A (2008) Arrangements in the modular evolution of proteins. Trends Biochem Sci 33(9):444–451
76. Grassi L, Fusco D, Sellerio A, Cora D, Bassetti B, Caselle M et al (2010) Identity and divergence of protein domain architectures after the yeast whole-genome duplication event. Mol Biosyst 6(11):2305–2315
77. Zhang X-C, Wang Z, Zhang X, Le MH, Sun J, Xu D et al (2012) Evolutionary dynamics of protein domain architecture in plants. BMC Evol Biol 12(1):6
78. Sharma M, Pandey GK (2016) Expansion and function of repeat domain proteins during stress and development in plants. Front Plant Sci 6:1218
79. Farris JS (1977) Phylogenetic analysis under Dollo's law. Syst Zool 26(1):77–88
80. Snel B, Bork P, Huynen M (2000) Genome evolution. Trends Genet 16(1):9–11
81. Kummerfeld SK, Teichmann SA (2005) Relative rates of gene fusion and fission in multidomain proteins. Trends Genet 21(1):25–30
82. Fong JH, Geer LY, Panchenko AR, Bryant SH (2007) Modeling the evolution of protein domain architectures using maximum parsimony. J Mol Biol 366(1):307–315
83. Wiedenhoeft J, Krause R, Eulenstein O (2010) Inferring evolutionary scenarios for protein domain compositions. In: Borodovsky M, Gogarten JP, Przytycka TM, Rajasekaran S (eds) Bioinforma. Res. Appl. 6th Int. Symp. ISBRA 2010 Storrs CT USA May 23-26 2010 Proc. Springer, Berlin, pp 179–190
84. Yang S, Bourne PE (2009) The evolutionary history of protein domains viewed by species phylogeny. PLoS One 4(12):e8378
85. Wu Y-C, Rasmussen MD, Kellis M (2012) Evolution at the subgene level: domain rearrangements in the drosophila phylogeny. Mol Biol Evol 29(2):689–705
86. Stolzer M, Siewert K, Lai H, Xu M, Durand D (2015) Event inference in multidomain families with phylogenetic reconciliation. BMC Bioinformatics 16(14):S8
87. Zmasek CM, Godzik A (2012) This Déjà Vu Feeling—analysis of multidomain protein evolution in eukaryotic genomes. PLoS Comput Biol 8(11):e1002701
88. Leclère L, Rentzsch F (2012) Repeated evolution of identical domain architecture in metazoan netrin domain-containing proteins. Genome Biol Evol 4(9):883–899
89. Nagy A, Bányai L, Patthy L (2011) Reassessing domain architecture evolution of metazoan proteins: major impact of errors caused by confusing paralogs and epaktologs. Genes. 2(3):516–561
90. Nagy A, Patthy L (2011) Reassessing domain architecture evolution of metazoan proteins: the contribution of different evolutionary mechanisms. Genes 2(3):578–598
91. Geer LY, Domrachev M, Lipman DJ, Bryant SH (2002) CDART: protein homology by domain architecture. Genome Res 12 (10):1619–1623
92. Moore AD, Held A, Terrapon N, Weiner J III, Bornberg-Bauer E (2014) DoMosaics: software for domain arrangement visualization and domain-centric analysis of proteins. Bioinformatics 30(2):282–283
93. Koestler T, von Haeseler A, Ebersberger I (2010) FACT: functional annotation transfer between proteins with similar feature architectures. BMC Bioinformatics 11(1):417
94. Hollich V, Sonnhammer ELL (2007) PfamAlyzer: domain-centric homology search. Bioinformatics 23(24):3382–3383
95. Terrapon N, Weiner J, Grath S, Moore AD, Bornberg-Bauer E (2014) Rapid similarity search of proteins using alignments of domain arrangements. Bioinformatics 30(2):274–281
96. Haider C, Kavic M, Sonnhammer ELL (2016) TreeDom: a graphical web tool for analysing domain architecture evolution. Bioinformatics 32(15):2384–2385
97. Vera-Parra N, Gutiérrez-Ramirez M, Lopez-Sarmiento D (2016) Automatic construction and graph-making of functional domain architectures. Adv Nat Appl Sci 10 (12):99–106

98. Doğan T, MacDougall A, Saidi R, Poggioli D, Bateman A, O'Donovan C et al (2016) UniProt-DAAC: domain architecture alignment and classification, a new method for automatic functional annotation in UniProtKB. Bioinformatics 32(15):2264–2271
99. Lee B, Lee D (2009) Protein comparison at the domain architecture level. BMC Bioinformatics 10(15):S5
100. Syamaladevi DP, Joshi A, Sowdhamini R (2013) An alignment-free domain architecture similarity search (ADASS) algorithm for inferring homology between multi-domain proteins. Bioinformation 9(10):491–499
101. Dohmen E, Kremer LPM, Bornberg-Bauer E, Kemena C (2016) DOGMA: domain-based transcriptome and proteome quality assessment. Bioinformatics 32(17):2577–2581

Chapter 16

New Insights on the Evolution of Genome Content: Population Dynamics of Transposable Elements in Flies and Humans

Lain Guio and Josefa González

Abstract

Understanding the abundance, diversity, and distribution of TEs in genomes is crucial to understand genome structure, function, and evolution. Advances in whole-genome sequencing techniques, as well as in bioinformatics tools, have increased our ability to detect and analyze the transposable element content in genomes. In addition to reference genomes, we now have access to population datasets in which multiple individuals within a species are sequenced. In this chapter, we highlight the recent advances in the study of TE population dynamics focusing on fruit flies and humans, which represent two extremes in terms of TE abundance, diversity, and activity. We review the most recent methodological approaches applied to the study of TE dynamics as well as the new knowledge on host factors involved in the regulation of TE activity. In addition to transposition rates, we also focus on TE deletion rates and on the selective forces that affect the dynamics of TEs in genomes.

Key words Long-read sequencing, Transposition rates, Self-regulation, Effective population size, Adaptation, Horizontal transfer

1 Transposable Elements Are Abundant and Active Genome Denizens

Transposable elements (TEs) are short DNA sequences, typically from a few hundred bp to ~10 kb long, which have the ability to move around in the genome by generating new copies of themselves. In addition to active autonomous elements, genomes also contained nonautonomous elements that can be mobilized by the enzymatic machinery of active TEs from the same family. Additionally, genomes contain TEs that cannot be mobilized anymore due to accumulation of mutations in their sequences [1]. TEs are an ancient, extremely diverse, and exceptionally active component of genomes. TEs have been found in virtually all organisms studied so far including bacteria, archaea, fungi, protists, plants, and animals [2–5]. The main TE groups, class I and class II, are present in all kingdoms, revealing their persistence over evolutionary time

Maria Anisimova (ed.), *Evolutionary Genomics: Statistical and Computational Methods*, Methods in Molecular Biology, vol. 1910, https://doi.org/10.1007/978-1-4939-9074-0_16, © The Author(s) 2019

[2]. These two classes of TEs differ in their transposition intermediates: while class I TEs transpose through RNA intermediates, class II TEs transpose directly as DNA. TEs within each class are further classified into (1) different orders, based on their insertion mechanism, structure, and encoded proteins; (2) different superfamilies, based on their replication strategy and on presence and size of target site duplications; and (3) different families, based on sequence conservation [2, 3]. Piegu et al. [1] criticized the current classification system, which accounts for sequence homology, structural features, and target site duplications, because it does not always take into account the evolutionary origins of the TEs [1–3]. As a consequence, phylogenetically unrelated classes or subclasses of TEs are grouped [1]. Piegu et al. [1] also suggested that a more inclusive classification that includes prokaryotic and eukaryotic TE classes should be considered. Recently, Arkhipova [6] proposed a TE classification system based on the replicative, integrative, and structural components of TEs, which integrates different aspects of all the existing classification systems [6].

TEs constitute a substantial albeit variable (from ~1% to almost 90%) proportion of genomes [7, 8] (Fig. 1). The identification methods, as well as the sequencing and assembly methods, have an important effect in the TE content estimation [4, 9–11]. In some cases, the TE-generated fraction of genomes is likely to be underestimated because methods for detecting TEs in genomic sequences are necessarily biased toward younger and more easily recognizable TEs. Indeed, new tools developed in recent years are able to identify TEs that remained hidden until now [4, 11]. As an example, when the human genome was first sequenced, ~40–45% of the genome was identifiable TEs, 5% was genes and other functional sequences (functional RNAs or regulatory regions), and the remaining ~50% of the genome had no identifiable origin [12]. de Koning et al. [13] using a highly sensitive new strategy named P-cloud found that at least 66–69% of the human genome is identifiable as repetitive sequences, most of them derived from TEs [13]. In *Drosophila melanogaster*, third-generation sequencing techniques (3GS) have allowed the detection of 37% more TE insertions in chromosome 2L compared to previously available short-read sequencing estimates (see below) [14]. In other *Drosophila* species such as *D. buzzatii*, the TE content has also been updated from 6% to 11%, thanks to the recent availability of whole-genome sequences [15].

As mentioned above, TEs are extremely active genomic denizens that are able to generate mutations of a great diversity of types [16–21]. TE-induced mutations range from subtle regulatory mutations to gross genomic rearrangements and often have phenotypic effects of a complexity that is not achievable by point mutations (Fig. 2). Among others, TEs can affect the expression of nearby genes by adding new splice sites, adenylation signals,

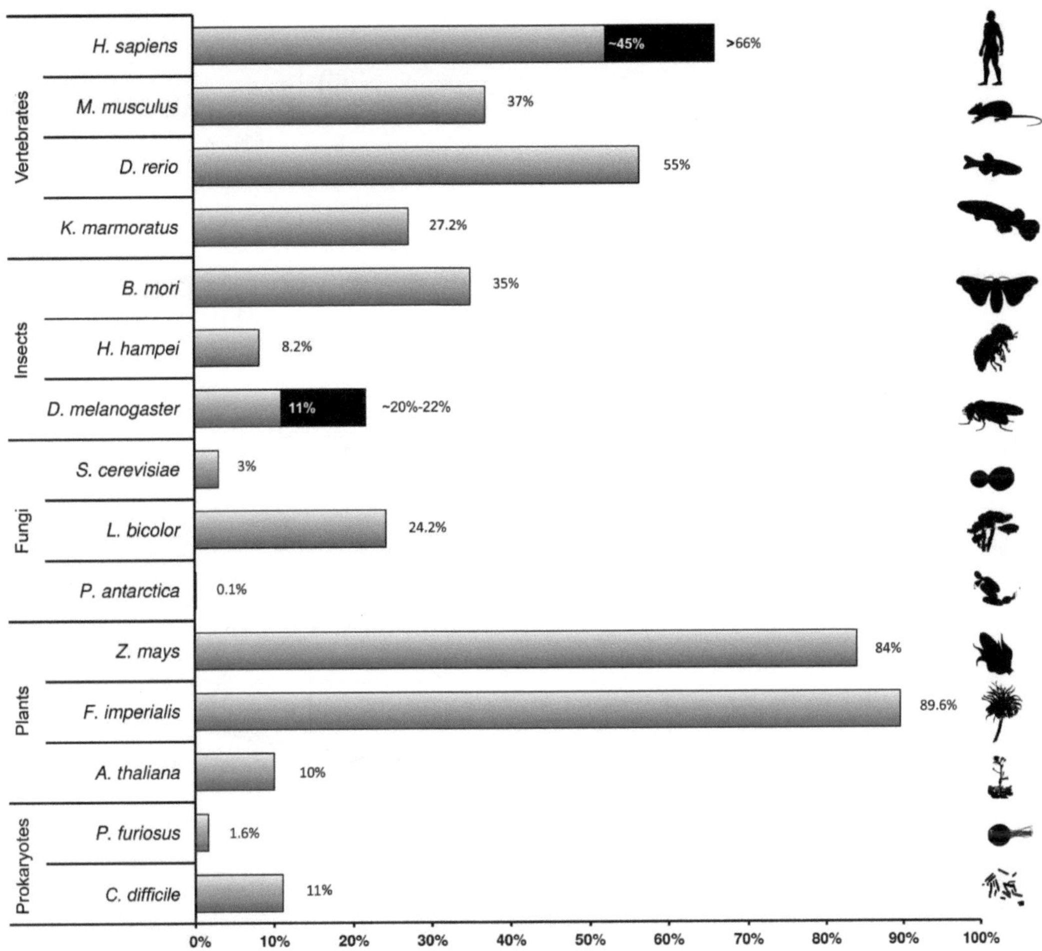

Fig. 1 TE content in the genome of different organisms expressed as percentage of the genome: *Homo sapiens* (~45% [12], >66% [13]) *Mus musculus* [143], *Saccharomyces cerevisiae* [144], *Arabidopsis thaliana* [145], *Pyrococcus furiosus* [146], *Clostridium difficile* [147], *Danio rerio* [133], *Kryptolebias marmoratus* [148], *Bombyx mori* [149], *Hypothenemus hampei* [150], *Drosophila melanogaster* (11%, [68], ~20% [69]), *Pseudozyma antarctica*, and *Laccaria bicolor* [151]. *Zea mays* [152] and *Fritillaria imperialis* [8]. All estimates were obtained with homology-based methods except [13] that uses P-cloud and [69] that uses de novo approaches

promoters, or transcription factor binding sites [22–24]. TEs can also be targets of epigenetic histone modifications that spread into adjacent genes affecting their expression [25, 26]. In addition to transcriptional changes, TEs have been shown to affect translation regulation when they are transcribed within a mRNA [27–29], to contribute to protein-coding regions both at the transcript and at the protein level [30–35], and TE-encoded proteins have been domesticated and are part of host genes [17, 36–40]. TE excision can lead to DNA deletions [41], and TE insertion can result in adding DNA through 3′ and, less frequently, through 5′ transduction [42, 43]. Finally, ectopic recombination between TEs causes

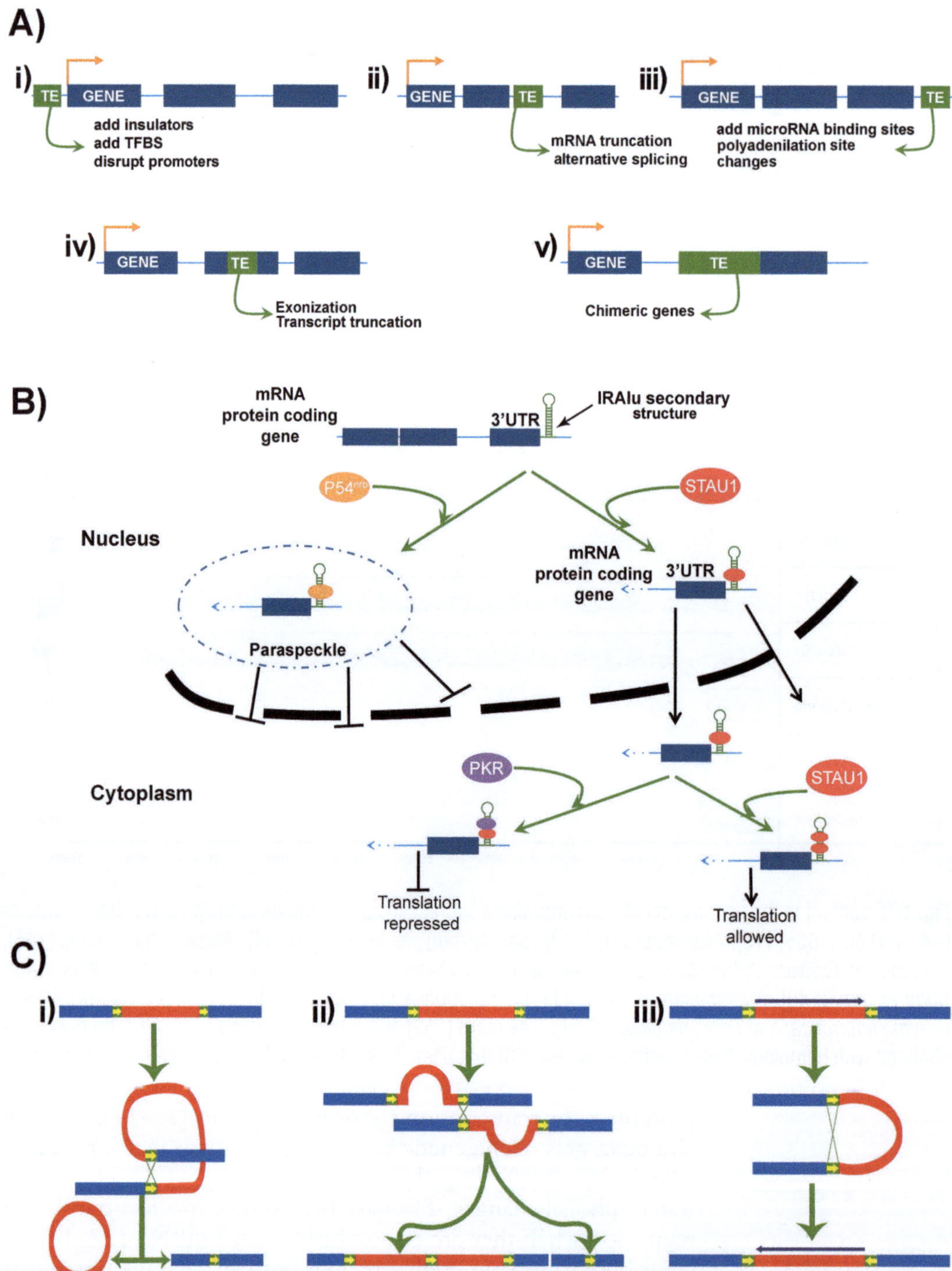

Fig. 2 Effects of TEs on the host genome. (**a**) TEs can affect the expression and/or structure of genes. Exons are represented as blue boxes and TEs as green boxes. (1) A TE inserted in the upstream region of a gene can add insulator sequences, transcription factor-binding site (TFBS), or can disrupt an existing promoter gene; (2) A TE inserted in an intron can truncate the mRNA or induce alternative splicing; (3) A TE inserted in the downstream region of a gene can add microRNA binding sites or alter the polyadenylation site; (4) A TE

deletions, duplications, and sequence rearrangements. Two recent studies in the human genome identified 516 chromosome rearrangements potentially generated by LINE-LINE nonallelic homologous recombination and 78 HERV-mediated rearrangements [44, 45]. Both studies used the annotations of LINEs and HERVs in the reference genome and look for evidence of rearrangements induced by these TEs using clinical databases of copy number variants containing information from thousands of patients. In addition to being associated with diseases [24, 46–49], the number of TE-induced mutations associated with positive effects on fitness-related traits also continues to increase both in humans and in *Drosophila* [50–63].

Overall, recent advances in sequencing technologies and in TE detection methods showed that, as expected, the TE content is higher than previously estimated. These new data also provided further evidence for the impact of TEs in genome function and genome structure. Thus, it is still indisputable that a thorough understanding of TE population dynamics is essential for the understanding of the eukaryotic genome structure, function, and evolution.

2 *Drosophila* and Humans: Two Extremes in TE Diversity and Population Dynamics

Much of the detailed information on TE evolution still comes from two species with the best-studied genomes: fruit flies (*D. melanogaster*) and humans. Fortunately, these two genomes represent two extremes in terms of TE diversity and population dynamics and thus give a reasonably diverse picture of the TE evolution and dynamics. For the rest of this chapter, we focus

Fig. 2 (continued) inserted in the exon of a gene can lead to exonization of the TE or to transcript truncation; (5) the whole domain of a TE protein could insert in the coding region of a gene generating a chimeric gene with host and TE domains [5, 21]. In addition to these changes that depend on where the TE is inserted and on the sequences that the TE is adding, TEs can also alter the posttranslational modifications of histones. (**b**) TEs could also induce translation repression by generating secondary structure in the 3′ UTR of genes that leads to changes in the localization of the mRNA. This secondary structure could bind to one of the protein components of *paraspeckle* (*P54^{nrb}*) and translocate to paraspeckle, a group of subnuclear bodies, avoiding moving out of the nucleus. However, the same secondary structure could bind to the dsRNA-binding protein *Staufen 1* (*STAU1*) and in this case translocate to cytoplasm. Once in the cytoplasm, the secondary structure could bind to *STAU1* again allowing translation, but under some situations mRNA could bind to the *ds-RNA- dependent protein kinase* (*PKR*) repressing translation [23]. (**c**) Ectopic recombination between TE copies (green boxes with yellow arrows) in the same orientation can lead to deletions when recombination takes place between copies located on the same chromatid (1) or deletions and duplications when recombination takes place between copies in different chromosomes (2) (recombination between two nonhomologous chromosomes should lead to a translocation). Ectopic recombination between TE copies in opposite orientation leads to inversion of the DNA between the two TEs (3)

primarily on these two genomes and will highlight the similarities and differences observed between them.

As mentioned above, the human reference genome has millions of TE copies, with 66–69% of the genome mostly derived from TE sequences [13]. Two human retrotransposable element (class I) families, LINE1 (L1, long interspersed nuclear element 1) and Alu, account for 60% of all interspersed repeat sequences. The vast majority of the TEs in the human genome are fixed, and most families are inactive. However, some elements of the main families of human endogenous retrovirus (HERV-K) and LINE1 elements show autonomous transposition. Meanwhile, elements of Alu and the hybrid SVA elements formed by SINEs (short interspersed nuclear elements), VNTRs (variable number tandem repeat), and Alus show nonautonomous activity [64–66].

In contrast, the fruit fly *D. melanogaster* reference genome contains only thousands of individual TE copies (5416 TE copies in FlyBase R6.04) accounting for only ~5.5% of the euchromatin [67]. If the missing percentage of TEs detected in chromosome 2L is similar in other chromosomes, the euchromatin TE content might be higher (~ 8.7%) [14]. If heterochromatin is also included, TEs account for 11–20% of the *D. melanogaster* genome [68, 69]. *D. melanogaster* TEs belong to approximately 100 diverse families of both class I and class II elements [69, 70]. Each family consists of 1–304 copies with no dominant family corresponding to the majority of TEs. The only exception is INE-1 family that contains ~2000 copies and has been inactive for the past ~3--4.6 million years [71–73]. The majority of TE families are considered to be active in *Drosophila*: individual TE copies are generally polymorphic in the population and show a high sequence similarity [69, 70, 74, 75]. Indeed, there is experimental evidence showing that *Gypsy* and *ZAM* elements are active [76, 77]. Besides, there is indirect evidence for the activity of 24 *D. melanogaster* superfamilies based on a whole-genome sequencing experiment of mutation accumulation lines [75] (Table 1).

Why do these two genomes differ so profoundly in content, diversity, and activity of TEs? The answer must lie in different aspects of TE population dynamics within genomes and forces that lead to varying rates of TE family birth and extinction. In the rest of this review, we focus on the state of knowledge of different aspects of TE population dynamics and discuss aspects of TE family evolution. Specifically, we focus on rates of TE transposition, fixation, or loss in human and *D. melanogaster* populations due to stochastic forces and natural selection for or against TE insertions and forces that affect coexistence of multiple TE families and the standing diversity of TE types (Fig. 3).

Table 1
Summary of recent TE population dynamic studies

Objectives	Findings	Relevance for TE dynamics	References
Overview of new discoveries about TEs in 75 basidiomycete fungi genomes	TE content varies among species displaying different lifestyles from 0.1% to 45.2%. The correlation between TE content and genome size is not strong. TEs seem essential for chromosomal architecture. A large battery of mechanisms to avoid transposition is present	The result of most TE activity is likely neutral as they often insert in intergenic regions. However, TEs play an important role in the evolution of plant pathogens and probably in symbiotic species	[151]
Characterization of TE content in the only selfing hermaphroditic vertebrate: the mangrove killifish *Kryptolebias marmoratus*	TE content is 27%. There is a great diversity of families with a pronounce abundance of Helitrons compared to its closest phylogenetic relatives. TE sequence divergence is also higher in *K. marmoratus* compared to close species	Against expectations, the number and composition of TEs in these selfing organisms is comparable to that of many other fish with outcrossing mating systems. The high Helitron content is one of the factors that could explain the high genetic diversity observed in this selfing killifish	[148]
Testing whether genome size equilibrium observed in 10 mammals and 24 birds species is due to covariation between DNA gain by transposition and DNA loss by deletion	DNA gain varies by more than sixfold across mammals and 30-fold across birds. DNA loss varies by twofold in mammals and threefold in birds. Neither DNA gain nor loss can solely explain variation in genome size. DNA loss exceeded gain in all but two lineages. Midsize deletions (31 bp to 10 kb) play a larger role than microdeletions (1–30 bp) in DNA loss	Genome size equilibrium is maintained through DNA loss counteracting DNA gains through TE expansions. DNA loss has probably been driven by large deletions (>10 kb). Genome expansion via transposition could promote genome contraction through TE-mediated deletions	[134]
Understanding the differences in abundance and diversity of L1 elements across vertebrates	Vertebrate L1s differ in the length of the 5′ UTR, 3′ UTR, and intergenic regions. They also differ in base composition with mammals and lizards showing a stronger A bias on the positive strand than frog and fish	Mammals show very little 5′ UTR homology due to the frequent acquisition of novel nonhomologous 5′ UTR during evolution. This seems not to occur in other groups of vertebrates since the relative conservation of the 5′ UTR and ORF1 suggests that the host do not repress transposition in a sequence-specific way	[153]

(continued)

Table 1 (continued)

Objectives	Findings	Relevance for TE dynamics	References
Understanding the role of TEs in *D. melanogaster* genome evolution, by estimating their insertion and deletion rates	24 TE superfamilies are active in mutation accumulation lines. TE activity is background dependent. There is an association between activity of some TE families and chromatin state, as well as a week correlation between insertion activity and GC content, and a negative correlation between deletion activity and exon content	Insertion rate is higher than deletion rate which helps explain the relative stability of TE numbers and genome size in *Drosophila* in the face of previously reported deletion bias. Heterochromatin may play a bigger role than recombination in shaping TE accumulation	[75]
Characterization and description of TEs in the coffee berry borer *Hypothenemus hampei* genome	8.3% of the genome are TEs (880 TE sequences): 49.24% of the TEs are MITEs. Several new families described: Hypo belonging to *Gypsy* superfamily, *Hamp* a new non-LTR family and *rosa* a new DNA TE family	Low TE content, compared with other insects, could be related to the reproductive characteristics and the population size of this species. Males have a chromosome set not transmitted to the next generation like asexual populations. The colonization of America probably produced a founder effect	[150]
To develop a comprehensive assessment of transposition activity at the *A. thaliana* species level	The analysis includes 211 samples collected all over the world. 165 of the 326 families annotated in *A. thaliana* showed recent transposition activity at the species level. TE composition and activity are strongly affected both by environmental and genetic factors	TEs have pervasive effects on the expression and methylation status of nearby genes which are likely deleterious and could help explain why bursts of transposition were not detected. Its self-fertilizing mating system should also lead to accelerated elimination of deleterious TE insertions. TEs are also involved in the generation of large-effect alleles at adaptive trait loci	[154]

(continued)

Table 1 (continued)

Objectives	Findings	Relevance for TE dynamics	References
Characterization of TE presence/absence in 216 *A. thaliana* accessions with respect to the reference genome	TE deletions were biased toward pericentromeric regions, while TE insertions had a more uniform distribution over chromosomes. TE variants associated with changes in nearby gene expression and local and distal methylation patterns	TEs are a significant source of genetic variation. Most TEs present at low frequencies. TEs likely play a role in facilitating epigenomic and transcriptional differences between *A. thaliana* accessions	[155]
To understand the role of TE in genome evolution of the sweet potato *Ipomoea batatas*	1405 TEs described based on transcriptomic data. 417 TEs are expressed in one or more tissues and 107 in the seven tissues analyzed	TE activity is tissue- and background-specific. Although several TEs are expressed in all the tissues and strains analyzed, some of them are active only in one specific strain and/or tissue. Authors suggest that TEs may play a role in environmental adaptation	[156]

3 Methodology Used to Study TE Population Dynamics

TE dynamics continues to be studied using three main approaches: mathematical modeling, computer simulations, and the analysis of empirical data. Often a combination of these approaches is used to better understand TE abundance, diversity, and distribution (Table 2). Le Rouzic et al. [78] applied the statistical framework originally developed to infer speciation and extinction dynamics in species phylogenies to reconstruct the evolutionary history of TEs [78]. The model allows to estimate and to interpret the pattern of transposition activity that results in different TE copy number distributions [78]. The authors also performed computer simulations to provide reference dynamics that aid in the interpretation of the results obtained (Table 2).

Traditionally, mathematical models considered the relationship between the host and a homogenous group of active TEs. However, the TE content of any genome is a mixed of autonomous and nonautonomous insertions. Xue and Goldenfeld [79] proposed a mathematical model that considers the relationship between nonautonomous and autonomous TEs as a predator-prey dynamic. Unlike previous models that also use the analogy to ecological models, Xue and Goldenfeld model takes into account the

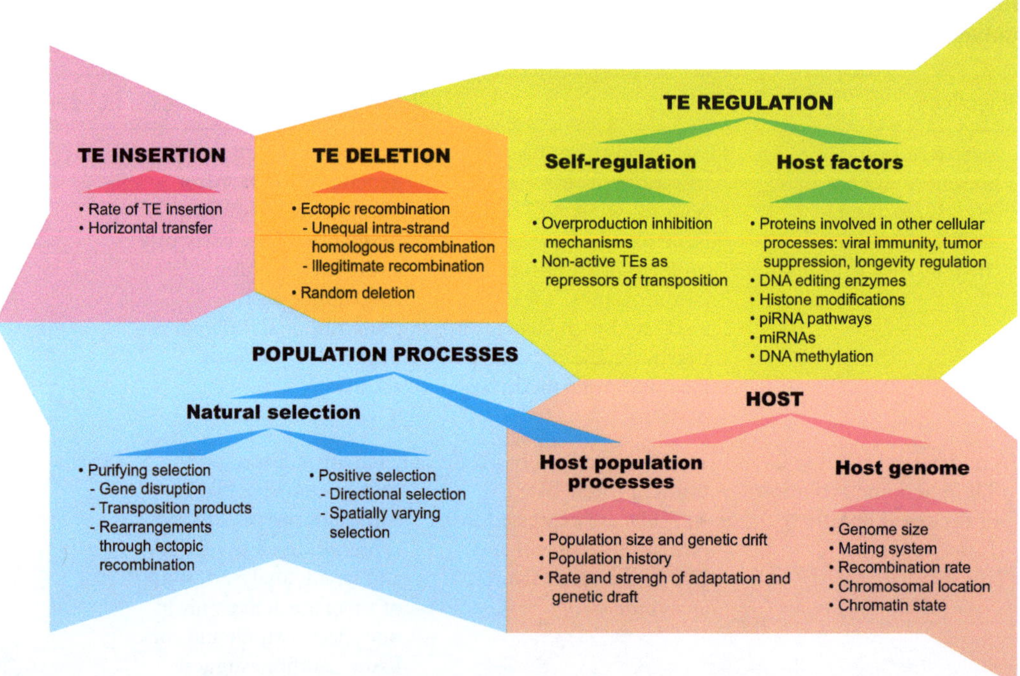

Fig. 3 Factors that influence the population and evolutionary dynamics of TEs. Our understanding of TE population and evolutionary dynamics is still incomplete. The different factors that affect TE population and evolutionary dynamics are interrelated, new factors have been identified in recent years, and future research is still likely to reveal existence of additional factors

molecular level interactions between transposable elements and the small copy number of the active transposons. The model predicts oscillations in the number of TEs in a time scale much longer than the cell replication time, suggesting that the genome stores the predator-prey state during successive generations [79].

TE dynamics have also been analyzed in variable environments [80, 81] (Table 2). Gogolesky et al. [81] proposed a stochastic computational model to analyze the dynamics of active TEs in genomes of sexual diploid organisms under environmental stress. They based their model in the Fisher geometrical model of fitness landscapes. Overall, the authors conclude that the presence of inactive copies of TEs is necessary for the transposition-selection equilibrium of autonomous copies and that the mutator capacity of TEs might be important when host populations face rapid environmental changes [81].

Other recently developed methods analyzed the influence of the mating system in TE dynamics, different modes of selection, or applied branching models for studying the propagation of particular TE classes [82–84] (Table 2).

In addition to mathematical modeling and simulations, multiple computational tools have been developed to analyze TEs in

Table 2
Summary of recent mathematical models and computer simulations applied to the study of TE dynamics

Model description	TEs modelled	Conclusions	References
The model quantifies the transposition activity over time based on the distribution of transposition events in the phylogenetic tree and the tree topology	*Fot* subfamilies from *Fusarium oxysporum*	The four subfamilies analyzed are still active with two of them showing clear changes in their transposition dynamics. The results obtained showed that regulation of transposition by the number of copies is not strong enough to maintain stable transposition-deletion equilibrium	[78]
Considering the genome as an ecosystem, the model analyzes the interaction between nonautonomous and autonomous TEs as a predator-prey relationship in individual cells	*L1* and *Alus* from *Homo sapiens*	The model predicts oscillations in the number of TEs in a time scale much longer than the cell replication time. Thus, the genome stores the predator-prey state during successive generations	[79]
The model, based in the Fisher geometric model, analyzes TE dynamics under changing environments in clonal organisms	Autonomous and nonautonomous TEs in asexual population	The model predicts that when nonautonomous TE copies are present, the transposition activity is lost and thus the stability of the host-TE system is compromised. Changes in the environment may induce bursts of transposition activity associated with faster adaptation. However, it is unlikely that the transposition activity is maintained in the long term	[80]
The model, based on the Fisher geometrical model, analyzes TEs dynamics in sexual diploid organisms under environmental changes	TEs in sexual diploid populations	The model suggests that the presence of inactive copies of TEs is necessary for the transposition-selection equilibrium of active copies and that the mutagenic role of TEs is crucial when host populations face rapid environmental changes	[81]
The model, based in the selfish DNA theory, analyzes the invasion dynamics of active TEs during the first stages of an experimental evolution experiment	*Mos1* and *peach*, *mariner* family from *Drosophila melanogaster*	The model predicts lower invasion frequencies than the ones observed experimentally. A substantial rate of replicative transposition during the initial invasion of the element was	[102]

(continued)

Table 2 (continued)

Model description	TEs modelled	Conclusions	References
		inferred from the discrepancy between observed and theoretical copy numbers	
The model analyzes the impact of intermediate selfing rates on TE dynamics and the influence of the mating system on the evolutionary properties of TEs	Active TEs in a diploid hermaphrodite population	The model predicts that the efficiency of TEs as genomic parasites decreases with the selfing rate, although rare TE invasions can still occur even in populations with 90% selfers. The model predicts TE extinction if populations change from sexual to asexual reproduction, although empirical data does not strongly support this result	[82]
The model studies the evolutionary behavior of TE copy number and the molecular evolution of their DNA sequences	TEs in sexual diploid populations	The model predicts that weak selection allows high copy numbers of TEs most of them inactive copies, while strong selection reduces the number of TEs but increases the proportion of active copies. Regarding TE sequences, the model shows that the phylogeny of these sequences allows distinguishing active copies from non- and less active copies	[83]
The model analyzes the propagation of LTR TEs by taking into account the TE position in the chromosome, the degradation level of the TEs, and the duplication rate that varies with the degradation level	*roo*, *Gypsy* and *DM412*, TEs of LTR family from *Drosophila melanogaster*	The simulation estimates several parameters affecting the propagation of TEs and identifies the initial copy from which three LTR families have spread on the euchromatin part of the 3L chromosome	[84]

sequenced genomes in the last 5 years. While some of these tools aimed at assessing the global abundance and diversity of TEs in the genome, such as dnaPipeTE, or to annotate TEs in assembled genomes, such as REPET, most of them are focused on discovering and/or genotyping individual copies of TEs in the genome using next-generation sequencing (NGS) data [11, 64, 85–90]. The diversity of methods available makes it difficult to choose the most appropriate one for the analyses of a given genome. To try to overcome this limitation, Nelson et al. [91] developed an integrated pipeline named McClintock that incorporates six complementary TE detection methods. McClintock generates standardized output for the different TE detection methods, thus facilitating the comparison of the results obtained with the different pipelines, as well as facilitating their installation and use [91]. This and other studies that compared the performance of several tools arrived to the same conclusion: several computational tools should be combined to increase the accuracy of TE analysis [64, 86, 91].

The availability of third-generation sequencing techniques (3GS) should help improve the detection and genotyping of TE insertions. Although 3GS was developed before 2010 [92], it has only been in the last few years when this technique has started to be used [14, 93]. Chakraborty et al. [14] reported the assembly of a *D. melanogaster* genome from a Zimbabwe strain using long-read single molecule real-time sequencing with 147X coverage. Among several novel structural variants described, they identified 37% additional TE insertions in the 2L chromosome compared with a previous study that used 70X coverage of short reads [14, 94]. 3GS technologies have also been applied to the sequencing of human genomes, although a detailed analysis of TE content based on long-read data has not been performed yet [95–97].

Recently, Disdero and Filée [98] introduced the first tool that uses long-read sequences to identify TE insertions in the *D. melanogaster* genome: LoRTE [98]. The authors argue that available software based on short reads fail to correctly identify TEs that are present in highly repetitive regions of the genome, while long-read technologies should allow us to identify all TEs in a given genome. LoRTE, developed in Python, verifies presence and/or absence of previously annotated TEs and can also detect new insertions not previously annotated in the reference genome. LoRTE is able to work with low-coverage sequences (<10X) providing an efficient accurate TE annotation in a cost-effective manner [98].

4 Rates of Transposition

4.1 Empirical Estimates of the Rates of Transposition in *Drosophila* and Humans

Transposition rates in *D. melanogaster* have been traditionally estimated empirically by in situ hybridization and by using PCR approaches. The activation of TEs following intra- and interspecific hybridization has been studied in different *Drosophila* species [99–101]. For example, Vela et al. [100] estimated transpositions rates in *D. buzzatii-D. koepferae* interspecific hybrid flies by in situ hybridization [100]. They found that hybrids showed at least one order of magnitude higher transposition rates than parental lines for at least three TE families [100]. Robillard et al. [102] estimated transposition rates by qPCR in an experimental evolution study in which a TE insertion was introduced in a strain lacking insertions from that particular family [102]. In the first generations after the introduction of the TE insertion, the transposition rate was 0.33–0.45 per copy per generation, while in the following generations, transposition rates were reduced at least one order of magnitude per copy per generation. These values represent the first steps in the invasion of a TE in a genome that is faster than the rate of transposition when measured in natural populations [102].

In the first edition of this chapter [103], we anticipated that NGS would allow studying transposition rates in a deeper and more accurate way. Indeed, recent studies have taken advantage of NGS data to estimate transposition rates in *D. melanogaster*. Rahman et al. [89] estimated using NGS data the transposition rate in the reference strain by comparing two available genomes that were sequenced with ~15 years difference. The average transposition rate for TEs belonging to different families was 7×10^{-5}, which is on the same order of magnitude as the previously reported rates (~10^{-4}–10^{-5}). Furthermore, they confirmed the prediction of increased transposition rate in inbred lines: they estimated a higher average number of TE insertions in lab strains inbred for more generations compared with strains inbred for a smaller number of generations [89]. Adrion et al. [75] estimated spontaneous insertion and deletion rates in *D. melanogaster* mutation accumulation lines [75]. The authors identified 24 active superfamilies and estimated genome-wide insertion rates to be higher than deletion rates: 2.11×10^{-9} vs. 1.37×10^{-10} per site per generation, respectively. Superfamily-specific rates of insertion varied from 0 to 5.13×10^{-3} insertions per copy per generation and were within the range of previously estimated rates [75] (Table 1).

In humans, previous studies estimated the transposition rate as in 1 in 95 to 1 in 250 births for L1, 1 in 20 births for Alu insertions, and 1 in 916 births for SVA retrotransposons [104–107]. Although there are several recent studies that estimate transposition rate in humans using NGS data, they all focused on somatic transposition in the brain or in tumor samples [47, 48, 90].

4.2 Transposition Control Mechanisms

Understanding the mechanisms controlling the transposition of TEs is central to our understanding of TE dynamics. Many different mechanisms of TE regulation have been described [43, 108, 109]. In this section, we will highlight recent advances in both TE self-regulation and regulation by host factors.

4.2.1 TE Self-Regulation

Self-regulation of transposition was first described in prokaryotes and soon after in TEs involved in hybrid dysgenesis in *Drosophila* [110]. Recent studies have cast some doubt on one of the self-regulation mechanisms described: transposase overproduction inhibition. The transposase overproduction inhibition mechanism regulates the transposition of *IS630-Tc1-mariner piggyBac* and *hobo-AC-Tam* (*hAT*) superfamilies [111, 112]. However, several studies reported contradictory results suggesting that transposase inhibition by overproduction does not always happen [113]. Bire et al. [113] suggested that some works failed to detect transposase inhibition because cellular cofactors are necessary to execute this regulation system, and as such it can only be detected in in vivo experiments [113]. However, Woodard et al. [114] showed that aggregation of transposase proteins produces filamentous structures (rodlets) in the nucleus in a host independent manner [114]. The authors further showed that a decline in transposition occurs after transposase concentrations are high enough for filamentous structures to be visible [114]. Thus, it is still not clear why some in vitro experiments failed to detect transposase overproduction inhibition [114].

4.2.2 Regulation by Host Factors

Small RNAs, such as small-interfering RNAs (siRNAs) and piwi-interacting RNAs (piRNAs), are well-known to play an essential role in silencing TEs and preventing transposition. Several recent reviews highlight the monumental progress in this field [115–119]. In addition to posttranscriptional regulation of TEs, small RNAs are involved in transcriptional regulation as well. In mouse, piRNAs are required for de novo methylation and silencing of TEs [120]. In *Drosophila*, Piwi proteins repress transcription and correlate with an increase in repressive chromatin marks at loci targeted by piRNAs [121].

While the role of siRNAs and piRNAs has been established for several years, a role of micro RNAs (miRs) in suppressing the mobility of retrotransposons was only recently described [122]. The authors showed that mir-128 binds to L1 RNA and represses its integration in humans [122].

New studies have also provided evidence for the role in TE repression of proteins previously known for their roles in other cellular processes such as interferon-stimulated proteins, the tumor suppressor *p53*, and the longevity regulating protein *SIRT6*. Several interferon-stimulated genes, such as the *Moloney leukemia virus 10* (*MOV10*), the *zinc-finger antiviral protein* (*ZAP*), and the *3′ repair exonuclease 1* (*TREX1*), which are associated with virus response, have been recently involved in the inhibition of L1 activity [66, 123]. Recently, it has also been shown that the *p53* transcription factor, which is involved in stress response networks and acts to restrict oncogenesis, also restricts retrotransposon activity in zebra fish, flies, and humans [124]. The authors showed that *p53* interacts with components of the piwi-interacting RNA to suppress retrotransposition [124]. Finally, the longevity regulating protein *SIRT6* is also involved in retrotransposon repression by coordinating their packaging into transcriptionally repressive heterochromatin. *SIRT6* binds to the 5′ UTR region of retrotransposons and mono-ADP ribosylates the *Krüppel-associated protein 1* (*KAP1*) facilitating the interaction of *KAP1* with the *heterochromatin protein 1α* (*HP1α*) leading to chromatin compaction [125].

5 Rate of Fixation and Frequency Distribution

5.1 Natural Selection Against TE Insertions

Natural selection and stochastic processes influence both the rate of fixation and the frequency distribution of TEs in populations. The efficiency of selection depends on the effective population size, which largely differs between *Drosophila* and humans: $>10^8$ and $\sim 10^4$, respectively [126, 127]. Thus, while in *Drosophila* the high efficiency of selection should led to the removal of slightly deleterious TE insertions, in humans, these insertions may accumulate in the genome. Indeed most of the TE sequences in the human genome are remnants of ancient insertions [12].

A review by Barrón et al. [128] explored the latest insights on the nature of selection acting against the deleterious effects of TEs in *D. melanogaster* populations [128]. More recently, Kofler et al. [129] analyzed intraspecific TE dynamics between *D. melanogaster* and *D. simulans* populations to shed light on the long-term evolution of TEs [129]. They confirmed that most of the TEs are present at low frequencies in *D. melanogaster* and showed that the same pattern is present in *D. simulans.* Based on computer simulations showing that 50% of the TE families have temporally heterogeneous transposition rates, and on the differences in TE composition between populations of the same species, the authors suggested that TE activity has recently increased in the two species. They proposed that the demographic history of both species, with a recent colonization of different environments, could be the cause of the high TE activity detected [129].

In humans, a recent study took advantage of the 1000 Genome Project data that reports 16,192 polymorphic TEs to perform the most complete TE dynamics analysis to date [130]. Most of the polymorphic TEs were found to be present at very low frequencies: >93% of TEs showed <5% allele frequency in 26 human populations. These results confirm that overall polymorphic TE insertions are deleterious in humans as was previously suggested with smaller family-specific datasets [131].

5.2 TE-Induced Adaptations

Several recent reviews have compiled results that showcase the adaptive role of TEs [19, 24, 50, 59, 128]. We would like to highlight the recent discovery of a TE in a fish-like marine chordate that encodes RAG-like proteins with endonuclease-transposase activity [39]. This discovery provides evidence that supports the TE origin hypothesis for the adaptive immune system in jawed vertebrates [39]. Two other recent publications provide experimental evidence for a role of TEs as providers of functional transcription factor binding sites (TFBS) involved in immune response and in cell pluripotency [50, 132]. A recent study linked ERV elements in humans with the interferon response pathway [50]. The authors showed that ERVs carrying enhancers have been co-opted to activate different genes involve in inflammatory response activated by interferon. This example shows how the exaptation of one family of TEs could shape a transcriptional network to activate different genes with one trigger system [50]. Sundaram et al. [132] reported mouse-specific TEs that contain multiple transcription factor binding sites for pluripotency transcription factors. The majority of the TEs were experimentally shown to exhibit enhancer activity in mouse embryonic stem cells including an in silico reconstructed ancestral TE. This latter result suggests that ancestral TEs already had transcriptional regulatory sites [132].

In *Drosophila*, the adaptive role of several TEs has also been identified. Most of the TEs characterized so far are involved in stress response: viral infection and xenobiotics (*Doc1420*, [60, 61]), oxidative stress (*FBti0018880*, [53]), xenobiotic stress (*Accord*, [62, 63], and *FBti0019627*, [52]), cold stress (*FBti0019985*, [55]), and heavy metal stress (*FBti0019170*, [56]), while *FBti0019386* insertion was associated with faster developmental time [54]. Some of these adaptive insertions have been shown to affect gene expression through different molecular mechanisms, such as affecting the polyadenylation site choice [52], and adding TFBS [53], while others have been associated with gene duplication [60, 62].

6 Rate of Loss

A recent study estimated genome-wide and superfamily-specific TE deletion rates in *D. melanogaster* inbred lines [75]. The authors found that most of the deletions involved retrotransposon elements suggesting that the deletions were due to ectopic recombination instead of excision. Deletion rates were smaller than insertion rates estimated in the same inbred lines [75].

In vertebrates, lineage-specific differences in TE deletion rates have been reported [133]. A possible explanation for this observation is that the success of some families results in a competition for the genome resources leading to the elimination of other TE families [133].

In addition to TE deletion rates, DNA loss rates should also be considered. In the human linage, estimates of DNA loss are smaller than estimates of DNA gain, 650 Mb vs. 815 Mb [134], while in *D. melanogaster*, the rate of DNA loss is higher than the rate of DNA gain [135–137].

7 Horizontal Transfer of TE Insertions

In addition to parent to offspring transmission, TEs can also be horizontally transferred [138–141]. By combining simulation and analytical approaches, Groth and Blumenstiel [142] suggested that exposure rate to new TE families through horizontal transfer can be an important determinant of TE genomic content when the effects of drift in a population are weak [142]. Thus, larger populations are expected to carry a higher TE content if population exposure rate is proportional to population size [142]. So far, most of the evidence for TE horizontal transfer comes from closely related and geographically close species [140]. There are several examples of horizontal transfer of TEs in *Drosophila* species, while so far horizontal transfer of TEs has not been described in humans [138].

8 Conclusion

Recent years have seen an increase in the number of reference genome sequences available as well as of population genome datasets. The availability of all these genome sequences and the development of new bioinformatics tools have allowed us to update our previous estimates of genomic TE content that have increased both in humans and in *D. melanogaster*. These data has also allowed us to gather more evidence for the functional impact, both detrimental and beneficial, of TE insertions. Thus, it is still indisputable that understanding TE population dynamics is essential to understand genome structure, genome function, and genome evolution.

New methods developed to analyze the dynamics of TEs in populations have shed light on the interplay between autonomous and nonautonomous TE copies, TE invasion dynamics, and how the mating system influences the dynamics of TEs in genomes. We have also considerably advanced our knowledge on the host factors that regulate TE activity as well as in the genome features that influence TE dynamics (Fig. 3). Finally, differences in effective population sizes that affect the efficiency of selection against new TE insertions and differences in the rates of TE loss between humans and *D. melanogaster* can still be considered two important factors that contribute to the different abundance, diversity, and activity of TEs in this two species [103].

9 Questions

How differences in the rate of DNA loss can affect the evolutionary dynamics of TEs?

Why host regulation of transposition is relevant for TE dynamics?

Which is the most important factor explaining the differences in TE content, diversity, and activity between humans and *Drosophila*?

Have the next-generation sequencing (NGS) technologies allowed us to identify all the TEs in a given genome?

How does the interaction between active and inactive copies of TEs affect TE dynamics?

Acknowledgment

We thank the reviewers for providing constructive comments on a previous version of this manuscript. This work has received funding from the European Research Council (ERC) under the European Union's Horizon 2020 research and innovation program (H2020-ERC-2014-CoG-647900) and from the Spanish Ministry of Economy and Competitiveness/FEDER (BFU2014-57779-P).

References

1. Piegu B, Bire S, Arensburger P, Bigot Y (2015) A survey of transposable element classification systems--a call for a fundamental update to meet the challenge of their diversity and complexity. Mol Phylogenet Evol 86:90–109
2. Wicker T, Sabot F, Hua-Van A, Bennetzen JL, Capy P, Chalhoub B, Flavell A, Leroy P, Morgante M, Panaud O, Paux E, SanMiguel P, Schulman AH (2007) A unified classification system for eukaryotic transposable elements. Nat Rev Genet 8(12):973–982
3. Kapitonov VV, Jurka J (2008) A universal classification of eukaryotic transposable elements implemented in Repbase. Nat Rev Genet 9(5):411–412. author reply 414
4. Bao W, Kojima KK, Kohany O (2015) Repbase Update, a database of repetitive elements in eukaryotic genomes. Mob DNA 6:11
5. Hua-Van A, Le Rouzic A, Boutin TS, Filee J, Capy P (2011) The struggle for life of the genome's selfish architects. Biol Direct 6:19
6. Arkhipova IR (2017) Using bioinformatic and phylogenetic approaches to classify transposable elements and understand their complex evolutionary histories. Mob DNA 8:19
7. Touchon M, Rocha EP (2007) Causes of insertion sequences abundance in prokaryotic genomes. Mol Biol Evol 24(4):969–981
8. Ambrozova K, Mandakova T, Bures P, Neumann P, Leitch IJ, Koblizkova A, Macas J, Lysak MA (2011) Diverse retrotransposon families and an AT-rich satellite DNA revealed in giant genomes of Fritillaria lilies. Ann Bot 107(2):255–268
9. Chaisson MJ, Wilson RK, Eichler EE (2015) Genetic variation and the de novo assembly of human genomes. Nat Rev Genet 16 (11):627–640
10. Treangen TJ, Salzberg SL (2011) Repetitive DNA and next-generation sequencing: computational challenges and solutions. Nat Rev Genet 13(1):36–46
11. Flutre T, Duprat E, Feuillet C, Quesneville H (2011) Considering transposable element diversification in de novo annotation approaches. PLoS One 6(1):e16526
12. Human Genome Sequencing Consortium (2001) Initial sequencing and analysis of the human genome. Nature 409(6822):860–921
13. de Koning AP, Gu W, Castoe TA, Batzer MA, Pollock DD (2011) Repetitive elements may comprise over two-thirds of the human genome. PLoS Genet 7(12):e1002384
14. Chakraborty M, VanKuren NW, Zhao R, Zhang X, Kalsow S, Emerson JJ (2018) Hidden genetic variation shapes the structure of functional elements in Drosophila. Nat Genet 50(1):20–25
15. Rius N, Guillen Y, Delprat A, Kapusta A, Feschotte C, Ruiz A (2016) Exploration of the *Drosophila buzzatii* transposable element content suggests underestimation of repeats

in Drosophila genomes. BMC Genomics 17:344

16. Kidwell MG, Lisch DR (2000) Transposable elements and host genome evolution. Trends Ecol Evol 15(3):95–99
17. Feschotte C, Pritham EJ (2007) DNA transposons and the evolution of eukaryotic genomes. Annu Rev Genet 41:331–368
18. Cowley M, Oakey RJ (2013) Transposable elements re-wire and fine-tune the transcriptome. PLoS Genet 9(1):e1003234
19. Casacuberta E, Gonzalez J (2013) The impact of transposable elements in environmental adaptation. Mol Ecol 22(6):1503–1517
20. Belyayev A (2014) Bursts of transposable elements as an evolutionary driving force. J Evol Biol 27(12):2573–2584
21. Rebollo R, Romanish MT, Mager DL (2012) Transposable elements: an abundant and natural source of regulatory sequences for host genes. Annu Rev Genet 46:21–42
22. Feschotte C (2008) Transposable elements and the evolution of regulatory networks. Nat Rev Genet 9(5):397–405
23. Elbarbary RA, Lucas BA, Maquat LE (2016) Retrotransposons as regulators of gene expression. Science 351(6274):aac7247
24. Chuong EB, Elde NC, Feschotte C (2017) Regulatory activities of transposable elements: from conflicts to benefits. Nat Rev Genet 18 (2):71–86
25. Lippman Z, Gendrel AV, Black M, Vaughn MW, Dedhia N, McCombie WR, Lavine K, Mittal V, May B, Kasschau KD, Carrington JC, Doerge RW, Colot V, Martienssen R (2004) Role of transposable elements in heterochromatin and epigenetic control. Nature 430(6998):471–476
26. Sentmanat MF, Elgin SC (2012) Ectopic assembly of heterochromatin in *Drosophila melanogaster* triggered by transposable elements. Proc Natl Acad Sci U S A 109 (35):14104–14109
27. Capshew CR, Dusenbury KL, Hundley HA (2012) Inverted Alu dsRNA structures do not affect localization but can alter translation efficiency of human mRNAs independent of RNA editing. Nucleic Acids Res 40 (17):8637–8645
28. Fitzpatrick T, Huang S (2012) 3-′-UTR-located inverted Alu repeats facilitate mRNA translational repression and stress granule accumulation. Nucleus 3(4):359–369
29. Liu WM, Chu WM, Choudary PV, Schmid CW (1995) Cell stress and translational inhibitors transiently increase the abundance of mammalian SINE transcripts. Nucleic Acids Res 23(10):1758–1765
30. Makalowski W, Mitchell GA, Labuda D (1994) Alu sequences in the coding regions of mRNA: a source of protein variability. Trends Genet 10(6):188–193
31. Gotea V, Makalowski W (2006) Do transposable elements really contribute to proteomes? Trends Genet 22(5):260–267
32. Wu M, Li L, Sun Z (2007) Transposable element fragments in protein-coding regions and their contributions to human functional proteins. Gene 401(1-2):165–171
33. Charng YC, Liu LD (2013) The extent of Ds1 transposon to enrich transcriptomes and proteomes by exonization. Bot Stud 54(1):14
34. Mandal AK, Pandey R, Jha V, Mukerji M (2013) Transcriptome-wide expansion of non-coding regulatory switches: evidence from co-occurrence of Alu exonization, antisense and editing. Nucleic Acids Res 41 (4):2121–2137
35. Hoen DR, Bureau TE (2015) Discovery of novel genes derived from transposable elements using integrative genomic analysis. Mol Biol Evol 32(6):1487–1506
36. Huda A, Bushel PR (2013) Widespread exonization of transposable elements in human coding sequences is associated with epigenetic regulation of transcription. Transcr Open Access 1(1)
37. Abascal F, Tress ML, Valencia A (2015) Alternative splicing and co-option of transposable elements: the case of TMPO/LAP2alpha and ZNF451 in mammals. Bioinformatics 31 (14):2257–2261
38. Lin L, Jiang P, Park JW, Wang J, Lu ZX, Lam MP, Ping P, Xing Y (2016) The contribution of Alu exons to the human proteome. Genome Biol 17:15
39. Huang S, Tao X, Yuan S, Zhang Y, Li P, Beilinson HA, Zhang Y, Yu W, Pontarotti P, Escriva H, Le Petillon Y, Liu X, Chen S, Schatz DG, Xu A (2016) Discovery of an active RAG transposon illuminates the origins of V(D)J recombination. Cell 166 (1):102–114
40. Shaheen M, Williamson E, Nickoloff J, Lee SH, Hromas R (2010) Metnase/SETMAR: a domesticated primate transposase that enhances DNA repair, replication, and decatenation. Genetica 138(5):559–566
41. Nordborg M, Walbot V (1995) Estimating allelic diversity generated by excision of different transposon types. Theor Appl Genet 90 (6):771–775

42. Moran JV, DeBerardinis RJ, Kazazian HH Jr (1999) Exon shuffling by L1 retrotransposition. Science 283(5407):1530–1534
43. Goodier JL, Kazazian HH Jr (2008) Retrotransposons revisited: the restraint and rehabilitation of parasites. Cell 135(1):23–35
44. Campbell IM, Gambin T, Dittwald P, Beck CR, Shuvarikov A, Hixson P, Patel A, Gambin A, Shaw CA, Rosenfeld JA, Stankiewicz P (2014) Human endogenous retroviral elements promote genome instability via non-allelic homologous recombination. BMC Biol 12:74
45. Startek M, Szafranski P, Gambin T, Campbell IM, Hixson P, Shaw CA, Stankiewicz P, Gambin A (2015) Genome-wide analyses of LINE-LINE-mediated nonallelic homologous recombination. Nucleic Acids Res 43 (4):2188–2198
46. Hancks DC, Kazazian HH Jr (2012) Active human retrotransposons: variation and disease. Curr Opin Genet Dev 22(3):191–203
47. Helman E, Lawrence MS, Stewart C, Sougnez C, Getz G, Meyerson M (2014) Somatic retrotransposition in human cancer revealed by whole-genome and exome sequencing. Genome Res 24(7):1053–1063
48. Evrony GD, Lee E, Park PJ, Walsh CA (2016) Resolving rates of mutation in the brain using single-neuron genomics. Elife 5:e12966
49. Payer LM, Steranka JP, Yang WR, Kryatova M, Medabalimi S, Ardeljan D, Liu C, Boeke JD, Avramopoulos D, Burns KH (2017) Structural variants caused by Alu insertions are associated with risks for many human diseases. Proc Natl Acad Sci U S A 114 (20):E3984–E3992
50. Chuong EB, Elde NC, Feschotte C (2016) Regulatory evolution of innate immunity through co-option of endogenous retroviruses. Science 351(6277):1083–1087
51. Gonzalez J, Lenkov K, Lipatov M, Macpherson JM, Petrov DA (2008) High rate of recent transposable element-induced adaptation in *Drosophila melanogaster*. PLoS Biol 6 (10):e251
52. Mateo L, Ullastres A, Gonzalez J (2014) A transposable element insertion confers xenobiotic resistance in Drosophila. PLoS Genet 10(8):e1004560
53. Guio L, Barron MG, Gonzalez J (2014) The transposable element Bari-Jheh mediates oxidative stress response in Drosophila. Mol Ecol 23(8):2020–2030
54. Ullastres A, Petit N, Gonzalez J (2015) Exploring the phenotypic space and the evolutionary history of a natural mutation in *Drosophila melanogaster*. Mol Biol Evol 32 (7):1800–1814
55. Merenciano M, Ullastres A, de Cara MA, Barron MG, Gonzalez J (2016) Multiple independent retroelement insertions in the promoter of a stress response gene have variable molecular and functional effects in drosophila. PLoS Genet 12(8):e1006249
56. Le Manh H, Guio L, Merenciano M, Rovira Q, Barron MG, Gonzalez J (2017) Natural and laboratory mutations in *kuzbanian* are associated with zinc stress phenotypes in *Drosophila melanogaster*. Sci Rep 7:42663
57. McCue AD, Nuthikattu S, Reeder SH, Slotkin RK (2012) Gene expression and stress response mediated by the epigenetic regulation of a transposable element small RNA. PLoS Genet 8(2):e1002474
58. Schrader L, Kim JW, Ence D, Zimin A, Klein A, Wyschetzki K, Weichselgartner T, Kemena C, Stokl J, Schultner E, Wurm Y, Smith CD, Yandell M, Heinze J, Gadau J, Oettler J (2014) Transposable element islands facilitate adaptation to novel environments in an invasive species. Nat Commun 5:5495
59. Shapiro JA (2017) Exploring the read-write genome: mobile DNA and mammalian adaptation. Crit Rev Biochem Mol Biol 52 (1):1–17
60. Magwire MM, Bayer F, Webster CL, Cao C, Jiggins FM (2011) Successive increases in the resistance of Drosophila to viral infection through a transposon insertion followed by a duplication. PLoS Genet 7(10):e1002337
61. Aminetzach YT, Macpherson JM, Petrov DA (2005) Pesticide resistance via transposition-mediated adaptive gene truncation in Drosophila. Science 309(5735):764–767
62. Schmidt JM, Good RT, Appleton B, Sherrard J, Raymant GC, Bogwitz MR, Martin J, Daborn PJ, Goddard ME, Batterham P, Robin C (2010) Copy number variation and transposable elements feature in recent, ongoing adaptation at the *Cyp6g1* locus. PLoS Genet 6(6):e1000998
63. Daborn PJ, Yen JL, Bogwitz MR, Le Goff G, Feil E, Jeffers S, Tijet N, Perry T, Heckel D, Batterham P, Feyereisen R, Wilson TG, ffrench-Constant RH (2002) A single p450 allele associated with insecticide resistance in Drosophila. Science 297(5590):2253–2256
64. Rishishwar L, Marino-Ramirez L, Jordan IK (2017) Benchmarking computational tools for polymorphic transposable element detection. Brief Bioinform 18:908

65. Rishishwar L, Wang L, Clayton EA, Marino-Ramirez L, McDonald JF, Jordan IK (2017) Population and clinical genetics of human transposable elements in the (post) genomic era. Mob Genet Elements 7(1):1–20
66. Goodier JL (2016) Restricting retrotransposons: a review. Mob DNA 7:16
67. Bergman CM, Quesneville H, Anxolabehere D, Ashburner M (2006) Recurrent insertion and duplication generate networks of transposable element sequences in the *Drosophila melanogaster* genome. Genome Biol 7(11):R112
68. Sessegolo C, Burlet N, Haudry A (2016) Strong phylogenetic inertia on genome size and transposable element content among 26 species of flies. Biol Lett 12(8):20160407
69. Quesneville H, Bergman CM, Andrieu O, Autard D, Nouaud D, Ashburner M, Anxolabehere D (2005) Combined evidence annotation of transposable elements in genome sequences. PLoS Comput Biol 1(2):166–175
70. Kaminker JS, Bergman CM, Kronmiller B, Carlson J, Svirskas R, Patel S, Frise E, Wheeler DA, Lewis SE, Rubin GM, Ashburner M, Celniker SE (2002) The transposable elements of the *Drosophila melanogaster* euchromatin: a genomics perspective. Genome Biol 3(12): RESEARCH0084
71. Kapitonov VV, Jurka J (2003) Molecular paleontology of transposable elements in the Drosophila melanogaster genome. Proc Natl Acad Sci U S A 100(11):6569–6574
72. Singh ND, Petrov DA (2004) Rapid sequence turnover at an intergenic locus in Drosophila. Mol Biol Evol 21(4):670–680
73. Yang HP, Barbash DA (2008) Abundant and species-specific DINE-1 transposable elements in 12 Drosophila genomes. Genome Biol 9(2):R39
74. Petrov DA, Fiston-Lavier AS, Lipatov M, Lenkov K, Gonzalez J (2011) Population genomics of transposable elements in *Drosophila melanogaster*. Mol Biol Evol 28 (5):1633–1644
75. Adrion JR, Song MJ, Schrider DR, Hahn MW, Schaack S (2017) Genome-wide estimates of transposable element insertion and deletion rates in *Drosophila melanogaster*. Genome Biol Evol 9(5):1329–1340
76. Kim A, Terzian C, Santamaria P, Pelisson A, Purd'homme N, Bucheton A (1994) Retroviruses in invertebrates: the *gypsy* retrotransposon is apparently an infectious retrovirus of *Drosophila melanogaster*. Proc Natl Acad Sci U S A 91(4):1285–1289
77. Leblanc P, Desset S, Giorgi F, Taddei AR, Fausto AM, Mazzini M, Dastugue B, Vaury C (2000) Life cycle of an endogenous retrovirus, ZAM, in *Drosophila melanogaster*. J Virol 74(22):10658–10669
78. Le Rouzic A, Payen T, Hua-Van A (2013) Reconstructing the evolutionary history of transposable elements. Genome Biol Evol 5 (1):77–86
79. Xue C, Goldenfeld N (2016) Stochastic predator-prey dynamics of transposons in the human genome. Phys Rev Lett 117 (20):208101
80. Startek M, Le Rouzic A, Capy P, Grzebelus D, Gambin A (2013) Genomic parasites or symbionts? Modeling the effects of environmental pressure on transposition activity in asexual populations. Theor Popul Biol 90:145–151
81. Gogolesky K, Startek A, Gambin A, Le Rouzic A (2016) Modelling the proliferation of transposable elements in populations under environmental stress. arXiv. arXiv:1611.04812
82. Boutin TS, Le Rouzic A, Capy P (2012) How does selfing affect the dynamics of selfish transposable elements? Mob DNA 3:5
83. Kijima TE, Innan H (2013) Population genetics and molecular evolution of DNA sequences in transposable elements. I. A simulation framework. Genetics 195(3):957–967
84. Moulin S, Seux N, Chretien S, Guyeux C, Lerat E (2017) Simulation-based estimation of branching models for LTR retrotransposons. Bioinformatics 33(3):320–326
85. Goubert C, Modolo L, Vieira C, ValienteMoro C, Mavingui P, Boulesteix M (2015) De novo assembly and annotation of the Asian tiger mosquito (*Aedes albopictus*) repeatome with dnaPipeTE from raw genomic reads and comparative analysis with the yellow fever mosquito (*Aedes aegypti*). Genome Biol Evol 7(4):1192–1205
86. Ewing AD (2015) Transposable element detection from whole genome sequence data. Mob DNA 6:24
87. Kofler R, Gomez-Sanchez D, Schlotterer C (2016) PoPoolationTE2: comparative population genomics of transposable elements using pool-seq. Mol Biol Evol 33 (10):2759–2764
88. Fiston-Lavier AS, Barron MG, Petrov DA, Gonzalez J (2015) T-lex2: genotyping, frequency estimation and re-annotation of transposable elements using single or pooled next-generation sequencing data. Nucleic Acids Res 43(4):e22

89. Rahman R, Chirn GW, Kanodia A, Sytnikova YA, Brembs B, Bergman CM, Lau NC (2015) Unique transposon landscapes are pervasive across Drosophila melanogaster genomes. Nucleic Acids Res 43(22):10655–10672
90. Treiber CD, Waddell S (2017) Resolving the prevalence of somatic transposition in Drosophila. Elife 6:e28297
91. Nelson MG, Linheiro RS, Bergman CM (2017) McClintock: an integrated pipeline for detecting transposable element insertions in whole genome shotgun sequencing data. G3 (Bethesda) 7:2763
92. McCarthy A (2010) Third generation DNA sequencing: Pacific Biosciences' single molecule real time technology. Chem Biol 17 (7):675–676
93. McCoy RC, Taylor RW, Blauwkamp TA, Kelley JL, Kertesz M, Pushkarev D, Petrov DA, Fiston-Lavier AS (2014) Illumina TruSeq synthetic long-reads empower de novo assembly and resolve complex, highly-repetitive transposable elements. PLoS One 9(9): e106689
94. Cridland JM, Macdonald SJ, Long AD, Thornton KR (2013) Abundance and distribution of transposable elements in two Drosophila QTL mapping resources. Mol Biol Evol 30(10):2311–2327
95. Sudmant PH, Rausch T, Gardner EJ, Handsaker RE, Abyzov A, Huddleston J, Zhang Y, Ye K, Jun G, Fritz MH, Konkel MK, Malhotra A, Stutz AM, Shi X, Casale FP, Chen J, Hormozdiari F, Dayama G, Chen K, Malig M, Chaisson MJP, Walter K, Meiers S, Kashin S, Garrison E, Auton A, Lam HYK, Mu XJ, Alkan C, Antaki D, Bae T, Cerveira E, Chines P, Chong Z, Clarke L, Dal E, Ding L, Emery S, Fan X, Gujral M, Kahveci F, Kidd JM, Kong Y, Lameijer EW, McCarthy S, Flicek P, Gibbs RA, Marth G, Mason CE, Menelaou A, Muzny DM, Nelson BJ, Noor A, Parrish NF, Pendleton M, Quitadamo A, Raeder B, Schadt EE, Romanovitch M, Schlattl A, Sebra R, Shabalin AA, Untergasser A, Walker JA, Wang M, Yu F, Zhang C, Zhang J, Zheng-Bradley X, Zhou W, Zichner T, Sebat J, Batzer MA, McCarroll SA, Genomes Project C, Mills RE, Gerstein MB, Bashir A, Stegle O, Devine SE, Lee C, Eichler EE, Korbel JO (2015) An integrated map of structural variation in 2,504 human genomes. Nature 526 (7571):75–81
96. Huddleston J, Chaisson MJP, Steinberg KM, Warren W, Hoekzema K, Gordon D, Graves-Lindsay TA, Munson KM, Kronenberg ZN, Vives L, Peluso P, Boitano M, Chin CS, Korlach J, Wilson RK, Eichler EE (2017) Discovery and genotyping of structural variation from long-read haploid genome sequence data. Genome Res 27(5):677–685
97. Pendleton M, Sebra R, Pang AW, Ummat A, Franzen O, Rausch T, Stutz AM, Stedman W, Anantharaman T, Hastie A, Dai H, Fritz MH, Cao H, Cohain A, Deikus G, Durrett RE, Blanchard SC, Altman R, Chin CS, Guo Y, Paxinos EE, Korbel JO, Darnell RB, McCombie WR, Kwok PY, Mason CE, Schadt EE, Bashir A (2015) Assembly and diploid architecture of an individual human genome via single-molecule technologies. Nat Methods 12(8):780–786
98. Disdero E, Filee J (2017) LoRTE: detecting transposon-induced genomic variants using low coverage PacBio long read sequences. Mob DNA 8:5
99. Kelleher ES, Edelman NB, Barbash DA (2012) Drosophila interspecific hybrids phenocopy piRNA-pathway mutants. PLoS Biol 10(11):e1001428
100. Vela D, Fontdevila A, Vieira C, Garcia Guerreiro MP (2014) A genome-wide survey of genetic instability by transposition in Drosophila hybrids. PLoS One 9(2):e88992
101. Romero-Soriano V, Modolo L, Lopez-Maestre H, Mugat B, Pessia E, Chambeyron S, Vieira C, Garcia Guerreiro MP (2017) Transposable element misregulation is linked to the divergence between parental piRNA pathways in drosophila hybrids. Genome Biol Evol 9(6):1450–1470
102. Robillard E, Le Rouzic A, Zhang Z, Capy P, Hua-Van A (2016) Experimental evolution reveals hyperparasitic interactions among transposable elements. Proc Natl Acad Sci U S A 113(51):14763–14768
103. Gonzalez J, Petrov DA (2012) Evolution of genome content: population dynamics of transposable elements in flies and humans. Methods Mol Biol 855:361–383
104. Cordaux R, Hedges DJ, Herke SW, Batzer MA (2006) Estimating the retrotransposition rate of human Alu elements. Gene 373:134–137
105. Ewing AD, Kazazian HH Jr (2010) High-throughput sequencing reveals extensive variation in human-specific L1 content in individual human genomes. Genome Res 20 (9):1262–1270
106. Huang CR, Schneider AM, Lu Y, Niranjan T, Shen P, Robinson MA, Steranka JP, Valle D, Civin CI, Wang T, Wheelan SJ, Ji H, Boeke JD, Burns KH (2010) Mobile interspersed

repeats are major structural variants in the human genome. Cell 141(7):1171–1182

107. Xing J, Zhang Y, Han K, Salem AH, Sen SK, Huff CD, Zhou Q, Kirkness EF, Levy S, Batzer MA, Jorde LB (2009) Mobile elements create structural variation: analysis of a complete human genome. Genome Res 19 (9):1516–1526
108. Ernst C, Odom DT, Kutter C (2017) The emergence of piRNAs against transposon invasion to preserve mammalian genome integrity. Nat Commun 8(1):1411
109. McCullers TJ, Steiniger M (2017) Transposable elements in Drosophila. Mob Genet Elements 7(3):1–18
110. Charlesworth B, Charlesworth D (1983) The population dynamics of transposable elements. Genet Res 42(1):1–27
111. Lohe AR, Hartl DL (1996) Autoregulation of mariner transposase activity by overproduction and dominant-negative complementation. Mol Biol Evol 13(4):549–555
112. Grabundzija I, Irgang M, Mates L, Belay E, Matrai J, Gogol-Doring A, Kawakami K, Chen W, Ruiz P, Chuah MK, VandenDriessche T, Izsvak Z, Ivics Z (2010) Comparative analysis of transposable element vector systems in human cells. Mol Ther 18 (6):1200–1209
113. Bire S, Casteret S, Arnaoty A, Piegu B, Lecomte T, Bigot Y (2013) Transposase concentration controls transposition activity: myth or reality? Gene 530(2):165–171
114. Woodard LE, Downes LM, Lee YC, Kaja A, Terefe ES, Wilson MH (2017) Temporal self-regulation of transposition through host-independent transposase rodlet formation. Nucleic Acids Res 45(1):353–366
115. Wheeler BS (2013) Small RNAs, big impact: small RNA pathways in transposon control and their effect on the host stress response. Chromosome Res 21(6-7):587–600
116. Clark JP, Lau NC (2014) Piwi proteins and piRNAs step onto the systems biology stage. Adv Exp Med Biol 825:159–197
117. Toth KF, Pezic D, Stuwe E, Webster A (2016) The piRNA pathway guards the germline genome against transposable elements. Adv Exp Med Biol 886:51–77
118. Yang F, Xi R (2017) Silencing transposable elements in the Drosophila germline. Cell Mol Life Sci 74(3):435–448
119. Luo S, Lu J (2017) Silencing of transposable elements by piRNAs in drosophila: an evolutionary perspective. Genomics Proteomics Bioinformatics 15(3):164–176
120. Aravin AA, Sachidanandam R, Bourc'his D, Schaefer C, Pezic D, Toth KF, Bestor T, Hannon GJ (2008) A piRNA pathway primed by individual transposons is linked to de novo DNA methylation in mice. Mol Cell 31 (6):785–799
121. Le Thomas A, Rogers AK, Webster A, Marinov GK, Liao SE, Perkins EM, Hur JK, Aravin AA, Toth KF (2013) Piwi induces piRNA-guided transcriptional silencing and establishment of a repressive chromatin state. Genes Dev 27(4):390–399
122. Hamdorf M, Idica A, Zisoulis DG, Gamelin L, Martin C, Sanders KJ, Pedersen IM (2015) miR-128 represses L1 retrotransposition by binding directly to L1 RNA. Nat Struct Mol Biol 22(10):824–831
123. Ariumi Y (2016) Guardian of the human genome: host defense mechanisms against LINE-1 retrotransposition. Front Chem 4:28
124. Wylie A, Jones AE, D'Brot A, Lu WJ, Kurtz P, Moran JV, Rakheja D, Chen KS, Hammer RE, Comerford SA, Amatruda JF, Abrams JM (2016) p53 genes function to restrain mobile elements. Genes Dev 30(1):64–77
125. Van Meter M, Kashyap M, Rezazadeh S, Geneva AJ, Morello TD, Seluanov A, Gorbunova V (2014) SIRT6 represses LINE1 retrotransposons by ribosylating KAP1 but this repression fails with stress and age. Nat Commun 5:5011
126. Karasov T, Messer PW, Petrov DA (2010) Evidence that adaptation in Drosophila is not limited by mutation at single sites. PLoS Genet 6(6):e1000924
127. Park L (2011) Effective population size of current human population. Genet Res (Camb) 93(2):105–114
128. Barron MG, Fiston-Lavier AS, Petrov DA, Gonzalez J (2014) Population genomics of transposable elements in Drosophila. Annu Rev Genet 48:561–581
129. Kofler R, Nolte V, Schlotterer C (2015) Tempo and mode of transposable element activity in drosophila. PLoS Genet 11(7): e1005406
130. Rishishwar L, Tellez Villa CE, Jordan IK (2015) Transposable element polymorphisms recapitulate human evolution. Mob DNA 6:21
131. Boissinot S, Davis J, Entezam A, Petrov D, Furano AV (2006) Fitness cost of LINE-1 (L1) activity in humans. Proc Natl Acad Sci U S A 103(25):9590–9594
132. Sundaram V, Choudhary MN, Pehrsson E, Xing X, Fiore C, Pandey M, Maricque B,

Udawatta M, Ngo D, Chen Y, Paguntalan A, Ray T, Hughes A, Cohen BA, Wang T (2017) Functional cis-regulatory modules encoded by mouse-specific endogenous retrovirus. Nat Commun 8:14550

133. Chalopin D, Naville M, Plard F, Galiana D, Volff JN (2015) Comparative analysis of transposable elements highlights mobilome diversity and evolution in vertebrates. Genome Biol Evol 7(2):567–580
134. Kapusta A, Suh A, Feschotte C (2017) Dynamics of genome size evolution in birds and mammals. Proc Natl Acad Sci U S A 114 (8):E1460–E1469
135. Leushkin EV, Bazykin GA, Kondrashov AS (2013) Strong mutational bias toward deletions in the Drosophila melanogaster genome is compensated by selection. Genome Biol Evol 5(3):514–524
136. Petrov DA, Lozovskaya ER, Hartl DL (1996) High intrinsic rate of DNA loss in Drosophila. Nature 384(6607):346–349
137. Petrov DA, Hartl DL (1998) High rate of DNA loss in the Drosophila melanogaster and *Drosophila virilis* species groups. Mol Biol Evol 15(3):293–302
138. Loreto EL, Carareto CM, Capy P (2008) Revisiting horizontal transfer of transposable elements in Drosophila. Heredity (Edinb) 100(6):545–554
139. Schaack S, Gilbert C, Feschotte C (2010) Promiscuous DNA: horizontal transfer of transposable elements and why it matters for eukaryotic evolution. Trends Ecol Evol 25 (9):537–546
140. Peccoud J, Loiseau V, Cordaux R, Gilbert C (2017) Massive horizontal transfer of transposable elements in insects. Proc Natl Acad Sci U S A 114(18):4721–4726
141. Peccoud J, Cordaux R, Gilbert C (2018) Analyzing horizontal transfer of transposable elements on a large scale: challenges and prospects. Bioessays 40(2)
142. Groth SB, Blumenstiel JP (2017) Horizontal transfer can drive a greater transposable element load in large populations. J Hered 108 (1):36–44
143. Mouse Genome Sequencing Consortium (2002) Initial sequencing and comparative analysis of the mouse genome. Nature 420 (6915):520–562
144. Kim JM, Vanguri S, Boeke JD, Gabriel A, Voytas DF (1998) Transposable elements and genome organization: a comprehensive survey of retrotransposons revealed by the complete *Saccharomyces cerevisiae* genome sequence. Genome Res 8(5):464–478
145. Arabidopsis Genome Initiative (2000) Analysis of the genome sequence of the flowering plant *Arabidopsis thaliana*. Nature 408 (6814):796–815
146. Filee J, Siguier P, Chandler M (2007) Insertion sequence diversity in archaea. Microbiol Mol Biol Rev 71(1):121–157
147. Sebaihia M, Peck MW, Minton NP, Thomson NR, Holden MT, Mitchell WJ, Carter AT, Bentley SD, Mason DR, Crossman L, Paul CJ, Ivens A, Wells-Bennik MH, Davis IJ, Cerdeno-Tarraga AM, Churcher C, Quail MA, Chillingworth T, Feltwell T, Fraser A, Goodhead I, Hance Z, Jagels K, Larke N, Maddison M, Moule S, Mungall K, Norbertczak H, Rabbinowitsch E, Sanders M, Simmonds M, White B, Whithead S, Parkhill J (2007) Genome sequence of a proteolytic (Group I) *Clostridium botulinum* strain Hall A and comparative analysis of the clostridial genomes. Genome Res 17(7):1082–1092
148. Rhee JS, Choi BS, Kim J, Kim BM, Lee YM, Kim IC, Kanamori A, Choi IY, Schartl M, Lee JS (2017) Diversity, distribution, and significance of transposable elements in the genome of the only selfing hermaphroditic vertebrate *Kryptolebias marmoratus*. Sci Rep 7:40121
149. Osanai-Futahashi M, Suetsugu Y, Mita K, Fujiwara H (2008) Genome-wide screening and characterization of transposable elements and their distribution analysis in the silkworm, *Bombyx mori*. Insect Biochem Mol Biol 38(12):1046–1057
150. Hernandez-Hernandez EM, Fernandez-Medina RD, Navarro-Escalante L, Nunez J, Benavides-Machado P, Carareto CMA (2017) Genome-wide analysis of transposable elements in the coffee berry borer *Hypothenemus hampei* (Coleoptera: Curculionidae): description of novel families. Mol Genet Genomics 292(3):565–583
151. Castanera R, Borgognone A, Pisabarro AG, Ramirez L (2017) Biology, dynamics, and applications of transposable elements in basidiomycete fungi. Appl Microbiol Biotechnol 101(4):1337–1350
152. Tenaillon MI, Hufford MB, Gaut BS, Ross-Ibarra J (2011) Genome size and transposable element content as determined by high-throughput sequencing in maize and *Zea luxurians*. Genome Biol Evol 3:219–229

153. Boissinot S, Sookdeo A (2016) The evolution of LINE-1 in vertebrates. Genome Biol Evol 8(12):3485–3507
154. Quadrana L, Bortolini Silveira A, Mayhew GF, LeBlanc C, Martienssen RA, Jeddeloh JA, Colot V (2016) The *Arabidopsis thaliana* mobilome and its impact at the species level. Elife 5:e15716
155. Stuart T, Eichten SR, Cahn J, Karpievitch YV, Borevitz JO, Lister R (2016) Population scale mapping of transposable element diversity reveals links to gene regulation and epigenomic variation. Elife 5:e20777
156. Yan L, Gu YH, Tao X, Lai XJ, Zhang YZ, Tan XM, Wang H (2014) Scanning of transposable elements and analyzing expression of transposase genes of sweet potato *Ipomoea batatas*. PLoS One 9(3):e90895

Part V

Population Genomics and Omics in Light of Disease and Evolution

Chapter 17

Association Mapping and Disease: Evolutionary Perspectives

Søren Besenbacher, Thomas Mailund, Bjarni J. Vilhjálmsson, and Mikkel H. Schierup

Abstract

In this chapter, we give a short introduction to the genetics of complex diseases emphasizing evolutionary models for disease genes and the effect of different models on the genetic architecture, and we give a survey of the state-of-the-art of genome-wide association studies (GWASs).

Key words Complex diseases, Association mapping, Genome-wide association studies, Common disease/common variant

1 Introduction

A combination of genes and environment determines our phenotype. The degree to which genotype or environment influences our phenotype—the balance of nature versus nurture—varies from trait to trait, with some traits independent of genotype and determined by the environment alone and others determined by the genotype alone and independent of the environment.

A measure quantifying the importance of genotype compared to the environment is the so-called heritability. It is the fraction of the total phenotypic variation in the population explained by variation in the genotype within the population [1]. A trait of interest, say a common disease, which exhibits a nontrivial heritability, tells us that genes are important for understanding this trait and that it is worthwhile to identify the specific genetic polymorphisms influencing the trait. The first step toward this is *association mapping*: searching for genetic polymorphisms that, statistically, associate with the trait. Polymorphisms associated with a given phenotype need not influence that phenotype directly, but it is among those associated genetic polymorphisms that we will find the causal ones.

Maria Anisimova (ed.), *Evolutionary Genomics: Statistical and Computational Methods*, Methods in Molecular Biology, vol. 1910, https://doi.org/10.1007/978-1-4939-9074-0_17, © The Author(s) 2019

Genetic variants are correlated, a phenomenon called *linkage disequilibrium* (LD), so by examining the trait association of a few variants, we learn about the association of many others. Examining the association between a phenotypic trait and a few hundred thousand to a million genetic variants suffices to capture how most of the common variation in the entire genome associates with the trait [2–4]. When we find a genetic variant associated with the trait, we have not necessarily located a variant that has any functional effect on the trait, but we have located a genomic region containing genetic variation that does. LD is predominantly a local phenomenon, so correlated genetic variants tend to be physically near each other on the genome. If we observe an association between the phenotype and a variant, and the variant is not causally affecting the trait but is merely in LD with a causal variant, the causal variant is likely nearby. Further examination of the region might reveal which variants affect the trait, and how, but that often involves functional characterization and is beyond association mapping. With association mapping, we merely seek to identify genetic variation that associates with a trait.

2 The Allelic Architecture of Genetic Determinants for Disease

Many complex diseases show a high heritability, typically ranging between 20% and 80%. Each genetic variant that increases the risk of disease contributes to the measured heritability of the disease and thus explains some fraction of the estimated total heritability of the trait. For most diseases investigated, many variants contribute, and the fraction of the heritability explained for each is therefore low. The number of contributing variants, their individual effects on the disease probability, their selection coefficient, and their dominance relations can be collectively termed the genetic architecture of a common disease. Insights into this architecture are slowly emerging and reveal differences between diseases [5].

Below we first consider two proposed genetic architectures based on theoretical arguments: the common disease common variant (CDCV) architecture and the common disease rare variant (CDRV) architecture. CDCV states that most of the heritability can be explained by a few high-frequency variants with moderate effects, while CDRV states that most of the heritability can be explained by moderate- or low-frequency variants with large effects. We present population genetic arguments for the two architectures and the consequences of the two architectures for association mapping. Later, in Subheading 5.1, we present empirical knowledge we have obtained about the genetic architectures of common diseases.

2.1 Theoretical Models for the Allelic Architecture of Common Diseases

Understanding the distribution of the number and frequency of genetic variants in a population is the purview of population genetics. Using diffusion approximations we can derive the expected frequency distribution of independent mutations under mutation-drift-selection balance in a stable population (*see*, e.g., Wright [6]). Central parameters are the mutation rate, u, and the selection for or against an allele, measured by s, scaled with the effective population size, N. Mutations enter a population with a rate determined by Nu, and subsequently, their frequencies change in a stochastic manner. If a mutant allele is not subject to natural selection, for example, if it does not lead to any change in function, it is selectively neutral. Its frequency then rises and falls with equal probability. If the allele is under selection, it has a higher likelihood of increasing in frequency than decreasing if it is under positive selection ($s > 0$) and conversely for negative selection ($s < 0$).

At very high or very low frequencies, selection has an insignificant effect on the change in frequency, and the system evolves essentially entirely stochastic (genetic drift). At moderate frequencies, however, the effect of selection is more pronounced, and given sufficiently strong selection (of an order $Ns \gg 1$), the direction of changes in the allele frequency is almost deterministically determined by the direction of selection. An allele subject to a sufficiently strong selection that happens to reach moderate frequencies either halts its increase if selection works against it, and drifts back to a low frequency, or if selection favors it, it rapidly rises to high frequencies, where eventually the stochastic effects again dominate (*see* Fig. 1).

The range of frequencies, where drift dominates, or selection dominates, is determined by the strength of selection (Ns) and the genotypic characteristics of selection, as, e.g., dominance relations between alleles. For strong selection or in large populations, the process is predominantly deterministic for most frequencies, while for weak selection or a small population, the process is highly stochastic for most frequencies. The time an allele can spend at moderate frequencies is also determined by Ns and selection characteristics.

Pritchard and Cox [7, 8] used diffusion arguments to show that common diseases are expected to be caused by a large number of distinct mutations. This implies that genes commonly involved in susceptibility exert their effect through multiple independent mutations rather than a single mutation identical by descent in all carriers (*see* Fig. 2). Each mutation, if under weak purifying selection, is unlikely to reach moderate frequencies, and since the population will only have few carriers of each disease allele, each can only explain little of the heritability. The accumulated frequency of several alleles, each kept to low frequency by selection, can, however, reach moderate frequencies. So the heritability can be

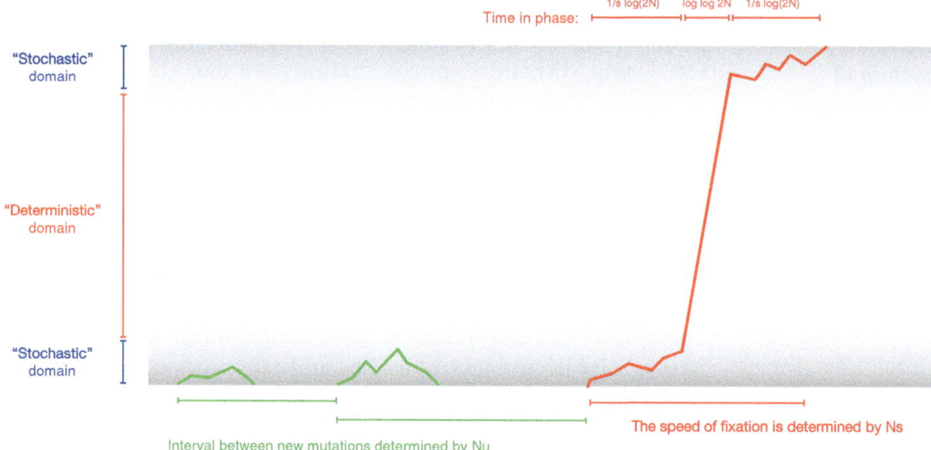

Fig. 1 Mutation, drift, and selection. New mutations enter a population at stochastic intervals, determined by the mutation rate, *u*, and the effective population size, *N*. For low or high frequencies, where the range of such frequencies is determined by the selection factor, *s*, and the effective population size, the frequency of a mutant allele changes stochastically. At medium frequencies, on the other hand, the frequency of the allele changes up or down, depending on *s*, in a practically deterministic fashion. If a positively selected allele reaches moderate frequency, it will quickly be brought to high frequency, at a speed also determined by *s* and *N*

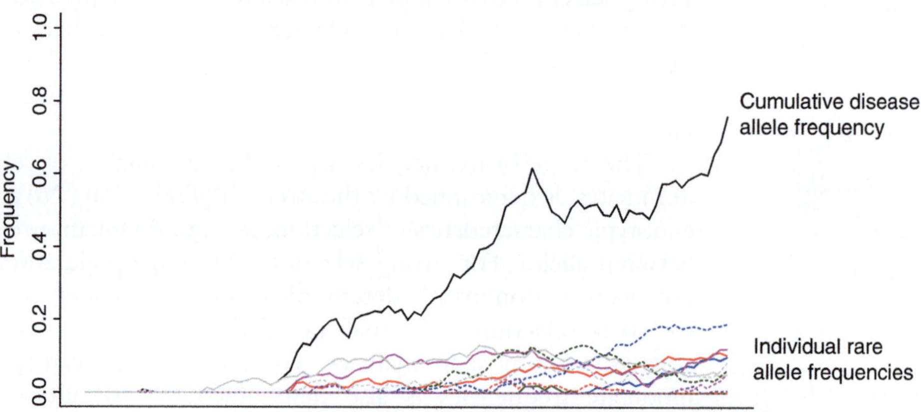

Fig. 2 Accumulation of several rare frequencies. If selection works against a set of alleles, each will be kept at a low frequency. Their accumulated frequency, however, can be high in the population

explained either by many recurrent mutations or many independent loci affecting the disease: the CDRV architecture.

Implicitly, this model assumes a population in mutation-selection equilibrium, and this does not necessarily match human populations. Humans have recently expanded considerably in numbers, and changes in our lifestyle, e.g., from hunter-gatherers to farmers might have changed the adaptive landscape driving selection of our genes.

The frequency range where drift, rather than deterministic selection, dominates is larger with a smaller population than with a larger population. We can think of the drift process as a birth-–death process operating on individual copies of genes, which is highly stochastic. Only when we consider a large number of these processes do we get an almost deterministic process. At low allele frequencies, the process is stochastic because we only have a few copies of the allele to consider. At higher frequencies, we have many copies, so we get the deterministic behavior. The same number of copies, however, constitutes a higher frequency of a small population than of a larger population. Consequently, selection is effective at much lower frequencies in a large population than it is in a small population; the absolute number of copies of a deleterious allele might be the same in a small and a large population, but they constitute a smaller fraction of the large population. In large populations, we expect to see deleterious mutations to be found at small frequencies unless, as is the case for most human populations, the large population size is a consequence of recent dramatic growth [9]. This effect is illustrated as the "transient period" in Fig. 3, where common genetic variants may contribute much more to disease than under stable demographic conditions. Following expansion, alleles that would otherwise be held at low frequency by selection may be at moderate frequencies and thus contribute a larger part of the heritability: the CDCV architecture.

Similarly, a recent change in the adaptive landscape of a population might cause an allele that was previously held at low frequency to be under positive selection and now rise in frequency [10]. In this transition period, an allele may be at a moderate frequency and therefore contributes significantly to the heritability of disease susceptibility (*see* Fig. 4).

Depending on which architecture underlies a given disease, different strategies are needed to discover the genetic variants

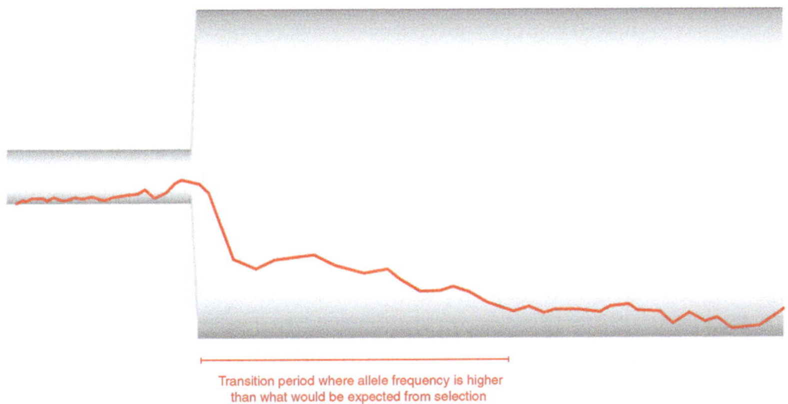

Fig. 3 A population out of equilibrium following an expansion. In a transition period following a population expansion, the allele frequency patterns are different from the patterns in a stable population

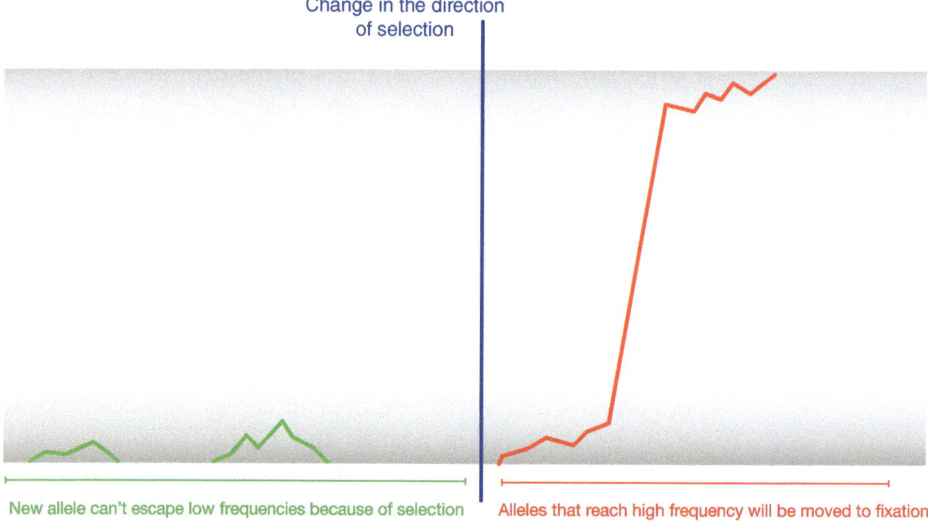

Fig. 4 A population out of equilibrium following changes in the selective landscape. If the selection of an allele changes direction, so the positively selected allele becomes negatively selected and vice versa, it will eventually move through moderate frequencies. Following a change in the selective landscape, it is thus possible to find alleles at moderate frequencies that would not otherwise be found

involved. When genome-wide association mapping was proposed as a strategy for discovering disease variants, the proposal was based on the hypothesis that, at least for some common diseases, the CDCV architecture underlies them. GWAS relies on the CDCV hypothesis for two practical reasons. The first is that the LD patterns across the genome greatly restrict examination to only a small fraction of the total possible variation. It is feasible to probe the common variants of a genome from a small selection of representative variants, but the association with rare variants is far less detectable. Second, statistical analysis of the association between polymorphism and disease is rather straightforward for moderate-frequency alleles but has far less power to detect association with low-frequency alleles.

While the GWAS approach is only practical as an approach for variant discovery for common alleles, it was necessary to hypothesize that the CDCV architecture would be underlying diseases of interest. The actual genetic architecture behind common diseases was unknown, but there were no alternative methods aimed at CDRV, so GWAS was the only show in town.

2.2 The Allelic Frequency Spectrum in Humans

The vast majority of human nucleotide variation is very rare because of our history of population bottlenecks followed by rapid growth. For instance, in the 2500 individuals of the 1000 genomes study, 64 million SNVs have frequency $<0.5\%$, and 20 million SNVs have frequency $>0.5\%$ [11]. Nevertheless the majority of heterozygous variants observed within a single individual are not rare [11]. The rare variants are most often very recent and therefore specific to populations, and they are also more often deleterious because selection has not yet acted on them [12]. This is particularly clear for loss-of-function variants and other protein-coding variants. A study of 2636 Icelanders found that the fraction of variants with a minor allele frequency (MAF) below 0.1% was 62% for protein-truncating variants, 46% for missense variants, and 38% for synonymous variants [13].

The strong recent population expansions have also allowed variants to increase in frequency by surfing on the population expansion wave front even if they would be selected against in a population with stable size. Thus, rare variants with large effects on disease may exist. The GWAS studies so far have been successful in identifying a large set of common variants associated with disease, so common variants contributing to disease do exist. It is likely that rare variants with large phenotypic effects also contribute to the heritability of many common diseases, but the extend is likely to be disease specific.

3 The Basic GWAS

The first GWASs were published around 2006 [14, 15] when Illumina and Affymetrix first introduced genotyping chips that made it possible to test hundreds of thousands of SNPs quickly and inexpensively. The GWASs' approach to find susceptibility variants for diseases boils down to testing approximately 0.3–-2 million SNPs (depending on chip type) for differences in allele frequencies between cases and controls, adjusting for the high number of multiple tests. This approach is a wonderfully simple procedure that requires no complicated statistics or algorithms but only well-known statistical tests and a minimum of computing power. Despite the simplicity, some issues remain, such as faulty genotype data and confounding factors that can result in erroneous findings if not handled properly. The most important aspects of any GWAS are, therefore, thorough quality control of the data used and measures to avoid and reduce the effect of confounding factors.

3.1 Statistical Tests

The primary analysis in an association study is usually testing each variant separately under the assumption of an additive or multiplicative model. One way of doing that is by creating a 2×2 allelic contingency table as shown in Table 1 by summing the number of A and B alleles seen in all case individuals and all control individuals. Be aware that we are counting alleles and not individuals in this contingency table, so N_{cases} will be equal to two times the number of case individuals because each individual carries two copies of each variant unless we are looking at non-autosomal DNA. If there is no association between the variant and the disease in question, we

Table 1
Contingency table for allele counts in case/control data

	Allele A	Allele B	
Case	$N_{\text{case,A}}$	$N_{\text{case,B}}$	N_{cases}
Control	$N_{\text{control,A}}$	$N_{\text{control,B}}$	N_{controls}
	N_{A}	N_{B}	N

Table 2
Expected allele counts in case/control data

	Allele A	Allele B	
Case	$(N_{\text{cases}} \cdot N_{\text{A}})/N$	$(N_{\text{cases}} \cdot N_{\text{B}})/N$	N_{cases}
Control	$(N_{\text{controls}} \cdot N_{\text{A}})/N$	$(N_{\text{controls}} \cdot N_{\text{B}})/N$	N_{controls}
	N_{A}	N_{B}	N

would expect the fraction of cases that have a particular allele to match the fraction of controls that have that allele. In that case, the expected allele count (EN) would be as shown in Table 2. To test whether the difference between the observed allele counts (in Table 1) and the expected allele counts (in Table 2) is significant, a Pearson χ^2 statistic can be calculated:

$$X^2 = \Sigma_{\text{Phenotype}}\Sigma_{\text{Allele}}\left(N_{\text{Phenotype, Allele}} - \text{EN}_{\text{Phenotype, Allele}}\right)^2/\text{EN}_{\text{Phenotype, Allele}}$$

This statistic approximates a χ^2 distribution with 1 degree of freedom, but if the expected allele counts are very low (<10), the approximation breaks down. This means that if the MAF is very low or if the total sample size, N, is small, an exact test, such as the Fisher's exact test, should be applied. An alternative to the tests that use the 2×2 allelic contingency table and thereby assumes a multiplicative model is the Cochran–Armitage trend test that assumes an additive risk model [16]. This test is preferred by some since it does not require an assumption of Hardy–Weinberg equilibrium in cases and controls combined [17].

While a 1 degree of freedom test that assumes an additive or multiplicative model is usually the first analysis, some studies also perform a test that would be better at picking up associations following a dominant or recessive pattern, for instance, by performing a 2 degrees of freedom test of the null hypothesis of no association between rows and columns in the 2×3 contingency table that counts genotypes instead of alleles.

3.2 Effect Estimates

A commonly used way of measuring the effect size of an association is the allelic odds ratio (OR), which is the ratio of the odds of being a case given that you carry n copies of alleles A to the odds of being a case if you carry $n-1$ copies of allele A. Assuming a multiplicative model, this can be calculated as:

$$\begin{aligned}\mathrm{OR} &= \left(N_{\mathrm{case,A}}/N_{\mathrm{control,A}}\right)/\left(N_{\mathrm{case,B}}/N_{\mathrm{control,B}}\right)\\ &= N_{\mathrm{case,A}}N_{\mathrm{control,B}}/N_{\mathrm{case,B}}N_{\mathrm{control,A}}\end{aligned}$$

Another measure of effect size that is perhaps more intuitive is the relative risk (RR), which is the disease risk in carriers divided by the disease risk in noncarriers. This measure, however, suffers from the weakness that it is harder to estimate. If our cases and controls were sampled from the population in an unbiased way, the allelic RR could be calculated as:

$$\mathrm{RR} = \left(N_{\mathrm{case,A}}/N_A\right)/\left(N_{\mathrm{case,B}}/N_B\right)$$

but it is very rare to have an unbiased population sample in association studies because the studies are generally designed to deliberately oversample the cases to increase the power. This oversampling affects the RR as calculated by the formula above but not the OR which is one of the reasons why the OR is usually reported in association studies instead of the RR.

3.3 Quality Control

Data quality problems can be either variant specific or individual specific, and inspection usually results in the removal of both problematic individuals and problematic variants from the data set.

Individual-specific problems can be caused by low DNA quality or contamination by foreign DNA. A sample of low DNA quality results in a high rate of missing data, where particular variants cannot be called, and there is a higher risk of miscalling variants. It is, therefore, recommended that individuals lacking calls in more than 2–3% of the variants are removed from the analysis. Excess heterozygosity is an indicator of sample contamination, and individuals displaying that should also be disregarded. Sex checks and other kinds of phenotype tests might also be applied to remove individuals, where the genotype information does not match the phenotype information due to a sample mix-up [18].

For a given variant, the data from an individual can be suspicious in two ways: it can fail to be called by the genotype-calling program or it can be miscalled. Typically, a conservative cutoff value is used in the calling process securing that most problems show up as missing data rather than miscalls. Most problematic variants, therefore, reveal a high fraction of missing data, and variants missing calls above a given threshold (typically, 1–5%) are removed. Miscalls typically occur when the homozygotes are hard to distinguish from the heterozygotes, and some of the heterozygotes are being misclassified as homozygotes or vice versa. Both biases manifest as deviation from Hardy–Weinberg equilibrium, and SNPs that show large deviations from Hardy–Weinberg equilibrium within the controls should be removed [19].

3.4 Confounding Factors

Confounding in GWAS can arise if there are genotyping batch effects or if there is population or family structure in the sample. For example, if cases and controls in GWAS are predominantly collected from geographically distinct areas, association signals could arise due to genetic differences caused by geographic variation, and most of such genetic signals are unlikely to be causal. Such confounding due to population structure typically occurs when samples have different genetic ancestry, e.g., if the sample contains individuals of both European and Asian ancestry. Population structure confounding can also happen when the population structure is more subtle, especially for large sample sizes. Methods for inferring population substructure, such as principal components analysis, are useful for detecting outliers we can remove from the data [20]. However, this approach is not suitable when dealing with subtle structure, as a small bias can become significant in a large enough sample of individuals of similar genetic ancestry.

Confounding in GWAS can be detected as inflation of the test statistics, beyond what is expected due to truly causal variants. A useful way of visualizing such inflation of test statistics is the so-called quantile–quantile (QQ) plot. In this plot, ranked values of the test statistic are plotted against their expected distribution under the null hypothesis. In the case of no true positives and no inflation of the test statistic due to population structure or cryptic relatedness, the points of the plot lie on the $x = y$ line (*see* Fig. 5a). True positives show an increase in values above the line in the right tail of the distribution but do not affect the rest of the points since

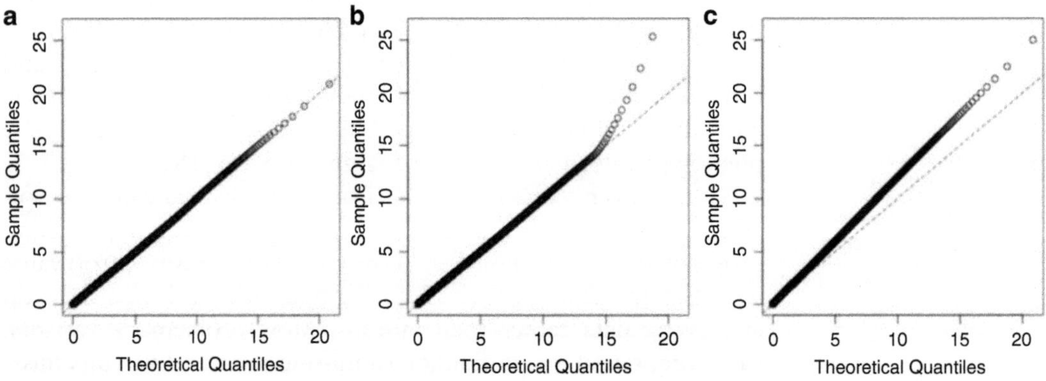

Fig. 5 QQ plots from a χ^2 distribution. (**a**) A QQ plot, where the observation follows the expected distribution. (**b**) A QQ plot, where the majority of observations follow the expected distribution, but where some have unexpectedly high values, i.e., are statistically significant. (**c**) A QQ plot, where the observations all seem to be higher than expected, which is an indication that the observations are not following the expected distribution

only a small fraction of the SNPs is expected to be true positives (Fig. 5b). Cryptic relatedness and population stratification lead to a deviation from the null distribution across the whole distribution and can, thus, be seen in the QQ plot as a line with a slope larger than 1 (Fig. 5c).

Several approaches accounting for population structure in GWAS have been proposed. Devlin and Roeder [21, 22] proposed *genomic control*, i.e., to shrink the observed χ^2 test statistic to make the median coincide with the expected value under the null model. However, studies by Yang et al. [23] and Bulik-Sullivan et al. [24] pointed out that the median and mean χ^2 statistic is expected to be inflated for polygenic traits, even when there is no population structure confounding. With that in mind, we recommend adjusting for the confounders in the statistical model instead of performing genomic control. One such approach is to include covariates that capture the relevant structure in the model. Price et al. [25] proposed including the largest principal components as covariates in the model to adjust for population structure. This approach has proved to be effective in most cases. However, if the sample includes related individuals or if it is very large, controlling for the top PCs may not be able to capture subtle structure. An alternative approach is to use mixed models [26, 27], where the expected genetic relatedness between the individuals is included in the model. Advances in computational efficiency of mixed models [28] now enable analysis of very large and complex data sets, such as the UK biobank data set [29].

Besides population structure, family structure or cryptic relatedness can also confound the analyses. Here one can identify closely related individuals by calculating a genetic relatedness matrix and prune the data so that it does not contain any close relatives. Lastly, sequencing batch effects due to incomplete randomizations can lead to structure, unrelated to genetics, which confounds the analysis. A study on polygenic prediction of longevity by Sebastiani et al. [30] serves as a warning. The researchers applied two different kinds of chips and failed to remove several SNPs that exhibited bad quality on only one of the chips [31]. If the fraction of the two different kinds of chips had been the same in both cases and controls that would probably not have resulted in false signals, unfortunately, the chip with the bad SNPs was used in twice as many cases as controls. When this genotyping batch effect was discovered, the authors had to retract their publication from Science. Type and frequency of errors that may happen during sample preparation and SNP calling are likely to vary through time and space, so case and control samples should be completely randomized as early as possible in the procedure of genotypic typing. Failure to carefully plan this aspect of an investigation introduces errors in the data that are hard, if not impossible, to disclose, and they may reduce interesting findings to mere artifacts.

3.5 Meta-analysis of GWAS

The statistical power to detect association depends directly on the sample size used, all other things being equal. This fact has driven researchers to collaborate across institutions and countries in GWAS consortia, where they combine multiple cohorts in one large analysis. However, for logistic and legal reasons, it may not be possible to share individual-level genotypes, which are required for all of the GWAS approaches covered so far. Meta-analyses of GWASs performed in each cohort are a solution to this problem. These require coordination between the researchers, where they share GWAS summary statistics instead of individual-level genotypes. These summary statistics are then meta-analyzed using statistical approaches that either assume a constant effect across cohorts or not. In recent years many large-scale GWAS meta-analyses have been published, and the resulting summary statistics of these are often made public, providing a treasure trove for understanding genetics of common diseases and traits [32].

3.6 Replication

The best way to make sure that a finding is real is to replicate it. If the same signal is found in an independent set of cases and controls, it means that the association is unlikely to be the result of a confounding factor specific to the original data. Likewise, if the association persists after typing the markers using another genotyping method, it means that it is not a false positive due to some artifact of the genotyping method used.

When trying to replicate a finding, the best strategy is to try to replicate it in a population of similar ancestry. A marker that correlates with a true causal variant in one population might not be correlated with the same variant in a population of different ethnicity, where the LD structure can be different. This is especially problematic when trying to replicate an association found in a non-African population in an African population [33]. A marker might easily have 20 completely correlated markers in a European population, but no good correlates in an African population. To replicate a finding in the European population of one of these variants, it does not suffice to test one of the variants in an African population; all 20 variants must be tested. This, however, also offers a way to fine map the signal and possibly find the causative variant [34].

Before spending time and effort to replicate an association signal in a foreign cohort, it is a good idea to search for the existing partial replication of the marker within the data. Usually, a marker is surrounded by several correlated markers on the genotyping chip, and if one marker shows a significant association, then the correlated markers should show an association too. If a marker is significantly associated with a disease, but no other marker in the region is, then it should be viewed as suspicious.

4 Imputation: Squeezing More Information Out of Your Data

The current generation of SNP chip types includes only 0.3–-2 million of the nine to ten million common SNPs in the human (i.e., SNPs with a MAF of more than 5%). Because of the correlation between SNPs in LD, however, the SNP chips can still claim to assay most of the common variants in the genome (in European populations at least). Although the Illumina HumanHap300 chip only directly tests about 3% of the ten million common SNPs, it still covers 77% of the SNPs in HapMap with a squared correlation coefficient (r^2) of at least 0.8 in a population of European ancestry [35]. The corresponding fraction in a population of African ancestry is only 33%, however.

These numbers expose two limitations of the basic GWAS strategy. First, there is a substantial fraction of the common SNPs that are not well covered by the SNP chips even in European populations (23% in the case of the HumanHap300 chip). Second, we rely on tagging to test a large fraction of the common SNPs, and this diluted signal from correlated SNPs inevitably causes us to overlook true associations in many instances. An efficient way of alleviating these limitations is genotype imputation, where genotypes that are not directly assayed are predicted using information from a reference data set that contains data from a large number of variants. Such imputation improves the GWAS in multiple ways: It boosts the power to detect associations, gives a more precise location of an association, and makes it possible to do meta-analyses between studies that used different SNP chips [36].

4.1 Selection of Reference Data Set

The two important choices when performing imputation are the reference data set to use and the software to use. Usually, a publicly available reference data set, such as the 1000 Genomes Project [11] or the large Haplotype Reference Consortium [37], is used. Alternatively, researchers sequence a part of their study cohort and thus create their own reference data set. The latter strategy has the advantage that one can be certain that the ancestry of the reference data matches the ancestry of the study cohort. It is important that the reference data be from a population that is similar to the study population. If the reference population is too distantly related to the study population, the reliability of the imputed data will be reduced. The quality and nature of the reference data also limit the quality of the imputed data in other ways. A reference data set consisting of only a small number of individuals is not able to reliably estimate the frequency of rare variants and that in turn means that the imputation of rare variants lacks in accuracy. This means that there is a natural limit to how low a frequency a variant can have and still be reliably imputed.

The largest publicly available reference data set is the Haplotype Reference Consortium (HRC) that combines whole-genome sequence data from 20 studies of predominantly European ancestry [37]. The first release of this reference panel has data from 32,611 samples at 39,235,157 SNPs. The large sample size means that variants with minor allele frequencies as low as 0.1% can correctly be imputed using this data set.

The use of imputation methods does not only offer the possibility of increased SNP coverage, but, given the right reference data, also eases the analysis of common non-SNP variation, such as indels and copy number variations (CNVs). So far some reference panels have, however, only include SNVs and disregarded indels and structural variants. The increasing quality of whole-genome sequencing and software for calling structural variants means that better data sets that include structural variants should soon become available. Imputation will then make it possible to use the SNP chips to test many indels and structural variants that are not being (routinely) tested today [38].

4.2 Imputation Software

The commonly applied genotype imputation methods, such as IMPUTE2 [39], BIMBAM [40], MaCH-Admix [41], and minimac3 [42], are all based on hidden Markov models (HMMs). Comparisons of these software packages have shown that they produce data of broadly similar quality but that they are superior to imputation software based on other methodological approaches [36, 43]. The basic HMMs used in these programs are similar to earlier HMMs developed to model LD patterns and estimate recombination rates.

When the sample size is large, imputation using these HMM-based methods imposes a high computational burden. One possible way of decreasing this burden is to pre-phase the samples so that resolved haplotypes are used as input for the imputation software instead of genotypes [44]. But even with pre-phasing, the computational task is far from trivial, and whole-genome imputation is not a task that can be performed on a single computer. This computational problem can be solved by using one of the two free imputation services that have recently been launched (https://imputationserver.sph.umich.edu, https://imputation.sanger.ac.uk). These services allow users to upload their data through a web interface and choose between a set of reference panels. The data set will then be imputed on a High Performance Computing Cluster, and the user will receive an email when the imputed data is ready for download.

4.3 Testing Imputed Variants

Since imputation is based on probabilistic models, the output is merely a probability for each genotype for the unknown variants in a given individual. That is, instead of reporting the genotype of an individual as AG, say, the program reports that the probability of the genotype being AA is 5%, that of being AG is 93%, and that of being GG is 2%. This nature of the output data challenges the GWAS. The simplest way of analyzing the imputed data is to use the "best guess" genotype, i.e., assume the genotype with the highest probability and ignore the others. In the example above, the individual would be given the genotype AG at the SNP in question, and usually, an individual's genotype would be considered as missing if none of the genotypes have a probability larger than a certain threshold (e.g., 90%). The use of "best guess" genotype is problematic since it does not take the uncertainty of the imputed genotypes into account, may introduce a systematic bias, and lead to false positives and false negatives. A better way is to report a logistic regression on the expected allele count—in the example above, the expected allele count for allele A would be 1.03 ($2p_{AA} + p_{AG}$). This method has proved to be surprisingly robust at least when the effect of the risk allele is small [45], which is the case for most of the variants found through GWAS. An even better solution is to use methods that fully account for the uncertainty of the imputed genotypes [45–47].

5 Current Status

After the first GWAS saw publication in 2005, it was followed by many more studies, and today almost 4000 such studies of human diseases or traits have been published (Fig. 6a). The first GWASs had moderate sample sizes with hundreds of samples, but over the years the sample sizes and thereby the power of the studies have gradually been increasing (Fig. 6b). Imputation and later also next-generation sequencing have resulted in a rapid increase in the

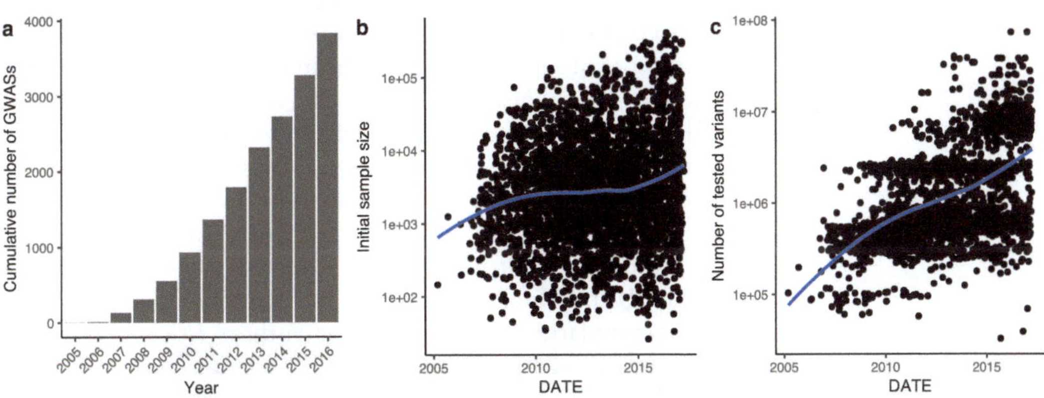

Fig. 6 GWAS statistics from the NHGRI-EBI GWAS Catalog [63] (accessed June 2017). (**a**) The cumulative number of GWASs published since 2005. (**b**) The initial sample sizes of the GWASs. For dichotomous traits the combined number of cases and controls is shown. Replication samples are not counted. (**c**) The number of tested variants in each study

number of variants that are tested in a GWAS (Fig. 6c). All these GWASs published in the last decade have increased our knowledge about the genetic architecture of common diseases a lot. In this section, we will go through some of the insights that have been revealed by these studies.

5.1 Polygenic Architecture of Common Diseases

GWASs have consistently shown that most complex traits and diseases have very polygenic architectures with a large number of causal variants with small effects. The small effect sizes mean that enormous sample sizes are needed to detect the associated variants and that each variant only explains a small fraction of the heritability. Even though large sample sizes have led to the discovery of many loci affecting common diseases, the aggregated effect of all these loci still only explains a small fraction of the heritability.

A good example is type 2 diabetes where researchers by 2012 had identified 63 associated loci that collectively only explained 5.7% of the liability-scale variance [48]. Such results led to much discussion about the possible source of the remaining "missing heritability" [49, 50]. A significant contribution to this debate was when researchers in 2010 started using mixed linear models to estimate the heritability explained by all common variants not only those that surpass a conservative significance threshold. These studies showed that a significant fraction of the so-called missing heritability was not truly missing from the GWAS data sets but only hidden due to small effect sizes. This was first illustrated in height where 180 statistically significant SNPs could only explain 10% of the heritability, but this fraction increased to 45% when all genotyped variants were considered [51].

For common diseases, such analyses have typically shown that around half of the heritability can be explained by considering all common variants. Given the small individual contribution of each of the discovered variants and that the individual contribution of the yet to be found variants will be even smaller, it is likely that the actual number of causal variants will be much more than a thousand for many common diseases. Recent data shows that in many diseases these causal variants are relatively uniformly distributed along the genome. It has, for instance, been estimated that 71–100% of 1 MB windows in the genome contribute to the heritability of schizophrenia [52]. Another article recently estimated that most 100 kB windows contribute to the variation of height and that more than 100,000 markers have an independent effect on height. This strikingly large number leads the authors to propose a new "omnigenic" model in which most genes expressed in a cell type that is relevant for a given disease have a nonzero contribution to the heritability of that disease [53].

5.2 Pleiotropy

The variants that have been discovered by GWASs so far reveal numerous examples where one genetic locus affects multiple often seemingly unrelated traits [54, 55]. One explanation for such a shared association between a pair of traits is mediation where the shared locus affects the risk of one of the traits, and that trait is causal for the other. Another possible explanation is pleiotropy where the shared locus is independently causal for both traits. It is possible to distinguish between mediation and true pleiotropy by adjusting or stratifying for one trait while testing the other. In the case of mediation, it is also possible to determine the direction of the causation. In general, it is difficult to make such causal inference from observational data, but Mendelian randomization, which uses significantly associated variants as instrumental variables, can in some circumstances be used to assess a causal relationship between a potential risk factor and a disease. For instance, Voight and colleagues used SNPs associated with lipoprotein levels to assess whether the correlation between different forms of lipoprotein and myocardial infarction risk was causal [56]. They found that while low-density lipoprotein (LDL) had a causal effect on disease risk, high-density lipoprotein (HDL) did not.

The fact that pleiotropy is widespread has several implications. One is that variants that have already been found to affect one trait can be prioritized in other studies since they are more likely also to affect another trait than a random variant is. Another implication is that we cannot always examine the effect of selection by studying one trait in isolation. There are multiple examples of antagonistic pleiotropy where a variant increases the risk of one disease while decreasing the risk of another.

5.3 Differences Between Diseases

Because of differences in age of onset and severity, we do not expect identical allelic architectures in all common diseases. Using the currently available GWAS data sets, we can now start to identify these differences in the allelic architectures, but because of the significant differences in samples sizes and the number of tested variants, this is not an easy task.

The data available to date show that the degree of polygenicity differs between diseases with schizophrenia, for example, having more predicted loci than immune disorders [57] and hypertension [52]. Results also show that rare variants play a larger role in some diseases compared to others. Rare variants, for example, have a greater role in amyotrophic lateral sclerosis than in schizophrenia [58] and are even less important in lifestyle-dependent diseases such as type 2 diabetes [59].

6 Perspectives

The price of whole-genome sequencing is still declining, and it is not unreasonable to expect that at some point in the future, a majority of people will get their genomes sequenced. At that point the availability of genetic data will no longer be a limiting factor in studies of common human diseases. In order to make the most of such huge data sets, the genetic information needs to be combined with high-quality phenotypic and environmental information. If that is achieved, we will be able to explain most—if not all—of the additive genetic variance for the common human diseases. Having large population data sets where genetic data is combined with extensive phenotypic data including information about lifestyle, diet and other environmental risk factors will also enable much better studies of pleiotropy and gene–environment interactions. A few large population data sets are already available now with the UK Biobank [29]—a prospective study of 500,000 individuals—being the best example.

While GWASs have found a lot of loci that are associated with common diseases, the actual causal variant and the functional mechanism driving the causation are still unknown for a large fraction of the loci. In order to understand the functional mechanism of a specific locus, it is necessary to combine sequence data with other types of data. This includes gene expression data (from the correct tissue) and epigenetic data such as methylation. Such data sets are fortunately also becoming cheaper to produce and thus more abundant as a result of falling sequencing costs. Furthermore large consortium data sets such as GTEx [60], ENCODE [61], and Roadmap Epigenomics [62] mean that each lab studying these mechanisms will not have to produce all the data themselves but can in part rely on these public data sets. It is thus likely that we in the future not only will find many more GWAS loci for each common disease but we will also have a much better understanding of how each of these loci affects the disease.

7 Questions

1. How can you distinguish causal variants from other variants when all variants have been typed? Is there any statistical way of distinguishing between correlation and causality just from genotype data? Could you use functional annotations?
2. Consider a GWAS data set, where in the top ten ranked statistics you have five markers that are close together and the remaining five scattered across the genome. Would you consider the five close markers more or less likely to be a true positive? Why? If one of them is a false positive, what would you think about the others?
3. Why is the RR but not the OR estimate affected by a biased case/control sample?
4. How would you test for, e.g., dominant or recessive effects in a contingency table?

References

1. Visscher PM, Hill WG, Wray NR (2008) Heritability in the genomics era |[mdash]| concepts and misconceptions. Nat Rev Genet 9:255. https://doi.org/10.1038/nrg2322
2. Cardon LR, Abecasis GR (2003) Using haplotype blocks to map human complex trait loci. Trends Genet 19:135
3. de Bakker PIW, Yelensky R, Pe'er I et al (2005) Efficiency and power in genetic association studies. Nat Genet 37:1217–1223. https://doi.org/10.1038/ng1669
4. Daly MJ, Rioux JD, Schaffner SF et al (2001) High-resolution haplotype structure in the human genome. Nat Genet 29:229–232. https://doi.org/10.1038/ng1001-229
5. Shi H, Kichaev G, Paşaniuc B (2016) Contrasting the genetic architecture of 30 complex traits from summary association data. Am J Hum Genet 99:139–153. https://doi.org/10.1016/j.ajhg.2016.05.013
6. Wright S (1931) Evolution in mendelian populations. Genetics 16:97–159
7. Pritchard JK (2001) Are rare variants responsible for susceptibility to complex diseases? Am J Hum Genet 69:124–137. https://doi.org/10.1086/321272
8. Pritchard JK, Cox NJ (2002) The allelic architecture of human disease genes: common disease–common variant… or not? Hum Mol Genet 11:2417
9. Reich DE, Lander ES (2001) On the allelic spectrum of human disease. Trends Genet 17:502
10. Di Rienzo A, Hudson RR (2005) An evolutionary framework for common diseases: the ancestral-susceptibility model. Trends Genet 21:596–601. https://doi.org/10.1016/j.tig.2005.08.007
11. 1000 Genomes Project Consortium, Auton A, Brooks LD et al (2015) A global reference for human genetic variation. Nature 526:68–74. https://doi.org/10.1038/nature15393
12. Quintana-Murci L (2016) Understanding rare and common diseases in the context of human evolution. Genome Biol 17:225. https://doi.org/10.1186/s13059-016-1093-y
13. Gudbjartsson DF, Helgason H, Gudjonsson SA et al (2015) Large-scale whole-genome sequencing of the Icelandic population. Nat Genet 47:435. https://doi.org/10.1038/ng.3247
14. Klein RJ, Zeiss C, Chew EY et al (2005) Complement factor H polymorphism in age-related macular degeneration. Science 308:385–389. https://doi.org/10.1126/science.1109557
15. Duerr RH, Taylor KD, Brant SR et al (2006) A genome-wide association study identifies IL23R as an inflammatory bowel disease gene. Science 314:1461–1463. https://doi.org/10.1126/science.1135245
16. Lewis CM (2002) Genetic association studies: design, analysis and interpretation. Brief Bioinform 3:146–153
17. Balding DJ (2006) A tutorial on statistical methods for population association studies. Nat Rev Genet 7:781–791. https://doi.org/10.1038/nrg1916
18. Wellcome Trust Case-Control Consortium (2007) Genome-wide association study of 14,000 cases of seven common diseases and 3,000 shared controls. Nature 447:661–678. https://doi.org/10.1038/nature05911
19. Mccarthy MI, Abecasis GCAR, Cardon LR et al (2008) Genome-wide association studies for complex traits: consensus, uncertainty and challenges. Nat Rev Genet 9:356. https://doi.org/10.1038/nrg2344
20. Patterson N, Price AL, Reich D (2006) Population structure and eigenanalysis. PLoS Genet 2:e190. https://doi.org/10.1371/journal.pgen.0020190
21. Devlin B, Roeder K (1999) Genomic control for association studies. Biometrics 55:997–1004
22. Devlin B, Roeder K, Wasserman L (2001) Genomic control, a new approach to genetic-based association studies. Theor Popul Biol 60:155–166. https://doi.org/10.1006/tpbi.2001.1542
23. Yang J, Weedon MN, Purcell S et al (2011) Genomic inflation factors under polygenic inheritance. Eur J Hum Genet 19:807–812. https://doi.org/10.1038/ejhg.2011.39
24. Bulik-Sullivan BK, Loh P-R, Finucane HK et al (2015) LD Score regression distinguishes confounding from polygenicity in genome-wide association studies. Nat Genet 47:291–295. https://doi.org/10.1038/ng.3211
25. Price AL, Patterson NJ, Plenge RM et al (2006) Principal components analysis corrects for stratification in genome-wide association studies. Nat Genet 38:904–909. https://doi.org/10.1038/ng1847
26. Yu J, Pressoir G, Briggs WH et al (2006) A unified mixed-model method for association mapping that accounts for multiple levels of relatedness. Nat Genet 38:203–208. https://doi.org/10.1038/ng1702
27. Price AL, Zaitlen NA, Reich D, Patterson N (2010) New approaches to population

stratification in genome-wide association studies. Nat Rev Genet 11:459–463. https://doi.org/10.1038/nrg2813
28. Loh P-R, Tucker G, Bulik-Sullivan BK et al (2015) Efficient Bayesian mixed-model analysis increases association power in large cohorts. Nat Genet 47:284–290. https://doi.org/10.1038/ng.3190
29. Sudlow C, Gallacher J, Allen N et al (2015) UK biobank: an open access resource for identifying the causes of a wide range of complex diseases of middle and old age. PLoS Med 12: e1001779. https://doi.org/10.1371/journal.pmed.1001779
30. Sebastiani P, Solovieff N, Puca A et al (2010) Genetic signatures of exceptional longevity in humans. Science. https://doi.org/10.1126/science.1190532
31. Alberts B (2010) Editorial expression of concern. Science 330:912. https://doi.org/10.1126/science.330.6006.912-b
32. Zheng J, Erzurumluoglu AM, Elsworth BL et al (2017) LD Hub: a centralized database and web interface to perform LD score regression that maximizes the potential of summary level GWAS data for SNP heritability and genetic correlation analysis. Bioinformatics 33:272–279. https://doi.org/10.1093/bioinformatics/btw613
33. Teo Y-Y, Small KS, Kwiatkowski DP (2010) Methodological challenges of genome-wide association analysis in Africa. Nat Rev Genet 11:149–160. https://doi.org/10.1038/nrg2731
34. Zaitlen N, Paşaniuc B, Gur T et al (2010) Leveraging genetic variability across populations for the identification of causal variants. Am J Hum Genet 86:23–33. https://doi.org/10.1016/j.ajhg.2009.11.016
35. International HapMap Consortium, Frazer KA, Ballinger DG et al (2007) A second generation human haplotype map of over 3.1 million SNPs. Nature 449:851–861. https://doi.org/10.1038/nature06258
36. Marchini J, Howie B (2010) Genotype imputation for genome-wide association studies. Nat Rev Genet 11:499–511. https://doi.org/10.1038/nrg2796
37. McCarthy S, Das S, Kretzschmar W et al (2016) A reference panel of 64,976 haplotypes for genotype imputation. Nat Genet 48:1279–1283. https://doi.org/10.1038/ng.3643
38. Sudmant PH, Kitzman JO, Antonacci F et al (2010) Diversity of human copy number variation and multicopy genes. Science 330:641–646. https://doi.org/10.1126/science.1197005
39. Howie BN, Donnelly P, Marchini J (2009) A flexible and accurate genotype imputation method for the next generation of genome-wide association studies. PLoS Genet 5: e1000529. https://doi.org/10.1371/journal.pgen.1000529
40. Servin B, Stephens M (2007) Imputation-based analysis of association studies: candidate regions and quantitative traits. PLoS Genet 3: e114. https://doi.org/10.1371/journal.pgen.0030114
41. Liu EY, Li M, Wang W, Li Y (2013) MaCH-admix: genotype imputation for admixed populations. Genet Epidemiol 37:25–37. https://doi.org/10.1002/gepi.21690
42. Das S, Forer L, Schönherr S et al (2016) Next-generation genotype imputation service and methods. Nat Genet 48:1284–1287. https://doi.org/10.1038/ng.3656
43. Nothnagel M, Ellinghaus D, Schreiber S et al (2009) A comprehensive evaluation of SNP genotype imputation. Hum Genet 125:163–171. https://doi.org/10.1007/s00439-008-0606-5
44. Howie B, Fuchsberger C, Stephens M et al (2012) Fast and accurate genotype imputation in genome-wide association studies through pre-phasing. Nat Genet 44:955–959. https://doi.org/10.1038/ng.2354
45. Guan Y, Stephens M (2008) Practical issues in imputation-based association mapping. PLoS Genet 4:e1000279. https://doi.org/10.1371/journal.pgen.1000279
46. Marchini J, Howie B, Myers S et al (2007) A new multipoint method for genome-wide association studies by imputation of genotypes. Nat Genet 39:906–913. https://doi.org/10.1038/ng2088
47. Stephens M, Balding DJ (2009) Bayesian statistical methods for genetic association studies. Nat Rev Genet 10:681–690. https://doi.org/10.1038/nrg2615
48. Morris AP, Voight BF, Teslovich TM, Ferreira T et al (2012) Large-scale association analysis provides insights into the genetic architecture and pathophysiology of type 2 diabetes. Nat Genet 44:981
49. Maher B (2008) Personal genomes: the case of the missing heritability. Nature 456:18–21. https://doi.org/10.1038/456018a
50. Manolio TA, Collins FS, Cox NJ, Goldstein DB (2009) Finding the missing heritability of complex diseases. Nature 461:747

51. Yang J, Benyamin B, McEvoy BP, Gordon S (2010) Common SNPs explain a large proportion of the heritability for human height. Nat Genet 42:565
52. Loh P-R, Bhatia G, Gusev A et al (2015) Contrasting genetic architectures of schizophrenia and other complex diseases using fast variance-components analysis. Nat Genet 47:1385–1392. https://doi.org/10.1038/ng.3431
53. Boyle EA, Li YI, Pritchard JK (2017) An expanded view of complex traits: from polygenic to omnigenic. Cell 169:1177
54. Price AL, Spencer CCA, Donnelly P (2015) Progress and promise in understanding the genetic basis of common diseases. Proc R Soc B 282:20151684. https://doi.org/10.1098/rspb.2015.1684
55. Pickrell JK, Berisa T, Liu JZ et al (2016) Detection and interpretation of shared genetic influences on 42 human traits. Nat Genet 48:709–717. https://doi.org/10.1038/ng.3570
56. Voight BF, Peloso GM, Orho-Melander M (2012) Plasma HDL cholesterol and risk of myocardial infarction: a mendelian randomisation study. Lancet 380:572
57. Ripke S, O'Dushlaine C, Chambert K et al (2013) Genome-wide association analysis identifies 13 new risk loci for schizophrenia. Nat Genet 45:1150–1159. https://doi.org/10.1038/ng.2742
58. Van Rheenen W, Shatunov A, Dekker AM (2016) Genome-wide association analyses identify new risk variants and the genetic architecture of amyotrophic lateral sclerosis. Nat Genet 48:1043
59. Gorlov IP, Gorlova OY, Amos CI (2015) Allelic spectra of risk SNPs are different for environment/lifestyle dependent versus independent diseases. PLoS Genet 11:e1005371
60. GTEx Consortium (2013) The Genotype-Tissue Expression (GTEx) project. Nat Genet 45:580–585. https://doi.org/10.1038/ng.2653
61. ENCODE Project Consortium, Bernstein BE, Birney E et al (2012) An integrated encyclopedia of DNA elements in the human genome. Nature 489:57–74. https://doi.org/10.1038/nature11247
62. Roadmap Epigenomics Consortium, Kundaje A, Meuleman W et al (2015) Integrative analysis of 111 reference human epigenomes. Nature 518:317–330. https://doi.org/10.1038/nature14248
63. MacArthur J, Bowler E, Cerezo M et al (2017) The new NHGRI-EBI Catalog of published genome-wide association studies (GWAS Catalog). Nucleic Acids Res 45:D896–D901. https://doi.org/10.1093/nar/gkw1133

Chapter 18

Ancestral Population Genomics

Julien Y. Dutheil and Asger Hobolth

Abstract

Borrowing both from population genetics and phylogenetics, the field of population genomics emerged as full genomes of several closely related species were available. Providing we can properly model sequence evolution within populations undergoing speciation events, this resource enables us to estimate key population genetics parameters such as ancestral population sizes and split times. Furthermore we can enhance our understanding of the recombination process and investigate various selective forces. With the advent of resequencing technologies, genome-wide patterns of diversity in extant populations have now come to complement this picture, offering an increasing power to study more recent genetic history.

We discuss the basic models of genomes in populations, including speciation models for closely related species. A major point in our discussion is that only a few complete genomes contain much information about the whole population. The reason being that recombination unlinks genomic regions, and therefore a few genomes contain many segments with distinct histories. The challenge of population genomics is to decode this mosaic of histories in order to infer scenarios of demography and selection. We survey modeling strategies for understanding genetic variation in ancestral populations and species. The underlying models build on the coalescent with recombination process and introduce further assumptions to scale the analyses to genomic data sets.

Key words Ancestral population, Coalescence, Demography, Divergence, Markov model, Migration, Recombination, Selection, Speciation

1 Introduction

We are in the population genomics era where data sets from the 1000 human genomes project [1], the great apes project [2], and the 1001 arabidopsis genomes project [3] are available. The underlying data sets contain genotypic information for thousands of individuals in one or several species, in the form of de novo sequenced genomes or variation compared to an available "reference" genome (a.k.a. *resequencing*). By comparing genomes from several individuals of the same species or closely related species, we can obtain information about split times, population sizes, recombination events, and selection in contemporary and ancestral

Maria Anisimova (ed.), *Evolutionary Genomics: Statistical and Computational Methods*, Methods in Molecular Biology, vol. 1910,
https://doi.org/10.1007/978-1-4939-9074-0_18,

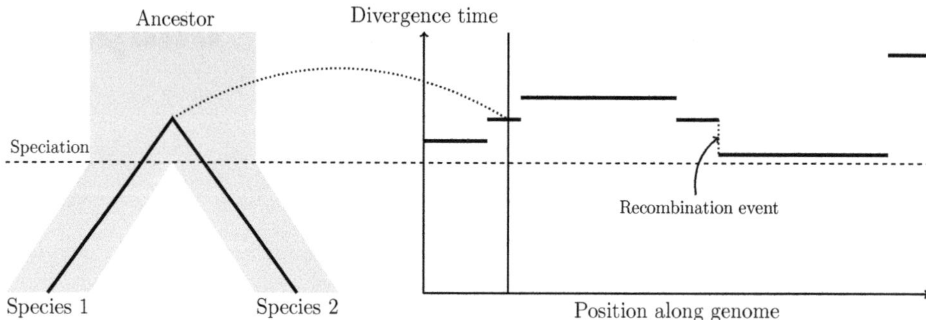

Fig. 1 Left: Isolation model of two species. Right: The coalescent process along the genomes of the two species. By comparing the two genomes we obtain information about the split time of the species and the ancestral population size. Furthermore the breakpoints along the genomes correspond to recombination events, so we also have information about the recombination process

species (*see* Fig. 1). In this chapter we discuss various models for obtaining this information.

Comparing homologous sequences available for a given locus to infer their degree of relatedness enables the discovery of the parental relationships of the sequences, depicted as a tree thereby named *genealogy*. When one sequence sampled from one individual of one species is compared with sequences from other species, the resulting genealogy contains information about the history of species, the so-called phylogeny. The phylogeny summarizes the relationship and the divergence times between the species.

Conversely, when sequences from several individuals within a species are sampled, we have access to the genetic variation in contemporary populations. The evolutionary forces that shape genetic variation within a species are genetic drift, mutation, recombination, and selection and are the subject of population genetics. The key modeling tool in population genetics is coalescent theory. Classical coalescent theory describes the genetic ancestry of a sample of homologous DNA sequences from the same species. This genealogical description includes times to common ancestry, which is measured back into the past.

Molecular phylogenetics and population genetics have accumulated 50 years of methodological developments. The convergence of these two fields and their key mathematical and statistical tools is needed in order to fully understand genomic sequence alignments, because comparing genealogies and phylogenies is at the heart of the study of the speciation process [4].

We describe the interplay between population genetics and phylogenetics by reviewing the methods and models that have been developed to understand evolutionary history from genomic data (*see* Table 1 for a comparative summary of all methods).

Table 1
Methods comparison

Principle	ARG Approx.	Spec.	Parameters estimated	Rate variation/ sequencing errors	Data set	Reference
Infer genealogy from independent loci, use distribution of inferred divergence and topology counts to estimate parameters	Independent loci	I	T, N_A	—	Primates: 53 "random" autosomal intergenic non-repetitive DNA segments of 2–20 kb	[5]
Count alignment patterns, fit EM model to infer parameters	—	I	T_1, N_A	Correction with outgroup	Primates	[6]
Likelihood calculation under a demographic model, numerical integration over genealogies	Independent loci	I	T_1, T_2, N_A	Independent estimate of rate	Primates	[7]
	Independent loci	IM	T_1, T_2, N_1, N_2, N_A, $m_{1\to2}$, $m_{2\to1}$	RAS	Drosophila	[8]
	Independent loci	IM	T_1, T_2, N_1, N_2, N_A, $m_{1\to2}$, $m_{2\to1}$	Independent estimate of rate	Primates: same data as [9] restricted to human, chimpanzee, gorilla, and orangutan	[10]
Bayesian inference	Independent loci	I	T_1, T_2, T_3, T_4, N_{A1}, N_{A2}, N_{A3}	RAS + branch-specific departure from molecular clock	Primates: 15,000 neutral loci (7.4 Mb)	[9]
Integrating over a subset of candidate genealogies using a hidden Markov model	Markov process	I	T_1, T_2, N_{A1}, N_{A2}	—	Primates: 1 Mb alignment	CoalHMM [11]
	Markov process	I	T_1, T_2, N_{A1}, N_{A2}, r	RAS	Primates: 1 Mb alignment	CoalHMM [12]
Integrating over the discretized distribution of divergence for a pair of genomes	Markov process	I	T, N_A, r	—	Orangutans: two full genomes	CoalHMM [13]

(continued)

Table 1
(continued)

Principle	ARG Approx.	Spec.	Parameters estimated	Rate variation/ sequencing errors	Data set	Reference
Integrating over the discretized distribution of divergence for a pair of haploid genomes in a population	Markov process	—	$Ne_{1...n}$, r	—	Human diploid genomes	PSMC [14]
Integrating over the discretized distribution of divergence of the most recent coalescence event with multiple haploid genomes in a population	Markov process	—	$Ne_{1...n}$, r fixed	—	Human diploid genomes	MSMC [15]
Use the conditional sampling distribution to approximate the integration over the discretized distribution of divergence coalescence events within multiple haploid genomes in one or more population(s)	Markov process + CSD	—	$Ne_{1...n}$, r fixed	—	Human diploid genomes	diCal [16–18]
Extension of the pairwise sequentially Markov coalescent with site frequency spectrum based on many individuals	Markov process + Poisson random field	—	$Ne_{1...n}$	—	Unphased human genomes	SMC++ [19]
Bayesian sampling of ARG, using a discretized distribution of divergence time conditioned on multiple haploid genomes in a population	Markov process + "threading"	—	$Ne_{1...n}$, r	—	Human diploid genomes	ARGweaver [20]

This table summarizes and compares existing ancestral population genomics methods. Parameters correspond to the one in Fig. 4. RAS: Rate across site model, assuming an a priori distribution of evolutionary rate (usually a discretized gamma distribution) over alignment positions

2 Coalescent Theory and Speciation

We start by describing the standard coalescent model within one population. The coalescent model describes the shape of the genealogy of several sequences sampled from a single population. For more information on the coalescent, we refer to [21, 22] and [23]. This section describes the coalescent process as a chronological process. In the next section, we will see how it can be modeled as a spatial process along the genome. In subsequent sections we extend the standard model to include two or more populations. In the cases where multiple populations are present we describe both the isolation model and the isolation-with-migration model.

2.1 The Standard Coalescent Model

The standard coalescent model is a continuous-time approximation of the neutral Wright–Fisher model. In the Wright–Fisher model the number of chromosomes $2N$ (we consider diploid organisms) is fixed in each non-overlapping generation. Each chromosome in a new generation chooses its ancestor uniformly at random from the previous generation.

Consider two chromosomes. The probability of the two chromosomes choosing the same ancestor is $1/(2N)$ and the probability of the two chromosomes not finding a common ancestor is $1 - 1/(2N)$. Let R_2 denote the number of generations back in time when the two individuals find a most recent common ancestor (MRCA). By repeating the argument above, the probability of the two chromosomes not finding a common ancestor r generations back in time is

$$P(R_2 > r) = \left(1 - \frac{1}{2N}\right)^r.$$

If we scale time t in units of $2N$, i.e., set $r = 2Nt$, we get

$$P(R_2 > r) = \left(1 - \frac{1}{2N}\right)^r = \left(1 - \frac{1}{2N}\right)^{2Nt} \approx e^{-t},$$

where the approximation is valid for large N. In coalescent time units the waiting time $T_2 = R_2/(2N)$ before coalescence of two individuals is therefore exponentially distributed with mean one.

These considerations can be extended to multiple individuals. In general the time T_n before two of n individuals coalesce is exponentially distributed with rate $\binom{n}{2}$.

The waiting time W_n for a sample of n individuals to find the most recent common ancestor (MRCA) is given by

$$W_n = T_n + T_{n-1} + \cdots + T_2,$$

where T_k are independent exponential random variables with parameter $\binom{k}{2}$; *see* Fig. 2 for an illustration. It follows that the mean of W_n is

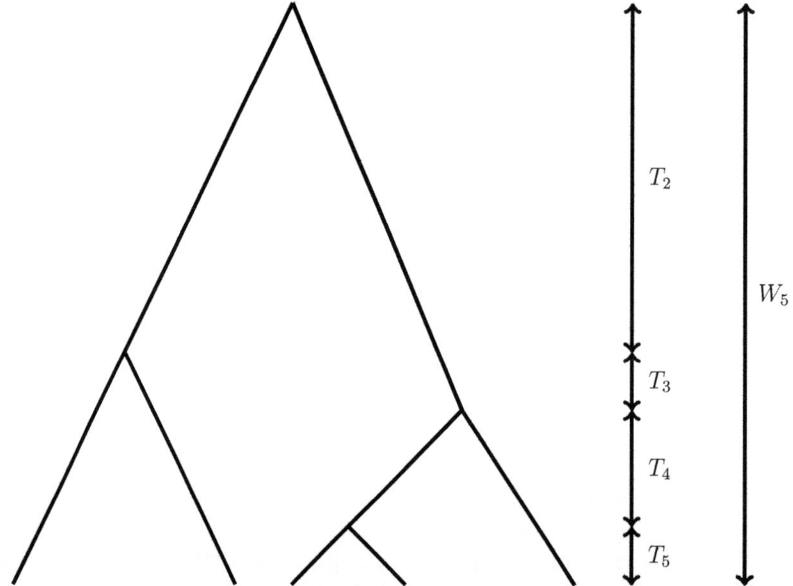

Fig. 2 Illustration of the coalescent process. The waiting time before two out of n individuals coalesce is T_n and the time before a sample of n individuals find common ancestry is W_n

$$E[W_n] = \sum_{k=2}^{n} E[T_k] = \sum_{k=2}^{n} \frac{2}{k(k-1)} = 2\sum_{k=2}^{n}\left(\frac{1}{k-1} - \frac{1}{k}\right)$$
$$= 2\left(1 - \frac{1}{n}\right).$$

Note that $\lim_{n\to\infty} E[W_n] = 2$.

The variance of W_n is

$$\text{Var}[W_n] = \sum_{k=2}^{n} \text{Var}[T_k] = \sum_{k=2}^{n} \binom{k}{2}^{-2}$$
$$= 8\sum_{k=1}^{n-1} \frac{1}{k^2} - 4\left(1 - \frac{1}{n}\right)\left(3 + \frac{1}{n}\right).$$

Note that $\lim_{n\to\infty}\text{Var}[W_n] = (\frac{8\pi^2}{6} - 12) = 1.16$.

The consequences of these calculations are that when we only sample within a population we are limited to relatively recent events. The expected time for a large sample to find their MRCA is approximately $2 \times (2N) = 4N$ generations with standard deviation $\sqrt{1.16} \times (2N) = 2.15N$ generations. As a consequence, a neutral sample within a population contains little information beyond $6N$ generations.

Humans have a generation time of approximately 20 years and an effective population size of approximately $N = 10,000$ (*see* [21, p. 251]), and therefore $6N$ generations correspond to approximately 1.2 million years (My) for humans. Therefore human diversity at neutral loci contains little demographic information beyond 1.2 My.

2.2 Adding Mutations to the Standard Coalescent Model

Now suppose mutations occur at a rate u per locus per generation. In a lineage of r generations, we then expect ru mutations or in the coalescent time units with $r = 2Nt$ we expect $2Ntu$ mutations. We let $\theta = 4Nu$ be the mutation rate parameter. Since u is small we can make a Poisson approximation of the binomial number of mutations in a lineage of r generations

$$\text{Bin}(r,u) = \text{Bin}(2Nt,\theta/(2 \cdot 2N)) \approx \text{Pois}(t\theta/2).$$

We have thus arrived at the following two-step process for simulating samples under the coalescent: (a) simulate the genealogy by merging lineages uniformly at random and with waiting times exponentially distributed with rate $\binom{n}{2}$ when n lineages are present; (b) on each lineage in the tree add mutations according to a Poisson process with rate $\theta/2$.

Another possibility is to scale the coalescent process such that one mutation is expected in one time unit. In this case the exponentially distributed waiting times in (a) have rate $\binom{n}{2}(2/\theta)$, and in (b) the mutations are added with unit rate. We use the latter version of the coalescent-with-mutations process below.

2.3 Taking Recombination into Account

For species where recombination occurs, different parts of the genome come from distinct ancestors, and therefore have a distinct history. Figure 3 exemplifies this phenomenon for two species. It displays the genealogical relationships for two sequences which underwent a single recombination event. In the presence of recombination, each position of a genome alignment therefore has a specific genealogy, and close positions are more likely to share the same one (recall Fig. 1). The genome alignment can therefore be described as an ordered series of genealogies, spanning a variable amount of sites, and then changing because of a recombination event [4]. The genealogy is therefore depicted as a complex graph with nodes representing both coalescence and recombination events, the ancestral recombination graph (ARG, Fig. 3c). A single genome thus contains different samples from the distribution of the age of the MRCA, and the distribution contains information about the ancestral population size and speciation time. The coalescent with recombination serves as a basis for modeling genome-wide genealogy, a point that we will further develop in Subheading 4.

3 Adding Genetic Barriers and Gene Flow to the Picture: The Structured Coalescent

In this section we extend the standard coalescent model. We consider coalescent models with multiple species and introduce population splits or speciation events. The models that we describe are

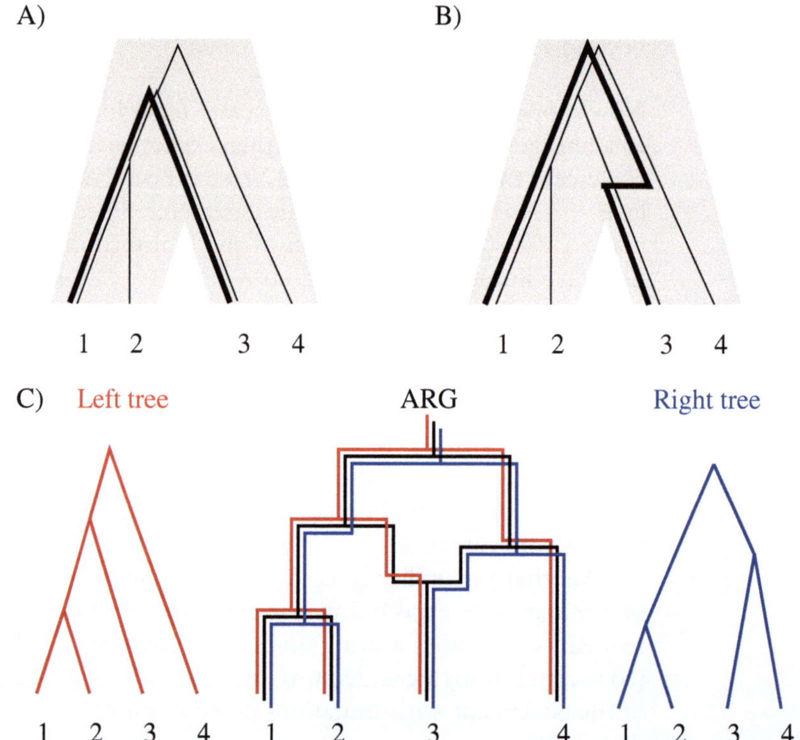

Fig. 3 Ancestral recombination graph for two species. (**a**) Genealogy of four sampled sequences from two species. The bold line shows the divergence of two sequences of interest. (**b**) A single recombination event happened between the lineages of sequences 3 and 4 (horizontal line), so that in a part of the sequences, the genealogy is as depicted by the bold line and therefore displays an older divergence. (**c**) The corresponding ancestral recombination graph (in black) with the trees of each side of the recombination break point superimposed (red: left tree; blue: right tree). When going backward in time, a split corresponds to a recombination event and a merger to a coalescence event

shown in Fig. 4 (see also Table 1) and include: (a) The two species isolation model; (b) The two species isolation-with-migration models; (c) The three species isolation model (and incomplete lineage sorting); and (d) The three species isolation-with-migration model. We also discuss the general multiple species isolation-with-migration model. The two species isolation model was introduced in [24] and the isolation-with-migration model was introduced in [25].

3.1 Isolation Model with Two Species

If the sequences are sampled from two distinct species that have diverged a time T ago (*see* Fig. 4a), then the distribution of the age of the MRCA is shifted to the right with the amount T, resulting in the distribution

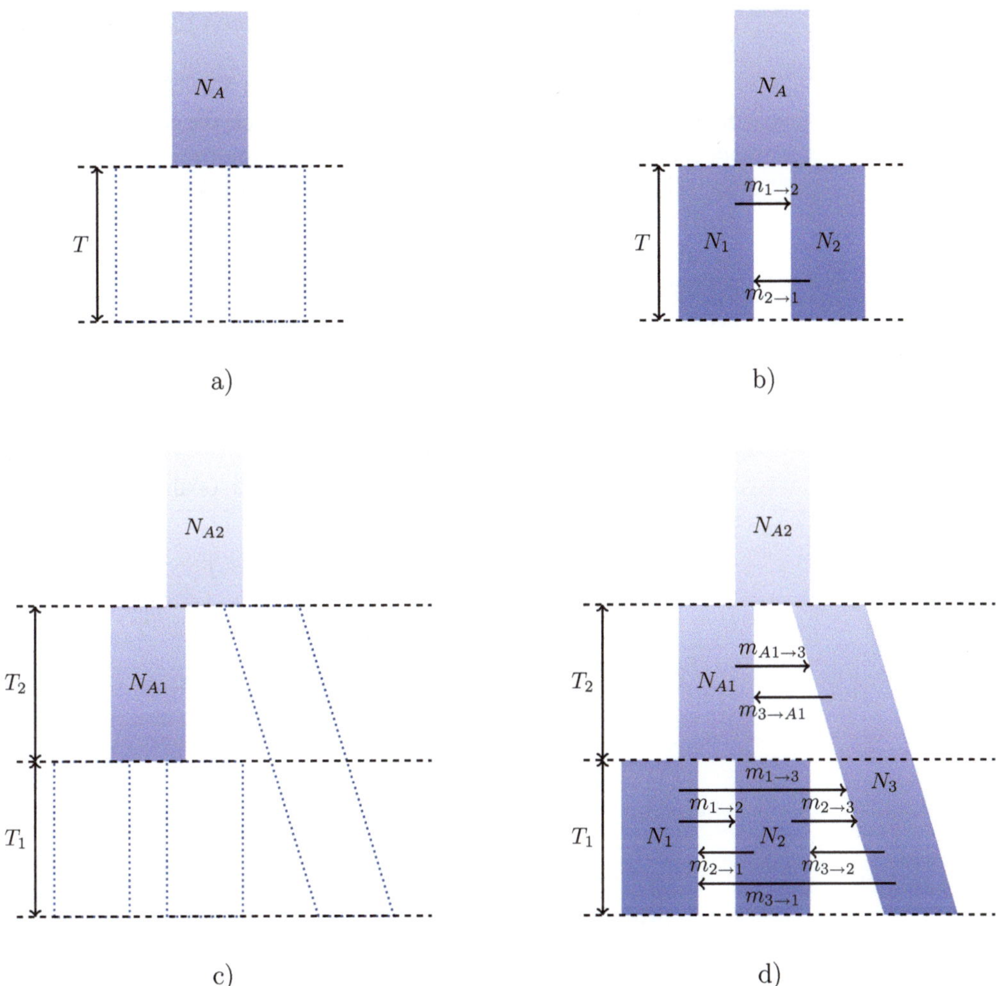

Fig. 4 Speciation models and associated parameters. In all exemplified models effective population size is constant between speciation events, represented by dash lines. The timing of the speciation events, noted T are parameters of the models, together with ancestral effective population sizes, noted N_A. In some cases, contemporary population sizes can also be estimated, and are noted N_i, where i is the index of the population. Models with post-divergence genetic exchanges have additional migration parameters labeled $m_{\text{from}\to\text{to}}$. The number of putative migration rates increases with the number of contemporary populations under study, and some models might consider some of them to be equal or eventually null to reduce complexity. (**a**) Isolation model with two species. (**b**) Isolation-migration model with two species. (**c**) Isolation model with three species. (**d**) Isolation-Migration model with three species

$$f_{T_2}(t) = \begin{cases} 0 & \text{if } t < T \\ \dfrac{2}{\theta_A} e^{-2(t-T)/\theta_A} & \text{if } t > T \end{cases},$$

where $\theta_A = 4N_A \cdot u$ is the ancestral mutation rate. The mean time to coalescent is $E[T_2] = T + \theta_A/2$ and the average divergence time between two sequences is twice this quantity, that is, $2T + \theta_A$. Since $\theta_A = 4N_A u$ it follows that the larger the size of the ancestral population, the bigger the difference between the speciation time and the divergence time.

The variance of the divergence time is $\text{Var}[T_2] = \theta_A^2/4$. With access to the distribution of divergence times, we could estimate the speciation time and population size from the mean and variance of the distribution. Unfortunately we do not know the complete distribution of divergence times and it is not immediately available to us, because long regions are needed for precise divergence estimation but have experienced one or more recombination events.

3.2 Isolation Model with Three or More Species and Incomplete Lineage Sorting

Now consider the isolation model with three species depicted in Fig. 4c. Such a model is often used for the human–chimpanzee--gorilla (HCG) triplet (e.g., [10–12]).

The density function for the time to coalescence between sample 1 and sample 2 is given by

$$f_{T_2}(t) = \begin{cases} 0 & \text{if } t < T_1 \\ \dfrac{2}{\theta_{A1}} e^{-2(t-T_1)/\theta_{A1}} & \text{if } T_1 < t < T_{12} \\ P_{12} \dfrac{2}{\theta_{A2}} e^{-2(t-T_{12})/\theta_{A2}} & \text{if } t > T_{12}, \end{cases} \tag{1}$$

where

$$T_{12} = T_1 + T_2 \quad \text{and} \quad P_{12} = e^{-2(T_{12}-T_1)/\theta_{A1}}$$

is the probability of the two samples *not* coalescing in the ancestral population of sample 1 and sample 2. In the upper right corner of Fig. 5 we plot the density (Eq. 1) with parameters that resemble the HCG triplet.

If sample 1 and sample 2 do not coalesce in the ancestral population of sample 1 and sample 2, then the three trees ((1,2),3), ((1,3),2), and ((2,3),1) are equally likely. The probability of the gene tree being different from the species tree is thus

$$\Pr(\text{incongruence}) = \frac{2}{3} P_{12} = \frac{2}{3} e^{-2(T_{12}-T_1)/\theta_{A1}}. \tag{2}$$

The event that the gene tree is different from the species tree is called incomplete lineage sorting (ILS). ILS is important because species tree incongruence often manifests itself as a relatively clear signal in a sequence alignment and thereby allows for accurate estimation of population parameters. In Fig. 6 we show the (in) congruence probability Eq. 2. We also refer to Exercise 1 (*see* Subheading 8.1) and Exercise 2 (*see* Subheading 8.2) for more discussion of ILS.

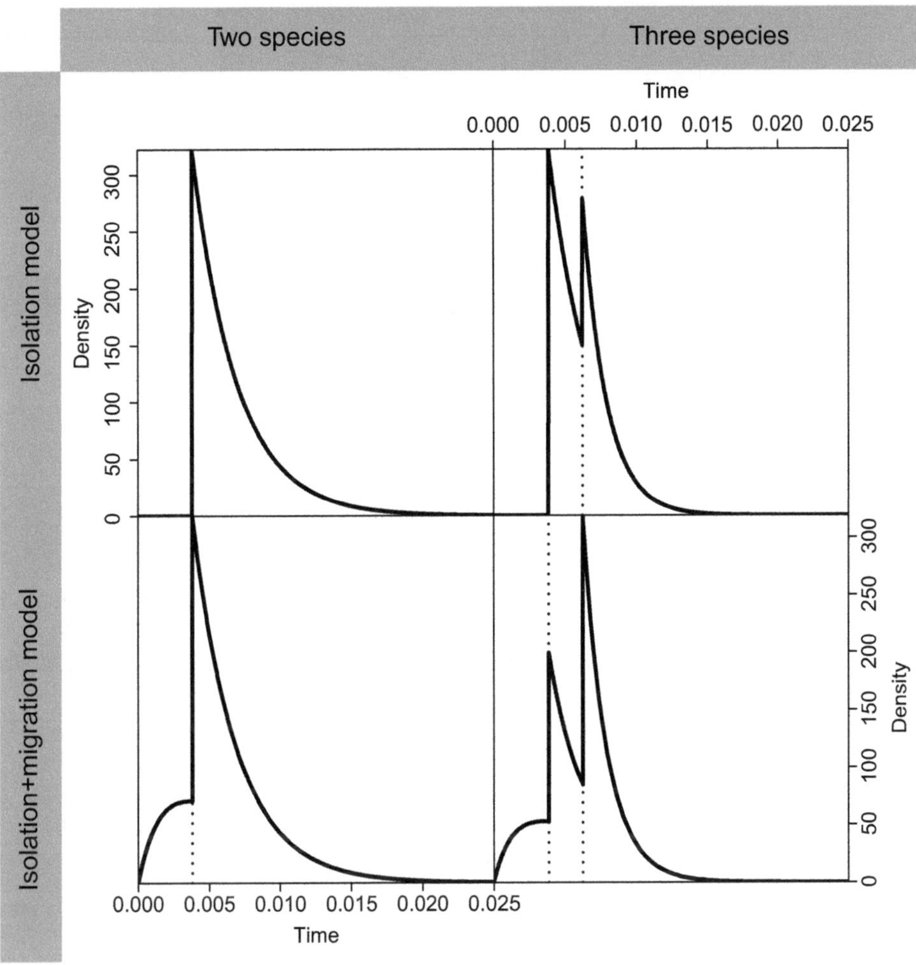

Fig. 5 Illustration of the density for coalescent in various models and data layout. The curves are the probability density functions. In the most simple case with two species, a constant ancestral population size and a punctual speciation (top left panel), more genomic regions find a common ancestor close to the species split (the vertical line), while a few regions have a more ancient common ancestor, distributed in an exponential manner (*see* Eq. 1). If speciation is not punctual and migration occurred after isolation of the species, then some sequences have a common ancestor which is more recent than the species split and the distribution in the ancestor becomes more complex (bottom left panel, *see* Eqs. 4 and 6). When a third species is added (right panel), then another discontinuity appears and all distributions depend on additional parameters, particularly when migration is allowed. We use $\theta_{A1} = 0.0062$, $\theta_{A2} = 0.0033$ and $\tau_1 = 0.0038$ (the first vertical line), $\tau_2 = 0.0062$ (the second vertical line) corresponding to the HCG triplet. Ancestral population sizes are taken from the simulation study in Table 6 in Wang and Hey [8]: $\theta_1 = 0.005$ and $\theta_2 = 0.003$. Migration parameters are all set to 50

In the three species isolation model the mean coalescent time for a sample from population 1 and a sample from population 2 is given by

$$E[T_2] = T_1 + (1 - P_{12})\frac{\theta_{A1}}{2} + P_{12}\frac{\theta_{A2}}{2}. \tag{3}$$

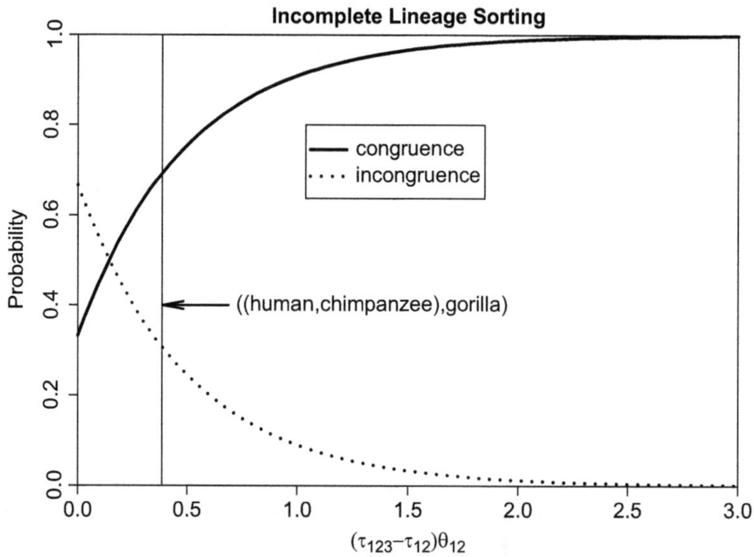

Fig. 6 Probability (Eq. 2) of gene tree and species tree being incongruent. In case of the HCG triplet we obtain $(T_{12} - T_1)/\theta_{A1} = (0.0062 - 0.0038)/0.0062 = 0.39$ which corresponds to an incongruence probability of 30%

Burgess and Yang [9] describe the speciation process for human, chimpanzee, gorilla, orangutan (O), and macaques (M) using an isolation model with five species. The HCGOM model contains four ancestral parameters θ_{HC}, θ_{HCG}, θ_{HCGO}, and θ_{HCGOM}. In this case (Eq. 3) extends to

$$\begin{aligned} E[T_2] = T_{\mathrm{HC}} &+ (1 - P_{\mathrm{HC}})\frac{\theta_{\mathrm{HC}}}{2} P_{\mathrm{HC}}(1 - P_{\mathrm{HCG}})\frac{\theta_{\mathrm{HCG}}}{2} \\ &+ P_{\mathrm{HC}}P_{\mathrm{HCG}}(1 - P_{\mathrm{HCGO}})\frac{\theta_{\mathrm{HCGO}}}{2} \\ &+ P_{\mathrm{HC}}P_{\mathrm{HCG}}P_{\mathrm{HCGO}}(1 - P_{\mathrm{HCGOM}})\frac{\theta_{\mathrm{HCGOM}}}{2}. \end{aligned}$$

3.3 Isolation-with-Migration Model with Two Species and Two Samples

The isolation-with-migration (IM) model with two species is shown in Fig. 4b. The IM-model has six parameters: The mutation rates θ_1, θ_2, and θ_A, the migration rates m_1 and m_2, and the speciation time T. We let $\Theta = (\theta_1, \theta_2, \theta_A, m_1, m_2, T)$ be the vector of parameters.

Wang and Hey [8] consider a situation with two genes. Before time T the system is in one of the following five states:

S_{11} : Both genes are in population 1.
S_{22} : Both genes are in population 2.
S_{12} : One gene is in population 1 and the other is in population 2.
S_1 : The genes have coalesced and the single gene is in population 1.
S_2 : The genes have coalesced and the single gene is in population 2.

The instantaneous rate matrix Q is given by

	S_{11}	S_{12}	S_{22}	S_1	S_2
S_{11}	$\cdot$	$2m_2$	0	$2/\theta_1$	0
S_{12}	m_1	$\cdot$	m_2	0	0
S_{22}	0	$2m_1$	$\cdot$	0	$2/\theta_2$
S_1	0	0	0	$\cdot$	m_2
S_2	0	0	0	m_1	$\cdot$

Starting in state a, the density for coalescent in population 1 at time $t < T$ is given by [26]

$$f_1(t) = (e^{Qt})_{aS_{11}}(2/\theta_1), \tag{4}$$

the density for coalescent in population 2 at time $t < T$ is

$$f_2(t) = (e^{Qt})_{aS_{22}}(2/\theta_2), \tag{5}$$

and the total density for a coalescent at time $t < T$ is

$$f(t) = f_1(t) + f_2(t). \tag{6}$$

Here $e^A = \sum_{i=0}^{\infty} A^i/(i!)$ is the matrix exponential of the matrix A and $(e^A)_{ij}$ is entry (i, j) in the matrix exponential.

After time T the system only has two states: S_{AA} corresponding to two genes in the ancestral population and S_A corresponding to one single gene in the ancestral population. The rate of going from state S_{AA} to state S_A is $2/\theta_A$. The density for coalescent in the ancestral population at time $t > T$ is therefore

$$f(t) = \left[(e^{QT})_{aS_{11}} + (e^{QT})_{aS_{12}} + (e^{QT})_{aS_{22}}\right]\frac{2}{\theta_A}e^{-(2/\theta_A)(t-T)}. \tag{7}$$

In Fig. 5 we illustrate the coalescent density in the two species isolation-with-migration model.

The likelihood for a pair of homologous sequences X is given by

$$P(X|\Theta) = L(\Theta|X) = \int_0^{\infty} P(X|t)f(t|\Theta)dt \tag{8}$$

where $f(t) = f(t|\Theta)$ given by Eqs. 6 and 7 is the density of the two sequences finding a MRCA at time t and $P(X|t)$ is the probability of the two sequences given that they find a MRCA at time t. The latter term is calculated using a distance-based method. One possibility is to use the infinite sites model where it is assumed that substitutions happen at unique sites, i.e., there are no recurrent substitutions. In this case the number of differences between the two sequences follows a Poisson distribution with rate 1.

For an application of the isolation-with-migration model with two sequences, we refer to [8]; a discussion of their approach can be found in [27].

3.4 Isolation-with-Migration Model with Three or More Species and Three or More Samples

Hey [28] considered the multipopulation isolation-with-migration (IM) model. Recall from Fig. 4b that the two-population IM model has six parameters: two present population sizes, one ancestral population size, one speciation time, and two migration rates. The three-population IM model in Fig. 4d has fifteen parameters: three present population sizes, two ancestral population sizes, two speciation times, and eight migration rates. In general a k-population IM model has $3k - 2 + 2(k - 1)^2$ parameters:

- k present population sizes,
- $(k - 1)$ ancestral population sizes,
- $(k - 1)$ speciation times, and
- $2(k - 1)^2$ migration rates.

See Fig. 5 for an example of divergence distribution with three species and migration and Exercise 3 (*see* Subheading 8.3) for a derivation of the number of migration rates in the general k-population model. For $k = 5$, 6, and 7 we obtain 45, 66, and 91 parameters. Because the number of parameters becomes very large even for small k, Hey [28] suggests adding constraints to the migration rates, e.g., setting some rates to zero or introducing symmetry conditions where rates between populations are the same.

4 Approximating the Coalescent with Recombination Along Genomes

Before the genomic era, multilocus population genetics models were addressing a small fraction of the complete ancestral recombination graph (ARG) by considering independent loci. As sequencing technologies evolved and allowed access to larger samples of genomic diversity, this independence assumption had to be relaxed and more explicit modeling of the ARG was required. Yet the complexity of the coalescent with recombination process makes its application to genome-scale data sets very challenging. Two directions of analysis methods have emerged: simulation-based or spatial approximations along the genome. In this chapter we focus on the latter and refer to Kelleher et al. [29] and Staab et al. [30] for the former. Simonsen and Churchill [31] described the first model of the joint distribution of genealogies at two loci for two genomes. Wiuf and Hein [32] extended this approach and described the coalescent as a spatial process along the genome. McVean and

Cardin [33] further approximated the description with a Markov process. In this section we describe and discuss these types of approximations.

4.1 The Independent Loci Approach: Free Recombination Between, No Recombination Within

The simplest way to handle issues relating to the ancestral recombination graph is to divide the data into presumably independent loci. Such analyses are therefore restricted to candidate regions that are not too large (to avoid including a recombination point) and not too close (to ensure several recombination events happened between loci). Each region can then be described by a single underlying tree, reducing the analytical and computational load.

Using 15,000 loci distant from 10 kb totaling 7.4 Mb and the isolation model introduced above, Burgess and Yang [9] (Table 2, model (b) sequencing errors) find the following ancestral population sizes and speciation times estimates for human (H), chimpanzee (C), gorilla (G), orangutan (O), and macaque (M) ancestors: $\theta_{\mathrm{HC}} = 0.0062$, $\theta_{\mathrm{HCG}} = 0.0033$, $\theta_{\mathrm{HCGO}} = 0.0061$, $\theta_{\mathrm{HCGOM}} = 0.0118$ and $T_{\mathrm{HC}} = 0.0038$, $T_{\mathrm{HCG}} = 0.0062$, $T_{\mathrm{HCGO}} = 0.0137$, $T_{\mathrm{HCGOM}} = 0.0260$. Converting these estimates into time units requires an estimate of the substitution rate, either absolute or deduced from a scaling point. Using $u = 10^{-9}$ as an estimate for substitutions per year, this leads to an estimate of 3.8 My for the human–chimpanzee speciation, a very recent estimate. Using the same data, Yang [10] showed that the isolation-with-migration model was preferred. Yang finds a more ancient speciation time $T_{\mathrm{HC}} = 0.0053$ (5.3 My with $u = 10^{-9}$) when migration is accounted for.

4.2 State-Space Model: Simonsen–Churchill Framework

The coalescent with recombination for two loci and two sequences is originally described in Simonsen and Churchill [31] as a continuous-time Markov chain backward in time with eight states as shown in Fig. 7. This Markov chain is given a careful treatment in the textbooks by Durrett [34, Section 3.1.1] and Wakeley [21, Section 7.2.4], and we therefore only briefly explain the basic properties of the model here.

A single sequence is either linked (●—●, ×—●, ●—×, or ×—× meaning that it contains material ancestral to the sample at both loci, or it is unlinked (●–, –●, –×, or ×–) when it contains material ancestral to the sample at only one locus. The coalescent rate is one for any two sequences, and the recombination rate is $\rho/2$ for any linked sequence. The chain begins at time zero in state 1 with two linked sequences. After an exponential waiting time with rate $1 + \rho$ the chain enters state 8 with probability $1/(1 + \rho)$ or state 2 with probability $\rho/(1 + \rho)$. The transition from state 1 to state 8 is a coalescent event, and the left and right tree heights are identical. The transition from state 1 to state 2 is a recombination event that breaks apart one of the two sequences. All other transitions have similar interpretations. Common ancestry for a locus is marked

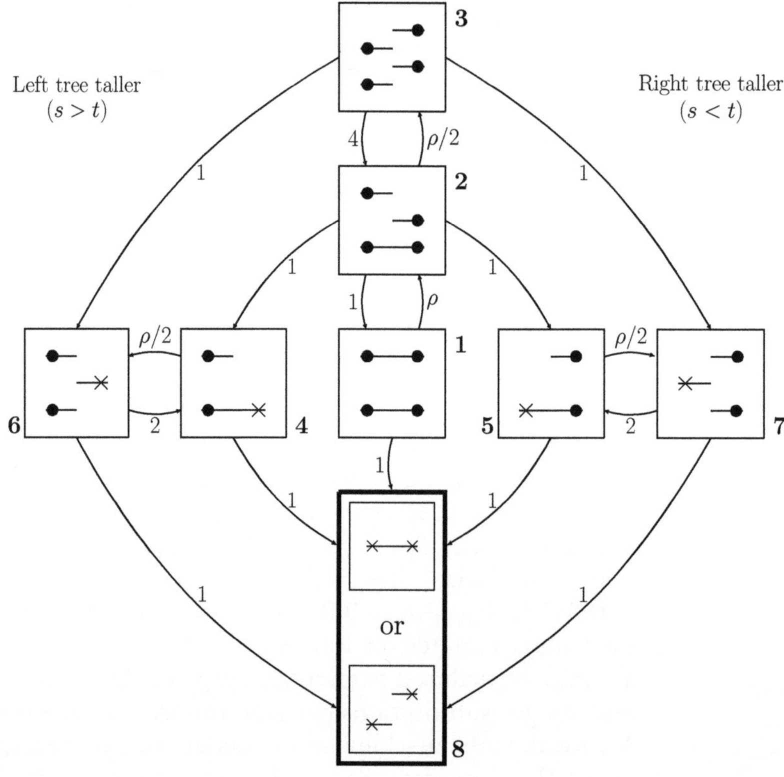

Fig. 7 State transition diagram for two loci and two sequences described as a continuous-time Markov chain backward in time. The figure is adapted from Figure 7.7 in Wakeley [21]. A line with a bullet or a cross at both ends is a linked sequence (ancestral material to the sample at both loci), whereas a line with a bullet or a cross at one end only is a sequence with ancestral material at one locus only. A cross denotes common ancestry. *s* and *t* denote the heights of the left and right trees, respectively

with a $\times$, so the transition from, e.g., state 1 to state 8 is a transition to the state $\times\!\!-\!\!\times$.

The height S of the left tree is the first time at which the process enters one of the states 5, 7, or 8 (states with a left $\times$), and the height T of the right tree is the first time at which one of the states 4, 6, or 8 is entered (states with a right $\times$). When state 8 is entered from state 1 the two tree heights are identical. State 8 is absorbing because only the tree heights are of interest.

The two key ingredients for the state-space model are the conditional probability for staying in a state $P(T=s|S=s)$ and the conditional density $q(t|s)$ of a new tree height t conditional on a change and a previous tree height s. Hobolth and Jensen [35] show that the conditional probability of no change from the left to the right tree is

$$P(T=s|S=s)=e^{s}[e^{As}]_{11}, \tag{9}$$

and the conditional density $q(t|s)$ of T given $S=s$ and given $T\neq S$ is

$$q(t|s) = \begin{cases} e^{-(s-t)} \dfrac{[e^{\Lambda t}]_{12} + [e^{\Lambda t}]_{13}}{e^{-s} - [e^{\Lambda s}]_{11}} & t < s, \\ e^{-(t-s)} \dfrac{[e^{\Lambda s}]_{12} + [e^{\Lambda s}]_{13}}{e^{-s} - [e^{\Lambda s}]_{11}} & t > s, \end{cases} \tag{10}$$

where Λ denotes the 8×8 rate matrix from Fig. 7.

Wakeley [21, Section 7.2.4] noted that the transitions between state 4 and 6 and the transitions between state 5 and 7 can be removed from the chain if we are only interested in the tree heights. Actually, even more transitions can be removed from the chain. Note from Eqs. 9 and 10 that we only need the entries $(1,1), (1,2)$, and $(1, 3)$ in $e^{\Lambda t}$ for calculating the probability of the same tree height in the next position and the transition density conditional on a change. These entries can be found from a reduced rate matrix where states 4, 5, 6, and 7 are removed and the rate from states 2 and 3 to a new absorbing state equals 2. In other words, define the reduced rate matrix

$$\tilde{\Lambda} = \begin{pmatrix} -(1+\rho) & \rho & 0 & 1 \\ 1 & -(3+\rho/2) & \rho/2 & 2 \\ 0 & 4 & -6 & 2 \\ 0 & 0 & 0 & 0 \end{pmatrix},$$

where states are numbered 1, 2, 3, and 4. The holding time and transition density for the model are now given by Eqs. 9 and 10 with Λ substituted by $\tilde{\Lambda}$.

In the left plot in Fig. 8 we illustrate the probability (Eq. 9) of the same tree height in the left and right loci conditional on the tree height in the left locus and different recombination rates.

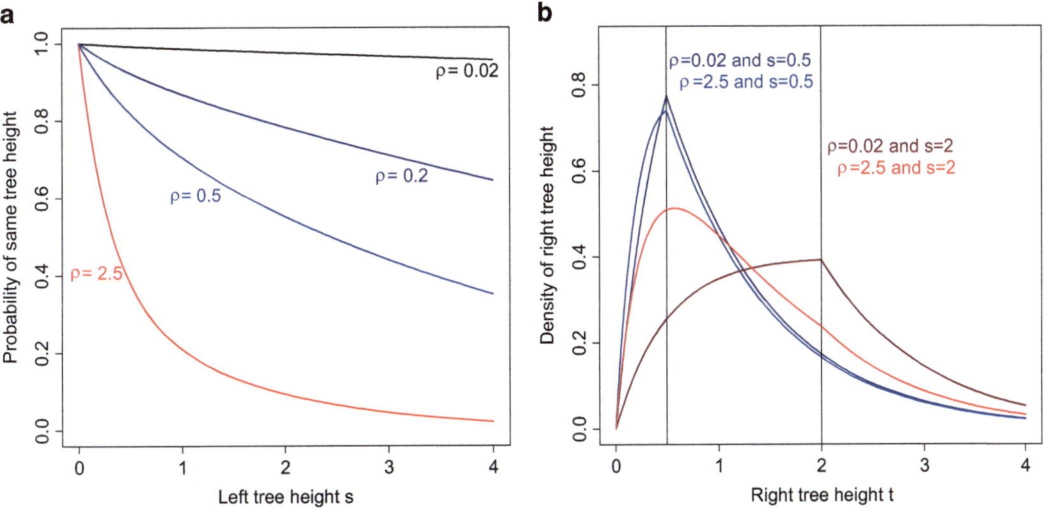

Fig. 8 (**a**) Probability of same tree height. (**b**) Density for right tree height conditional on the left tree height being equal to s and a recombination rate equal to ρ

As expected the probability for identical tree heights decreases with the height of the left tree and with the recombination rate.

In the right plot in Fig. 8 we illustrate the density (Eq. 10) of the right tree height conditional on the left tree height and a change in tree height. When the recombination rate increases, the density for the right tree height moves toward smaller tree heights. The reason is that at least one recombination is needed for having a change in tree height. We also observe that the density is continuous but not differentiable in the position of the left tree height.

4.3 Time Discretization: Setting Up the Finite State HMM

Li and Durbin [14] and Mailund et al. [13] analyze pairs of sequences using a hidden Markov model (HMM). The hidden states are tree heights (times to the most recent common ancestor), and the tree height is discretized to obtain a finite hidden state space. The observed states of the HMM are alignment columns, with probabilities corresponding to a substitution process on the tree (*see* Fig. 9). In the Li and Durbin model, an infinite site model is assumed and observed states are converted to binary data, telling whether the site is heterozygous (one mutation) or homozygous (no mutation).

We now describe how we discretize time for the case of two sequences considered in the previous section. The discrete version of the Markov process is used to build a finite Markov chain along the two sequences. When the finite Markov chain is combined with a substitution process, we obtain an HMM as in Li and Durbin [14].

Let the discrete time points (backward in time) of the Markov chain be $d_0 = 0 < d_1 < d_2 < \cdots < d_{M-1} < d_M = \infty$ and denote the corresponding states by $1, 2, \ldots, M$. State m ($m \in \{1, \ldots, M\}$) then corresponds to a tree height in the interval between d_{m-1} and d_m. The continuous stationary distribution is $\pi(t) = \exp(-t)$, and therefore the discrete times are chosen such that

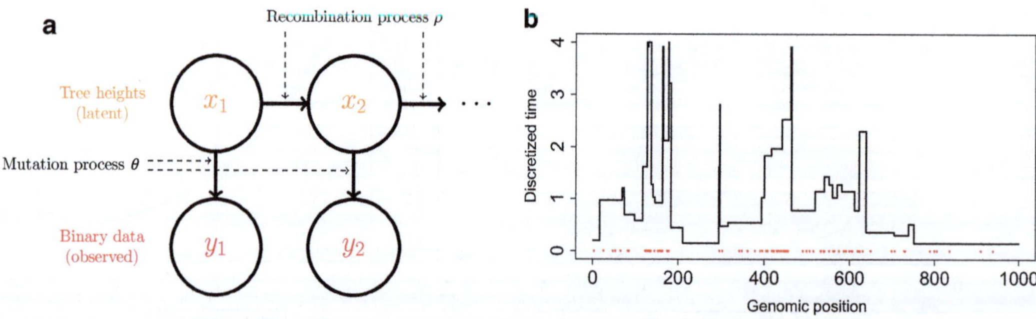

Fig. 9 (**a**) Graphical structure of the hidden Markov Model. (**b**) Simulation from the hidden Markov model

$1 - \exp(-d_m) = m/M$, or $d_m = -\log(1 - m/M)$, where we define $\log(0) = -\infty$.

We now get for $1 \leq \ell, r \leq M$ the joint density

$$P(L = \ell, R = r)$$
$$= \begin{cases} \sum_{k \in \{5,7\}} \sum_{j \in \{5,7\}} \sum_{i \in \{1,2,3\}} [e^{\Lambda d_{\ell-1}}]_{1i} \\ [e^{\Lambda(d_\ell - d_{\ell-1})}]_{ij} [e^{\Lambda(d_{r-1} - d_\ell)}]_{jk} [e^{\Lambda(d_r - d_{r-1})}]_{k8} & \text{if } \ell < r \\ \sum_{i \in \{1,2,3\}} [e^{\Lambda d_{\ell-1}}]_{0i} [e^{\Lambda(d_\ell - d_{\ell-1})}]_{i8} & \text{if } \ell = r \\ P(L = r, R = \ell) & \text{if } \ell > r. \end{cases} \quad (11)$$

The reason for the first case is that in order for the left tree height to be in state $\ell < r$, it must be in state 1, 2, or 3 at time $d_{\ell-1}$ and in state 5 or 7 at time d_ℓ (i.e., there have been no coalescent events before time $d_{\ell-1}$ and a left coalescent event between time $d_{\ell-1}$ and d_ℓ), and similarly it must still be in state 5 or 7 at time d_{r-1} and in state 8 at time d_r (i.e., there have been no coalescent events between time d_ℓ and time d_{r-1} and a right coalescent event between time d_{r-1} and time d_r). The next case corresponds to no coalescent events before time $d_{\ell-1}$ and both a left and a right coalescent event between time $d_{\ell-1}$ and d_ℓ. The last case is due to symmetry of the chain.

From the joint tree states (ℓ, r) we easily get the conditional tree states

$$P_{(\ell,r)} = P(r|\ell) = P(R = r|L = \ell) = \frac{P(L = \ell, R = r)}{P(L = \ell)},$$

where $P(L = \ell) = \Sigma_r P(R = r, L = \ell)$. These probabilities are used in the HMM.

4.4 Careful Treatment of Mutation Process

A careful treatment of the mutation process allows for a more coarse binning procedure and is needed to avoid biasing the results. In continuous time the probability for a mutation given a tree height t is given by $\mu(t) = 1 - \exp(-\theta t)$, and the stationary tree height distribution is $\pi(t) = \exp(-t)$. The probability of a mutation conditionally on the hidden state m becomes

$$\begin{aligned}\mu_m &= p(y_i = 1|x_i = m)\\ &= p(y_i = 1|t\in(d_{m-1}, d_m)) = \frac{p(y_i = 1|t\in(d_{m-1}, d_m))}{p(t\in(d_{m-1}, d_m))}\\ &= \frac{\int_{d_{m-1}}^{d_m} p(y_i = 1|t)\pi(t)dt}{\int_{d_{m-1}}^{d_m} \pi(t)dt} = \frac{\int_{d_{m-1}}^{d_m} (1 - e^{-\theta t})e^{-t}dt}{\int_{d_{m-1}}^{d_m} e^{-t}dt}\\ &= 1 - e^{-\theta d_{m-1}} \frac{\left(1 - e^{-(1+\theta)(d_m - d_{m-1})}\right)}{(1+\theta)\left(1 - e^{-(d_m - d_{m-1})}\right)}.\end{aligned} \tag{12}$$

Note that with a fine discretization we have that the interval $d_m - d_{m-1}$ is small and the first-order Taylor expansion $\exp(-az) \approx 1 - az$ for z small gives

$$p(y_i = 1|x_i = m) \approx 1 - e^{-\theta d_{m-1}},$$

as perhaps expected. We are, however, discretizing the interval $[0, \infty[$, so it is not possible to avoid one or more large bins. Generally we have found that a careful treatment of the mutation process is crucial for accurate inference [36].

4.5 Statistical Inference of Population Parameters from Sequences

Here we choose to focus on three inference methods for estimating the recombination rate. The first method is based on the full likelihood obtained from the classical forward (or backward) algorithm for HMMs. The second is based on the distribution of the distance between segregating sites. This summary statistics was used in Harris and Nielsen [37] for demographic inference. It is sometimes also described as the distribution of the distance between heterozygote sites, runs of homozygosity, or the nearest-neighbor distribution. The third summary statistics is the probability that two sites at certain distance apart are both heterozygote sites. This probability is closely related to the pair correlation function from spatial statistics [36] and to the zygosity correlation introduced in [38].

4.5.1 Summary Statistics: Runs of Homozygosity and Pair Correlation

Recall that in continuous time the probability for a mutation given a tree height t is given by $\mu(t) = 1 - \exp(-\theta t)$, and the stationary tree height distribution is $\pi(t) = \exp(-t)$. The marginal probability for a mutation is therefore given by

$$\int_0^{\infty} \mu(t)\pi(t)dt = \theta/(1+\theta). \tag{13}$$

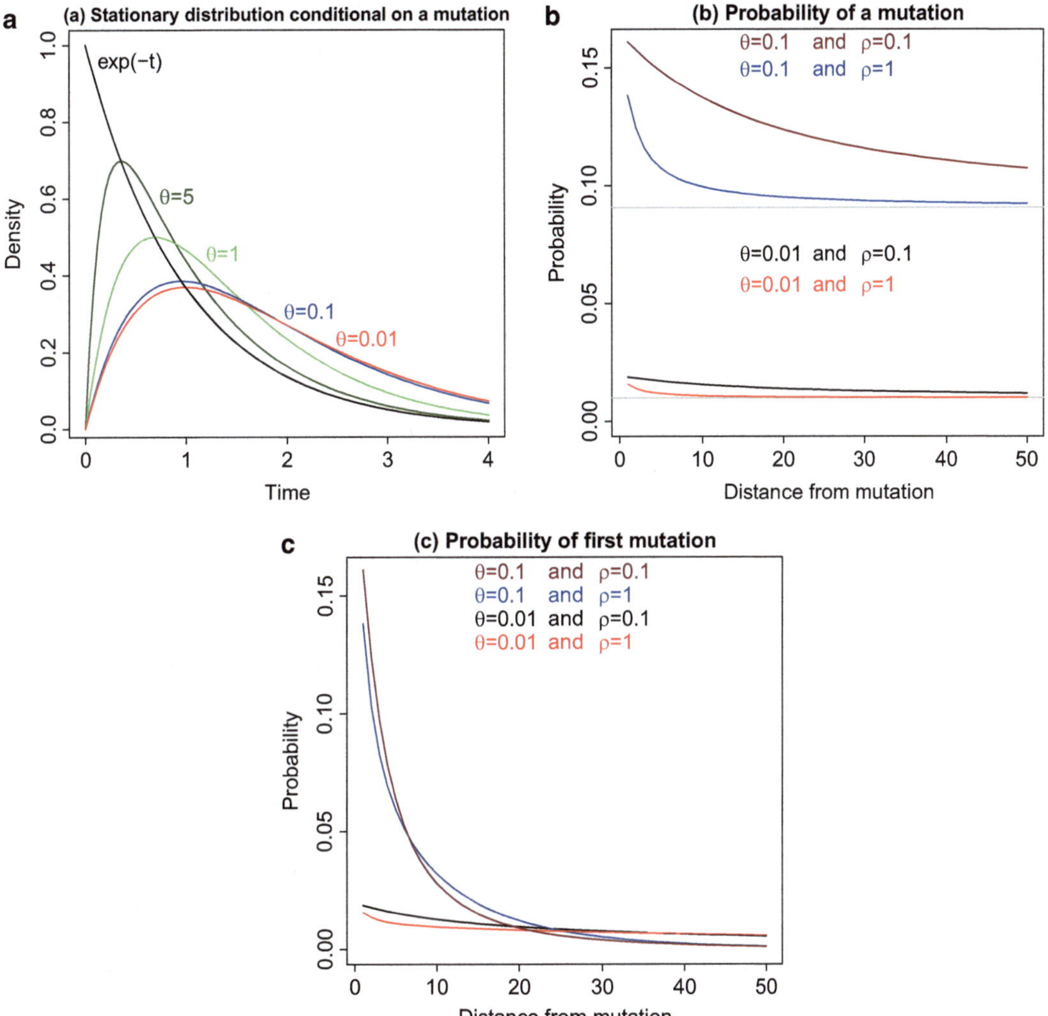

Fig. 10 (**a**) Stationary distribution of tree height conditional on a mutation. (**b**) Probability of a mutation at various distances away from a mutation. (**c**) Probability of the first mutation at various distances away from a mutation

We also get the stationary distribution

$$\phi(t) = \frac{\mu(t)\pi(t)}{\int_0^\infty \mu(t)\pi(t)dt} = \frac{1+\theta}{\theta} e^{-t}\left(1 - e^{-\theta t}\right)$$

for a tree height t conditional on a mutation. Figure 10a shows $\phi(t)$ for different values of θ. Note that small mutation rates imply a higher tree height when we condition on a mutation. In discrete time the probability for a mutation given a tree height m was given by Eq. 12. Let $\boldsymbol{\mu} = (\mu_1, \ldots, \mu_M)$ be the vector of mutation probabilities. The stationary distribution $\boldsymbol{\phi} = (\phi_1, \ldots, \phi_M)$ for a state m conditional on a mutation is given by

$$\phi_\ell = \frac{\mu_\ell \pi_\ell}{\sum_{m=1}^{M} \mu_m \pi_m},$$

where $\pi_m = 1/M$ because this is how the time discretization was chosen.

The probability for a mutation at a distance r from a typical mutation is then given by

$$\kappa(r) = \boldsymbol{\phi}' P^r \mu,$$

where $'$ denotes vector transpose. In Fig. 10b we show $\kappa(r)$ as a function of ρ and θ. Note that the curves converge to $\theta/(1 + \theta)$ and that the behavior for small r is determined by the recombination rate.

The distribution of runs of homozygosity is given by

$$\nu(r) = \boldsymbol{\phi}'[P\text{diag}(e - \mu)]^{r-1} P\boldsymbol{\mu}.$$

Here $e = (1, \ldots, 1)$ is the vector of length M with 1 in every entry and diag$(e - \mu)$ is the diagonal matrix with $e - \mu$ on the diagonal. In Fig. 10c we show $\nu(r)$ as a function of ρ and θ.

4.5.2 Parameter Estimation

We estimate the mutation rate using an estimating equation based on the marginal probability for a mutation (Eq. 13). If the observed frequency of a mutation is $\hat{p}$, then the mutation rate is $\hat{\theta} = \hat{p}/(1 - \hat{p})$ (see left plot in Fig. 11). The recombination rate is estimated using maximum likelihood for the HMM and goodness of fit for the pair correlation (see middle plot in Fig. 11) and runs of homozygosity (see right plot in Fig. 11).

We simulated 50 sequences of length 20,000 base pairs and with mutation rate $\theta = 0.1$ and recombination rate $\rho = 0.1$. We estimated the mutation rate using the estimating equation and the recombination rate using maximum likelihood and the HMM, and goodness of fit for the pair correlation and nearest neighbor (Fig. 12) [35]. As expected the HMM procedure shows the best results because here we are using all the available information. It seems, however, that we are not losing too much power when applying the pair correlation function. This is in contrast to the nearest-neighbor summary statistics that perform much worse than the other two methods.

We have provided a detailed treatment of the main components involved in an analysis of pair of DNA sequences based on an HMM derived from coalescent theory. Pairwise sequentially Markov coalescent (PSMC) models have been extensively applied to various organisms, see, for instance [39–43].

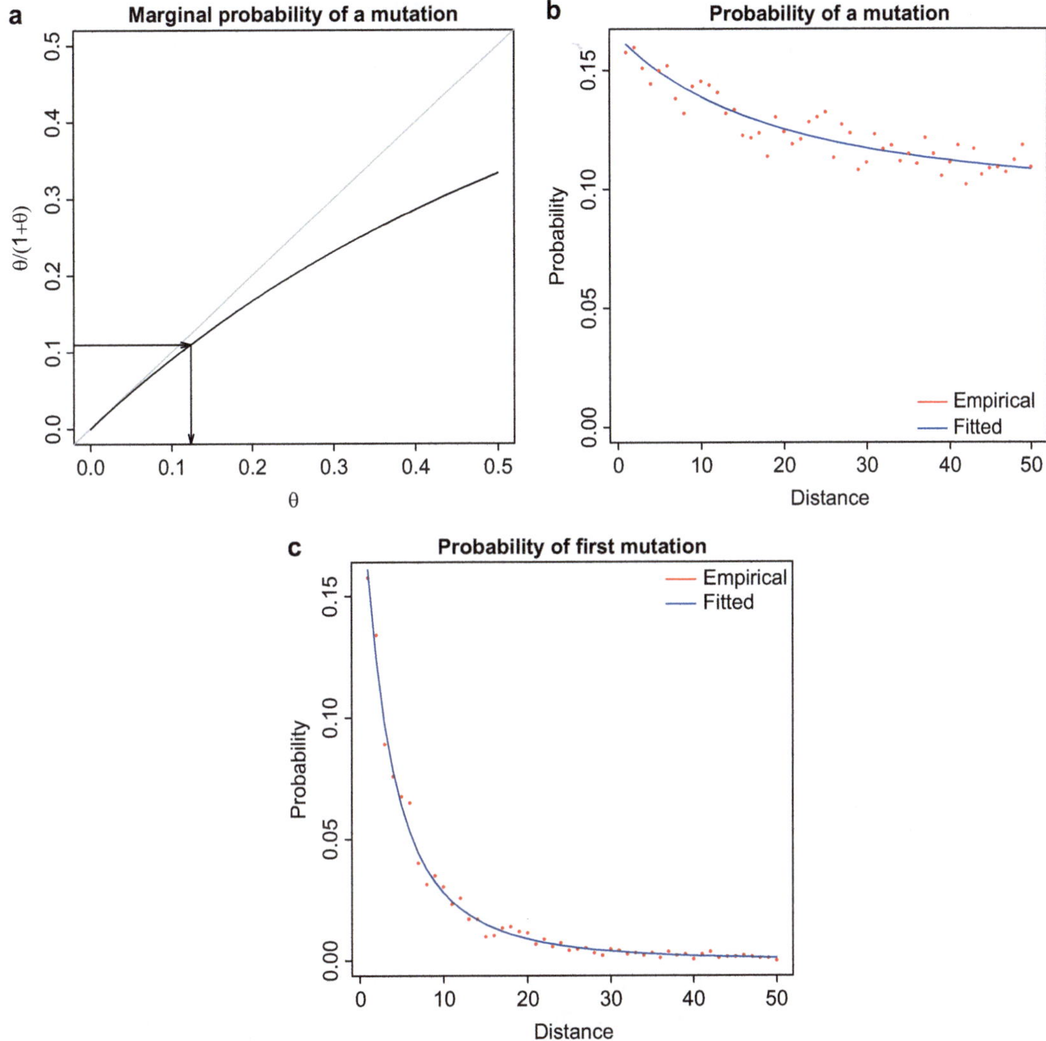

Fig. 11 Parameter estimation for summary statistics. (**a**) The mutation rate θ is estimated from the observed number of mutations and length of the region. (**b**) The recombination rate ρ is estimated using the empirical distribution of a mutation at various distances from a mutation. (**c**) The recombination rate is estimated using the empirical distribution of the first mutation from a mutation

5 Extending the Pairwise Sequentially Markov Coalescent

Extending the SMC to more than two genomes has proved to be challenging. The number of hidden states becomes prohibitive, as several divergence times have to be modeled and combined with distinct possible topologies. Further simplifications are therefore needed to account for an increasing number of genomes.

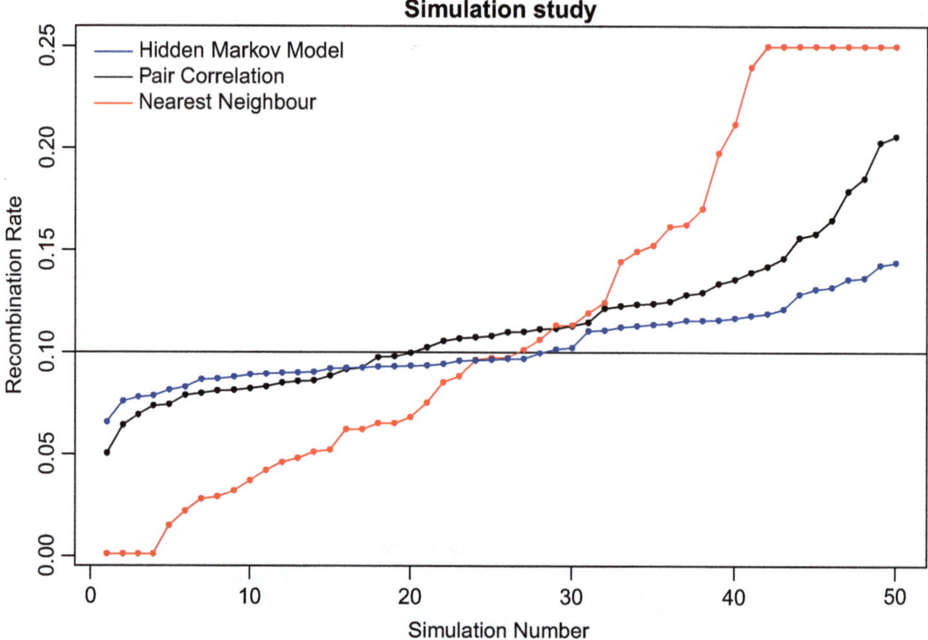

Fig. 12 Results of parameter estimation for simulation study. The pair correlation summary performs rather well compared to the full HMM data analysis. Nearest neighbor is a poor summary statistics

5.1 From 2 to n Genomes

5.1.1 The Multiple Sequentially Markov Coalescent (MSMC)

Schiffels and Durbin [15] proposed to extend the PSMC model [14] to more than two haploid genomes by modeling the most recent coalescence event in the sample. In this framework, the hidden states of the model are a combination of divergence times, taken from a discretized distribution, and identity of the corresponding haplotypes involved. The rationale for such simplification was that the PSMC showed poor resolution in the recent past [14], and considering more genomes would bring additional signal. The drawback of this implementation is that the more genomes are considered, the more "shifted" toward the present is the timeframe where population parameters can be inferred. As a result, the authors reported that with more than 8 diploid individuals (16 haploid genomes), parameters can virtually not be estimated (see also [44] for an illustration of this effect with simulations). Another consequence of this approach is that the recombination rate parameter cannot be reliably estimated [15]. The MSMC was used to infer the recent history of human population. In particular, the authors introduced the possibility to label individuals and look at cross-coalescence rate between groups, a way to get a fine-tuned view of population divergence [15, 45].

5.1.2 The Demographic Inference with Composite Approximate Likelihood (diCal)

An alternative approach was introduced by Song and colleagues [16–18]. The demographic inference with composite approximate likelihood (diCal) approach is based on the conditional sampling distribution, which computes the likelihood of one genome conditioned on the observation of others. Using the so-called composite likelihood formula, it is therefore possible to compute the likelihood of the data for n genomes as the product of the likelihood of one genome given the $n - 1$ other ones and the likelihood the remaining $n - 1$ genomes:

$$P(D_{1\ldots n}|\Theta) = \Pr(D_1|D_{2\ldots n}, \Theta) \times P(D_{2\ldots n|\Theta}),$$

where Θ is the set of model parameters and $D_{1\ldots n}$ denotes the data set with n genomes. By further noting that

$$P(D_{2\ldots n}|\Theta) = P(D_2|D_{3\ldots n}, \Theta) \times P(D_{3\ldots n-1}|\Theta)$$

the likelihood of the full data set can be computed by recursion. The terms $P(D_i|D_{i+1\ldots n})$ form the conditional sampling distribution (CSD). Paul et al. [16] proposed a way to compute the CSD at the cost of introducing several additional hypotheses: (a) the haplotypes upon which the sample is conditioned are considered independent, that is, no coalescence events involving these haplotypes are allowed and (b) mutations can only occur once in any lineage (infinite site hypothesis). The likelihood resulting from this approximated CSD is therefore not exact. This approach was introduced by Li and Stephens [46] and is referred to as the product of approximate conditionals (PAC) model. Under the PAC model, the likelihood depends on the order by which the data is conditioned, which can be circumvented with permutation procedures. While the CSD-based SMC does not have the same drawbacks as the MSMC of Schiffels and Durbin [15], its computational efficiency decreases as the number of haplotypes considered increases and becomes impractical for more than 10 genomes [19]. An elegant feature of the diCal approach is that it can be extended to more complex demographic models, including population structure and gene flow [18, 45]. Such extension is of interest as the SMC approximation has been shown to be sensitive to strong population structure [47].

5.1.3 Extending the SMC with Conditional Site Frequency Spectra (CSFS)

In order to use the large amount of data available in "1000 genomes" projects, Terhorst et al. [19] extended the PSMC in a different direction. Instead of modeling the genealogy of the complete sample, the authors proposed to model the divergence of two haplotypes (the PSMC model) as *hidden states*, yet considering the full set of genomes as *observed states*. In this approach, the transition probabilities of the coalescent HMM are similar to the PSMC (or to be more precise, similar to the MSMC with two haplotypes, as the original PSMC uses the SMC of McVean and Cardin [33] and not the SMC' of Marjoram and Wall [48]), but the emission probabilities are extended to account for the full site frequency spectrum of hundreds of genomes. This *conditional site frequency spectrum* (CSFS) is computed using coalescence theory, offering a generalization of the Poisson random field (PRF) model introduced by Sawyer and Hartl [49]. Just like the original PRF, however, the CSFS ignores linkage of observed states, only linkage between the two conditioned haplotypes is modeled via the SMC. Additional data reduction steps are therefore required to ensure that the independence condition of sampled sites is met.

5.1.4 Explicit Reconstruction of the Ancestral Recombination Graph

While the ARG contains all historical information about a sample of genomes, genomes themselves contain very little information regarding the underlying ARG. As a result, in most statistical inference methods is the ARG treated as a variable accounted for, but not directly inferred. In the SMC models presented above, this is taken care of by the hidden Markov methodology, which computes a likelihood for a given sample by summing over all possible ARG (via the so-called *forward* algorithm). The Viterbi algorithm and the posterior decoding procedure are HMM algorithms that allow to reconstruct a posteriori the most likely ARG for a sample, such procedures are notably used for the inference of patterns of incomplete lineage sorting along genomes [11, 12, 50, 51]. Yet the variance in such estimation is typically very large [12].

Rasmussen et al. [20] proposed a different approach: they developed a Bayesian sampler of ARGs conditioned on a set of genome sequences. Similar in principle to the PAC and CSD approaches, the authors proposed to generate the ARG of n genomes conditioned on the ARG of $n - 1$ genomes, a procedure they refer to as *threading*. The generated ARGs can then be used to infer evolutionary processes of interest. Palacios et al. [52] developed a non-parametric method that allows to estimate the variation in time of the effective population size based on such reconstructed ARG. Rasmussen et al. further showed that while the model used for inference is purely neutral, the a posteriori inferred ARG contains signature of selection, visible for instance as a decrease of the time of the most common ancestor of two samples in the data close to coding sequences. Such approaches offer promising avenues for the development of new statistical methods to detect genomic regions with unusual history.

5.2 The Case of Multiple Species

Hobolth et al. [11] developed a hidden Markov model (HMM) to infer the ancestral recombination graph between three closely related species. Because this model only contains one haploid genome per species, it only allows to infer population parameters in the ancestral species. Dutheil et al. [12] reparametrized this model in the context of the sequentially Markov coalescent. In contrast to the previous approaches, only four hidden states were

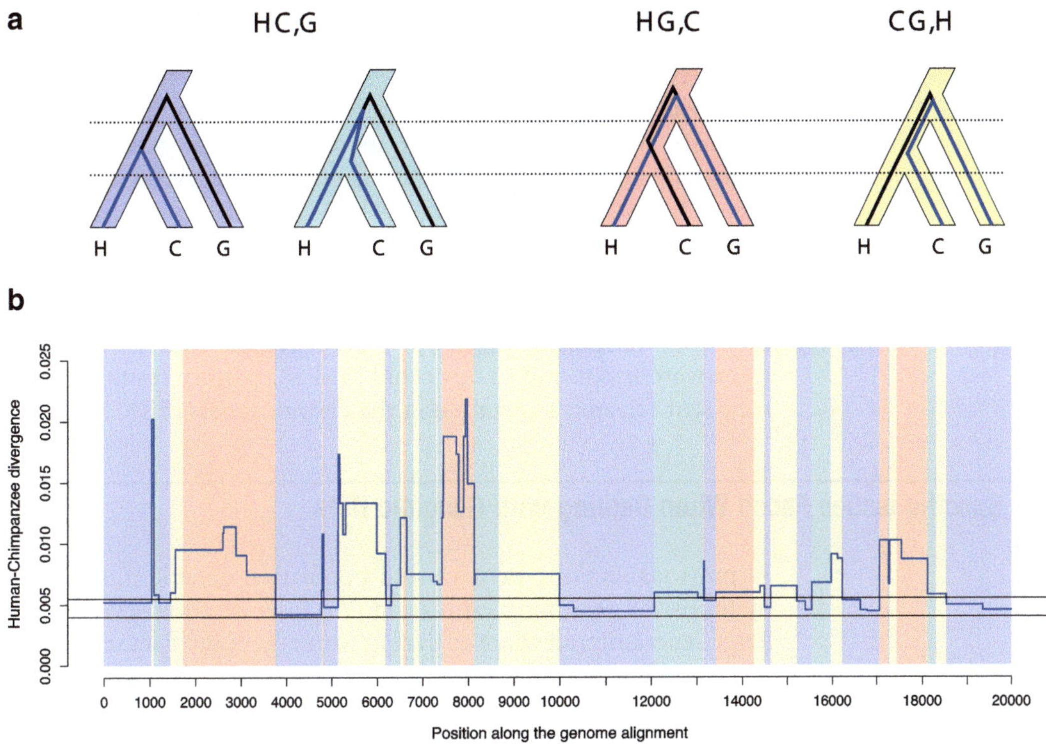

Fig. 13 The coalescent process along genomes of three closely related species. (**a**) Four archetypes of coalescence scenarios with three species, exemplified with human, chimpanzee, and gorilla. In the first scenario, human and chimpanzee coalesce within the human–chimpanzee common ancestor. In the three other scenarios, all sequences coalesce within the common ancestor of all species, with probability 1/3 depending on which two sequences coalesce first. (**b**) Example of genealogical changes along a piece of an alignment. The alignment was simulated using the true coalescent process and parameters corresponding to the human–chimpanzee–orangutan history. The blue line depicts the variation along the genome of the human–chimpanzee divergence. The background colors depict the change in topology, red and yellow corresponding to incomplete lineage sorting. Each change in color or break of the blue line is the result of a recombination event

considered, corresponding to four alternative scenarios of lineage segregation (Fig. 13). In states 1 and 2, the genealogy is consistent with the phylogeny and lineages segregate in the same order as the species. In states 2, 3, and 4, allele divergence predates the first speciation event and ancestral polymorphism persists between the two speciation events, leading to incomplete lineage sorting. The scenarios depicted by states 2, 3 and 4 are equally likely, and in the case of states 3 and 4, the resulting topology is inconsistent with the phylogenetic tree. This model therefore does not rely directly on divergence variation along the genome alignment but uses patterns of topology variation instead to compute the speciation times and ancestral population sizes.

Using this approach, Hobolth et al. estimated a speciation time between human and chimpanzee around 4.1 My and a large ancestral effective population size of 60,000 for the human–chimpanzee ancestor. Dutheil et al. [12] found similar estimates with the same data set while accounting for substitution rate variation across sites and estimated an average recombination rate of 1.7 cM/Mb. With sequencing of more great ape genomes, this approach allowed to estimate population size in several ape ancestors ([27, 50, 53], reviewed in [54]). As ILS is a proxy for ancestral effective population size, a major result of these studies is that the distribution of ILS is not uniform along the genome. For instance, it is reduced in proximity of genes, a pattern that can be explained by background selection [27, 50]. Large regions of the X chromosome were also found to be devoid of ILS, a pattern resulting from recurrent selective sweeps along the chromosomes [55].

6 Specific Issues Faced When Dealing with Genomic Data

In previous sections we discussed population genetic models and methods for parameter estimation. We now describe several challenges encountered when analyzing whole-genome data sets, at the intra- and interspecific levels.

6.1 Sequencing Errors and Rate Variation

Sequencing errors are a well-described source of bias in population genetics analyses, resulting in an excess of singletons [56]. At both the intra- and interspecific/populational level, such error therefore leads to incorrect estimates of local divergence, in particular for recent times. When more divergent sequences are compared, for instance, from distinct species, the issue becomes more complex as the error rate differs between and within sequences due to coverage variation, but also properties of the genome (base composition, repeated elements, etc.). Such errors result in a departure from the molecular clock hypothesis, thus potentially leading to biases in parameter estimates, such as asymmetries in genealogy frequencies [57, 58]. In this respect, data preprocessing becomes a crucial step in any genomic analysis. Methods would also benefit in many cases of inclusion of a proper modeling of such errors. Burgess and Yang noticed that sequencing errors can be seen as a contemporary acceleration in external branches, resulting in an extra branch length [9]. Such an extra length can be easily accommodated in many models. It has to be noted that only a differential in error rates between lineages results in a departure from molecular clock, and in such approaches, one still has to consider that at least one sequence is error-free. In addition, as noted by the authors, assuming a constant error rate over all genomic positions may also turn out to be inappropriate, and better models should allow this rate to vary across the sequence. Such approaches still have to be explored. Moreover, sequencing errors are not distinguishable from lineage-specific acceleration (or deceleration in another species). In that respect, sequence quality scores can be a valuable source of information. They are currently used to preprocess the data by removing doubtful regions, but can ultimately be used in the modeling framework.

The substitution rate also varies along the genome, which potentially affects the reconstruction of sequence genealogy, a phenomenon well known by phylogeneticists. In such case the tools developed for phylogenetic analysis can be applied with a reasonable cost. This generally consists in assuming a prior distribution of the site-specific rate and integrating the likelihood over all possible rates [8, 9, 12]. Alternatively, one can also use one or more outgroup sequences to calibrate the rate, as in [6, 7].

6.2 Diploid Data and Phasing

While sequencing of diploid individuals allows to infer the two alleles present at heterozygous positions, establishing how these alleles are combined on each homologous chromosome requires an additional, error-prone step calling *phasing*. Analyses based on the comparison of individuals from distinct species do not require such information, as the coalescence time of two alleles from the same species is expected to have happened much after the speciation time of the compared species. In such case alleles at each heterozygous position can be sampled randomly [13] in order to build a composite haploid genome. The same rationale applies with respect to the use of the human reference genome, a composite genome obtained from multiple individuals. Conversely, inferences at the population level typically rely on the modeling of haploid genomes and therefore require phased data. A notable exception is the PSMC [14], as well as its extension SMC++ [19], which, when applied to one diploid individual, only requires the knowledge of the position of heterozygous positions.

6.3 Structural Variation and Genome Alignment

Genome data are intrinsically fragmented, firstly because of chromosomal organization, but also because of rearrangements that prevent molecule-to-molecule alignment from one species to another. A genome data set is therefore a set of distinct alignments, one per synteny block. Synteny information can only be extracted when individual genomes are available, which is typically not the case for most "re-sequencing" data sets. At the population level, however, such large-scale variation is considered negligible (but see, for instance, [59] for an exception), while it becomes more prominent when genomes from distinct species are compared. In such cases, a genome alignment is constructed with potential errors ultimately leading to the comparison of nonhomologous regions. So far, the only way to deal with such errors is to restrict the analysis on regions where orthology can be unambiguous resolved, mostly by removing short synteny blocks and regions that contain a high proportion of repeated elements, gaps, and duplications.

7 Discussion

Studying the speciation process with genome data implies new modeling challenges, as the basic configuration of a population genetics data set is drastically changed: instead of having a few loci sequenced in several individuals, we have an (almost) exhaustive set of loci sequenced in several individuals for multiple closely related species. The change involves the spatial dimension, but also time, as the process under study occurred much further back in time than the ones that are commonly studied with a "standard" population genetics data set. The use of the spatial signal has a major consequence, namely, that recombination has to be taken into account, even if it is not directly modeled.

Apart from these considerations, ancestral population genomics, as population genetics, heavily relies on the study of sequence genealogy, its shape, but also its variation. The underlying models build on existing intraspecies population modeling, as they only need to add the species divergence process, that is, a moment in time where two populations stop exchanging genetic material and evolve fully independently. The simplest isolation model assumes that the speciation is instantaneous, while the isolation-with-migration model assumes that the two neo-species can still exchange some material, at least for a certain time after the split. Such a model is not different from a pure isolation model where the ancestral population is structured into two subpopulations: in the first case the speciation time is defined as the time of the split, while in the second case it is the time of the last genetic exchange. Recent work on primates [10] suggests that the speciation of human and chimpanzee was not instantaneous. If the average divergence of the human and chimpanzee is a bit more than 6 My (using widely accepted mutation rate), then the split of the two species initiated around 5.5 My ago, and the last genetic exchange can be dated around 4 My.

The fact that we sample a large number of positions in the genome thus appears to have the power to counterbalance the reduced sampling of individuals within population, allowing the estimation of demographic parameters in the ancestor. Nonetheless, complexity limits are rapidly reached, when considering, for example, three closely related species that can exchange migrants. More complex demographic scenarios, incorporating, for instance, variation in population sizes, will also add additional parameters that might not all be identifiable.

If the ancient speciation processes have left signatures in the contemporary genomes, we do not know yet how far back in time this is true. Intuitively, the signal is maximal when the variation in divergence due to polymorphism is large enough compared to the total divergence. The divergence due to polymorphism is proportional to the ancestral population size, while the divergence of species is only dependent on the time when it happened. So the further back in time we are looking at, the bigger the population sizes need to be so that the ancient polymorphism leaves a signature in the total divergence time. In addition to this, one has to take into consideration sequence saturation due to the too large number of substitutions that accumulated since ancient splits, and the fact that demographic scenarios complexity increases with time. For instance, when considering the evolution of a species over several millions of generations, the probability that a bottleneck, resetting the signal from past events, occurred once is not negligible.

We are in the population genomics era. Data sets are available that allow us to understand the evolutionary processes that are associated with the formation and evolution of species. Analyzing such data sets with the current methodologies however offers major challenges: (1) developing the appropriate computational tools able to handle such data sets with current machines (both in terms of processor speed and memory usage) and (2) design realistic models with enough complexity to capture the most important historical events while remaining computationally tractable.

8 Exercises

8.1 ILS in Primates

Assuming that there are 5 My between the speciation times of human with the gorilla and the orangutan, that the HG ancestral effective population size was 50,000, what is the expected amount of ILS between human, gorilla, and orangutan? Assuming that another 2.5 My separates the speciations of human with chimpanzee and gorilla, with an HC effective ancestral population size of 50,000, what is the expected amount of ILS between human, chimpanzee, and orangutan? We assume a generation time of 20 years for all extent and ancestral primates.

8.2 Estimating Ancestral Population Size from the Observed Amount of ILS

Given that 30% of incomplete lineage sorting is observed between human, chimpanzee, and gorilla and assuming a generation time of 20 years and a that 2.5 My separate the splits between human/chimpanzee and human—chimpanzee/gorilla, what is the effective ancestral population size compatible with this observed amount? Using Burgess and Yang's method [9], a researcher finds a higher estimate of Ne than expected. What could explain this discrepancy?

8.3 Number of Migration Rates in the General k-Population IM Model

In this exercise we show that a k-population IM model has $2(k-1)^2$ migration rates.

1. Starting at the bottom of the k-population IM model argue that the number of migration rates at the level of k populations is $k(k-1)$.

2. Moving up to the next level where $(k - 1)$ populations are present (one of them being an ancestral population, we assume that there two speciation events are never simultaneous) argue that the new ancestral population introduces $2(k - 1)$ new migration rates.
3. Moving up yet another level where $(k - 2)$ populations are present argue that the new ancestral population introduces 2 $(k - 2)$ new migration rates.
4. Show that the total number of migration rates is $2(k - 1)^2$.

Acknowledgements

We thank an anonymous reviewer for constructive comments and detailed suggestions on the manuscript.

References

1. 1000 Genomes Project Consortium, Auton A, Brooks LD, Durbin RM, Garrison EP, Kang HM, Korbel JO, Marchini JL, McCarthy S, McVean GA, Abecasis GR (2015) A global reference for human genetic variation. Nature 526(7571):68–74
2. Prado-Martinez J, Sudmant PH, Kidd JM, Li H, Kelley JL, Lorente-Galdos B, Veeramah KR, Woerner AE, O'Connor TD, Santpere G, Cagan A, Theunert C, Casals F, Laayouni H, Munch K, Hobolth A, Halager AE, Malig M, Hernandez-Rodriguez J, Hernando-Herraez I, Prüfer K, Pybus M, Johnstone L, Lachmann M, Alkan C, Twigg D, Petit N, Baker C, Hormozdiari F, Fernandez-Callejo M, Dabad M, Wilson ML, Stevison L, Camprubí C, Carvalho T, Ruiz-Herrera A, Vives L, Mele M, Abello T, Kondova I, Bontrop RE, Pusey A, Lankester F, Kiyang JA, Bergl RA, Lonsdorf E, Myers S, Ventura M, Gagneux P, Comas D, Siegismund H, Blanc J, Agueda-Calpena L, Gut M, Fulton L, Tishkoff SA, Mullikin JC, Wilson RK, Gut IG, Gonder MK, Ryder OA, Hahn BH, Navarro A, Akey JM, Bertranpetit J, Reich D, Mailund T, Schierup MH, Hvilsom C, Andrés AM, Wall JD, Bustamante CD, Hammer MF, Eichler EE, Marques-Bonet T (2013) Great ape genetic diversity and population history. Nature 499 (7459):471–475
3. Weigel D, Mott R (2009) The 1001 genomes project for Arabidopsis thaliana. Genome Biol 10(5):107
4. Siepel A (2009) Phylogenomics of primates and their ancestral populations. Genome Res 19(11):1929–1941
5. Chen FC, Li WH (2001) Genomic divergences between humans and other hominoids and the effective population size of the common ancestor of humans and chimpanzees. Am J Hum Genet 68(2):444–456
6. Patterson N, Richter DJ, Gnerre S, Lander ES, Reich D (2006) Genetic evidence for complex speciation of humans and chimpanzees. Nature 441(7097):1103–1108
7. Yang Z (2002) Likelihood and Bayes estimation of ancestral population sizes in hominoids using data from multiple loci. Genetics 162 (4):1811–1823
8. Wang Y, Hey J (2010) Estimating divergence parameters with small samples from a large number of loci. Genetics 184(2):363–379
9. Burgess R, Yang Z (2008) Estimation of hominoid ancestral population sizes under Bayesian coalescent models incorporating mutation rate variation and sequencing errors. Mol Biol Evol 25(9):1979–1994
10. Yang Z (2010) A likelihood ratio test of speciation with gene flow using genomic sequence data. Genome Biol Evol 2:200–211
11. Hobolth A, Christensen OF, Mailund T, Schierup MH (2007) Genomic relationships and speciation times of human, chimpanzee, and gorilla inferred from a coalescent hidden Markov model. PLoS Genet 3(2):e7

12. Dutheil JY, Ganapathy G, Hobolth A, Mailund T, Uyenoyama MK, Schierup MH (2009) Ancestral population genomics: the coalescent hidden Markov model approach. Genetics 183(1):259–274
13. Mailund T, Dutheil JY, Hobolth A, Lunter G, Schierup MH (2011) Estimating speciation time and ancestral effective population size of Bornean and Sumatran orangutan subspecies using a coalescent hidden Markov model. PLoS Genet 7(3):e1001319
14. Li H, Durbin R (2011) Inference of human population history from individual whole-genome sequences. Nature 475 (7357):493–496
15. Schiffels S, Durbin R (2014) Inferring human population size and separation history from multiple genome sequences. Nat Genet 46 (8):919–925. http://www.nature.com/ng/journal/v46/n8/full/ng.3015.html
16. Paul JS, Steinrücken M, Song YS (2011) An accurate sequentially Markov conditional sampling distribution for the coalescent with recombination. Genetics 187(4):1115–1128
17. Sheehan S, Harris K, Song YS (2013) Estimating variable effective population sizes from multiple genomes: a sequentially Markov conditional sampling distribution approach. Genetics 194(3):647–662. https://doi.org/10.1534/genetics.112.149096
18. Steinrücken M, Paul JS, Song YS (2013) A sequentially Markov conditional sampling distribution for structured populations with migration and recombination. Theor Popul Biol 87:51–61
19. Terhorst J, Kamm JA, Song YS (2017) Robust and scalable inference of population history from hundreds of unphased whole genomes. Nat Genet 49(2):303–309
20. Rasmussen MD, Hubisz MJ, Gronau I, Siepel A (2014) Genome-wide inference of ancestral recombination graphs. PLoS Genet 10(5): e1004342
21. Wakeley J (2008) Coalescent theory: an introduction, 1st edn. Roberts and Company Publishers, Arapahoe County
22. Hein J, Schierup MH, Wiuf C (2005) Gene genealogies, variation and evolution: a primer in coalescent theory. Oxford University Press, Oxford
23. Tavaré S (2004) Ancestral inference in population genetics, vol 1837. Springer, New York, pp 1–188
24. Takahata N, Nei M (1985) Gene genealogy and variance of interpopulational nucleotide differences. Genetics 110(2):325–344
25. Nielsen R, Wakeley J (2001) Distinguishing migration from isolation: a Markov chain Monte Carlo approach. Genetics 158 (2):885–896
26. Tavaré S (1979) A note on finite homogeneous continuous-time Markov chains. Biometrics 35:831–834
27. Hobolth A, Andersen LN, Mailund T (2011) On computing the coalescence time density in an isolation-with-migration model with few samples. Genetics 187(4):1241–1243
28. Hey J (2010) Isolation-with-migration models for more than two populations. Mol Biol Evol 27(4):905–920
29. Kelleher J, Etheridge AM, McVean G (2016) Efficient coalescent simulation and genealogical analysis for large sample sizes. PLoS Comput Biol 12(5):e1004842
30. Staab PR, Zhu S, Metzler D, Lunter G (2015) Scrm: efficiently simulating long sequences using the approximated coalescent with recombination. Bioinformatics 31(10):1680–1682
31. Simonsen N, Churchill N (1997) A Markov chain model of coalescence with recombination. Theor Popul Biol 52(1):43–59
32. Wiuf C, Hein J (1999) Recombination as a point process along sequences. Theor Popul Biol 55(3):248–259
33. McVean GAT, Cardin NJ (2005) Approximating the coalescent with recombination. Philos Trans R Soc Lond B Biol Sci 360 (1459):1387–1393
34. Durrett R (2008) Probability models for DNA sequence evolution. Probability and its applications. Springer, New York
35. Hobolth A, Jensen JL (2014) Markovian approximation to the finite loci coalescent with recombination along multiple sequences. Theor Popul Biol 98:48–58. https://doi.org/10.1016/j.tpb.2014.01.002
36. Nielsen SV, Simonsen S, Hobolth A (2016) Inferring population genetic parameters: particle filtering, HMM, ripleys K-function or runs of homozygosity? In: Algorithms in bioinformatics. Lecture notes in computer science. Springer, Cham, pp 234–245
37. Harris K, Nielsen R (2013) Inferring demographic history from a spectrum of shared haplotype lengths. PLoS Genet 9(6):e1003521. http://journals.plos.org/plosgenetics/article?id=10.1371/journal.pgen.1003521
38. Lynch M, Xu S, Maruki T, Jiang X, Pfaffelhuber P, Haubold B (2014) Genome-wide linkage-disequilibrium profiles from single individuals. Genetics 198(1):269–281. https://doi.org/10.1534/genetics.114.166843

39. Nadachowska-Brzyska K, Burri R, Smeds L, Ellegren H (2016) PSMC analysis of effective population sizes in molecular ecology and its application to black-and-white Ficedula flycatchers. Mol Ecol 25(5):1058–1072. https://doi.org/10.1111/mec.13540

40. Deinum EE, Halligan DL, Ness RW, Zhang YH, Cong L, Zhang JX, Keightley PD (2015) Recent evolution in Rattus norvegicus is shaped by declining effective population size. Mol Biol Evol 32(10):2547–2558. https://doi.org/10.1093/molbev/msv126

41. Thomas CG, Wang W, Jovelin R, Ghosh R, Lomasko T, Trinh Q, Kruglyak L, Stein LD, Cutter AD (2015) Full-genome evolutionary histories of selfing, splitting, and selection in Caenorhabditis. Genome Res 25(5):667–678. https://doi.org/10.1101/gr.187237.114

42. Nadachowska-Brzyska K, Li C, Smeds L, Zhang G, Ellegren H (2015) Temporal dynamics of avian populations during pleistocene revealed by whole-genome sequences. Curr Biol 25(10):1375–1380. https://doi.org/10.1016/j.cub.2015.03.047

43. Wallberg A, Han F, Wellhagen G, Dahle B, Kawata M, Haddad N, Simões ZLP, Allsopp MH, Kandemir I, De la Rúa P, Pirk CW, Webster MT (2014) A worldwide survey of genome sequence variation provides insight into the evolutionary history of the honeybee Apis mellifera. Nat Genet 46(10):1081–1088. http://www.nature.com/ng/journal/v46/n10/full/ng.3077.html

44. Dutheil JY (2017) Hidden Markov models in population genomics. Methods Mol Biol 1552:149–164

45. Raghavan M, Steinrücken M, Harris K, Schiffels S, Rasmussen S, DeGiorgio M, Albrechtsen A, Valdiosera C, Ávila Arcos MC, Malaspinas AS, Eriksson A, Moltke I, Metspalu M, Homburger JR, Wall J, Cornejo OE, Moreno-Mayar JV, Korneliussen TS, Pierre T, Rasmussen M, Campos PF, Damgaard PDB, Allentoft ME, Lindo J, Metspalu E, Rodríguez-Varela R, Mansilla J, Henrickson C, Seguin-Orlando A, Malmström H, Stafford T, Shringarpure SS, Moreno-Estrada A, Karmin M, Tambets K, Bergström A, Xue Y, Warmuth V, Friend AD, Singarayer J, Valdes P, Balloux F, Leboreiro I, Vera JL, Rangel-Villalobos H, Pettener D, Luiselli D, Davis LG, Heyer E, Zollikofer CPE, Ponce de León MS, Smith CI, Grimes V, Pike KA, Deal M, Fuller BT, Arriaza B, Standen V, Luz MF, Ricaut F, Guidon N, Osipova L, Voevoda MI, Posukh OL, Balanovsky O, Lavryashina M, Bogunov Y, Khusnutdinova E, Gubina M, Balanovska E, Fedorova S, Litvinov S, Malyarchuk B, Derenko M, Mosher MJ, Archer D, Cybulski J, Petzelt B, Mitchell J, Worl R, Norman PJ, Parham P, Kemp BM, Kivisild T, Tyler-Smith C, Sandhu MS, Crawford M, Villems R, Smith DG, Waters MR, Goebel T, Johnson JR, Malhi RS, Jakobsson M, Meltzer DJ, Manica A, Durbin R, Bustamante CD, Song YS, Nielsen R, Willerslev E (2015) Population genetics. Genomic evidence for the Pleistocene and recent population history of native Americans. Science 349(6250):aab3884

46. Li N, Stephens M (2003) Modeling linkage disequilibrium and identifying recombination hotspots using single-nucleotide polymorphism data. Genetics 165(4):2213–2233

47. Eriksson A, Mahjani B, Mehlig B (2009) Sequential Markov coalescent algorithms for population models with demographic structure. Theor Popul Biol 76(2):84–91

48. Marjoram P, Wall JD (2006) Fast "coalescent" simulation. BMC Genet 7(1):16

49. Sawyer SA, Hartl DL (1992) Population genetics of polymorphism and divergence. Genetics 132(4):1161–1176. http://www.genetics.org/content/132/4/1161

50. Scally A, Dutheil JY, Hillier LW, Jordan GE, Goodhead I, Herrero J, Hobolth A, Lappalainen T, Mailund T, Marques-Bonet T, McCarthy S, Montgomery SH, Schwalie PC, Tang YA, Ward MC, Xue Y, Yngvadottir B, Alkan C, Andersen LN, Ayub Q, Ball EV, Beal K, Bradley BJ, Chen Y, Clee CM, Fitzgerald S, Graves TA, Gu Y, Heath P, Heger A, Karakoc E, Kolb-Kokocinski A, Laird GK, Lunter G, Meader S, Mort M, Mullikin JC, Munch K, O'Connor TD, Phillips AD, Prado-Martinez J, Rogers AS, Sajjadian S, Schmidt D, Shaw K, Simpson JT, Stenson PD, Turner DJ, Vigilant L, Vilella AJ, Whitener W, Zhu B, Cooper DN, de Jong P, Dermitzakis ET, Eichler EE, Flicek P, Goldman N, Mundy NI, Ning Z, Odom DT, Ponting CP, Quail MA, Ryder OA, Searle SM, Warren WC, Wilson RK, Schierup MH, Rogers J, Tyler-Smith C, Durbin R (2012) Insights into hominid evolution from the gorilla genome sequence. Nature 483 (7388):169–175

51. Munch K, Mailund T, Dutheil JY, Schierup MH (2014) A fine-scale recombination map of the human-chimpanzee ancestor reveals faster change in humans than in chimpanzees and a strong impact of GC-biased gene conversion. Genome Res 24(3):467–474. https://doi.org/10.1101/gr.158469.113

52. Palacios JA, Wakeley J, Ramachandran S (2015) Bayesian nonparametric inference of population size changes from sequential genealogies. Genetics 201(1):281–304
53. Prüfer K, Munch K, Hellmann I, Akagi K, Miller JR, Walenz B, Koren S, Sutton G, Kodira C, Winer R, Knight JR, Mullikin JC, Meader SJ, Ponting CP, Lunter G, Higashino S, Hobolth A, Dutheil J, Karakoç E, Alkan C, Sajjadian S, Catacchio CR, Ventura M, Marques-Bonet T, Eichler EE, André C, Atencia R, Mugisha L, Junhold J, Patterson N, Siebauer M, Good JM, Fischer A, Ptak SE, Lachmann M, Symer DE, Mailund T, Schierup MH, Andrés AM, Kelso J, Pääbo S (2012) The bonobo genome compared with the chimpanzee and human genomes. Nature 486(7404):527–531
54. Mailund T, Munch K, Schierup MH (2014) Lineage sorting in apes. Annu Rev Genet 48:519–535
55. Dutheil JY, Munch K, Nam K, Mailund T, Schierup MH (2015) Strong selective sweeps on the X chromosome in the human-chimpanzee ancestor explain its low divergence. PLoS Genet 11(8):e1005451
56. Achaz G (2008) Testing for neutrality in samples with sequencing errors. Genetics 179 (3):1409–1424
57. Slatkin M, Pollack JLL (2008) Subdivision in an ancestral species creates asymmetry in gene trees. Mol Biol Evol 25(10):2241–2246
58. Hobolth A, Dutheil JY, Hawks J, Schierup MH, Mailund T (2011) Incomplete lineage sorting patterns among human, chimpanzee and orangutan suggest recent orangutan speciation and widespread natural selection. Genome Res 21(3):349–356
59. Stukenbrock EH, Jørgensen FG, Zala M, Hansen TT, McDonald BA, Schierup MH (2010) Whole-genome and chromosome evolution associated with host adaptation and speciation of the wheat pathogen Mycosphaerella graminicola. PLoS Genet 6(12):e1001189

Chapter 19

Introduction to the Analysis of Environmental Sequences: Metagenomics with MEGAN

Caner Bağcı, Sina Beier, Anna Górska, and Daniel H. Huson

Abstract

Metagenomics has become a part of the standard toolkit for scientists interested in studying microbes in the environment. Compared to 16S rDNA sequencing, which allows coarse taxonomic profiling of samples, shotgun metagenomic sequencing provides a more detailed analysis of the taxonomic and functional content of samples. Long read technologies, such as developed by Pacific Biosciences or Oxford Nanopore, produce much longer stretches of informative sequence, greatly simplifying the difficult and time-consuming process of metagenomic assembly. MEGAN6 provides a wide range of analysis and visualization methods for the analysis of short and long read metagenomic data. A simple and efficient analysis pipeline for metagenomic analysis consists of the DIAMOND alignment tool on short reads, or the LAST alignment tool on long reads, followed by MEGAN. This approach performs taxonomic and functional abundance analysis, supports comparative analysis of large-scale experiments, and allows one to involve experimental metadata in the analysis.

Key words Metagenomics, Software, MEGAN, Taxonomic analysis, Functional analysis, Long reads

1 Introduction

Metagenomics is the study of microbiome samples, such as obtained from ocean water, soil, plant matter, or feces, say, using high-throughput DNA sequencing [1]. Metagenomic sequencing allows the study of microorganisms found in environmental samples without relying on culturing methods or prior knowledge of the composition of the community. With metagenomics, one can determine the taxonomic and functional content of samples.

While most metagenomic projects to date have used short read sequencing (next-generation sequencing), there is increasing interest in using long read sequencing technologies in this area. Long read technologies have been considered too expensive, difficult, or error-prone for application in metagenomics. However, this is changing and computational analysis methods designed for processing short reads now need to be modified to work well on long

Maria Anisimova (ed.), *Evolutionary Genomics: Statistical and Computational Methods*, Methods in Molecular Biology, vol. 1910, https://doi.org/10.1007/978-1-4939-9074-0_19,

reads, so as to make good use of the ability of long reads to cover multiple genes.

A major computational challenge in metagenomics is the alignment of sequencing reads against a comprehensive reference database. Billions of reads can be aligned against a large protein reference database in reasonable time using high-throughput alignment tools such as DIAMOND [2]. Long reads require frame-shift aware alignment tools, such as LAST [3, 4], because insertions or deletions due to sequencing errors impact long reads, as discussed in Subheading 2.

In the following, we will first discuss how to perform basic alignment and analysis of short reads in Subheading 2.1 and long reads in Subheading 2.2. We will then show, in Subheading 3, how to compare large numbers of samples in MEGAN6 [5] and perform basic statistical analysis of the samples and their metadata. In Subheading 4 we briefly discuss the challenges we will have to face to further improve the analysis of data from environmental samples. Finally, in Subheading 4.1 we describe some additional resources available for using MEGAN 6.

2 Workflows for Metagenomic Analysis with MEGAN

The basic workflow for using MEGAN consists of two main steps: read alignment against a reference database and then import an analysis of the alignments in MEGAN. The aim of pipeline is to perform taxonomic and functional binning of the input reads.

The alignment can be performed using a number of different tools depending on the type of sequencing data and on the chosen database, its sequence type, size, and available computer power. For smaller databases more sensitive tools can be chosen such as MALT [6] or even BLAST [7]. These tools generally offer higher sensitivity at the cost of a longer runtime. For large datasets and databases, it is more suitable to choose an alignment tool such as DIAMOND or LAST. We use the NCBI NR database [8] with both of the latter tools, because it is the largest and most comprehensive protein database available today. NCBI NR contains 144.5 million protein sequences (August 2017).

2.1 Short Read Pipeline

We describe here the basic short read analysis pipeline as shown in Fig. 1. By default, we use DIAMOND to align reads against the full NCBI NR database.

Before running the pipeline, one can optionally perform preprocessing, that is, quality control, trimming, and filtering, of the raw reads. However, these steps usually have little impact on the results of the alignment-based analysis described in this document.

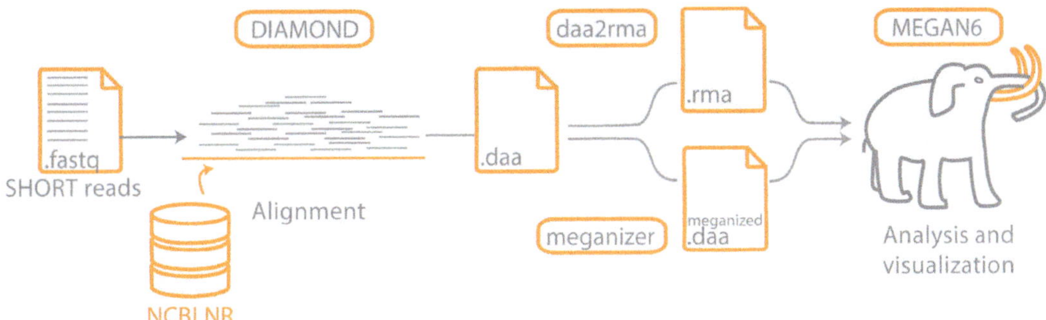

Fig. 1 Basic pipeline for short read analysis

2.1.1 Read Alignment with DIAMOND

DIAMOND uses double indexed alignment, which means both the reference database and the query are indexed for comparison. This leads to a large speedup especially for large queries and databases. Like BLASTX, DIAMOND uses the "seed and extend" method to find all matches between a query and the database. To further increase speed, DIAMOND utilizes spaced seeds, which are long seeds where only some positions are used for matching the seed. This leads to another increase of speed without decreasing sensitivity.

DIAMOND can be run either in fast or sensitive mode. Fast mode will run around 20,000 times faster than BLASTX on short reads and will be able to find 75–90% of all relevant matches that one would find with BLASTX, while sensitive mode provides a speedup of 2500× while recovering up to 94% of significant matches.

2.1.2 Taxonomic and Functional Classification with MEGAN6

DIAMOND can save alignments in a compressed format called DAA (DIAMOND alignment archive) format. DAA files can be imported into MEGAN6 in multiple ways. A small number of small DAA files can easily be imported interactively using menu items provided in MEGAN. For larger datasets and or many files, one should use the command-line tools provided with MEGAN. These include `daa2rma`, which will generate a RMA file as used by MEGAN from one or two (for paired reads) DIAMOND files and `daa-meganizer`, which analyzes a DAA file and then appends the result to the end of the file. Such "meganized" DAA files can then be opened directly in MEGAN. The latter approach is much faster and is more space efficient. However, to use paired reads all alignments have to be in the same file.

One can use the program `blast2rma` to process the output of a range of different alignment programs, such as BLAST.

During the processing of alignments for MEGAN, the reads will be assigned to nodes in the NCBI taxonomy and any functional classifications that have been configured in the import dialog or on the command-line. Taxonomic binning of each read is done separately, by assigning it to the lowest common ancestor (LCA) of its significant matches. Matches can be filtered by multiple parameters, for example, e-value and bit-score, as well as sequence identity. Only matches passing those filters will be used to determine the LCA. It is also important to choose the minimum support (or minimum support percentage), the number or percentage of reads that must be assigned to a single taxon before it will be part of the final result. Reads assigned to a taxon that does not pass the minimum support filter will be pushed up the taxonomy until a taxon is found that passes the filter.

Functional binning is performed by mapping the NCBI database accessions for the matches of a read to identifiers of the selected functional classification. Mapping files are currently available for InterPro2GO [9, 10] (InterPro families embedded in a GO-based hierarchy), eggNOG [11], KEGG [12], and SEED [13].

2.1.3 Investigation of the Results

The resulting files can be opened and interactively investigated using the MEGAN6 graphical user interface. The first view when opening a file is always a hierarchical representation of the taxonomic composition of the sample. Selecting different nodes of this tree, the user can uncover further information on the reads mapped to the represented taxon. Selecting `Inspect Reads` on a node will open the Inspector Window, which displays the reads assigned to that node, as well as their alignments. This functionality can be used both in the Taxonomy Viewer, where nodes represent taxa, and in any of the Functional Viewers. Figure 2a shows an example of the Inspector Window.

Instead of just viewing a listing of the matches and alignments, it is also possible to select `Show Alignments`. This will open the Alignment Viewer (Fig. 2b), where for each of the database references with matches from the reads assigned to the selected node it is

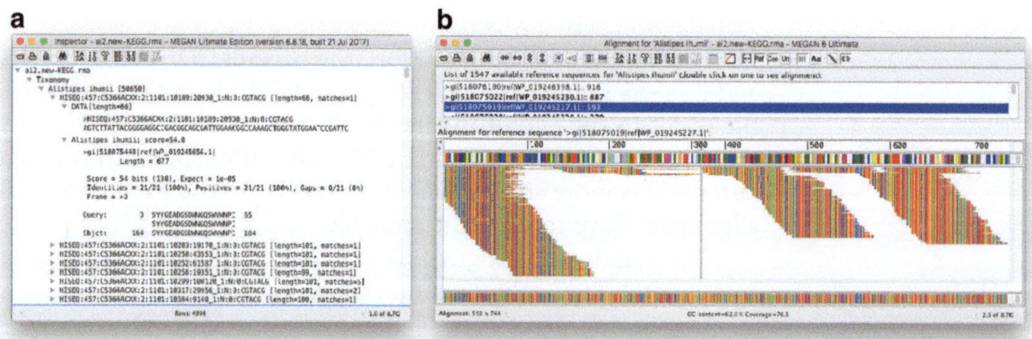

Fig. 2 (**a**) The Inspector Viewer showing some reads that have been assigned to *Alistipes ihumii*. (**b**) The Alignment Viewer showing reads aligned to a reference sequence

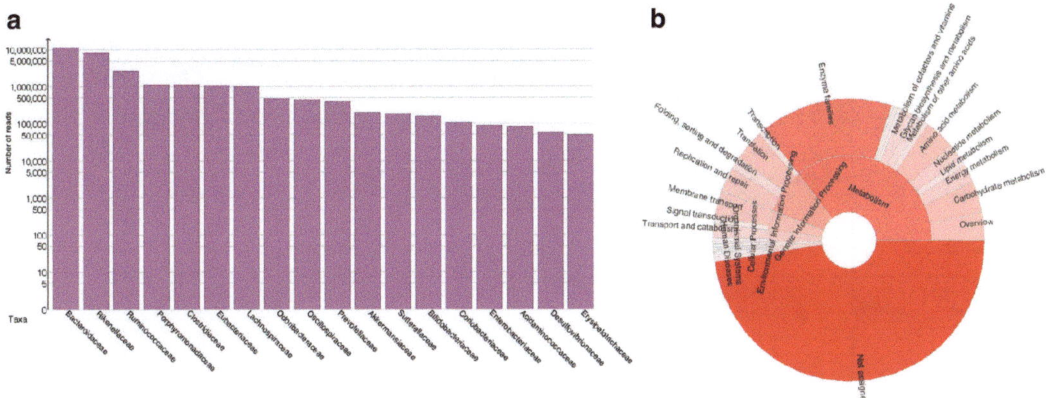

Fig. 3 (**a**) Bar chart of taxonomic assignments on family level, sorted by abundance. (**b**) Radial chart of functional assignments to KEGG for the same sample from [14]

possible to show the alignment of all of those reads on the reference. This can be useful, say, to determine how much of a reference gene is covered by reads.

Apart from being able to investigating taxonomic diversity, the advantage of using metagenomic sequencing to study an environmental sample is the ability to study the functional potential of the community. MEGAN currently provides four different functional classification systems for this purpose: InterPro & GO, eggNOG, KEGG, and SEED.

Each functional classification is displayed as a tree. The nodes of the tree can be investigated very much like the nodes of the taxonomic tree. Abundances can be visualized using different visualization options from simple bar charts over box plots and heat maps to radial tree charts drawn based on the abundances of the selected nodes. Two examples show charts that are shown in Fig. 3.

Alignments or reads matching a selected function can be exported to a text file or extracted to a new MEGAN document. This makes it possible to study only a part of a microbial community that is of particular interest. For example, if you select nodes associated with antibiotic resistance genes, you can determine which taxonomic assignment the reads assigned to antibiotic resistance genes have. An example of this is shown in Fig. 4.

If you want to study the full gene sequence of proteins found in your samples and be able to compare variants of those genes, it can be helpful to use gene-centric assembly [15]. Gene-centric assembly uses the alignments to reference proteins to assemble the matching reads. One can thus obtain the gene sequences from different organisms found in a sample for further analysis steps.

We will introduce more possibilities for studying the taxonomic and functional diversity of multiple samples in comparison in Subheading 3.

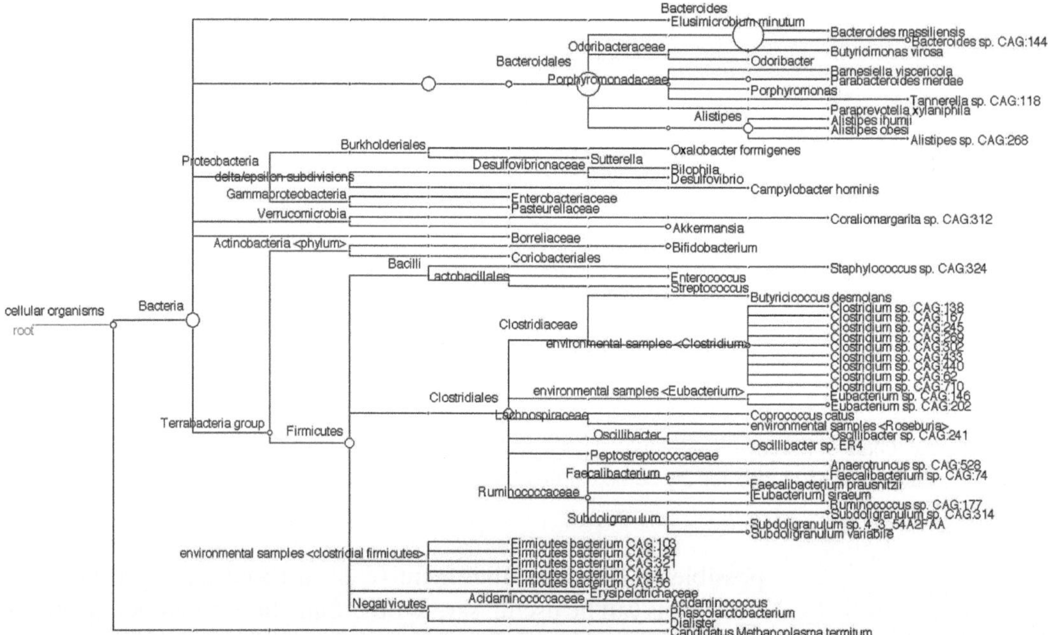

Fig. 4 Taxonomic assignment of reads from the day 0 sample for "Alice" from the ASARI [14] dataset which have been assigned to "resistance of fluoroquinolones" in the SEED hierarchy

2.2 Long Read Pipeline

As presented in the previous section, using metagenomic short reads, one can assembly gene sequences and obtain variants of a single gene using a gene-centric assembly, or of course use other assembly techniques. However, using short read data, it is very difficult to establish whether different genes are present in the same organism. We can connect the genes if they are found on a single DNA molecule with long sequencing reads, provided by third generation sequencing technologies such as PacBio [16] or Oxford Nanopore [17].

The PacBio and Nanopore devices can produce reads that are hundreds of thousands of bases long, with error rates of around 10%, say [17]. In contrast to short reads, which each can be safely assumed to overlap with only a single gene, long read will usually overlap or contain multiple genes. Hence, many popular short read alignment and analysis algorithms may require modification so as to take into account that a given read can align to multiple genes.

2.2.1 Long Read Analysis Pipeline

The basic long read analysis pipeline is analogous to the above described short read pipeline, and consists of the alignment and MEGAN analysis steps (Fig 5), but the details of the analysis pipeline as well as some components of MEGAN6 differ from the short read solution.

As described in the following, for long reads alignment is performed using LAST, processing of the alignments requires an additional step and MEGAN provides some modified algorithms for processing and visualizing long reads.

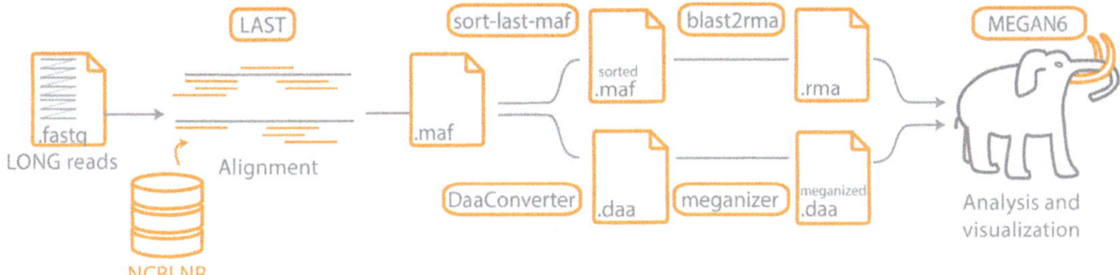

Fig. 5 Basic pipeline for long read analysis

```
Score = 86 bits (159), Expect = 7e-13
Identities = 34/37 (92%), Positives = 34/37 (92%), Gaps = 2/37 (5%)
Frame = -1
Query:  1080    EAVMVLSLDAEA\LVGYRE/KFPAWMDADRFEIKPRK   976
                EAVMVLSLDAEA LV YRE KFPAWMDADRFEIKPRK
Sbjct:  232     EAVMVLSLDAEA-LVRYRE-KFPAWMDADRFEIKPRK   266
```

Fig. 6 A frame-shift aware DNA-to-protein alignment produced by LAST

2.2.2 Alignment Using LAST

Third generation sequencing technologies produce much longer reads, with a higher error rate (approximately 10%, mostly insertions and deletions). Most DNA-to-protein aligners (such as BLASTX [7] or DIAMOND) translate the complete DNA query sequence in all six reading frames and then align the translated sequences against the protein database. Insertions or deletions in long reads cause a frame-shift and break translation-based alignments. LAST is a frame-shift aware aligner that incorporates single-base insertions or deletions into the alignment calculation. These are represented as "\" for forward-shifts and "/" for reverse-shifts, as shown in Fig. 6.

LAST, when used with large databases, such as NCBI-nr, splits the database into several volumes and indexes them individually. Similarly the large input files are loaded in separate volumes, and each volume of input is searched against each volume of the database. LAST, by default, generates output in MAF, "Multiple Alignment Format."

2.2.3 Taxonomic and Functional Classification of Long Reads

Because of processing both the query and database in different volumes and writing the output as soon as it is generated, the alignments for a single read appear in different parts of the MAF output of LAST. MEGAN processes alignment files line-by-line, identifies all alignments of a single read, and then assigns that read to a taxonomic and/or functional class. The unordered structure of LAST output prevents MEGAN from doing this. Thus, MAF files produced by LAST must be sorted before they are imported to MEGAN. For this task, MEGAN provides a command-line script, called *sort-last-maf*.

Alternatively, the user can use *DAA_Converter* (available at http://github.com/BenjaminAlbrecht84/DAA_Converter), which converts a given MAF file to a DAA file. This has several advantages, including space compression and faster processing. Additionally, the output of LAST can directly be piped into *DAA_Converter* which will then convert the output into a DAA file as LAST continues to operate. The trade-off when using *DAA_Converter* currently is that the alignments are filtered out with the default settings in MEGAN6 and resulting DAA file only has the alignments that would pass the filter, making it impossible to change filtration parameters without running LAST again once the conversion is done.

Similar to short reads, these long read MAF and DAA can then be imported into MEGAN and each read will get assigned to a taxon and/or functional class(es) of any provided functional hierarchy. The filtration based on bit-score of alignments work differently for long reads. In case of short reads, the alignments are filtered globally—only those that are within top 10% (by default) of the best-scoring alignment are taken into account. For long reads, this filtration is applied to each "gene" separately, as one long read can contain many different genes along its length. The alignments that overlap significantly (>90% by default) are grouped into *segments*, denoting different genes, and each interval is then processed individually in the filtering step.

The LCA algorithm to assign reads to taxonomic classes is also modified for long reads. As there are multiple genes on a single long read, and each of them may be conserved in different clades of the taxonomic tree, the *naïve LCA* is usually uninformative. Instead long reads are assigned to the most specific taxon that covers more than a fixed percentage (>80% by default) of every base pair that has an alignment. This algorithm assigns reads specifically to lower levels of taxonomy as long as they cover a gene which has low level conservation, other taxa gets lower percentages of coverage. Functional classification of long reads does not necessarily assign each read into one functional class, instead reads are assigned to the functional class of best-scoring alignment in each *segment*, thus each segment is assigned to one function and one read can be assigned to multiple different functional classes.

2.2.4 Investigation of the Results

The first view the user gets when a long read dataset is loaded in to MEGAN6 is identical to that of a short read dataset; however, there are some underlying differences and several investigation modes designed specifically for long reads.

Due to a large variability of read length of long reads [18], it is impractical for MEGAN to report number of reads assigned to class as a mean of abundance. Using the raw read length is also not feasible for Nanopore technology as reads tend to have "head"

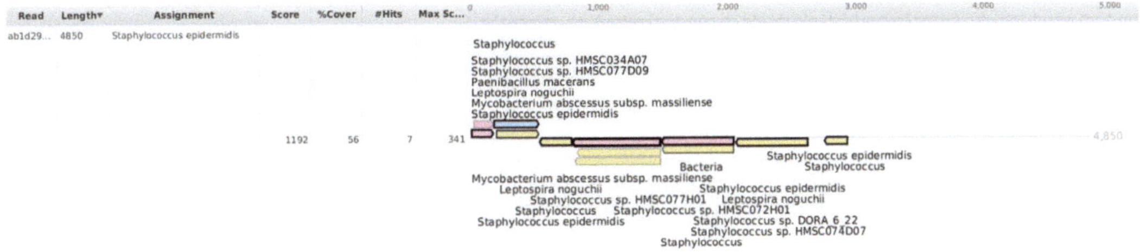

Fig. 7 Long Read Inspector in MEGAN6. The read is drawn as a line in the middle and the protein alignments are drawn as arrows on their corresponding positions and strands on the read

and "tail" regions composed of random bases [19] (Fig. 7 shows a read whose tail region has no significant alignment to any protein in the database). Thus, the default mean of reporting the abundance for a particular taxon or functional class in long read pipeline is the number of aligned bases.

The number of alignments on a long read can easily exceed hundreds and complicates the Alignment Viewer and the Inspector features of MEGAN6. In order to simplify the investigation of alignments on the reads, MEGAN6 offers a *Long Read Inspector* window (Fig. 7), accessible via right-click on any of the nodes in the main view. This inspector draws reads as horizontal lines and alignments as arrows on their corresponding positions. The names of taxa or functional classes are also linked to these alignment arrows.

The Inspector Window helps particularly in the case of suspicious assignments. Figure 8a shows the inspector view for a read that was assigned to *Trichuris trichiura*, a human parasitic whipworm, in a sample of known mixture of microorganisms [20]. A closer inspection to Fig. 8a lets us see that, although the read is spanned by several alignments from *Escherichia coli*, it is assigned to *T. trichiura* because the total length of alignments to *T. trichiura* is longer than 80% whereas it is below that for *E. coli* and all other competing taxa.

For further analysis of such suspicious assignments, MEGAN6 offers a remote BLAST function, in which selected reads are aligned against a selected database (such as the nucleotide collection—NCBI nt) on the NCBI website and the resulting assignments are captured, processed, and presented in a new MEGAN document. In Fig. 8b, we see that our "suspicious" read is assigned to *E. coli*, which was in the known mixture of microorganisms, based on remote NCBI-BLAST against NCBI nt.

Similar to exporting alignments and reads as explained in the previous section, these can also be exported in general feature format (GFF) for downstream analysis. This provides a simple way of obtaining the annotation, especially for long reads and contigs. The annotations exported to the GFF files contain the accessions of

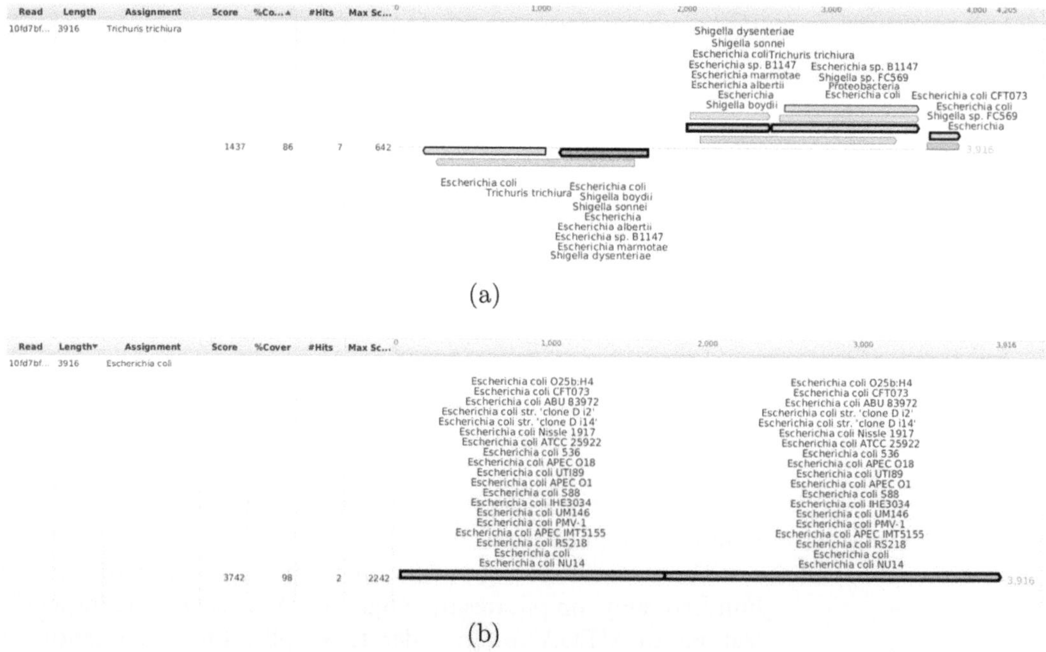

Fig. 8 MEGAN6 offers a remote BLAST functionality, namely "BLAST on NCBI," which can be used for suspicious assignments. (**a**) Long Read Inspector view for a read assigned to *Trichuris trichiura*, based on protein alignments against NCBI nr. (**b**) Long Read Inspector view for the same read as in (**a**), assigned to *Escherichia coli*, after searching it against nucleotide collection of NCBI using the remote BLAST functionality of MEGAN6

references and their corresponding taxonomic and/or functional mappings depending on which mapping files were used during importing the dataset into MEGAN.

3 Comparison of Multiple Samples

Most modern metagenomics experiments include the collection and analysis of multiple samples to compare different groups with controls or study the dynamic changes of a microbial community over time. Hence, a very important feature of MEGAN is the ability to load multiple datasets into a single "comparison document" (megan file). This is a light-weight file that does not contain the original reads and alignments, but allows one to compare the taxonomic and functional diversity of multiple samples.

To be able to easily compare groups of samples and relate findings to features attached to samples, it is helpful to import metadata. Metadata should be provided in tabular format and connect the sample IDs to attributes whose values can be text, numeric, or boolean values. Using this information you can group samples in different visualizations. For example, this allows easier interpretation of the principal component analysis (PCoA) plots in MEGAN. Principal components can be calculated using different distance measures including Bray–Curtis or simple Euclidean distances. MEGAN can include bi-plots and tri-plot vectors into the PCoA plot, which represent the top taxonomic or functional classes and metadata features, respectively, that correlate most with the differences between samples. Figure 9 shows multiple examples of PCoA plots including bi-plot and tri-plot vectors.

MEGAN can also calculate and visualize co-occurrence and correlation plots. For correlation there are two options. The first is useful for time series analysis, because it calculates correlations between different taxa. This can be used to determine how changes in abundance of one taxon influence changes in another, which makes it possible to detect potential interactions between taxa. To distinguish the effect of interactions between taxa from it being caused by an external influence, it is useful to check out the other attribute correlation plot, which calculates correlations between taxa and metadata. So, if, for example, two taxa are correlated to each other and correlated to the same external influence from the metadata, then they might be less likely to be influencing each other, but are perhaps both influenced by the same attribute of the metadata. An example of an attribute correlation plot is shown in Fig. 10.

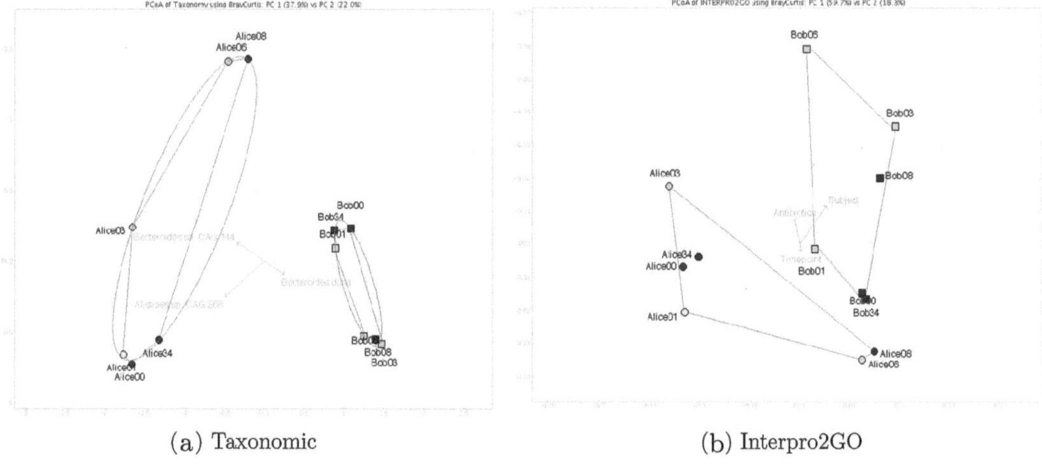

(a) Taxonomic (b) Interpro2GO

Fig. 9 PCoA analysis of 12 samples associated with "Alice" (round shapes) and "Bob" (square shapes), from [14]. Time points of antibiotic intake are colored light blue, time points before and after antibiotic intake dark red. (**a**) A PCoA plot based on Bray–Curtis distances as calculated by MEGAN using the taxonomic abundances for the samples. The green vectors represent the bi-plot vectors. The samples are grouped by individual, showing the convex hulls of the groups as well as ellipses. (**b**) is based on the same data but using the abundances of GO terms in the InterPro2GO hierarchy and only showing the convex hulls of the group. Here the orange vectors are the tri-plot vectors, showing the relation of metadata values to the principal components

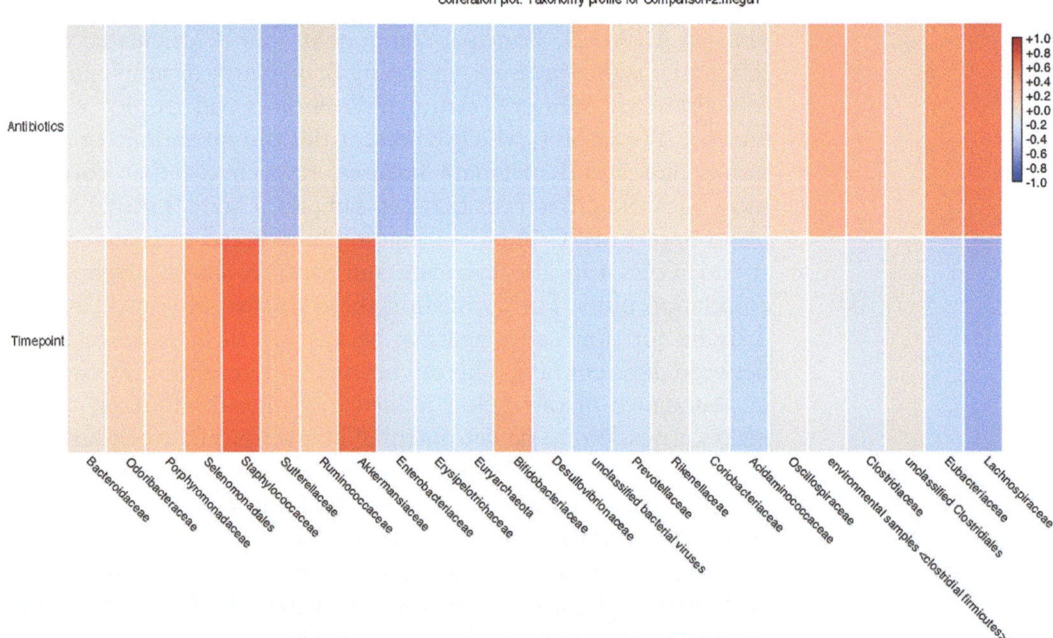

Fig. 10 Attribute correlation plot for the data from [14] for two healthy individuals taking antibiotics for 6 days (day 1–6). Correlation is shown as a heat map with red marking positive correlation between the attribute and the taxon and blue marking negative correlation. Correlations are shown for antibiotics intake (boolean) and time (day 0, 1, 3, 6, 8, and 34)

4 Outlook

It goes without saying that the quality and quantity of the input sequencing data limits the reliability of the output analysis. More directly, quality of the MEGAN hierarchy assignments is determined by the quality of the read alignment, which, in turn, depends on the chosen database and alignment tool. On the one hand, the database needs to be well annotated and comprehensive, as it is only possible to analyze the organisms or entities present in it. On the other hand, the alignment tool needs to be sensitive in order to identify the matching sequence. It is especially difficult to deal with sets of very similar sequences. Currently, for the human gut microbiome sequencing data analyzed with the basic short read pipeline, as much as 30% of reads are not assigned to any node in the course of the taxonomic analysis.

In order to avoid the bias introduced by the database one can also use one of the database-free strategies, e.g., k-mer counting. They are good for tracking the global changes in the data, but it is difficult to correct for possible contaminations. Although MEGAN does not support this type of analysis, it enables global comparisons with PCoA based on the profiles computed for each of the samples.

Another approach is assembly based analysis. In brief, the reads are assembled and then the scaffolds or contigs are annotated and investigated. This approach provides some information on gene co-localization at a cost of data loss in the form of unassembled reads and short contigs. Full metagenomic read assembly [21] is a very complex and computationally expensive task that MEGAN does not address.

Application of the long read sequencing technologies opens new perspective for metagenomics analysis. Long reads provide information on gene co-location on a single DNA molecule, and make assembly much easier. But, long reads also pose new algorithmic challenges in aspects of the protein alignment, hierarchy assignment, and abundance computation. As long read technologies continue to evolve, so, too, must the corresponding analysis algorithms.

MEGAN is a powerful visual analytics tool that provides a wide range of the algorithms for analysis of metagenomics sequencing data. MEGAN can run on hundreds of samples along with hundreds of metadata columns. It is the main workhorse of the Tubiom project where metagenomics profiles of 10,000 volunteers are collected and mined for correlations with the vast metadata (www.tuebiom.de).

4.1 MEGAN Resources

MEGAN Community software is freely available on the website: ab.inf.uni-tuebingen.de/data/software/megan6, together with the current mapping files for taxonomic and functional analysis.

Short read datasets presented in this chapter and used for visualizations are publicly accessible in MEGAN via MeganServer. The dataset used in the Long Read Pipeline section was downloaded from the supplementary material of Brown et al. [20]. Instructions for use of MEGAN and user support can be found on the MEGAN community website (megan.informatik.uni-tuebingen.de).

References

1. Handelsman J (2004) Metagenomics: application of genomics to uncultured microorganisms. Microbiol Mol Biol Rev 68(4):669–685
2. Buchfink B, Xie C, Huson DH (2015) Fast and sensitive protein alignment using DIAMOND. Nat Methods 12:59–60
3. Kiełbasa SM, Wan R, Sato K, Horton P, Frith MC (2011) Adaptive seeds tame genomic sequence comparison. Genome Res 21 (3):487–493
4. Sheetlin SL, Park Y, Frith MC, Spouge JL (2014) Frameshift alignment: statistics and post-genomic applications. Bioinformatics 30 (24):3575–3582
5. Huson DH, Beier S, Flade I, Górska A, El-Hadidi M, Mitra S, Ruscheweyh HJ, Tappu R (2016) MEGAN community edition–interactive exploration and analysis of large-scale microbiome sequencing data. PLoS Comput Biol 12(6):e1004957. https://doi.org/10.1371/journal.pcbi.1004957
6. Herbig A, Maixner F, Bos KI, Zink A, Krause J, Huson DH (2016) MALT: fast alignment and analysis of metagenomic DNA sequence data applied to the Tyrolean Iceman. bioRxiv 050559. https://doi.org/10.1101/050559
7. Altschul SF, Gish W, Miller W, Myers EW, Lipman DJ (1990) Basic local alignment search tool. J Mol Biol 215(3):403–410

8. Geer LY, Marchler-bauer A, Geer RC, Han L, He J, He S, Liu C, Shi W, Bryant SH (2010) The NCBI biosystems database. Nucleic Acids Res 38(2009):492–496
9. Mitchell A, Chang HY, Daugherty L, Fraser M, Hunter S, Lopez R, McAnulla C, McMenamin C, Nuka G, Pesseat S, Sangrador-Vegas A, Scheremetjew M, Rato C, Yong SY, Bateman A, Punta M, Attwood TK, Sigrist CJ, Redaschi N, Rivoire C, Xenarios I, Kahn D, Guyot D, Bork P, Letunic I, Gough J, Oates M, Haft D, Huang H, Natale DA, Wu CH, Orengo C, Sillitoe I, Mi H, Thomas PD, Finn RD (2014) The InterPro protein families database: the classification resource after 15 years. Nucleic Acids Res 43(D1):D213–D221
10. Hunter S, Corbett M, Denise H, Fraser M, Gonzalez-Beltran A, Hunter C, Jones P, Leinonen R, McAnulla C, Maguire E, Maslen J, Mitchell A, Nuka G, Oisel A, Pesseat S, Radhakrishnan R, Rocca-Serra P, Scheremetjew M, Sterk P, Vaughan D, Cochrane G, Field D, Sansone SA. EBI metagenomics–a new resource for the analysis and archiving of metagenomic data. Nucleic Acids Res 42(D1):D600–D606
11. Powell S, Szklarczyk D, Trachana K, Roth A, Kuhn M, Muller J, Arnold R, Rattei T, Letunic I, Doerks T, Jensen LJ, von Mering C, Bork P (2012) eggNOG v3.0: orthologous groups covering 1133 organisms at 41 different taxonomic ranges. Nucleic Acids Res 40(D1):D284–D289
12. Kanehisa M, Goto S (2000) KEGG: Kyoto encyclopedia of genes and genomes. Nucleic Acids Res 28(1):27–30
13. Overbeek R, Olson R, Pusch GD, Olsen GJ, Davis JJ, Disz T, Edwards RA, Gerdes S, Parrello B, Shukla M, Vonstein V, Wattam AR, Xia F, Stevens R The seed and the rapid annotation of microbial genomes using subsystems technology (RAST). Nucleic Acids Res 42 (D1):D206–D214
14. Willmann M, El Hadidi M, Huson DH, *et al* (2015) Antibiotic selection pressure determination through sequence-based metagenomics. Antimicrob Agents Chemother 59 (12):7335–7345
15. Huson DH, Tappu R, Bazinet AL, Xie C, Cummings MP, Nieselt K, Williams R (2017) Fast and simple protein-alignment-guided assembly of orthologous gene families from microbiome sequencing reads. Microbiome 5 (1):11
16. Eid J, Fehr A, Gray J, Luong K, Lyle J, Otto G, Peluso P, Rank D, Baybayan P, Bettman B, Bibillo A, Bjornson K, Chaudhuri B, Christians F, Cicero R, Clark S, Dalal R, Dixon J, Foquet M, Gaertner A, Hardenbol P, Heiner C, Hester K, Holden D, Kearns G, Kong X, Kuse R, Lacroix Y, Lin S, Lundquist P, Ma X, Marks P, Maxham M, Murphy D, Park I, Pham T, Phillips M, Roy J, Sebra R, Shen G, Sorenson J, Tomaney A, Travers K, Trulson M, Vieceli J, Wegener J, Wu D, Yang A, Zaccarin D, Zhao P, Zhong F, Korlach J, Turner S (2009) Real-time DNA sequencing from single polymerase molecules. Science 323(5910):133–138
17. Mikheyev AS, Tin MY (2014) A first look at the Oxford Nanopore MinION sequencer. Mol Ecol Resour 14(6):1097–1102
18. Laver T, Harrison J, ONeill PA, Moore K, Farbos A, Paszkiewicz K, Studholme DJ (2015) Assessing the performance of the Oxford Nanopore Technologies MinION. Biomol Detect Quant 3:1–8
19. Yang C, Chu J, Warren RL, Birol I (2017) NanoSim: nanopore sequence read simulator based on statistical characterization. GigaScience 6(4):1–6
20. Brown BL, Watson M, Minot SS, Rivera MC, Franklin RB (2017) MinION nanopore sequencing of environmental metagenomes: a synthetic approach. GigaScience 6(3):1–10
21. Medvedev P, Georgiou K, Myers G, Brudno M (2007) Computability of models for sequence assembly. Gene 4645:289–301

Chapter 20

Multiple Data Analyses and Statistical Approaches for Analyzing Data from Metagenomic Studies and Clinical Trials

Suparna Mitra

Abstract

Metagenomics, also known as environmental genomics, is the study of the genomic content of a sample of organisms (microbes) obtained from a common habitat. Metagenomics and other "omics" disciplines have captured the attention of researchers for several decades. The effect of microbes in our body is a relevant concern for health studies. There are plenty of studies using metagenomics which examine microorganisms that inhabit niches in the human body, sometimes causing disease, and are often correlated with multiple treatment conditions. No matter from which environment it comes, the analyses are often aimed at determining either the presence or absence of specific species of interest in a given metagenome or comparing the biological diversity and the functional activity of a wider range of microorganisms within their communities. The importance increases for comparison within different environments such as multiple patients with different conditions, multiple drugs, and multiple time points of same treatment or same patient. Thus, no matter how many hypotheses we have, we need a good understanding of genomics, bioinformatics, and statistics to work together to analyze and interpret these datasets in a meaningful way. This chapter provides an overview of different data analyses and statistical approaches (with example scenarios) to analyze metagenomics samples from different medical projects or clinical trials.

Key words Metagenomics, Metatranscriptomics, Microbiome, Clinical trials, Comparative metagenomics

1 Introduction

The diversity of species on earth is high, and most of them are microorganisms. Their ubiquitous presence makes it extremely difficult to identify and classify all microbes in a laboratory environment. Standard genomics tries to enrich pure cultures and study them: for example, the taxonomy, the genome, the genes, and the pathways. However, only a miniscule fraction of all microbes can be cultured because of their complex symbiosis and nutrient requirements in other organisms. The scientific community is now equipped with the development of new sequencing techniques

Maria Anisimova (ed.), *Evolutionary Genomics: Statistical and Computational Methods*, Methods in Molecular Biology, vol. 1910, https://doi.org/10.1007/978-1-4939-9074-0_20,

and high-throughput analysis. The study of the genomic content of a sample of microorganisms obtained from a common habitat is made possible with the field of metagenomics, also known as environmental genomics [1]. Instead of taking the DNA for sequencing from isolated cultures it is obtained directly from the environment. Therefore, the analysis of microbes that are deemed unculturable (which means current laboratory culturing techniques are unable to grow them) with standard laboratory techniques becomes possible. Two main approaches commonly used in metagenomic studies: *marker gene-based metagenomics* (e.g., 16S amplicon sequencing) and *metagenomic shotgun sequencing*. In the first approach, DNA is used as the template for PCR to amplify a segment of the conserved 16S ribosomal RNA (rRNA) gene sequence. Universal primers complementary to conserved regions are used so that the region can be amplified from any bacteria. After purification of PCR products, sequencing of the 16S rRNA gene is performed [2]. In the second approach, shotgun sequencing, DNA is broken up randomly into multiple small segments, which are sequenced using the chain termination method to obtain reads. Multiple overlapping reads for the target DNA are obtained by performing several rounds of this fragmentation and sequencing. Computer programs then use the overlapping ends of different reads to assemble them into a continuous sequence [3].There are several publications discussing the differences in microbial biodiversity discovery between 16S amplicon and shotgun sequencing, for example see [4]. In a recent study using water samples from Brazil's major river floodplain systems, authors showed shotgun sequencing outdone by amplicon [5]. Here, the authors ascribed the poor performance of shotgun sequencing mainly to the weakness of the database used in the study, as compared to databases for the 16S rRNA gene. This study can be used as a caution for people working with rare environments (*See article by Catherine Offord in The Scientist*[1]). Comparisons of the two methods in well-studied systems such as the gut microbiome have generally found that shotgun sequencing identifies more microbial diversity [6].

Further recent advancement of culturomics approach is shedding light on multiple high-throughput culture conditions [7, 8]. As the samples used in metagenomics do not contain the genome of just one but many different microorganisms, the possibility of analyzing their functional and metabolic interplay arises. Next-generation sequencing technology (NGS) has effectively transformed infectious disease research throughout the last decade, fuelling the growth in genetic data and providing huge number of DNA reads at an affordable cost. Many studies use these

[1] https://www.the-scientist.com/?articles.view/articleNo/50044/title/Shotgun-Sequencing-Outdone-by-Amplicon/.

techniques, which examine microorganisms that inhabit niches in the human body, sometimes causing disease, and researchers often try to correlate these microorganisms and their change with multiple treatment conditions (e.g., *see* [9]). Gene annotations in these studies support the association of specific genes or metabolic pathways with health and with specific diseases. In a recent article authors discussed how host gene–microbial interactions are major determinants for the development of multifactorial chronic disorders and thus for the relationship between genotype and phenotype [10]. There are many other reports based on the application of metagenomics in understanding oral health and disease [11–13]. As recently described by Forbes et al., metagenomics and other "omics" disciplines could provide the solution to a cultureless future in clinical microbiology, food safety, and public health [14].

No matter from which environment it comes, the analysis of datasets from such studies are similar to some extent. Most projects aim at determining either the presence or absence of specific species of interest, or to obtain an overview of the taxa represented in a given metagenome and comparing the biological diversity and the functional activity of a wider range of microorganisms within their communities. The importance increases for comparison of different datasets, as researchers will need to determine and understand the similarities and dissimilarities within the metagenomes of different environments. These environments can be multiple patients with different conditions, multiple drugs, or multiple time points of same treatment or same patient. Further, sometimes researchers also may compare different environments for example to study antibiotic resistance genes (ARG) and understand which environments are more prone to such ARGs. Thus, no matter how many hypotheses we have, we need a good understanding of genomics, bioinformatics, and statistics to work together to analyze and interpret these datasets in a meaningful way.

This chapter provides an overview of different data analyses and statistical approaches to analyze metagenomics samples from a number of clinically derived datasets. The methodological description of this chapter will be guided by three main scenarios. The first one is a published data set from human atherosclerotic plaque samples (Scenario 1) [15]; the second one is a clinical trial example comparing the effects of two omega-3 polyunsaturated fatty acids (PUFAs) supplements on healthy volunteers (Scenario 2) [16]; and the third one is another clinical trial example comparing the efficacy of two drugs for an infectious disease (Scenario 3).

The Scenarios 3 came from an ongoing unpublished project; therefore, the real datasets are not provided. This chapter is mainly focused on multiple data analyses/annotation and statistical approaches that can be used in similar situations, but any biological finding of the example scenarios is not explained here. Although all of these scenarios are derived from medical projects, the analyses approach can be adapted to environmental samples as well. On this occasion, I must emphasize the importance to have good metadata, that is, a detailed description of each parameter like health status or sampling site or age or any similar information relating to specific samples that may be important for the analyses. Good metadata are key to good analyses and noise reduction in data analysis processes.

2 Description of Example Studies

2.1 Scenario 1: Metagenomic Analyses of Human Atherosclerotic Plaque Samples

To investigate microbiome diversity within human atherosclerotic tissue samples high-throughput metagenomic analysis was employed on (1) atherosclerotic plaques obtained from a group of patients who underwent endarterectomy due to recent transient cerebral ischemia or stroke and (2) presumed stabile atherosclerotic plaques obtained from autopsy from a control group of patients who all died from causes not related to cardiovascular disease. Our data provides evidence that suggest a wide range of microbial agents in atherosclerotic plaques, and an intriguing new observation that shows this microbiota displayed differences between symptomatic and asymptomatic plaques, as judged from the taxonomic profiles in these two groups of patients. Additionally, functional annotations reveal significant differences in basic metabolic and disease pathway signatures between these groups.

In this project, we demonstrate the feasibility of novel high-resolution techniques aimed at identification and characterization of microbial genomes in human atherosclerotic tissue samples. Our analysis suggests that distinct groups of microbial agents might play different roles during the development of atherosclerotic plaques. These findings may serve as a reference point for future studies in this area of research. The workflow in Fig. 1 provides a brief description of the sample processing and analyses pipeline for the study described in Scenario 1. If readers want to know more details of the methodology, please refer to (15). This scenario is an example of analyzing host-associated metagenome samples.

2.1.1 Methodology Details

For this study, we used atherosclerotic tissue samples from a group of 15 patients that underwent elective carotid endarterectomy following repeated transient ischemic attacks or minor strokes (samples from symptomatic atherosclerotic plaques as cases).[2] Further, we have asymptomatic atherosclerotic plaques from seven

[2] All methods and experimental manuals were approved by The National Committee on Health Research Ethics (Danish) and was granted by the Ethical Committee of the region of Copenhagen (H-3-2011-013).

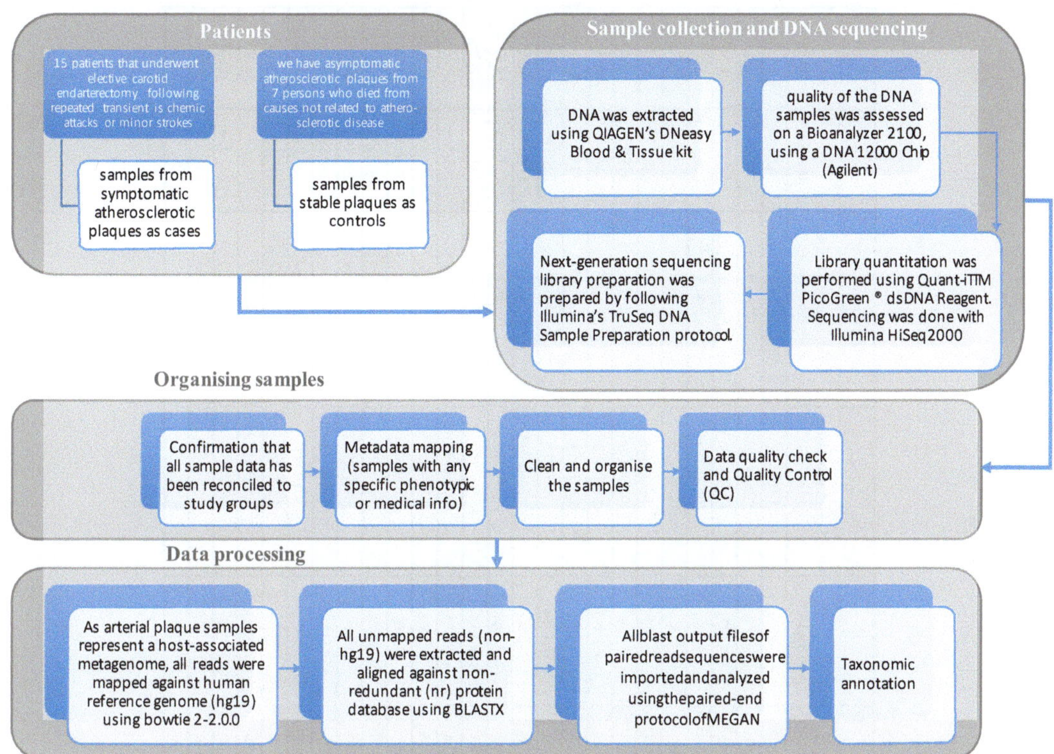

Fig. 1 Analysis pipeline for the study of human atherosclerotic plaque samples. Interested readers may refer to the full study here [15]

persons who died from causes not related to atherosclerotic disease (samples from stable plaques as controls).[3]

All 22 arterial plaque samples resulted in 2,610,268,774 shotgun sequencing reads. After mapping these reads against Hg19 using bowtie 2 [17] with "very-sensitive" parameters to filter all human-like sequences from our samples. The average amount of non-Hg19 reads is 884,727,044 (average 33.89% per sample, Table 1). These non-Hg19 reads were extracted and aligned against nonredundant (nr) protein database (version 30.07.2012) [18] using BLASTX (ncbi-blast-2.2.25+; Max e-value 10e−3) [19]. After performing the BLASTX alignment, all output files of paired read sequences were imported and analyzed using the paired-end protocol of MEGAN5 [20]. For all non-Hg19 annotated reads, 2–16% (mean 4.6%) were assigned as bacteria in different samples. The rest of reads were assigned to Eukaryota. Table 1 provides details of sequencing read statistics and assignments of reads after different stages of data processing. R statistical

[3] These samples originated from the tissue bank at the Department of Forensic Medicine (Approval No. 1501230).

Table 1
Sample statistics and read assignments

Patients	Platform	SampleID-DNA	Tissue Sample info.	Library preparation date	Raw reads	Non-Hg19 reads	Non-Hg19%	RapSearch processed (as in MEGAN)	Assigned in MEGAN (MSc:50 MSp25 MinCompl:0.44 and paired protocol)	Bacteria	% Bacteria
Cases	Hiseq 2000	48	20/10-11 gDNA CP	3/14/2012	93,124,682	31,504,036	33.83%	8,404,029	3,763,197	243,336	2.90%
		49	10/11-11 gDNA CP	3/14/2012	101,840,018	53,068,718	52.11%	12,405,108	6,772,304	436,463	3.52%
		50	9/1-12 gDNA CP	3/14/2012	94,765,328	43,109,046	45.49%	11,643,905	6,419,975	505,239	4.34%
		51	11/1-12 gDNA CP	3/14/2012	112,426,390	50,653,838	45.06%	10,711,699	4,899,567	216,511	2.02%
		52	18/1-12 gDNA CP	3/14/2012	88,328,470	36,795,952	41.66%	10,574,678	5,766,665	492,013	4.65%
		53	20/1-12 gDNA CP	3/14/2012	124,000,764	41,858,084	33.76%	10,057,354	4,676,002	243,775	2.42%
		54	10/1-12 gDNA CP	3/14/2012	93,124,682	82,366,266	88.45%	19,199,623	9,229,157	365,993	1.91%
		237	31/1-12 tissue DNA	10/18/2012	141,334,092	58,853,792	41.64%	6,946,473	3,854,397	249,369	3.59%
		238	1/2-12 tissue DNA	10/18/2012	153,302,968	36,976,828	24.12%	5,216,154	3,002,923	835,412	16.02%
		239	7/2-12 tissue DNA	10/18/2012	154,652,444	48,177,200	31.15%	5,643,923	3,045,461	222,363	3.94%
		240	27/2-12 tissue DNA	10/18/2012	101,591,496	46,000,276	45.28%	5,166,877	2,640,033	110,186	2.13%
		241	5/3-12 tissue DNA	10/18/2012	99,927,824	42,032,952	42.06%	4,607,355	2,471,259	153,425	3.33%
		242	13/3-12 tissue DNA	10/18/2012	82,850,664	46,721,284	56.39%	5,002,916	2,716,756	135,617	2.71%
		243	25/4-12 tissue DNA	10/18/2012	104,094,892	52,853,464	50.77%	5,726,930	3,095,542	193,819	3.38%
	Hiseq 2500	977 (P0613)	13/06-12 tissue DNA	7/11/2013	111,699,176	13,191,406	11.81%	2,344,220	955,602	74,646	3.18%
	Miseq V1	P0613 repeat		7/11/2013	7,184,184	980,004	13.64%	759,262	228,262	31,889	4.20%
		977+APD1			118,883,360	14,171,410	11.92%	3,103,482	1,183,864	106,535	3.43%
Controls	Hiseq 2000	55	AP26 gDNA (Control)	3/14/2012	105,779,932	29,019,712	27.43%	8,911,853	4,040,359	293,160	3.29%
		56	AP25 gDNA (Control)	3/20/2012	128,471,814	10,767,822	8.38%	4,680,735	1,922,688	204,397	4.37%
		232	AP 21 tissue DNA (Control)	3/18/2012	127,173,774	25,793,632	20.72%	3,267,969	1,726,714	164,711	5.04%
		233	AP24 tissue DNA (Control)	3/18/2012	166,547,592	29,282,304	17.58%	4,559,468	2,634,338	649,226	14.24%
		234	AP27 tissue DNA (Control)	3/18/2012	114,673,124	37,683,568	32.86%	4,550,457	2,477,740	193,659	4.26%
		235	AP28 tissue DNA (Control)	3/18/2012	151,195,284	34,767,160	22.99%	4,633,661	2,550,975	231,926	5.01%
		236	AP29 tissue DNA (Control)	3/18/2012	152,179,180	32,269,700	21.21%	4,254,572	2,319,403	204,778	4.81%
					2,610,268,774	685,143,146	26.25%				4.60%

programming language [21] was used for multivariate statistics. Later in Subheading 3, we will describe few of the analysis approaches revisiting this study.

In this study our data provided evidence that suggest a wide range of microbial agents (some pathogens) in atherosclerotic plaques, and these microbes displayed differences between symptomatic and asymptomatic plaques as judged from the taxonomic profiles in these two groups of patients. Further, fluorescence in situ hybridization (FISH) was performed to validate the presence of biofilm-like structures of few pathogens (which have been previously predicted from taxonomic analyses) in the symptomatic atherosclerotic plague samples. FISH staining demonstrates the presence of live bacteria; thus, this is a very good approach for cross-validation of any computational finding in the lab.

There are also potentials of using this data for not only taxonomic annotation but also to reveal the functional profiles through partial assembly of specific members and their functional annotations. Functional annotations reveal significant differences in basic metabolic and disease pathway signatures between these groups. Here, we will not provide details of the whole study, but interested readers may refer to [15].

On this occasion, it is necessary to mention that in any similar project in future, for alignment purpose, we would have used DIAMOND [22] which uses improved algorithms and additional heuristics and works much faster compared to available other aligners. Scenario 1 is an example of analyzing shotgun sequence datasets obtained from tissue samples or host-associated metagenome. In case readers have shotgun sequence datasets from environmental samples or from fecal samples, they do not need to perform alignment step to get rid of the host-associated sequences, unless there is any doubt of contamination. Normally we suggest to have control or blank samples in two wells per 96-well plate to address any issue with contaminations.

2.2 Scenario 2: The Effect of Omega-3 Polyunsaturated Fatty Acid Supplements on the Human Intestinal Microbiota

2.2.1 Study Design

A randomized, open-label, crossover trial of 8 weeks' treatment with 4 g mixed eicosapentaenoic acid (EPA)/docosahexaenoic acid (DHA) in two formulations (soft-gel capsules and drinks) with a 12-week "washout" period [16] is chosen. Healthy volunteers aged greater than 50 years of both genders were included in this study. Participants were randomized to take two types of EPA and DHA compositions (Fig. 2):

1. Two 200 mL drinks per day (providing approximately as the triglyceride daily) at any suitable time of day, or
2. Four soft-gel capsules (each containing 250 mg EPA and 250 mg DHA as the ethyl ester) twice daily with meals (providing 2000 mg EPA and 2000 mg DHA per day), both for 8 weeks.

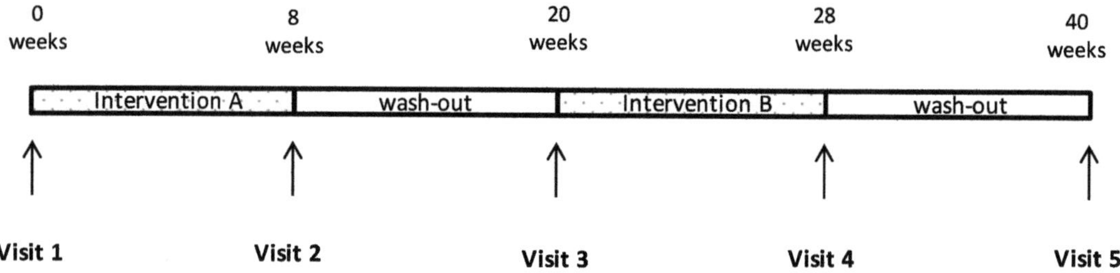

Fig. 2 Schedule of visits for the study to understand the effect of omega-3 polyunsaturated fatty acid supplements on the human intestinal microbiota

After a 12-week "washout" period, participants took the second intervention for 8 weeks. We also included a final study visit after a second 12-week "washout" period (V5; Fig. 2). Fecal samples were collected at five time-points for microbiome analysis by 16S rRNA PCR and Illumina MiSeq sequencing. Parallel red blood cell (RBC) fatty acid analysis was performed by liquid chromatography–tandem mass spectrometry.

2.2.2 Sample Preparation and Sequencing

Microbial DNA extractions were performed based on the method of Yu and Morrison, [23] with slight modifications. DNA was extracted from approximately 250 mg feces using the QIAamp DNA Stool Mini Kit (Qiagen, Germany) with bead beating. DNA Library Prep Kit for Illumina, NEBNext Singleplex Oligos for Illumina (New England Biolabs, UK), and unique in-house-designed index primers (Integrated DNA Technologies, UK) were used to allow for multiplexing of samples. Twelve cycles of enrichment PCR were performed, and final libraries were cleaned with AMPure Beads (Beckman Coulter, UK). Successful libraries were confirmed by DNA 1000 bioanalyzer chips or DNA Analysis screen tapes (Agilent, UK). Quantification was performed with the Quant-iT dsDNA Assay Kit, broad range. A total of 30 ng of each library was pooled and sequenced on an Illumina MiSeq (2 × 250 bp) [24]. The variable region (V4) of the 16S rRNA gene was sequenced for these samples.

2.2.3 Data Analyses

Demultiplexed FASTQ files were trimmed of adapter sequences using cutadapt [25]. Paired reads were merged using fastq-join [26] under default settings and then converted to FASTA format. Consensus sequences were removed if they contained any ambiguous base calls, two contiguous bases with a PHRED quality score lower than 33, or a length more than 2 bp different from the expected length of 240 bp. Further analysis was performed using QIIME [27]. Operational taxonomy units (OTUs) were picked using usearch [28] and aligned to the Greengenes reference database using PyNAST [29]. Taxonomy was assigned using the RDP 2.2 classifier [30]. The resulting OTU BIOM files from the above analyses were imported in MEGAN for detailed group-specific analyses, annotations, and plots [31]. R statistical programming language [21] was used for multivariate statistics and other plots.

This dataset and method pipeline are purely described as an example for similar analyses; thus, we will not explain the results here, but interested, readers may see [16]. Scenario 2 is a typical example of analyzing 16S sequence data. In Subheading 3, we will describe few of the analysis approaches using data from this study.

2.3 Scenario 3: Comparing Effects of Two Drug Treatments for an Infectious Disease

In a given situation suppose we need to compare treatment effect of two drugs (e.g., X and Y) or more, where we have time series data, that is, patient samples from multiple time points of the treatment course for both drugs. This time series data can be either collected every day of the treatment period or in intervals. Furthermore, for practical reasons we might not be able to obtain data at a desired day but $\pm 1/2$ days. It is important to select an error threshold and be consistent with that throughout the project. For example, we need to have a similar depth of sequencing reads or need to follow subsample comparison as detailed later, and, also, we need to discard samples with very low number of reads. Further during alignment to reference database and during mapping to taxonomy similar scores and thresholds should be used for all samples (please check best parameter selections in individual websites while using specific tools). Additionally, there can be multiple fundamental factors in patient samples such as age, gender, and geography that may not contribute in a similar manner to resiliency. Figure 3 shows a schematic of the metadata structure, which may help to understand the complexity of a typical clinical trial.

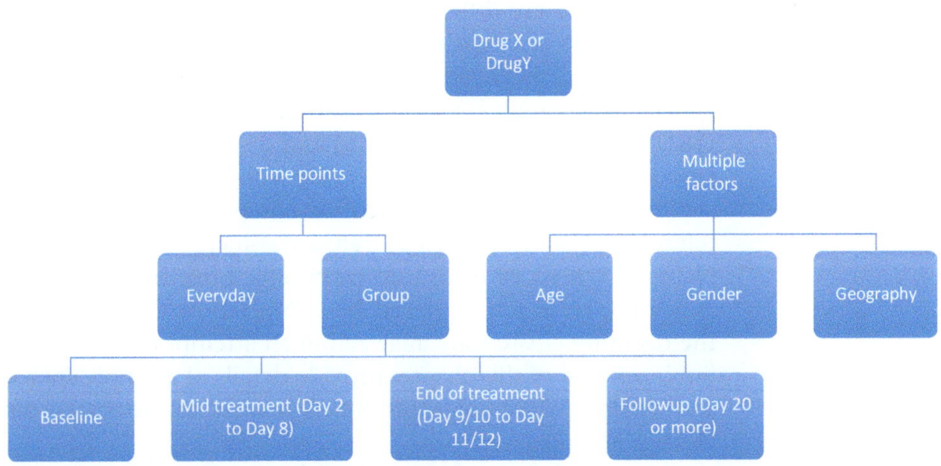

Fig. 3 Schematic diagram of multiple factors in a clinical study

2.3.1 Sample Preparation and Sequencing and Data Analyses

In a clinically relevant setting this type of study wants to know which drug works better for a similar group of patients. Patients are randomized between drug arms to control any selection bias. Usually in this type of projects as we want to compare several factors, we need many samples to start with. Readers are advised to seek statistics help to do power calculation to obtain the preferred sample size. In general, as we end up having hundreds of samples, we usually go for 16S sequencing as a cost-effective solution. However, some projects can also use shotgun sequencing. Similar to previous examples, we assume that we have sequenced (either 16S or shotgun sequencing) our samples and performed further analysis process as outlined earlier to obtain taxonomic profile (following data analyses methods as described in previous scenarios) for each patient at each time point. Besides analyzing time series of each individual separately, we have also grouped them in certain time points such as baseline, mid-treatment, end of treatment, and follow-up. Besides treatment groups, patients are also compared based on multiple factors such as age, gender, and geography.

3 General Methods for Annotation and Statistical Analyses

Broadening our focus beyond these studies, additional analysis techniques are explained below which are used in these studies and also can be used in similar projects.

3.1 Taxonomic and Functional Annotation

Taxonomic annotation addresses the question, *'Who is out there?'* or in other words tries to obtain information regarding the species composition of a given metagenome. On the other hand, functional annotation attempts to answer the question, *'What are they doing?'* There are different approaches for metagenome analyses, among which one type of approach is to use phylogenetic markers to distinguish between different species in a sample. The most widely used marker is the small subunit ribosomal ribonucleic acid (SSU rRNA) gene (16S or 18S) and a second type of method is based on analyzing the nucleotide composition of reads. In a supervised approach the nucleotide composition of a collection of reference genomes is used to train a classifier, which is then used to place a given set of reads into taxonomic bins. In an unsupervised approach, reads are clustered by composition similarity and then the resulting clusters are analyzed in an attempt to place the reads. Subheading 4 of this chapter provide details of multiple approaches and available different tools which readers can use according to their preferences.

In general, for annotating 16S rRNA sequences we use QIIME [27] and for shotgun sequencing we use MEGAN [31] which can also be used for 16S. MEGAN is a highly efficient program for interactive analysis and comparison of microbiome data, allowing one to explore hundreds of samples and billions of reads. While taxonomic profiling is performed based on the NCBI taxonomy, MEGAN also provides a number of different functional profiling approaches. MEGAN Community Edition also supports the use of metadata in the context of principal coordinate analysis and clustering analysis [31]. In all the three scenarios explained in this chapter, MEGAN is used as primary tool for annotations. For more details on MEGAN tool, *see* Chapter 23.

If we have shotgun sequencing then we have good option for functional annotation, but with 16S sequences we can only perform taxonomic analyses with confidence although there are few tools which might predict metagenome functional content from marker genes [32, 33]. Most shotgun annotation pipelines (such as MEGAN [31], MG-RAST [34], IMG/MER [35], EBI Metagenomics [36]) support functional annotations and they often use databases such as KEGG [37], SEED [38], eggNOG [39], and COG/KOG [40], as well as protein domain databases such as TIGRFAM [41] and PFAM [42].

3.2 Metagenome Assembly

Similar in nature to the genomic assembly, which is the reconstruction of genomes from the sequenced DNA segments (or *reads*), metagenome assembly is more complex. The main goal is to stitch together the fragments of the reads that could be from the same genome. Here the reads consist of mixture of DNA from different organisms and also may have widely different levels of abundance. Few recent reviews discussed new challenges and opportunities as well as assessed the most common and freely available metagenome assembly tools with respect to their output statistics, their sensitivity for low-abundance community members and variability in resulting community profiles as well as their ease of use. Interested readers please refer to reviews [43, 44].

3.3 Rarefaction Curves

Rarefaction curves represent a powerful method for comparing species richness among habitats on an equal-effort basis based on the construction of the so-called rarefaction curves [45]. This is a very useful tool for statistical data analyses that helps us to Correct for bias in species number due to unequal sample sizes by standardization to the number of species expected in a sample if it had the same total size as the smallest sample. As an example, we have two sample groups, first having 50 individuals and second 30 individuals with multiple number of species obtained from their taxonomic analyses. Rarefaction helps us to compare the situation, if we would have same number of individuals in two sample groups. Rarefaction curves are used differently in case of 16S and shotgun metagenomics. Ni and colleagues have described methods for estimating a reasonable and practical amount for SSU rRNA gene sequencing and explained how much metagenomic sequencing is enough to

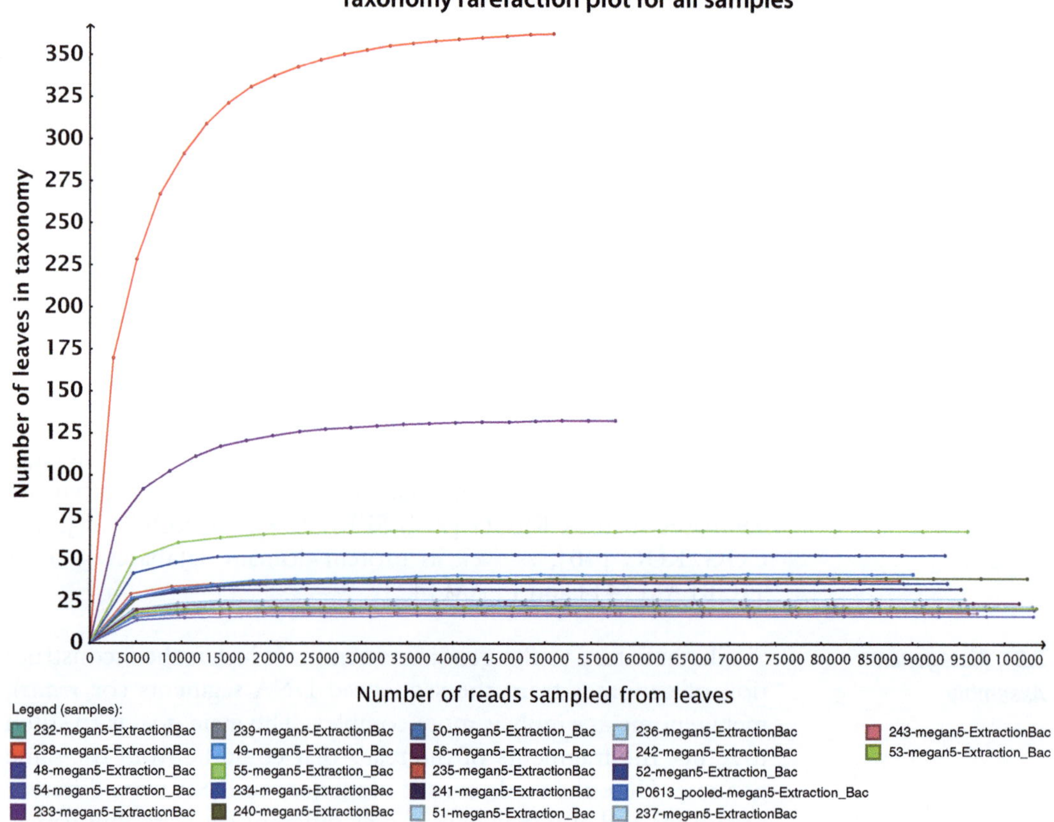

Fig. 4 Rarefaction. Rarefaction plot using annotated species profile for all 22 (unstable and stable) atherosclerotic plaque samples. These curves show the number of nodes that would be present if based on 10%, 20%, and up to 90% of the reads

achieve a given goal [46]. In metagenomic shotgun sequencing, the fraction of the metagenome represented in the data set is termed coverage, which can be assessed through rarefaction curve. Interested readers may refer to a recent publication which has advocated for the estimation of the average coverage obtained in metagenomic studies, and briefly presented the advantages of different approaches [47].

In Scenario 1, for comparing case and control groups from human atherosclerotic plaque samples, we computed rarefaction curves from the normalized profile of 22 samples using the bacterial reads, showing the number of nodes that would be present in the analysis if based from 10% to 90% of the reads (Fig. 4). From sequence statistics (Table 1) and the rarefaction curve (Fig. 4), it is apparent that 2 (sample 233 and 238) of the 22 samples had much higher sequencing depth than the other samples. Later in the study we therefore omitted these two samples from merged case vs. control analyses.

Similarly, in Scenario 2 also, rarefaction was performed at various levels to compare diversity for different sample groupings. All groups were rarefied to the lowest read number, and the diversity calculated using weighted and unweighted UniFrac as well as the non-phylogenetic Bray–Curtis dissimilarity measure.

3.4 Subsample Comparison

In situations like Fig. 3, where two samples have much higher sequencing depth, another option can be subsample comparison. In this process without excluding high-depth samples from further study, another approach is to simulate subsample of lowest sample size (of other samples in the study) for sufficient number of times. And then take a median of the subsamples to generate a pseudo profile, which can serve as a good comparable sample for the group. For example, if in a study for most of the samples sequence reads are in a range of 200,000–300,000. However, only few samples have approx. 1 million reads, in those cases we simulate subsample of 200,000 reads from them for large number of times (say 1000) and we take median of the profiles, which we can then compare with other samples.

3.5 Comparative Visualization

Comparative visualization includes different types of plots and charts (pie charts, histograms, and many other kinds of plots) which can help us to draw basic conclusions regarding our data. For example, Fig. 5 depicts basic comparison of patients in two drug treatment groups for certain time points such as baseline, mid-treatment, end of treatment and follow up (from Scenario 3).

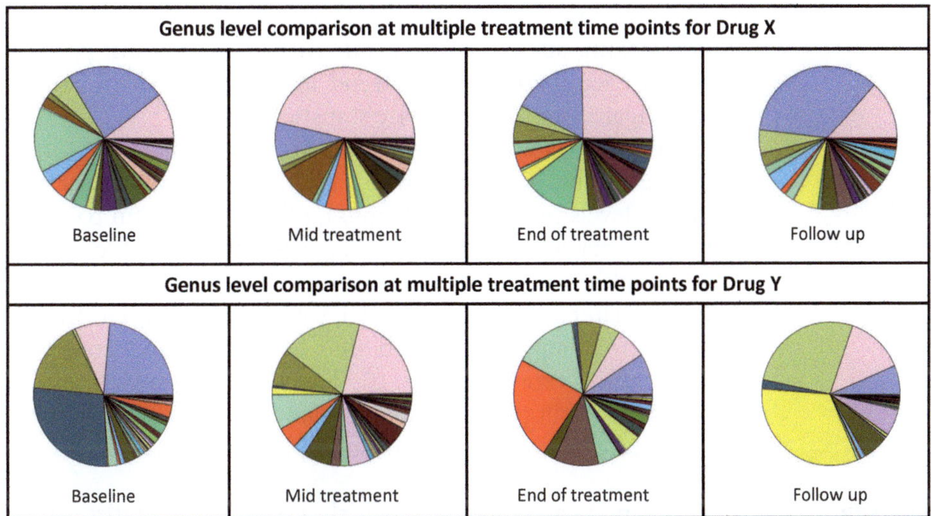

Fig. 5 Genus level taxonomic comparison of patients' microbiome (median of each time point group) in two drug treatment groups for certain time points such as baseline, mid-treatment, end of treatment and follow up. Here different colors indicate different genera and the size of each color in the pie reflects the percentage of those genus in median microbiome for each time point group and for each drug

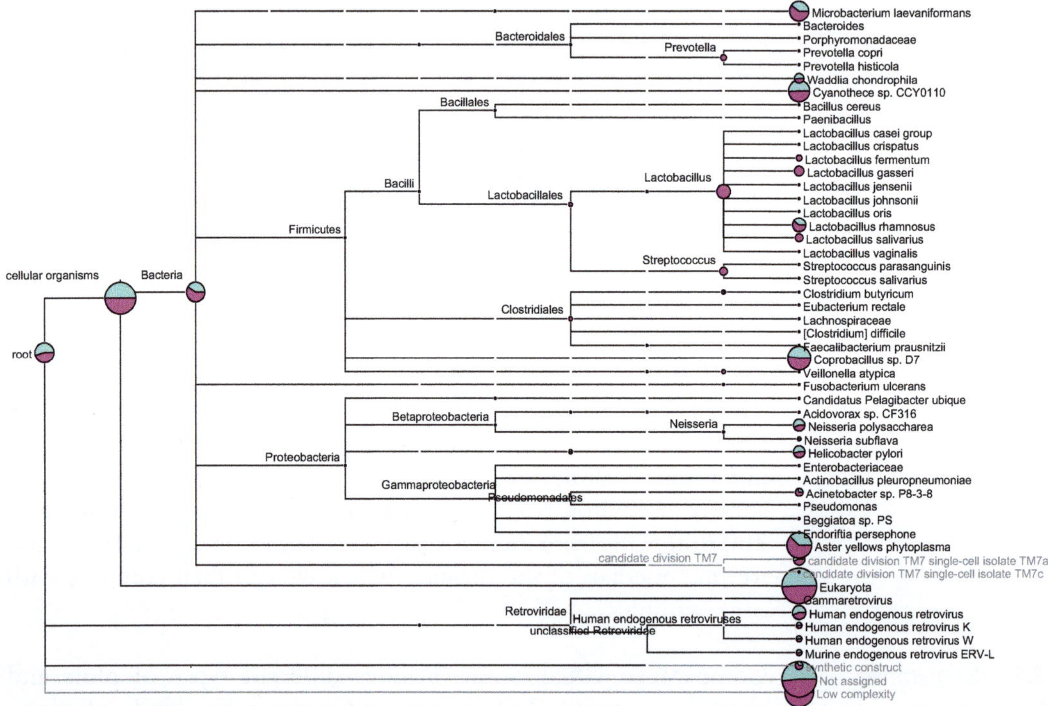

Fig. 6 Tree view at "family" level taxonomy comparing merged data from cases and control samples using data from Scenario 1

Form this figure we can easily see that the microbiome pattern in drug X over treatment period is more consistent (or more stable over the time) than in drug Y. Here with visual comparison we are not making any conclusion, but with these types of plots we can start to see if there is any trend in our data, which can later be investigated with appropriate statistical tests.

Further as metagenomic data are often hierarchical in nature, besides doing basic plots which can be done only at certain taxonomic levels (e.g., family/genus), often it is helpful to display the whole data as comparative tree view. For example in Scenario 1, samples from cases and controls have grouped closely (as can be seen later in Subheading 3.9), we can explore their broad differences by comparing total biome from cases and controls using comparative tree view (Fig. 6). This kind of tree view also help us to assess multiple time point samples from single patient or grouped data comparison for multiple factors (e.g., in Scenario 3).

3.6 Diversity Analyses

Diversity analyses is one of the prominent statistical analysis approaches that address some of the downstream analysis steps associated with metagenomic studies. Species abundance estimates in the community are used to make inference about diversity on the whole community. The terms *alpha*, *beta*, and *gamma diversity* were all introduced by R. H. Whittaker to describe the spatial component of biodiversity [48]. *Alpha diversity* is just the diversity of each site (samples in each group). *Beta diversity* represents the differences in species composition among sites. *Gamma diversity* is the diversity of the entire landscape of different sites (all species pool from multiple samples). A *diversity index* measures how many different types (such as species) are there in a dataset (a community) and simultaneously takes into account how evenly the basic entities (such as individuals) are distributed among these types. Three commonly used measures of diversity, Simpson's index, Shannon's entropy, and the total number of species, are related to Renyi's definition of a generalized entropy, and are well explained and compared by Hill [49]. Interested readers may also refer to [50] for consistent terminology for quantifying species diversity. Many other publications also explain this topic very well.

3.7 Comparison Using Distance Matrices

Another common technique to compare metagenomic datasets is using distance matrices. First, a taxonomic profile is computed for each data set. Second, a matrix of pairwise distances is determined using one of several possible ecological indices. Finally, the distances are represented using an appropriate visualization technique. Mitra et al. [51] explained multiple distance matrices (such as Bray–Curtis, Kulczynski, χ^2, Hellinger, and Goodall) in the context of multiple metagenome comparison. In addition to these *UniFrac* is another distance metric used for comparing biological communities. It differs from dissimilarity measures such as Bray–Curtis by incorporating information on the relative relatedness of community members by incorporating phylogenetic distances between observed organisms in the computation [52–54]. Both weighted (quantitative) and unweighted (qualitative) variants of UniFrac are often used in microbial ecology, where the former accounts for abundance of observed organisms, while the latter only considers their presence or absence.

3.8 Boxplots

In descriptive statistics, "boxplot" or alternatively called "box and whisker plot," is an important and one of the most informative tools that is used for graphically depicting groups of numerical data through their quartiles [55]. The boxplot is a quick way of examining multiple groups of data graphically, which easily provides information regarding quartiles, range, variation, and even outliers and enables us to compare within and between group samples. For example, Fig. 7 shows distribution of samples in multiple time point for both drugs (example data in Scenario 3). From this plot we can clearly gather the idea that diversity with drug X is consistently higher than that with drug Y. Further in Fig. 5 we have already seen that microbiome pattern in drug X showed less disruption, thus from these two figures we can hypothesize that drug Y

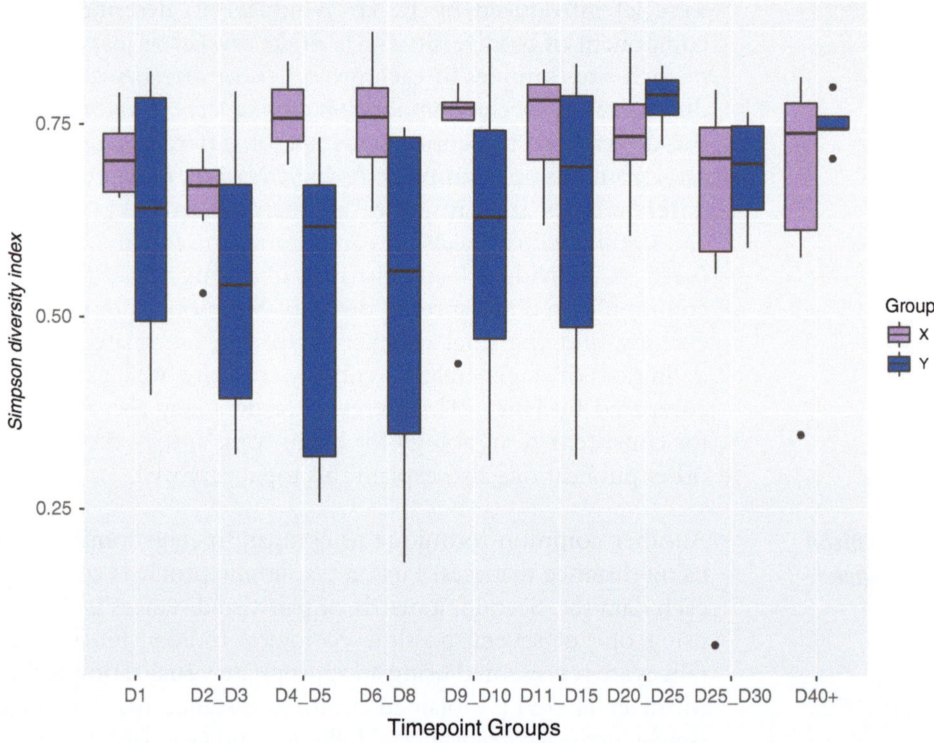

Fig. 7 Boxplot showing Simpson diversity indices for samples from each time point and for both the drugs X and Y

being more disruptive to the microbiome. Such hypotheses can help us in further statistical analyses.

3.9 Hierarchical Clustering

Cluster analysis, especially hierarchical clustering [56, 57], is an important tool for the exploratory and unsupervised analysis (where we do not need a training dataset to feed the programme) of high dimensional datasets and often used in genomics and other fields for their ability to simultaneously uncover multiple layers of clustering structure. In our example, Fig. 8 depicts a hierarchical clustering result of family level taxonomic comparison data for all 22 samples. Interestingly, samples 238 and P0613 were mostly different, and among the other samples, all unstable plaques clustered together, apart from all stable plaque controls that clustered separately.

Interestingly, the asymptomatic atherosclerotic plaques have more abundance of host microbiome-associated microbial families such as *Porphyromonadaceae*, *Bacteroidaceae*, *Micrococcaceae*, and *Streptococcaceae* than the symptomatic atherosclerotic plaques. In contrast, the symptomatic atherosclerotic plaques have more abundance of pathogenic microbial families such as *Helicobacteraceae*, *Neisseriaceae*, and sulfur-consuming families such as sulfur-

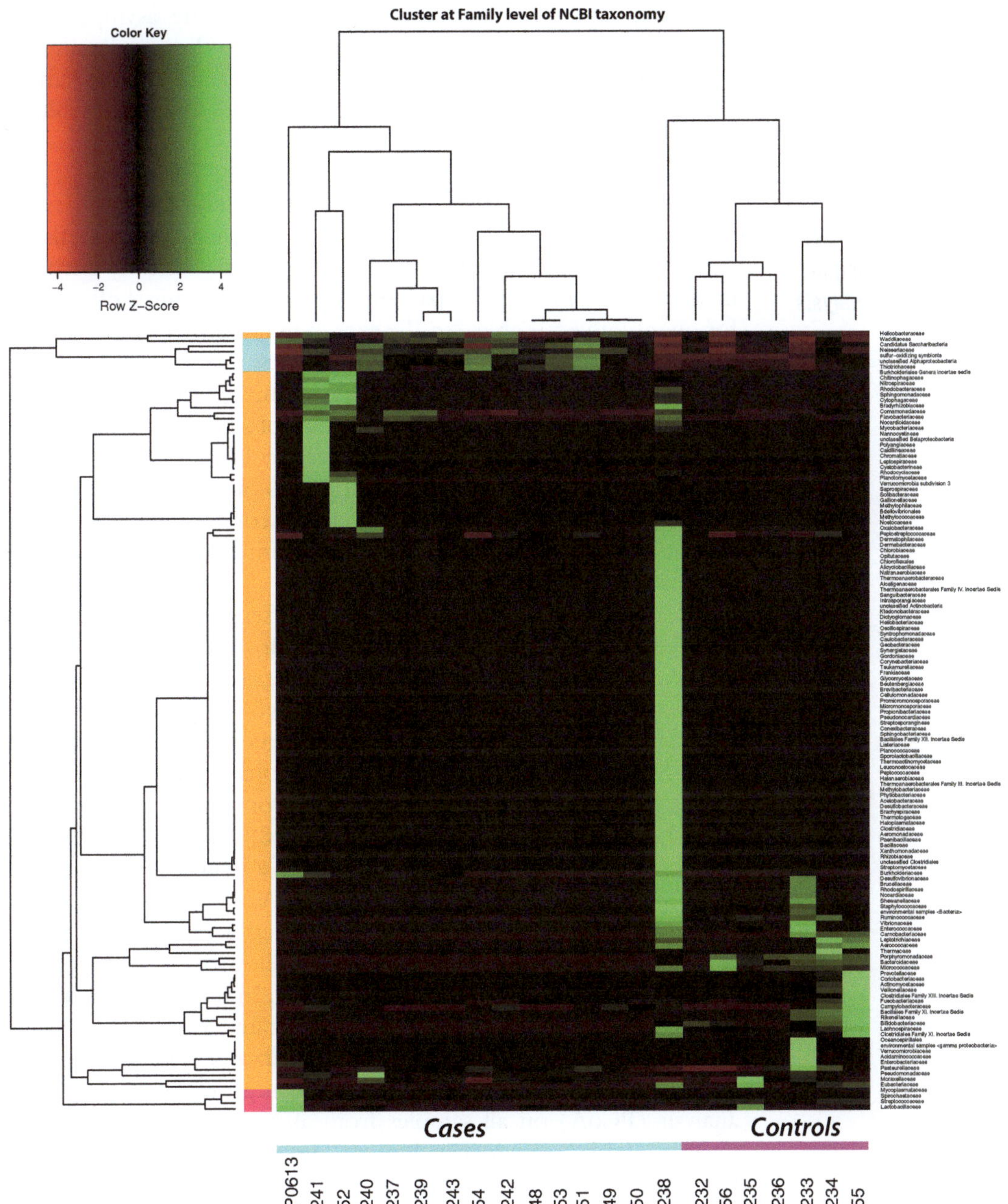

Fig. 8 Taxonomic comparison of all DNA samples. Hierarchical clustering result of "family" level taxonomic comparisons of data from Scenario 1: unstable atherosclerotic plaques from 15 patients with symptomatic atherosclerotic disease (unstable plaques) and stable plaques from a control group of seven patients that died from other causes than atherosclerosis (controls). Red indicates downregulation, green indicates upregulation, and black indicates no change in read abundance level comparing to all samples. Hierarchical clustering was computed with average linkage, whereas Pearson correlation was used for clustering the families (rows) and Spearman correlation was used for clustering the datasets (columns), respectively

oxidizing symbionts and *Thiotrichaceae* than the asymptomatic atherosclerotic plaques (Fig. 8). For P0613, the species profile appeared very different from all other samples. Thus, this sample also treated as an outlier in further analyses (*see* [15] if interested in actual study).

3.10 Principal Component Analysis (PCA) and Principal Coordinates Analysis (PCoA)

PCA and PCoA are tools for multivariate analysis. PCA uses an orthogonal transformation to convert a set of observations of possibly correlated variables into a set of values of linearly uncorrelated variables called principal components [58]. This is often used for quantitative variables, so the axes in graphic have a quantitative weight, and the positions of the samples are in relation with those weight. On the other hand, PCoA or multidimensional scaling (MDS) is a means of visualizing the level of similarity of individual cases of a dataset [59]. PCoA is similar to Polar ordination (PO; [60]) arranges samples between endpoints or 'poles' according to the distance matrix maximizing the linear correlation between the distances in the distance matrix. If further interested in these methods please see [61].

For multiple sample comparison we often use PCoA and PCA, these are among the best tools available for multivariate analysis. These can give us powerful information of similarities and dissimilarities within samples. When coupled with phenotypic data or metadata (using colors and symbols etc.), these can be very helpful tools to understand within group variations. As an example, we have used PCoA on 22 plaque samples from Scenario 1 (Fig. 9). Here we can see that sample 238 and 238 being very different possibly due to high sequence depth (as also seen in Fig. 4).

Biplots: In addition to PCA or PCoA, variables can also be plotted on the same diagram (this is called a *biplot*). The biplot provides a useful tool of data analysis and allows the visual appraisal of the structure of large data matrices [62]. In our examples, where taxa are variables, biplot can show important taxa which helps in determining relatedness represented as arrows. For example, in Scenario 2, β diversity was compared using principal coordinate analysis (PCoA) on all samples from all visits, where biplots are displayed with green arrows (Fig. 10). From this PCoA with biplot, we interpret that samples from volunteers 8, 13, and 16 are different than the other volunteers and that they have higher abundance of *Succinivibrionaceae*, *Gammaproteobacteria*, *Aeromonadales*, etc.

3.11 Canonical-Correlation Analysis (CCA) and Canonical-Correspondence Analysis (CCA)

CCA (correlation) seeks to find the linear combination of the X_i and Y_j that have the greatest correlation with each other where $X = (X_1, \ldots, X_n)$ and $Y = (Y_1, \ldots, Y_m)$ of random variables thus it is often used as a dimension–reduction method. The method was first introduced by Harold Hotelling [63]. On the other hand, CCA (correspondence) is a multivariate method to elucidate the relationships between biological assemblages of species and their

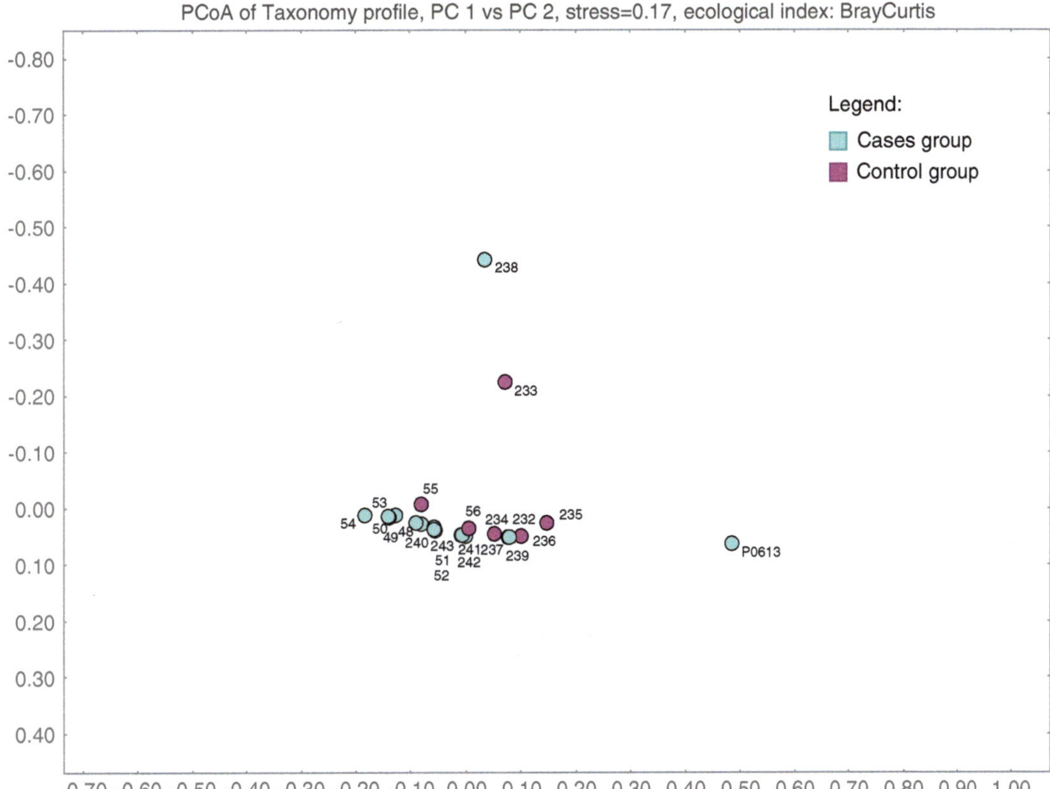

Fig. 9 principal coordinate analyses (PCoA) of "family" level taxonomic comparisons of data from Scenario 1: unstable atherosclerotic plaques from 15 patients with symptomatic atherosclerotic disease (cases: cyan) and stable plaques from a control group of seven patients that died from other causes than atherosclerosis (controls: magenta)

environment. This method by Cajo J. F. ter Braak involves a canonical correlation analysis and a direct gradient analysis [64]. By environment we mean any kind of metadata, such as some physicochemical parameters obtained from same group where the species data is obtained. The idea is to relate the prevalence of a set of species to a collection of environmental variables. Biplots are often used in CCA (correspondence) for visualization purpose. For example, in our Scenario 2, a typical illustration of correlation and correspondence analyses between the microbiome and RBC fatty acid data is displayed in Fig. 11.

In this occasion it is important to note that CCA does not perform variable selection. Further, when the number of variables exceeds the number of observations (or sample size), CCA cannot be applied directly due to singularity of the covariance matrix. In a recent study [65] the authors have discussed this problem and a few existing solutions. Additionally, they developed a method for structure-constrained sparse canonical correlation analysis

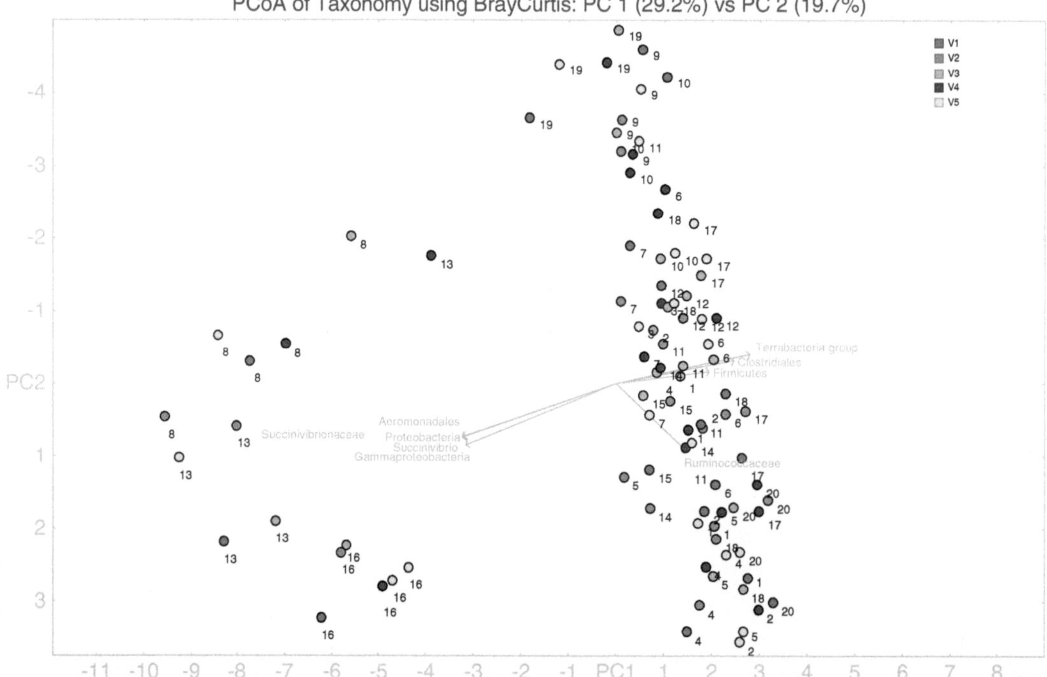

Fig. 10 principal coordinate analyses (PCoA) of level taxonomic comparisons of data from Scenario 2: all samples (V1–V5) for all participants, where biplots are displayed with green arrows. Each visit is denoted by a different color

(ssCCA) in a high-dimensional setting. ssCCA takes into account the phylogenetic relationships among bacteria, which provides important prior knowledge on evolutionary relationships among bacterial taxa (*see* [65] if interested).

3.12 Multivariate Analyses

Multivariate data analysis refers to any statistical approach used to analyze data with more than one variable. For example, as described in Scenario 3 we have multiple factors. The key to identifying important microbial taxa associated with two treatments is that the large datasets from each patient are compared within groups, and then the metadata from the patients' groups are compared against each other. Analysis of multivariate data in response to factors, groups, or treatments in an experimental design needs sophisticated methods.

To achieve this, we can use PERMANOVA (permutational multivariate analysis of variance) [66] to test the homogeneity of multivariate dispersions within groups, on the basis of any resemblance measure. PERMANOVA is a better approach than ANOVA (Analysis of variance)/MANOVA (Multivariate analysis of variance) for our study as PERMANOVA works with any distance measure that is appropriate to the data, and uses permutations to make it

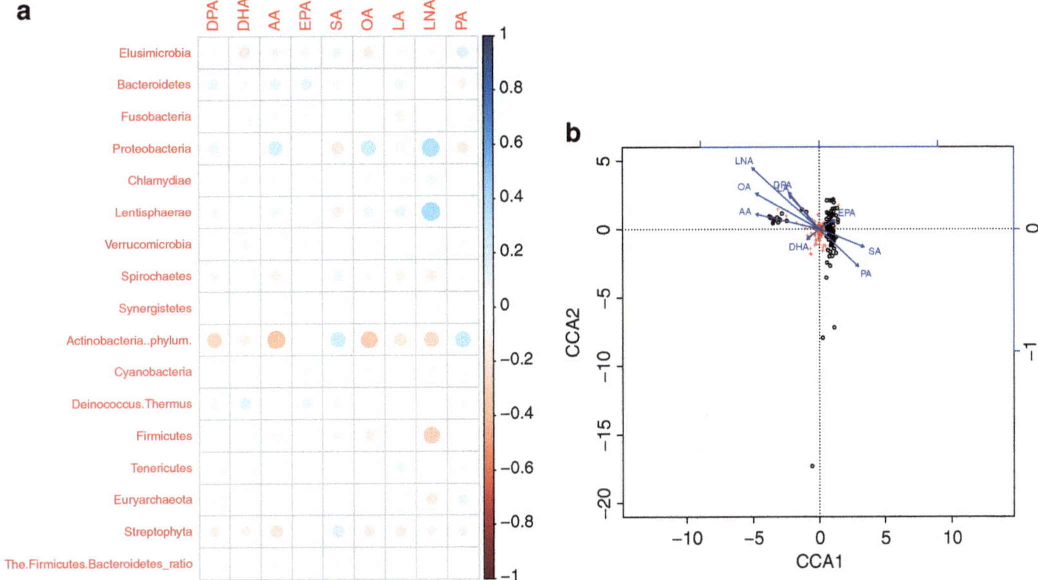

Fig. 11 (**a**) Pearson correlation between genus level microbiome and RBC fatty acid data. (**b**) Canonical correspondence analysis of microbiome (genus level taxonomy) distribution in relation to blood parameters (biplot: represented by blue arrows). Red crosses represent taxa and black circles represents individual samples

distribution free, unlike assuming normal distributions. Finally, in addition to the above multiple comparisons, we can examine if there is consistency of microbiota changes and patterns across the geographical locales of treatment subjects; as our samples are from different countries. We are not showing the details of multivariate analyses, but there are multiple available packages for such analyses with good tutorials. Interested readers may visit these packages and websites as detailed below.

The Primer-E package [67] is commonly used by microbial ecologists and allows for multiple multivariate statistical analyses. We often use R statistical programming language [21] for multivariate statistics. Moreover R is used for several types of graphical representations. Particular packages provide in-built functions and libraries (within R environment) specially for metagenomic datasets such as Bioconductor [68], vegan [69], and phyloseq [70].

4 Tools and Packages Commonly Used in Metagenomic Studies

A list of multiple tools is provided below for analyzing metagenomic data from raw sequence reads to final comparisons and statistical analyses. Discussion of all these tools are beyond the scope of this chapter, but interested readers can see recent review articles [71–74] and it must be noted that there can be other tools as well outside this list.

1. *Processing of raw sequence reads and quality control (QC):*
   (a) FastQC (https://www.bioinformatics.babraham.ac.uk/projects/fastqc/).
   (b) Fastx_toolkit (http://hannonlab.cshl.edu/fastx_toolkit/).
   (c) Cut-adapt (both adapter trimming and quality trim) [25].
   (d) BBTools (http://jgi.doe.gov/data-and-tools/bbtools/).
   (e) Condetri (Read trimmer for Illumina data) [75].
   (f) Trimmomatic (allows multiple threads) [76].
   (g) SolexaQA [77].
   (h) PRINSEQ [78].
2. *Alignment tool:*
   (a) BLAST [18].
   (b) USEARCH [28].
   (c) DIAMOND [22].
   (d) Rapsearch [79].
   (e) PyNAST [29].
3. *Analyses for 16S projects: OTU clustering, picking, and taxonomic assignment.*
   (a) QIIME [27].
   (b) USEARCH [28].
   (c) RDP classifier [30].
   (d) SILVA (for 16S + 18S) [80].
   (e) Mothur [81].
   (f) SILVAngs (https://www.arb-silva.de/documentation/silvangs/).
   (g) MEGAN [31].
   (h) AmpliconNoise [82].
   (i) Open reading frame (ORF) prediction, for example, with MG-DOTUR [83].
4. *Assembly of shotgun metagenomics data.*
   (a) Reference-based assembly.
      - MIRA 4 [84].
      - MetaAMOS (https://www.cbcb.umd.edu/software/metamos).
   (b) De novo assembly.
      - Newbler (Roche).
      - iAssembler [85].
      - EULER [86].

- Velvet [87].
- SOAP [88].
- Abyss [89].

(c) The next generation of assembly tools.

- MetaVelvet-SL [90].
- Meta-IDBA [91].
- InteMAP [92].
- SAT-Assembler [93].
- IDBA-UD [94].

5. *Removing near-exact matches by maping to specific genomes.*

(a) Bowtie 2 [17].

6. *Binning tools for metagenomes.*

(a) Composition-based binning algorithms.

- S-GSOM [95].
- PhylopythiaS [96].
- TACAO [97].
- PCAHIER [98].
- ESOM [95].
- ClaMS [99].

(b) Similarity-based binning software include tools.

- MEGAN [31].
- IMG/MER 4 [35].
- MG-RAST [34].
- CARMA [100].
- MetaPhyler [101].

(c) Unsupervised binning.

- PhylopythiaS+ [102].
- PhymmBL [103].
- ESOMs [104].
- VizBin [105].
- IFCM (fuzzy c-means method) [106].

7. *Binning of metagenome contigs for reconstructing single genomes.*

(a) ICoVeR [107].

(b) MyCC [108].

(c) MetaBAT [109].

(d) GroopM [110].

   (e) MaxBin2 [111].
   (f) CONCOCT [112].
8. *Identification of genes within the reads/assembled contigs or "gene calling".*
   (a) MetaGeneMark [113].
   (b) Prodigal [114].
   (c) Orphelia [115].
   (d) FragGeneScan [116].
9. *Predict for clustered regularly interspaced short palindromic repeats (CRISPRs).*
   (a) CRT [117].
   (b) PILER-CR [118].
   (c) IMG/MER [35].
10. *Annotation pipelines.*
   (a) MEGAN [31].
   (b) QIIME for 16S projects [27].
   (c) Galaxy platform.
   (d) MG-RAST [34].
   (e) IMG/MER [35].
   (f) Primer-E package [67].
   (g) Several packages built within R [21].
      - Vegan [69].
      - Phyloseq [70].
      - Bioconductor [68].
11. *Prediction of functional content from metagenomics.*
   (a) PICRUSt [33].
   (b) Tax4Fun [32].
12. *Statistical computing.*
   (a) R [21].
   (b) Many other tools can be used for statistical analyses.
13. *Web service for the analysis of metagenomic data.*
   (a) The EBI Metagenomics service [36].
   (b) European Nucleotide Archive (ENA).
   (c) MG-RAST [34].
   (d) METAGENassist [119].
   (e) BusyBee Web [120].
   (f) Meta4 [121].

5 Concluding Remarks

This chapter has illustrated multiple data analyses and annotation techniques in metagenomic studies with three case studies. This is not a chapter about any new method development but a description of optimized pipelines using various available tools. With these example scenarios, the use of multiple pipelines has been demonstrated to analyze and interpret the data starting from very raw sequence to the final statistical outputs. Example scenarios describe some of the tools that we have used for analyzing the projects selected for demonstration, but besides these there are plenty of other available tools for metagenomics, most of which are listed in Subheading 4. This chapter does not provide the details of the tools or describe their pros and cons but this can be a good starting point for the readers to explore available options to analyze and interpret their datasets. From this chapter readers shall get an idea of current research projects in medical studies and multiple approaches used to analyze the data originating from these projects, although readers should keep in mind that this is not an exclusive list of possible pipelines for analyzing metagenomic samples. There might be other approaches as well. While step-by-step instructions of all the tools is beyond the scope of this chapter, the methods outline here might be useful to researchers to plan, analyze, and interpret their research projects successfully.

References

1. Riesenfeld CS, Schloss PD, Handelsman J (2004) Metagenomics: genomic analysis of microbial communities. Annu Rev Genet 38:525–552
2. Clarridge JE (2004) Impact of 16S rRNA gene sequence analysis for identification of bacteria on clinical microbiology and infectious diseases. Clin Microbiol Rev 17 (4):840–862
3. Staden R (1979) A strategy of DNA sequencing employing computer programs. Nucleic Acids Res 6(7):2601–2610
4. Ranjan R, Rani A, Metwally A, McGee HS, Perkins DL (2016) Analysis of the microbiome: advantages of whole genome shotgun versus 16S amplicon sequencing. Biochem Biophys Res Commun 469(4):967–977
5. Tessler M, Neumann JS, Afshinnekoo E, Pineda M, Hersch R, Velho LFM et al (2017) Large-scale differences in microbial biodiversity discovery between 16S amplicon and shotgun sequencing. Sci Rep 7(1):6589
6. Jovel J, Patterson J, Wang W, Hotte N, O'Keefe S, Mitchel T et al (2016) Characterization of the gut microbiome using 16S or shotgun metagenomics. Front Microbiol 7:459
7. Greub G (2012) Culturomics: a new approach to study the human microbiome. Clin Microbiol Infect 18(12):1157–1159
8. Lagier JC, Khelaifia S, Alou MT, Ndongo S, Dione N, Hugon P et al (2016) Culture of previously uncultured members of the human gut microbiota by culturomics. Nat Microbiol 1(12):8
9. Peterson J, Garges S, Giovanni M, McInnes P, Wang L, Schloss JA et al (2009) The NIH human microbiome project. Genome Res 19 (12):2317–2323
10. Virgin HW, Todd JA (2011) Metagenomics and personalized medicine. Cell 147 (1):44–56
11. Wang J, Qi J, Zhao H, He S, Zhang Y, Wei S et al (2013) Metagenomic sequencing reveals microbiota and its functional potential associated with periodontal disease. Sci Rep 3:1843

12. Xu P, Gunsolley J (2014) Application of metagenomics in understanding oral health and disease. Virulence 5(3):424–432
13. Ai D, Huang R, Wen J, Li C, Zhu J, Xia LC (2017) Integrated metagenomic data analysis demonstrates that a loss of diversity in oral microbiota is associated with periodontitis. BMC Genomics 18(Suppl 1):1041
14. Forbes JD, Knox NC, Ronholm J, Pagotto F, Reimer A (2017) Metagenomics: the next culture-independent game changer. Front Microbiol 8:1069
15. Mitra S, Drautz-Moses DI, Alhede M, Maw MT, Liu Y, Purbojati RW et al (2015) In silico analyses of metagenomes from human atherosclerotic plaque samples. Microbiome 3:14
16. Watson H, Mitra S, Croden FC, Taylor M, Wood HM, Perry SL et al (2018) A randomised trial of the effect of omega-3 polyunsaturated fatty acid supplements on the human intestinal microbiota. Gut 67 (11):1974–1983
17. Langmead B, Salzberg SL (2012) Fast gapped-read alignment with Bowtie 2. Nat Methods 9(4):357–U54
18. Benson DA, Karsch-Mizrachi I, Clark K, Lipman DJ, Ostell J, Sayers EW (2012) GenBank. Nucleic Acids Res 40(D1):D48–D53
19. Altschul SF, Gish W, Miller W, Myers EW, Lipman DJ (1990) Basic local alignment search tool. J Mol Biol 215(3):403–410
20. Huson DH, Mitra S, Ruscheweyh HJ, Weber N, Schuster SC (2011) Integrative analysis of environmental sequences using MEGAN4. Genome Res 21(9):1552–1560
21. R Development Core Team (2008) R: A language and environment for statistical computing. R Foundation for Statistical Computing, Vienna
22. Buchfink B, Xie C, Huson DH (2015) Fast and sensitive protein alignment using DIAMOND. Nat Methods 12(1):59–60
23. Yu ZT, Morrison M (2004) Improved extraction of PCR-quality community DNA from digesta and fecal samples. BioTechniques 36 (5):808–812
24. Taylor M, Wood HM, Halloran SP, Quirke P (2017) Examining the potential use and long-term stability of guaiac faecal occult blood test cards for microbial DNA 16S rRNA sequencing. J Clin Pathol 70(7):600–606
25. Martin M (2011) Cutadapt removes adapter sequences from high-throughput sequencing reads. Genomics 98(1):152–153
26. Aronesty E (2011) ea-utils: "Command-line tools for processing biological sequencing data". https://github.com/ExpressionAnalysis/ea-utils
27. Caporaso JG, Kuczynski J, Stombaugh J, Bittinger K, Bushman FD, Costello EK et al (2010) QIIME allows analysis of high-throughput community sequencing data. Nat Methods 7(5):335–336
28. Edgar RC (2010) Search and clustering orders of magnitude faster than BLAST. Bioinformatics 26(19):2460–2461
29. Caporaso JG, Bittinger K, Bushman FD, DeSantis TZ, Andersen GL, Knight R (2010) PyNAST: a flexible tool for aligning sequences to a template alignment. Bioinformatics 26(2):266–267
30. Wang Q, Garrity GM, Tiedje JM, Cole JR (2007) Naive Bayesian classifier for rapid assignment of rRNA sequences into the new bacterial taxonomy. Appl Environ Microbiol 73(16):5261–5267
31. Huson DH, Beier S, Flade I, Gorska A, El-Hadidi M, Mitra S et al (2016) MEGAN community edition—interactive exploration and analysis of large-scale microbiome sequencing data. PLoS Comput Biol 12(6):12
32. Aßhauer KP, Wemheuer B, Daniel R, Meinicke P (2015) Tax4Fun: predicting functional profiles from metagenomic 16S rRNA data. Bioinformatics 31(17):2882–2884
33. Langille MGI, Zaneveld J, Caporaso JG, McDonald D, Knights D, Reyes JA et al (2013) Predictive functional profiling of microbial communities using 16S rRNA marker gene sequences. Nat Biotechnol 31 (9):814–821
34. Wilke A, Bischof J, Gerlach W, Glass E, Harrison T, Keegan KP et al (2016) The MG-RAST metagenomics database and portal in 2015. Nucleic Acids Res 44(Database issue):D590–D5D4
35. Markowitz VM, Chen IMA, Palaniappan K, Chu K, Szeto E, Pillay M et al (2014) IMG 4 version of the integrated microbial genomes comparative analysis system. Nucleic Acids Res 42(Database issue):D560–D5D7
36. Hunter S, Corbett M, Denise H, Fraser M, Gonzalez-Beltran A, Hunter C et al (2014) EBI metagenomics—a new resource for the analysis and archiving of metagenomic data. Nucleic Acids Res 42(Database issue): D600–D6D6
37. Du JL, Yuan ZF, Ma ZW, Song JZ, Xie XL, Chen YL (2014) KEGG-PATH: Kyoto encyclopedia of genes and genomes-based pathway analysis using a path analysis model. Mol BioSyst 10(9):2441–2447

38. Overbeek R, Begley T, Butler RM, Choudhuri JV, Chuang HY, Cohoon M et al (2005) The subsystems approach to genome annotation and its use in the project to annotate 1000 genomes. Nucleic Acids Res 33 (17):5691–5702
39. Powell S, Forslund K, Szklarczyk D, Trachana K, Roth A, Huerta-Cepas J et al (2014) eggNOG v4.0: nested orthology inference across 3686 organisms. Nucleic Acids Res 42(D1):D231–D2D9
40. Tatusov RL, Galperin MY, Natale DA, Koonin EV (2000) The COG database: a tool for genome-scale analysis of protein functions and evolution. Nucleic Acids Res 28 (1):33–36
41. Haft DH, Selengut JD, White O (2003) The TIGRFAMs database of protein families. Nucleic Acids Res 31(1):371–373
42. Finn RD, Coggill P, Eberhardt RY, Eddy SR, Mistry J, Mitchell AL et al (2016) The Pfam protein families database: towards a more sustainable future. Nucleic Acids Res 44(D1): D279–DD85
43. Vollmers J, Wiegand S, Kaster AK (2017) Comparing and evaluating Metagenome assembly tools from a microbiologist's perspective—not only size matters! PLoS One 12(1):e0169662
44. Ghurye JS, Cepeda-Espinoza V, Pop M (2016) Metagenomic assembly: overview, challenges and applications. Yale J Biol Med 89(3):353–362
45. Bacaro G, Rocchini D, Ghisla A, Marcantonio M, Neteler M, Chiarucci A (2012) The spatial domain matters: spatially constrained species rarefaction in a free and open source environment. Ecol Complex 12:63–69
46. Ni J, Yan Q, Yu Y (2013) How much metagenomic sequencing is enough to achieve a given goal? Sci Rep 3:1968
47. Rodriguez RL, Konstantinidis KT (2014) Estimating coverage in metagenomic data sets and why it matters. ISME J 8 (11):2349–2351
48. Whittaker RH (1972) Evolution and measurement of species diversity. Taxon 21 (2/3):213–251
49. Hill MO (1973) Diversity and evenness: a unifying notation and its consequences. Ecology 54(2):427–432
50. Tuomisto H (2010) A consistent terminology for quantifying species diversity? Yes, it does exist. Oecologia 164(4):853–860
51. Mitra S, Gilbert JA, Field D, Huson DH (2010) Comparison of multiple metagenomes using phylogenetic networks based on ecological indices. ISME J 4(10):1236–1242
52. Lozupone C, Knight R (2005) UniFrac: a new phylogenetic method for comparing microbial communities. Appl Environ Microbiol 71(12):8228–8235
53. Hamady M, Lozupone C, Knight R (2010) Fast UniFrac: facilitating high-throughput phylogenetic analyses of microbial communities including analysis of pyrosequencing and PhyloChip data. ISME J 4(1):17–27
54. Lozupone C, Lladser ME, Knights D, Stombaugh J, Knight R (2011) UniFrac: an effective distance metric for microbial community comparison. ISME J 5(2):169–172
55. McGill R, Tukey JW, Larsen WA (1978) Variations of box plots. Am Stat 32(1):12–16
56. Kaufman L, Rousseeuw PJ (1990) Finding groups in data: an introduction to cluster analysis, 1st edn. John Wiley, New York
57. Hastie T, Tibshirani R, Friedman J (2009) Hierarchical clustering. In: The elements of statistical learning, 2nd edn. Springer, New York
58. Pearson K (1901) On lines and planes of closest fit to systems of points in space. Philos Mag 2(7–12):559–572
59. Borg I, Groenen PJ (2005) Modern multidimensional scaling: theory and applications. Springer Science & Business Media, New York
60. Bray JR, Curtis JT (1957) An ordination of the upland forest communities of southern Wisconsin. Ecol Monogr 27(4):326–349
61. Michael PW Ordination methods—an overview. http://ordination.okstate.edu/overview.htm
62. Gabriel KR (1971) Biplot graphic display of matrices with application to principal component analysis. Biometrika 58(3):453–467
63. Hotelling H (1936) Relations between two sets of variates. Biometrika 28:321–377
64. Terbraak CJF (1986) Canonical correspondence-analysis—a new eigenvector technique for multivariate direct gradient analysis. Ecology 67(5):1167–1179
65. Chen J, Bushman FD, Lewis JD, Wu GD, Li H (2013) Structure-constrained sparse canonical correlation analysis with an application to microbiome data analysis. Biostatistics 14(2):244–258
66. Anderson MJ (2001) A new method for non-parametric multivariate analysis of variance. Austral Ecol 26(1):32–46
67. Clarke KG, Gorley RN (2006) PRIMER v6: user manual/tutorial. PRIMER-E, Plymouth

68. Gentleman RC, Carey VJ, Bates DM, Bolstad B, Dettling M, Dudoit S et al (2004) Bioconductor: open software development for computational biology and bioinformatics. Genome Biol 5(10):16
69. Oksanen J, Blanchet FG, Friendly M, Kindt R, Legendre P, McGlinn D, et al (2008) The vegan package. https://cran.r-project.org/web/packages/vegan/vegan.pdf
70. McMurdie PJ, Holmes S (2013) Phyloseq: an R package for reproducible interactive analysis and graphics of microbiome census data. PLoS One 8(4):11
71. Teeling H, Glockner FO (2012) Current opportunities and challenges in microbial metagenome analysis-a bioinformatic perspective. Brief Bioinform 13(6):728–742
72. Oulas A, Pavloudi C, Polymenakou P, Pavlopoulos GA, Papanikolaou N, Kotoulas G et al (2015) Metagenomics: tools and insights for analyzing next-generation sequencing data derived from biodiversity studies. Bioinform Biol Insights 9:75–88
73. Pavlopoulos GA, Oulas A, Iacucci E, Sifrim A, Moreau Y, Schneider R et al (2013) Unraveling genomic variation from next generation sequencing data. BioData Min 6:13
74. Lindgreen S, Adair KL, Gardner PP (2016) An evaluation of the accuracy and speed of metagenome analysis tools. Sci Rep 6:19233
75. Smeds L, Künstner A (2011) ConDeTri—a content dependent read trimmer for Illumina data. PLoS One 6(10):e26314
76. Bolger AM, Lohse M, Usadel B (2014) Trimmomatic: a flexible trimmer for illumina sequence data. Bioinformatics 30 (15):2114–2120
77. Cox MP, Peterson DA, Biggs PJ (2010) SolexaQA: at-a-glance quality assessment of illumina second-generation sequencing data. BMC Bioinformatics 11:485
78. Schmieder R, Edwards R (2011) Quality control and preprocessing of metagenomic datasets. Bioinformatics 27(6):863–864
79. Ye Y, Choi J-H, Tang H (2011) RAPSearch: a fast protein similarity search tool for short reads. BMC Bioinformatics 12:159
80. Quast C, Pruesse E, Yilmaz P, Gerken J, Schweer T, Yarza P et al (2013) The SILVA ribosomal RNA gene database project: improved data processing and web-based tools. Nucleic Acids Res 41(D1): D590–D5D6
81. Schloss PD, Westcott SL, Ryabin T, Hall JR, Hartmann M, Hollister EB et al (2009) Introducing mothur: open-source, platform-independent, community-supported software for describing and comparing microbial communities. Appl Environ Microbiol 75 (23):7537–7541
82. Quince C, Lanzen A, Davenport RJ, Turnbaugh PJ (2011) Removing noise from Pyrosequenced amplicons. BMC Bioinformatics 12:38
83. Schloss PD, Handelsman J (2008) A statistical toolbox for metagenomics: assessing functional diversity in microbial communities. BMC Bioinformatics 9:34
84. Chevreux B, Pfisterer T, Drescher B, Driesel AJ, Müller WEG, Wetter T et al (2004) Using the miraEST assembler for reliable and automated mRNA transcript assembly and SNP detection in sequenced ESTs. Genome Res 14(6):1147–1159
85. Zheng Y, Zhao LJ, Gao JP, Fei ZJ (2011) iAssembler: a package for de novo assembly of Roche-454/sanger transcriptome sequences. Bmc Bioinformatics 12:8
86. Pevzner PA, Tang H, Waterman MS (2001) An Eulerian path approach to DNA fragment assembly. Proc Natl Acad Sci U S A 98 (17):9748–9753
87. Zerbino DR, McEwen GK, Margulies EH, Birney E (2009) Pebble and rock band: heuristic resolution of repeats and scaffolding in the velvet short-read de novo assembler. PLoS One 4(12):9
88. Luo R, Liu B, Xie Y, Li Z, Huang W, Yuan J et al (2012) SOAPdenovo2: an empirically improved memory-efficient short-read de novo assembler. GigaScience 1:18
89. Jackman SD, Vandervalk BP, Mohamadi H, Chu J, Yeo S, Hammond SA et al (2017) ABySS 2.0: resource-efficient assembly of large genomes using a bloom filter. Genome Res 27(5):768–777
90. Afiahayati, Sato K, Sakakibara Y (2015) MetaVelvet-SL: an extension of the velvet assembler to a de novo metagenomic assembler utilizing supervised learning. DNA Res 22(1):69–77
91. Peng Y, Leung HCM, Yiu SM, Chin FYL (2011) Meta-IDBA: a de Novo assembler for metagenomic data. Bioinformatics 27(13): i94–i101
92. Lai B, Wang F, Wang X, Duan L, Zhu H (2015) InteMAP: integrated metagenomic assembly pipeline for NGS short reads. BMC Bioinformatics 16:244
93. Zhang Y, Sun Y, Cole JR (2014) A scalable and accurate targeted gene assembly tool (SAT-assembler) for next-generation

sequencing data. PLoS Comput Biol 10(8): e1003737

94. Peng Y, Leung HCM, Yiu SM, Chin FYL (2012) IDBA-UD: a de novo assembler for single-cell and metagenomic sequencing data with highly uneven depth. Bioinformatics 28 (11):1420–1428
95. Chan C-KK, Hsu AL, Tang S-L, Halgamuge SK (2008) Using growing self-Organising maps to improve the binning process in environmental whole-genome shotgun sequencing. J Biomed Biotechnol 2008:513701
96. Patil KR, Roune L, McHardy AC (2012) The PhyloPythiaS web server for taxonomic assignment of metagenome sequences. PLoS One 7(6):e38581
97. Diaz NN, Krause L, Goesmann A, Niehaus K, Nattkemper TW (2009) TACOA—taxonomic classification of environmental genomic fragments using a kernelized nearest neighbor approach. BMC Bioinformatics 10:56
98. Zheng H, Wu H (2010) Short prokaryotic DNA fragment binning using a hierarchical classifier based on linear discriminant analysis and principal component analysis. J Bioinforma Comput Biol 8(6):995–1011
99. Pati A, Heath LS, Kyrpides NC, Ivanova N (2011) ClaMS: a classifier for metagenomic sequences. Stand Genomic Sci 5(2):248–253
100. Krause L, Diaz NN, Goesmann A, Kelley S, Nattkemper TW, Rohwer F et al (2008) Phylogenetic classification of short environmental DNA fragments. Nucleic Acids Res 36 (7):2230–2239
101. Liu B, Gibbons T, Ghodsi M, Treangen T, Pop M (2011) Accurate and fast estimation of taxonomic profiles from metagenomic shotgun sequences. BMC Genomics 12(2):S4
102. Gregor I, Dröge J, Schirmer M, Quince C, McHardy AC (2016) PhyloPythiaS+: a self-training method for the rapid reconstruction of low-ranking taxonomic bins from metagenomes. PeerJ 4:e1603
103. Brady A, Salzberg S (2011) PhymmBL expanded: confidence scores, custom databases, parallelization and more. Nat Methods 8(5):367
104. Dick GJ, Andersson AF, Baker BJ, Simmons SL, Thomas BC, Yelton AP et al (2009) Community-wide analysis of microbial genome sequence signatures. Genome Biol 10(8):R85
105. Laczny CC, Sternal T, Plugaru V, Gawron P, Atashpendar A, Margossian HH et al (2015) VizBin—an application for reference-independent visualization and human-augmented binning of metagenomic data. Microbiome 3:1
106. Liu Y, Hou T, Kang B, Liu F (2017) Unsupervised binning of metagenomic assembled Contigs using improved fuzzy C-means method. IEEE/ACM Trans Comput Biol Bioinform 14(6):1459–1467
107. Broeksema B, Calusinska M, McGee F, Winter K, Bongiovanni F, Goux X et al (2017) ICoVeR—an interactive visualization tool for verification and refinement of metagenomic bins. BMC Bioinformatics 18:233
108. Lin H-H, Liao Y-C (2016) Accurate binning of metagenomic contigs via automated clustering sequences using information of genomic signatures and marker genes. Sci Rep 6:24175
109. Kang DD, Froula J, Egan R, Wang Z (2015) MetaBAT, an efficient tool for accurately reconstructing single genomes from complex microbial communities. PeerJ. 3:e1165
110. Imelfort M, Parks D, Woodcroft BJ, Dennis P, Hugenholtz P, Tyson GW (2014) GroopM: an automated tool for the recovery of population genomes from related metagenomes. PeerJ 2:e603
111. Wu Y-W, Tang Y-H, Tringe SG, Simmons BA, Singer SW (2014) MaxBin: an automated binning method to recover individual genomes from metagenomes using an expectation-maximization algorithm. Microbiome 2:26
112. Alneberg J, Bjarnason BS, de Bruijn I, Schirmer M, Quick J, Ijaz UZ et al (2014) Binning metagenomic contigs by coverage and composition. Nat Methods 11 (11):1144–1146
113. Zhu W, Lomsadze A, Borodovsky M (2010) Ab initio gene identification in metagenomic sequences. Nucleic Acids Res 38(12):e132
114. Hyatt D, Chen G-L, LoCascio PF, Land ML, Larimer FW, Hauser LJ (2010) Prodigal: prokaryotic gene recognition and translation initiation site identification. BMC Bioinformatics 11:119
115. Hoff KJ, Lingner T, Meinicke P, Tech M (2009) Orphelia: predicting genes in metagenomic sequencing reads. Nucleic Acids Res 37(Web Server issue):W101–W1W5
116. Rho M, Tang H, Ye Y (2010) FragGeneScan: predicting genes in short and error-prone reads. Nucleic Acids Res 38(20):e191
117. Bland C, Ramsey TL, Sabree F, Lowe M, Brown K, Kyrpides NC et al (2007) CRISPR recognition tool (CRT): a tool for automatic detection of clustered regularly interspaced

palindromic repeats. BMC Bioinformatics 8:209

118. Edgar RC (2007) PILER-CR: fast and accurate identification of CRISPR repeats. BMC Bioinformatics 8:18
119. Arndt D, Xia J, Liu Y, Zhou Y, Guo AC, Cruz JA et al (2012) METAGENassist: a comprehensive web server for comparative metagenomics. Nucleic Acids Res 40(Web Server issue):W88–W95
120. Laczny CC, Kiefer C, Galata V, Fehlmann T, Backes C, Keller A (2017) BusyBee web: metagenomic data analysis by bootstrapped supervised binning and annotation. Nucleic Acids Res 45(W1):W171–W1W9
121. Richardson EJ, Escalettes F, Fotheringham I, Wallace RJ, Watson M (2013) Meta4: a web application for sharing and annotating metagenomic gene predictions using web services. Front Genet 4:168

Chapter 21

Systems Genetics for Evolutionary Studies

Pjotr Prins, Geert Smant, Danny Arends, Megan K. Mulligan, Rob W. Williams, and Ritsert C. Jansen

Abstract

Systems genetics combines high-throughput genomic data with genetic analysis. In this chapter, we review and discuss application of systems genetics in the context of evolutionary studies, in which high-throughput molecular technologies are being combined with quantitative trait locus (QTL) analysis in segregating populations.

The recent explosion of high-throughput data—measuring thousands of RNAs, proteins, and metabolites, using deep sequencing, mass spectrometry, chromatin, methyl-DNA immunoprecipitation, etc.—allows the dissection of causes of genetic variation underlying quantitative phenotypes of all types. To deal with the sheer amount of data, powerful statistical tools are needed to analyze multidimensional relationships and to extract valuable information and new modes and mechanisms of changes both within and between species. In the context of evolutionary computational biology, a well-designed experiment and the right population can help dissect complex traits likely to be under selection using proven statistical methods for associating phenotypic variation with chromosomal locations.

Recent evolutionary expression QTL (*e*QTL) studies focus on gene expression adaptations, mapping the gene expression landscape, and, tentatively, define networks of transcripts and proteins that are jointly modulated sets of *e*QTL networks. Here, we discuss the possibility of introducing an evolutionary "prior" in the form of gene families displaying evidence of positive selection, and using that prior in the context of an *e*QTL experiment for elucidating host-pathogen protein-protein interactions.

Here we review one exemplar evolutionairy eQTL experiment and discuss experimental design, choice of platforms, analysis methods, scope, and interpretation of results. In brief we highlight how *e*QTL are defined; how they are used to assemble interacting and causally connected networks of RNAs, proteins, and metabolites; and how some QTLs can be efficiently converted to reasonably well-defined sequence variants.

Key words Systems genetics, Genetical genomics, QTL, eQTL, xQTL, R-genes, Evolution, R/qtl, LMM, GEMMA, NGS, Genomics, Metabolomics, Network inference, GeneNetwork

1 Introduction

Genetics concerns the study of heritably quantitative or complex traits. Many agricultural traits of interest, such as milk production in cattle and response to fertilizer in crops and most human, animal, and plant diseases, are complex traits. Associating, or linking,

Maria Anisimova (ed.), *Evolutionary Genomics: Statistical and Computational Methods*, Methods in Molecular Biology, vol. 1910, https://doi.org/10.1007/978-1-4939-9074-0_21,

complex traits with certain positions on the genome is achieved through the mapping of the so-called quantitative trait loci (QTL).

Mapping QTL in experimental populations is possible when linkage and/or association information is available. When we have a population of individuals with known genotypes, it may be possible to link a phenotype with a certain genotype. To genotype individuals, first marker maps are created. A marker is a known genomic location, where the genotype of an individual can be determined. In the early days, the genotype was determined by visible chromosome features, later with restriction fragment length polymorphism (RFLP) and amplified fragment length polymorphism (AFLP, *see* also [1–3]), and, increasingly, with SNP/haplotype data [4]. When all individuals with genotype A at a marker location somewhere on the genome are susceptible to a disease and all other individuals with genotype B are not, there is linkage/association or a QTL. If it is clear cut, i.e., single QTL explains all phenotype variance, it is likely to be a single gene effect. Often it is not clear cut, and we need statistics to determine the strength of association between phenotype and genotype.

It is also possible to use linkage disequilibrium (LD) to map QTL in outbred and natural populations. LD occurs when certain stretches of the genome (haplotypes) show nonrandom behavior based on allele frequencies and recombination. Associating haplotype frequencies with phenotypes potentially renders QTL. Kim et al. describe the genome-wide pattern of LD in a sample of 19 *Arabidopsis thaliana* accessions using SNP microarrays [5]. LD is tested, for example, by Dixon et al., to globally map the effect of polymorphism on gene expression in 400 children from families recruited through a proband with asthma [6].

The use of terms “association” and “linkage” can be confusing, even in literature. In this text we use **association** with haplotypes in natural populations of unrelated individuals and **linkage** with markers in families and groups of families, often termed experimental populations. Note some genetic studies are hybrids of both methods, such as Dixon et al. [6], and individuals are related, i.e., some within-family linkage information is available for 400 children from 206 families which should be accounted for in the analysis.

Statistical power can be increased by using experimental crosses instead of natural populations. For example, each individual line in a set of recombinant inbred lines (RILs) is homozygous across the genome, doubling the genetic variance, simplifying genetic models, and increasing statistical power. For model organisms, such as *A. thaliana*, *Caenorhabditis elegans*, *Drosophila melanogaster*, and *Mus musculus*, genotyped and even fully sequenced experimental crosses are available; i.e., for these species it is not necessary to generate a new cross, and for these crosses comprehensive SNP and sequence data may be available. One of the features of inbred model organisms is that they are “immortal” which means that experiments conducted more than 10, even 30, years ago can still be compared with those today. Databases, such as GeneNetwork [7, 8], contain thousands of studies conducted on the same individual mouse strains.

Systems genetics combines genetics with high-throughput molecular technologies. Combining gene expression, as measured by microarray probes or RNA sequencing, with linkage leads to gene expression QTL (*e*QTL). Such *e*QTL studies elucidate how genotypic variation underlies, for example, morphological phenotypes, by using gene expression levels as intermediate molecular phenotypes. In other words, the expression level, as measured by a microarray probe or probe set, is treated as a phenotype, i.e., a gene expression trait. This phenotype is associated with the genome in the form of one or more *e*QTL. With microarrays, the genomic location of the probe is usually known. Therefore, expression phenotype and probe connect two types of genomic information: *e*QTL location(s) and gene location. It is usually assumed that *e*QTL loci represent *cis*- or *trans*-transcription regulators of the target gene [9]. If the *e*QTL is located close to the gene on the genome, the *e*QTL may point to a *cis*-regulator. If the *e*QTL is located far from the gene on the genome, the *e*QTL may point to a *trans*-regulator of a single gene or even *e*QTL *trans*-bands that regulate multiple genes (*see* Fig. 1a and [10, 11]).

In a similar fashion, proteins and metabolites can be measured to map protein QTL (*p*QTL) and metabolite QTL (*m*QTL). A remarkable study published in 1994 used two-dimensional protein electrophoresis and a restriction fragment length polymorphism map (RFLP) [12]. Deep sequencing, chromatin, and methyl-DNA immunoprecipitation are just a few of the latest technologies that add to the arsenal of tools available for the study of the genetic variation underlying quantitative phenotypes. Together, *e*QTL, *m*QTL, and *p*QTL are referred to as *x*QTL. Different *x*QTL appear to confirm each other, for example, with the *A. thaliana* glucosinolate pathway where *e*QTL, *m*QTL, and *p*QTL were mapped together and used to infer the underlying pathways [13]. Such causal inference can lead to dissecting pathways and gene networks which is an active field of research, e.g., [14–16] (*see* also Fig. 1).

1.1 Evolutionary xQTL Studies

From the perspective of evolutionary biology, systems genetics has been applied to elucidate evolutionary adaptations of transcript regulation. For example, Fraser et al. introduced a test for lineage-specific selection on gene expression and analyzed the directionality of microarray *e*QTL for 112 haploid segregants of a genetic cross between two strains of the budding yeast *Saccharomyces cerevisiae*, reanalyzing the two-color cDNA microarray data of Brem and Kruglyak [17]. They found that hundreds of gene expression levels have been subjected to lineage-specific selection. Comparing these findings with independent population genetic

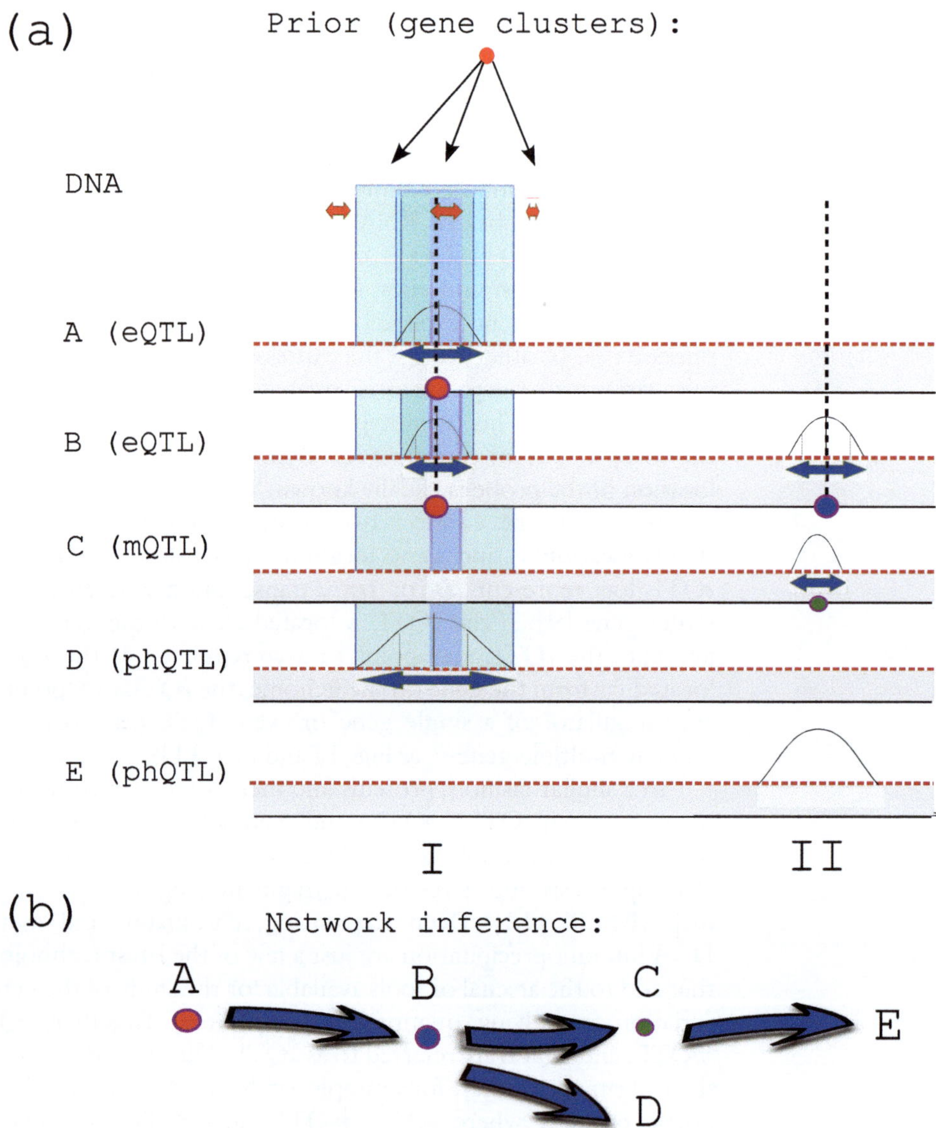

Fig. 1 In this hypothetical and schematic example related to mapped locations on a chromosome, prior information is combined with multiple phenotype-genotype QTL mappings to zoom in on genomic areas and to reason about causal relations between different layers of information. (**a**) The prior (red area on the chromosome) points out that certain sections are of interest; these sections consist of related genes with high homology showing evidence of positive selection, as discussed in the main text. The blue double arrow points out the confidence interval for each QTL, above the significance threshold (red dotted line). The accumulated evidence (light-blue areas) leads to a narrowed down section on the genome, where in this case the prior information is the most specific. In addition, expression phenotypes A and B point to exact gene locations (dotted line, based on exact probe information). (**b**) To infer causal relationships, network inference is possible. On the left (vertical I), traits A, B, and D map to one hot spot, where A may be a regulator of B because one QTL is shared. B causes metabolite phenotype C, again a shared QTL. Phenotype D matches A and B, and phenotype E matches A, B, and C. These causal relationships are drawn by arrows. The figure suggests that, even if individual QTL are not very informative, the accumulated evidence starts to paint a picture

evidence of selective sweeps suggests that this lineage-specific selection has resulted in recent sweeps at over a hundred genes, most of which led to increased transcript levels. Fraser et al. suggest that adaptive evolution of gene expression is common in yeast, that regulatory adaptation can occur at the level of entire pathways, and that similar genome-wide scans may be possible in other species, including human [18].

In another *S. cerevisiae* study, Zou et al., by reanalyzing the same two-color cDNA microarray data, uncovered genetic regulatory network divergence between duplicate genes. They found evidence that the regulation of the ancestral gene diverged due to gene duplication [19].

Li et al. studied plasticity of gene expression in *C. elegans*, using a set of 80 RILs generated from a cross of N2 (Bristol) and CB4856 (Hawaii), representing two genetic and ecological extremes of *C. elegans*. While the overall level of polymorphism among wild isolates of *C. elegans* is relatively low, the genetic distance between N2 and CB4856 is high, representing millions of years of genetic drift. Differential expression induced in a RIL population by temperatures of 16 °C and 24 °C has a strong genetic component. With a group of transgenes, there was prominent evidence for a common master regulator: an *e*QTL trans-band of 66 coregulated genes appeared at 24 °C. The results suggest widespread genetic variation of differential expression responses to environmental impacts and demonstrate the potential of systems genetics for mapping the molecular determinants of phenotypic plasticity [11], leading to a more generalized systems genetics, where value is added from environmental perturbation [20].

Hager et al. determined that genetic architecture supports mosaic brain evolution and independent brain-body size regulation by a quantitative genetic approach involving over 10,000 BXD mouse RILs. The BXD family consists of over 100 lines derived from parental strains that differ at five million single nucleotide polymorphisms (SNPs), indels, transposons, and copy-number variants. This model system harbors naturally occurring genetic variation at a level approximating that of human populations. The study utilizes a high-density linkage analysis to map loci modulating phenotypic variation in overall brain size, body size, and the size of seven major brain parts: neocortex, cerebellum, striatum, olfactory bulb, hippocampus, lateral geniculate nucleus, and basolateral complex of the amygdala. Under the mosaic evolutionary hypothesis, the size of different systems evolves independently due to differential selective pressures associated with different tasks. They identified independent loci for size variation in seven key parts of the brain and observe that brain parts show low or no phenotypic correlation, as is predicted by a mosaic scenario. They also demonstrate that variation in brain size is independently regulated from body size [21].

Kliebenstein et al. detected significant gene network variation in 148 RILs originating from a cross between two *A. thaliana* accessions, Bay-0 and Shahdara. They were able to identify *e*QTL controlling network responses for 18 out of 20 *a priori* defined gene networks, representing 239 genes [22].

According to Gilad, *e*QTL studies show that (1) variation in gene expression levels is both widespread and highly heritable; (2) gene expression levels are highly amenable to genetic mapping; and (3) most strong *e*QTL are found near the target gene, suggesting that variation in cis-regulatory elements underlies much of the observed variation in gene expression levels [23]. Meanwhile, Alberts et al. suggest that sequence polymorphisms influencing the binding of microarray probes may cause many false cis *e*QTL, which should be accounted for [24].

1.2 Adding a Prior

QTL mapping links complex traits with one or more locations on the genome (*see* Fig. 1). Such a location is a wide measure because a QTL is a statistical estimate and rarely a precise indicator. On the genome, a single QTL may represent tens, hundreds, and even thousands of real genes. Combining the QTL with high-throughput technologies, such as microarrays, can add information. To zoom in on the genes underlying QTL, information from other sources has to be utilized. Such *a priori* knowledge (prior) could consist of results from traditional linkage studies or association studies of, for example, human disease. That way one can assign a specific regulatory role to polymorphic sites in a genomic region known to be associated with disease [23]. Other useful priors can be derived from existing information on gene ontology terms, metabolic pathways, and protein-protein interactions, which can be used to identify genes and pathways [25], provided these databases are sufficiently informative.

Zou et al., for example, used gene ontology as a **prior** and concluded that *trans*-acting *e*QTL divergence between duplicate pairs of genes is related to a fitness defect under treatment conditions, but not with fitness under normal condition [19].

Chen et al. identified strong candidate genes for resistance to leaf rust in barley and on the general pathogen response pathway using a custom barley microarray on 144 doubled haploid lines of the St/Mx population [26]. Fifteen thousand six hundred and eighty-five *e*QTL were mapped from 9557 genes. Correlation analysis identified 128 genes that were correlated with resistance, of which 89 had *e*QTL colocating with the phenotypic QTL (phQTL) or classic QTL. Transcript abundance in the parents and conservation of synteny with rice prioritized six genes as candidates for Rphq11, the phQTL of largest effect [26].

In this chapter we discuss the steps needed to design an *x*QTL experiment to make use of systems genetics in evolutionary studies more concrete. As the prior we add information on plant host genes showing evidence of positive selection.

2 Designing an Evolutionary *x*QTL Experiment

An experimental design based on systems genetics can highlight sections of the genome showing correlation with an evolutionary trait. One such evolutionary trait of interest is plant resistance against pathogens. Plants have developed mechanisms to defend themselves against pests. When a pathogen, such as potato blight *Phytophthora infestans*, or a nematode, such as *Meloidogyne hapla*, infects a plant, it uses a battery of so-called effectors to help invade the plant. Some of these effector molecules act to dissolve cellulose [27]. Intriguingly, other molecules are involved in actively reprogramming plant cells. Such plant-pathogen effectors have been shown to mimic plant transcription factors [28] and switch on genes that help the pathogen [29]. A susceptible plant allows the pathogen to suppress defense mechanisms and to change cell configuration. For example, the nematodes *M. hapla* and *Globodera rostochiensis* transform plant cells, so they become elaborate feeding structures. The genetics of this plant-pathogen interaction is potentially even relevant for human medicine, as an increased understanding of host-pathogen relationships may help understand the workings of the innate immune system and nematode immunomodulation [30, 31]. The innate immune system, through plant resistance genes (R-genes, *see* Box 1), influences susceptibility to infections in all multicellular organisms and is a much older evolutionary mechanism than the advanced adaptive immune system found in higher organisms.

Box 1: Adaptive evolution in R-genes

Plant resistance genes (R-genes) are a homologous family of genes, formed by gene duplication events and hypothesized to be involved in an evolutionary arms race with pathogen effectors. R-genes are involved in recognizing specific pathogens with cognate avirulence genes and initiating defense signaling that results in disease resistance [32]. R-genes are characterized by a molecular gene-for-gene interaction [33] in which a specific allele of a disease resistance gene recognizes an avirulence protein or pathogen allele. This specificity is often encoded, at least in part, in a relatively fast-evolving leucine-rich repeat (LRR) region [34], which consists of a varying number of LRR modules. Activation of at least some

(continued)

Box 1: (continued)
of these proteins is regulated in trans, as has been shown for RPM1 and RPS2 [35].

A single *A. thaliana* plant has about 150 R-genes, representing a subset of R-genes in the overall population. The protein products of R-genes are involved in molecular interactions. They generally have a recognition site which can dock against, i.e., recognize, one or more specific molecule(s). The proteins encoded by the largest class of R-genes carry a nucleotide-binding site LRR domain (NB-LRR, also referred to as NB-ARC-LRR and NBS-LRR). NB-LRR R-genes can be further subdivided based on their N-terminal structural features into TIR-NB-LRR, which have homology to the *Drosophila* Toll and mammalian interleukin-1 receptors and CC-NB-LRR, which contain a putative coiled-coil motif [36]. The LRR domain appears to mediate specificity in pathogen recognition, while the N-terminal TIR, or coiled-coil motif, is likely to play a role in downstream signaling [34]. When a molecule is docked, the R-protein is able to activate pathways in the cell, resulting in, for example, a hypersensitive response causing apoptosis and preventing spread of infection.

Meanwhile, one single R-protein only recognizes one type of invading molecule. Therefore, through its R-genes, one individual plant only recognizes a limited number of strains of invading pathogens, as the individual pathogens have variation in effectors too. When a pathogen evolves to use nonrecognized effectors, the plant becomes susceptible. The success of plant defense is determined by both evolution and the variation of specificity in a population. Unlike the evolved mammal immune system, which can change in a living organism and learn about invasions "on the fly" [37], plant R-genes depend on the variation inside a gene pool to provide the resistance against a pathogen; *see*, for example, Holub et al. [38]. Even so, many genes involved in pathogen recognition undergo rapid adaptive evolution [39], and studies have found that *A. thaliana* R-genes show evidence of positive selection, e.g., [40–42].

In this chapter we do not limit ourselves to (known) R-genes. Plants have evolved a complex array of chemical and enzymatic defenses, both constitutive and inducible, that are not involved in pathogen detection but whose effectiveness influences pathogenesis and disease resistance. The genes underlying these defenses comprise a substantial portion of the host genome. Based on genomic sequencing, it is estimated that some 14% of the 21,000 genes in *A. thaliana* are related to defense against pathogens [43]. Most of these genes are not involved in direct pathogen detection, but their protein products interact directly with pathogen proteins or protein products at the molecular level. Among these proteins, for example, are chitinases and endoglucanases that attack and degrade the cell walls of pathogens and which pathogens counterattack with inhibitors. Such systems of antagonistically interacting proteins provide the opportunity for molecular coevolution of individual systems of attack and resistance [39].

In this chapter we design an experiment to look for all gene families showing evidence of positive selection. This evidence of positive selection is the prior for *e*QTL analysis: combining known genomic locations of gene families with *e*QTL locations derived from gene expression variation in a host-pathogen interaction experiment, which hopefully results in zooming in on gene families involved in plant resistance. The prior adds statistical power in locating putative gene families involved in host-pathogen coevolution (Fig. 1). Note that, in this chapter, the term "interaction" is used in two ways. The first is for QTL interaction, where two QTL on the genome interact **statistically**. The second is for host-pathogen gene-for-gene interaction, where gene products from different species interact **physically**.

2.1 Create a Prior with PAML

To create the prior, we use Ziheng Yang's codeml implementation of phylogenetic analysis by maximum likelihood (PAML) [44]. PAML can find amino acid sites which show evidence of positive selection using d_N/d_S ratios, which is the ratio of non-synonymous over synonymous substitution (ω, *see* [44]). The calculation of maximum likelihood for multiple evolutionary models is computationally expensive, and executing PAML over an alignment of a hundred sequences may take hours, sometimes days, on a PC. The software for generating the prior is prepackaged and makes up the workflow in Chap. 25, which includes BLAST [45], Clustal Omega [46], pal2nal [47], PAML [44], and BioRuby [48].

It is possible to find nonoverlapping large gene families by using BLASTCLUST, a tool that is part of the BLAST tool set [45]. After fetching the *A. thaliana* cDNA sequences from the Arabidopsis Information Resource (TAIR) [49], convert the sequences to a protein BLAST database format. Based on a homology criterion, the identity score and genes are clustered into putative gene families by running BLASTCLUST with 70% amino acid sequence identity. Note that the percentage identity may not render all families and will leave out a number of genes. It is used here for demonstration purposes only. For *A. thaliana* such a genome-wide search finds at least 60 gene families, including some R-gene families.

After aligning all family sequences, use PAML's codeml to find evidence of positive selection in the gene families. Clustal Omega is used to align the amino acid sequences and create a phylogenetic tree. Next, pal2nal creates codon alignments, which can be used by PAML. Finally run PAML's codeml M0-M3 (one ratio vs. nearly neutral) tests and M7-M8 (beta vs. beta + ω) tests in a computing cluster environment as shown in Chap. 25.

An M0-M3 $\chi 2$ test finds that 43 gene families (out of 60) show significant evidence of positive selection. M7-M8, meanwhile, finds 35 gene families. Therefore, based on the described procedure, approximately half the families show significant evidence of positive selection and can be considered candidate gene families involved in host-pathogen interactions. Note that this number contains false positives because the evolutionary model may be too simplistic; *see* also [50]. Nevertheless, these candidate gene families can be used as an effective filter for further research.

When a gene family displays evidence of positive selection, the genome locations can be used as a prior for systems genetics (*see* Fig. 1). With the full genome sequence of *A. thaliana* available, the location of gene families showing evidence of positive selection is known. For example, in the *Columbia* (Col-0) ecotype, the majority of the 149 R-genes are combined in clusters spreading 2–9 loci; the remaining 40 are isolated. Clusters are organized in so-called superclusters [36, 51]. Phylogenetic analysis shows that such clusters are the result of both old segmental duplications and recent chromosome rearrangements [36, 52].

2.2 Select a Suitable Experimental Population

To select a suitable experimental population, the choice of parents is key. Because we want a descriptive evolutionary prior based on gene families with known genome locations, we also need a sequenced genome, from one parent and ideally from both of the parental strains. The choice of parents for QTL analysis is normally based on large (classical) phenotypic differences. For testing pathogen resistance, the choice would ideally be one susceptible parent and one resistant (nonsusceptible) parent. For *e*QTL, phylogenetic distance can be used, when there is no obvious phenotype. In general, it is a good idea to choose one or both parents from common library strains based on, for example, *Columbia* (Col-0), *Landsberg erecta* (Ler-0), *Wassilewskija* (Ws-0), or *Kashmir* (Kas-1). This is because a great number of experimental resources and online information will be available. In addition, a reference genetic background is provided in this way, which allows the comparison of the effects of QTL and mutant alleles [53]. A number of RIL populations can be found through TAIR, a model organism database providing a centralized, curated gateway to *Arabidopsis* biology, research materials, and community [49].

2.3 Select an xQTL Technology

A large part of published *x*QTL studies is based on gene expression *e*QTL partly because gene expression probe provides a direct genomic link. When it comes to selecting single-color or two-color arrays, one consideration may be that two-color arrays have higher efficiency when using a distant pair design [54].

Deep sequencing technology (RNA-seq, [55]) is affordable for *e*QTL studies. The main advantage over microarrays is improved signal-to-noise ratios and possibly improved coverage depending on the reference genome. Microarrays are noisy partly due to cross hybridization, e.g., [56], and have limited signal on low-abundance transcripts or expressors; both facts are detrimental to significance. Deep sequencing is no *panacea*, however, since it accentuates the high expressors. High expressors are expressed thousands of times higher than low expressors. Low expressors may lack significance for differential expression. Worse because deep sequencing is stochastic, many low expressors may even be absent. Another point to consider is that currently at least 1 in 1000 nucleotide base pairs is misread, which makes it harder to disentangle error from genetic variation. Only when a sequence polymorphism is measured many times (say 20×), it can be considered to represent genetic variation.

Also a choice for a certain *e*QTL technology should take into account that, when looking at differential gene expression analysis, different microarray platforms agree with each other, but overlap between microarray and deep sequencing is much lower, suggesting a technical bias [57].

For an example of a metabolite *m*QTL study, *see* Keurentjes et al. [58] and Fu et al. [59]. For a study integrating *e*QTL, *p*QTL, *m*QTL, and classical phenotypic QTL, *see* Fu et al. [60] and Jansen et al. [13].

2.4 Sizing the Experimental Population

The size of the experimental population should be large enough to give informative results. For classical QTL analysis, the sizing may be assisted using estimates of total environmental variance and the total genetic variance derived from the accessions, selected as parents. Roughly, population sizes of 200 RILs, without replications, will allow detection of large-effect QTL with an explained variance of 10% in confidence intervals of 10–20 cM. Detection of small-effect QTL or mapping accuracy below 5% requires increasing the population size to at least 300 RILs [53]. It is important to note that QTL mapping accuracy is a function of marker density and population size. The number of strains to use differs between inbred lines. The promise of extreme dense marker maps, such as delivered by SNPs, does not automatically translate to higher accuracy. It is the number of recombination events in the population for a particular genomic region that limits QTL interval size. In fact, current marker maps, in the order of thousands of (evenly spread) markers per genome, suite population sizes of a few hundred RILs.

It is a fallacy, for example, to expect higher mapping power when combining an ultradense SNP map with just 20 individuals.

For high-throughput *x*QTL, the experimental population should be sized against an acceptable false discovery rate (FDR), minimizing for type I and type II errors. This can be achieved using a permutation strategy to assess statistical significance, maintaining the correlation of the expression traits while destroying any genetic linkages or associations in natural populations: marker data is permuted while keeping the correlation structure in the trait data, such as presented by Breitling et al. [61]. Unfortunately, this information differs for every experiment and is only available afterward. Analyzing a similar experiment, using the same tissue and data acquisition technology, may give an indication [60], but when no such material is available, a crude estimate may be had by taking the thresholds of a (classic) single-trait QTL experiment and adjusting that for multiple testing by the Bonferroni correction (minimize type I errors) or Benjamini-Hochberg correction (minimize type II errors). Note that Bonferroni results in a very conservative estimate.

2.5 Analyzing the xQTL Experiment with R/qtl

R/qtl is extensible, interactive free software for the mapping of *x*QTL in experimental crosses. It is implemented as an add-on package for the widely used statistical language/software R. Since its introduction, R/qtl has become a reference implementation with an extensive guide on QTL mapping [62].

R/qtl includes multiple QTL mapping (MQM), as described in [10], an automated procedure, which combines the strengths of generalized linear model regression with those of interval mapping. MQM can handle missing data by analyzing probable genotypes. MQM selects important marker cofactors by multiple regression and backward elimination. QTL are moved along the chromosomes using these preselected markers as cofactors. QTL are interval mapped using the most informative model through maximum likelihood. MQM for R/qtl brings the following advantages to QTL mapping: (1) higher power, as long as the QTL explain a reasonable amount of variation; (2) protection against overfitting, because MQM fixes the residual variance from the full model; (3) prevention of ghost QTL detection (between two QTL in coupling phase); and (4) detection of negating QTL (QTL in repulsion phase) [10].

MQM for R/qtl brings additional advantages to systems genetics data sets with hundreds to millions of traits: (5) a pragmatic permutation strategy for control of the FDR and prevention of locating false QTL hot spots, as discussed above; (6) high-performance computing by scaling on multi-CPU computers, as well as clustered computers, by calculating phenotypes in parallel, through the message passing interface (MPI) of the parallel package for R; and (7) visualizations for exploring interactions in a genomic circle plot and cis- and trans-regulation. MQM comes with a 40-page tutorial for MQM and is part of the software distribution of R/qtl [10, 63].

2.6 Matching the Prior

After detecting *e*QTL, we have a map of gene regulation in the form of a cis-trans map. When taking *a priori* information into account, i.e., genomic locations derived through other methods, we can potentially match the genomic locations of genes and gene families with the *e*QTL cis-trans map. Until now, there has been no combined QTL and evolutionary study, involving PAML, for host-pathogen relationships in plants, though they have been conducted separately.

2.7 Combining xQTL Results: Causality and Network Inference

In addition to identifying *e*QTL or *x*QTL, it is possible to think in terms of grouping related traits by correlations. Molecular and phenotypic traits can be informative for inferring underlying molecular networks. When two independent non-correlated traits share multiple QTL, inference of a functional relationship is possible (Fig. 1b). Thus, distinguishing trait causality, reactivity, or independence can be based upon logic involving underlying QTL. This was the basic idea in Jansen and Nap 2001 [64]. Later, people started to use biological variation as an extra source for reasoning because if A affects B, biological variation in trait A is propagated to B and not vice versa. This assumes there is no hidden trait C affecting both A and B; *see* also Li et al. [15].

Mapping QTL for thousands of molecular phenotypes is the first step in attempting to reconstruct gene networks. Not only can network reconstruction be used within a particular layer, say within *e*QTL analysis, i.e., transcript data only, but also across layers. Such interlevel (system) analysis integrates transcript *e*QTL, protein *p*QTL, metabolite *m*QTL, and classical QTL [13].

The examination of pairwise correlation between traits can lead to the hypothesis of a functional relationship when that correlation is high. Beyond the detected QTL, the correlation between residuals among traits, after accounting for QTL effects, or correlations between traits conditional on other traits is further evidence for a network connection. To infer directional effects, it is necessary to analyze the correlations among pairs of traits in detail. If trait A maps to a subset of the QTL of trait B, then the common QTL can be taken as evidence for their network connection, while the distinct QTL can be used to infer the direction (Fig. 1b), unless all the common QTL have widespread pleiotropic effects, which is when a single gene influences multiple traits. If traits A and B have common QTL, without QTL that are distinct, then the inference is more complicated, and further analysis is needed to discriminate pleiotropy from any of the possible orderings among traits [13, 15].

Li et al. [15] point out that, despite the exciting possibilities of correlation analysis, extreme caution is advised, especially in intralevel analyses, owing to the potential impact of correlated measurement error (leading to false-positive connections). By introducing a prior, however, causal inference becomes feasible for realistic population sizes [15]. The outcome of a causal inference on two traits sharing a common QTL may be either that one is causal for the other or that they are independent. In the first case, QTL-induced variation is propagated from one trait to the other, while in the latter case, the two traits may even be regulated by different genes or polymorphisms within the QTL region, and their apparent relationship (correlation) is explained by linkage disequilibrium and not by a shared biological pathway [15].

3 Discussion

A QTL is a statistical property connecting genotype with phenotype. In this chapter, we reviewed studies which, with various degrees of success, combine some type of prior information with *x*QTL. We propose that a search for genome-wide evidence of positive selection can produce a valid and interesting prior for *x*QTL analysis. This is achieved by combining information of genomic locations of putative gene families, possibly involved in plant-pathogen interactions, with QTL locations derived from a systems genetics experiment. Both the *e*QTL example and the search for genome-wide evidence of positive selection pressure are essentially exploratory and result in a list of putative genes, or gene families, with known genomic locations. The combined information yields candidate genes and pathways that are under positive selection pressure and, potentially, involved in host-pathogen interactions. We explain that it is possible to design an *e*QTL experiment using existing experimental populations, e.g., using an *A. thaliana* RIL population, and analyze results with existing free and open-source software, such as the R/qtl tool set.

Systems genetics bridges the study of quantitative traits with molecular biology and gives new momentum to QTL population studies. Genetic variation at multiple loci in combination with environmental factors can induce molecular or phenotypic variation. Variation may manifest itself as linear patterns among traits at different levels that can be deconstructed. Correlations can be attributed to detectable QTL and a logical framework based on common and distinct QTL and propagation of biological variation, which can be used to infer network causality, reactivity, or independence [15]. Unexplained biological variation can be used to infer direction between traits that share a common QTL and have no distinct QTL, though it may be difficult to separate biological from technical variation. Prior knowledge and complementary experiments, such as deletion mapping followed by independent gene expression studies between parental lines, may validate or disprove implicated network connections [65].

Evolutionary systems genetics can help dissect the underlying genetics of pathogen susceptibility in plants. Where "evolutionary genetics" describes how evolutionary forces shape biodiversity, as observed in nature, "evolutionary systems genetics" describes how phenotype variation in a population is formed by genotype variation between, for example, host and pathogen involved in an evolutionary arms race.

For purpose of online analysis we created GeneNetwork.org (GN) [7], a free and open-source (FOSS) framework for web-based genetics that can be deployed anywhere. GN allows biologists to upload high-throughput experimental data, such as expression data from microarrays and RNA-seq, and also classical phenotypes, such as disease phenotypes. These phenotypes can be mapped interactively against genotypes using embedded tools, such as R/QTL [10] for model organisms and FaST-LMM [66] and GEMMA [67] which are suitable for human populations and outbred crosses, such as the mouse diversity outcross. Interactive D3 graphics are included from R/qtl charts, and presentation-ready figures can be generated. Recently we have added functionality for phenotype correlation [68], correlation trait loci [16], and network analysis [14]. For examples on using GeneNetwork, *see* also Mulligan et al. [8].

If you want to know more about *e*QTL, we suggest the review by Gilad et al. [23], which also discusses *e*QTL in genome-wide association studies (GWAS), useful in situations where experimental crosses are not available (such as with many pathogens and humans). For further reading on R-gene evolution, we recommend Bakker et al. [34]. For R/qtl analysis, we recommend the R/qtl guide [62] and our MQM tutorial online [63]. For integrating different *x*QTL methods and causal inference, we recommend Li et al. [15] and Jansen et al. [13].

4 Questions

1. What is an *e*QTL, and why does it present two genomic locations?
2. Can a prior, as used here, really add statistical power, or is it no more than circumstantial evidence?
3. When designing an evolutionary systems genetics experiment, what are the steps to consider?
4. How can causality be inferred from QTL networks?

References

1. Qin L, Prins P, Jones JT et al (2001) Genest, a powerful bidirectional link between cdna sequence data and gene expression profiles generated by cdna-aflp. Nucleic Acids Res 29 (7):1616–1622
2. Mckeown PC, Laouielle-duprat S, Prins P et al (2011) Identification of imprinted genes subject to parent-of-origin specific expression in arabidopsis thaliana seeds. BMC Plant Biol 11:113
3. Nandi S, Subudhi PK, Senadhira D et al (1997) Mapping QTLs for submergence tolerance in rice by AFLP analysis and selective genotyping. Mol Gen Genet 255(1):1–8
4. Meaburn E, Butcher LM, Schalkwyk LC, Plomin R (2006) Genotyping pooled DNA using 100K SNP microarrays: a step towards genomewide association scans. Nucleic Acids Res 34 (4):e27
5. Kim S, Plagnol V, Hu TT et al (2007) Recombination and linkage disequilibrium in arabidopsis thaliana. Nat Genet 39(9):1151–1155
6. Dixon AL, Liang L, Moffatt MF et al (2007) A genome-wide association study of global gene expression. Nat Genet 39(10):1202–1207
7. Sloan Z, Arends D, Broman KW et al (2016) Genenetwork: framework for web-based genetics. JOSS 1(2):25
8. Mulligan MK, Mozhui K, Prins P, Williams RW (2017) Genenetwork: a toolbox for systems genetics. Methods Mol Biol 1488:75–120
9. Gibson G, Weir B (2005) The quantitative genetics of transcription. Trends Genet 21 (11):616–623
10. Arends D, Prins P, Jansen RC, Broman KW (2010) R/qtl: high-throughput multiple QTL mapping. Bioinformatics 26 (23):2990–2992
11. Li Y, Alvarez OA, Gutteling EW et al (2006) Mapping determinants of gene expression plasticity by genetical genomics in C. elegans. PLoS Genet 2(12):e222
12. Damerval C, Maurice A, Josse JM, De Vienne D (1994) Quantitative trait loci underlying gene product variation: a novel perspective for analyzing regulation of genome expression. Genetics 137(1):289–301
13. Jansen RC, Tesson BM, Fu J, Yang Y, Mcintyre LM (2009) Defining gene and QTL networks. Curr Opin Plant Biol 12(2):241–246
14. Langfelder P, Horvath S (2008) Wgcna: an r package for weighted correlation network analysis. BMC Bioinformatics 9:559
15. Li Y, Tesson BM, Churchill GA, Jansen RC (2010) Critical reasoning on causal inference in genome-wide linkage and association studies. Trends Genet 26(12):493–498
16. Arends D, Li Y, Brockmann G et al (2016) Correlation trait loci (ctl) mapping: phenotype network inference subject to genotype. JOSS. 1 (6):87
17. Brem RB, Kruglyak L (2005) The landscape of genetic complexity across 5,700 gene expression traits in yeast. Proc Natl Acad Sci U S A 102(5):1572–1577
18. Fraser HB, Moses AM, Schadt EE (2010) Evidence for widespread adaptive evolution of gene expression in budding yeast. Proc Natl Acad Sci U S A 107(7):2977–2982
19. Zou Y, Su Z, Yang J, Zeng Y, Gu X (2009) Uncovering genetic regulatory network divergence between duplicate genes using yeast eqtl landscape. J Exp Zool B Mol Dev Evol 312 (7):722–733
20. Li Y, Breitling R, Jansen RC (2008) Generalizing genetical genomics: getting added value from environmental perturbation. Trends Genet 24(10):518–524
21. Hager R, Lu L, Rosen GD, Williams RW (2012) Genetic architecture supports mosaic brain evolution and independent brain-body size regulation. Nat Commun 3:1079
22. Kliebenstein DJ, West MA, Van leeuwen H et al (2006) Identification of QTLs controlling gene expression networks defined a priori. BMC Bioinformatics 7:308
23. Gilad Y, Rifkin SA, Pritchard JK (2008) Revealing the architecture of gene regulation: the promise of eQTL studies. Trends Genet 24 (8):408–415
24. Alberts R, Terpstra P, Li Y et al (2007) Sequence polymorphisms cause many false cis eqtls. PLoS One 2(7):e622
25. Franke L, Bakel van H, Fokkens L et al (2006) Reconstruction of a functional human gene network, with an application for prioritizing positional candidate genes. Am J Hum Genet 78(6):1011–1025
26. Chen X, Hackett CA, Niks RE et al (2010) An eqtl analysis of partial resistance to puccinia hordei in barley. PLoS One 5(1):e8598
27. Qin L, Kudla U, Roze EH et al (2004) Plant degradation: a nematode expansin acting on plants. Nature 427(6969):30
28. Saijo Y, Schulze-lefert P (2008) Manipulation of the eukaryotic transcriptional machinery by

bacterial pathogens. Cell Host Microbe 4 (2):96–99
29. Chen LQ, Hou BH, Lalonde S et al (2010) Sugar transporters for intercellular exchange and nutrition of pathogens. Nature 468 (7323):527–532
30. Hewitson JP, Grainger JR, Maizels RM (2009) Helminth immunoregulation: the role of parasite secreted proteins in modulating host immunity. Mol Biochem Parasitol 167(1):1–11
31. Bird PI, Trapani JA, Villadangos JA (2009) Endolysosomal proteases and their inhibitors in immunity. Nat Rev Immunol 9 (12):871–882
32. Dangl JL, Jones JD (2001) Plant pathogens and integrated defence responses to infection. Nature 411(6839):826–833
33. Flor H (1956) The complementary genic systems in flax and flax rust*. Adv Genet 8:29–54
34. Bakker EG, Toomajian C, Kreitman M, Bergelson J (2006) A genome-wide survey of R gene polymorphisms in Arabidopsis. Plant Cell 18 (8):1803–1818
35. Mackey D, Belkhadir Y, Alonso JM, Ecker JR, Dangl JL (2003) Arabidopsis rin4 is a target of the type iii virulence effector avrrpt2 and modulates rps2-mediated resistance. Cell 112 (3):379–389
36. Richly E, Kurth J, Leister D (2002) Mode of amplification and reorganization of resistance genes during recent arabidopsis thaliana evolution. Mol Biol Evol 19(1):76–84
37. Medzhitov R, Janeway CA Jr (1997) Innate immunity: impact on the adaptive immune response. Curr Opin Immunol 9(1):4–9
38. Holub EB (2001) The arms race is ancient history in Arabidopsis, the wildflower. Nat Rev Genet 2(7):516–527
39. Bishop JG, Dean AM, Mitchell-olds T (2000) Rapid evolution in plant chitinases: molecular targets of selection in plant-pathogen coevolution. Proc Natl Acad Sci U S A 97 (10):5322–5327
40. Xiao S, Emerson B, Ratanasut K et al (2004) Origin and maintenance of a broad-spectrum disease resistance locus in Arabidopsis. Mol Biol Evol 21(9):1661–1672
41. Mondragon-palomino M, Meyers BC, Michelmore RW, Gaut BS (2002) Patterns of positive selection in the complete nbs-lrr gene family of Arabidopsis thaliana. Genome Res 12 (9):1305–1315
42. Sun X, Cao Y, Wang S (2006) Point mutations with positive selection were a major force during the evolution of a receptor-kinase resistance gene family of rice. Plant Physiol 140 (3):998–1008
43. Bevan M, Bancroft I, Bent E, Chalwatzis N (1998) Analysis of 1.9 Mb of contiguous sequence from chromosome 4 of Arabidopsis thaliana. Nature 391(6666):485–488
44. Yang Z (1997) PAML: a program package for phylogenetic analysis by maximum likelihood. Comput Appl Biosci 13(5):555–556
45. Altschul SF, Madden TL, Schaffer AA et al (1997) Gapped blast and psi-blast: a new generation of protein database search programs. Nucleic Acids Res 25(17):3389–3402
46. Sievers F, Wilm A, Dineen D et al (2011) Fast, scalable generation of high-quality protein multiple sequence alignments using clustal omega. Mol Syst Biol 7:539
47. Suyama M, Torrents D, Bork P (2006) Pal2nal: robust conversion of protein sequence alignments into the corresponding codon alignments. Nucleic Acids Res 34(Web Server): W609–W612
48. Goto N, Prins P, Nakao M et al (2010) BioRuby: bioinformatics software for the Ruby programming language. Bioinformatics 26 (20):2617–2619
49. Rhee SY, Beavis W, Berardini TZ et al (2003) The Arabidopsis Information Resource (TAIR): a model organism database providing a centralized, curated gateway to Arabidopsis biology, research materials and community. Nucleic Acids Res 31(1):224–228
50. Anisimova M, Nielsen R, Yang Z (2003) Effect of recombination on the accuracy of the likelihood method for detecting positive selection at amino acid sites. Genetics 164(3):1229–1236
51. Arabidopsis Genome Initiative (2000) Analysis of the genome sequence of the flowering plant arabidopsis thaliana. Nature 408 (6814):796–815
52. Michelmore RW, Meyers BC (1998) Clusters of resistance genes in plants evolve by divergent selection and a birth-and-death process. Genome Res 8(11):1113–1130
53. Salinas J, Sanchez-serrano J (2006) Arabidopsis protocols. Humana Press Inc, Totowa, NJ
54. Fu J, Jansen RC (2006) Optimal design and analysis of genetic studies on gene expression. Genetics 172(3):1993–1999
55. Mortazavi A, Williams BA, Mccue K, Schaeffer L, Wold B (2008) Mapping and quantifying mammalian transcriptomes by rna-seq. Nat Methods 5(7):621–628
56. Eklund AC, Turner LR, Chen P et al (2006) Replacing cRNA targets with cDNA reduces microarray cross-hybridization. Nat Biotechnol 24(9):1071–1073
57. Hoen PA, Ariyurek Y, Thygesen HH et al (2008) Deep sequencing-based expression

analysis shows major advances in robustness, resolution and inter-lab portability over five microarray platforms. Nucleic Acids Res 36 (21):e141

58. Keurentjes JJ, Sulpice R, Gibon Y et al (2008) Integrative analyses of genetic variation in enzyme activities of primary carbohydrate metabolism reveal distinct modes of regulation in Arabidopsis thaliana. Genome Biol 9(8): R129
59. Fu J, Swertz MA, Keurentjes JJ, Jansen RC (2007) Metanetwork: a computational protocol for the genetic study of metabolic networks. Nat Protoc 2(3):685–694
60. Fu J, Keurentjes JJ, Bouwmeester H et al (2009) System-wide molecular evidence for phenotypic buffering in Arabidopsis. Nat Genet 41(2):166–167
61. Breitling R, Li Y, Tesson BM et al (2008) Genetical genomics: spotlight on QTL hotspots. PLoS Genet 4(10):e1000232
62. Broman K, Sen S (2009) A guide to QTL mapping with R/qtl. Springer, New York, NY
63. Arends D, Prins P, Broman KW, Jansen RC (2010) Tutorial - multiple-QTL mapping (MQM) analysis. http://www.rqtl.org/tutorials/MQM-tour.pdf
64. Jansen RC, Nap JP (2001) Genetical genomics: the added value from segregation. Trends Genet 17(7):388–391
65. Wayne ML, Mcintyre LM (2002) Combining mapping and arraying: an approach to candidate gene identification. Proc Natl Acad Sci U S A 99(23):14903–14906
66. Lippert C, Listgarten J, Liu Y et al (2011) Fast linear mixed models for genome-wide association studies. Nat Methods 8(10):833–835
67. Zhou X, Stephens M (2012) Genome-wide efficient mixed-model analysis for association studies. Nat Genet 44(7):821–824
68. Wang X, Pandey AK, Mulligan MK et al (2016) Joint mouse-human phenome-wide association to test gene function and disease risk. Nat Commun 7:10464

Part VI

Handling Genomic Data: Resources and Computation

Chapter 22

Semantic Integration and Enrichment of Heterogeneous Biological Databases

Ana Claudia Sima, Kurt Stockinger, Tarcisio Mendes de Farias, and Manuel Gil

Abstract

Biological databases are growing at an exponential rate, currently being among the major producers of Big Data, almost on par with commercial generators, such as YouTube or Twitter. While traditionally biological databases evolved as independent silos, each purposely built by a different research group in order to answer specific research questions; more recently significant efforts have been made toward integrating these heterogeneous sources into unified data access systems or interoperable systems using the FAIR principles of data sharing. Semantic Web technologies have been key enablers in this process, opening the path for new insights into the unified data, which were not visible at the level of each independent database. In this chapter, we first provide an introduction into two of the most used database models for biological data: relational databases and RDF stores. Next, we discuss ontology-based data integration, which serves to unify and enrich heterogeneous data sources. We present an extensive timeline of milestones in data integration based on Semantic Web technologies in the field of life sciences. Finally, we discuss some of the remaining challenges in making ontology-based data access (OBDA) systems easily accessible to a larger audience. In particular, we introduce natural language search interfaces, which alleviate the need for database users to be familiar with technical query languages. We illustrate the main theoretical concepts of data integration through concrete examples, using two well-known biological databases: a gene expression database, Bgee, and an orthology database, OMA.

Key words Data integration, Ontology-based data access, Knowledge representation, Query processing, Keyword search, Relational databases, RDF stores

Abbreviations

ABox	Assertional box
Bgee	data*B*ase for *G*ene *E*xpression *E*volution, https://bgee.org/
FK	Foreign key in a relational database
HBB	Hemoglobin unit beta gene
IRI	Internationalized Resource Identifier
OBDA	Ontology-based data access

Maria Anisimova (ed.), *Evolutionary Genomics: Statistical and Computational Methods*, Methods in Molecular Biology, vol. 1910, https://doi.org/10.1007/978-1-4939-9074-0_22,

OMA	*O*rthologous *Ma*trix, a database for the inference of orthologs among complete genomes.—https://omabrowser.org, SPARQL endpoint: https://sparql.omabrowser.org/sparql
PK	Primary key in a relational database
PK-FK	Primary key-foreign key relationship; enables joining two tables in a relational database
RDB	Relational database
RDF	Resource Description Framework
SODA	Search Over Relational Databases [21]
SQL	Structured Query Language
SPARQL	SPARQL Protocol and RDF Query Language
TBox	Terminological box
URI	Uniform Resource Identifier

1 Introduction

Biological databases have grown exponentially in recent decades, both in number and in size, owing primarily to modern high-throughput sequencing techniques [1]. Today, the field of genomics is almost on par with the major commercial generators of Big Data, such as YouTube or Twitter, with the total amount of genome data doubling approximately every 7 months [2]. While most biological databases have initially evolved as independent silos, each purposely built by a different research group in order to collect data and respond to a specific research question, more recently significant efforts have been made toward *integrating* the different data sources, with the aim of enabling more powerful insights from the aggregated data, which would not be visible at the level of individual databases.

Let us consider the following example. An evolutionary biologist might want to answer the question "What are the human-rat orthologs, expressed in the liver, that are associated with leukemia?". Getting an answer for this type of question usually requires information from at least three different sources: an orthology database (e.g., OMA [3], OrthoDB [4], or EggNog [5]); a gene expression database, such as Bgee [6]; and a proteomics database containing disease associations (e.g., UniProt [7]). In the lack of a unified access to the three data sources, obtaining this information is a largely manual and time-consuming process. First, the biologist needs to know *which* databases to search through. Second, depending on the interface provided by these databases, he or she might need to be familiar with a technical query language, such as SQL or SPARQL (note: a list of acronyms is provided at the beginning of this chapter). At the very least, the biologist is required to know the specific identifiers (IDs) and names used by the research group that created the database, in order to search for relevant entries. An integrated view, however, would allow the user to obtain this information automatically, without knowing any of the details regarding the structure of the underlying data sources—nor the type of storage these databases use—and eventually not even specific IDs (such as protein or gene names).

Biological databases are generally characterized by a large heterogeneity, not only in the type of information they store but also in the model of the underlying data store they use—examples include relational databases, file-based stores, graph based, etc. Examples of databases considered fundamental to research in the life sciences can be found in the ELIXIR Europe's Core Data Resources, available online at https://www.elixir-europe.org/platforms/data. In this chapter we will mainly discuss two types of database models: the relational model (i.e., relational databases) and a graph-based data model, RDF (the Resource Description Framework).

Database systems have been around since arguably the same time as computers themselves, serving initially as "digitized" copies of tabular paper forms, for example, in the financial sector, or for managing airline reservations. Relational databases, as well as the mathematical formalism underlying them, namely, the relational algebra, were formalized in the 1970s by E.F. Codd, in a foundational paper that now has surpassed 10,000 citations [8]. The relational model is designed to structure data into so-called tuples, according to a predefined schema. Tuples are stored as rows in tables (also called "relations"). Each table usually defines an entity, such as an object, a class, or a concept, whose instances (the tuples) share the same attributes. Examples of relations are "Gene", "Protein", "Species", etc. The attributes of the relation will represent the columns of the table, for example, "gene name." Furthermore, each row has a unique identifier. The column (or combination of columns) that stores the unique identifier is called a primary key and can be used not only to uniquely identify rows *within* a table but also to connect data *between* multiple tables, through a Primary Key-Foreign key relationship. Doing such a connection is called a join. In fact, a join is only one of the operations defined by relational algebra. Other common operations include projection, selection, and others. The operands of relational algebra are the database tables, as well as their attributes, while the operations are expressed through the Structured Query Language (SQL). For a more in-depth discussion on relational algebra, we refer the reader to the original paper by E.F. Codd [8].

This chapter is structured as follows. In Sect. 2, we give a brief introduction to relational databases, through the concrete example of the Bgee gene expression database. We introduce the basics of Semantic Web technologies in Sect. 3. Readers who are already familiar with the Semantic Web stack might skip Sect. 3 and jump directly to Sect. 4, which presents an applied use case of Semantic Web technologies in the life sciences: modeling the Bgee and OMA databases. Section 5 represents the core of this chapter. Here, we present ontology-based data integration (Sect. 5.1) and illustrate it through the concrete example of a unified ontology for Bgee and OMA (Sect. 5.2), as well as the mechanisms required to further extend the integrated system with other heterogeneous sources such as the UniProt protein knowledge base (Sect. 5.3). We introduce natural language interfaces, which enable easy data access even for nontechnical users, in Sect. 5.4. We present an extensive timeline of milestones in data integration based on Semantic Web technologies in the field of life sciences in Sect. 6. Finally, we conclude in Sect. 7.

2 Modeling a Biological Database with Relational Database Technology

In this section we will demonstrate how to model a biological database with relational database technology.

Figure 1 illustrates the data model of a sample extracted from the Bgee database. The sample contains five tables and their relationships, shown as arrows, where the direction of the arrow is oriented from the foreign key of one table to the primary key of a related one. For example, the Primary Key (PK) of the *Species* table is the *SpeciesID*. Following the relationships highlighted in bold, we see that the *SpeciesID* also appears in the two tables connected to *Species*: *GlobalCond* and *Gene*. In these tables, the attribute plays the

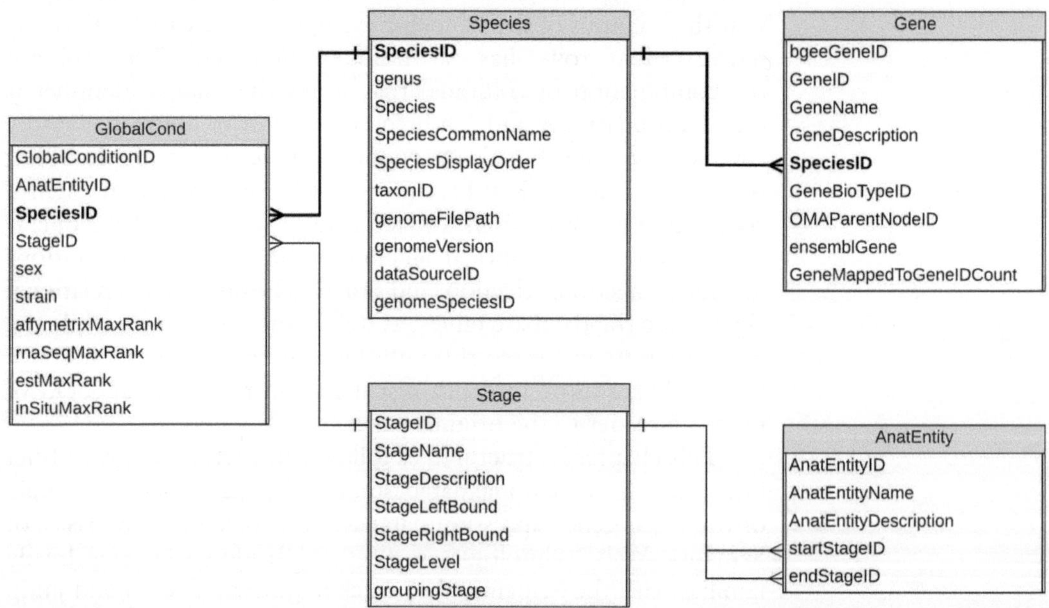

Fig. 1 Sample relational database (extracted from the gene expression database Bgee)

role of a Foreign Key (FK). The PK-FK relationships allow combining or aggregating data from related tables. For example, by joining *Species* and *Gene*, through the *SpeciesID*, we can find to which species a gene belongs. Concretely, let's assume we want to find the species where the gene "HBB" can be found. Given that this information is stored in the *SpeciesCommonName* attribute, we can retrieve it through the following SQL query:

```
SELECT SpeciesCommonName from Species JOIN Gene
WHERE Gene.GeneName = 'HBB' and Species.SpeciesID = Gene.
SpeciesID
```

This query enables retrieving (via the "SELECT" keyword) the attribute corresponding to the species name (*SpeciesCommonName*) by joining the Species and Gene tables, based on their primary key-foreign key relationship, namely, via the *SpeciesID*, on the condition that the *GeneName* exactly matches "HBB." For a more detailed introduction to the syntax and usage of SQL, we refer the reader to an online introductory tutorial [9], as well as the more comprehensive textbooks [10, 11].

Taking this a step further, we can imagine the case where a second relational database also stores information about genes, but perhaps with some additional data, such as associations with diseases. Can we still combine information across these distinct databases? Indeed, as long as there is a common point between the tables in the two databases, such as the *GeneID* or the *SpeciesID*, it is usually possible to combine them into a single, federated database and use SQL to query it through federated joins. An example of using federated databases for biomedical data is presented in [12].

2.1 Limitations of Relational Databases and Emerging Solutions for Data Integration

So far, we have seen that relational databases are a mature, highly optimized technology for storing and querying structured data. Also, combined with a powerful and expressive query language, SQL, they allow users to federate (join) data even from different databases.

However, there are certain relationships that are not natural for relational databases. Let us consider the relationship "hasOrtholog". Both the domain and the range of this relationship, as defined in the Orthology Ontology [13], are the same—a gene. For example, the hemoglobin (HBB) gene in human has the Hbb-bt orthologous gene in the mouse (expressed via the relation *hasOrtholog*). In the relational database world, this translates into a so-called self-join. As the name suggests, this requires joining one table—in this case, *Gene*—with itself, in order to retrieve the answer. These types of "self-join" relations, while frequent in the real world (e.g., a manager of an employee is also an employee, a friend of a person is also a person, etc.), are inefficient in the context of relational databases. While there are sometimes ways to avoid self-joins, these require even more advanced SQL fluency on the part of the programmer [14].

Moreover, relational databases are typically not well-suited for applications that require frequent schema changes. Hence, NoSQL stores have gained widespread popularity as an alternative to traditional relational database management systems [15–17]. These systems do not impose a strict schema on the data and are therefore more flexible than relational databases in the cases where the structure of the data is likely to change over time. In particular, graph databases, such as Virtuoso [18], are very well suited for data integration, as they allow easily combining multiple data sources into a single graph. We discuss this in more detail in Sect. 3.

These and other considerations have led to the vision of the Semantic Web, formalized in 2001 by Tim Berners Lee et al. [19]. At a high-level, the Semantic Web allows representing the semantics of data in a structured, easy to interlink, machine-readable way, typically by use of the Resource Description Framework (RDF)—a graph-based data model. The gradual adoption of RDF stores, although widespread in the Web context and in the life sciences in particular, did not replace relational databases altogether, which lead to a new challenge: how will these heterogeneous data sources now be integrated?

Initial integration approaches in the field of biological databases have been largely manual: first, many of them (either relational or graph-based) have included cross-references to other sources. For example, UniProt contains links to more than 160 other databases. However, this raises a question for the user: which of the provided links should be followed in order to find relevant connections? While a user can be assumed to know the contents of a few related databases, we can hardly expect anyone to be familiar with more than 160 of them! To avoid this problem, other databases have chosen an orthogonal approach: instead of referencing links to other sources, simply copy the relevant data from those sources into the database. This approach also has a few drawbacks. First, it generates redundant data (which might result in significant storage space consumption), and, most importantly, it might lead to the use of stale, outdated results. Moreover, this approach is contradictory to best practices of data warehousing used widely across various domains in industry. For a discussion on this, we refer the reader to [20].

Databases such as UniProt are highly comprehensive, with new results being added to each release, results that may sometimes even contradict previous results. Duplication of this data into another database can quickly lead to missing out the most recent information or to high maintenance efforts required to keep up with the new changes. In the following sections, we discuss an alternative approach: integrating heterogeneous data sources through the use of a unifying data integration layer, namely, an integrative ontology, that aligns, but also enriches the existing data, with the purpose of facilitating knowledge discovery.

Throughout the remainder of this chapter, we will combine theoretical aspects of data integration with concrete examples, based on our SODA project [21], as well as from our ongoing research project, Bio-SODA [22], where we are currently building an integrated data access system for biological databases (starting with OMA and Bgee), using a natural language search interface. In the context of this project, Semantic Web technologies, such as RDF, are used to enhance interoperability among heterogeneous databases at the semantic level (e.g., RDF graphs with predefined semantics). Moreover, currently, several life science and biomedical databases such as OMA [3], UniProt [7], neXtProt [22], the European Bioinformatics Institute (EMBL-EBI) RDF data [24], and the WorldWide Protein Data Bank [25] already provide RDF data access, which also justifies an RDF-based approach to enable further integration efforts to include these databases. A recent initiative for (biological) data sharing is based on the FAIR principles [26], aiming to make data *f*indable, *a*ccessible, *i*nteroperable, and *re*-usable.

3 Semantic Web Technologies

The Semantic Web, as its name shows, emerged mainly as a means to attach semantics (meaning) to data on the Web [19]. In contrast to relational databases, Semantic Web technologies rely on a graph data model, in order to enable interlinking data from disparate sources available on the Web. Although the vision of the Semantic Web still remains an ideal, many large datasets are currently published based on the Linked Data principles [27] using Semantic Web technologies (e.g., RDF). The Linked Open Data Cloud illustrates a collection of a large number of different resources including DBPedia, UniProt, and many others.

In this section, we will describe the Semantic Web (SW) stack, focusing on the technologies that enhance data integration and enrichment. For a more complete description of the SW stack, we refer the reader to the comprehensive introductions in [28–30].

The Semantic Web stack is presented in Fig. 2. We will focus on the following standards or layers of the stack: URI, the syntax layer (e.g., Turtle (TTL), an RDF serialization format), RDF, OWL, RDFS, and SPARQL. These layers are highlighted in gray in Fig. 2.

3.1 Unique Resource Identifier (URI)

A Uniform Resource Identifier (URI) is a character sequence that identifies an abstract or physical resource. A URI is classified as a locator, a name, or both. The Uniform Resource Locators (URLs) are a subset of URIs that, in addition to identifying a resource, provide a means of locating the resource by describing its primary

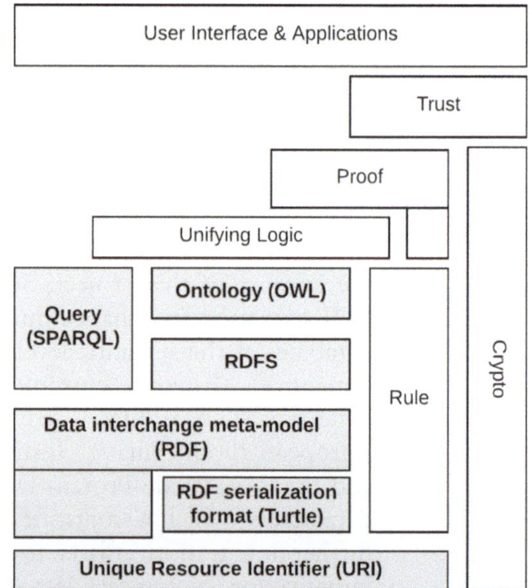

Fig. 2 The Semantic Web stack modified from [31]

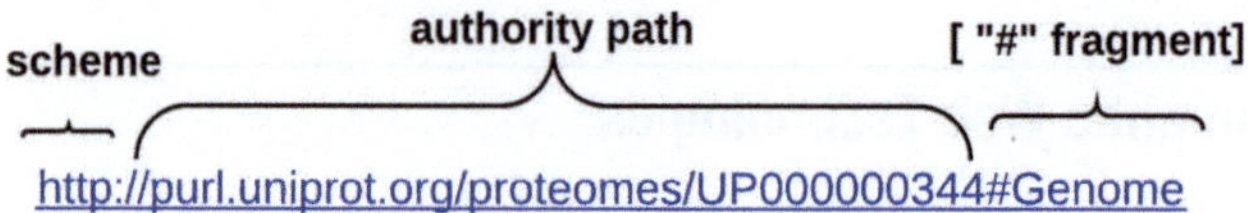

Fig. 3 An example of a UniProt URI with a fragment

access or network "location." For example, https://bgee.org is a URI that identifies a resource (i.e., the Bgee gene expression website), and it implies solely a representation of this resource (i.e., an HTML Web page). This resource is accessible through the HTTPS protocol.

The Uniform Resource Name (URN) is also a URI that refers to both the "urn" scheme [32], which are URIs required to remain globally unique and persistent *even* when the resource does not exist anymore or becomes unavailable, and to any other URI with the properties of a name. For example, the URN urn:isbn:978-1-61779-581-7 is a URI that refers to a previous edition of this book by using the International Standard Book Number (ISBN). However, no information about the location and how to get this resource (book) is provided.

The URI syntax consists of a hierarchical sequence of components referred to as the scheme, authority, path, query, and fragment [33]. Figure 3 describes a UniProt URI that includes these components.

An individual scheme does not have to be classified as being just one of "name" or "locator." Instances of URIs from any given scheme may have the characteristics of names (URN) or locators

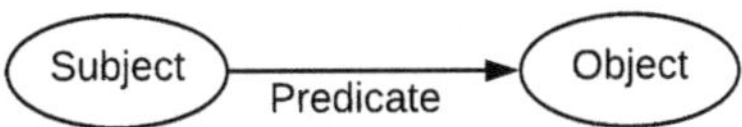

Fig. 4 An RDF graph with two nodes (subject and object) and an edge connecting them (predicate)

(URL) or both (URN + URL). Further examples of URIs with variations in their syntax components are:

- ftp://ftp.bgee.org/current/download/calls/expr_calls/Sus_scrofa_expr_simple_development.tsv.zip
- http://www.ensembl.org/Multi/Search/Results?q=BRCA2
- mailto:Bgee@sib.swiss
- urn:miriam:pubmed:26615188
- https://www.ncbi.nlm.nih.gov/pubmed/26615188

3.2 Resource Description Framework (RDF)

The Resource Description Framework (RDF) is a framework for describing information about resources in the World Wide Web, which are identified with URIs. In the previous section, we have seen that data in relational databases is organized into tables, according to some predefined schema. In contrast, in RDF stores, data is mainly organized into triples, namely, *<subject, predicate, object>*, similarly to how sentences in natural language are structured. An informal example would be: *<Bob, isFriendOf, Alice>*. A primer on triples and the RDF data model, using this simple example, is available online [34]. Figure 4 illustrates the RDF triple: the subject represents the resource being described, the predicate is a property of that resource, and finally the object is the value of the property (i.e., an attribute of the subject).

Triples can be defined using the RDF. The data store for RDF data is also called a "triple store." Moreover, in analogy to the data model (or the schema) of a relational database, the high-level structure of data in a triple store can be described using an *ontology*. According to Studer et al. [35], an ontology is a formal, explicit specification of a shared conceptualization. "Formal" refers to the fact that the expressions must be machine readable: hence, natural language is excluded. In this context, we can mention description logic (DL)-based languages [36], such as OWL 2 DL (*see* Sect. 3.3 for further details) to define ontologies. A DL ontology is the equivalent of a knowledge base (KB). A KB is mainly composed of two components that describe different statements in ontologies: the terminological box (TBox, i.e., the schema) and the assertional box (ABox, i.e., the data). Therefore, the conceptual statements form the set of TBox axioms, whereas the instance level statements form the set of ABox assertions. To exemplify this, we can mention the following DL axioms: $Man \equiv Human \sqcap Male$

(a TBox axiom that states a man is a human and male) and *john: Man* (an ABox assertion that states john is an instance of man).

Given that one of the goals of the Semantic Web is to *assign unambiguous names to resources* (URIs), an ontology should be more than a simple description of data in a *particular* triple store. Rather, it should more generally serve as a description of a *domain*, for instance, genomics (*see* Gene Ontology [37]) or orthology (*see* Orth Ontology [13]). Different *instantiations* of this domain, for example, by different research groups, should reuse and extend this ontology. Therefore, constructing good ontologies requires careful consideration and agreement between domain specialists, with the goal of formally representing knowledge in their field. As a consequence, ontologies are usually defined in the scope of consortiums—such as the Gene Ontology Consortium [38] or the Quest for Orthologs Consortium [39]. A notable collaborative effort is the Open Biological and Biomedical Ontology (OBO) Foundry [40]. It established principles for ontology development and evolution, with the aim of maximizing cross-ontology coordination and interoperability, and provides a repository of life science ontologies, currently, including about 140 ontologies.

To give an example of RDF data in a concrete life sciences use case, let us consider the following RDF triples, which illustrate a few of the assertions used in the OMA orthology database to describe the human hemoglobin protein ("HBB"), using the first version of the ORTH ontology [13]:

```
oma:PROTEIN_HUMAN04027 rdf:type orth:Protein.
oma:PROTEIN_HUMAN04027 oma:geneName "HBB".
oma:PROTEIN_HUMAN04027 biositemap:description "Hemoglobin
subunit beta".
oma:PROTEIN_HUMAN04027 obo:RO_0002162 <http://www.uniprot.
org/taxonomy/9606>.
```

This simple example already illustrates most of the basics of RDF. The instance that is being defined—the HBB protein in human—has the following URI in the OMA RDF store: http://omabrowser.org/ontology/oma#PROTEIN_HUMAN04027

The URI is composed of the OMA prefix, http://omabrowser.org/ontology/oma# (abbreviated here as "oma:"), and a fragment identifier, *PROTEIN_HUMAN04027*. The first triple describes the *type* of this resource—namely, an orth:Protein—based on the Orthology Ontology, prefixed here as "orth:," http://purl.org/net/orth#. As mentioned previously, this is a higher-level ontology, which OMA reuses and instantiates. It is important to note that other ontologies are used as well in the remaining assertions: for example, the last triple references the UniProt taxonomy ID *9606*. This is based on the National Center for Biotechnology Information (NCBI) organismal taxonomy [41]. If we follow the link in a

Web browser, we see that it identifies the "Homo sapiens" species, while the property obo:*RO_0002162* (i.e., http://purl.obolibrary.org/obo/RO_0002162) simply denotes "in taxon" in OBO [40]. Lastly, the concept also has a human-readable description, "Hemoglobin subunit beta."

3.3 RDF Schema (RDFS)

RDF Schema (RDFS) provides a vocabulary for modeling RDF data and is a semantic extension of RDF. It provides mechanisms for describing groups (i.e., classes) of related resources and the relationships between these resources. The RDFS is defined in RDF. The RDFS terms are used to define attributes of other resources such as the domains (*rdfs:domain*) and ranges (*rdfs:range*) of properties. Moreover, the RDFS core vocabulary is defined in a namespace informally called *rdfs* here, and it is conventionally associated with the prefix *rdfs:*. That namespace is identified by the URI http://www.w3.org/2000/01/rdf-schema#.

In this section, we will mostly focus on the RDF and RDFS terms used in this chapter. Further information about RDF/RDFS terms is available in [42].

- Classes
  - **rdfs:Resource**—all things described by RDF are called *resources*, which are instances of the class *rdfs:Resource* (i.e., *rdfs:Resource* is an instance of *rdfs:Class*).
  - **rdfs:Class** is the class of resources that are RDF classes. Resources that have properties (attributes) in common may be divided into classes. The members of a class are instances.
  - **rdf:Property** is a relation between subject and object resources, i.e., a predicate. It is the class of RDF properties.
  - **rdfs:Literal** is the class of literal values such as textual strings and integers. *rdfs:Literal* is a subclass of *rdfs:Resource.*
- Properties
  - **rdfs:range** is an instance of *rdf:Property.* It is used to state that the values of a property are instances of one or more classes. For example, *orth:hasHomolog rdfs:range orth:SequenceUnit* (*see* Fig. 5a). This statement means that the values of *orth:hasHomolog* property can only be instances of *orth:SequenceUnit* class.
  - **rdfs:domain** is an instance of *rdf:Property.* It is used to state that any resource that has a given property is an instance of one or more classes. For example, *orth:hasHomolog rdfs:domain orth:SequenceUnit* (*see* Fig. 5b). This statement means that resources that assert the *orth:hasHomolog* property must be instances of *orth:SequenceUnit* class.
  - **rdf:type** is an *rdf:Property* that is used to state that a resource is an instance of a class.

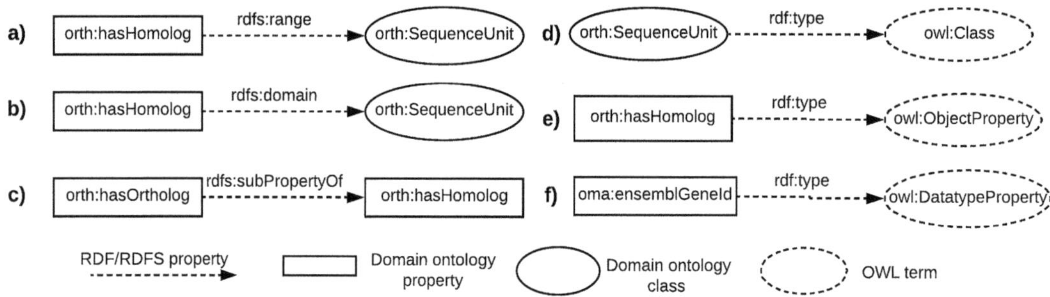

Fig. 5 Examples of RDF/RDFS statements

- **rdfs:subClassOf** is an *rdf:Property* to assert that all instances of one class are instances of another. For example, if *C1 rdfs: subClassOf C2* then an instance of *C1* is also an instance of *C2* but not vice versa.
- **rdfs:subPropertyOf** is used to state that all resources related by one property (i.e., the subject of rdfs:subPropertyOf) are also related by another (i.e., the object of rdfs:subPropertyOf, the "super-property"). For example, all orthologous relations are also homologous relations. Because of this, in the latest release candidate of the Orthology Ontology [13], it is stated that *orth:hasOrtholog* is a sub-property of *orth: hasHomolog*. Figure 5c illustrates this statement.

3.4 Web Ontology Language (OWL)

The first level above RDF/RDFS in the Semantic Web stack (*see* Fig. 2) is an ontology language that can formally describe the meaning of resources. If machines are expected to perform useful reasoning tasks on RDF data, the language must go beyond the basic semantics of RDF Schema [43]. Because of this, OWL and OWL 2 (i.e., Web Ontology languages) include more terms for describing properties and classes, such as relations between classes (e.g., disjointness, *owl:disjointWith*), cardinality (e.g., "exactly 2," *owl:cardinality*), equality (i.e., *owl:equivalentClass*), richer typing of properties, characteristics of properties (e.g., symmetry, *owl: SymmetricProperty*), and enumerated classes (i.e., *owl:oneOf*). The owl: prefix replaces the following URI namespace: http://www.w3.org/2002/07/owl#.

As a full description of OWL and OWL 2 is beyond the scope of this chapter, we refer the interested reader to [44, 45]. In the following, we focus solely on some essential modeling features that the OWL languages offer in addition to RDF/RDFS vocabularies.

- **owl:Class** is a subclass of rdfs:Class. Like rdfs:Class, an owl:Class groups instances that share common properties. However, this new OWL term is defined due to the restrictions on DL-based

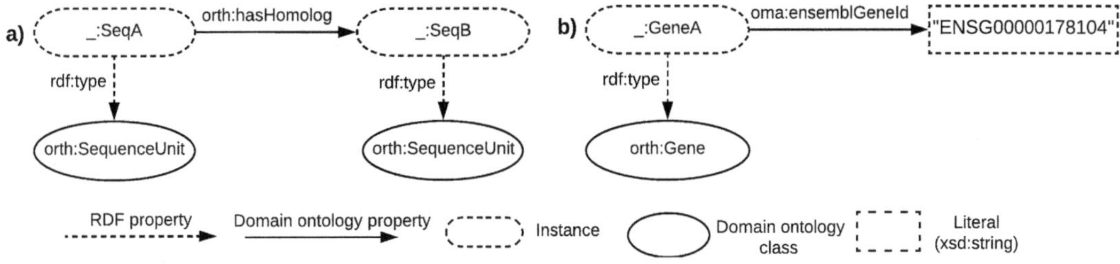

Fig. 6 Examples of instances of orth:SequenceUnit and orth:Gene and object and datatype property assertions

OWL languages (e.g., OWL DL and OWL Lite; OWL 2 DL and its syntactic fragments EL, QL, and RL). These restrictions imply that not all RDFS classes are legal OWL DL/OWL 2 DL classes. For example, the *orth:SequenceUnit* entity in the ORTH ontology is stated as an OWL class (i.e., *orth:SequenceUnit rdf:type owl:Class*—Fig. 5d illustrates this axiom). Therefore, *orth:SequenceUnit* is also an RDFS class since owl:Class is a subclass of rdfs:Class.

- **owl:ObjectProperty** is a subclass of rdf:Property. The instances of *owl:ObjectProperty* are *object properties* that link individuals to individuals (i.e., members of an *owl:Class*). For example, the *orth:hasHomolog* object property (*see* Fig. 5e) relates one *orth:SequenceUnit* individual to another one. Figure 5a illustrates this example.
- **owl:DatatypeProperty** is a subclass of rdf:Property. The instances of *owl:DatatypeProperty* are *datatype properties* that link individuals to data values. To illustrate a datatype property, we can mention the *oma:ensemblGeneId* (*see* Figs. 5f and 6b). This property asserts a gene identifier to an instance of an *orth:Gene.*

Further information about OWL languages are available as World Wide Web Consortium (W3C) recommendations in [46] and [47].

3.5 RDF Serialization Formats

RDF is a graph-based data model which provides a grammar for its syntax. Using this grammar, RDF syntax can be written in various concrete formats which are called RDF serialization formats. For example, we can mention the following formats: Turtle [48], RDF/XML (an XML syntax for RDF) [49], and JSON-LD (a JSON syntax for RDF) [50]. In this section, we will solely focus on the Turtle format.

Turtle language (TTL) allows for writing an RDF graph in a compact textual form. To exemplify this serialization format, let us consider the following turtle document that defines the homologous and orthologous relations:

```
@prefix rdfs: <http://www.w3.org/2000/01/rdf-schema#> .
@prefix rdf: <http://www.w3.org/1999/02/22-rdf-syntax-ns#> .
@prefix owl: <http://www.w3.org/2002/07/owl#> .
@prefix orth: <http://purl.org/net/orth#> .

# http://purl.org/net/orth#SequenceUnit
orth:SequenceUnit rdf:type owl:Class .

orth:hasHomolog rdf:type owl:ObjectProperty ;
                rdf:type owl:SymmetricProperty ;
                rdfs:domain orth:SequenceUnit ;
                rdfs:range orth:SequenceUnit .

orth:hasOrtholog rdf:type owl:ObjectProperty ;
                 rdfs:subPropertyOf orth:hasHomolog .
```

This example introduces many of features of the Turtle language: `@prefix` and prefixed names (e.g., `@prefix rdfs: http://www.w3.org/2000/01/rdf-schema#`), predicate lists separated by ";" (e.g., `orth:hasOrtholog rdf:type owl:ObjectProperty; rdfs:subPropertyOf orth:hasHomolog.`), comments prefixed with "#" (e.g., # http://purl.org/net/orth#SequenceUnit), and a simple triple where the subject, predicate, and object are separated by white spaces and ended with a "." (e.g., `orth:SequenceUnit rdf:type owl:Class`).

Further details about TTL serialization are available as a W3C recommendation in [48]

3.6 Querying the Semantic Web with SPARQL

Once we have defined the knowledge base (TBox and ABox), how can we use it to retrieve relevant data? Similar to SQL for relational databases, data in RDF stores can be accessed by using a query language. One of the main RDF query languages, especially used in the field of life sciences, is SPARQL [51]. A SPARQL query essentially consists of a graph *pattern*, namely, conjunctive RDF triples, where the values that should be retrieved (the unknowns—either subjects, predicates, or objects) are replaced by variable names, prefixed by "?". Looking again at the previous example, if we want to get the description of the "HBB" protein from OMA, we would simply use a graph pattern, where the value of the "description"—the one we want to retrieve—is replaced by a variable as follows:

```
SELECT ?description WHERE {
        ?protein oma:geneName "HBB".
        ?protein biositemap:description ?description.
}
```

The choice of variable name itself is not important (we could have used "?*x*", "?var", etc., albeit with a loss of readability).

Essentially, we are interested in the description of a protein about which we only know a name—“HBB.”

In order to get a sense of how large bioinformatics databases currently are, but also to get a hands-on introduction into how they can be queried using SPARQL, we propose to retrieve the total number of proteins in UniProt in Exercise A at the end of this chapter. Furthermore, Exercise C will allow trying out and refining the OMA query introduced above, but also writing a new one, using the OMA SPARQL endpoint.

4 Modeling Biological Databases with Semantic Web Technologies

In this section we show a concrete example of how we can use Semantic Web technologies to model the two biology databases Bgee and OMA.

Figure 7 illustrates a fragment of a candidate ontology describing the relational database sample from Bgee (*see* Fig. 1). The ellipses illustrate classes of the ontology, either specific to the Bgee ontology, such as *AnatomicEntity* (the equivalent of the *anatEntity* table in the relational view), or classes from imported ontologies, such as the *Taxon* class (the prefix “up:” denoting the UniProt ontology, http://purl.uniprot.org/core/). The advantage of using external (i.e., imported) classes is that integration with other databases which also instantiate these classes will be much simpler. For example, we will see that the class *Gene* serves as the “join point” between OMA and Bgee. Arrows define properties of the ontology: either *datatype properties* (similar to attributes of a table in the relational world), such as the *speciesName* or the *stageName*, or *object properties*, which are similar to primary key-foreign key relationships, given that they link instances of one class to those of another. If we compare Fig. 7 (the ontology view) against Fig. 1 (the relational view), we notice that the object properties *isExpressedIn* and *isAbsentIn* only appear explicitly in the ontology. This is because the values of these properties will actually be calculated on-the-fly, from *multiple* attributes in the relational database. Given that Bgee is mainly used to query gene expressions, these properties are *exposed* as new semantic properties in the domain ontology, namely, expression or absence of expression of a gene in a particular anatomic entity. This is one of the means through which the semantic layer can not only describe but also *enrich* the data available in the underlying layers (in this case, in the relational database). The domain of both the *isExpressedIn* and *isAbsentIn* properties is in this case a gene, while the range is an anatomic entity, such that triples that instantiate this relationship will have the structure: <*Gene, isExpressedIn, AnatomicEntity*>.

Given that the OMA ontology is significantly larger than the one for Bgee, we only show here the class hierarchy in Fig. 8. The

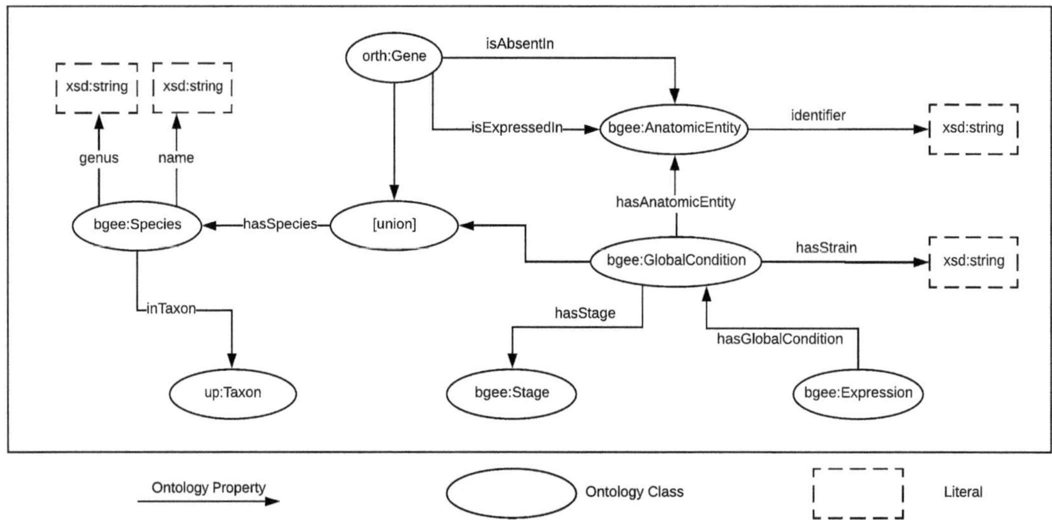

Fig. 7 A portion of the ontology defined over the relational database sample from Bgee. For readability purposes, we omitted the namespace ("bgee:") for the ontology properties

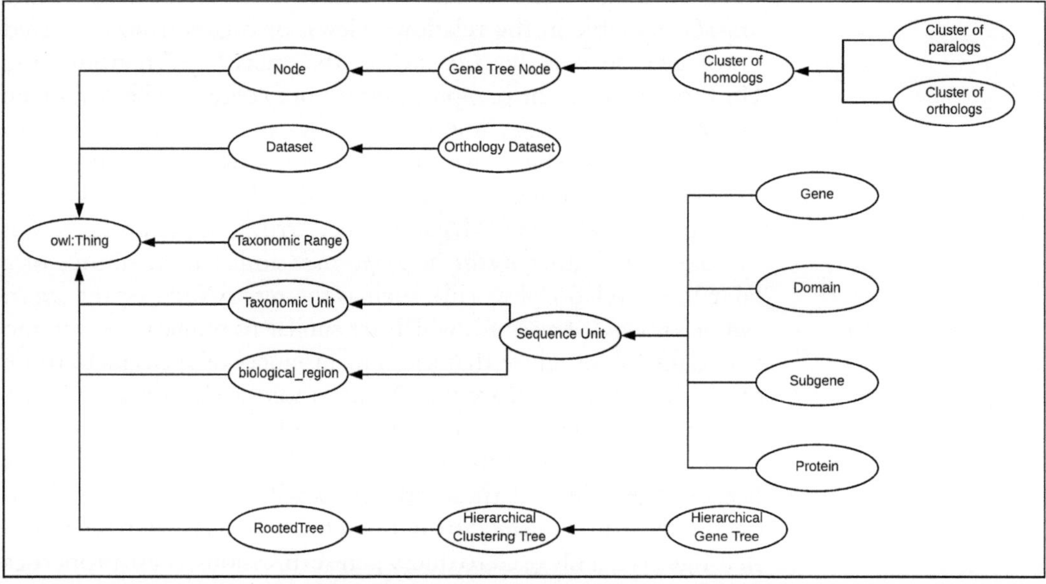

Fig. 8 The class hierarchy of the OMA ontology. Ellipses indicate class labels, while arrows indicate the "rdfs: subClassOf " property. Further details are available in [13]

most important concepts in the ontology are shown in the top right corner, namely, the cluster of orthologs and the cluster of paralogs, which store information about gene orthology (or paralogy) in a hierarchical tree structure (the gene-tree node). Similarly to the Bgee ontology, the *Gene* class in OMA is external. Arrows indicate the "rdfs:subClassOf" relationship—for example, both the "Cluster of Orthologs" and the "Cluster of Paralogs" classes—are subclasses of the "Cluster of Homologs" class. For a description of the ontology, as well as a discussion regarding its design within the Quest for Orthologs Consortium, we point the reader to [13]. Furthermore, the ontology can be explored or visualized in WebVOWL [52] using the Web page of the OMA SPARQL endpoint [53] available online at https://sparql.omabrowser.org/sparql.

Until here we have explored a few relatively simple examples in order to get familiar with the basics of Semantic Web technologies (URIs, RDF triples, and SPARQL). However, we can now introduce a more complex query that will better illustrate the expressivity of the SPARQL query language for accessing RDF stores—that is, for integrating and joining data *across* different databases.

Since all RDF stores structure data using the same standard model for data interchange, the main requirements in order to efficiently join multiple sources are:

1. That they each expose data through a SPARQL endpoint that supports federation (SPARQL 1.1)
2. That the sources share URIs or ontologies

This is the reason why already today we can jointly query, for example, OMA and UniProt—essentially, integrating the two databases by means of executing a federated SPARQL query.

To illustrate this, let us consider the following example: what are the human genes available in the OMA database that have a known association with leukemia? OMA does *not* contain any information related to diseases, however, UniProt does. In this case, since OMA already cross-references UniProt with the *oma: xrefUniprot* property, we can write the following federated SPARQL query, which will be running at the OMA SPARQL endpoint:

```
select distinct ?proteinOMA ?proteinUniProt
where {
     service <http://sparql.uniprot.org/sparql> {
           ?proteinUniProt a up:Protein .
           ?proteinUniProt up:organism taxon:9606 .  # Homo Sapiens
           ?proteinUniProt up:annotation ?annotation . # annotations of this protein
           entry
           ?annotation rdfs:comment ?text
           filter( regex(str(?text), "leukemia") )      # only those containing the
           text "leukemia"
     }

     ?proteinOMA a orth:Protein.
     ?proteinOMA oma:xrefUniprot ?proteinUniProt.
}
```

We skip the details regarding the prefixes used in the example and focus on the new elements in the query. The main part to point out is the "*service <http://sparql.uniprot.org/sparql>*" block, delimited between the inner brackets. This enables using the SPARQL endpoint of UniProt remotely, as a service. Through this mechanism, the query will first fetch from UniProt all instances of proteins that are annotated with a text that contains "leukemia" (this is achieved by the *filter* keyword in the *service* block). Then, using the cross-reference *oma:xrefUniprot* property, the query will return all the *equivalent* entries from OMA. From here, the user can explore, either in the OMA browser or by further refining the SPARQL query, other properties of these proteins: for example, their orthologs in a given species available in the database. In Exercise D at the end of this chapter, we encourage the reader to try this out in the OMA SPARQL endpoint. Note that the same results can be obtained by writing this query in the UniProt SPARQL endpoint and referencing the OMA one as a service. For an overview of federation techniques for RDF data, we refer the reader to the survey [54].

The mechanisms illustrated so far, while indeed powerful for federating distinct databases, have a major drawback: they require the user to know the schema of the databases (otherwise, how would we know which properties to query in the previous examples?), and, more importantly, they require all users to be familiar with a technical query language, such as SPARQL. While very expressive, formulating such queries can quickly become overwhelming for non-programmer users. In the following, we will look at techniques that aim to overcome these limitations.

5 Ontology-Based Integration of Heterogeneous Data Stores

So far we have seen some of the alternatives available for storing biological data—relational databases and triple stores. In this section, we look at how these heterogeneous sources can be integrated and accessed in a unified, user-friendly manner that does not require knowledge of the location or structure of the underlying data nor of the technical language (SQL or SPARQL) used to retrieve the data. The architecture we present is inspired by work presented in [21], which focused strictly on keyword search in relational databases.

5.1 A System's Perspective

We start with a bottom-up description of the layers that make up an integrated data access system, followed by a concrete example using the two bioinformatics databases introduced above: the orthology database OMA and the gene expression database Bgee.

The main four layers of an integrated data access system, as shown in Fig. 9, are:

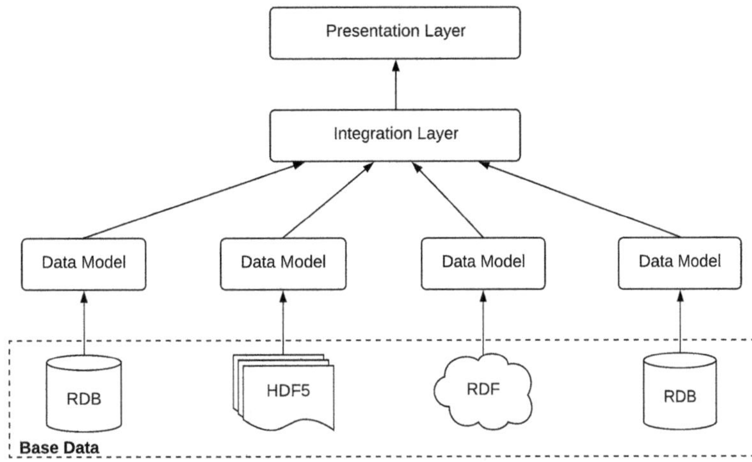

Fig. 9 Integrated data access system

5.1.1 Base Data Layer

This represents the physical storage layer, where all the actual data, for example, experimental results, annotations, etc., are kept. Figure 9 illustrates only a few of the possible storage types, namely, relational databases, hierarchical data stores (e.g., HDF5), and RDF stores. At this low-level layer, the data are usually structured so as to optimize machine parameters, such as storage space, complexity of joins required to answer physical queries, etc. Therefore, it is not designed for human readability. Furthermore, tables, column names, or even IDs may not match any real terms. For example, the Bgee relational database uses the table name "anatEntity" to refer to the term "anatomic entity," while others may be even further away from the original terms.

5.1.2 Data Model Layer

This layer is used to describe, at a higher level of abstraction, the data contained in the physical storage. Here, for example, original names for terms are recovered while also creating a mapping between these higher-level terms ("Anatomical Entity") and their corresponding physical layer location (table "anatEntity" in schema Bgee). The data model layer can be viewed as the first *semantic layer* in the system, as it allows representing the actual terms referred to in the underlying physical storage while abstracting away the details of the actual structure of the physical storage. The data model layer can be understood as an ontology, however, only applicable to the level of an individual database.

5.1.3 Integration Layer

The integration layer performs a similar task to the data model layer, in that it defines a mapping between high-level concepts ("Anatomical Entity") and *all the occurrences* where these concepts can be found in the physical storage (table "anatEntity" in schema Bgee, class "Anatomic Entity" in UniProt, etc.). In doing so, the integration layer also *aligns* the different data models, by defining which identifiers from one data model correspond to which ones from the others. In the case of biological databases, this is usually done by taking into account *cross-references*, which already exist between most databases, as we have seen in the SPARQL query in Sect. 5.

While the data model layer can be seen as a *local* ontology, the integration layer will serve as a *global* ontology. The integration layer can be queried using, for example, SPARQL. However, in order to get the results from the underlying sources, the SPARQL query needs be translated in the native query languages of the underlying sources (e.g., SQL for relational databases). This is achieved by using the *mappings* defined in the global ontology. For example, the keyword "expressed in" does not have a direct correspondence in Bgee, but it can be translated into an SQL procedure (in technical terms, it represents an SQL *view* of the data). Without going into details, at a high level, the property "gene A *expressed in* anatomic entity B" will be computed by looking at the number of experiments stored in the database, showing the expression of A in B. It is conceivable that in another database, which could also form part of the integrated system, this information is available explicitly. In this case the mapping would simply be a 1-to-1 correspondence to the property value stored in the database. The role of the integration layer is to capture *all* the occurrences where a certain concept (entity or property) can be found, along with a *mapping* for each of the occurrences, defining *how* information about this concept can be computed from the base data.

To summarize, the integration layer abstracts away the *location and structure* of data in the underlying sources, providing users a unified access through a global ontology. One of the drawbacks of this approach is that, in the lack of a presentation layer, such as a user-friendly query interface (e.g., a visual query builder or a keyword-based search interface), the data represented in the global ontology is accessible mainly through a technical query language, such as SPARQL. Therefore, in order to be able to access the data, users are required to become fluent in the respective query language.

It is worth at this point mentioning that most data integration systems available at the time of this writing only offer the three layers presented so far. Examples of such systems, generically denoted as ontology-based data access (OBDA) systems, are Ontop [55], Ultrawrap [56], or D2RQ [57].

5.1.4 Presentation Layer

The three layers presented so far already achieve data integration, but with a significant drawback, which is that the user is required to know a technical query language, such as SPARQL. The role of the presentation layer is to expose data from all integrated resources in an easy to access, user-friendly manner. The presentation layer abstracts away the structure of the integration layer and exposes data through a search interface that users (including non-programmers) are familiar with, such as keyword search [21, 58] or even full natural language search [59, 60].

The challenges in building the presentation layer are manyfold: first, human language is inherently ambiguous. As an example, let us assume a user asks: "Is the HBB gene expressed in the blood?" What does the user mean? The hemoglobin gene (HBB) in general? Or just in the human? The system should be proactive in helping the user clarify the semantics or intents of the question, before trying to compute the underlying SPARQL query. Second, the presentation layer should provide not only raw results but also an explanation—for example, what sources were queried, how many items from each source have been processed in order to generate the response, etc. This enables the user to validate the generated results or to otherwise continue refining the question. Third, the presentation layer must also rank the results according to some relevance metric, similarly to how search results are scored in Web search engines. Given that the number of results retrieved from the underlying sources can easily become overwhelming (e.g., searching for "HBB" in Bgee returns over 200 results), it is important that the most relevant ones are shown first.

From a technical point of view, the presentation layer maintains an index (i.e., the vocabulary) of all keywords stored in the lower layers, both data and metadata (descriptions, labels, etc.), such that each keyword in a user query can be mapped to existing data in the lower layers. An important observation is that the presentation layer highly relies on the quality of the *annotations* available in the lower layers. In the lack of human-readable labels and descriptions in the global ontology, the vocabulary collected by the presentation layer will miss useful terms that the user might search for. One way to detect and fix this problem is to always log user queries and improve the quality of the annotations "on demand," whenever the queries cannot be solved due to missing items in the vocabulary. For a more extended discussion on the topic of labels and their role in the Semantic Web, refer to [61].

Finally, it is worth noting that none of these layers need to be centralized—indeed, even in the case of the integration layer, although its role is to build a *common* view of all data in the physical storage, it can be distributed across multiple machines, just as long as the presentation layer knows which machine holds which part of the unified view.

5.2 A Concrete Example: A Global Ontology to Unify OMA and Bgee

So far we have seen an abstract view of a system for data integration across heterogeneous databases. It is time to look at how this translates into a real-world example, using the Bgee relational database and the OMA RDF database.

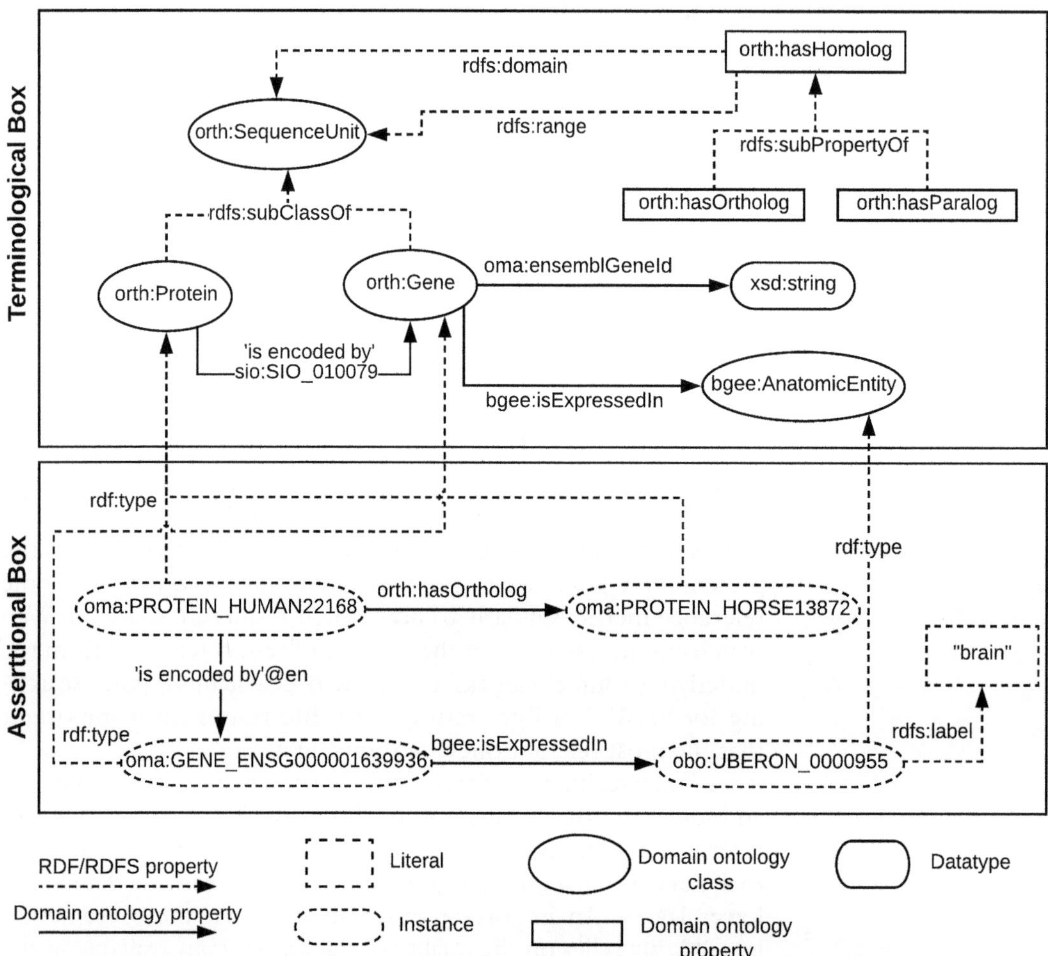

Fig. 10 A sample global ontology for integrating OMA and Bgee and an example assertion

The top part of Fig. 10, the *terminological box*, illustrates part of the global ontology (layer 3, integration layer) for the two databases, with most of the terms being part of OMA, except for *Anatomic Entity*, which is specific to Bgee. As mentioned previously, OMA extends the ORTH ontology, which is why the corresponding terms in the ontology are prefixed with "*orth:*." The *Gene* concept can actually be found in both Bgee and OMA; therefore the global ontology will define mappings to both sources. As we can see in the ontology, the *Gene* is the common point that joins together OMA and Bgee. The gene IDs used in both databases are Ensembl IDs [62], stored in the *ensemblGeneId* string property. For example, the human hemoglobin gene, "HBB," which we previously showed as an example entry in OMA, corresponds to the *ENSG00000244734* Ensemble ID and can also be found in Bgee.

The lower part of Fig. 10, the *assertional box*, illustrates an example assertion—in this case, that the protein *HUMAN22168* in OMA is orthologous to the protein *HORSE13872* and that, furthermore, this protein is encoded by the gene with the Ensemble ID *ENSG000001639936*. Moreover, this gene is expressed in the brain (the Uberon ID for this being "*UBERON:0000955*"). The human-readable description is stored in the String literal *label*—as, for example, the name of the anatomic entity, "brain," shown in the bottom-right corner in the figure. Without labels, much of the available data would not be easily searchable by a human user nor by an information retrieval system.

Note that with this sample ontology, we can already answer questions related to orthology *and* gene expression jointly, such as the first part of our introductory query: "What are the human-rat orthologs, expressed in the liver...?". This question essentially refers to pairs of orthologous *Genes* (those in human and rat) and their expression in a given *Anatomic Entity* (the liver). Apart from the *Species* class, which is not explicitly shown, all of the information is already captured by the ontology in Fig. 10. A similar mechanism can be used to further extend this to UniProt (for instance, based again on gene IDs as the "join point," or by using existing cross-references, as we have shown in the previous section), therefore enabling users to ask even more complex queries.

5.3 How to Link a Database with an Ontology?

One of the main challenges in implementing technologies for the Semantic Web was recognized from early on (*see* the study published in 2001 by Calvanese et al. [63]) to be the problem of integrating *heterogeneous* sources. In particular, one of the observations made was that integrating legacy data will not be feasible through a simple 1-to-1 mapping of the underlying sources into an integrative ontology (e.g., mapping all attributes of tables in relational databases to properties of classes in an ontology), but rather through more complex transformations, that map *views* of the data into elements of the global ontology [63].

To illustrate this with a concrete example, let us consider again the unified ontology for OMA and Bgee that we introduced in the previous section. Although Figure 10 shows properties such as "gene *isExpressedIn*" or "gene *hasOrtholog*," this data is actually not *explicitly* stored in the underlying databases but rather needs to be computed on-the-fly based on the available data. For example, the "isExpressedIn" property can be computed based on the number of experiments which show the expression of a gene in a certain anatomic entity in Bgee. Deciding the exact threshold for when a gene is considered as "expressed" according to the data available is not straightforward and needs to be agreed upon by domain specialists. Therefore, the integration layer will also serve to *enrich* the data available in the underlying layers, by defining new concepts based on this data (e.g., the presence or absence of gene expression in an anatomic entity).

At this point it is worth clarifying an important question: why are mappings necessary? Why is it not enough to replicate the data in the different underlying formats into a single, uniform way (e.g., translate all RDB data into RDF)? The answer is that not only would such a translation require a lot of engineering effort, but more importantly, it would transform the data from a format that is highly optimized for data access, into a format that is optimized for different purposes (data integration and reasoning). Querying *relational databases* still is, today, the most efficient means of accessing very large quantities of structured data. Transforming all of it into RDF would in many cases mean *downgrading* the overall performance of the system. In some cases storing RDF data in the relational format was proven to be more efficient [64].

So how are mappings then created? One of the main mechanisms to achieve this is currently the W3C standard R2RML, available as a W3C recommendation online [65]. R2RML enables mapping relational data to the RDF model, as chosen by the programmer. For a concrete example of how mappings can be defined and what are the advantages of this approach, we refer the reader to [66]. A mapping essentially defines a *view* of the data, which is a query (in this case, an SQL query) that allows retrieving a relevant portion of the underlying data, in order to answer a higher-level question (e.g., what is "expressed in"?). The materialization of this query (the answer) will be returned in RDF format, on demand, according to the mapping. This avoids duplicating or translating data in advance from the underlying relational database into RDF until it is really needed, in order to answer a user query.

For a discussion regarding the limitations of R2RML and alternative approaches to define mappings from relational data to RDF, we refer the reader to the survey [67].

5.4 Putting Things Together

So far we have seen how individual sources can be represented into a single, unified ontology, and we had a high-level view of a data access system that enables users to ask queries and get responses in a unified way, without knowledge of where data is located or how it is structured. In this section we finally look at how all of these components can work together in answering natural language queries on biological databases. Although there are multiple alternatives to natural language interfaces, including visual query interfaces or keyword-based search interfaces, it has been shown that natural language interfaces are the most appropriate means to query Semantic Web data for non-technical end-users [68]. As a consequence, natural language querying, based on Semantic Web technologies, is currently one of the active areas of research, examples of recent systems implementing an ontology-based natural language interface including the Athena [59] and TRDiscover [60] systems.

First, recall the user question we formulated in the beginning of this chapter: "What are the human-rat orthologs, expressed in the liver, that are associated with leukemia?" Let us assume the resources at hand to answer this question are the biological databases OMA, Bgee, and UniProt. The four main steps required to translate the natural language question into the underlying query languages of OMA, Bgee, and UniProt will be:

(a) Identify entities in the query

This is the natural language processing step that extracts the main concepts the user is interested in, based on the keywords of the input query: *orthologs*, *human*, *rat*, *expressed*, *liver*, *associated*, and *leukemia*.

(b) Identify matches of the entities in the integrative ontology

The extracted keywords will be searched for in the vocabulary of the presentation layer, resulting in one or multiple URIs, given that a keyword can match multiple concepts. For example, the keyword "orthologs" can match either the entity "OrthologCluster" or the property "hasOrtholog" of a gene in OMA. The index of the presentation layer will also return the location the URI originates from (OMA or Bgee or UniProt).

(c) Construct subqueries for each of the matches

The extracted URIs will be used to construct subqueries on each of the underlying data sources. This step requires translating the original query into the native language of each underlying database, with specific mechanisms for each type of database (relational or triple store). At a high level, the translation process involves finding the minimal sub-schema (or subgraph in the case of RDF data) that covers all the keywords matched from the input query. Taking the example previously shown in Fig. 10, the minimal subgraph that contains "orthologs" and "expressed" will essentially contain only two nodes of the entire graph: *Gene* (which is both the domain and the range of the "hasOrtholog" property in the Orthology Ontology) and *AnatomicEntity* (which is the range of the "isExpressedIn" property in the Bgee ontology). All the unknowns of the query (e.g., which ortholog genes) are replaced by variables. The final subqueries for OMA and Bgee might therefore (*informally*) look like this:

```
OMA: select ?gene1 ?gene2 where {
             ?protein1 a Protein.
             ?protein1 inTaxon "Homo sapiens".
             ?protein1 isEncodedBy ?gene1.
             ?protein1 hasOrtholog ?protein2.
             ?protein2 inTaxon "Rattus norvegicus".
             ?protein2 isEncodedBy ?gene2.
     }
```

Note that we have simplified the actual query for readability purposes (using the literals "Homo sapiens" and "Rattus norvegicus" instead of their corresponding URIs). This subquery will cover the keywords: *ortholog*, *human*, and *rat*. Notice that the query should return *genes*, not proteins, because the join point between OMA and Bgee is the *Gene* class.

```
Bgee: select ?gene where {
                    ?gene a Gene.
                    ?gene isExpressedIn ?anatomicEntity.
                    ?anatomicEntity rdfs:label "liver".
  }
```

This subquery will therefore cover the *expressed* and *liver* keywords. The final step will be then to get the similar subquery for UniProt (which we omit here for brevity) and to compute the joint result, namely, the intersection between all the sets returned by the subqueries.

(d) Join the results from each of the subqueries

This final step is essential in keeping the performance of the system to an acceptable level. Joining (federating) the results of several subqueries into a unified result is not an easy task and requires a careful ordering of the operations from all subqueries. To understand this problem, let us consider again our example and try to see how many results each of the subqueries will return. First, if we take a look at the OMA browser and try to find all orthologs between human and rat, this will amount to more than 21,000 results. However, is the user really interested in all of them? Certainly not, as the input query shows—the user is only interested in a small fraction of the orthologs, namely, those that are expressed in the liver and have an association with leukemia (according to the data stored in Bgee and UniProt). How many are these? If we now refer to UniProt and look for the disease *leukemia*, we will find that there are only 20 entries which illustrate the association with this disease. Clearly, getting only the orthologs of these 20 entries will be much more efficient than retrieving all 21,000 pairs from OMA first and then removing most of them to only keep relevant ones.

However, note that in this case, we only know this information because we constructed the queries and tried them out by hand first. How should the system estimate the number of results (i.e., the cardinality of each subquery) in advance? This question has been an active area of research for a long time. Some of the methods used to tackle this problem are either to precompute statistics regarding the number of results available in different tables of the underlying sources [69] or to use statistics regarding previously asked queries to optimize the new ones, for example, via statistical machine learning [70]. In the first case, we would, for instance, store the individual counts of different orthologous pairs while also keeping statistics about diseases if we expect these types of questions to be asked frequently, whereas in the second case, we would simply look at the number of results similar subqueries generated in the past, to optimize which results to fetch first. For a recent study of optimization methods for federated SPARQL queries, *see* [71].

(e) Present the user the final results

Finally, the joined results are returned to the user, along with an explanation regarding the constructed query and the entities that were matched in order to construct it. In this way, the user has the opportunity to validate the correctness of the answer or otherwise to further refine the question.

For a more in-depth discussion regarding natural language query interfaces in ontology-based data access systems, we refer the reader to Athena [59] and TRDiscover [60].

6 Timeline of Semantic Web Technologies and Ontology-Based Data Integration in Life Sciences

The field of life sciences has been an early adopter of Semantic Web technologies, due to the need of interoperability and integration of biological data spread across different databases. In this section, we provide a brief timeline (*see* Fig. 11), including the example ontologies introduced in this chapter.

- 1995: Davidson et al. [72] suggest basic steps to integrate bioinformatics data (common data model, match semantically related objects, schema integration, transform data into federated database, match semantically equivalent data).
- 2000: **TAMBIS (Transparent Access to Multiple Bioinformatics Information Sources)** [73] proposes a unified ontology covering many aspects of the bioinformatics knowledge space.
- 2000: The **"Gene Ontology—a tool for the unification of biology"** [37] is the first significant milestone in unifying diverse biological databases, focusing on gene functions. Even before the publication of the Semantic Web paper by Tim Berners Lee (in the following year), the GO highlighted the benefits of controlled vocabularies and standardized naming, both precursors of Semantic Web technologies, which were adopted in the GO in the year 2002 [74]. Today it is, arguably, the most

1995	Davidson et al. suggest **basic steps to integrate bioinformatics data**
2000	**TAMBIS**: Transparent Access to Multiple Bioinformatics Information Sources
	Gene Ontology
2001	**BioMoby:** a unified registry of web services for life scientists
2003	**Integrating biological databases** in Nature Reviews Genetics
	UniProt: The Universal Protein Knowledge
2004	First **International Workshop on Data Integration in the Life Sciences**
2005	**HCLS IG**: Semantic Web Health Care and Life Sciences Interest Group
2006	**OBO Foundry**: Open Biological and Biomedical Ontology Foundry
	OLS: Ontology Lookup Service
2007	**National Center for Biomedical Ontology (NCBO) BioPortal**: a web portal to biomedical ontologies
2008	**BioMoby**: interoperable access to over 1400 bioinformatics resources
	BioGateway: a semantic systems biology tool for the life sciences
	Special issue **Database Integration in Life Sciences** in Briefings in Bioinformatics
2009	Review on **Ontologies and Semantic Web Technologies** in Briefings in Bioinformatics
2010	**NCBO** launches a **SPARQL endpoint**
2012	**Semantic Web meets Integrative Biology**
2016	**Orthology Ontology**

Fig. 11 A selective timeline of data integration efforts in life sciences

comprehensive resource of computable knowledge regarding gene functions and products.

- 2001: Launch of the **BioMoby project** [75] providing a unified registry of Web services for life scientists using a consensus-driven approach. It listed, for instance, all services converting gene names to GO terms or all databases accepting GO terms. The registry is currently no longer maintained.
- 2003: A **Nature Reviews Genetics** article on **Integrating Biological Databases** [76] highlights the "database-surfing" problem (i.e., the time-consuming process of manually visiting multiple databases to answer complex biological research questions) and argues for standardized naming of biological objects to overcome the problem. Link integration, view integration, and data warehousing are proposed for data integration. Arguably, link integration has since become the most adopted solution.
- 2003: Launch of **UniProt** [77] by the UniProt Consortium, a collaboration between the Swiss Institute of Bioinformatics (SIB), the European Bioinformatics Institute (EBI), and the Protein Information Resource (PIR). UniProt is the world's most comprehensive freely accessible resource on protein sequences and functional annotation. Since 2008 the data is

published in RDF, and since 2013 a SPARQL endpoint is provided [78].

- 2004: The first **International Workshop on Data Integration in the Life Sciences**, held in Leipzig, promotes "a Bioinformatics Semantic Web" and highlights solutions for heterogeneous data integration. The workshop continues to be held every year, and its proceedings (e.g., [79]) provide a good overview of advances in the field.
- 2005: The W3C Consortium launches the **Semantic Web Health Care and Life Sciences Interest Group** (HCLS IG) to develop the use of Semantic Web technologies to improve health care and life sciences research. Today, the HCLS Linked Data Guide [80] provides best practices for publication of biological Linked Data on the Web.
- 2006: The **OBO Foundry** [40] establishes principles for ontology development and evolution to support biomedical data integration through a suite of orthogonal interoperable reference ontologies.
- 2006: Publication of the **Ontology Lookup Service** (OLS), a repository for biomedical ontologies with the aim to provide a single point of access (with controlled vocabulary queries) to the latest ontology versions. It allows interactive browsing, as well as programmatic access [81].
- 2007: Launch of the **National Center for Biomedical Ontology (NCBO) BioPortal** [82], a web portal to biomedical ontologies. OBO ontologies are a central component. The portal started with 50 ontologies; to date it is the most comprehensive repository with currently 852 biomedical ontologies and more than eight million classes.
- 2008: Launch of the **BioMoby Consortium** [83] and the first **release of the BioMoby Semantic Web Service**, at the time providing interoperable access to over 1400 bioinformatics resources worldwide.
- 2008: **BioGateway** [84] provides a single SPARQL entry point to all OBO candidate ontologies, the GO annotation files, the SWISS-PROT protein set, the NCBI taxonomy, and several in-house ontologies.
- 2008: The ***Briefings in Bioinformatics*** **journal launches a special issue** dedicated to **Database Integration in Life Sciences** [85], acknowledging the major challenge of integrating data scattered over millions of publications and thousands of heterogeneous databases.
- 2008: **Bio2RDF** [86] applies Semantic Web technology to various publicly available databases (converting them into RDF format and linking with normalized URIs and a common

ontology). Updates continue to be provided for increased interoperability among bioinformatics databases [87, 88].

- 2009: *Briefings in Bioinformatics* publishes a review on **Biological Knowledge Management** [89], highlighting the transforming role of ontologies and Semantic Web technologies in enabling knowledge representation and extraction from heterogeneous bioinformatics databases.
- 2010: **NCBO launches a SPARQL endpoint**, available at http://sparql.bioontology.org/.
- 2012: Publication of a survey highlighting the benefits of integration using Semantic Web technologies in the field of **Integrative Biology** [90].
- 2016: Publication of the **Orthology Ontology** [13].

7 Conclusions and Outlook

Data integration is arguably one of the most important enablers of new scientific discoveries, given that research data is currently growing at an unprecedented rate. This is especially true in the case of biological databases. While data integration poses many challenges, the emergence of standards, integrative ontologies, as well as the availability of cross-references between many of the biological databases make the problem easier to tackle. This chapter has provided a brief introduction to the methods that can be used to integrate heterogeneous databases using Semantic Web technologies while also providing a concrete example of achieving this goal for three well-known existing biological databases: OMA, Bgee, and UniProt.

Although there would be many more aspects to cover and much of the work for achieving wide-scale data integration still remains to be done, we would like to end this chapter by reinforcing the following conclusion, extracted from a study of Biological Ontologies for Biodiversity Knowledge Discovery [91]:

> We hope that current work will spur interest and feedback from scientists and bioinformaticians who see data integration, interoperability, and reuse as the solution to bringing the past 300 years of biological exploration of the planet into currency for science and society.

8 Exercises

A. Querying UniProt with SPARQL

The goal of this warm-up exercise is to get familiar with a SPARQL endpoint and to write your first SPARQL query. For this purpose, open the link to the UniProt SPARQL endpoint, http://sparql.

uniprot.org/ in a Web browser. How many entries do you think are available in UniProt? To find out, simply check the bottom-left corner of the Web page—you will notice that the total number of triples is always kept up to date there. How many of these entries describe proteins? To find out, try running the following SPARQL query that counts all instances of the database that belong to the protein class. What is the result?

```
PREFIX up:<http://purl.uniprot.org/core/>
SELECT (count(?protein) as ?count)
WHERE
{ ?protein a up:Protein. }
```

Notice that the UniProt SPARQL web page includes many examples on the right-hand side—in order to get more familiar with UniProt and SPARQL, try further some of the sample queries provided there.

B. Exploring Biological Ontologies Through Keyword Search in the Ontology Lookup Service

We have seen in Sect. 3.6 an example assertion about the "HBB" gene in the human, including the following triple:

```
oma:PROTEIN_HUMAN04027 obo:RO_0002162 <http://www.uniprot.
org/taxonomy/9606> .
```

This triple essentially asserts that the gene is located in the *Homo sapiens* taxon. However, as a regular user, how could you know what the URIs for "in taxon" and *Homo sapiens* are? One of the possible ways to get these identifiers is by searching for the keywords of interest in the *Ontology Lookup Service* (OLS). To do this, go to the Web page of the service https://www.ebi.ac.uk/ols/index, and try to enter first "in taxon". What is the result? Try also *Homo sapiens.* What about "human"?

C. Querying OMA with SPARQL

Recall from Sect. 3.6 the sample query we presented for retrieving the description of the human hemoglobin gene from OMA. We provide it in a more explicit form here:

```
SELECT ?description WHERE {
        ?protein oma:geneName "HBB".
        ?protein <http://biontology.org/ontologies/biositemap.owl#description> ?de-
        scription.
}
```

First try to think about possible information that is missing from this query. For example, is this query guaranteed to return a single result (remember we are using an *orthology* database)?

Try to look again at how the human "HBB" protein is defined in Sect. 3. Then, try to run the SPARQL query as-is in the OMA SPARQL endpoint: https://sparql.omabrowser.org/sparql. What do you get? What is the reason? Try to print out more information about the protein, not just its description. For example, add another triple pattern to capture the *oma:hasOMAId* property value as well (don't forget to add it to the selected variables in the first line!), perhaps also the taxon ID in UniProt. What can you deduce? Can you correct the query so that it only gets the description we were originally interested in?

D. Federated Queries Using SPARQL (OMA and UniProt)

In Sect. 4 we presented an example Federated Query using the SPARQL endpoint of OMA and the remote SPARQL endpoint of UniProt, as a service. We recall the query here:

```
prefix up:<http://purl.uniprot.org/core/>
prefix taxon:<http://purl.uniprot.org/taxonomy/>
select distinct ?proteinOMA ?proteinUniProt
where {
     service <http://sparql.uniprot.org/sparql> {
             ?proteinUniProt a up:Protein .
             ?proteinUniProt up:organism taxon:9606 .    # Homo Sapiens
             ?proteinUniProt up:annotation ?annotation . # annotations of this
              protein entry
             ?annotation rdfs:comment ?text
             filter( regex(str(?text), "leukemia") )     # only those containing
             the text "leukemia"
     }

     ?proteinOMA a orth:Protein.
     ?proteinOMA oma:xrefUniprot ?proteinUniProt.
}
```

Try running this query in the OMA SPARQL endpoint, https://sparql.omabrowser.org/sparql. You might need to wait a couple of minutes to get the remote results. Next, try to look at the examples provided in the right side of the page to see how to get more properties of the *proteinOMA* variable—for example, try getting the description or the OMA ID. Next, try modifying this query so that it can run in the UniProt SPARQL endpoint, invoking the OMA one as a service. Remember to get the relevant prefixes and define them in the header of the query first ("oma," "orth"). You can get these by looking at "Namespace prefixes" in the OMA SPARQL Web page. Finally, test your modifications using UniProt, http://sparql.uniprot.org/.

Acknowledgments

We would like to thank all the members of the Bio-SODA SNF project at ZHAW Zurich University of Applied Sciences, at the University of Lausanne (UNIL) and at the SIB Swiss Institute of Bioinformatics for their valuable feedback: Adrian Altenhoff (UNIL, SIB), Maria Anisimova (ZHAW, SIB), Christophe Dessimoz (UNIL, SIB), Frederic Bastian (UNIL, SIB), Heinz Stockinger (SIB), Marc Robinson-Rechavi (UNIL, SIB), and Erich Zbinden (ZHAW, SIB). This work was supported by the Swiss National Science Foundation (SNSF) funded under grant NRP 75 Big Data.

References

1. Mole B (2004) The gene sequencing future is here. http://www.sciencenews.org/article/gene-sequencing-future-here. Accessed 15 Feb 2018
2. Stephens ZD, Lee SY, Faghri F et al (2015) Big data: astronomical or genomical? PLoS Biol 13 (7):e1002195. https://doi.org/10.1371/journal.pbio.1002195
3. Altenhoff AM, Škunca N, Glover N et al (2014) The OMA orthology database in 2015: function predictions, better plant support, synteny view and other improvements. Nucleic Acids Res 43(D1):D240–D249. https://doi.org/10.1093/nar/gku1158
4. Waterhouse RM, Tegenfeldt F, Li J et al (2012) OrthoDB: a hierarchical catalog of animal, fungal and bacterial orthologs. Nucleic Acids Res 41(D1):D358–D365. https://doi.org/10.1093/nar/gks1116
5. Powell S, Szklarczyk D, Trachana K et al (2011) eggNOG v3. 0: orthologous groups covering 1133 organisms at 41 different taxonomic ranges. Nucleic Acids Res 40(D1): D284–D289
6. Bastian F, Parmentier G, Roux J et al (2008) Bgee: integrating and comparing heterogeneous transcriptome data among species. In: International Workshop on Data Integration in the Life Sciences. Springer, Berlin, pp 124–131
7. UniProt Consortium (2014) UniProt: a hub for protein information. Nucleic Acids Res 43 (D1):D204–D212. https://doi.org/10.1093/nar/gku989
8. Codd EF (1970) A relational model of data for large shared data banks. Commun ACM 13 (6):377–387. https://doi.org/10.1145/362384.362685
9. W3Schools. Online SQL Tutorial. https://www.w3schools.com/sql/sql_intro.asp. Accessed 15 Feb 2018
10. Beaulieu A (2009) Learning SQL: master SQL fundamentals. O'Reilly Media, Inc, Sebastopol, CA
11. Fehily C (2014) SQL (database programming). Questing Vole Press, Pacific Grove, CA. (2015 Edition)
12. Teodoro D, Pasche E, Wipfli R et al (2009) Integration of biomedical data using federated databases. Swiss Medical Informatics, Muttenz
13. Fernández-Breis JT, Chiba H, del Carmen L-GM et al (2016) The orthology ontology: development and applications. J Biomed Semant 7(1):34. https://doi.org/10.1186/s13326-016-0077-x
14. Self-join incurs more I/O activities and increases locking overhead (2013) http://sqltouch.blogspot.ch/2013/07/self-join-incurs-more-io-activities-and.html. Accessed 15 Feb 2018
15. Sadalage PJ, Fowler M (2012) NoSQL distilled: a brief guide to the emerging world of polyglot persistence. Pearson Education, Upper Saddle River, NJ
16. Hunger M, Boyd R, Lyon W (2016) RDBMS & Graphs: why relational databases aren't always enough. https://neo4j.com/blog/rdbms-graphs-why-relational-databases-arent-enough/. Accessed 15 Feb 2018
17. Stockinger K, Bödi R, Heitz J et al (2017) ZNS-Efficient query processing with Zurich-NoSQL. Data Know Eng 112:38–54
18. Erling O, Mikhailov I (2009) RDF support in the virtuoso DBMS. In: Networked knowledge-networked media. Springer, Berlin, pp 7–24. https://doi.org/10.1007/978-3-642-02184-8_2

19. Berners-Lee T, Hendler J, Lassila O (2001) The semantic web. Sci Am 284(5):34–43
20. Kimball R, Ross M, Mundy J et al (2015) The kimball group reader: relentlessly practical tools for data warehousing and business intelligence remastered collection. John Wiley & Sons, New York, NY
21. Blunschi L, Jossen C, Kossmann D et al (2012) Soda: generating sql for business users. Proc VLDB Endow 5(10):932–943
22. Bio-SODA: enabling complex, semantic queries to bioinformatics databases through intuitive searching over data (2017) https://www.zhaw.ch/no_cache/en/research/people-publications-projects/detail-view-project/projekt/3066/. Accessed 15 Feb 2018
23. Lane L, Argoud-Puy G, Britan A et al (2011) neXtProt: a knowledge platform for human proteins. Nucleic Acids Res 40(D1): D76–D83. https://doi.org/10.1093/nar/gkr1179. Sparql endpoint. Available at: https://sparql.nextprot.org/
24. Li W, Cowley A, Uludag M et al (2015) The EMBL-EBI bioinformatics web and programmatic tools framework. Nucleic Acids Res 43 (W1):W580–W584. https://doi.org/10.1093/nar/gkv279. RDF data. Available at: https://www.ebi.ac.uk/rdf
25. Berman H, Henrick K, Nakamura H (2003) Announcing the worldwide protein data bank. Nat Struct Mol Biol 10(12):980. RDF data. Available at https://pdbj.org/help/rdf
26. Wilkinson MD, Dumontier M, Aalbersberg IJ et al (2016) The FAIR guiding principles for scientific data management and stewardship. Sci Data 3:160018. https://doi.org/10.1038/sdata.2016.18
27. Bizer C, Heath T, Berners-Lee T (2009) Linked data-the story so far. Int J Semant Web Inf Syst 5(3):1–22
28. Domingue J, Fensel D, Hendler JA (eds) (2011) Handbook of semantic web technologies. Springer Science & Business Media, New York, NY
29. Hitzler P, Krotzsch M, Rudolph S (2009) Foundations of semantic web technologies. CRC press, Boca Raton, FL
30. Patel-Schneider PF (2005) A revised architecture for semantic web reasoning. In: International Workshop on Principles and Practice of Semantic Web Reasoning. Springer, Berlin, pp 32–36
31. Bratt S (2007) Semantic Web, and Other Technologies to Watch. https://www.w3.org/2007/Talks/0130-sb-W3CTechSemWeb/#(24). Accessed 15 Feb 2018
32. URN syntax, RFC2141 (1997) https://tools.ietf.org/html/rfc2141. Accessed 15 Feb 2018
33. URI Syntax, RFC3986 (2005) https://tools.ietf.org/html/rfc3986. Accessed 15 Feb 2018
34. RDF 1.1 Primer (2014) https://www.w3.org/TR/rdf11-primer/. Accessed 15 Feb 2018
35. Studer R, Benjamins VR, Fensel D (1998) Knowledge engineering: principles and methods. Data Know Eng 25(1-2):161–197
36. Baader F (ed) (2003) The description logic handbook: theory, implementation and applications. Cambridge university press, Cambridge
37. Ashburner M, Ball CA, Blake JA (2000) Gene ontology: tool for the unification of biology. Nat Genet 25(1):25. https://doi.org/10.1038/75556
38. Gene Ontology Consortium (2014) Gene ontology consortium: going forward. Nucleic Acids Res 43(D1):D1049–D1056. https://doi.org/10.1093/nar/gku1179
39. Dessimoz C, Gabaldón T, Roos DS et al (2012) Toward community standards in the quest for orthologs. Bioinformatics 28 (6):900–904. https://doi.org/10.1093/bioinformatics/bts050
40. Smith B, Ashburner M, Rosse C et al (2007) The OBO Foundry: coordinated evolution of ontologies to support biomedical data integration. Nat Biotechnol 25(11):1251
41. Federhen S (2011) The NCBI taxonomy database. Nucleic Acids Res 40(D1):D136–D143. https://doi.org/10.1093/nar/gkr1178
42. McBride B (2004) The resource description framework (RDF) and its vocabulary description language RDFS. In: Handbook on ontologies. Springer, Berlin, pp 51–65
43. OWL Web Ontology Language Overview (2004) https://www.w3.org/TR/owl-features/. Accessed 15 Feb 2018
44. Motik B, Patel-Schneider PF, Parsia B, Bock C, Fokoue A, Haase P, Hoekstra R, Horrocks I, Ruttenberg A, Sattler U, Smith M (2009) OWL 2 web ontology language: structural specification and functional-style syntax. W3C Recomm 27(65):159
45. OWL Web Ontology Language semantics and abstract syntax (2004) W3C Recommendation, 10. https://www.w3.org/TR/owl-semantics/. Accessed 15 Feb 2018
46. W3C Owl Working Group (2012) OWL 2 web ontology language document overview. https://www.w3.org/TR/owl2-overview/. Accessed 15 Feb 2018

47. Dean M, Schreiber G, Bechhoffer S, et al (2004) OWL web ontology language reference. W3C Recommendation. https://www.w3.org/TR/owl-ref/. Accessed 15 Feb 2018
48. Prud'hommeaux E, Carothers G (2014) RDF 1.1 Turtle: terse RDF triple language. W3C recommendation. http://www.w3.org/TR/2014/REC-turtle-20140225/. The latest edition is available at http://www.w3.org/TR/turtle/
49. World Wide Web Consortium (2014) RDF 1.1 XML Syntax. The latest edition is available at https://www.w3.org/TR/rdf-syntax-grammar/
50. World Wide Web Consortium (2014) JSON-LD 1.0: a JSON-based serialization for linked data. The latest edition is available at https://www.w3.org/TR/json-ld/
51. SPARQL Query Language for RDF (2008) https://www.w3.org/TR/rdf-sparql-query/. Accessed 15 Feb 2018
52. Lohmann S, Link V, Marbach E et al (2014) WebVOWL: web-based visualization of ontologies. In: International Conference on Knowledge Engineering and Knowledge Management. Springer, Cham, pp 154–158
53. Altenhoff AM, Glover NM, Train CM et al (2017) The OMA orthology database in 2018: retrieving evolutionary relationships among all domains of life through richer web and programmatic interfaces. Nucleic Acids Res 46(D1):D477–D485
54. Rakhmawati NA, Umbrich J, Karnstedt M et al (2013) A comparison of federation over SPARQL endpoints frameworks. In: International Conference on Knowledge Engineering and the Semantic Web. Springer, Berlin, pp 132–146
55. Calvanese D, Cogrel B, Komla-Ebri S et al (2017) Ontop: answering SPARQL queries over relational databases. Semant Web 8 (3):471–487
56. Sequeda JF, Miranker DP (2013) Ultrawrap: SPARQL execution on relational data. Web Semant 22:19–39
57. Bizer C, Seaborne A (2004) D2RQ-treating non-RDF databases as virtual RDF graphs. In: Proceedings of the 3rd international semantic web conference (ISWC2004). Springer, New York, NY
58. Gasteiger E, Gattiker A, Hoogland C et al (2003) ExPASy: the proteomics server for in-depth protein knowledge and analysis. Nucleic Acids Res 31(13):3784–3788. https://doi.org/10.1093/nar/gkg563
59. Saha D, Floratou A, Sankaranarayanan K et al (2016) Athena: an ontology-driven system for natural language querying over relational data stores. Proc VLDB Endow 9(12):1209–1220. https://doi.org/10.14778/2994509.2994536
60. Song D, Schilder F, Smiley C et al (2015) TR Discover: a natural language interface for querying and analyzing interlinked datasets. In: International Semantic Web Conference, 2. Springer, Cham
61. Ell B, Vrandečić D, Simperl E (2011) Labels in the web of data. Semant Web 2011:162–176. https://doi.org/10.1007/978-3-642-25073-6_11
62. Kinsella RJ, Kähäri A, Haider S et al (2011) Ensembl BioMarts: a hub for data retrieval across taxonomic space. Database 2011: bar030. https://doi.org/10.1093/database/bar030
63. Calvanese D, De Giacomo G, Lenzerini M (2001) Ontology of integration and integration of ontologies. Descr Logics 49(10-19):30
64. Bornea MA, Dolby J, Kementsietsidis A et al (2013) Building an efficient RDF store over a relational database. In: Proceedings of the 2013 ACM SIGMOD International Conference on Management of Data. ACM, Washington, DC, pp 121–132
65. R2RML: RDB to RDF mapping language (2012) https://www.w3.org/TR/r2rml/. Accessed 15 Feb 2018
66. Sequeda JF (2016) Integrating relational databases with the Semantic Web. IOS Press, Amsterdam
67. Michel F, Montagnat J, Zucker CF (2013) A survey of RDB to RDF translation approaches and tools. Dissertation. Inria Sophia Antipolis
68. Kaufmann E, Bernstein A (2007) How useful are natural language interfaces to the semantic web for casual end-users? In: The Semantic Web. Springer, Berlin, pp 281–294
69. Leis V, Gubichev A, Mirchev A et al (2015) How good are query optimizers, really? Proc VLDB Endow 9(3):204–215
70. Wu W, Chi Y, Zhu S et al (2013) Predicting query execution time: are optimizer cost models really unusable? In: Data Engineering (ICDE), 2013 IEEE 29th International Conference on 2013 Apr 8. IEEE, New York, NY, pp 1081–1092
71. Montoya G, Skaf-Molli H, Hose K (2017) The Odyssey approach for optimizing federated SPARQL queries. In: International Semantic Web Conference. Springer, Cham, pp 471–489
72. Davidson SB, Overton C, Buneman P (1995) Challenges in integrating biological data sources. J Comput Biol 2(4):557–572. https://doi.org/10.1089/cmb.1995.2.557

73. Stevens R, Baker P, Bechhofer S et al (2000) TAMBIS: transparent access to multiple bioinformatics information sources. Bioinformatics 16(2):184–186
74. Magkanaraki A, Alexaki S, Christophides V et al (2002) Benchmarking rdf schemas for the semantic web. In: International Semantic Web Conference. Springer, Berlin, pp 132–146
75. Wilkinson MD, Links M (2002) BioMOBY: an open source biological web services proposal. Brief Bioinform 3(4):331–341. https://doi.org/10.1093/bib/3.4.331
76. Stein LD (2003) Integrating biological databases. Nat Rev Genet 4(5):337. https://doi.org/10.1038/nrg1065
77. Apweiler R, Bairoch A, Wu CH (2004) Protein sequence databases. Curr Opin Chem Biol 8 (1):76–80. https://doi.org/10.1016/j.cbpa.2003.12.004
78. UniProt Consortium (2012) Update on activities at the Universal Protein Resource (UniProt) in 2013. Nucleic Acids Res 41(D1): D43–D47. https://doi.org/10.1093/nar/gks1068
79. Da Silveira M, Pruski C, Schneider R (2017) Data integration in the life sciences. Springer International Publishing AG, Cham
80. HCLS Linked Data Guide (2012) https://www.w3.org/2001/sw/hcls/notes/hcls-rdf-guide/. Accessed 15 Feb 2018
81. Côté RG, Jones P, Apweiler R et al (2006) The ontology lookup service, a lightweight cross-platform tool for controlled vocabulary queries. BMC Bioinformatics 7(1):97. https://doi.org/10.1186/1471-2105-7-97
82. Salvadores M, Alexander PR, Musen MA et al (2013) BioPortal as a dataset of linked biomedical ontologies and terminologies in RDF. Semant Web 4(3):277–284. https://doi.org/10.3233/SW-2012-0086
83. BioMoby Consortium (2008) Interoperability with Moby 1.0—it's better than sharing your toothbrush! Brief Bioinform 9(3):220–231. https://doi.org/10.1093/bib/bbn003
84. Antezana E, Blondé W, Egaña M et al (2009) BioGateway: a semantic systems biology tool for the life sciences. BMC Bioinformatics 10 (10):S11. https://doi.org/10.1186/1471-2105-10-S10-S11
85. Database Integration in the Life Sciences (2008) Briefings in Bioinformatics, Special Issue 9(6). https://academic.oup.com/bib/issue/9/6
86. Belleau F, Nolin MA, Tourigny N et al (2008) Bio2RDF: towards a mashup to build bioinformatics knowledge systems. J Biomed Inform 41(5):706–716. https://doi.org/10.1016/j.jbi.2008.03.004
87. Callahan A, Cruz-Toledo J, Ansell P et al (2013) Bio2RDF release 2: improved coverage, interoperability and provenance of life science linked data. In: Extended Semantic Web Conference. Springer, Berlin, pp 200–212. https://doi.org/10.1007/978-3-642-38288-8_14
88. Dumontier M, Callahan A, Cruz-Toledo J et al (2014) Bio2RDF release 3: a larger connected network of linked data for the life sciences. In: Proceedings of the 2014 International Conference on Posters & Demonstrations Track
89. Antezana E, Kuiper M, Mironov V (2009) Biological knowledge management: the emerging role of the Semantic Web technologies. Brief Bioinform 10(4):392–407. https://doi.org/10.1093/bib/bbp024
90. Chen H, Yu T, Chen JY (2012) Semantic web meets integrative biology: a survey. Brief Bioinform 14(1):109–125. https://doi.org/10.1093/bib/bbs014
91. Walls RL, Deck J, Guralnick R et al (2014) Semantics in support of biodiversity knowledge discovery: an introduction to the biological collections ontology and related ontologies. PLoS One 9(3):e89606

Chapter 23

High-Performance Computing in Bayesian Phylogenetics and Phylodynamics Using BEAGLE

Guy Baele, Daniel L. Ayres, Andrew Rambaut, Marc A. Suchard, and Philippe Lemey

Abstract

In this chapter, we focus on the computational challenges associated with statistical phylogenomics and how use of the broad-platform evolutionary analysis general likelihood evaluator (BEAGLE), a high-performance library for likelihood computation, can help to substantially reduce computation time in phylogenomic and phylodynamic analyses. We discuss computational improvements brought about by the BEAGLE library on a variety of state-of-the-art multicore hardware, and for a range of commonly used evolutionary models. For data sets of varying dimensions, we specifically focus on comparing performance in the Bayesian evolutionary analysis by sampling trees (BEAST) software between multicore central processing units (CPUs) and a wide range of graphics processing cards (GPUs). We put special emphasis on computational benchmarks from the field of phylodynamics, which combines the challenges of phylogenomics with those of modelling trait data associated with the observed sequence data. In conclusion, we show that for increasingly large molecular sequence data sets, GPUs can offer tremendous computational advancements through the use of the BEAGLE library, which is available for software packages for both Bayesian inference and maximum-likelihood frameworks.

Key words Adaptive Markov chain Monte Carlo, Multipartite data, Generalized linear model, High-performance computing, BEAGLE, BEAST, Pathogen phylodynamics, Data integration, Bayesian phylogenetics, Phylogenomics

1 Introduction

Phylogenomics, a term coined by Eisen and Fraser [13], explores the intersection of evolutionary studies and genomic analyses. Accurate phylogenetic reconstruction using genomic data has important repercussions for answering particular questions in genome analysis, as phylogenomic analyses often involve estimating the underlying evolutionary history of sequences either as an intermediate goal or as an end point. The availability of more and more complete genomes can help to correct for phylogenetic reconstruction artifacts and contradictory results that often appeared in

Maria Anisimova (ed.), *Evolutionary Genomics: Statistical and Computational Methods*, Methods in Molecular Biology, vol. 1910, https://doi.org/10.1007/978-1-4939-9074-0_23,

molecular phylogenies based on a single or few orthologous genes [21]. Expanding the number of characters that can be used in phylogenetic reconstruction from a few thousand to tens of thousands, these large quantities of data lead to reduced estimation errors associated with site sampling, to very high power in the rejection of simple evolutionary hypotheses and to high confidence in estimated phylogenetic patterns [4].

Among the phylogenetic reconstruction approaches that have attained widespread recognition, Bayesian inference has become increasingly popular, in large part due to the availability of open-source software packages such as the Bayesian evolutionary analysis by sampling trees (BEAST) software [11] and MrBayes [29]. Bayesian phylogenetic inference is based on a quantity called the posterior distribution of trees, which involves a summation over all trees and, for each tree, integration over all possible combinations of branch length and substitution model parameter values [20]. Analytical evaluation of this distribution is practically infeasible, and hence needs to be approximated using a numerical method, the most common being Markov chain Monte Carlo (MCMC). The basic idea is to construct a Markov chain that has as its state space the parameters of the statistical model and a stationary distribution that is the posterior distribution of the parameters (including the tree) [20]. While MCMC integration has revolutionized the field of phylogenetics [34], the continuously increasing size of data sets is pushing the field of statistical phylogenetics to its limits.

While promising approaches to improve MCMC efficiency have emerged recently from the field of computational statistics, such as sequential Monte Carlo (SMC; see, e.g., Doucet [10]) and Hamiltonian Monte Carlo (HMC; see, e.g., Neal [27]), these approaches do not yet find widespread use in phylogenetics. The primary difficulty in this adoption centers around the tree that encompasses both continuous and discrete random variables. Instead, considerable attention is being meted on techniques for parallelization [32] to improve phylogenetic software run-times. Obtaining sufficient samples from a Markov chain may take many iterations, due to the large number of trees that may describe the relationships of a group of species and high autocorrelation between the samples. It is therefore of critical importance to perform each iteration in a computationally efficient manner, making optimal use of the available hardware. High-performance computational libraries, such as the broad-platform evolutionary analysis general likelihood evaluator (BEAGLE) [3], can be useful tools to enable efficient use of multicore computer hardware (or even special-purpose hardware), while at the same time requiring minimal knowledge from the software user(s).

In this chapter, we first introduce the BEAGLE software library and its primary purpose, characteristics, and typical usages in Subheading 2, along with the hardware specifications of the devices used for benchmarks in this chapter. In Subheading 3, we present computational benchmarks on the different hardware devices for a collection of data sets that are typically analyzed with models of varying complexity. Subheading 4 presents a brief overview of studies for which GPU computing capabilities were critical to analyze the data in a timely fashion. Given the increasing capabilities over hardware devices, we present an interesting avenue for further research in Subheading 5, in the form of adaptive MCMC.

2 The BEAGLE Library

BEAGLE [3] is a high-performance likelihood-calculation platform for phylogenetic applications. BEAGLE defines a uniform application programming interface (API) and includes a collection of efficient implementations for evaluating likelihoods under a wide range of evolutionary models, on graphics processing units (GPUs) as well as on multicore central processing units (CPUs). The BEAGLE library can be installed as a shared resource, to be used by any software aimed at phylogenetic reconstruction that supports the library. This approach allows developers of phylogenetic software to share any optimizations of the core calculations, and any program that uses BEAGLE will automatically benefit from the improvements to the library. For researchers, this centralization provides a single installation to take advantage of new hardware and parallelization techniques.

The BEAGLE project has been very successful in bringing hardware acceleration to phylogenetics. The library has been integrated into popular phylogenetics software including BEAST [11], MrBayes [29], PhyML [19], and GARLI [35] and has been widely used across a diverse range of evolutionary studies. The BEAGLE library is free, open-source software licensed under the Lesser GPL and available at https://beagle-dev.github.io.

2.1 Principles

2.1.1 Computing Observed Data Likelihoods

The most effective methods for phylogenetic inference involve computing the probability of observed character data for a set of taxa given an evolutionary model and phylogenetic tree, which is often referred to as the (observed data) likelihood of that tree. Felsenstein demonstrated an algorithm to calculate this probability [16], and his algorithm recursively computes partial likelihoods via simple sums and products. These partial likelihoods track the probability of the observed data descended from an internal node conditional on a particular state at that internal node.

The partial likelihood calculations apply to a subtree comprising a parent node, two child nodes, and connecting branches. It is repeated for each unique site pattern in the data (in the form of a multiple sequence alignment), for each possible character of the state space (e.g., nucleotide, amino acid, or codon), and for each internal node in the proposed tree. The computational complexity of the likelihood calculation for a given tree is $O(p \times s^2 \times n)$, where p is the number of unique site patterns in the sequence (typically on the order of 10^2–10^6), s is the number of states each character in the sequence can assume (typically 4 for a nucleotide model, 20 for an amino-acid model, or 61 for a codon model), and n is the number of operational taxonomic units (e.g., species and alleles).

Additionally, the tree space is very large; the number of unrooted topologies possible for n operational taxonomic units is given by the double factorial function $(2n - 5)!!$ [15]. Thus, to explore even a fraction of the tree space, a very large number of topologies need to be evaluated, and hence a very great number of likelihood calculations have to be performed. This leads to analyses that can take days, weeks, or even months to run. Further compounding the issue, rapid advances in the collection of DNA sequence data have made the limitation for biological understanding of these data an increasingly computational problem. For phylogenetic inferences, the computation bottleneck is most often the calculation of the likelihoods on a tree. Hence, speeding up the calculation of the likelihood function is key to increasing the performance of these analyses.

2.1.2 Parallel Computation

Advances in computer hardware, specifically in parallel architectures, such as many-core GPUs, multicore CPUs, and CPU intrinsics (e.g., SSE and AVX), have created opportunities for new approaches to computationally intensive methods. The structure of the likelihood calculation, involving large numbers of positions and multiple states, as well as other characteristics, makes it a very appealing computational fit to these modern parallel processors, especially to GPUs.

BEAGLE exploits GPUs via fine-grained parallelization of functions necessary for computing the likelihood on a (phylogenetic) tree. Phylogenetic inference programs typically explore tree space in a sequential manner (Fig. 1, *tree space*) or with only a small number of sampling chains, offering limited opportunity for task-level parallelization. In contrast, the crucial computation of partial likelihood arrays at each node of a proposed tree presents an excellent opportunity for fine-grained data parallelism, which GPUs are especially suited for. The use of many lightweight execution threads incurs very low overhead on GPUs, enabling efficient parallelism at this level.

In order to calculate the overall likelihood of a proposed tree, phylogenetic inference programs perform a post-order traversal, evaluating a partial likelihood array at each node. When using BEAGLE, the evaluation of these multidimensional arrays is offloaded to the library. While each partial likelihood array is still

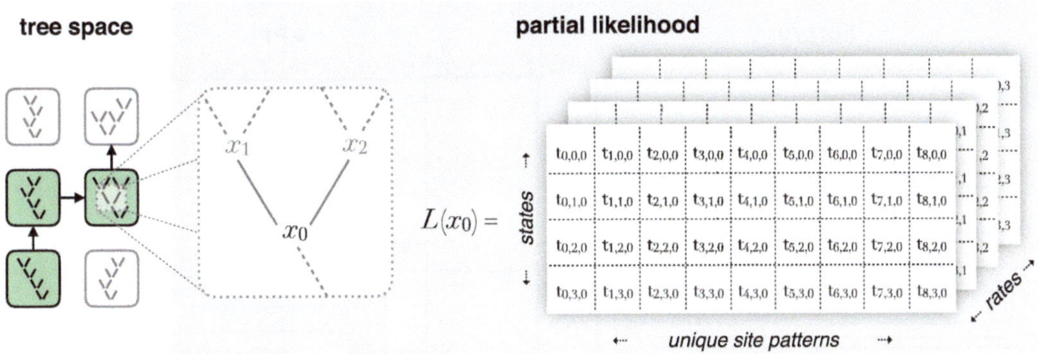

Fig. 1 Diagrammatic example of the tree sampling process and fine-grained parallel computation of phylogenetic partial likelihoods using BEAGLE for a nucleotide model problem with five taxa, nine site patterns, and four evolutionary rate categories. Each entry in a partial likelihood array L is assigned to a separate GPU thread t. In this example, 144 GPU threads are created to enable parallel evaluation of each entry of the partial likelihood array $L(x_0)$

evaluated in sequence, BEAGLE assigns the calculation of the array entries to separate GPU threads, for computation in parallel (Fig. 1, *partial likelihood*). Further, BEAGLE uses GPUs to parallelize other functions necessary for computing the overall tree likelihood, thus minimizing data transfers between the CPU and GPU. These additional functions include those necessary for computing branch transition probabilities, for integrating root and edge likelihoods, and for summing site likelihoods.

Multicore CPU parallelization through BEAGLE can only be done via multiple instances of the library, such that each instance computes a different data partition. Multiple CPU threads can be used (e.g., one for each partition) if the application program (BEAST, for the remainder of this chapter) creates the BEAGLE instances in separate computation threads, which will be the case when using BEAST. This approach suits the trend of increasingly large molecular sequence data sets, which are often heavily partitioned in order to better model the underlying evolutionary processes. BEAGLE itself does not employ any kind of load balancing nor are the site columns computed in individual threads. Each BEAGLE instance only parallelizes computation on CPUs via SSE vectorization.

BEAGLE can also use GPUs to perform partitioned analyses, however for problem sizes that are insufficiently large to saturate the capacity of one device, efficient computation requires multiple GPUs. Recent progress has been made in parallelizing the computation of multiple data subsets on one GPU [1], and future releases of BEAGLE will include this capability.

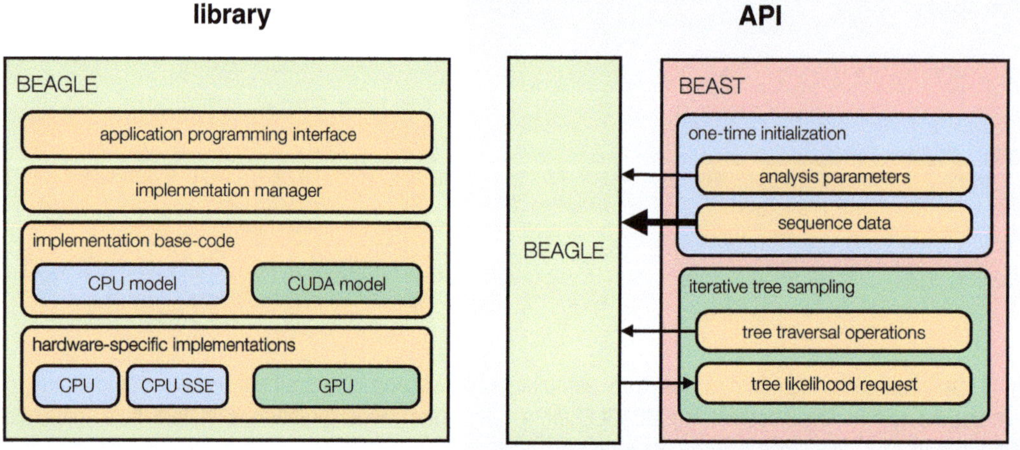

Fig. 2 Layer diagram depicting BEAGLE library organization, and illustration of API use. Arrows indicate direction and relative size of data transfers between the client program and library

2.2 Design

2.2.1 Library

The general structure of the BEAGLE library can be conceptualized as layers (Fig. 2, *library*), the upper most of which is the application programming interface. Underlying this API is an implementation management layer, which loads the available implementations, makes them available to the client program, and passes API commands to the selected implementation.

The design of BEAGLE allows for new implementations to be developed without the need to alter the core library code or how client programs interface with the library. This architecture also includes a plugin system, which allows implementation-specific code (via shared libraries) to be loaded at runtime when the required dependencies are present. Consequently, new frameworks and hardware platforms can more easily be made available to programs that use the library, and ultimately to users performing phylogenetic analyses.

Currently, the implementations in BEAGLE derive from two general models. One is a serial CPU implementation model, which does not directly use external frameworks. Under this model, there is a standard CPU implementation, and one with added SSE intrinsics, which uses vector processing extensions present in many CPUs to parallelize computation across character state values. The other implementation model involves an explicit parallel accelerator programming model, which uses the CUDA external computing framework to exploit NVIDIA GPUs. It implements fine-grained parallelism for evaluating likelihoods under arbitrary molecular evolutionary models, and thus harnessing the large number of processing cores to efficiently perform calculations [3, 32].

Recent progress has been made in developing new implementations for BEAGLE, beyond those described here, thus expanding the range of hardware that can be used. Upcoming releases of the library will include additional support for CPU parallelism via a multi-threaded implementation and will support the OpenCL standard, enabling the use of AMD GPUs [2].

2.2.2 Application Programming Interface

The BEAGLE API was designed to increase performance via fine-scale parallelization while reducing data transfer and memory copy overhead to an external hardware accelerator device (e.g., GPU). Client programs, such as BEAST [11], use the API to offload the evaluation of tree likelihoods to the BEAGLE library (Fig. 2, *API*). API functions can be subdivided into two categories: those which are only executed once per inference run and those which are repeatedly called as part of the iterative sampling process. As part of the one-time initialization process, client programs use the API to indicate analysis parameters such as tree size and sequence length, as well as specifying the type of evolutionary model and hardware resource(s) to be used. This allows BEAGLE to allocate the appropriate number and size of data buffers on device memory. Additionally at this initialization stage, the sequence data is specified and transferred to device memory. This costly memory operation is only performed once, thus minimizing its impact.

During the iterative tree sampling procedure, client programs use the API to specify changes to the evolutionary model and instruct a series of partial likelihood operations that traverse the proposed tree in order to find its overall likelihood. BEAGLE efficiently computes these operations and makes the overall tree likelihood as well as per-site likelihoods available via another API call.

2.3 Performance

Peak performance with BEAGLE is achieved when using a high-end GPU; however, the relative gain over using a CPU depends on model type and problem size as more demanding analyses allow for better utilization of GPU cores. Figure 3 shows speedups relative to serial CPU code when using BEAGLE with an NVIDIA P100 GPU for the critical partial likelihood function, with increasing unique site pattern counts and for two model types. Computing these likelihoods typically accounts for over 90% of the total execution time for phylogenetic inference programs and the relationship between speedups and problem size observed here primarily matches what would be observed for a full analysis.

Figure 3 includes performance results for computing partial likelihoods under both nucleotide and codon models. The vertical axis labels show the speedup relative to the average performance of a baseline serial, single threaded and non-vectorized, CPU implementation. This nonparallel CPU implementation provides a consistent performance level across different problem sizes and

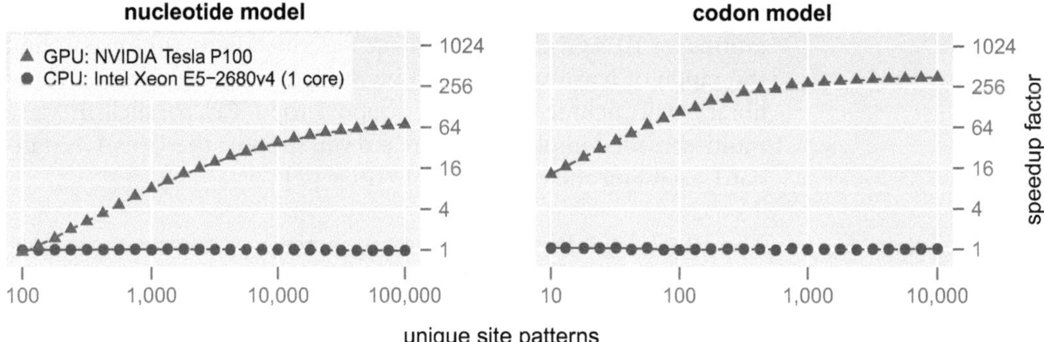

Fig. 3 Plots showing BEAGLE partial likelihood computation performance on the GPU relative to serial CPU code, under nucleotide and codon models and for an increasing number of unique site patterns. Speedup factors are on a log-scale

provides a relevant point of comparison as most phylogenetic inference software packages use serial code as their standard.

Using a nucleotide model, relative GPU performance over the CPU strongly scales with the number of site patterns. For very small numbers of patterns, the GPU exhibits poor performance due to greater execution overhead relative to overall problem size. GPU performance improves quickly as the number of unique site patterns is increased and by 10,000 patterns it is closer to a saturation point, continuing to increase but more slowly. At 100,000 nucleotide patterns, the GPU is approximately 64 times faster than the serial CPU implementation.

For codon-based models, GPU performance is less sensitive to the number of unique site patterns. This is due to the better parallelization opportunity afforded by the 61 biologically meaningful states that can be encoded by a codon. The higher state count of codon data compared to nucleotide data increases the ratio of computation to data transfer, resulting in increased GPU performance for codon-based analyses. For a problem size with 10,000 codon patterns, the GPU is over 256 times faster than the serial CPU implementation.

2.4 Memory Usage

When assessing the suitability of a phylogenetic analysis for GPU acceleration via BEAGLE, it is also important to consider if the GPU has sufficient on-board memory for the analysis to be performed. GPUs typically have less memory than what is available to CPUs, and the high transfer cost of moving data from CPU to GPU memory prevents direct use of CPU memory for GPU acceleration.

Figure 4 shows how much memory is required for problems of different sizes when running nucleotide and codon-model analyses in BEAST with BEAGLE GPU acceleration. Note that when multiple GPUs are available, BEAST can partition a data set into

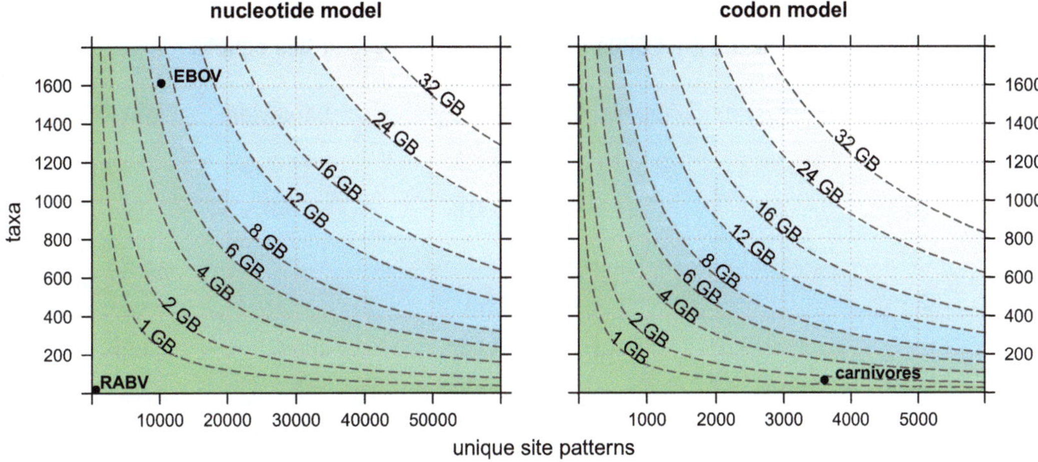

Fig. 4 Contour plots depicting BEAGLE memory usage on GPUs for BEAST nucleotide and codon-model analyses with 4 rate categories in double-precision floating-point format, for a range of problem sizes with different numbers of taxa and of unique site patterns. Memory requirements shown here assume an unpartitioned dataset. Partitioned analyses and more sophisticated models that use multiple BEAGLE instances incur memory overhead per additional library instance. Black dots indicate memory usage requirements for the unpartitioned version of three data sets subsequently described in this chapter

separate BEAGLE instances, one for each GPU. Thus, each GPU will only require as much memory as necessary for the data subset assigned to it. Typical PC-gaming GPUs have 8 GB of memory or less, while GPUs dedicated to high-performance computing, such as the NVIDIA Tesla series, may have as much as 24 GB of memory.

2.5 Hardware

Highly parallel computing technologies such as GPUs have overtaken traditional CPUs in peak performance potential and continue to advance at a faster pace. Additionally, the memory bandwidth available to the processor is especially relevant to data-intensive computations, such as the evaluation of nucleotide model likelihoods. In this measure as well, high-end GPUs significantly outperform equivalently positioned CPUs.

BEAGLE was designed to take advantage of this trend of increasingly advanced GPUs and uses runtime compilation methods to optimize code for whichever generation of hardware is being used. Table 1 lists hardware specifications for the processors used in this chapter. We note that further advancements in the GPU market for scientific computing are on its way, with NVIDIA preparing the launch (at the time of writing) of the Tesla V100 in Q3 of 2017. The new NVIDIA Tesla V100 features a total of 5120 CUDA cores and comes equipped with 32 GB of on-board memory with 900 GB/s of bandwidth. As such, it seems to have the potential to reach 7.5 TFLOPs in double-precision peak performance (DP PP), a roughly 50% increase over their current flagship, the Tesla P100.

Table 1
Hardware specifications for the Intel CPUs and NVIDIA GPUs used in this chapter

Hardware	Year	Cores	Memory	Bandwidth	DP PP
Xeon E5-2680v2	2013	2 × 10	64 GB	60 GB/s	0.45 TFLOPS
Xeon E5-2680v3	2014	2 × 12	64 GB	68 GB/s	0.96 TFLOPS
GTX 590	2011	2 × 512	2 × 1.5 GB	164 GB/s	0.31 TFLOPS
Tesla K20X	2012	2688	6 GB	250 GB/s	1.31 TFLOPS
Tesla K40	2013	2880	12 GB	288 GB/s	1.43 TFLOPS
Quadro P5000	2016	2560	16 GB	288 GB/s	NA
Tesla P100	2016	3584	16 GB	720 GB/s	4.70 TFLOPS

Estimated performance in double-precision peak performance (DP PP) taken from the manufacturer's website. Note that the Quadro P5000 GPU only lists performance in single precision and we hence list its double-precision performance as not available (NA)

3 Results

In this section, we compare the performance of various typical Bayesian phylogenetic, phylogenomic, and phylodynamic analyses on different multicore architectures. In Subheading 3.1, we analyze a data set of mitochondrial genomes [32] using a high-dimensional model of codon substitution which, albeit low in number of parameters, is particularly challenging in phylogenetic analyses specifically because of the high-dimensional state space. In Subheading 3.2, we analyze the largest Ebola virus data set at the time of publication [12] using a collection of nucleotide substitution models, i.e., one per codon position and an extra one for analyzing the intergenic regions, where the large number of taxa and unique site patterns offer an interesting test case for the comparison between CPU and GPU performance. Finally, Subheading 3.3 reports on the performance of analyzing data sets that complement sequence data with discrete trait data (typically host data or geographic data), for which transition rates between (a potential large number of) discrete trait states are parameterized as a generalized linear model (GLM). All performance evaluations in this results section were run for 100,000 iterations (which is usually insufficient to achieve convergence) in BEAST v1.8.4 [11], using double precision (both on CPU and GPU) and in conjunction with BEAGLE v2.1.2 [3]. By default, BEAST—through BEAGLE—uses SSE2 (Streaming SIMD Extensions 2), an SIMD instruction set extension to the x86 architecture, when performing calculations on CPU.

3.1 Carnivores

Selection is a key evolutionary process in shaping genetic diversity and a major focus of phylogenomics investigations [23]. Researchers frequently evaluate the strength of selection operating on genes or even individual codons in the entire phylogeny or in a subset of branches using statistical methods. Codon substitution models have been particularly useful for this purpose because they allow estimating the ratio of non-synonymous and synonymous substitution rates (dN/dS) in a phylogenetic framework. Goldman and Yang [18] and Muse and Gaut [26] developed the first codon-based evolutionary models (GY and MG, respectively), i.e., models that have codons as their states, incorporating biologically meaningful parameters such as transition/transversion bias, variability of a gene, and amino acid differences.

Full codon substitution models are computationally expensive compared to standard nucleotide substitution models due to their large state space. Compared to nucleotide models (4 states) and amino acid models (20 states), a full vertebrate mitochondrial codon model has 60 states (ignoring the four nonsense or stop codons). We restrict ourselves to the standard GY codon substitution model implementation in BEAST [11], employ the standard assumption that mutations occur independently at the three codon positions and therefore only consider substitutions that involve a single-nucleotide substitution, and assume that codons evolve independently from one another. Additionally, we allow for substitution rate heterogeneity among codons using a discrete gamma distribution (i.e., each codon is allowed to evolve at a different substitution rate) [33], which increases the computational demands of such an analysis fourfold (given that we allow for the standard assumption of four discrete rate categories).

As a first application of using state-of-the-art hardware in statistical phylogenetics, we reevaluate the performance of a full codon model on a set of mitochondrial genomes from extant carnivores and a pangolin outgroup [4, 32]. This genomic sequence alignment contains 10,869 nt columns that code for 12 mitochondrial proteins and when translated into a single 60-state vertebrate mitochondrial codon model, yields a total of 3623 alignment columns, of which 3601 site patterns are unique [32]. Figure 5 shows a comparison of the computational throughput between various CPU and GPU computing platforms. To this end, we make use of an option in BEAST [11] to split an alignment into two or more pieces of equal length, with each resulting alignment being evaluated on a separate processor core or computing device for optimal performance. Figure 5 shows that the analysis scales remarkably well on CPU, where the use of each additional processor core results in a performance increase. This can be attributed to the use of full codon models, which invokes a higher workload when evaluating each likelihood and hence more concurrent evaluation compared to thread communication. As the evaluation of the total

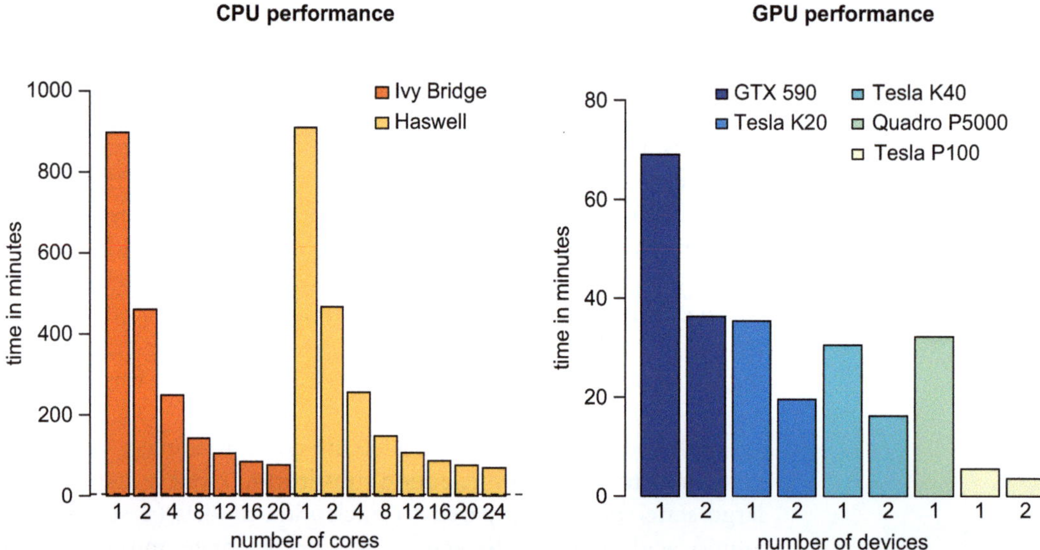

Fig. 5 Performance evaluation of different CPU and GPU configurations when estimating a single full codon substitution model on a full genome mitochondrial data set. Left: performance comparison on multicore CPUs (with the dashed line indicating optimal GPU performance). Right: performance comparison on GPUs. The numbers below the bars indicate the number of partitions/threads used in each analysis. The GTX 590, a GPU released in 2011 for PC gaming, performs equally well as a 24-core CPU configuration. However, GPUs from the Tesla generation, aimed at scientific computing, drastically outperform the CPU system, with the recently released P100 showcasing the impressive improvements in the GPU market, running 166 times faster than a single-core CPU setup

amount of unique site patterns is split over more processor cores, the workload per core decreases and the communication overhead increases, resulting in smaller relative performance increases.

The performance of a 24-core CPU setup is easily matched by a single GPU (the GTX 590) that was originally aimed at the market of PC gaming. However, subsequent improvements in GPU cards for scientific computing have yielded impressive performance gains, with a single Tesla K20 GPU outperforming 2 GTX 590 GPUs. Whereas the advent of the Tesla K40 offered further performance increases, it was mainly welcomed for having twice the amount of on-board memory, allowing for much larger data sets to be analyzed on GPU. The recent introduction of the Tesla P100 GPU promised and delivered astonishing results, as shown in Fig. 5, with a single Tesla P100 GPU delivering six times the performance of a Tesla K40 GPU on these high-dimensional full codon models. We conclude that the use of a high-performance computational library such as BEAGLE, in combination with a powerful GPU, has significantly facilitated the evaluation of and phylogenetic inference with full codon models.

3.2 Ebola Virus

The original developments within the BEAGLE library offered considerable computational speedup when evaluating codon models—up to a 52-fold increase when employing three GPU cards—and nucleotide models—up to a 15-fold increase when using three GPU cards—in double precision [32]. This may have resulted in the perception that GPU cards are mainly useful when evaluating codon models, but that the benefit for fitting models was not sufficiently substantial to warrant GPUs. To offer an objective assessment of the usefulness of GPUs in such cases, we analyze the use of various CPU and GPU configurations on a full genome Ebola virus data set, consisting of 1610 publicly available genomes sampled over the course of the 2013–2016 Ebola virus disease epidemic in West Africa [12] (we discuss this study in more detail in Subheading 4). This data set encompassing 18,992 nt columns is modelled with four partitions: one for each codon position and one additional partition for the intergenic region (which consists of several noncoding regions interspersed in the genome). The three codon partitions contain, respectively, 2366, 2328, and 2731 unique site patterns, while the intergenic partition contains 2785 unique site patterns. We model among-site rate variation [33] in each partition independently, which confronts us with a computationally demanding analysis for this large number of taxa and unique site patterns.

Figure 6 shows how the performance of such a large nucleotide data set scales with the available CPU and GPU resources. Contrary to the carnivores data set analysis in Subheading 3.1, this analysis does not scale particularly well with the number of CPU cores available, as the main benefit lies with splitting each partition into two subpartitions and only limited performance gains can be observed when using additional partitions or threads. Popular single GPU cards for scientific computing—such as the Tesla K40—match the optimal performance brought about by using 16 CPU cores, and may provide a useful alternative to multicore CPU systems. However, the decreasing cost for increasingly parallel multicore CPU systems makes this a difficult matchup for slightly older GPUs. More recently introduced GPU cards, such as the Tesla P100, are able to deliver a substantial performance improvement over a multicore CPU setup, with two Tesla P100 GPUs running in parallel offering over twice the performance of a 16-core CPU setup. We note that the GTX 590 cards, as well as a single Tesla K20 card, do not contain sufficient on-board memory to hold the full data set and as such, these benchmarks could not be run on those resources.

3.3 Phylogeography

As shown in the results in Subheadings 3.1 and 3.2, different partitions of the aligned sequence data can contain a large number of unique site patterns, rendering phylogenomic inference challenging. However, other data types are also included more

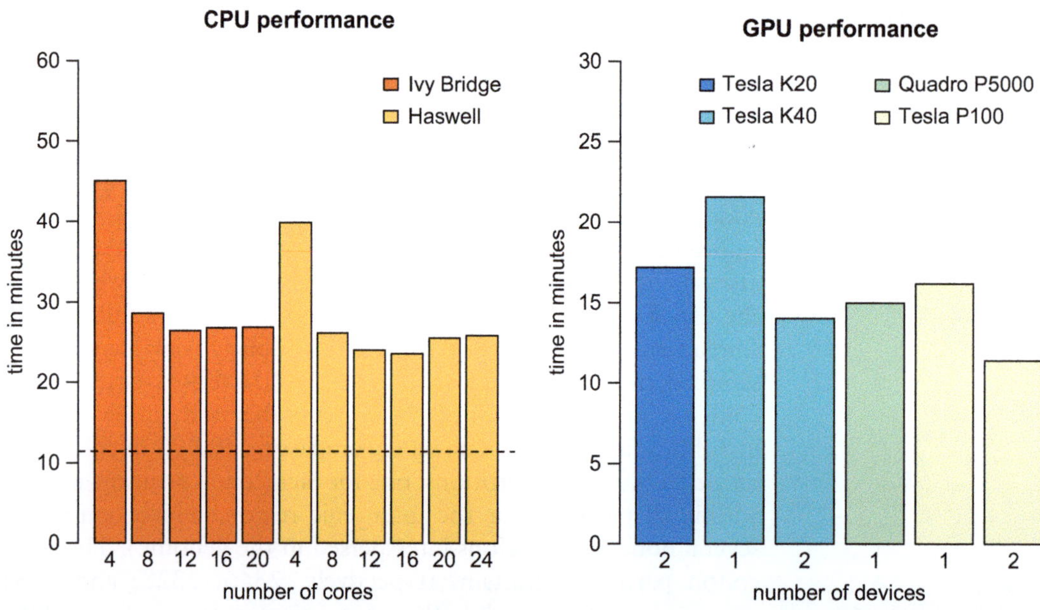

Fig. 6 Performance evaluation of different CPU and GPU configurations when estimating a four-partition nucleotide substitution model on a full genome Ebola virus data set. Left: performance comparison on multicore CPUs (with the dashed line indicating optimal GPU performance). Right: performance comparison on GPUs. The numbers below the bars indicate the number of partitions/threads used in each analysis. Because of the lower dimensionality compared to full codon models, nucleotide substitution models currently take less advantage of the large amounts of cores present on GPUs. The fastest GPU configuration—consisting of two Tesla P100 GPUs—outperforms by 107% the fastest CPU configuration that employs 16 threads on a 24-core Haswell CPU

frequently, such as trait data to be analyzed alongside the sequence data and hence potentially influencing the outcome of such an analysis (for an overview, see, e.g., Baele et al. [6]).

Arguably, the most frequently considered traits in phylodynamics, and molecular evolution in general, are spatial locations. The interest in spatial dispersal has developed into its own research field referred to as phylogeography, with Bayesian inference of discrete phylogenetic diffusion processes being adopted in the field of biogeography [30]. Jointly estimating the phylogeny and the trait evolutionary process, Lemey et al. [23] implemented a similar Bayesian full probabilistic connection between sequences and traits in BEAST [11], with applications focusing on spatiotemporal reconstructions of viral spread. These approaches offer extensive modelling flexibility at the expense of a quadratic growth in number of instantaneous rate parameters in the continuous-time Markov chain (CTMC) model as a function of the state dimensionality of the trait. This can be seen in Fig. 7, which shows two maps with different numbers of (discrete) locations and the corresponding CTMC models that describe the instantaneous rates of transition between these locations.

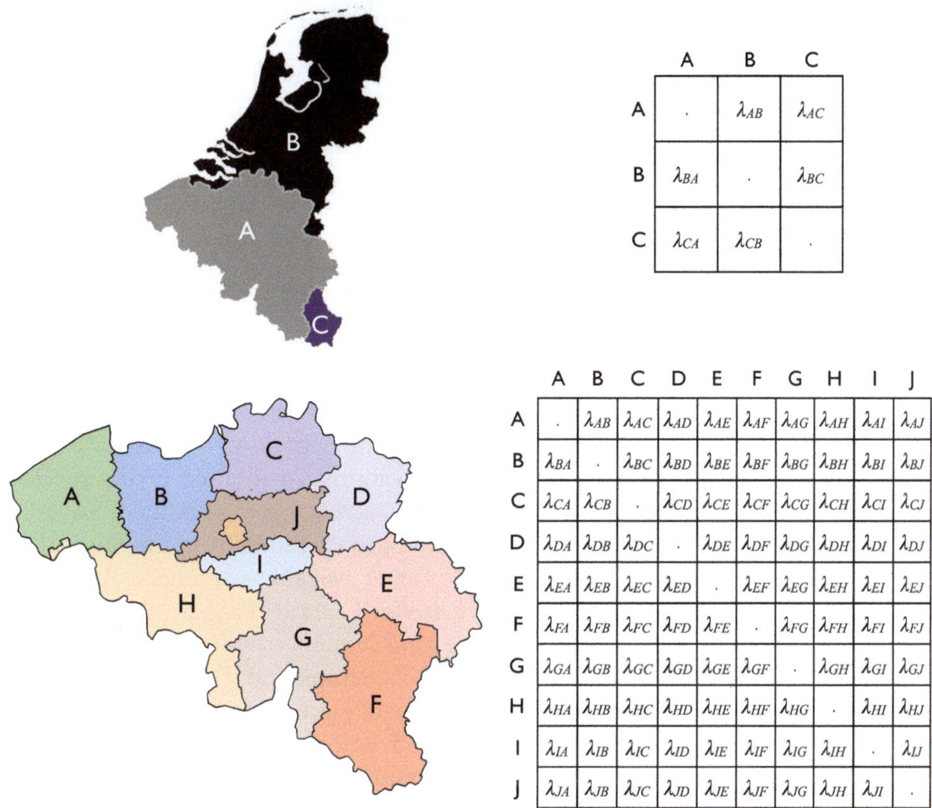

	A	B	C
A	.	λ_{AB}	λ_{AC}
B	λ_{BA}	.	λ_{BC}
C	λ_{CA}	λ_{CB}	.

	A	B	C	D	E	F	G	H	I	J
A	.	λ_{AB}	λ_{AC}	λ_{AD}	λ_{AE}	λ_{AF}	λ_{AG}	λ_{AH}	λ_{AI}	λ_{AJ}
B	λ_{BA}	.	λ_{BC}	λ_{BD}	λ_{BE}	λ_{BF}	λ_{BG}	λ_{BH}	λ_{BI}	λ_{BJ}
C	λ_{CA}	λ_{CB}	.	λ_{CD}	λ_{CE}	λ_{CF}	λ_{CG}	λ_{CH}	λ_{CI}	λ_{CJ}
D	λ_{DA}	λ_{DB}	λ_{DC}	.	λ_{DE}	λ_{DF}	λ_{DG}	λ_{DH}	λ_{DI}	λ_{DJ}
E	λ_{EA}	λ_{EB}	λ_{EC}	λ_{ED}	.	λ_{EF}	λ_{EG}	λ_{EH}	λ_{EI}	λ_{EJ}
F	λ_{FA}	λ_{FB}	λ_{FC}	λ_{FD}	λ_{FE}	.	λ_{FG}	λ_{FH}	λ_{FI}	λ_{FJ}
G	λ_{GA}	λ_{GB}	λ_{GC}	λ_{GD}	λ_{GE}	λ_{GF}	.	λ_{GH}	λ_{GI}	λ_{GJ}
H	λ_{HA}	λ_{HB}	λ_{HC}	λ_{HD}	λ_{HE}	λ_{HF}	λ_{HG}	.	λ_{HI}	λ_{HJ}
I	λ_{IA}	λ_{IB}	λ_{IC}	λ_{ID}	λ_{IE}	λ_{IF}	λ_{IG}	λ_{IH}	.	λ_{IJ}
J	λ_{JA}	λ_{JB}	λ_{JC}	λ_{JD}	λ_{JE}	λ_{JF}	λ_{JG}	λ_{JH}	λ_{JI}	.

Fig. 7 Graphical depiction of the Benelux countries (top) and the provinces of Belgium (bottom). When these countries and provinces are used as discrete location states in a discrete trait model, this yields, respectively, a 3 × 3 and a 10 × 10 CTMC model with instantaneous rates of transition between each pair of locations. Such models are subject to the same restrictions as those used in popular substitution models, i.e., the rows sum to 0. As such, these CTMC models consist of, respectively, 6 and 90 free parameters to be estimated

Many phylodynamic hypotheses can be addressed through the combination of genetic and trait data, but additional data in the form of covariates can help explain the evolutionary or epidemiological process. Such covariates can be used in a GLM formulation on a matrix of transition rate parameters between locations defining a CTMC process. Lemey et al. [25] developed an approach to simultaneously reconstruct spatiotemporal history and identify which combination of covariates associates with the pattern of spatial spread. This approach involves parameterizing each rate of among-location movement, typically denoted as the *ij*th elements (λ_{ij}) of the CTMC transition rate matrix, in the phylogeographic model as a log linear function of various potential covariates:

$$\log\lambda_{ij} = \beta_1\delta_1 x_{i,j,1} + \beta_2\delta_2 x_{i,j,2} + \cdots + \beta_N\delta_N x_{i,j,N}, \tag{1}$$

where β_i is the estimated effect size of covariate x_i, δ_i is a binary indicator that tracks the posterior probability of the inclusion of covariate x_i in the model, and N equals the number of covariates; further, in the case of Fig. 7: $i, j \in \{A, B, C\}$ with $N = 3$ (top), and $i, j \in \{A, B, C, D, E, F, G, H, I, J\}$ with $N = 10$ (bottom). Priors and posteriors for the inclusion probabilities (δ) can be used to express the support for each predictor in terms of Bayes factors (for more information, *see* Baele et al. [6], Lemey et al. [25]). We discuss examples of such possible predictors in a phylogeographic setting in Subheading 4 but focus here on performance benchmarks for such generalized linear models.

3.3.1 Bat Rabies

We here assess the performance of a phylodynamic setup aimed at reconstructing the spatial dispersal and cross-species dynamics of rabies virus (RABV) in North American bat populations based on a set of 372 nucleoprotein gene sequences (nucleotide positions: 594–1353). The data set comprises a total of 17 bat species sampled between 1997 and 2006 across 14 states in the USA [31]. Two additional species that had been excluded from the original analysis owing to a limited amount of available sequences, *Myotis austroriparius* (Ma) and *Parastrellus hesperus* (Ph), are also included here [14]. We also include a viral sequence with an unknown sampling date (accession no. TX5275, sampled in Texas from *Lasiurus borealis*), which will be adequately accommodated in our inference. This leads to a total of 548 unique site patterns. Following Faria et al. [14], we employ two GLM-diffusion models for this analysis, one on the discrete set of 17 bat species and another on the discrete set of 14 location states.

Figure 8 shows the performance of various multicore platforms on the bat rabies Bayesian phylodynamic analysis. In contrast to previous examples, the low number of sites (and hence unique site patterns) in the alignment does not offer many options for splitting the observed data likelihood over additional threads. While four CPU cores offer the optimal performance across our CPU platforms, using more threads for the analysis causes serious communication overhead, slowing down the analysis. Comparing the CPU results with the GPU results shows that, across all multicore platforms tested, a 4-core CPU offers the best performance.

Nonetheless, this scenario provides a very interesting use case for employing multiple graphics cards for scientific computing. Even though the (relatively small) dimensions of this particular example do not allow for a performance increase, it will be beneficial for higher-dimensional cases to compute each diffusion model on a separate GPU. When assuming independent diffusion processes that only depend on the underlying phylogeny, each of the trait diffusion models can be computed on a different GPU, whereas the data alignment can be split into two subpartitions of equal complexity (i.e., with an equal number of unique site patterns) and hence also be computed in parallel over the two GPUs. However, the limited sequence data size and the relatively restricted

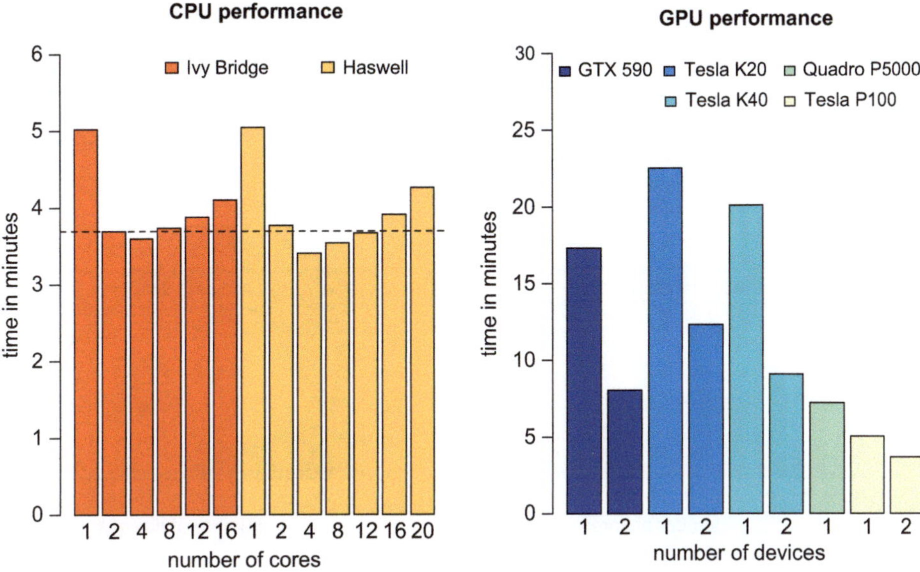

Fig. 8 Performance evaluation of different CPU and GPU configurations when estimating a single-partition nucleotide substitution model on a rabies virus data set, along with two GLMs. Left: performance comparison on multicore CPUs (with the dashed line indicating optimal GPU performance). Right: performance comparison on GPUs. Given the short length of the alignment, communication overhead becomes quite severe when adding a large number of CPU processor cores. Even two state-of-the-art Tesla P100 GPUs fail to outperform four CPU cores, given the low number of sequences and the low number of states for the GLMs

number of discrete locations make this data set less suited for illustrating performance increases using GPUs.

3.3.2 Ebola Virus

We here assess the performance of a similar setup as in the previous section, but using the data from the 2013–2016 West African epidemic caused by the Ebola virus [12]. We hence use the nucleotide data from the previous Ebola example (*see* Subheading 3.2) and augment it with location states. Using a phylogeographic GLM that integrates covariates of spatial spread, we have examined which features influenced the spread of EBOV among administrative regions at the district (Sierra Leone), prefecture (Guinea), and country (Liberia) levels. This resulted in a GLM parameterizing transition rates among 56 discrete location states according to 25 potential covariates (*see* Dudas et al. [12] for more information), resulting in a computationally challenging analysis.

As shown in Fig. 9, we have evaluated the performance of this challenging data set on our different multicore platforms. By comparing these benchmarks with those in Fig. 6, it's clear that the addition of a high-dimensional discrete trait model is much harder to process for any multicore CPU configuration. Adding more CPU cores to the analysis does not improve performance by much, indicative of the discrete trait model being the main

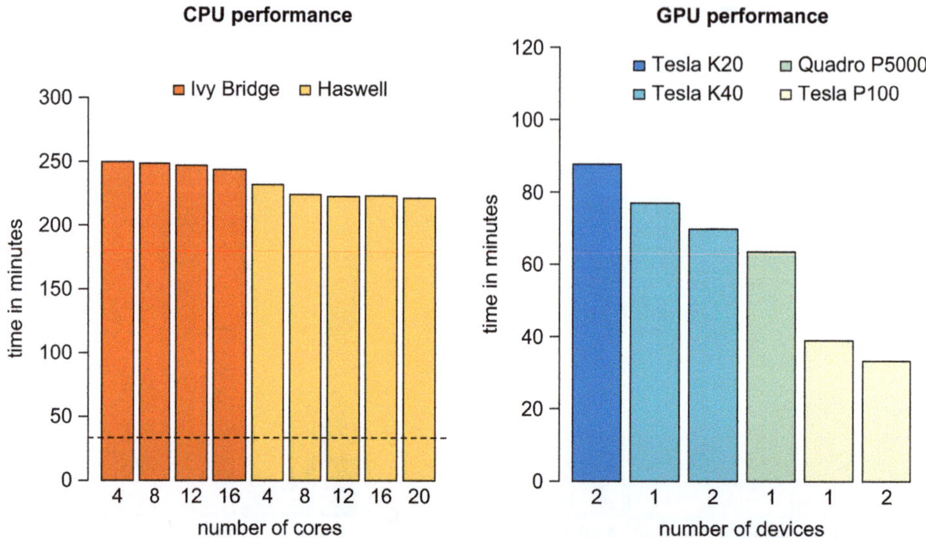

Fig. 9 Performance evaluation of different CPU and GPU configurations when estimating a four-partition nucleotide substitution model on a full genome Ebola virus data set along with a GLM-diffusion model for the location traits. Left: performance comparison on multicore CPUs (with the dashed line indicating optimal GPU performance). Right: performance comparison on GPUs. While a previous comparison of solely the nucleotide data resulted in a performance increase of over 100% for GPUs over CPUs, we observe a far more pronounced benefit for GPU in this case with two Tesla P100 GPUs outperforming a 12-core CPU setup by 568%

bottleneck in this analysis. This can be attributed to the high dimension of the discrete phylogeographic model [23] and the fact that this model describes a single column of characters, which cannot be split into multiple partitions. Relative to the computational complexity of modelling the location states, splitting the observed sequence data over multiple partitions/threads yields relatively small performance improvements.

Some of the (older) GPUs cannot fit the full data set in memory (such as the GTX 590 and a single Tesla K20), but those that are able to vastly outperform any CPU setup. Further, as these GPUs are better equipped to handle high-dimensional models, splitting the observed sequence data over multiple physical cards still yields noticeable performance gains. In contrast to Fig. 8, where two discrete phylogeographic models were used that could each be computed on different GPUs, the fact that this example only considers a single trait observation explains why less performance gains can be obtained by adding an additional GPU to the analysis.

4 Examples

In this section, we highlight examples of large sequence data sets that are augmented with trait data in the form of discrete geographic locations, for which BEAGLE [3] offers impressive computational benefits, specifically when running these analyses on powerful graphics cards for scientific computing. Further, discrete phylogeographic models can be equipped with generalized linear models to identify predictors of pathogen spread. Both inclusion probabilities and conditional effect sizes for these predictors are estimated in order to determine support for such explanatory variables of (pathogen) spread.

4.1 Human Influenza H3N2

A potentially powerful predictor for the behavior of influenza and other infectious diseases comes in the form of information on global human movement patterns, of which the worldwide air transportation network is by far the best studied system of global mobility in the context of human infectious diseases [8].

Lemey et al. [25] use a discrete phylogeographic model equipped with a GLM to show that the global dynamics of influenza H3N2 are driven by air passenger flows, whereas at more local scales spread is also determined by processes that correlate with geographic distance. For a data set that encompasses 1441 time-stamped hemagglutinin sequences (sampled between 2002 and 2007) and up to 26 locations to be used in a discrete phylogeographic model equipped with a GLM, BEAGLE can offer substantial performance gains. A snapshot of a visual reconstruction through geographic space is presented in Fig. 10, which includes a summary of the support for the collection of covariates in the

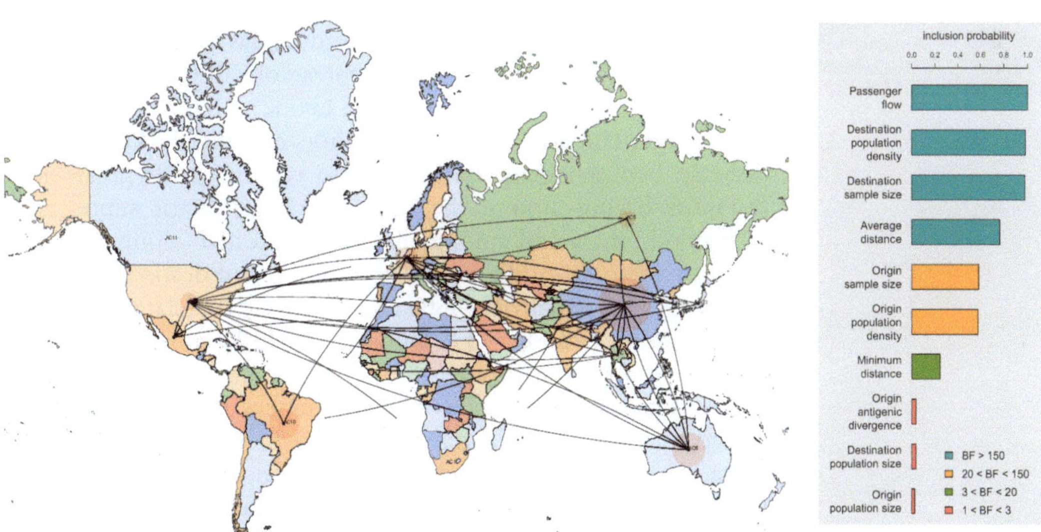

Fig. 10 Snapshot of the geographic spread of human influenza subtype H3N2, based on 1441 hemagglutinin sequences sampled between 2002 and 2007 [25]. A discrete phylogeographic approach was used, allocating the sequence data into a discrete number of locations and employing a generalized linear model on the parameters that model geographic spread. Inclusion probabilities and Bayes factor support are shown for the most prominent predictors of H3N2 geographic spread. D3 visualization is made using SpreaD3 [7], with circular polygon areas proportional to the number of tree lineages maintaining that location at that time

GLM that offer the strongest contribution to spatial spread among those tested. As illustrated in Fig. 10 (but *see* Lemey et al. [25] for additional data), there is strong evidence that air passenger flow is among the most dominant drivers of the global dissemination of H3N2 influenza viruses. Further, geographic spread is found to be inversely associated (data not shown; but *see* Lemey et al. [25]) with geographical distance between locations and with origin and destination population densities, which may seem counterintuitive. As the authors state, this negative association of population density with viral movement may suggest that commuting is less likely, per capita, to occur out of, or into, dense subpopulations.

4.2 Ebola Virus

During the two and a half years Ebola virus (EBOV) circulated in West Africa, it caused at least 28,646 cases and 11,323 deaths. As mentioned in Subheading 3.3.2, Dudas et al. [12] used 1610 genome sequences collected throughout the epidemic, representing over 5% of recorded Ebola virus disease (EVD) cases to reconstruct a detailed phylogenetic history of the movement of EBOV within and between the three most affected countries. This study considers a massive time-stamped data set that allows to uncover regional patterns and drivers of the epidemic across its entire duration, whereas individual studies had previously focused on either limited geographical areas or time periods. The authors use the phylogeographic GLM to test which features were important in shaping the spatial dynamics of EVD during the West African epidemic (*see* Fig. 11).

The phylogeographic GLM allowed Dudas et al. [12] to determine the factors that influenced the spread of EBOV among administrative regions at the district (Sierra Leone), prefecture (Guinea), and country (Liberia) levels. The authors find that EBOV tends to disperse between geographically close regions, with great circle distances having among the strongest Bayes factor support for inclusion in the GLM among all covariates tested (along with four other predictors). Additionally, both origin and destination population sizes are equally strongly and positively correlated with viral dissemination (*see* Fig. 11). Dudas et al. [12] conclude that the combination of the positive effect of population sizes with the inverse effect of geographic distance implies that the epidemic's spread followed a classic gravity-model dynamic, with intense dispersal between larger and closer populations. Finally, the authors found a significant propensity for virus dispersal to occur within each country, relative to internationally, suggesting that country borders may have provided a barrier for the geographic spread of EBOV.

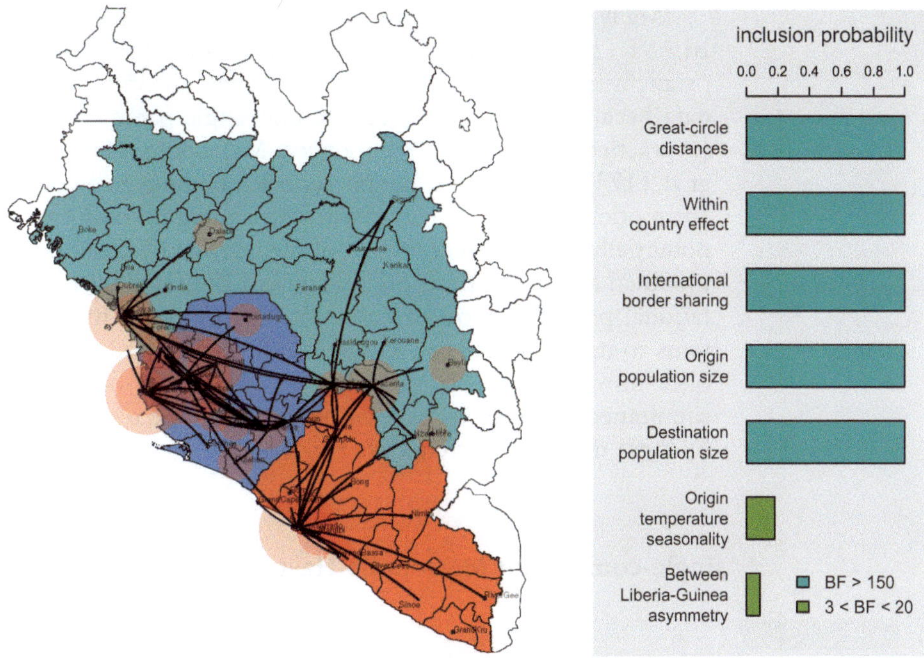

Fig. 11 Snapshot of the geographic spread of the Ebola virus during the 2013–2016 West African epidemic, based on 1610 whole genome sequences [12]. A discrete phylogeographic approach was used, allocating the sequence data into a discrete number of locations and employing a generalized linear model on the parameters that model geographic spread. Inclusion probabilities and Bayes factor support are shown for the most prominent predictors of Ebola virus geographic spread. D3 visualization is made using SpreaD3 [7], with circular polygon areas proportional to the number of tree lineages maintaining that location at that time

5 Adaptive MCMC

The various data sets described in this chapter so far have shown the use and computational performance of a wide range of models in phylogenetic, phylogenomic, and phylodynamic research. Whether employing full codon models (e.g., the carnivores data set), codon partition models (e.g., the Ebola data set), or discrete phylogeographic models, the number of parameters of a typical Bayesian phylogenetic analysis has increased drastically over the years. This is exacerbated by the use of partitioning strategies, resulting also in a potentially large array of likelihoods that need to be evaluated simultaneously, increasing run times for most phylogenetic analyses. In a similar fashion, computational resources available to researchers have also markedly increased, both in the form of multicore CPU technology and increasingly powerful graphics cards targeted towards scientific computing. The ubiquitous availability of multiprocessor and multicore computers practically has motivated the design of novel parallel algorithms to make efficient use of these machines [22, 32].

Many Bayesian phylogenetics software packages, such as BEAST [11] and MrBayes [29], do not fully exploit the inherent parallelism of such multicore systems when analyzing partitioned data because they typically update one single parameter at a time (a practice called single-component Metropolis–Hastings; Gilks et al. [17]). Such a single parameter often belongs to an evolutionary model for a single data partition, leading to only one of the potentially large collection of (observed) data likelihoods to be modified at any one time. Such a strategy does not use the computational power of modern-day multiprocessor and multicore systems to its full advantage. Updating all the models' parameters at once however leads to multiple data likelihoods being modified simultaneously, thereby making better use of the resources offered by these multicore systems (*see* Fig. 12).

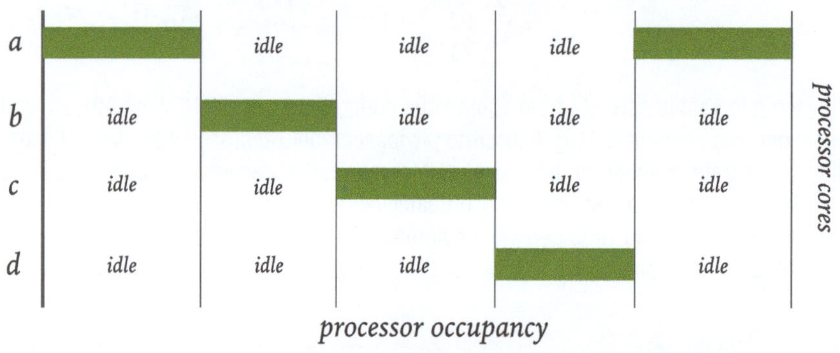

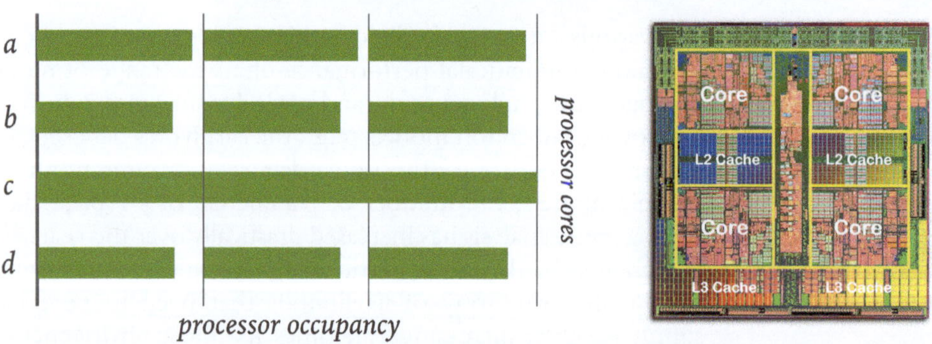

Fig. 12 Conceptual visualization on the potential benefits of an adaptive MCMC algorithm over single-component Metropolis–Hastings (green bars indicate that a processor is computing a specific likelihood). In Bayesian phylogenetics, the common practice of updating a single parameter (i.e., either *a*, *b*, *c*, or *d*) at a time leaves many CPU cores idle, underusing the computational performance of such architecture. Adaptive MCMC allows to update a collection of continuous parameters simultaneously (i.e., *a*, *b*, *c*, and *d*), putting many cores (in this case: 4) to work in a parallel fashion. Quad-Core AMD Opteron processor silicon die is shown, courtesy of Advanced Micro Devices, Inc. (AMD), obtained from Wikimedia Commons

In recent work, Baele et al. [5] propose to use multivariate components to update blocks of parameters, leading to acceptance or rejection for all of those parameters simultaneously, rather than updating all the parameters one by one in a sequential fashion using low-dimensional or scalar components [17]. To this end, the authors developed an adaptable variance multivariate normal (AVMVN) transition kernel for use in Bayesian phylogenetics, based on the work of Roberts and Rosenthal [28], to simultaneously estimate a large number of partition-specific parameters. Baele et al. [5] implemented this adaptive MCMC approach in the popular open-source BEAST software package [11], which enables this transition kernel to exploit the computational routines within the BEAGLE library [3]. The authors applied this transition kernel to a collection of clock model parameters, speciation model parameters, coalescent model parameters, and partition-specific evolutionary model parameters (which include substitution model parameters, varying rates across sites parameters, and relative rate parameters), although this kernel may find its use on parameters in many additional models.

Baele et al. [5] show that such an AVMVN transition kernel tremendously increases estimation performance over a standard set of single-parameter transition kernels. Importantly, the use of an AVMVN transition kernel requires a paradigm shift in assessing performance of transition kernels in MCMC. It is common to judge the performance of Bayesian phylogenetic software packages strictly by the time they take to evaluate proposed parameter values, often expressed in time per number of states or iterations. However, comparing transition kernels that only require a single processor core to evaluate a proposed value against transition kernels that require a collection of processor cores to evaluate a collection of proposed values simultaneously is unfair, as the latter will logically take up more time as this involves more (computational) work (on multiple processor cores). Hence, a fair comparison involves calculating the effective sample size (ESS) per time unit, as this takes into account differences in execution speed while still reporting a main statistic of interest.

We note that the approach of Baele et al. [5] has been shown to yield performance increases on CPU, but that it still needs to be tested on GPU. This is due to the specific design of the BEAGLE library [3], which evaluates a collection of BEAGLE likelihoods/instances sequentially on GPU. Current work is underway to simplify the run time process of the BEAGLE library on GPU, allowing for simultaneous evaluation of such a collection of instances.

6 Conclusions

In this chapter, we have focused on the computational challenges associated with typical analyses in the fields of phylogenetics, phylogenomics, and phylodynamics. We have provided a detailed description of how the BEAGLE library can employ multicore hardware to perform efficient likelihood evaluations and have focused on its interaction with the BEAST software package. Using benchmarks collected on a range of multicore hardware, both from the CPU and GPU market, we have shown that employing the BEAGLE high-performance computational library can considerably decrease computation time on these different systems and this for data sets with different characteristics in terms of size and complexity. The BEAGLE library allows to simultaneously compute the likelihoods for different data partitions on different CPU cores or even on different hardware devices, such as multiple GPU cards. In addition, existing data partitions can be split into multiple subpartitions to be computed in parallel across multicore hardware, yielding potentially drastic performance increases as shown in the benchmarks discussed in this chapter.

Having employed the BEAGLE library on state-of-the-art multicore hardware for a range of commonly used evolutionary models, we conclude that the combination of using BEAGLE and running analyses on powerful graphics cards aimed at the scientific computing market allows for massive performance gains for many challenging data sets. Given that sequence data sets keep growing in size and are being complemented with associated trait data, we have paid particular attention to a popular discrete trait model that parameterizes the transition rates between its states as a GLM, to allow for the inclusion of covariates to help explain transitions in the trait data. Graphics cards can be particularly useful when dealing with such models, as shown in the benchmarks presented, and we have hence presented a number of examples from the literature in which such a setup was used to perform the analyses.

Discrete phylogeography approaches (or discrete trait analyses), as the ones presented in this chapter, treat the sampling locations of the sequences as informative data, rather than uninformative auxiliary variables [9, 23, 25]. As such, the posterior distribution of the parameters given the data contains not only the likelihood of the sequences given the genealogy and substitution model but also the likelihood of the sampling locations given the genealogy and migration matrix, calculated by integrating over all possible discrete state transition histories using Felsenstein's pruning algorithm [16]. What makes this computationally demanding is a potential large number (equal to the number of branches in the phylogeny) of potentially high-dimensional (depending on the number of sampling locations) matrices, which can be parallelized across a large number of computing cores such as those found on a GPU.

Similarly, structured coalescent approaches also contain the likelihood of the sequences given the genealogy and substitution model in their posterior distribution of the parameters given the data. The use of BEAGLE will yield equal benefits to both approaches when it comes to the computation of the likelihood of the sequences given the genealogy and substitution model. However, rather than the likelihood of the sampling locations, structured coalescent approaches require computation of the probability density of the genealogy and migration history under the structured coalescent given the migration matrix and effective population sizes. To compute this density, a product of exponentials—one for each of the time intervals between successive events (coalescence, sampling, or migration)—needs to be calculated. If the number of demes is sufficiently large, a GPU implementation of the probability density of the genealogy and migration history under the structured coalescent may be able to compute the contribution to this density for each of those time intervals in a highly parallel manner.

Approximations to the structured coalescent include, for example, BASTA [9], which aims to compute the probability density of the genealogy under the structured coalescent, integrated over migration histories. The computational bottleneck of this approach lies with calculating and updating the probability distribution of lineages among demes, over all lineages and over all coalescent events. This involves computing the matrix exponential of the product of each time interval duration with the backwards-in-time migration rate matrix, of which the diagonal elements are defined such that the rows sum to zero. BEAGLE is equipped with a parallel thread block design for computing such finite-time transition probabilities, and to construct the finite-time transition probabilities in parallel across all lineages, and therefore has the potential to provide performance increases for structured coalescent approximations such as BASTA. However, the application software that calls upon BEAGLE needs to be implemented to rely on BEAGLE's API in order to achieve the corresponding performance increases.

Graphics cards aimed at the scientific computing market have traditionally offered roughly three times the single-precision performance compared to their double-precision performance (Tesla K40, K20X and K20). Previous generation cards, such as the Tesla K10, offered poor double-precision performance and focused solely on single-precision performance (up to 24 times their double-precision performance). The latest generation of GPUs, specifically the Tesla P100, offers tremendous double-precision performance, while single-precision performance is still twice as high. We therefore expect a doubling in performance for the computations described in this chapter if we would run them in single precision on GPU, provided that the decrease in accuracy would not lead to rescaling issues which would slow down the evaluations.

In theory, single-precision likelihood evaluations will be twice as fast as double-precision likelihood evaluations on CPU as well, but with more rescaling issues hampering performance. However, the influence of switching to single precision is more difficult to assess for CPUs, as there are a number of other factors to consider. Single-precision floating points are half the size compared to double-precision floating points and hence they may fit into a lower level of cache with a lower latency, which potentially frees up cache space to cache more (or other) data. Additionally, they require half the memory bandwidth, which frees up that bandwidth for other operations to be performed. Nevertheless, the total overall bandwidth will still be limited compared to that of powerful graphics cards and this will not suffice to bridge the performance differences between CPUs and GPUs in phylogenetics.

Finally, we have presented an interesting avenue for further increasing computational performance on multicore hardware, in the form of a new adaptive MCMC transition kernel. Traditional MCMC transition kernels generally update single parameters in a serial fashion triggering sequential likelihood evaluations on single cores. The adaptive transition kernel however updates a collection of continuous parameters simultaneously, triggering multiple likelihood evaluations in parallel on multiple cores and hence allowing for potentially large improvements in computational efficiency. Further research into this area is needed to continuously advance MCMC kernels and keep computation time manageable for a wide range of models in Bayesian phylogenetics.

7 Notes

1. We have showcased the potentially impressive performance gains brought about by using BEAGLE in conjunction with powerful graphics cards. However, users sometimes complain about the poor performance gains they experience when using a GPU for their analyses, which may have to do with their GPU being not particularly suited for scientific computing. We urge readers to be cautious as to which GPU they invest in, as there is an important distinction between graphics cards aimed at the gaming market and those aimed at the scientific computing market. Computer gaming cards mainly offer tremendous single-precision performance, but typically weak double-precision performance. We hence advise to invest in GPUs aimed at scientific computing, offering increased accuracy and performance in double precision. As a rule, computer gaming

cards have a much reduced cost compared to scientific computing cards, but we advise readers to check the technical specifications of the card before purchase.

2. While 32-bit operating systems are no longer the norm, such systems are still being used from time to time, and problems have been reported in the use and/or installation of BEAGLE on such systems. We strongly advise to install and run BEAST together with BEAGLE on a 64-bit operating system and, in the case of problems, urge users to check that their Java installation is a proper 64-bit installation as well (i.e., avoid 32-bit or mixed mode software installations). The BEAGLE website, hosted at https://github.com/beagle-dev/beagle-lib, contains installation instructions for Windows, Linux/Unix, and Mac systems.
3. While powerful GPUs can be purchased and installed in desktop computers for immediate use, high-performance computing (HPC) centers or computing clusters can also be equipped with GPUs. These systems typically run a job scheduler that allows users to submit BEAST analyses to either CPU or GPU nodes. In case the requested resources are not immediately available, the submitted job is placed in a queue until those resources become available, which may take some time. We hence strongly advise users (especially those who manually compose their input files) to first test their BEAST XML files on a local desktop machine with BEAGLE installed, in order to not have wasted precious time in a job queue only to find out the BEAST XML cannot be run properly.

8 Exercises

1. An important aspect to getting computations—such as those discussed in this chapter—up and running, is defining which hardware is available on your computer or server. This can easily be checked using BEAST once BEAGLE has been installed. To check this when using the BEAST graphical user interface (GUI), simply check the box that says "Show list of available BEAGLE resources and Quit"; alternatively, when using the command-line interface using a BEAST Java Archive (or JAR) file—which can usually be found in the `lib` directory within the BEAST folder—you can simply type:

```
java -jar beast.jar -beagle_info
```

If the path to the BEAGLE library hasn't been set up automatically, be sure to add its location to the command by adding:

```
java -Djava.library.path=/usr/local/lib ...
```

On a typical desktop system equipped with a GPU fit for scientific computing, this will yield the following output to screen:

```
Using BEAGLE library v2.1.2 for accelerated, parallel
likelihood evaluation. 2009-2013, BEAGLE Working Group.
Citation: Ayres et al (2012) Systematic Biology 61: 170-173

BEAGLE resources available:
0 : CPU
    Flags: PRECISION_SINGLE PRECISION_DOUBLE ...

1 : Intel(R) HD Graphics 530
    Global memory (MB): 1536
    Clock speed (Ghz): 1.05
    Number of compute units: 24
    Flags: PRECISION_SINGLE COMPUTATION_SYNCH ...

2 : Tesla K40c
    Global memory (MB): 11520
    Clock speed (Ghz): 0.74
    Number of cores: 2880
    Flags: PRECISION_SINGLE PRECISION_DOUBLE ...
```

In order to determine which resource to use for your computations, it's important to look into the specifications of the GPU as listed by the hardware vendor. For example, certain GPUs will be equipped with a large number of cores and yet they're aimed at the computer gaming market, which will result in poor double-precision performance. As we have shown in Table 1, the Tesla brand is typically well suited for GPU computing, but other cards may be appropriate as well if they deliver adequate double-precision peak performance. In the output printed above, it's quite obvious that we'll be interested in running our analyses on resource 2, i.e., a GPU equipped with thousands of computing cores (resource 1 is an integrated graphics unit, mainly fit for delivering graphics output to screen).

2. Once you have located a GPU fit for scientific computing on your desktop computer or server, try to perform your analysis both on the system's CPU and GPU to compare performance. Using BEAST's GUI, the default option is to run on CPU; if you'd like to run your analysis on a suitable GPU, use BEAST's GUI to select "GPU" where it says "Prefer use of:." However, most desktop computers don't come equipped with powerful

graphics cards, and most often servers aimed at high-performance computing (HPC) will be used for performing these types of computations. As such servers are typically instructed using a command-line interface, BEAST offers the possibility to assign computations to one or more specific GPUs. Using the system described here, it would hence make sense to run your analysis on resource 2, which can be done as follows:

```
java -jar beast.jar -beagle_gpu -beagle_order 2 data.xml
```

Note that not specifying `-beagle_order` will result in the analysis being run on the system's CPU, i.e., resource 0. Additionally, when employing a GPU for your analyses, adding the `-beagle_gpu` argument is highly advised. Many different combinations of using resources arise when your data set is partitioned into multiple subsets, for example if your data is partitioned according to gene and/or codon position. In such cases, it may be beneficial to split those partitions onto multiple resources by using the `-beagle_order` command-line option. For example, the Ebola virus data set (without trait data) has four partitions; it may be useful (although this depends on the actual hardware and needs to be tested) to compute the likelihood of one partition on the CPU (i.e., resource 0) and the other three likelihoods on the GPU (i.e., resource 2). This can be done as follows:

```
java -jar beast.jar -beagle_gpu -beagle_order 0,2,2,2
ebola.xml
```

3. In some cases, such as for example the carnivores data set analyzed in this chapter, only one (sequence) data partition is available. On CPU, drastic performance improvements can still be achieved by using a BEAGLE feature that allows to split up a data partition into multiple subsets, as can be seen in Fig. 5. This approach will lead to performance increases on most CPU systems, as many laptops now come equipped with 4-core processors; this can hence easily be tested on the system you're currently using. To split a (sequence) data partition into two subsets, you can use the following command:

```
java -jar beast.jar -beagle_instances 2 carnivores.xml
```

To generate the results in Fig. 5, we have used this approach to split the data set into 2, 4, 8, 12, 16, 20, and 24 subpartitions, increasing performance every step of the way. Note that on GPU, this approach will only lead to an increased overhead and hence worse performance compared to keeping the single data partition, as all the likelihood calculations end up on the same (GPU) device. With multiple GPUs in a system however, this can also lead to drastic performance improvements, as shown throughout this chapter.

Acknowledgements

The authors acknowledge support from the European 2020 Union Seventh Framework Programme for research, technological development, and demonstration under Grant Agreement no. 278433-PREDEMICS. The VIROGENESIS project receives funding from the European Union's Horizon 2020 research and innovation programme under grant agreement No. 634650. The COMPARE project receives funding from the European Union's Horizon 2020 research and innovation programme under grant agreement No. 643476. This work is supported by National Science Foundation (NSF) [DMS 1264153 to M.A.S.; DBI 0755048 and DBI 1356562 to Michael P. Cummings who supported D.L.A. and is also acknowledged here] and National Institutes of Health [R01 AI117011 and R01 HG006139 to M.A.S.]. G.B. acknowledges support from the Interne Fondsen KU Leuven/Internal Funds KU Leuven and of a Research Grant of the Research Foundation—Flanders (FWO; Fonds Wetenschappelijk Onderzoek—Vlaanderen). We acknowledge funding of the "Bijzonder Onderzoeksfonds," KU Leuven (BOF, No. OT/14/115), and the Research Foundation—Flanders (FWO, G0D5117N and G0B9317N). We gratefully acknowledge support from the NVIDIA Corporation with the donation of parallel computing resources used for this research. The computational resources and services used in this work were provided by the VSC (Flemish Supercomputer Center), funded by the Research Foundation—Flanders (FWO) and the Flemish Government—Department EWI.

References

1. Ayres DL, Cummings MP (2017) Configuring concurrent computation of phylogenetic partial likelihoods: accelerating analyses using the BEAGLE library. In: 17th International conference on algorithms and architectures for parallel processing: ICA3PP 2017, Collocated Workshops
2. Ayres DL, Cummings MP (2017) Heterogeneous hardware support in BEAGLE, a high-performance computing library for statistical phylogenetics. In: 46th International conference on parallel processing workshops (ICPPW 2017)
3. Ayres DL, Darling A, Zwickl DJ, Beerli P, Holder MT, Lewis PO, Huelsenbeck JP, Ronquist F, Swofford DL, Cummings MP, Rambaut A, Suchard MA (2012) BEAGLE: an application programming interface and high-performance computing library for statistical phylogenetics. Syst Biol 61(1):170–173
4. Baele G, Lemey P (2013) Bayesian evolutionary model testing in the phylogenomics era: matching model complexity with computational efficiency. Bioinformatics 29 (16):1970–1979

5. Baele G, Lemey P, Rambaut A, Suchard MA (2017) Adaptive MCMC in Bayesian phylogenetics: an application to analyzing partitioned data in BEAST. Bioinformatics 33 (12):1798–1805
6. Baele G, Suchard MA, Rambaut A, Lemey P (2017) Emerging concepts of data integration in pathogen phylodynamics. Syst Biol 66(1): e47–e65
7. Bielejec F, Baele G, Vrancken B, Suchard MA, Rambaut A, Lemey P (2016) Spread3: interactive visualization of spatiotemporal history and trait evolutionary processes. Mol Biol Evol 33 (8):2167–2169. https://doi.org/10.1093/molbev/msw082
8. Brockmann D, David V, Gallardo AM (2009) Human mobility and spatial disease dynamics. In: Diffusion fundamentals III. Leipziger Universitätsverlag, Leipzig, pp 55–81
9. De Maio N, Wu CH, O'Reilly KM, Wilson D (2015) New routes to phylogeography: a Bayesian structured coalescent approximation. PLoS Genet 11(8):e1005421
10. Doucet A, De Freitas N, Gordon N (eds) (2001) Sequential Monte Carlo methods in practice. Springer, New York
11. Drummond AJ, Suchard MA, Xie D, Rambaut A (2012) Bayesian phylogenetics with BEAUti and the BEAST 1.7. Mol Biol Evol 29 (8):1969–1973
12. Dudas G, Carvalho LM, Bedford T, Tatem AJ, Baele G, Faria NR, Park DJ, Ladner JT, Arias A, Asogun D, Bielejec F, Caddy SL, Cotten M, D'Ambrozio J, Dellicour S, Caro AD, Diclaro II JD, Durrafour S, Elmore MJ, Fakoli III LS, Faye O, Gilbert ML, Gevao SM, Gire S, Gladden-Young A, Gnirke A, Goba A, Grant DS, Haagmans BL, Hiscox JA, Jah U, Kargbo B, Kugelman JR, Liu D, Lu J, Malboeuf CM, Mate S, Matthews DA, Matranga CB, Meredith LW, Qu J, Quick J, Pas SD, Phan MVT, Pollakis G, Reusken CB, Sanchez-Lockhart M, Schaffner SF, Schieffelin JS, Sealfon RS, Simon-Loriere E, Smits SL, Stoecker K, Thorne L, Tobin EA, Vandi MA, Watson SJ, West K, Whitmer S, Wiley MR, Winnicki SM, Wohl S, Wölfel R, Yozwiak NL, Andersen KG, Blyden SO, Bolay F, Carroll MW, Dahn B, Diallo B, Formenty P, Fraser C, Gao GF, Garry RF, Goodfellow I, Günther S, Happi CT, Holmes EC, Keïta S, Kellam P, Koopmans MPG, Kuhn JH, Loman NJ, Magassouba N, Naidoo D, Nichol ST, Nyenswah T, Palacios G, Pybus OG, Sabeti PC, Sall A, Ströher U, Wurie I, Suchard MA, Lemey P, Rambaut A (2017) Virus genomes reveal factors that spread and sustained the Ebola epidemic. Nature 544(7650):309–315
13. Eisen JA, Fraser CM (2003) Phylogenomics: intersection of evolution and genomics. Science 300(5626):1706–1707
14. Faria NR, Suchard MA, Rambaut A, Streicker DG, Lemey P (2013) Simultaneously reconstructing viral cross-species transmission history and identifying the underlying constraints. Philos Trans R Soc B 368 (1614):20120196
15. Felsenstein J (1978) The number of evolutionary trees. Syst Biol 27(1):27–33
16. Felsenstein J (1981) Evolutionary trees from DNA sequences: a maximum likelihood approach. J Mol Evol 17(6):368–376
17. Gilks WR, Richardson S, Spiegelhalter DJ (1996) Markov chain Monte Carlo in practice. Chapman and Hall, London
18. Goldman N, Yang Z (1994) A codon-based model of nucleotide substitution for protein-coding DNA sequences. Mol Biol Evol 11 (5):725–736. http://mbe.oxfordjournals.org/content/11/5/725.abstract, http://mbe.oxfordjournals.org/content/11/5/725.full.pdf+html
19. Guindon S, Dufayard JF, Lefort V, Anisimova M, Hordijk W, Gascuel O (2010) New algorithms and methods to estimate maximum-likelihood phylogenies: assessing the performance of PhyML 3.0. Syst Biol 59 (3):307–321
20. Huelsenbeck JP, Ronquist F, Nielsen R, Bollback JP (2001) Bayesian inference of phylogeny and its impact on evolutionary biology. Science 294(5550):2310–2314
21. Jeffroy O, Brinkmann H, Delsuc F, Philippe H (2006) Phylogenomics: the beginning of incongruence? Trends Genet 22(4):225–231
22. Kobert K, Flouri T, Aberer A, Stamatakis A (2014) The divisible load balance problem and its application to phylogenetic inference. In: International workshop on algorithms in bioinformatics, pp 204–216
23. Kumar S, Filipski AJ, Battistuzzi FU, Pond SLK, Tamura K (2012) Statistics and truth in phylogenomics. Mol Biol Evol 29(2):457–472
24. Lemey P, Rambaut A, Drummond AJ, Suchard MA (2009) Bayesian phylogeography finding its roots. PLoS Comput Biol 5(9):e1000520
25. Lemey P, Rambaut A, Bedford T, Faria N, Bielejec F, Baele G, Russell CA, Smith DJ, Pybus OG, Brockmann D, Suchard MA (2014) Unifying viral genetics and human transportation data to predict the global transmission dynamics of human influenza H3N2. PLoS Pathog 10(2):e1003932
26. Muse SV, Gaut BS (1994) A likelihood approach for comparing synonymous and

nonsynonymous nucleotide substitution rates, with application to the chloroplast genome. Mol Biol Evol 11(5):715–724
27. Neal RM (2010) MCMC using Hamiltonian dynamics. In: Handbook of Markov chain Monte Carlo, vol 54. CRC Press, Boca Raton, pp 113–162
28. Roberts GO, Rosenthal JS (2009) Examples of adaptive MCMC. J Comput Graph Stat 18:349–367
29. Ronquist F, Teslenko M, van der Mark P, Ayres DL, Darling A, Höhna S, Larget B, Liu L, Suchard MA, Huelsenbeck JP (2012) MrBayes 3.2: efficient Bayesian phylogenetic inference and model choice across a large model space. Syst Biol 61(3):539–542
30. Sanmartín I, van der Mark P, Ronquist F (2008) Inferring dispersal: a Bayesian approach to phylogeny-based island biogeography, with special reference to the Canary Islands. J Biogeogr 35:428–449
31. Streicker DG, Turmelle AS, Vonhof MJ, Kuzmin IV, McCracken GF, Rupprecht CE (2010) Host phylogeny constrains cross-species emergence and establishment of rabies virus in bats. Science 329(5992):676–679
32. Suchard MA, Rambaut A (2009) Many-core algorithms for statistical phylogenetics. Bioinformatics 25:1370–1376
33. Yang Z (1996) Among-site rate variation and its impact on phylogenetic analyses. Trends Ecol Evol 11(9):367–372
34. Yang Z, Rannala B (1997) Bayesian phylogenetic inference using DNA sequences: a Markov chain Monte Carlo method. Mol Biol Evol 14(7):717–724
35. Zwickl DJ (2006) Genetic algorithm approaches for the phylogenetic analysis of large biological sequence datasets under the maximum likelihood criterion. PhD thesis, The University of Texas at Austin

Chapter 24

Scalable Workflows and Reproducible Data Analysis for Genomics

Francesco Strozzi, Roel Janssen, Ricardo Wurmus, Michael R. Crusoe, George Githinji, Paolo Di Tommaso, Dominique Belhachemi, Steffen Möller, Geert Smant, Joep de Ligt, and Pjotr Prins

Abstract

Biological, clinical, and pharmacological research now often involves analyses of genomes, transcriptomes, proteomes, and interactomes, within and between individuals and across species. Due to large volumes, the analysis and integration of data generated by such high-throughput technologies have become computationally intensive, and analysis can no longer happen on a typical desktop computer.

In this chapter we show how to describe and execute the same analysis using a number of workflow systems and how these follow different approaches to tackle execution and reproducibility issues. We show how any researcher can create a reusable and reproducible bioinformatics pipeline that can be deployed and run anywhere. We show how to create a scalable, reusable, and shareable workflow using four different workflow engines: the Common Workflow Language (CWL), Guix Workflow Language (GWL), Snakemake, and Nextflow. Each of which can be run in parallel.

We show how to bundle a number of tools used in evolutionary biology by using Debian, GNU Guix, and Bioconda software distributions, along with the use of container systems, such as Docker, GNU Guix, and Singularity. Together these distributions represent the overall majority of software packages relevant for biology, including PAML, Muscle, MAFFT, MrBayes, and BLAST. By bundling software in lightweight containers, they can be deployed on a desktop, in the cloud, and, increasingly, on compute clusters.

By bundling software through these public software distributions, and by creating reproducible and shareable pipelines using these workflow engines, not only do bioinformaticians have to spend less time reinventing the wheel but also do we get closer to the ideal of making science reproducible. The examples in this chapter allow a quick comparison of different solutions.

Key words Bioinformatics, Evolutionary biology, Big data, Parallelization, MPI, Cloud computing, Cluster computing, Virtual machine, MrBayes, Debian Linux, GNU Guix, Bioconda, CWL, Common Workflow Language, Guix Workflow Language, Snakemake, Nextflow

---

Availability: All included software, scripts, and Docker images are based on free and open-source software and can be found at https://github.com/EvolutionaryGenomics/scalability-reproducibility-chapter.

Maria Anisimova (ed.), *Evolutionary Genomics: Statistical and Computational Methods*, Methods in Molecular Biology, vol. 1910, https://doi.org/10.1007/978-1-4939-9074-0_24, © The Author(s) 2019

1 Introduction

1.1 Overview

In this chapter, we show how to create a bioinformatics pipeline using four workflow systems: CWL, GWL, Snakemake, and Nextflow. We show how to put them together, so you can adapt it for your own purposes while discussing in the process the different approaches. All scripts and source code can be found on GitHub. The online material allows a direct comparison of how such workflows are assembled with their syntax.

Due to large volumes, the analysis and integration of data generated by high-throughput technologies have become computationally intensive, and analysis can no longer happen on a typical desktop computer. Researchers therefore are faced with the need to scale analyses efficiently by using high-performance compute clusters or cloud platforms. At the same time, they have to make sure that these analyses run in a reproducible manner. And in a clinical setting, time becomes an additional constraint, with motivation to generate actionable results within hours.

In the case of evolutionary genomics, lengthy computations are often multidimensional. Examples of such expensive calculations are Bayesian analyses, inference based on hidden Markov models, and maximum likelihood analysis, implemented, for example, by MrBayes [1], HMMER [2], and phylogenetic analysis by maximum likelihood (PAML) [3]. Genome-sized data, or Big Data [4, 5], such as produced by high-throughput sequencers, as well as growing sample size, such as from UK Biobank, the Million Veterans Program, and the other large genome-phenome projects, are exacerbating the computational challenges, e.g., [6].

In addition to being computationally expensive, many implementations of major algorithms and tools in bioinformatics do not scale well. One example of legacy software requiring lengthy computation is Ziheng Yang's CodeML implementation of PAML [3]. PAML finds amino acid sites that show evidence of positive selection using d_N/d_S ratios, i.e., the ratio of nonsynonymous and synonymous substitution rate. For further discussion see also Chapter. 12. Executing PAML over an alignment of 100 sequences may take hours, sometimes days, even on a fast computer. PAML (version 4.x) is designed as a single-threaded process and can only exploit one central processing unit (CPU) to complete a calculation. To test hundreds of alignments, e.g., different gene families, PAML is invoked hundreds of times in a serial fashion, possibly taking days on a single computer. Here, we use PAML as an example, but the idea holds for any software program that is CPU bound, i.e., the CPU speed determines program execution time. A CPU bound program will be at (close to) 100% CPU usage. Many legacy programs are CPU bound and do not scale by themselves.

Most bioinformatics (legacy) programs today do not make effective use of multi-core computers

The reason most bioinformatics software today does not make full use of multicore computers or GPUs is because writing such software is difficult. (See also the text box below for a further treatment of this topic; *see* Box 1.)

A common parallelization strategy in bioinformatics is to start with an existing nonparallel application and run it by dividing data into independent units of work or jobs which run in parallel and do not communicate with each other. This is also known as an "embar rassingly parallel" solution, and we will pursue this below.

1.2 Parallelization in the Cloud

Cloud computing allows the use of "on-demand" CPUs accessible via the Internet and is playing an increasingly important role in bioinformatics. Bioinformaticians and system administrators previously had to physically install and maintain large compute clusters to scale up computations, but now cloud computing makes it possible to rent and access CPUs, GPUs, and storage, thereby enabling a more flexible concept of on-demand computing [7]. The cloud scales and commoditizes cluster infrastructure and management and, in addition, allows users to run their own operating system, usually not true with existing cluster and GRID infrastructure (a GRID is a heterogeneous network of computers that act together). A so-called hypervisor sits between the host operating system and the guest operating system, and it makes sure they are clearly separated while virtualizing host hardware. This means many guests can share the same machine that appears to the users as a single machine on the network. This allows providers to efficiently allocate resources. Containers are another form of light virtualization that is now supported by all the main cloud providers, such as Google, Microsoft, Rackspace OpenStack, and Amazon (AWS). Note that only OpenStack is available as free and open-source software.

An interesting development is that of portable batch systems (PBS) in the cloud. PBS-like systems are ubiquitous in high-performance computing (HPC). Both Amazon EC2 and Microsoft Cloud offer batch computing services with powerful configuration options to run thousands of jobs in the cloud while transparently automating the creation and management of virtual machines and containers for the user. As an alternative, Arvados is an open-source product specifically aimed at bioinformatics applications that makes the cloud behave as if it is a local cluster of computers, e.g., [8].

At an even higher level, MapReduce is a framework for distributed processing of huge datasets, and it is well suited for problems using large number of computers [9]. The map step takes a dataset and splits it into parts and distributes them to worker nodes. Worker nodes can further split and distribute data. At the reduce step, data is combined into a result, i.e., it is an evolved scatter and gather approach. An API is provided that allows programmers to access functionality. The Apache Hadoop project includes a MapReduce implementation and a distributed file system [10] that can be used with multiple cloud providers and also on private computer clusters. Another similar example is the Apache Spark project based on resilient distributed datasets (RDD)—a fault-tolerant collection of elements that can be accessed and operated on in parallel.

The advantage of such higher-level systems is that they go well beyond hardware virtualization: not only the hardware infrastructure but also the operating system, the job scheduler, and resource orchestration are abstracted away. This simplifies data processing, parallelization, and the deployment of virtual servers and/or containers. The downside is that users have less control over the full software stack and often needs to program and interact with an application programmers interface (API).

Overall, in the last decade, both commercial and noncommercial software providers have made cloud computing possible. Bioinformaticians can exploit these services.

1.3 A Pipeline for the Cloud

To create a bioinformatics pipeline, it is possible to combine remote cloud instances with a local setup. Prepare virtual machines or containers using similar technologies on a local network, such as a few office computers or servers, and then use these for calculations in the cloud when an analysis takes too long. The cloud computing resources may, for instance, support a service at peak usage, while regular loads are met with local infrastructure (i.e., burst compute). New ideas can be developed and pre-evaluated using modest in-house setups and then scaled to match the most demanding work.

Cloud services can be used for burst computing – enabling local clusters to be much smaller – as small as a single computer

In the following sections, we will provide instructions to deploy applications, and we will show how the use of workflow systems and reproducible environments can greatly simplify running scalable workflows on different environments, including the cloud.

1.4 Parallelization of Applications Using a Workflow

In case of embarrassingly parallel applications, programs are run independently as separate processes which do not communicate with each other. This is also a scatter and gather approach, i.e., inputs split into several jobs are fed into each process by the user. Job outputs are collected and collated. In bioinformatics, such tasks are often combined into computational pipelines. With the PAML example, each single job can be based on one alignment, potentially giving linear speed improvements by distributing jobs across multiple CPUs and computers. In other words, the PAML software, by itself, does not allow calculations in parallel, but it is possible to parallelize multiple runs of PAML by splitting the dataset. The downside of this approach is the deployment and configuration of pipeline software, as well as the management and complexity of splitting inputs and the collecting and collating of outputs. Also, pipelines are potentially fragile, because there is no real interprocess communication. For example, it is hard to predict the consequences of a storage or network error in the middle of a week- or month-long calculation.

Even for multithreaded applications that make use of multiple CPUs, such as BLAST and MrBayes, it is possible to scale up calculations by using a workflow. For example, MrBayes-MPI version 3.1.2 does not provide between-machine parallelization and is therefore machine bound, i.e., the machine's performance determines the total run time. Still, if one needs to calculate thousands of phylogenetic trees, discrete jobs can be distributed across multiple machines. A similar approach is often used for large-scale BLAST analyses over hundreds of thousands of sequences.

A pipeline typically consists of linear components, where one software tool feeds into another, combined with a scattering of jobs across nodes and a gathering and collation of results.

In existing compute clusters, to distribute work across nodes, portable batch system (PBS) schedulers are used, such as Slurm [11]. Many pipelines in bioinformatics are created in the form of Bash, Perl, or Python scripts that submit jobs to these schedulers. Such scripted pipelines have the advantage that they are easy to write and adaptable to different needs. The downside is that they are hard to maintain and not very portable, since the description of the environment and the software packages are not part of these scripts, reducing or completing preventing the reproducibility of a certain analysis in a different context. This has led to the current state of affairs in bioinformatics that it is surprisingly hard to share pipelines and workflows. As a result much effort is spent reinventing the wheel.

Most existing bioinformatics pipelines cannot easily be shared and reproduced

In recent years, a number of efforts have started to address the problem of sharing workflows and making analyses reproducible. One example is the Common Workflow Language (CWL), a specification for describing analysis workflows and tools in a way that makes them portable and scalable across a variety of environments—from workstations to cluster, cloud, and HPC environments. CWL is a large bioinformatics community effort. Different platforms support CWL, including Arvados, Galaxy, and Seven Bridges [8].

A second workflow language is the Guix Workflow Language (GWL) built on top of the GNU Guix software deployment system. GWL aims to provide a deterministic and bit-reproducible analysis environment.

A third workflow language and orchestrator, Nextflow, allows scalable and reproducible scientific workflows to run seamlessly across multiple platforms from local computers to HPC clusters and the cloud, offering a concise and expressive DSL to describe complex workflows. Nextflow is routinely used in organizations and institutes, such as the Roche Sequencing, the Wellcome Trust Sanger Institute, and the Center for Genomic Regulation (CRG) Nextflow workshop.

Forth there is Snakemake, another widely used workflow manager system, written in Python and inspired by GNU Make. It allows for the composition of workflows based on a graph of rules whose execution is triggered by the presence, absence, or modification of expected files and directories.

It is interesting to note that all these workflow languages and systems originated in bioinformatics. It suggests that in this rapidly growing field, the increasing computational needs and moreover the diverse demands made more formal solutions a necessity. It also suggests that existing workflow engines used in astronomy and physics, for example, have different requirements.

Box 1: Understanding Parallelization

Parallel computing is related to concurrent computing. In parallelized computing, a computational task is typically broken down in several, often many, very similar subtasks that can be processed independently and whose results are combined afterward, upon completion, i.e., a simple scatter and gather approach. In contrast, in distributed computing, the various processes often do not address related tasks; or when they do, the separate tasks may have a varied nature and often require some interprocess communication during execution. The latter is also a hallmark of supercomputing where compute nodes have high-speed connections.

In the bioinformatics space, we usually discuss embarrassingly parallel computing which means similar tasks are distributed across multiple CPUs without interprocess communication. This can be among multiple cores within a single processor, a multiprocessor system, or a network of computers, a so-called compute cluster.

Even so, parallel multicore programming easily becomes complex. Typically, parallel programming has to deal with extra data and control flow; it has to deal with deadlocks,

(continued)

Box 1: (continued)
where depending tasks wait for each other forever and, with race conditions, where tasks try to modify a shared resource (e.g., a file) at the same time resulting in a loss of data or an undetermined condition. This introduces additional complexity in software development, bug hunting, and code maintenance. Typically it takes more time to debug such code than to write it.

Writing programs that fully utilize multi-core architectures is hard

Not only is parallel programming intrinsically complicated; programmers also have to deal with communication overheads between parallel threads. MrBayes, for example, a program for calculating phylogenetic trees based on Bayesian analysis, comes with MPI support. MPI is a message-based abstraction of parallelization, in the form of a binary communication protocol implemented in a C programming library [12]. In some cases the parallelized version is slower than the single CPU version. For example, the MPI version calculates each Markov chain in parallel, and the chains need to be synchronized with each other, in a "scatter and gather" pattern. The chains spend time waiting for each other in addition to the communication overheads introduced by MPI itself. Later MrBayes adopted a hybrid use of coarse-grained OpenMPI and fine-grained use of pthreads or OpenMP leading to improved scalability, e.g., [13].

Another example of communication overhead is with the statistical programming language R [14], which does not have native threading support built into the language. One possible option is to use an MPI-based library which only allows coarse-grained parallelization from R, as each parallelized R thread starts up an R instance, potentially introducing large overheads, both in communication time and memory footprint. For a parallelized program to be faster than its single-threaded counterpart, these communication overheads have to be dealt with.

Parallelization in R is coarse-grained with large overhead

The need for scaling up calculations on multi-CPU computers has increased the interest in a number of functional programming languages, such as Erlang [15], Haskell [16], Scala [17], and Julia [18]. These languages promise to ease writing parallel software by introducing a higher level of abstraction of parallelization, combined with immutable data, automatic garbage collection, and good debugging support [5, 19]. For example, Erlang and Scala rely on Actors as an abstraction of parallelization and make reasoning about fine-grained parallelization easier and therefore less error prone.

Actors were introduced and explored by Erlang, a computer language originally designed for highly parallelized telecommunications computing. To the human programmer, each Actor appears as a linear piece of programming and is parallelized without the complexity of locks, mutexes, and semaphores. Actors allow for parallelization in a manageable way, where lightweight threads are guaranteed to be independent and each has a message queue, similar to MPI. Actors, however, are much faster, more intuitive, and, therefore, probably, safer than MPI. Immutable data, when used on a single multi-CPU computer, allows fast passing of data by reference between Actors. When a computer language supports the concept of immutability, it guarantees data is not changed between parallel threads, again making programming less error prone and easier to structure. Actors with support for immutable data are implemented as an integral part of the programming language in Erlang, Haskell, Scala, Elixir, and D [20].

Another abstraction of parallelized programming is the introduction of goroutines, part of the Go programming language [21]. Where MPI and Actors are related to a concept of message passing and mail boxes, goroutines are more closely related to Unix named pipes. Goroutines also aim to make reasoning about parallelization easier, by providing a pipe where data goes in and results come out, and this processing happens concurrently without use of mutexes, making it easier to reason about linear code. Goroutines are related to communicating sequential processes (CSP), the original paper by Tony Hoare in 1978 [22]. Meanwhile, recent practical implementations are driven by the ubiquity of cheap multicore computers and the need for scaling up. A Java implementation of CSP exists, named JCSP [23], and a Scala alternative named CSO [24]. Go made goroutines intuitive and a central part of the strongly typed compiled language.

Erlang, Elixir, Haskell, Scala, Julia, Go and D are languages offering useful abstractions and tools for multi-core programming

It is important to note that the problems, ideas, and concepts of parallel programming are not recent. They have been an important part of computer science theory for decennia. We invite the reader interested in parallel programming to read up on the languages that have solid built-in support high-level parallelization abstractions, in particular, Scala [17], Go [21], and D [20].

1.4.1 GPU Programming

Another recent development is the introduction of GPU computing or "heterogeneous computing" for offloading computations. Most GPUs consist of an array of thousands of cores that can execute similar instructions at the same time. Having a few thousand GPU cores can speed up processing significantly. Programming GPUs, however, is a speciality requiring specialized compilers and communication protocols, and there are many considerations, not least the I/O bottleneck between the main memory and the GPU's dedicated RAM [5]. Even so, it is interesting to explore the use of GPUs in bioinformatics since they come with almost every computer today and clusters of GPU can increasingly be found in HPC infrastructure and in the cloud, alike. With the advent of "deep neural networks" and the general adoption of machine learning techniques for Big Data, GPUs have become a mainstream technology in data mining.

2 Package Software in a Container

Container technologies, such as Docker and Singularity, have gained popularity because they have less overhead than full virtual machines (VMs) and are smaller in size [24]. Containers are fully supported by the major cloud computing providers and play an important role for portability across different platforms.

Adoption of container solutions on HPC has been problematic, mostly because of security concerns. Singularity [26] offers a decentralized environment encapsulation that works in user space and that can be deployed in a simpler way since no root privileges are required to execute tools provided with Singularity. That is, Singularity containers can be created on a system with root privileges but run on a system without root privileges—though it requires some special kernel support. Docker containers can be imported directly in Singularity, so when we present how to build Docker container images in the following sections, the reader should be aware that the same images can also be used with Singularity. Singularity is slowly being introduced in HPC setups [27].

GNU Guix also has support for creating and running Linux containers. One interesting benefit is that, because the software packaging system is read-only and provides perfect isolation, containers automatically can share specific software running on the underlying system, making running containers even lighter and extremely fast.

In this section we discuss three popular software distribution systems for Linux: Debian GNU/Linux (Debian), GNU Guix, and Conda can be used together on a single system allowing access to most bioinformatics software packages in use today. In this section we bundle tools that can be deployed in a Docker image, which can run on a single multicore desktop computer and a compute cluster and in the cloud.

2.1 Debian Med

Debian (http://www.debian.org) is the oldest software distribution (started 1993) mentioned here with the largest body of software packages. Debian targets a wide range of architectures and includes a kernel plus a large body of other user software including graphical desktop environments, server software, and specialist software for scientific data processing. Overall Debian represents millions of users and targets most platforms in use today, even though it is not the only packaging system around (RPM being a notable alternative, for RedHat, Fedora, OpenSuSE, and CentOS).

Debian Med is a project within Debian that packages software for medical practice and biomedical research. The goal of Debian Med is a complete open system for all tasks in medical care and research [28]. With Debian Med over 400 precompiled bioinformatics software programs are available for Linux, as well as some 400 R packages. Proper free and open-source software (FOSS) can easily be packaged and distributed through Debian. Debian and its derivatives, such as Ubuntu and Mint, share the deb package format and have a long history of community support for bioinformatics packages [28, 29].

2.1.1 Create a Docker Image with Debian

Using the bio packages already present in Debian, it is straightforward to build a Docker container that includes all the necessary software to run the example workflows. Here is the code for creating the Docker image (*see* also [30]). We created a pre-built Docker image which is available on Docker Hub [31].

Essentially, write a Docker script:

```
FROM debian:buster
RUN apt-get update && apt-get -y install perl clustalo paml
ADD pal2nal.pl /usr/local/bin/pal2nal.pl
RUN chmod +x /usr/local/bin/pal2nal.pl
```

And build and run the container:

```
docker build -t scalability_debian -f Dockerfile.debian
```

2.2 GNU Guix

GNU Guix (https://www.gnu.org/software/guix/) is a package manager of the GNU project that can be installed on top of other Linux distributions and represents a rigorous approach toward dependency management [32]. GNU Guix software packages are uniquely isolated by a hash value computed over all inputs, including the source package, the configuration, and all dependencies. This means that it is possible to have multiple versions of the same software and even different variants or combinations of software, e.g., Apache web server with SSL and without SSL compiled on a single system.

As of November 2017, GNU Guix provides over 6500 soft ware packages, including a wide range of dedicated scientific software for bioinformatics, statistics, and machine learning

2.2.1 Create a Docker Image with GNU Guix

GNU Guix has native support for creating Docker images. Creating a Docker image with GNU Guix is a one liner:

```
guix pack -f docker -S /bin=bin paml clustal-omega
```

which creates a reproducible Docker image containing PAML and Clustal Omega [33], including all of their runtime dependencies. Guix makes it very easy to write new package definitions using the Guile language (a LISP). If you want to include the definition of your own packages (that are not in Guix main line), you can include them dynamically. This is how we add pal2nal [34] in below GWL workflow example (see Subheading 3.3 below).

2.3 Conda

Conda (https://conda.io/docs/) is a cross-platform package manager written in Python that can be used to install software written in any language. Conda allows the creation of separate environments to deploy multiple or conflicting packages versions, offering a means of isolation. Note that this isolation is not as rigorous as that provided by GNU Guix or containers. The Bioconda [35] (https://bioconda.github.io/) project provides immediate access to over 2900 software packages for bioinformatics, and it is maintained by an active community of more than 200 contributors.

2.3.1 Create a Docker Image with Bioconda

A Docker container can be created starting from the "Miniconda" image template, which is based on Debian. The Docker instructions are comparable to those of Debian above:

```
FROM conda/miniconda3
RUN conda config --add channels conda-forge
RUN conda install -y perl=5.22.0
RUN conda install -y -c bioconda paml=4.9 clustalo=1.2.4
wget=1.19.1
ADD pal2nal.pl /usr/local/bin/pal2nal.pl
RUN chmod +x /usr/local/bin/pal2nal.pl
```

Note that we provide the version numbering of the packages. If you want to build this container, you can use the Dockerfile provided in the GitHub repository [30] and then run:

```
docker build -t scalability .
```

We also added a pre-built container image on Docker Hub [31].

Conda can also be used outside any container system to install the software directly on a local computer or cluster. To do that first install the Miniconda package https://conda.io/miniconda.html, and then you can create a separate environment with the necessary software to run the workflows. Following is an example to set up a working environment:

```
conda create -n scalability
source activate scalability
conda config --add channels conda-forge
conda install -y perl=5.22.0
conda install -y -c bioconda paml=4.9 clustalo=1.2.4
wget=1.19.1
wget http://www.bork.embl.de/pal2nal/distribution/pal2nal.
v14.tar.gz
tar xzvf pal2nal.v14.tar.gz
sudo cp pal2nal.v14/pal2nal.pl /usr/local/bin
sudo chmod +x /usr/local/bin/pal2nal.pl
```

Note that we use Miniconda here to bootstrap Bioconda. Bioconda can be bootstrapped in other ways. One of them is GNU Guix which contains a Conda package.

2.4 A Note on Software Licenses

All above packaging systems use free and open-source software (FOSS) released under a permissible license, i.e., a license permitting the use, modification, and distribution of the source code for any purpose. This is important because it allows software distributions to distribute all included software freely. Software that is made available under more restrictive licenses, such as for "academic nonprofit use only," cannot be distributed in this way. An example is PAML that used to have such a license. Only when it was changed PAML got included into Debian, etc. Also, for this book chapter, we asked the author of pal2nal to add a proper license. After adding the GPLv2, it became part of the Debian distribution; *see* also https://tracker.debian.org/pkg/pal2nal. This means that above Docker scripts can be updated to install the pal2nal Debian package.

When you use scientific software, always check the type of license under which it is provided, to understand what you can or cannot do with it. When you publish software, add a license along with your code, so others can use it and distribute it.

Typical licenses used in bioinformatics are MIT (Expat) and BSD, which are considered very permissive, and also GPL and the Apache License, which are designed to grant additional protections with regard to derivative works and patentability. Whenever possible, free software licenses such as mentioned above are encouraged for scientific software. Check the guidelines of your employer and funding agencies.

3 Create a Scalable and Reusable Workflow

3.1 Example Workflow

We have created a number of examples to test a scalable and reproducible workflow, the full code, and examples that are available on GitHub [30]. In this case putative gene families of the oomycete *Phytophthora infestans* are tested for evidence of positive selection. *P. infestans* is a single-cell pathogen, which causes late blight of potato and tomato. Gene families under positive selection pressure may be involved in protein–protein interactions and are potentially of interest for fighting late blight disease.

As an example the *P. infestans* genome data [36] was fetched from http://www.broadinstitute.org/annotation/genome/phytophthora_infestans/MultiDownloads.html, and predicted genes were grouped by \name{blastclust} using 70% identity (see also Chapter. 21). This resulted in 72 putative gene families listed on the online repository on GitHub [30].

The example workflow aligns amino acid sequences using Clustal Omega, creates a neighbor joining tree, and runs CodeML from the PAML suite. The following is one example to look for evidence of positive selection in a specific group of alignments:

```
clustalo -i data/clusterXXXXX/aa.fa --guidetree-out=data/
clusterXXXXXX/aa.ph > data/clusterXXXXXX/aa.aln
pal2nal.pl -output paml data/clusterXXXXX/aa.aln data/clus-
terXXXXX/nt.fa > data/clusterXXXXX/alignment.phy
cd data/clusterXXXXX
Codeml ../paml0-3.ctl
```

First we align amino acid with Clustal Omega, followed by translation to a nucleotide alignment with pal2nal. Next we test for evidence of positive selection using PAML's \name{Codeml} with models M0–M3. Note that the tools and settings used here are merely chosen for educational purposes. The approach itself here may result in false positives, as explained by Schneider et al. [37]. Also, PAML is not the only software that can test for evidence of positive selection, for example, the HyPhy molecular evolution and statistical sequence analysis software package contains similar functionality and uses MPI to parallelize calculations [38]. PAML is used here because it is a reference implementation and is suitable as an example how a legacy single-threaded bioinformatics application can be parallelized in a workflow.

In the next section, different workflow systems are presented that can be used to run the described analysis: in a scalable and reproducible manner, locally on a desktop, on a computer cluster, or in the cloud. All the code and data to run these examples is available on GitHub [30]. To load the code on your desktop, clone the git repository locally. The examples can be executed from the repository tree:

```
git clone https://github.com/EvolutionaryGenomics/scalabil-
ity-reproducibility-chapter.git
```

3.2 Common Workflow Language

Common workflow language (CWL, http://www.commonwl.org/) is a standard for describing workflows that are portable across a variety of computing platforms [39]. CWL is a specification and not a software in itself though it comes with a reference implementation which can be run with Docker containers. CWL promotes an ecosystem of implementations and supporting systems to execute the workflows across multiple platforms. The promise is that when you write a workflow for, e.g., Arvados, it should also run on another implementation, e.g., Galaxy.

Given that CWL takes inspiration from previously developed tools and GNU Make in particular [40], the order of execution in a CWL workflow is based on dependencies between the required tasks. However unlike GNU Make, CWL tasks are defined to be isolated, and you must be explicit about inputs and outputs. The benefits of explicitness and isolation are flexibility, portability, and scalability: tools and workflows described with CWL can transparently leverage software deployment technologies, such as Docker, and can be used with CWL implementations from different vendors, and the language itself can be applied to describe large-scale workflows that run in HPC clusters, or the cloud, where tasks are scheduled in parallel across many nodes.

CWL workflows are written in JSON or YAML formats. A workflow consists of blocks of steps, where each step in turn is made up of a task description that includes the inputs and outputs of the task itself. The order of execution of the tasks is determined automatically by the implementation engine. In the GitHub repository, we show an example of a CWL workflow to describe the analysis over the protein alignments. To test the workflow, you will need the CWL reference runner implementation:

```
pip install cwlref-runner
```

and then to run the example from the repository tree:

```
CWL/workflow.cwl --clusters data
```

To run the CWL workflow on a grid or cloud multi-node system, we can install another CWL implementation, this one built upon the toil platform [41]:

```
pip install toil[cwl]
toil-cwl-runner CWL/workflow.cwl --clusters data
```

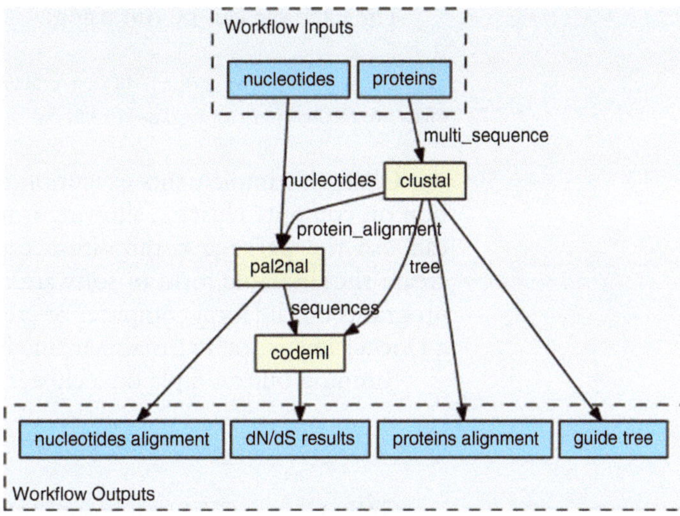

Fig. 1 Workflow automatically generated from the CWL schema displays how PAML's Codeml receives inputs from two sources and outputs the d_N/d_S information. A workflow engine figures out that it has to run clustal first, followed by pal2nal and Codeml as a linear sequence. For each input, the job can be executed in parallel

CWL comes with extra tooling, such as visualization of CWL workflows (Fig. 1). *See* view.commonwl.org for more examples.

3.3 Guix Workflow Language

The Guix Workflow Language (GWL) extends the functional package manager GNU Guix [32] with workflow management capabilities. GNU Guix provides an embedded domain-specific language (EDSL) for packages and package composition. GWL extends this EDSL with processes and process composition.

In GWL, a process describes the computation, for example, running the clustalo program. A workflow in the GWL describes how processes relate to each other. For example, the Codeml program can only run after both clustalo and pal2nal finished successfully.

The tight coupling of GWL and GNU Guix ascertains that not only the workflow is described rigorously but also the deployment of the programs on which the workflow depends.

To run the GWL example, you need to install GNU Guix (https://www.gnu.org/software/guix/manual/html_node/Binary-Installation.html) and the GWL installed on your computer. Once GNU Guix is available, installing GWL can be done using:

```
guix package -i gwl
```

The example can be run using:

```
cd scalability-reproducibility-chapter/GWL
guix workflow -r example-workflow
```

GWL also implements execution engines to offload computation on compute clusters, allowing it to scale. The process engines can use the package composition capabilities of GNU Guix to create the desirable form of software deployment—be it installing programs on the local computer or creating an application bundle, a Docker image, or a virtual machine image.

Running our example on a cluster that has Grid Engine:

```
guix workflow -r example-workflow -e grid-engine
```

GNU Guix + GWL can ensure full reproducibility of an analysis, including all software dependencies—all the way down to glibc. GNU Guix computes a unique string, a hash, on the complete set of inputs and the build procedure of a package. It can guarantee that a package is built with the same source code, dependency graph, and the same build procedure, and produces identical output. In GWL for each process and workflow, a hash is computed of the packages, the procedure, and the execution engine. By comparing hashes it is not only possible to compare whether the workflow is running using the exact same underlying software packages, and using the same procedures, but also the full graph of dependencies can be visualized. To obtain such an execution plot:

```
guix package -i graphviz
guix workflow -g example-workflow | dot -Tpdf > example-workflow.pdf
```

Note that, unlike the other workflow solutions discussed here, GWL does not use the time stamps of output files. The full dependency graph is set before running the tools, and it only needs to check whether a process returns an error state. This means that there are no issues around time stamps and output files do not have to be visible to the GWL engine.

3.4 Snakemake

Snakemake [42] is a workflow management system that takes inspiration from GNU Make [40], a tool to coordinate the compilation of large programs consisting of interdependent source files (https://snakemake.readthedocs.io/en/stable/).

Snakemake provides a DSL that allows the user to specify generator rules. A rule describes the steps that need to be performed to produce one or more output files, such as running a shell script. These output files may be used as inputs to other rules. The workflow is described as a graph in which the nodes are files (provided input files, generated intermediate files, or the desired output files) and the edges are inferred from the input/output interdependencies of connected rules.

When a user requests a certain file to be generated, Snakemake matches the file name against concrete or wildcard rules, traverses the graph from the target file upward, and begins processing the steps for every rule for which no new output file is available. Whether or not an output file is considered new depends on its time stamp relative to the time stamp of prerequisite input files. In doing so, Snakemake only performs work that has not yet been done or for which the results are out of date, just like GNU Make. Snakemake can be configured to distribute jobs to batch systems or to run jobs on the local system in parallel. The degree of parallelization depends on the dependencies between rules.

Snakemake is written in Python and allows users to import Python modules and use them in the definition of rules, for example. It also has special support for executing R scripts in rules, by exposing rule parameters (such as inputs, outputs, concrete values for wildcards, etc.) as an S4 object that can be referenced in the R script.

Snakemake provides native support for the Conda package manager. A rule may specify a Conda [35] environment file describing a software environment that should be active when the rule is executed. Snakemake will then invoke Conda to download the required packages as specified in the environment file. Alternatively, Snakemake can interface with an installation of the Singularity container system [26] and execute a rule within the context of a named application bundle, such as a Docker image.

To run the Snakemake workflow, you need to install Snakemake (example showed with Conda):

```
conda install -y -c bioconda snakemake=4.2.0
```

And then to run the example from the repository tree:

```
cd Snakemake
snakemake
```

3.5 Nextflow

Nextflow [43] is a framework and an orchestration tool that enables scalable and reproducible scientific workflows using software containers (https://www.nextflow.io/). It is written in the Groovy JVM programming language [44] and provides a domain-specific language (DSL) that simplifies writing and deploying complex workflows across different execution platforms in a portable manner.

A Nextflow pipeline is described as a series of processes, where each process can be written in any language that can be executed or interpreted on Unix-like operating systems (e.g., Bash, Perl, Ruby, Python, etc.). A key component of Nextflow is the dataflow programming model, which is a message-based abstraction for parallel programming similar to the CSP paradigm (*see* [23]). The main difference between CSP and dataflow is that in the former, processes communicate via synchronous messages, while in the latter, the messages are sent in an asynchronous manner. This approach is useful when deploying large distributed workloads because it has latency tolerance and error resilience. In practical term the dataflow paradigm uses a push model in which a process in the workflow sends its outputs over to the downstream processes that waits for the data to arrive before starting their computation. The communication between processes is performed through channels, which define inputs and outputs for each process. Branches in the workflow are also entirely possible and can be defined using conditions that specify if a certain process must be executed or not depending on the input data or on user defined parameters.

The dataflow paradigm is the closest representation of a pipeline idea where, after having opened the valve at the beginning, the flow progresses through the pipes. But Nextflow can handle this data flow in a parallel and asynchronous manner, so a process can operate on multiple inputs and emit multiple outputs at the same time. In a simple workflow where, for instance, there are 100 nucleotide sequences to be aligned with the NCBI NT database using BLAST, a first process can compute the alignment of the 100 sequences independently and in parallel, while a second process will wait to receive and collect each of the outputs from the 100 alignments to create a final results file. To allow workflow portability, Nextflow supports multiple container technologies such as Docker and Singularity and integrates natively with Git and popular code sharing platforms, such as GitHub. This makes it possible to precisely prototype self-contained computational workflows, tracking also all the modifications over time and ensuring the reproducibility of any former configuration. Nextflow allows executing workflows across different computing platforms by supporting several cluster schedulers (e.g., SLURM, PBS, LSF and SGE) and allowing direct execution on the Amazon cloud (AWS), using services, such as AWS Batch or automating the creation of a compute cluster in the cloud for the user.

To run the Nextflow example, you need to have Java 8 and a Docker engine (1.10 or higher) installed. Next install Nextflow with:

```
curl -s https://get.nextflow.io | bash
```

Run the example from the repository tree:

```
./nextflow run Nextflow/workflow.nf -with-docker evolutionar-
ygenomics/scalability
```

To save the graph of the executed workflow, it is sufficient to add the option "-with-dag workflow.pdf." The same example can also be run without Docker if the required packages have been installed locally following the Bioconda or Guix examples. In this case you can omit the "-with-docker" instruction. To run the example on a compute cluster or in the cloud, it is sufficient to specify a different executor (e.g., sge or awsbatch) in the Nextflow configuration file and ensure that those environments are configured to properly work with the Docker container.

4 Discussion

In this chapter we show how to describe and execute the same analysis using a number of workflow systems and how these follow different approaches to tackle execution and reproducibility issues. It is important to assess underlying design choices of these solutions and also to look at the examples we provide online. Even though it may look attractive to opt for the simplest choices, it may be that the associated maintenance burden may be cause for regret later.

The workflow tools introduced in this chapter offer direct integration of software packages. The overall advantage of the bundling software approach is that when software deployment and execution environment are controlled, the logic of the analysis pipeline can be developed separately using descriptive workflows. This separation allows communities to build best practice shareable pipelines without worrying too much about individual system architectures and the underlying environments. An example is the effort by the Global Alliance for Genomics and Health (GA4GH, https://www.ga4gh.org) to develop and share best practice analysis workflows with accompanying container images [45].

In this chapter we also discussed the scaling up of computations through parallelization. In bioinformatics, the common parallelization strategy is to take an existing nonparallel application and divide data into discrete units of work, or jobs, across multiple CPUs and clustered computers. Ideally, running jobs in parallel on a single multicore machine shows linear performance increase for every CPU added, but in reality it is less than linear [46]. Resource contention on the machine, e.g., disk or network I/O, may have processes wait for each other. Also, the last, and perhaps longest, running job causes total timing to show less than linear performance, as the already finished CPUs are idle. In addition to the resource contention on a single machine, the network introduces latencies when data is moved around.

Running the example workflow in the cloud has similar performance and scalability compared to running it on a local infrastructure, after adjusting for differences in hardware and network speeds. Cloud computing is an attractive proposition for scaling up calculation jobs and storing data. Cloud prices for virtual servers and data storage have decreased dramatically, and the possibility of using spot or preemptible instances (i.e., virtual servers that can be priced down to 70% or 80% the normal price but that can be shut down in any moment by the cloud provider) is making cloud computing solutions competitive for high-performance and scientific computing. Cloud essentially outsources hardware and related plumbing and maintenance. Sophisticated tooling allows any researcher to run software in the cloud. We predict an increasing number of groups and institutes will move from large-scale HPC clusters toward tight HPC cluster solutions that can handle continuous throughput with burst compute in the cloud.

Reproducibility is a prime concern in science. Today several solutions are available to address reproducibility concerns. Systems such as Docker and Singularity are built around bundling binary applications and executing them in a container context. Advanced package managers such as Conda or Guix allow the user to create separate software environments where different application versions can be deployed without collisions while ensuring control and traceability over changes and dependencies. All these solutions represent a different approach to address the reproducibility challenge while also offering a different user experience and requiring different setups to work properly. For instance, container-based systems such as Docker and Singularity are not always a viable option in HPC environments since they may require updates to the existing computing infrastructure. Also, HPC operating system installations may include kernel versions that do not allow for the so-called user namespaces, a fundamental component among the many kernel features that together allow an application to run in an isolated container environment. Another downside of containers is that it is hard to assess what is in them—they act like a black box. When creating containers with above Docker scripts, it depends on the time they are assembled what goes in. A Debian or Conda update between creating containers, for example, may include a different software version therefore a different dependency graph. Only GNU Guix containers provide a clear view on what is contained.

Containers provide isolation from the underlying operating system. On HPC environments it may be required to run software outside a container. While applications built with Guix or Conda can be run in isolation when container support is available, they do not require these features at runtime. As a package manager Conda, neither depends on container features nor on root privileges, but it pays for this convenience with a lack of both process isolation and bit-reproducibility [47]. GNU Guix, meanwhile, provides the most rigorous path to reproducible software deployment. In order to guarantee that packages are built in a bit-reproducible fashion and share binary packages, Guix requires to store packages in the directory /gnu/store. There are several work-arounds for this; one of them is by using containers, and another is by mounting /gnu/store from a host that has built privileges for that directory. A third option is to build packages targeted at a different directory, but this loses the bit-reproducibility and the convenience of binary installs. A fourth option is to provide relocatable binary installation packages that can be installed in a user available directory, similar to what Bioconda does. Such packages exist for sambamba, gemma, and the D-compiler.

Finally, each combination of these packaging and workflow solutions occupies a slightly different region in the solution space for the scalability and reproducibility challenge. Fortunately, the packaging tools can be used next to each other without interference, thereby providing a wealth of software packages for bioinformatics. Today, there is hardly ever a good reason to build software from source.

5 Questions

1. Using one of the packaging or container systems described (e.g., Conda, Guix, or Docker), prepare a working environment to run the examples. Now try to run the workflows using the tools presented and appreciate the different approaches to execute the same example.
2. Compare the different syntaxes used by the tools to define a workflow and explore how each tool describes the processes and the dependencies in a different way.
3. Use the Amazon EC2 calculation sheet, and calculate how much it would cost to store 100 GB in S3, and execute a calculation on 100 "large" nodes, each reading 20 GB of data. Do the same for another cloud provider.

References

1. Ronquist F, Huelsenbeck JP (2003) MrBayes 3: Bayesian phylogenetic inference under mixed models. Bioinformatics 19 (12):1572–1574
2. Eddy SR (2008) A probabilistic model of local sequence alignment that simplifies statistical significance estimation. PLoS Comput Biol 4 (5):e1000069
3. Yang Z (1997) PAML: a program package for phylogenetic analysis by maximum likelihood. Comput Appl Biosci 13(5):555–556
4. Doctorow C (2008) Big data: welcome to the petacentre. Nature 455(7209):16–21
5. Trelles O, Prins P, Snir M, Jansen RC (2011) Big data, but are we ready? Nat Rev Genet 12 (3):224

6. Durbin RM, Abecasis GR, Altshuler DL et al (2010) A map of human genome variation from population-scale sequencing. Nature 467(7319):1061–1073
7. Schadt EE, Linderman MD, Sorenson J, Lee L, Nolan GP (2010) Computational solutions to large-scale data management and analysis. Nat Rev Genet 11(9):647–657
8. Leipzig J (2017) A review of bioinformatic pipeline frameworks. Brief Bioinform 18 (3):530–536
9. Jeffrey D, Sanjay G (2004) Mapreduce: simplified data processing on large clusters
10. White T (2009) Hadoop: the definitive guide, 1st edn. O'Reilly, Sebastopol, CA
11. (2009) Slurm workload manager. https://slurm.schedmd.com
12. Graham RL, Woodall TS, Squyres JM (2005) Open MPI: a flexible high performance MPI
13. Stamatakis A, Ott M (2008) Exploiting fine-grained parallelism in the phylogenetic likelihood function with mpi, pthreads, and openmp: a performance study. Pattern Recognition in Bioinformatics. Springer, Berlin, pp 424–435
14. {r D e v e l o p m e n t C o r e T e a m} (2010) R: a language and environment for statistical computing
15. Cesarini F, Thompson S (2009) Erlang programming, 1st edn. O'Reilly Media, Inc., Sebastopol, CA
16. Hudak P, Peterson J, Fasel J (2000) A gentle introduction to haskell, version 98. http://haskell.org/tutorial/
17. Odersky M, Spoon L, Venners B (2008) Programming in scala. Artima, Walnut Creek CA
18. Bezanson J, Karpinski S, Shah VB, Edelman A (2012) Julia: a fast dynamic language for technical computing. CoRR. abs/1209.5145
19. Okasaki C (1998) Purely functional data structures. Cambridge University Press, Cambridge
20. Alexandrescu A (2010) The D programming language, 1st edn. Addison-Wesley Professional, Boston, MA. 460p
21. Griesemer R, Pike R, Thompson K (2009) The Go programming language. http://golang.org/
22. Hoare CAR (1978) Communicating sequential processes. Commun ACM 21:666–677
23. Welch P, Aldous J, Foster J (2002) Csp networking for java (jcsp. net). Comput Sci 2002:695–708
24. Sufrin B (2008) Communicating scala objects. Communicating Process Architectures, p 35
25. Di tommaso P, Palumbo E, Chatzou M et al (2015) The impact of docker containers on the performance of genomic pipelines. PeerJ 3: e1273
26. Kurtzer GM, Sochat V, Bauer MW (2017) Singularity: scientific containers for mobility of compute. PLoS One 12(5):e0177459
27. Sochat V (2017) Singularity registry: open source registry for singularity images. J Open Source Soft 2(18):426
28. Möller S, Krabbenhoft HN, Tille A et al (2010) Community-driven computational biology with debian linux. BMC Bioinformatics 11 (Suppl 12):S5
29. Möller S, Afgan E, Banck M et al (2014) Community-driven development for computational biology at sprints, hackathons and codefests. BMC Bioinformatics 15(14):S7
30. Strozzi F, Wurmus R, Roel J et al (2017) Data, workflow example and docker files for scalability and reproducibility chapter. https://github.com/EvolutionaryGenomics/scalability-reproducibility-chapter
31. Strozzi F, Wurmus R, Roel J et al (2017) Docker images for scalability and reproducibility chapter. https://hub.docker.com/u/evolutionarygenomics/
32. Courtès L (2013) Functional package management with guix. CoRR. abs/1305.4584
33. Sievers F, Wilm A, Dineen D et al (2011) Fast, scalable generation of high-quality protein multiple sequence alignments using clustal omega. Mol Syst Biol 7:539
34. Suyama M, Torrents D, Bork P (2006) Pal2nal: robust conversion of protein sequence alignments into the corresponding codon alignments. Nucleic Acids Res 34(Web Server Issue):W609–W612
35. Grüning B, Dale R, Sjödin A et al (2017) Bioconda: a sustainable and comprehensive software distribution for the life sciences. bioRxiv
36. Haas BJ, Kamoun S et al (2009) Genome sequence and analysis of the irish potato famine pathogen phytophthora infestans. Nature 461 (7262):393–398
37. Schneider A, Souvorov A, Sabath N et al (2009) Estimates of positive darwinian selection are inflated by errors in sequencing, annotation, and alignment. Genome Biol Evol 1:114–118
38. Pond SL, Frost SD, Muse SV (2005) HyPhy: hypothesis testing using phylogenies. Bioinformatics 21(5):676–679
39. Amstutz P, Crusoe MR, Tijanić N et al (2016) Common workflow language, v1.0
40. Stallman R, Mcgrath R (1989) Gnu make: a program for directing recompilation. Free Software Foundation, Boston, MA

41. Vivian J, Rao AA, Nothaft FA et al (2017) Toil enables reproducible, open source, big biomedical data analyses. Nat Biotechnol 35 (4):314–316
42. Köster J, Rahmann S (2012) Snakemake--a scalable bioinformatics workflow engine. Bioinformatics 28(19):2520–2522
43. Di tommaso P, Chatzou M, Floden EW et al (2017) Nextflow enables reproducible computational workflows. Nat Biotechnol 35 (4):316–319
44. Koenig D, Glover A, King P, Laforge G, Skeet J (2007) Groovy in action. Manning Publications Co. Greenwich, CT
45. (2017) Ga4gh platform for docker-based tools and workflows sharing. https://dockstore.org
46. Amdahl GM (1967) Validity of the single processor approach to achieving large scale computing capabilities. In: Proceedings of the April 18-20, 1967, Spring Joint Computer Conference. ACM, Washington, DC, pp 483–485
47. (2015) Reproducible builds https://reproducible-builds.org. https://reproducible-builds.org

Chapter 25

Sharing Programming Resources Between Bio* Projects

Raoul J. P. Bonnal, Andrew Yates, Naohisa Goto, Laurent Gautier, Scooter Willis, Christopher Fields, Toshiaki Katayama, and Pjotr Prins

Abstract

Open-source software encourages computer programmers to reuse software components written by others. In evolutionary bioinformatics, open-source software comes in a broad range of programming languages, including C/C++, Perl, Python, Ruby, Java, and R. To avoid writing the same functionality multiple times for different languages, it is possible to share components by bridging computer languages and Bio* projects, such as BioPerl, Biopython, BioRuby, BioJava, and R/Bioconductor.

In this chapter, we compare the three principal approaches for sharing software between different programming languages: by remote procedure call (RPC), by sharing a local "call stack," and by calling program to programs. RPC provides a language-independent protocol over a network interface; examples are SOAP and Rserve. The local call stack provides a between-language mapping, not over the network interface but directly in computer memory; examples are R bindings, RPy, and languages sharing the Java virtual machine stack. This functionality provides strategies for sharing of software between Bio* projects, which can be exploited more often.

Here, we present cross-language examples for sequence translation and measure throughput of the different options. We compare calling into R through native R, RSOAP, Rserve, and RPy interfaces, with the performance of native BioPerl, Biopython, BioJava, and BioRuby implementations and with call stack bindings to BioJava and the European Molecular Biology Open Software Suite (EMBOSS).

In general, call stack approaches outperform native Bio* implementations, and these, in turn, outperform "RPC"-based approaches. To test and compare strategies, we provide a downloadable Docker container with all examples, tools, and libraries included.

Key words Bioinformatics, R, Python, Ruby, Perl, Java, Web services, RPC, EMBOSS, PAML

1 Introduction

Bioinformatics has created its tower of Babel. The full set of functionality for bioinformatics, including statistical and computational methods for evolutionary biology, is implemented in a wide range of computer languages, e.g., Java, C/C++, Perl, Python, Ruby, and R. This comes as no surprise, as computer language design is the

Download: https://github.com/EvolutionaryGenomics/Cross-language-interfacing

Maria Anisimova (ed.), *Evolutionary Genomics: Statistical and Computational Methods*, Methods in Molecular Biology, vol. 1910, https://doi.org/10.1007/978-1-4939-9074-0_25, © The Author(s) 2019

result of multiple trade-offs, for example, in strictness, convenience, and performance. In this chapter we discuss strategies for combining solutions from different languages and look at performance implications of combining cross-language functionality. In the process we also highlight implications of such strategic choices.

Computer languages used in bioinformatics today typically fall into two groups: those compiled and those interpreted. Java, C++, and D, for example, are statically typed compiled languages, while R, Perl, Ruby, and Python are dynamically typed interpreted languages. In principle, a compiled language is converted into machine code once by a language compiler, and an interpreted language is compiled every time at runtime, the moment it is run by an interpreter. Static typing allows a compiler to optimize machine code for speed. Dynamic typing requires an interpreter and resolves variable and function types at runtime. Such design decisions cause Java, C++, and D to have stronger compile-time type checking and faster execution speed than R, Perl, Ruby, and Python. When comparing runtime performance of these languages, compiled statically typed languages, such as C++, D, and Java, generally outperform interpreted dynamically typed languages, such as Python, Perl, and R. For speed comparison between languages, see, for example, the benchmarks game.

Statically typed compiled languages tend to produce faster code at runtime

Runtime performance, however, is not the only criterion for selecting a computer language. R, Perl, Ruby, and Python offer sophisticated interactive analysis of data in an interpreted shell which is not directly possible with C++, D, or Java. Another important criterium may be conciseness. Interpreted languages generally allow functionality to be written in less lines of code. The number of lines matter, as it is often easier to grasp something expressed in a short and concise fashion, if done competently, leading to easier coding and maintenance of software and resulting in increased programmer productivity. In general, with R, Perl, Ruby, and Python, it takes less lines of code to write software than with C++, D, or Java; this is also visible from the examples in the benchmarks game.

Interpreted languages allow for concise code that is easier to read and results in increased programmer productivity

Based on the conciseness criterium, computer languages fall into these two groups. This suggests a trade-off between execution speed and conciseness/programmer productivity. Even so, strong typing may help later when refactoring code, perhaps regaining some of that lost productivity. The authors also note that in their experience, the more programming languages one masters, the easier it becomes mastering new languages (with the exception, perhaps, of Haskell). Learning new programming languages is important when writing software.

Logically, to fully utilize the potential of existing and future bioinformatics functionality, it is necessary to bridge between computer languages. Bioinformaticians cannot be expected to master every language, and it is inefficient to write the same functionality for every language. For example, R/Bioconductor contains unique and exhaustive functionalities for statistical methods, such as for gene expression analysis [1]. The singular implementation of this functionality in R has caused researchers to invest in learning the R language. Others, meanwhile, have worked on building bridges between languages. For example, RPy and Rserve allow accessing R functionality from Python [2], and JRI and Rserve allow accessing R functionality from Java [3, 4]. Other languages have similar bindings, such as RSRuby that allows accessing R from Ruby.

Discussing other important criteria for selecting a programming language, such as ease of understanding, productivity, portability, and the size and dynamics of the supporting Bio* project developer communities, is beyond the scope of this chapter. The authors, who have different individual preferences, wish to emphasize that every language has characteristics driven by language design and there is no single perfect all-purpose computer language. In practice, the choice of a computer language depends mainly on the individuals involved in a project, partly due to the investment it takes to master a language. Researchers and programmers have prior investments and personal preferences, which have resulted in a wide range of computer languages used in the bioinformatics community.

Contrasting with singular implementations, every mainstream Bio* project, such as BioPerl [5], Biopython [6], BioRuby [7], R/Bioconductor [1], BioJava [8], the European Molecular Biology Open Software Suite (EMBOSS) [9], and Bio++ [10], contains duplication of functionality. Every Bio* project consists of a group of volunteers collaborating at providing functionality for bioinformatics, genomics, and life science research under an open-source software (OSS) license. The BioPerl project does that for Perl, BioJava for Java, etc. Next to the language used, the total coverage of functionality, and perhaps quality of implementation, differs between projects. Not only is there duplication of effort, both in writing and testing code, but also there are differences in implementation, completeness, correctness, and performance. For example, implementations between projects differ even for something as straightforward as codon translation, e.g., in number of types of encoding and support for the translating of ambiguous nucleotides. EMBOSS, uniquely, attempts to predict the final amino acid in a sequence, even when there are only two nucleotides available for the last codon.

Whereas Chapter 25 discusses Internet data resources and how to share them, in this chapter, we discuss how to share functional resources by interfacing and bridging functionality between different computer languages. This is highly relevant to evolutionary biology as most classic phylogenetic resources were written in C, while nowadays phylogenetic routines are written in Java, Perl, Python, Ruby, and R. Especially for communities with relatively few software developers, we argue here that it is important to bridge these functional resources from multiple languages. For bridging, strategies are here discussed to invoke one program from another, use some form of remote procedure calls (RPC), or use a local call stack.

1.1 Bridging Functional Resources Calling from Program to Program

The most simple way of interfacing software is by invoking one program from another. This strategy is often used in Bio* projects, for example, for invoking external programs. A regular subset would be PAML [11], HMMER [12], ClustalW [13], MAFFT [14], Muscle [15], BLAST [16], and MrBayes [17]. The Bio* projects typically contain modules which invoke the external program and parse the results. The advantage of this approach is that it mimics running a program on the command line, so invocation is straightforward. Another advantage, in a web service context, is that if the called program crashes, it does not have to take the whole service down. There are also some downsides, however. Loading a new instance of a program every time incurs extra overhead. More importantly, nonstandard input and output makes the interface fragile, i.e., what happens when input or output differs between two versions of a program? A further downside is that external programs do not have fine-grained function access and have no support for advanced error handling and exceptions. What happens, for example, when the invoked program runs out of process memory? How to handle that gracefully? A final complication is that such a program is an external software deployment dependency, which may be hard to resolve for an end user.

1.2 Remote Procedure Call

In contrast to calling one program from another, true cross-language interfacing allows one language to access functions and/or objects in another language, as if they are native function calls. To achieve transparent function calls between different computer languages, there are two principal approaches. The first approach is for one language to call directly into another language's function or method over a network interface, the so-called remote procedure call (RPC). The second approach is to call into another language over a local "call stack."

In bioinformatics, cross-language RPC comes in the form of web services and binary network protocols. A web service application programming interface (API) is exposed, and a function call gets translated with its parameters into a language-independent format, a procedure called "marshalling." After calling the function on a server, the result is returned in, for example, XML and translated back through "unmarshalling." Examples of cross-language XML protocols are SOAP [18] and XML/RPC [19].

More techniques exist for web service-type cross-language RPC. For example, representational state transfer (REST), or ReSTful [20], is a straightforward HTTP protocol, often preferred over SOAP because of its simplicity. Another XML-based protocol is Resource Description Framework (RDF), as part of the semantic web specification. Both REST and RDF can be used for RPC solutions.

In addition, binary alternatives exist because XML-based protocols are not very efficient. XML is verbose, increasing the data load, and requires parsing at both marshalling and unmarshalling steps. In contrast, binary protocols are designed to reduce the data transfer load and increase speed. Examples of binary protocols are Rserve [3], which is specifically designed for R, and Google protocol buffers [21]. Another software framework based on a binary protocol is Thrift, by the Apache software foundation, designed for scalable cross-language service development [22]. Finally, also worth considering are very fast interoperable messaging-based paradigms, such as ZeroMQ [23], and high-level message-level optimizers, such as GraphQL.

1.3 Local Call Stack

The alternative to RPC is to create native local bindings from one language to another using a shared native call stack, essentially linking into code of a different computer language. With the call stack, function calls do not run over the network but over a stack implementation in shared computer memory. In a single virtual machine, such as the JVM and Erlang Beam, compiled code can share the same call stack, which can make cross-language calling efficient. For example, the languages Java, Jython, JRuby, Clojure, Groovy, and Scala can transparently call into each other when running on the same virtual machine using native speeds.

Native call stack sharing is also supported at the lowest level by the computer operating system through compiled shared libraries. These shared libraries have an extension .so on Linux, .dylib on OSX, and .dll on Windows. The shared libraries are designed so that they contain code and data that provide services to independent programs, which allows the sharing and changing of code and data in a modular fashion. Shared library interfaces are well defined at the operating system level, and languages have a way of binding them. Specialized interface bindings to shared libraries exist for every language, for example, R's C modules, the Java Native Interface (JNI) for the JVM, Foreign Function Interfaces (FFI) for Python and Ruby, the Parrot native compiler interface PerlXS for Perl.

With (dynamic) shared libraries, certain algorithms can be written in a low-level, high-performance compiled computer language, such as C/C++, D, or FORTRAN. And high-level languages, such as Perl, Python, Ruby, R, and even Java, can access these algorithms. This way, languages can be mixed to optimize solutions. Creating these shared library interfaces, however, can be a tedious exercise, which often calls for code generators. One such generator is the Simplified Wrapper and Interface Generator (SWIG) [24], which consists of a macro-language, a C header file parser, and the tools to bind low-level shared libraries to a wide range of languages. For C/C++, SWIG can parse the header files and generate the bindings for other languages, which, in turn, call into these shared libraries. The Boost project has similar facilities for mapping calls to SWIG. C FFI's that come with programming languages, such as Python's CFFI and Ruby's FFI, tend to be the easiest to work with.

Even though this extensive functionality for interfacing is available, the full potential of creating cross-language adapters is not fully exploited in bioinformatics. Rather than bridge two languages, researchers often opt to duplicate functionality. This is possibly due to a lack of information on the effort involved and the added complexity of creating a language bridge. Also, the impact on performance may be an unknown quantity. A further complication is the need to understand, to some degree, both sides of the equation, i.e., to provide an R function to Python requires some understanding of both R and Python, at least to the level of reading the documentation of the shared module and creating a working binding. Likewise, binding Python to C using a call stack approach requires some understanding of both Python and C. Sometimes, binding of complex functions can be daunting, and deployment may be a concern, e.g., when creating shared library bindings on Linux, they may not easily work on Windows or macOS.

1.4 Comparing Approaches

Here, we compare bridging code from one language to another using the RPC approach and the call stack approach. As a comparison we also provide a program-to-program approach and show how dependencies can be fixated. The comparison is done in the form of short experiments (scripts) which can be executed by the reader. To measure performance between different approaches, we use codon translation as an example of shared functionality between Bio* projects. Codon translation is a straightforward algorithm with table lookups. Such sequence translation is representative of many bioinformatics tasks that deal with genome-sized data and require many function calls with small-sized parameters.

In this chapter we first focus on comparing R and Python bindings. We include native Bio* implementations, i.e., Biopython, BioRuby, BioPerl, BioJava, and EMBOSS (C) for an absolute speed comparison. Next we try bindings on the JVM.

Examples and tests can in principle be experimented with a computer running Linux, macOS, or Windows. To ease trials, we have defined GNU Guix packages that contain the tools and their dependencies. From this we have created a downloadable Docker image that supports all interfaces and performance examples (GNU Guix and Docker are discussed in Chapter 25).

2 Results

2.1 Calling into R

R is a free and open-source environment for statistical computing and graphics [25]. R comes with a wide range of functionality, including modules for bioinformatics, such as bundled in R/Bioconductor [1]. R is treated as a special citizen in this chapter because the language is widely used and comes with statistical algorithms for evolutionary biology, such as Ape [26] and SeqinR [27], both available through the comprehensive R archive network (CRAN).

R defines a clear interface between the high-level language R and low-level highly optimized C and FORTRAN libraries, some of which have been around for a long time, such as the libraries for linear regression and linear algebra. In addition, the R environment successfully handles cross-platform packaging of C, C++, FORTRAN, and R code. The combination of features has resulted in R becoming the open-source language of choice in a number of communities, including statistics and some disciplines in biology. R/Bioconductor has gene expression analysis [1] and R/qtl [28] and R/qtlbim [29] for QTL mapping (see also QTL mapping in Chapter 21). Not all is lost, however, for those not comfortable with the R language itself. R can act as an intermediate between functionality and high-level languages. A number of libraries have been created that interface to R from other languages, either providing a form of RPC, through RSOAP or Rserve, or a call stack interface calling into the R-shared library and executing R commands, for example, RPy for Python, RSPerl for Perl, RSRuby for Ruby, and JRI for Java. Of the last call stack approaches, RPy currently has the most complete implementation; *see* also [2].

In this chapter, we compare different approaches for invoking full R functionality from another language. To test cross-language calling, we elected to demonstrate codon translation. Codon-to-protein amino acid translation is representative for a relatively simple computation that potentially happens thousands of times with genome-sized data. Every Bio* project includes such a translation function, so it is a fair way to test for language interoperability and performance. For data, we use a WormBase [30] C. *elegans* cDNA FASTA file (33 Mb), containing 24,652 nucleotide sequences, predicted to translate to protein (Fig. 1).

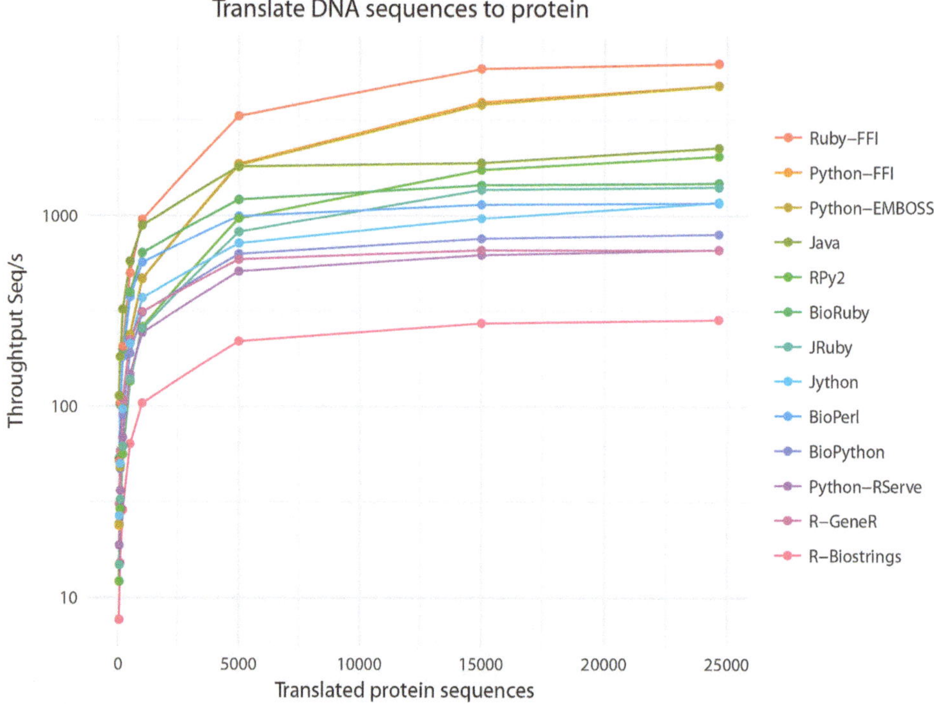

Fig. 1 Throughput of mRNA to protein translation using combinations of cross-language calling with a range of programming resources. WormBase *C. elegans* predicted protein coding DNA that was parsed in FASTA format and translated into amino acids. Tests were executed inside a container. Different file sizes were used containing 500, 1000, 5000, 15,000, and 25,000 sequences (*X*-axis) and the number of sequences processed per seconds (*Y*-axis $\log_{10}$ scale). Measurements were taken on an AMD Opteron(TM) 6128 8 cores at 2.0 GHz, 4 sockets × 8 cores, with 512 GB RAM DDR3 ECC, and an HDD SATA of 2 TB. Broadly the figure shows that sustained throughput is reached quickly and flattens out. R-Biostrings performs poorly at 285 Seq/s, while R-GeneR and Rserve (Python+Rserve+GeneR) perform at the level of native Bio* libraries, respectively, 658 Seq/s and 660 Seq/s. The cross-language Ruby-FFI at 6256 Seq/s calls EMBOSS C translation and outperforms all others

2.1.1 Using GeneR with Plain R

"The R/Bioconductor GeneR package [31] supports fast codon translation with the strTranslate function implemented in C." GeneR supports the eukaryotic code and other major encoding standards. R usage is:

```
library(GeneR)
strTranslate("atgtcaatggtaagaaatgtatcaaatcagagcgaaaaattg-
gaaattttgt")
[1] "MSMVRNVSNQSEKLEIL"
```

The \name{R+GeneR} script (also available here) reads:

```
fasta = 'dna.fa'
library(GeneR)
idx = indexFasta(fasta)
```

```
lines <-readLines( paste(fasta,'.ix',sep='') )
index <-read.table(paste(fasta,'.ix',sep='') )[,1]
n = 0
for (i in 1:times) {
  for (name in index) {
   readFasta (file=fasta, name = name)
   ntseq = getSeq(0)
   aaseq = strTranslate(ntseq)
   cat(">",name," (",n,")\n",aaseq,"\n",sep="")
   n = n+1
  }
}
```

and parses the nucleotide FASTA input and outputs amino acid FASTA. Run the script:

```
docker run --rm -v 'pwd'/tmp:/tmp -v 'pwd'/scripts:/scripts -e \
  BATCH_VARS=/tmp/test-dna-${i}.fa -t bionode bash -c "source
/etc/profile
cd /book-evolutionary-genomics
./scripts/create_test_files.rb
R -q --no-save --no-restore --no-readline --slave < src/R/
DNAtranslate_GeneR.R" > /dev/null
```

Used directly from R, the throughput of the GeneR module is about 658 sequences per second (Seq/s) on the test system, an AMD Opteron(TM) 6128 CPU at 2.00 GHz (*see* also Fig. 1). When checking the implementation by reading the source code, in the first edition, we found that the GeneR FASTA parser was a huge bottleneck. The FASTA parser implementation created an index on disk and reloaded the full index file from disk for each individual sequence, thereby incurring a large overhead for every single sequence.

To see if we could improve throughput, we replaced the slow FASTA parser with \name{R+Biostrings} which reads FASTA once into RAM using the R/Bioconductor BioStrings module and still uses GeneR to translate. At the time, this implementation was 1.6 times faster than GeneR. At this time GeneR is 3.2 on average faster than reading with Biostrings which had a throughput of 284.83 Seqs/s proving some work was done by the authors to improve GeneR. The second script can be found here.

2.1.2 Calling into R from Other Languages with RPC

One strategy for bridging between languages is to use R as a network server and invoke remote procedure calls (RPC) over the network.

1. SOAP

   SOAP allows processes to communicate using XML over HTTP in a client/server setup. SOAP is an operating system and computer language "agnostic," so it can be used to bridge between languages. In the previous edition of this chapter {Ref to Previous Edition, same chapter}, we wrote a R/SOAP [32] adapter for codon translation and invoked it from Python (a Python to R bridge). That client script can be found here. The SOAP bridge was dropped from this chapter because the SOAP packages are not maintained and it was by far the slowest method of cross-language interfacing we tried! The marshalling and unmarshalling of simple string objects using XML over a local network interface takes a lot of computational resources. We do not recommend using SOAP.

2. Rserve

Rserve [3] is a custom binary network protocol, more efficient than XML-based protocols [3]. R data types are converted into Rserve binary data types. Rserve was originally written for Java, but nowadays connectors exist for other languages. With Rserve, Python and R do not have to run on the same server. Furthermore, all data structures will automatically be converted from native R to native Python and numpy types and back.

With RServe fired up a Python example is:

```
import pyRserve
conn = pyRserve.connect()
conn.eval('library(GeneR)')
conn.eval('strTranslate("atgtcaatggtaagaaatgtatcaaatcagagc-
gaaaaattggaaattttgt")')
'MSMVRNVSNQSEKLEIL'
```

where Rserve+GeneR uses the GeneR translate function. In our test Biopython [6] is used for parsing FASTA, and at 797 Seq/s, even with this network bridge, Python+Rserve's speed is on par with that of R. The script can be found here.

2.1.3 Calling into R from Other Languages with the Call Stack Approach

Another strategy for bridging language is to use a native call stack, i.e., data does not get transferred over the network. RPy2 executes R code from within Python over a local call stack [2]. Invoking the same GeneR functions from Python:

```
import rpy2.robjects as robjects
from rpy2.robjects.packages import importr
importr('GeneR')
strTranslate=robjects.r['strTranslate']
strTranslate("atgtcaatggtaagaaatgtatcaaatcagagcgaaaaattg-
gaaattttgt")[0]
'MSMVRNVSNQSEKLEIL'
```

This example uses Biopython for parsing FASTA and invokes GeneR translation over a call stack handled by RPy2. At 2049 Seq/s, throughput is the highest of our calling into R examples. The Python implementation outperforms the other FASTA parsers, and GeneR is fast too when only the translation function is called (GeneR's strTranslate is actually written in C, not in R). Still, there are some overheads for bridging and transforming string objects from Python into R and back. The RPy2 call stack approach is efficient for passing data back and forth. The script can be found here.

2.2 Native Bio* Implementations

When dealing with cross-language transport comparisons, it is interesting to compare results with native language implementations. For example, Biopython [6] would be:

```
from Bio.Seq import Seq
from Bio.Alphabet import generic_dna
coding_dna = Seq("atgtcaatggtaagaaatgtatcaaatcagagcgaaaaattg-
gaaattttgt", generic_dna)
coding_dna.translate()
Seq('MSMVRNVSNQSEKLEIL', ExtendedIUPACProtein())
```

which runs at 797 Seq/s which is slower than the Python3+RPy2+GeneR version. This is because the translate function is written in Python and not in C. It is, however, still faster than R+GeneR. Ruby+BioRuby runs faster at 1481 Seq/s. Perl+BioPerl is in the middle with 1165 Seq/s. We can assume the Biopython, BioPerl, and BioRuby implementations are reasonably optimized for performance. Therefore, throughput reflects the performance of these interpreted languages (*see* Fig. 1).

Java is a statically typed compiled language. Java+BioJava [8] outperforms the interpreters and runs at 2266 Seq/s.

The source code for all examples can be found here in the {Biopython}, {BioRuby}, {BioPerl}, and {BioJava} subdirectories.

2.3 Using the JVM for Cross-Language Support

The Java virtual machine (JVM) is a "bytecode" standard that represents a form of computer intermediate language. This language conceptually represents the instruction set of a stack-oriented capability architecture. This intermediate language, or "bytecode," is not tied to Java specifically, and in the last 10 years, a number of languages have appeared which target the JVM, including JRuby (Ruby on the JVM), Jython (Python on the JVM), Groovy [33], Clojure [34], and Scala [35]. These languages also compile into bytecode and share the same JVM stack. The shared JVM stack allows transparent function calling between different languages.

An example of calling BioJava translation from a Scala program:

```
import org.biojava.nbio.core.sequence.transcription.TranscriptionEngine
import org.biojava.nbio.core.sequence._

val transcriber = TranscriptionEngine.getDefault()
val dna = new DNASequence("atgtcaatggtaagaaatgtatcaaatcagagcgaaaaattggaaattttgt")
val rna = dna.getRNASequence(transcriber)
rna.getProteinSequence(transcriber)
'MSMVRNVSNQSEKLEIL'
```

which uses the BioJava libraries.

A native Java function, such as getProteinSequence, is directly invoked from the other language without overheads (the passed-in transcriber object is passed by reference, just like in Java). In fact, Scala compiles to bytecode, which maps one to one to Java, including the class definitions. The produced bytecode is a native Java bytecode; therefore, the performance of calling BioJava from Scala or Java is exactly the same. This also holds for other languages on the JVM, such as Clojure and Groovy.

We have also included a JRuby example that calls into BioJava4 on the JVM and runs at 1413 Seq/s. JRuby is an interpreter on the JVM that still needs some translation calling into JVM functions. It is therefore slower than native calls.

2.4 Shared C Library Cross-Calling Using EMBOSS Codon Translation

EMBOSS is a free and OSS analysis package specially developed for the needs of the molecular biology user community, mostly written in C [9].

2.4.1 FFI

Using Foreign Function Interface (FFI), it is possible to load dynamic libraries at runtime, define classes to map composite data types, and bind functions for a later use inside your host programming language. We used FFI to bind the EMBOSS translation function to Python and Ruby. The Python example:

```
from ctypes import *
import os
emboss = cdll.LoadLibrary(os.path.join(os.path.dirname(os.path.abspath(__file__)),"emboss.so"))
trnTable = emboss.ajTrnNewI(1)
ajpseq = emboss.ajSeqNewNameC(b"atgtcaatggtaagaaatgtatcaaatcagagcgaaaaattggaaattttgt", b"Test sequence")
ajpseqt = emboss.ajTrnSeqOrig(trnTable,ajpseq,1)
seq = emboss.ajSeqGetSeqCopyC(ajpseqt)
seq = str(c_char_p(seq).value,'utf-8')
```

```
print(seq)
MSMVRNVSNQSEKLEILX
```

The Ruby example:

```
require 'ffi'

module Emboss
  extend FFI::Library
  ffi_lib "./emboss.so"
  attach_function :ajTrnNewI, [:int], :pointer
   attach_function :ajSeqNewNameC, [:pointer, :pointer], :
pointer
  attach_function :ajTrnSeqOrig, [:pointer, :pointer, :int], :
pointer
  attach_function :ajSeqGetSeqCopyC, [:pointer], :string
end

trnTable = Emboss.ajTrnNewI(1)
ajpseq = Emboss.ajSeqNewNameC("atgtcaatggtaagaaatgtatcaaatca-
gagcgaaaaattggaaattttgt", "Test sequence")
ajpseqt = Emboss.ajTrnSeqOrig(trnTable,ajpseq,1)
aa = Emboss.ajSeqGetSeqCopyC(ajpseqt)
print aa,"\n"
MSMVRNVSNQSEKLEILX
```

In both cases the advantage of FFI is that it does not require to compile any source code, just loading the shared library and binding what is needed. Python has a native library called ctypes, and more sophisticated libraries are available to help the programmer bind complex data structures and functions. Ruby has a dedicated gem called [ruby-ffi].

The Ruby and Python FFI outperforms all above methods at 6257 Seq/s and 4787 Seq/s, respectively (*see* Fig. 1). Plotting the time in seconds spent to translate the sequences, Ruby and Python FFI are the lowest (quickest) in the whole comparison (*see* Fig. 2). The high speed points out that (1) the invoked Biopython and BioRuby functions are reasonably efficient at parsing FASTA, (2) the FFI-generated call stack is efficient for moving data over the local call stack, and (3) the EMBOSS transeq DNA to protein translation is optimal C code.

2.5 Calling Program to Program

Calling program to program is far more common than you may think because even when you run a program in a shell, such as Bash, you are calling program to program. You can invoke EMBOSS from the command line:

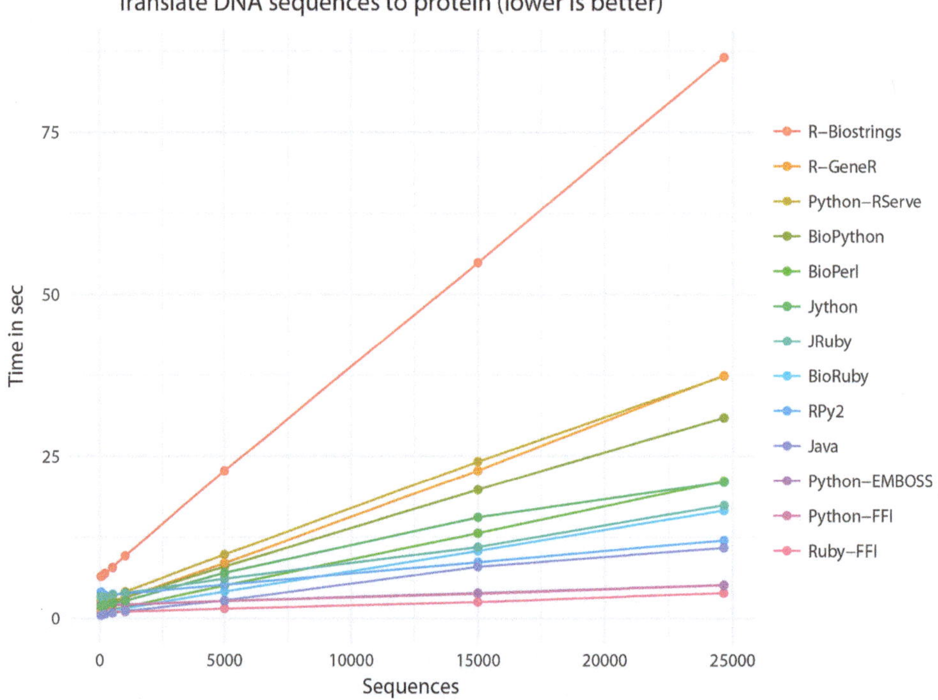

Fig. 2 Number of seconds needed for processing mRNA to protein translation using cross-language calling with a range of programming resources. *See* Fig. 1 for the setup. The figure shows that for all the implementations, the time increases linearly with the number of sequences in input. R-Biostrings performs poorly with an upstart of 6.50 s and the highest slope. The cross-language Ruby-FFI, Python FFI, and Python-EMBOSS with an upstart slightly higher than Java have a very minimal slope; Ruby-FFI has a nearly constant time

```
transeq test-dna.fa test.pep
```

transeq is written in C and runs at a very fast 23,478 Seq/s. Invoking above EMBOSS' transeq in Python looks like this:

```
os.system("transeq "+fn+" out.pep")
for seq_record in SeqIO.parse("out.pep", "fasta"):
    print(">",seq_record.id)
    seq = str(seq_record.seq)
    print(seq)
```

and this combination runs at 4768 Seq/s. That is close to Python FFI and a third of the speed of transeq on its own because of Python parsing the output. Every parsing step has a cost attached.

2.6 Web Services

A discussion on bridging languages would not be complete if we did not include web services, particularly using REST API's. Service like TogoWS and EBI web services which include EMBOSS transeq

(SOAP) offer functionality over http(s) and can be used from any programming language. Here a Ruby example of using TogoWS:

```
## Invoke irb by loading BioRuby
% irb -r bio

## Create a TogoWS object
>> togows = Bio::TogoWS::REST.new
=>  #<Bio::TogoWS::REST:0x007f840faab9d8 @pathbase="/",
@http=#<Net::HTTP togows.dbcls.jp:80 open=false>,
  @header={"User-Agent"=>"BioRuby/1.5.1"}, @debug=false>

## Search for UniProt entries by keywords
>> togows.search('uniprot', 'lung cancer')
=> "KKLC1_MACFA\nKKLC1_HUMAN\nDLEC1_HUMAN\n .....

## Retrieve one UniProt entry (or multiple entries if you like)
>> entry = togows.entry('uniprot', 'KKLC1_MACFA')

## See the entry content
>> puts entry
ID   KKLC1_MACFA              Reviewed;          114 AA.
AC   Q4R717;
 :

## Convert the retrieved UniProt entry into FASTA format
>> puts togows.convert(entry, 'uniprot', 'fasta')
>KKLC1_MACFA RecName: Full=Kita-kyushu lung cancer antigen
1 homolog;
MNVYLLLASGILCALMTVFWKYRRFQRNTGEMSSNSTALALVRPSSTGLINSNTDNNLSV
YDLSRDILNNFPHSIAMQKRILVNLTTVENKLVELEHILVSKGFRSASAHRKST
```

Web services can harness a lot of power because they use large databases and access up-to-date information. As an example, let's generate RDF from above entry:

```
## Retrieve PubMed entry and convert it into RDF/Turtle
(or JSON or XML if you like)
>> puts togows.entry('pubmed', '16381885', 'ttl')
@prefix dc: <http://purl.org/dc/elements/1.1/> .
@prefix dcterms: <http://purl.org/dc/terms/> .
@prefix rdfs: <http://www.w3.org/2000/01/rdf-schema#> .
@prefix prism: <http://prismstandard.org/namespaces/2.0/ba-
sic/> .
@prefix medline: <http://purl.jp/bio/10/pubmed/> .

<http://rdf.ncbi.nlm.nih.gov/pubmed/16381885>    medline:pmid
"16381885" ;
```

```
rdfs:label        "pmid:16381885" ;
dc:identifier     "16381885" ;
medline:own       "NLM" ;
```

Unfortunately, data centric web services can be slow, i.e., sending and retrieving data over the internet incurs large latency and throughput penalties. Sometimes they use powerful back ends, and it is possible to submit large batch jobs which compete with locally installed solutions. Examples are the BLAST service [16] and GeneNetwork [36].

3 Discussion

The half-life of bioinformatics software is 2 years—Pjotr Prins

In this chapter we show that there are many ways of bridging between computer languages. Cross-language interfacing is a topic of importance to evolutionary genomics (and beyond) because computational biologists need to provide tools that are capable of complex analysis and cope with the amount of biological data generated by the latest technologies. Cross-language interfacing allows sharing of code. This means computer software can be written in the computer language of choice for a particular purpose. Flexibility in choice of computer programming language allows optimizing of computational resources and, perhaps even more important, software developer resources, in bioinformatics.

When some functionality is needed that exists in a different computer language than the one used for a project, a developer has the following options: either rewrite the code in the preferred language, essentially a duplication of effort, or bridge from one language to the other. For bridging, there are essentially two technical methods that allow full programmatic access to functionality: through RPC or a local call stack. A third option may be available when functionality can be reached through the command line, as shown above with transeq.

RPC function invocation, over a network interface, has the advantage of being language agnostic and even machine independent. A function can run on a different machine or even over the Internet, which is the basis of web services and may be attractive even for running services locally. RPC XML-based technologies, however, are slow because of expensive parsing and high data load. Our metrics suggest that it may be worth experimenting with binary protocols, such as Rserve and Apache Thrift.

When performance is critical, e.g., when much data needs to be processed, or functions are invoked millions of times, a native call stack approach may be preferred over RPC. Metrics suggest that the EMBOSS C implementation performs well and that binding to the native C libraries with FFI is efficient (*see* Fig. 2). Alternatively, it is possible to use R as an intermediate to C libraries. Interestingly, calling R libraries, many of which are written in C, may give higher performance than calling into native Bio* implementations. For example, Python+RPy2+GeneR is faster that Biopython pure Python implementation of sequence translation, and it is also faster than R calling into GeneR directly—confirming a common complaint that R can be slow.

Even though RPC may perform less well than local stack-based approaches, RPC has some real advantages. For example, if you have a choice of calling a local BLAST library or call into a remote and ready NCBI RPC interface, the latter lacks the deployment complexity. Also the public resource may be more up to date than a copied server running locally. This holds for many curated services that involve large databases, such as PDB [37], Pfam [38], KEGG [39], and UniProt [40]. Chapter 25 gives a deeper treatment of these Internet resources.

From the examples given in this chapter, it may be clear that actual invocation of functions through the different technologies is similar, i.e., all listed Python scripts look similar, provided the underlying dependencies on tools and libraries have been resolved. The main difference between implementations is with deployment of software, rather than invocation of functionality. The JVM approach is of interest, because it makes bridging between supported languages transparent and deployment straightforward. Not only can languages be mixed, but also the advanced Java tool chain is available, including debuggers, profilers, load distributors, and build tools. Other shared virtual machines, such as .NET and Parrot, potentially offer similar advantages but are less used in bioinformatics.

In the first edition, we wrote that when striving for reliable and correct software solutions, the alternative strategy of calling computer programs as external units via the command line should be discouraged: not only is it less efficient that a program gets started every time a function gets called, but also a potential deployment nightmare is introduced. What happens when the program is not installed, or the interface changed between versions, or when there is some other error? With the full programmatic interfaces, discussed in this chapter, incompatibilities between functions get caught much earlier. In this edition of the chapter, we add that efficiency considerations still hold, and error handling can be problematic. When it comes to deployment, however, there now exist solutions that fixate versions of software and give control of the dependency graph, i.e., a tool like transeq can be coupled with its exact version against your software. To ascertain coupling: first there are containers, such as offered by Docker, that allow for bundling software binaries. Second, some recent software distributions allow for formal deployment solutions with reproducible dependency graphs. If you want to know more, check the GNU Guix and NixOS projects. It is possible to combine these deployment technologies. In fact, with this chapter, we provide tools and scripts defined as GNU Guix packages and hosted in a Docker container. These solutions are discussed in Chapter 25.

Choosing a computer language should not be based on runtime performance considerations alone. The maturity of the language and accompanying libraries, tools, and documentation should count heavily, as well as the activity of the community involved. The time saved by using a known language versus learning a new language should be factored in. The main point we are trying to make here is that it is possible to mix languages using different interfacing strategies. This allows leveraging existing functionality, as written by others, using a language of choice. Depending on one's needs, it is advisable to test possible alternatives for performance, as the different tests show that performance varies.

Whichever language and bridging technology is preferred, we think it important to test the performance of different ways of interfacing languages, as there is (1) a need for combining languages in bioinformatics and (2) it is not always clear what impact a choice of cross-language interface may have on performance. By testing different bridging technologies and functional implementations, the best solution should emerge for a specific scenario.

So far, we have focused on the performance of cross-language calling. In Chapter 25, scalability of computation is discussed by programming for multiple processors and machines.

4 Questions

1. Install the Docker container and run different tests. Can you replicate the differences of throughput statistics?
2. Why are network protocols such as Rserve slower than native call stack approaches?
3. What are possible advantages of using a virtual machine, such as the JVM?
4. If you were to bridge between your favorite language and an R library, what options do you have?

Acknowledgments

We thank all open-source software developers for creating such great tools and libraries for the scientific community.

References

1. Gentleman RC, Carey VJ, Bates DM et al (2004) Bioconductor: open software development for computational biology and bioinformatics. Genome Biol 5(10):R80
2. Gautier L (2010) An intuitive Python interface for Bioconductor libraries demonstrates the utility of language translators. BMC Bioinformatics 11(Suppl 12):S11
3. Urbanek S (2003) Rserve a fast way to provide R functionality to applications. In Proceedings of the 3rd International Workshop on Distributed Statistical Computing (DSC 2003), Vienna, Austria
4. Urbanek S (2009) How to talk to strangers: ways to leverage connectivity between R, Java and objective C. Comput Stat 24:303–311
5. Stajich JE, Block D, Boulez K et al (2002) The Bioperl toolkit: Perl modules for the life sciences. Genome Res 12(10):1611–1618
6. Cock PJ, Antao T, Chang JT et al (2009) Biopython: freely available Python tools for computational molecular biology and bioinformatics. Bioinformatics 25(11):1422–1423
7. Goto N, Prins P, Nakao M et al (2010) Bioruby: bioinformatics software for the Ruby programming language. Bioinformatics 26 (20):2617–2619
8. Holland RC, Down TA, Pocock M et al (2008) BioJava: an open-source framework for bioinformatics. Bioinformatics 24(18):2096–2097
9. Rice P, Longden I, Bleasby A (2000) EMBOSS: the european molecular biology open software suite. Trends Genet 16 (6):276–277
10. Dutheil J, Gaillard S, Bazin E et al (2006) Bio++: a set of C++ libraries for sequence analysis, phylogenetics, molecular evolution and population genetics. BMC Bioinformatics 7:188
11. Yang Z (1997) PAML: a program package for phylogenetic analysis by maximum likelihood. Comput Appl Biosci 13(5):555–556
12. Eddy SR (2008) A probabilistic model of local sequence alignment that simplifies statistical significance estimation. PLoS Comput Biol 4 (5):e1000069
13. Larkin MA, Blackshields G, Brown NP et al (2007) Clustal W and clustal X version 2.0. Bioinformatics 23(21):2947–2948
14. Katoh K, Kuma K, Toh H, Miyata T (2005) MAFFT version 5: improvement in accuracy of multiple sequence alignment. Nucleic Acids Res 33(2):511–518
15. Edgar RC (2004) MUSCLE: a multiple sequence alignment method with reduced time and space complexity. BMC Bioinformatics 5:113
16. Altschul SF, Madden TL, Schaffer AA et al (1997) Gapped BLAST and PSI-BLAST: a new generation of protein database search programs. Nucleic Acids Res 25(17):3389–3402
17. Ronquist F, Huelsenbeck JP (2003) MrBayes 3: Bayesian phylogenetic inference under mixed models. Bioinformatics 19 (12):1572–1574
18. Box D, Ehnebuske D, Kakivaya G et al (2000) Simple object access protocol (SOAP) 1.1
19. St Laurent S, Johnston J, Dumbill E (2001) Programming Web services with XML-RPC. pub-ORA, 213p
20. Richardson L, Ruby S (2007) Restful web services. pub-ORA, xxiv + 419p
21. Muller J, Lorenz M, Geller F, Zeier A, Plattner H (2010) Assessment of communication protocols in the EPC network-replacing textual SOAP and XML with binary google protocol buffers encoding. In: Industrial Engineering and Engineering Management (IE\&EM), 2010 IEEE 17Th International Conference on. IEEE, New York, NY, pp 404–409
22. Agarwal A, Slee M, Kwiatkowski M (2007) Thrift: scalable cross-language services implementation
23. Hintjens P (2013) Zeromq: messaging for many applications. O'Reilly Media, Sebastopol, CA, 516p
24. Beazley D (1996) SWIG: an easy to use tool for integrating scripting languages with C and C++. Proceedings of the 4th conference on USENIX Tcl/Tk Workshop, 1996-Volume 4, USENIX Association, 15p
25. Development Core Team R (2010) R: a language and environment for statistical computing
26. Paradis E, Claude J, Strimmer K (2004) APE: analyses of phylogenetics and evolution in R language. Bioinformatics 20(2):289–290
27. Charif D, Thioulouse J, Lobry JR, Perriere G (2005) Online synonymous codon usage analyses with the ade4 and seqinR packages. Bioinformatics 21(4):545–547
28. Arends D, Prins P, Jansen RC, Broman KW (2010) R/qtl: high-throughput multiple QTL mapping. Bioinformatics 26 (23):2990–2992

29. Yandell BS, Mehta T, Banerjee S et al (2007) R/qtlbim: QTL with Bayesian interval mapping in experimental crosses. Bioinformatics 23(5):641–643
30. Harris TW, Antoshechkin I, Bieri T et al (2010) WormBase: a comprehensive resource for nematode research. Nucleic Acids Res 38(Database issue):D463–D467
31. Cottret L, Lucas A, Marrakchi E et al GeneR: R for genes and sequences analysis
32. Warnes G (2004) RSOAP provides a SOAP interface for the open-source statistical package R
33. Koenig D, Glover A, King P, Laforge G, Skeet J (2007) Groovy in action. Manning Publications Co, Greenwich, CT
34. Halloway S (2009) Programming Clojure. Pragmatic Bookshelf, Raleigh, NC
35. Odersky M, Altherr P, Cremet V et al (2004) An overview of the Scala programming language. LAMP-EPFL. (IC/2004/64)
36. Sloan Z, Arends D, Broman KW et al (2016) Genenetwork: framework for web-based genetics. J Open Source Soft 1(25):2
37. Berman HM, Battistuz T, Bhat TN et al (2002) The protein data bank. Acta Crystallogr D Biol Crystallogr 58(Pt 6, 1):899–907
38. Finn RD, Mistry J, Tate J et al (2010) The Pfam protein families database. Nucleic Acids Res 38(Database Issue):D211–D222
39. Kanehisa M, Goto S (2000) KEGG: kyoto encyclopedia of genes and genomes. Nucleic Acids Res 28(1):27–30
40. Bairoch A, Apweiler R, Wu CH et al (2005) The universal protein resource (UniProt). Nucleic Acids Res 33(Database issue):D154–D159

Index

A

B

Maria Anisimova (ed.), *Evolutionary Genomics: Statistical and Computational Methods*, Methods in Molecular Biology, vol. 1910, https://doi.org/10.1007/978-1-4939-9074-0,

C

D

E

F

Q

R

S

T

U

MIX
Papier aus verantwortungsvollen Quellen
Paper from responsible sources
FSC® C105338

If you have any concerns about our products,
you can contact us on
ProductSafety@springernature.com

In case Publisher is established outside the EU,
the EU authorized representative is:
Springer Nature Customer Service Center GmbH
Europaplatz 3, 69115 Heidelberg, Germany

Printed by Libri Plureos GmbH
in Hamburg, Germany